Integration of Pharmaceutical Discovery and Development

Case Histories

Pharmaceutical Biotechnology

Series Editor: Ronald T. Borchardt
The University of Kansas
Lawrence, Kansas

Recent volumes in this series:

Volume 4 BIOLOGICAL BARRIERS TO PROTEIN DELIVERY
Edited by Kenneth L. Audus and Thomas J. Raub

Volume 5 STABILITY AND CHARACTERIZATION OF PROTEIN AND PEPTIDE DRUGS: Case Histories
Edited by Y. John Wang and Rodney Pearlman

Volume 6 VACCINE DESIGN: The Subunit and Adjuvant Approach
Edited by Michael F. Powell and Mark J. Newman

Volume 7 PHYSICAL METHODS TO CHARACTERIZE PHARMACEUTICAL PROTEINS
Edited by James N. Herron, Wim Jiskoot, and Daan J. A. Crommelin

Volume 8 MODELS FOR ASSESSING DRUG ABSORPTION AND METABOLISM
Edited by Ronald T. Borchardt, Philip L. Smith, and Glynn Wilson

Volume 9 FORMULATION, CHARACTERIZATION, AND STABILITY OF PROTEIN DRUGS: Case Histories
Edited by Rodney Pearlman and Y. John Wang

Volume 10 PROTEIN DELIVERY: Physical Systems
Edited by Lynda M. Sanders and R. Wayne Hendren

Volume 11 INTEGRATION OF PHARMACEUTICAL DISCOVERY AND DEVELOPMENT: Case Histories
Edited by Ronald T. Borchardt, Roger M. Freidinger, Tomi K. Sawyer, and Philip L. Smith

A Chronological Listing of Volumes in this series appears at the back of this volume

---

A Continuation Order Plan is available for this series. A continuation order will bring delivery of each new volume immediately upon publication. Volumes are billed only upon actual shipment. For further information please contact the publisher.

Integration of Pharmaceutical Discovery and Development

Case Histories

Edited by

Ronald T. Borchardt
The University of Kansas
Lawrence, Kansas

Roger M. Freidinger
Merck Research Laboratories
West Point, Pennsylvania

Tomi K. Sawyer
ARIAD Pharmaceuticals, Inc.
Cambridge, Massachusetts

and

Philip L. Smith
SmithKline Beecham
Collegeville, Pennsylvania

Plenum Press • New York and London

Library of Congress Cataloging-in-Publication Data

Integration of pharmaceutical discovery and development
: case histories / edited by Ronald T. Borchardt ...
[et al.].
p. cm. -- (Pharmaceutical biotechnology ; v.
11)
Includes bibliographical references and index.
ISBN 0-306-45743-1
1. Drug development. 2. Drugs--Design. 3. Drugs-
-Testing. I. Borchardt, Ronald T. II. Series.
[DNLM: 1. Drug Design. 2. Chemistry,
Pharmaceutical. 3. Drug Evaluation. QV 744I5923
1998]
RM301.25.I57 1998
615'.19--dc21
DNLM/DLC
for Library of Congress 98-27185
CIP

ISBN 0-306-45743-1

A Division of Plenum Publishing Corporation
233 Spring Street, New York, N.Y. 10013

http://www.plenum.com

10 9 8 7 6 5 4 3 2 1

Printed in the United States of America

Contributors

Wade J. Adams • Discovery Chemistry, Pharmacia & Upjohn, Inc., Kalamazoo, Michigan 49001-0199

Akwete L. Adjei • Abbott Laboratories, North Chicago, Illinois 60064

Kimberly K. Adkison • Glaxo Wellcome Research and Development, Research Triangle Park, North Carolina 27709

Fahad Al-Obeidi • Department of Chemistry, Selectide Research Center, Hoechst-Marion Roussel, Tucson, Arizona 85724

Robert C. Andrews • Glaxo Wellcome Research and Development, Research Triangle Park, North Carolina 27709

Paul A. Aristoff • Discovery Chemistry, Pharmacia & Upjohn, Inc., Kalamazoo, Michigan 49001-0199

Bruce J. Aungst • DuPont Merck Pharmaceutical Company, Experimental Station, Wilmington, Delaware 19880-0500

Wilfried Bauer • Novartis Pharma AG, Basel, Switzerland CH-4002

Judd Berman • Glaxo Wellcome Research and Development, Research Triangle Park, North Carolina 27709

Pradip K. Bhatnagar • Department of Medicinal Chemistry, SmithKline Beecham Pharmaceuticals, King of Prussia, Pennsylvania 19406-0939

Lawrence Birkemo • Glaxo Wellcome Research and Development, Research Triangle Park, North Carolina 27709

James Blanchard • Arizona Health Sciences Center, University of Arizona, Tucson, Arizona 85724

Steven G. Blanchard • Glaxo Wellcome Research and Development, Research Triangle Park, North Carolina 27709

David Bodmer • Novartis Pharma AG, Basel, Switzerland CH-4002

H. Neal Bramson • Glaxo Wellcome Research and Development, Research Triangle Park, North Carolina 27709

Ulrich Briner • Novartis Pharma AG, Basel, Switzerland CH-4002

Peter J. Brown • Glaxo Wellcome Research and Development, Research Triangle Park, North Carolina 27709

Christian Bruns • Novartis Pharma AG, Basel, Switzerland CH-4002

George Burton • SmithKline Beecham Pharmaceuticals, Collegeville, Pennsylvania 19426-0989

Eugene N. Bush • Abbott Laboratories, North Chicago, Illinois 60064-3500

David J. Carini • DuPont Merck Pharmaceutical Company, Experimental Station, Wilmington, Delaware 19880-0500

Kong Teck Chong • Pharmacia & Upjohn, Inc., Kalamazoo, Michigan 49007

David D. Christ • DuPont Merck Pharmaceutical Company, Experimental Station, Wilmington, Delaware 19880-0500

Brenda V. Dawson • Health Sciences, The University of Auckland, 92019 Auckland, New Zealand

George V. De Lucca • DuPont Merck Pharmaceutical Company, Experimental Station, Wilmington, Delaware 19880-0500

Annette M. Doherty • Department of Chemistry, Parke-Davis Pharmaceutical Research Division, Warner-Lambert Company, Ann Arbor, Michigan 48105

Robert T. Dorr • Arizona Cancer Center, University of Arizona, Tucson, Arizona 85724

David Drewry • Glaxo Wellcome Research and Development, Research Triangle Park, North Carolina 27709

John V. Duncia • DuPont Merck Pharmaceutical Company, Experimental Station, Wilmington, Delaware 19880-0500

Geneviève Durand-Cavagna • Merck Sharp & Dohme-Chibret Research Center, Riom, 63203 France

Harma M. Ellens • Department of Medicinal Chemistry, SmithKline Beecham Pharmaceuticals, King of Prussia, Pennsylvania 19406-0939

John D. Elliott • Department of Medicinal Chemistry, SmithKline Beecham Pharmaceuticals, King of Prussia, Pennsylvania 19406-0939

Susan Erickson-Viitanen • DuPont Merck Pharmaceutical Company, Experimental Station, Wilmington, Delaware 19880-0500

Stephen V. Frye • Glaxo Wellcome Research and Development, Research Triangle Park, North Carolina 27709

Kenneth W. Funk • Abbott Laboratories, North Chicago, Illinois 60064

Liang-Shang L. Gan • Glaxo Wellcome Research and Development, Research Triangle Park, North Carolina 27709

Paul D. Gesellchen • Lilly Research Laboratories, Eli Lilly and Company, Indianapolis, Indiana 46285

Jonathan Greer • Abbott Laboratories, North Chicago, Illinois 60064-3500

Mac E. Hadley • Department of Cell Biology and Anatomy, University of Arizona, Tucson, Arizona 85724

Kathy A. Halm • Glaxo Wellcome Research and Development, Research Triangle Park, North Carolina 27709

Fortuna Haviv • Abbott Laboratories, North Chicago, Illinois 60064-3500

David J. Hermann • Glaxo Wellcome Research and Development, Research Triangle Park, North Carolina 27709

Joanna P. Hinton • Department of Pharmacokinetics and Drug Metabolism, Parke-Davis Pharmaceutical Research, Warner-Lambert Company, Ann Arbor, Michigan 48105

Ralph Hirschmann • Department of Chemistry, University of Pennsylvania, Philadelphia, Pennsylvania 19104-6323

Victor J. Hruby • Department of Chemistry, University of Arizona, Tucson, Arizona 85724

William F. Huffman • Department of Medicinal Chemistry, SmithKline Beecham Pharmaceuticals, King of Prussia, Pennsylvania 19406-0939

Prabhakar K. Jadhav • DuPont Merck Pharmaceutical Company, Experimental Station, Wilmington, Delaware 19880-0500

Richard L. Jarvest • SmithKline Beecham Pharmaceuticals, Harlow, Essex CM19 5AW, England

Richard K. Jensen • Discovery Chemistry, Pharmacia & Upjohn, Inc., Kalamazoo, Michigan 49001-0199

Andrea Kay • Novartis Pharma Ltd., East Hanover, New Jersey 07936

Andrew G. King • Department of Molecular Virology and Host Defense, SmithKline Beecham Pharmaceuticals, Collegeville, Pennsylvania 19426

Hollis D. Kleinert • Abbott Laboratories, North Chicago, Illinois 60064

Judith Knittle • Abbott Laboratories, North Chicago, Illinois 60064-3500

Kenneth A. Koeplinger • Pharmacia & Upjohn, Inc., Kalamazoo, Michigan 49007

M. Amparo Lago • Department of Medicinal Chemistry, SmithKline Beecham Pharmaceuticals, King of Prussia, Pennsylvania 19406-0939

Patrick Y. S. Lam • DuPont Merck Pharmaceutical Company, Experimental Station, Wilmington, Delaware 19880-0500

Ioana Lancranjan • Novartis Pharma AG, Basel, Switzerland CH-4002

Frank W. Lee • Glaxo Wellcome Research and Development, Research Triangle Park, North Carolina 27709

Norman Levine • Department of Dermatology, University of Arizona, Tucson, Arizona 85724

Jiunn H. Lin • Drug Metabolism, Merck Research Laboratories, West Point, Pennsylvania 19486

Franco Lombardo • Central Research Division, Pfizer Inc., Groton, Connecticut 06340

Dagfinn Løvhaug • Nycomed Imaging AS, Bioreg Research, Oslo N0371, Norway

John A. Lowe III • Central Research Division, Pfizer Inc., Groton, Connecticut 06340

Peter Marbach • Novartis Pharma AG, Basel, Switzerland CH-4002

Linda Mizen • SmithKline Beecham Pharmaceuticals, Collegeville, Pennsylvania 19426-0989

Walter Morozowich • Discovery Chemistry, Pharmacia & Upjohn, Inc., Kalamazoo, Michigan 49001-0199

Eliot H. Ohlstein • Department of Medicinal Chemistry, SmithKline Beecham Pharmaceuticals, King of Prussia, Pennsylvania 19406-0939

Drazen Ostovic • Pharmaceutical Research and Development, Merck Research Laboratories, West Point, Pennsylvania 19486

Guy E. Padbury • Pharmacia & Upjohn, Inc., Kalamazoo, Michigan 49007

Arthur A. Patchett • Departments of Medicinal Chemistry and Biochemistry & Physiology, Merck Research Laboratories, Rahway, New Jersey 07065

Catherine E. Peishoff • Department of Medicinal Chemistry, SmithKline Beecham Pharmaceuticals, King of Prussia, Pennsylvania 19406-0939

Louis M. Pelus • Department of Molecular Virology and Host Defense, SmithKline Beecham Pharmaceuticals, Collegeville, Pennsylvania 19426

Michael E. Pierce • DuPont Merck Pharmaceutical Company, Experimental Station, Wilmington, Delaware 19880-0500

Bernard Plazonnet • Merck Sharp & Dohme-Chibret Research Center, Riom, 63203 France

Janos Pless • Novartis Pharma AG, Basel, Switzerland CH-4002

Gerald S. Ponticello • Merck Research Laboratories, West Point, Pennsylvania 19486

William M. Potts • Department of Drug Metabolism and Pharmacokinetics, SmithKline Beecham Pharmaceuticals, King of Prussia, Pennsylvania 19406-0939

Thomas J. Raub • Pharmacia & Upjohn, Inc., Kalamazoo, Michigan 49007

Friedrich Raulf • Novartis Pharma AG, Basel, Switzerland CH-4002

Rodney Robison • Novartis Pharma Ltd., East Hanover, New Jersey 07936

Donna L. Romero • Discovery Chemistry, Pharmacia & Upjohn, Inc., Kalamazoo, Michigan 49001-0199

Saul H. Rosenberg • Abbott Laboratories, North Chicago, Illinois 60064

Tomi K. Sawyer • Ariad Pharmaceuticals, Cambridge, Massachusetts 02139

William C. Schinzer • Discovery Chemistry, Pharmacia & Upjohn, Kalamazoo, Michigan 49001-0199

Francis J. Schwende • Pharmacia & Upjohn, Inc., Kalamazoo, Michigan 49007

Joel E. Shaffer • Glaxo Wellcome Research and Development, Research Triangle Park, North Carolina 27709

John Sharkey • Novartis Pharma Ltd., East Hanover, New Jersey 07936

Robert T. Shuman • Lilly Research Laboratories, Eli Lilly and Company, Indianapolis, Indiana 46285

Achintya K. Sinhababu • Glaxo Wellcome Research and Development, Research Triangle Park, North Carolina 27709

Philip L. Smith • Department of Drug Delivery, SmithKline Beecham Pharmaceuticals, Collegeville, Pennsylvania 19426

Roy G. Smith • Departments of Medicinal Chemistry and Biochemistry & Physiology, Merck Research Laboratories, Rahway, New Jersey 07065

Thomas Soranno • Novartis Pharma Ltd., East Hanover, New Jersey 07936

Barbara Stolz • Novartis Pharma AG, Basel, Switzerland CH-4002

Elizabeth E. Sugg • Glaxo Wellcome Research and Development, Research Triangle Park, North Carolina 27709

Michael F. Sugrue • Merck Research Laboratories, West Point, Pennsylvania 19486

David Sutton • SmithKline Beecham Pharmaceuticals, Harlow, Essex CM19 5AW, England

W. Gary Tarpley • Discovery Chemistry, Pharmacia & Upjohn, Inc., Kalamazoo, Michigan 49001-0199

Suvit Thaisrivongs • Pharmacia & Upjohn, Inc., Kalamazoo, Michigan 49007

Richard C. Thomas • Discovery Chemistry, Pharmacia & Upjohn, Inc., Kalamazoo, Michigan 49001-0199

Gaochao Tian • Glaxo Wellcome Research and Development, Research Triangle Park, North Carolina 27709

Timothy K. Tippin • Glaxo Wellcome Research and Development, Research Triangle Park, North Carolina 27709

Larry Tremaine • Central Research Division, Pfizer Inc., Groton, Connecticut 06340

Bharat K. Trivedi • Department of Medicinal Chemistry, Parke-Davis Pharmaceutical Research, Warner-Lambert Company, Ann Arbor, Michigan 48105

Andrew C. G. Uprichard • Department of Cardiac and Vascular Diseases, Parke-Davis Pharmaceutical Research Division, Warner-Lambert Company, Ann Arbor, Michigan 48105

Joseph P. Vacca • Medicinal Chemistry, Merck Research Laboratories, West Point, Pennsylvania 19486

R. Anthony Vere Hodge • SmithKline Beecham Pharmaceuticals, Harlow, Essex CM19 5AW, England

Peter Vit • Novartis Pharma AG, Basel, Switzerland CH-4002

Robert E. Waltermire • DuPont Merck Pharmaceutical Company, Experimental Station, Wilmington, Delaware 19880-0500

Gisbert Weckbecker • Novartis Pharma AG, Basel, Switzerland CH-4002

Thomas N. Wheeler • Glaxo Wellcome Research and Development, Research Triangle Park, North Carolina 27709

Steven M. Winter • Central Research Division, Pfizer Inc., Groton, Connecticut 06340

Matthew J. Wyvratt • Departments of Medicinal Chemistry and Biochemistry & Physiology, Merck Research Laboratories, Rahway, New Jersey 07065

Zhiyang Zhao • Pharmacia & Upjohn, Inc., Kalamazoo, Michigan 49007

Gail L. Zipp • Pharmacia & Upjohn, Inc., Kalamazoo, Michigan 49007

Preface

In the late 1980s, it became painfully evident to the pharmaceutical industry that the old paradigm of drug discovery, which involved highly segmented drug design and development activities, would not produce an acceptable success rate in the future. Therefore, in the early 1990s a paradigm shift occurred in which drug design and development activities became more highly integrated. This new strategy required medicinal chemists to design drug candidates with structural features that optimized pharmacological (e.g., high affinity and specificity for the target receptor), pharmaceutical (e.g., solubility and chemical stability), biopharmaceutical (e.g., cell membrane permeability), and metabolic/pharmacokinetic (e.g., metabolic stability, clearance, and protein binding) properties. Successful implementation of this strategy requires a multidisciplinary team effort, including scientists from drug design (e.g., medicinal chemists, cell biologists, enzymologists, pharmacologists) and drug development (e.g., analytical chemists, pharmaceutical scientists, physiologists, and molecular biologists representing the disciplines of pharmaceutics, biopharmaceutics, and pharmacokinetics/drug metabolism).

With this new, highly integrated approach to drug design now widely utilized by the pharmaceutical industry, the editors of this book have provided the scientific community with case histories to illustrate the nature of the interdisciplinary interactions necessary to successfully implement this new approach to drug discovery. In the first chapter, Ralph Hirschmann provides a historical perspective of why this paradigm shift in drug discovery has occurred. Subsequent chapters describe in detail the strategies used to discover the following drugs or drug candidates: renin inhibitors (Chapter 2, S. H. Rosenberg and H. D. Kleinert, Abbott Laboratories); angiotensin II antagonists (Chapter 3, D. Carini *et al.*, DuPont Merck); thrombin inhibitors (Chapter 4, R. T. Shuman and P. D. Gesellchen, Lilly Research Laboratories); endothelin receptor antagonists (Chapter 5, A. M. Doherty and A. C. G. Uprichard, Parke-Davis; Chapter 6, J. D. Elliott *et al.*, SmithKline

Beecham); LHRH antagonists (Chapter 7, F. Haviv *et al.,* Abbott Laboratories); LHRH agonists (Chapter 8, K. W. Funk *et al.,* Abbott Laboratories); somatostatin agonists (Chapter 9, P. Marbach *et al.,* Novartis); HIV protease inhibitors (Chapter 10, G. E. Padbury *et al.,* Pharmacia & Upjohn; Chapter 11, J. H. Lin *et al.,* Merck Research Laboratories; Chapter 12, G. V. De Lucca *et al.,* DuPont Merck); reverse transcriptase inhibitors (Chapter 13, W. J. Adams *et al.,* Pharmacia & Upjohn); antiherpesvirus agents (Chapter 14, R. L. Jarvest *et al.,* SmithKline Beecham); ester prodrugs of β-lactam antibiotics (Chapter 15, L. Mizen and G. Burton, SmithKline Beecham); hematoregulators (Chapter 16, P. K. Bhatnagar *et al.,* SmithKline Beecham), inhibitors of 5α-reductase (Chapter 17, S. V. Frye *et al.,* Glaxo Wellcome); α_{1A} receptor antagonists (Chapter 18, K. K. Adkison *et al.,* Glaxo Wellcome); inhibitors of secretory phospholipase A_2 (Chapter 19, S. G. Blanchard *et al.,* Glaxo Wellcome); CCK-B receptor antagonists (Chapter 20, F. Lombardo *et al.,* Pfizer; Chapter 21, B. K. Trivedi and J. P. Hinton, Parke-Davis); CCK-A agonists (Chapter 22, E. E. Sugg *et al.,* Glaxo Wellcome); growth hormone secretagogues (Chaper 23, A. A. Patchett *et al.,* Merck Research Laboratories); carbonic anhydrase inhibitors (Chapter 24, G. S. Ponticello *et al.,* Merck Research Laboratories); and melanotropic peptides (Chapter 25, M. E. Hadley *et al.,* University of Arizona, University of Auckland, Hoechst-Marion Roussel, and Ariad Pharmaceuticals).

Lastly, we thank all of the authors for their valuable and timely contributions. We hope that the case histories presented in this book will illustrate the benefits of this highly integrated approach to drug discovery and will facilitate the discovery of novel drugs in the future.

Contents

Chapter 1

Introduction 1
Ralph Hirschmann

Chapter 2

Renin Inhibitors
Saul H. Rosenberg and Hollis D. Kleinert

1. The Renin Angiotensin System (RAS) 7
2. *In Vitro* Assays 8
3. Renin Inhibitor Design 10
   3.1. Novel Transition-State Analogues 10
   3.2. Models to Evaluate Pharmacological Responses 12
   3.3. Molecular Weight, Proteolytic Stability, and Aqueous Solubility . . 13
   3.4. Renin Inhibitors with Oral Bioavailability 17
4. Conclusions 24
   References 25

Chapter 3

The Discovery and Development of Angiotensin II Antagonists
David J. Carini, David D. Christ, John V. Duncia, and Michael E. Pierce

1. Introduction 29
2. Development of a Tetrazole Derivative 30
   2.1. Chemical Stability and Potential Toxicity of Tetrazoles 30
   2.2. Metabolism of Tetrazoles 32

2.3. The Search for Tetrazole Replacements 33
2.4. Synthetic Availability of Biphenyltetrazoles 39
3. An Active Metabolite of Losartan 44
3.1. Identification of EXP3174 44
3.2. Should We Develop EXP3174? 45
3.3. The Search for a Superior EXP3174 Analogue 45
4. Early Evaluation of Losartan's Activity in Humans 47
5. Selective versus Balanced Angiotensin II Receptor Antagonists 48
6. Conclusion 51
References 52

Chapter 4

Development of an Orally Active Tripeptide Arginal Thrombin Inhibitor
Robert T. Shuman and Paul D. Gesellchen

1. Introduction 57
2. Identification of Lead Compounds 60
2.1. *In Vitro* Structure–Activity Relationships 61
2.2. *In Vivo* Structure–Activity Relationships 64
3. Development of Parenteral Clinical Candidate 69
3.1. Development of Licensed Compound (Efegatran) 69
3.2. Summary of Clinical Data on Efegatran 70
4. Development of an Oral Candidate 70
4.1. *In Vivo* Oral Bioavailability 71
4.2. Oral Dosing in Efficacy Models 73
4.3. Pharmacokinetics of Oral Candidate 73
4.4. Clinical Data for Oral Candidate 75
5. Conclusion 77
References 78

Chapter 5

Discovery and Development of an Endothelin A Receptor-Selective Antagonist PD 156707
Annette M. Doherty and Andrew C. G. Uprichard

1. Introduction 81
2. Discovery of PD 156707: Medicinal Chemistry, Pharmacology, and Pharmacokinetics 84

2.1. Identification of Lead Structures 84
2.2. Structure–Activity Relationships 86
2.3. Pharmacokinetics/Selection 87
2.4. Chemistry/Chemical Development 89
2.5. Biological Evaluation of PD 156707 90
2.6. Metabolism 92
2.7. Assay Development 93
3. Efficacy Studies: Which Disease States? 96
3.1. Hypertension 96
3.2. Heart Failure 99
3.3. Pulmonary Hypertension 100
3.4. Stroke 103
4. Future Plans 103
5. Summary 105
References 105

Chapter 6

Endothelin Receptor Antagonists
John D. Elliott, Eliot H. Ohlstein, Catherine E. Peishoff, Harma M. Ellens, and M. Amparo Lago

1. Introduction 113
2. Rational Design of SB 209670 115
3. Pharmacological, Drug Metabolism, and Pharmacokinetic Characterization of SB 209670 121
4. Selection of the Orally Bioavailable Candidate SB 217242 121
5. Conclusion 127
References 127

Chapter 7

LHRH Antagonists
Fortuna Haviv, Eugene N. Bush, Judith Knittle, and Jonathan Greer

1. Mechanism of Action of LHRH Agonists and Antagonists 131
2. Structural Differences of LHRH Agonists and Antagonists 133
2.1. Reduction of Size of LHRH Analogues 135

2.2. Enzymatic Stability of LHRH Analogues and Effect of *N*-methyl Substitution on Enzymatic Stability of LHRH Agonists 136
2.3. Effect of *N*-methyl Substitution on Water Solubility of LHRH Antagonists. Discovery of A-75998 137
3. Biological Testing Strategy 137
3.1. *In Vitro* Testing of A-75998: Receptor Binding, Inhibition of LH Release, and Histamine Release 138
3.2. *In Vivo* Studies of A-75998 in Rat, Dog, and Monkey 138
3.3. Pharmacokinetics of A-75998 in Rat, Dog, and Monkey 140
4. Aggregation and Formulation of A-75998 141
5. LHRH Antagonists in Clinical Evaluation 144
5.1. Clinical Study of A-75998 144
5.2. Current LHRH Antagonists in Clinical Studies 144
6. Summary 146
References 146

Chapter 8

LHRH Agonists

Kenneth W. Funk, Jonathan Greer, and Akwete L. Adjei

1. Introduction 151
1.1. Background 152
1.2. Drug Candidate Selection 153
2. Physical Chemistry and Chemical Characterization 157
2.1. Bulk Drug Synthesis 158
2.2. Manufacturing Controls 160
2.3. Physical Characteristics and Methods 161
2.4. Chemical Characterization and Methods 163
2.5. Moisture and Acetic Acid 163
2.6. Amino Acid Analysis 164
3. Formulation Chemistry of Leuprolide Acetate 165
3.1. *In Vitro* Studies 165
3.2. *In Vivo* Studies 166
4. Clinical Development 169
4.1. Standards and Controls 169
4.2. Physical and Chemical Characterization 171
4.3. Pathology and Toxicology 174
4.4. Clinical Pharmacokinetics and Pharmacodynamics 177
5. Conclusions 178
References 179

Chapter 9

Discovery and Development of Somatostatin Agonists
Peter Marbach, Wilfried Bauer, David Bodmer, Ulrich Briner, Christian Bruns, Andrea Kay, Ioana Lancranjan, Janos Pless, Friedrich Raulf, Rodney Robison, John Sharkey, Thomas Soranno, Barbara Stolz, Peter Vit, and Gisbert Weckbecker

1. Introduction 183
2. Somatostatin Receptors 184
2.1. Heterogeneity of Somatostatin Receptors 184
2.2. The Somatostatin Receptor Gene Family 184
2.3. Tissue Distribution 185
2.4. Pharmacology 186
3. Discovery and Development of Sandostatin® 186
3.1. Synthesis of Octreotide 189
3.2. Pharmacodynamic Tests 190
3.3. Pharmacokinetic Studies 191
3.4. Toxicology 191
3.5. Clinical Development 192
4. Development of Sandostatin® LAR® 193
4.1. Manufacture 194
4.2. Preclinical Studies 194
4.3. Clinical Studies 195
5. Oncolar™: Technical Development of a New LAR Formulation of Octreotide 196
5.1. Manufacture 196
5.2. Preclinical Studies 196
6. Antiproliferative Effects of Single-Agent Octreotide 197
6.1. Mechanism of Antiproliferative Action 198
6.2. Route of Administration and Plasma Levels 199
6.3. Octreotide as a Potentiator of Standard Anticancer Regimens 199
7. Development of Octreotide for Oncological Uses beyond the Control of Disease-Related Symptoms in GEP Tumors 201
7.1. Somatostatin Receptor Binding and Growth Factor Suppression 201
7.2. Clinical Trials 202
8. Radiolabeled Octreotide Analogues 202
8.1. Imaging of Tumors with OctreoScan® 203
8.2. Tumor Radiotherapy with SMT 487 203
9. Summary and Outlook 204
References 205

Chapter 10

Factors Impacting the Delivery of Therapeutic Levels of Pyrone-Based HIV Protease Inhibitors
Guy E. Padbury, Gail L. Zipp, Francis J. Schwende, Zhiyang Zhao, Kenneth A. Koeplinger, Kong Teck Chong, Thomas J. Raub, and Suvit Thaisrivongs

1. Introduction ... 211
1.1. HIV Protease as a Therapeutic Target ... 211
1.2. Pyrone-Based Inhibitors ... 213
1.3. Factors that Affect Drug Delivery ... 215
1.4. Life in a Perfect World ... 215
2. Efficacy ... 216
2.1. Effect/Importance of Protein Binding ... 216
2.2. Clinical Targets ... 219
3. Pharmacokinetics ... 220
3.1. Total versus Unbound Intrinsic Clearance ... 220
3.2. Factors Affecting Clearance ... 221
3.3. Absolute Oral Bioavailability versus Systemic Exposure ... 224
4. Life in the Real World ... 225
4.1. Selection of a Viable Chemical Template ... 225
4.2. Identification of a Final Clinical Candidate ... 227
References ... 229

Chapter 11

The Integration of Medicinal Chemistry, Drug Metabolism, and Pharmaceutical Research and Development in Drug Discovery and Development: The Story of Crixivan®, an HIV Protease Inhibitor
Jiunn H. Lin, Drazen Ostovic, and Joseph P. Vacca

1. Introduction ... 233
2. Discovery of L-735,524 (Crixivan®) ... 234
3. Improvement of Solubility ... 238
4. Physicochemical Properties of MK-639 (Indinavir) ... 241
5. pH-Dependent Oral Absorption ... 246
6. *In Vitro/In Vivo* Metabolism ... 248
7. Backup Compounds ... 249
8. Conclusion ... 252
References ... 254

Chapter 12

De Novo Design and Discovery of Cyclic HIV Protease Inhibitors Capable of Displacing the Active-Site Structural Water Molecule

George V. De Lucca, Prabhakar K. Jadhav, Robert E. Waltermire, Bruce J. Aungst, Susan Erickson-Viitanen, and Patrick Y. S. Lam

1. Introduction 257
2. Initiation of Program at DMPC 258
3. Design of Cyclic Ureas 259
   3.1. *De Novo* Design 259
   3.2. Confirmation of Design 262
   3.3. Molecular Recognition 265
4. First Clinical Candidate DMP 323 266
   4.1. Discovery and Optimization 266
   4.2. Chemistry and Process Development 267
   4.3. Clinical Study 270
5. Second Clinical Candidate DMP 450 270
   5.1. Discovery and Optimization 270
   5.2. Safety and Pharmacokinetics 271
   5.3. Chemistry and Process Development 272
   5.4. Clinical Study 273
6. Future Cyclic Ureas 274
   6.1. Potency 275
   6.2. Resistance Profile 276
   6.3. Pharmacokinetics 278
   6.4. Design and Physicochemical Properties 280
7. Conclusion 280
   References 281

Chapter 13

Discovery and Development of the BHAP Nonnucleoside Reverse Transcriptase Inhibitor Delavirdine Mesylate

Wade J. Adams, Paul A. Aristoff, Richard K. Jensen, Walter Morozowich, Donna L. Romero, William C. Schinzer, W. Gary Tarpley, and Richard C. Thomas

1. Introduction, Goals, and Strategy 285
2. Discovery of Initial Lead (PNU-80493E) 287

3. Selection of First-Generation Candidate (PNU-87201) 288
4. Development of PNU-87201E (Atevirdine Mesylate) 292
5. Goals for Second-Generation Candidate 292
6. Selection Process 293
7. Water-Soluble Compounds 294
8. Development of PNU-90152T (Delavirdine Mesylate) 300
   8.1. Pharmacology 300
   8.2. Formulation/Salt Selection/Crystal Form 301
   8.3. Absorption, Distribution, Metabolism, and Excretion 305
   8.4. Safety/Toxicokinetics 308
   8.5. Clinical Summary 309
9. Conclusions 310
   References 310

Chapter 14

Famciclovir: Discovery and Development of a Novel Antiherpesvirus Agent
Richard L. Jarvest, David Sutton, and R. Anthony Vere Hodge

1. Introduction 313
   1.1. Identification of Penciclovir as an Antiherpesvirus Agent 314
   1.2. Antiviral Activity and Spectrum of Activity 315
   1.3. Mechanism of Action 316
   1.4. Oral Bioavailability 321
2. Prodrug Forms of Penciclovir 321
   2.1. Strategy and Evaluation of Oral Bioavailability 321
   2.2. Evaluation of Metabolic Conversion in Human Body Fluids and Tissues 326
   2.3. Selection of Preferred Oral Candidate: Famciclovir 327
   2.4. Other Routes of Administration 327
3. Preclinical Evaluation of Famciclovir 329
   3.1. Animal Models of Infection 329
   3.2. Chirality of Metabolic Products from Famciclovir 330
   3.3. Identification of Enzymatic Oxidation in Humans 331
4. Clinical Evaluation 331
   4.1. Metabolism and Pharmacokinetics 331
   4.2. Efficacy 333
5. Conclusion 337
   References 338

Chapter 15

The Use of Esters as Prodrugs for Oral Delivery of β-Lactam Antibiotics
Linda Mizen and George Burton

1. Introduction ... 345
2. Chemical Overview ... 347
3. Animal Bioavailability Studies and Selection ... 350
   3.1. Penicillins, Penems, Trinem ... 351
   3.2. Cephalosporins ... 353
4. Hydrolysis Rates and Physicochemical Properties ... 357
   4.1. Hydrolysis by Liver ... 357
   4.2. Hydrolysis by Small Intestine ... 358
   4.3. Hydrolysis by Blood ... 358
   4.4. Physicochemical Properties ... 360
5. Dosing Vehicles and Formulations ... 361
6. Summary and Conclusions ... 361
   References ... 362

Chapter 16

Hematoregulators: A Case History of a Novel Hematoregulatory Peptide, SK&F 107647
Pradip K. Bhatnagar, William F. Huffman, Andrew G. King, Dagfinn Løvhaug, Louis M. Pelus, William M. Potts, Philip L. Smith

1. Introduction ... 367
2. Hematopoiesis, Endogenous Regulators, and Host Defense Mechanism ... 368
3. Unmet Needs ... 369
4. Nonproteinaceous Hematoregulators ... 371
   4.1. Polymeric Carbohydrate: Betafectin ... 371
   4.2. Low-Molecular-Weight Hematoregulators ... 371
5. SK&F 107647 and Analogues ... 375
   5.1. Structure–Activity Relationships of SK&F 107647 ... 376
   5.2. Mechanism of Action ... 379
   5.3. Colony Stimulating Activity Induction Assay ... 379
   5.4. Hematopoietic Synergistic Factor Assay ... 379
   5.5. Preclinical Studies ... 380
6. Conclusions ... 383
   References ... 384

Chapter 17

Discovery and Development of GG745, a Potent Inhibitor of Both Isozymes of 5α-Reductase
Stephen V. Frye, H. Neal Bramson, David J. Hermann, Frank W. Lee, Achintya K. Sinhababu, and Gaochao Tian

1. Introduction ... 393
   1.1. 5α-Reductases ... 393
   1.2. Pathophysiology of DHT ... 397
   1.3. Finasteride: Clinical Effects of a Type 2-Selective 5α-Reductase Inhibitor ... 398
   1.4. Potential Utility of a Dual 5α-Reductase Inhibitor ... 399
2. Enzymology of 5α-Reductases ... 399
   2.1. Time Dependence of Inhibition by Δ^1 4-Azasteroids ... 399
   2.2. Modeling of the Clinical Effect of Finasteride ... 404
3. Discovery of Dual 5α-Reductase Inhibitors: 6-Azasteroids ... 405
   3.1. Medicinal Chemistry ... 405
   3.2. Pharmacokinetic Studies: *In Vivo* and *in Vitro* Correlations ... 408
4. Discovery of GG745 ... 410
5. Initial Clinical Studies with GG745 ... 413
   5.1. Interspecies Scaling/Dose Selection ... 413
   5.2. Pharmacokinetic and Pharmacodynamic Results in Man ... 414
   References ... 417

Chapter 18

Discovery of a Potent and Selective α_{1A} Antagonist: Utilization of a Rapid Screening Method to Obtain Pharmacokinetic Parameters
Kimberly K. Adkison, Kathy A. Halm, Joel E. Shaffer, David Drewry, Achintya K. Sinhababu, and Judd Berman

1. Introduction ... 423
   1.1. Benign Prostatic Hyperplasia ... 423
   1.2. Therapeutic Use of α_{1A}-Selective Antagonists ... 424
   1.3. Project Goal ... 425
2. Research Strategy ... 425
   2.1. Compound Progression and Critical Path ... 425
   2.2. Discovery of α_{1A}-Selective Oxazole-Containing Antagonists ... 426

3. Pharmacokinetic/Pharmacodynamic Strategy 433
3.1. *In Vitro* Metabolism Screening Prior to Pharmacokinetic Studies 433
3.2. Improved Pharmacokinetic Throughput: Mixture Dosing Coupled with LC/MS Analysis 434
3.3. Pharmacokinetic Evaluation of Other Leads 439
3.4. Pharmacodynamics of the Lead Compound 440
4. Advancement of Compound **18** to Exploratory Development 442
References 442

Chapter 19

Discovery of Bioavailable Inhibitors of Secretory Phospholipase A_2

Steven G. Blanchard, Robert C. Andrews, Peter J. Brown, Liang-Shang L. Gan, Frank W. Lee, Achintya K. Sinhababu, and Thomas N. Wheeler

1. Introduction 445
1.1. Therapeutic Target 445
1.2. Program Objective 446
2. *In Vitro* Identification of Active-Site Inhibitors of $sPLA_2$ 446
2.1. "Dual Substrate" Strategy for Inhibitor Discovery 446
2.2. *In Vitro* Profile of Substrate Analogue PLA_2 Inhibitors 447
3. *In Vivo* Anti-inflammatory Activity of Initial Candidates 448
3.1. Choice of Animal Model 448
3.2. *In Vivo* Activity Is Dependent on Formulation of the Test Compound 449
3.3. Activity in the Rat Carrageenan Paw Edema Model 450
4. Pharmacokinetic and Metabolic Fate of Candidate Inhibitors 452
4.1. Plasma Levels and Metabolic Profiles after i.v. and p.o. Dosing ... 452
4.2. *In Vitro* Studies 453
4.3. Conclusions Based on Metabolism Studies 457
5. Preparation of Inhibitors Designed to Address the Observed Metabolic Instability 458
5.1. Synthesis and *in Vitro* Evaluation of Inhibitory Activity 458
5.2. Evaluation of *in Vitro* Stability 460
5.3. Pharmacokinetic Studies 460
5.4. *In Vivo* Activity of Inhibitors with Improved Metabolism and Pharmacokinetics 460

6. Summary and Conclusions 461
References 462

Chapter 20

The Anxieties of Drug Discovery and Development: CCK-B Receptor Antagonists
Franco Lombardo, Steven M. Winter, Larry Tremaine, and John A. Lowe III

1. Introduction 465
2. Chemistry 466
3. Initial Drug Metabolism Studies 468
4. Formulation Studies 471
5. A New Analogue with Improved Aqueous Solubility: CP-310,713 476
6. Lessons Learned 477
References 478

Chapter 21

CI-1015: An Orally Active CCK-B Receptor Antagonist with an Improved Pharmacokinetic Profile
Bharat K. Trivedi and Joanna P. Hinton

1. Introduction 481
1.1. First-Generation CCK-B Antagonists 482
1.2. CI-988 Pharmacokinetic Retrospective 483
1.3. Objectives of the Discovery Team 488
2. Discovery of CI-1015 488
2.1. Design Strategy 488
2.2. Structure–Activity Relationship Study 488
3. Preclinical Characterization of Backup Candidates 494
3.1. *In Vitro* and *in Vivo* Comparison 494
3.2. Pharmacokinetic Evaluations in Rat 494
3.3. Brain Penetration Studies 498
3.4. Evaluation of Potential for Gastric Acid Secretion 499
3.5. Pharmacokinetic Evaluation in Monkey 500
4. Conclusion 500
References 503

Chapter 22

Orally Active Nonpeptide CCK-A Agonists
Elizabeth E. Sugg, Lawrence Birkemo, Liang-Shang L. Gan, and Timothy K. Tippin

1. Introduction 507
2. *In Vivo* Profile of GW7854 508
3. Pharmaceutical Studies with GW7854 510
   3.1. Batch Variation 510
   3.2. Dosing Vehicle 510
4. Pharmacology Studies 510
   4.1. The Mouse Gallbladder Emptying Assay 510
   4.2. Alternate Species 511
   4.3. The Conditioned Feeder Rat Model 511
5. Pharmacokinetic Profile of GW7854 511
6. The Caco-2 Model for Intestinal Absorption 512
   6.1. Correlation with Rat Intestinal Absorption 513
   6.2. Structure–Transport Relationships 513
7. Bioavailability versus Bioactivity 516
8. Oral versus Intraduodenal Dosing 521
9. Discussion 521
10. Clinical Implications 522
    References 522

Chapter 23

Orally Active Growth Hormone Secretagogues
Arthur A. Patchett, Roy G. Smith, and Matthew J. Wyvratt

1. Introduction 525
2. Discovery of GHRP-6 Mimics: Benzolactam L-692,429 527
   2.1. Clinical Studies with L-692,429 528
   2.2. Structure–Activity–Bioavailability Relationships for the Benzolactams 529
3. New Structural Leads 534
   3.1. Privileged Structure Screening 534
   3.2. Discovery of MK-0677 536
4. Mechanism of Action of GH Secretagogues 544
   4.1. Biochemistry 544

4.2. Characterization of the GH Secretagogue Receptor (GHS-R) 545
4.3. Cloning the GH Secretagogue Receptor 546
4.4. GH Secretagogue Receptor and GH Pulsatility 546
5. Conclusion 547
References 549

Chapter 24

Dorzolamide, a 40-Year Wait: From an Oral to a Topical Carbonic Anhydrase Inhibitor for the Treatment of Glaucoma
Gerald S. Ponticello, Michael F. Sugrue, Bernard Plazonnet, and Geneviève Durand-Cavagna

1. Introduction 555
2. Benzothiazoles 557
3. Benzothiophenes 558
4. Thienothiopyrans 559
5. Dorzolamide 560
6. Pharmacology 563
6.1. *In Vitro* 563
6.2. *In Vivo* 564
7. Pharmaceutical Research and Development Studies 566
8. Safety Assessment Studies 567
9. Summary 571
References 572

Chapter 25

Discovery and Development of Novel Melanogenic Drugs: Melanotan-I and -II
Mac E. Hadley, Victor J. Hruby, James Blanchard, Robert T. Dorr, Norman Levine, Brenda V. Dawson, Fahad Al-Obeidi, and Tomi K. Sawyer

1. Introduction 575
2. The Melanocortin Peptides and Receptors 576
2.1. Melanocortin Peptides 576
2.2. Melanocortin Receptors 577
3. Discovery of the MT-I and MT-II as MSH Superagonists 579
3.1. Structure–Activity Studies of α-MSH 579
3.2. Design and Chemistry of MT-I and MT-II 580

3.3. *In Vitro* and *in Vivo* Pharmacology of MT-I and MT-II 582
4. Development of MT-I and MT-II as Novel Melanogenic Drugs 583
4.1. Stability, Pharmacokinetic, and Toxicological Studies 583
4.2. Drug Delivery and Clinical Studies 585
5. Summary and Future Directions 590
References 591

Index 597

Chapter 1

Introduction

Ralph Hirschmann

It is a pleasure to write an introduction to this timely book. As the reader is well aware, drug discovery has changed dramatically during the second half of this century. Let me cite a few examples. When I graduated from the University of Wisconsin in 1950, UV spectroscopy had established itself as a tool of the organic chemist, but IR spectroscopy was not available in Madison. By the time I started to work at Merck that year, IR capability was available. NMR, on the other hand, did not begin its tour de force until the end of the decade. Circular dichroism was also to emerge as an important tool during that decade. High-resolution mass spectrometry was not known, nor was it possible to use cocrystallization of macromolecules with ligands in order to interpret the X-ray structure of bioactive complexes. The modern era of biology was to be ushered in by Watson and Crick in 1953. Its impact on the discovery of oral drugs evolved slowly or rapidly, depending on one's expectations. As I pointed out elsewhere, rational drug design was known, at least since the 1940s, and continued to flourish in response to discoveries in synthetic organic chemistry, biochemistry, and pharmacology. Rational drug design was also advanced enormously by the advent in the 1970s of computerized molecular modeling. Screening continued to play a critical role in drug discovery in the 1950s. Although supposedly in disfavor in the allegedly new era of rational drug design, screening never really disappeared; indeed, it emerged stronger than ever after the concept of the screening of diverse libraries was almost universally embraced and greatly facilitated by automation. Taken together, these developments and other advances in chemistry, biology, and in physical measurements enabled the pharmaceutical industry to discover new leads and to bring them to can-

---

Ralph Hirschmann • Department of Chemistry, University of Pennsylvania, Philadelphia, Pennsylvania 19104-6323.

Integration of Pharmaceutical Discovery and Development: Case Studies, edited by Borchardt *et al.*, Plenum Press, New York, 1998.

didate product status with unprecedented speed. It is at this juncture that safety and efficacy, pharmaceutical formulations, production costs, pharmacokinetic and metabolic properties emerge as the remaining, formidable challenges.

The issues relating to safety and efficacy have been largely in the hand of the Gods, but the emerging field of proteomics may improve our ability to predict. The obstacles relating to pharmaceutical formulations and production costs are now generally solved by expert pharmaceutical chemists, process research chemists, and engineers. This leaves "only" pharmacokinetic and metabolism issues, especially oral bioavailability and biological half-lives, for what is often termed rather cavalierly the "endgame." This book is devoted to a discussion by experts of these pharmacokinetic and metabolism issues as they relate to optimization of a drug candidate's efficacy.

It is important to recognize that the conventional strategy employed in the industry had been flawed in that the medicinal chemist focused his or her attention almost exclusively on but two issues: potency and specificity. The oral bioavailability issues were left for the endgame. When it came time for the latter, the rules were straightforward. Make the compound more lipophilic, and if this does not bring the required results, make the compound more hydrophilic. (It was also permissible to carry out the exercise in the reverse order.) Unfortunately, very often this tactic failed and the endgame never reached the desired goal. To be sure, helpful concepts have been well understood by medicinal chemists for many years. These include consideration of the molecular weight and of the log P of the compound of interest. For both of these, acceptable limits had been set empirically. Conversely, it was well understood that it is desirable to avoid functionality that can facilitate elimination via conjugation. Further, the prodrug approach has in fact allowed several compounds to make important contributions to therapy. Drug latentiation, a refinement of the prodrug concept, has even made selective drug delivery possible. Further, ingenious pharmaceutical delivery systems have also had a significant impact. Finally, scientists in departments of drug metabolism and pharmacokinetics developed a better understanding of other factors that affect oral bioavailability and biological half-life such as rate of dissolution, acid and base stability, metabolism, protein binding, active transport, passage through channels, conformation, charge, glomerular filtration, brush border metabolism, first-pass metabolism in the liver, and so forth.

Although these advances represented invaluable additions to the armamentarium of the overall process, they were of little help to the medicinal chemists because the available research capabilities were restricted to compounds already in Development and were not available to guide the medicinal chemists, thereby forcing them to concentrate almost exclusively on potency and specificity in the search for a drug candidate. This was simply a consequence of the fact that compounds in Development invariably (and properly) have a higher priority than those in Basic Research. Only during the recent past has the industry assigned a group of individuals with pharmacokinetic and drug metabolism capability to serve exclusively Basic Research.

Consider a hypothetical case where a medicinal chemist seeks to optimize a lead that binds a highly hydrophobic pocket. He or she is likely to find that the more hydrophobic groups that can be incorporated into the lead structure, the greater potency of the analogue and the greater everyone's "exuberance." That the resulting candidate compound will have negligible water solubility and thus little or no oral bioavailability should not have come as a surprise, and yet this became a very common scenario. The fact is that if a compound is too hydrophobic, it will lack adequate solubility in water and therefore also oral bioavailability. Conversely, if the compound is too water soluble, the extraction from aqueous medium by the membrane, a requirement for transport, can become very difficult. Thus, generally there has to be a balance between hydrophobic properties and water solubility. From this perspective the above adage "to improve oral bioavailability make the compound more hydrophobic and if this fails, make it more hydrophilic," has a semirational basis.

I became involved with the search for an orally bioavailable peptide when Dan Veber, our colleagues, and I sought to find an orally bioavailable somatostatin-related drug in the early 1970s. We knew that whereas a peptide such as the cyclic tetradecapeptide somatostatin is rapidly degraded by proteases, small cyclic peptides such as cyclic hexapeptides are not. Importantly, it was mistakenly believed at that time that susceptibility to enzymatic cleavage is the only obstacle to oral bioavailability of peptides. Indeed, this erroneous notion is still expressed in the current literature. MK-678, a cyclic hexapeptide, emerged from this research and proved to be indeed stable to relevant proteases. To our chagrin, its oral bioavailability was, nevertheless, below 5%. This taught us that stability to proteases is a necessary condition for oral bioavailability, but not a sufficient one. We speculated that the culprit responsible for the lack of oral bioavailability of MK-678 was the secondary amide bonds. For this reason, Professors A. B. Smith III, K. C. Nicolaou, and I embarked on a research program to replace the secondary amide bonds by alternate scaffolds. The program with Smith led to the design and synthesis of HIV-1 protease inhibitors in which an NH-displaced pyrrolinone scaffold replaced the amide backbone. Pleasingly, assay results obtained at Merck suggested that these pyrrolinone-based enzyme inhibitors displayed better transport into the lipid bilayer of lymphocytes than did their peptide counterparts.

We proposed an explanation for the improved transport properties of the pyrrolinones vis-à-vis their peptide counterparts based on the observation by W. D. Stein in the late 1960s that cellular transport correlates inversely with solvation by water. Stein reasoned that extraction of a compound into a lipid bilayer requires desolvation and, therefore, energy. Shortly thereafter, Diamond and Wright showed that intramolecular hydrogen bonding permits 1,2-dihydroxycyclohexane to exhibit better transport properties than the isomeric 1,3-diol; they attributed this to the fact that the latter binds four molecules of water, but the former only two. We proposed that the intramolecular hydrogen bond between the carbonyl and the NH of two neighboring pyrrolinone rings may similarly reduce solvation by two molecules of water when compared with a conventional secondary amide bond. It is pleasing that the important studies by Conradi and his collaborators in Pharmaceutical Research and Development at Upjohn demonstrated convincingly in the early 1990s that increasing methylation of tetrapeptide amide nitrogens increases their passive transport across Caco-2 cell monolayers, and that charge and chain length are more important than lipophilicity in predicting flux across a rabbit intestine. These results not only were consistent with the observations of Stein, but also were pleasing to us because they provided experimental support for our speculation in the late 1970s that the secondary amide bonds (i.e., the peptide scaffold) were at least partly responsible for the poor oral bioavailability of MK-678.

The above-mentioned Caco-2 cells, developed by Borchardt and his collaborators at the University of Kansas and Per Artursson and his associates at Uppsala University, have become a widely used model for the direct assessment of cell transport. This epithelial cell line is transformed and thus immortal. Its value derives in part from the fact that the experiments require little of the test compound and the protocol can measure both apical-to-basolateral and basolateral-to-apical transport. Thus, it has become one of the important new tools available to pharmaceutical scientists interested in studying intestinal transport of drugs.

It had seemed paradoxical to me that while MK-678 had poor oral bioavailability, it was rapidly eliminated unchanged from the circulation after parenteral administration. Further, it had been shown by Karls and collaborators that clearance by liver and kidney is unaffected by desolvation energy. It is tempting to speculate that the resolution of the above paradox lies in the important role played by the efflux pumps (P-glycoproteins) that have assumed enormous importance in the phenomenon referred to as multiple drug resistance (MDR). Recent investigations, notably by L. Z. Benet, R. Borchardt, P. Burton, and others, have focused on the role of P-glycoprotein in drug transport. Taken together, these concepts suggest the possibility that the poor oral bioavailability of compounds such as MK-678 may be related, at least in part, to the fact that the drug, after extraction by the cell membranes, is pumped out of the cell back into the gut by P-glycoprotein. If this concept has validity, it provides an entirely new perspective for the oral bioavailability problem associated with, for example, peptides of low molecular weight that are stable to proteases. It would also explain why MK-678 is, on the one hand, poorly bioavailable, and, on the other hand, readily removed from circulation.

A very recent paper from the Netherlands Cancer Institute lends credence to this concept. These investigators used MDR 1a (−/−) mice, which lack functional P-glycoprotein in the intestine, to show that P-glycoprotein limits the oral uptake of paclitaxel and that this pump does indeed affect the direct elimination of taxol from the circulation. It is also relevant that the Caco-2 cells possess a polar-ized efflux system that is inhibited by compounds such as cyclosporin and verapamil, further enhancing the practical value of these cells. J. H. Lin and his colleagues in the Drug Metabolism Department at Merck recently studied the effect of verapamil, a potent P-glycoprotein inhibitor, on the intestinal absorption of MK-678 in rats and found that the absorption was not facilitated. These results suggest, but do not prove, that MK-678 is not a substrate of P-glycoprotein.

It is also known that P-glycoprotein is an important constituent of the blood–brain barrier. Thus, one might be tempted to conclude that inhibition of P-glycoproteins might greatly simplify the life of the medicinal chemist by facilitating oral bioavailability, reducing elimination from the circulation, and making it easier to get CNS-active drugs across the blood–brain barrier. Unfortunately, life is not that simple. Recently, I have used as the facetious title of several lectures the question, "Did God install the blood–brain barrier to punish the medicinal chemist?" The answer, of course, is no. That the barrier serves to protect the brain was convincingly shown by G. R. Lankas and his associates in the Safety Assessment Department at Merck when they demonstrated that P-glycoprotein deficiency in a subpopulation of CF-1 mice enhances avermectin-induced neurotoxicity. P-glycoprotein takes on even greater importance if—as suggested by Benet—it acts to facilitate drug metabolism in conjunction with cytochrome P450, both appropriately positioned in the gastrointestinal tract.

The chapters that follow give us every reason to be optimistic about the future impact of pharmacokinetic and drug metabolism research on the early stages of drug discovery.

Chapter 2

Renin Inhibitors

Saul H. Rosenberg and Hollis D. Kleinert

The search for renin inhibitors as an improved modality for antihypertensive therapy was a lively area of research in the 1980s. Incredibly, no fewer than 15 pharmaceutical companies were actively involved in the field during this time. It is of course impossible to summarize within the scope of a single chapter the myriad approaches taken by these various groups, subject matter that can be found in several comprehensive review articles (Greenlee, 1990; Wood *et al.,* 1994; Rosenberg, 1995). It is equally impossible to fully describe the scope of this multidisciplinary research effort as it occurred at our institution, for the discovery phase alone involved the synthesis (prior to the birth of combinatorial chemistry) and biological evaluation of over 4000 novel renin inhibitors. Instead, this chapter will focus on the strategies that we used to identify and then overcome the numerous barriers that are encountered during the discovery and development of a drug candidate. Some of these hurdles, such as the requirements for intrinsic efficacy and safety, are routine to all drug discovery projects. Others, including conferring oral activity to a peptidic molecule, were more specific to the renin inhibitor project and required new approaches that brought together many diverse disciplines.

1. THE RENIN ANGIOTENSIN SYSTEM (RAS)

The genesis of the project was the choice of renin inhibition as a biochemical target. Renin is the first and rate-limiting enzyme in one of the principal systems for the regulation of blood pressure, the well characterized renin–angiotensin

---

Saul H. Rosenberg and Hollis D. Kleinert • Abbott Laboratories, North Chicago, Illinois 60064.
Integration of Pharmaceutical Discovery and Development: Case Studies, edited by Borchardt *et al.,* Plenum Press, New York, 1998.

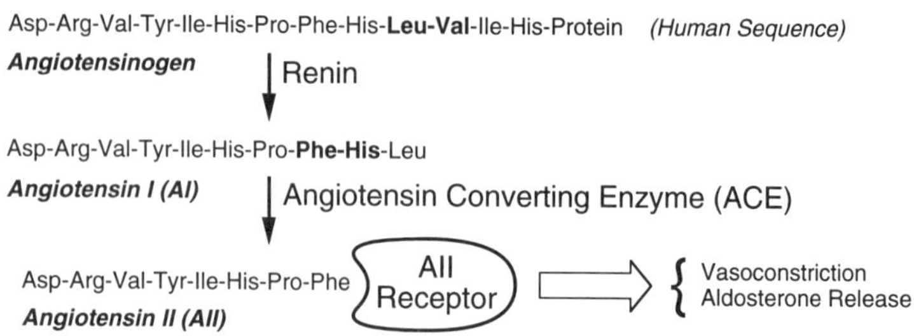

Figure 1. The renin–angiotensin system (RAS).

cascade (Fig. 1; Peach, 1977). Renin cleaves its natural substrate, angiotensinogen, at the Leu-Val scissile bond to produce the decapeptide angiotensin I (AI). AI has little intrinsic activity, but it is rapidly converted by angiotensin-converting enzyme (ACE), to the octapeptide angiotensin II (AII), one of the most potent known vasoconstrictors. AII also stimulates the release of aldosterone, which in turn promotes sodium retention and a secondary upregulation of blood pressure through an increase in vascular volume. At the inception of the project, ACE inhibitors were marketed drugs with proven antihypertensive activity that verified the concept of RAS blockade (Atkinson and Robertson, 1979). Certain side effects were associated with ACE inhibition, however, the most notable of which was a chronic cough in 6–14% of patients (Israili and Hall, 1992; Karlberg, 1993; Overlack, 1996). ACE is a nonselective enzyme. Among its multiple substrates are bradykinin, substance P, enkephalin, and other endogenous peptides (Erdös and Skidgel, 1986; Sunman and Sever, 1993), and its involvement in the bradykinin system has been implicated as the source of the ACE inhibitor-induced cough (Overlack, 1996; Fox *et al.,* 1996). We and others reasoned that inhibition of renin and antagonism of the AII receptor should conceptually afford antihypertensive activity equivalent to ACE inhibition but with an improved side effect profile. The latter approach has recently reached fruition following the discovery, through random screening, of nonpeptide ligands for the AII receptor (Steinberg *et al.,* 1993; Goa and Wagstaff, 1996). In contrast, the search for inhibitors of renin was an early exercise in rational drug design of peptidomimetic structures.

2. *IN VITRO* ASSAYS

Our primary biological tests were *in vitro* enzyme inhibition assays, and these were established early in the project. To provide an intrinsic measurement of the ability of an inhibitor to bind to renin, compounds were first tested against purified human renal renin at its pH optimum (pH 6.0) using human angiotensinogen as the enzyme substrate (Bolis *et al.,* 1987). Under physiologic conditions, how-

Table I
Inhibitory Potencies of Renin Inhibitors Incorporating Transition-State Mimics with Enhanced Affinity for the Active Site

	IC_{50} (nM)	
Compound	Purified[a]	Plasma[b]
H-261	0.7	—
1	2	—
2	1500	—
3	10	—
4	0.6	—
5a	0.55	3.3
5b	2	10
6	0.63	2.8
Enalkiren	0.78	14
A-65317	0.37	0.57

[a] Purified human renal renin, pH 6.0.
[b] Human plasma renin, pH 7.4.

ever, an inhibitor will not only encounter renin at a different pH (pH 7.4), but may also bind to plasma proteins thereby reducing its effective concentration and consequently its activity. We have found that both the pH of the assay and the presence of plasma proteins can profoundly affect measured potency, and that the magnitude of these effects varied with the structure of the inhibitor (Rosenberg, 1995; Table I). An *in vitro* assay employing the endogenous renin and angiotensinogen present in human plasma was therefore established as a secondary, more biologically relevant, measurement of inhibitory activity (Plattner *et al.*, 1988).

Because the rationale for renin inhibition was the prediction of an improved side effect profile compared with ACE inhibition, a renin inhibitor must not interact with other endogenous aspartic proteinases. To determine enzymatic specificity, we established assays for pepsin, cathepsin D, and gastricsin (Bolis *et al.*, 1987). In general, renin inhibitors did not exhibit significant activity against these related enzymes.

In addition to compound evaluation, there were two other applications for the *in vitro* renin inhibition assays. The first was the assessment of potential *in vivo* models. By establishing renin assays based on plasma from various species (monkey, dog, mouse, rat, ferret, hamster, hog, gerbil, guinea pig, and sheep), we could determine the relative sensitivity of renin from a given species to an inhibitor derived from the human angiotensinogen sequence. As expected, our renin inhibitors were most active against human and monkey renin. Our bioavailability determinations also benefited from an efficient *in vitro* assay. We used the degree to which plasma sample extracts inhibited renin, along with the previously determined inhibition curve, to calculate the amount of drug present in the original sample, a bioassay that allowed us to measure low circulating titers of drug in small-volume blood samples. When high plasma drug levels were achieved as a result of either longer-acting intravenous or orally active renin inhibitors, the gold standard (but more tedious) HPLC procedures were employed to analyze the pharmacokinetics.

3. RENIN INHIBITOR DESIGN

Renin belongs to the aspartic proteinase class of proteolytic enzymes: Two aspartic acid residues lie in an active-site cleft and catalyze the addition of water across the scissile amide bond to effect hydrolysis. By 1983, renin was well characterized. The proposal that the reaction proceeded through a tetrahedral transition state had been strengthened by crystal structures of related fungal enzymes (Bott *et al.*, 1982; James *et al.*, 1982). Additionally, Tewksbury *et al.* (1981) had shown that the cleavage site in human angiotensinogen lay between residues Leu-10 and Val-11, which differed from the leucine-leucine scissile bond sequence in nonprimate species. Szelke *et al.* (1983) used this information to design the hydroxyethylene isostere as a mimic of the tetrahedral transition state. This fragment was incorporated into the minimum substrate sequence (as a replacement for both the Leu-10 and Val-11 residues) to provide H-261 (Fig. 2), the first inhibitor reported to possess nanomolar potency against human renin (IC_{50} = 0.7 nM). Another early replacement for the Leu-Val dipeptide was the unusual γ-amino acid statine that is found in the naturally occurring peptide pepstatin (isovaleryl-Val-Val-Sta-Ala-Sta), an extremely potent inhibitor of the aspartic proteinase pepsin (Workman and Burkitt, 1979). Whereas pepstatin itself was a weak inhibitor of renin (IC_{50} = 22 μM), Boger *et al.* (1983) incorporated statine into the angiotensinogen sequence to provide statine-containing renin inhibitory peptide (SCRIP, Fig. 2; IC_{50} = 16 nM). These two compounds demonstrated that a transition-state mimic would bind tightly to the active site of renin and provided the foundation for subsequent inhibitor design.

3.1. Novel Transition-State Analogues

From the outset, we knew that the renin inhibitor project must overcome an immense hurdle, namely, to be competitive with existing antihypertensive therapies, oral activity was an absolute requirement. We also knew that there was no precedent for the oral absorption of a linear peptide of the size of SCRIP or H-261. Because the presence of peptide bonds renders compounds susceptible to degradation and high molecular weight limits intestinal absorption and enhances hepatic elimination (Plattner and Norbeck, 1990), our strategy was to design inhibitors in which both peptidic character and molecular weight were minimized. Large

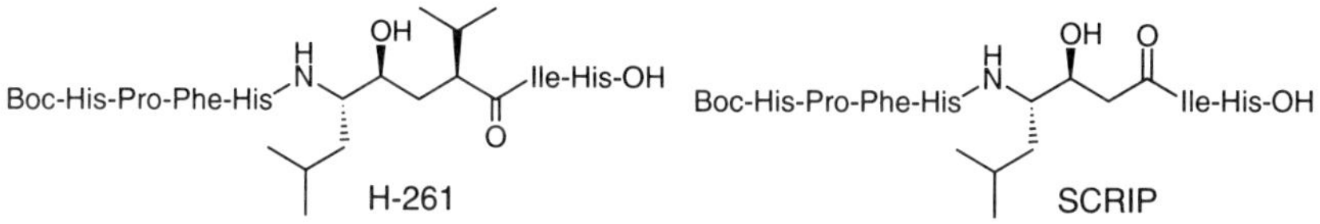

Boc-His-Pro-Phe-His-N(H)-CH(CH2CH(CH3)2)-CH(OH)-CH2-CH(iPr)-C(O)-Ile-His-OH
H-261

Boc-His-Pro-Phe-His-N(H)-CH(CH2CH(CH3)2)-CH(OH)-CH2-C(O)-Ile-His-OH
SCRIP

Figure 2. Renin inhibitors H-261 and SCRIP.

polypeptide renin inhibitors such as SCRIP incorporate multiple side chains, each of which can fit into a specific pocket within the enzyme thereby contributing to the overall binding energy. Some of these interactions would obviously be lost in smaller inhibitors in which residues have been eliminated. Therefore, to reduce molecular weight without sacrificing potency, we sought to design transition-state mimics with enhanced affinity to renin so as to compensate for any lost inhibitor–enzyme interactions. This approach would also provide novel structures that would guarantee a solid patent position.

We employed numerous strategies to discover proprietary and potent transition-state mimics (Greenlee, 1990; Rosenberg, 1995). The most successful of these, as measured by both activity and synthetic accessibility, is outlined in Fig. 3 (Luly *et al.*, 1988). Sequential deletions had demonstrated that a protected Phe-His dipeptide at the N-terminus was the minimum sequence compatible with good inhibitory potency (Plattner *et al.*, 1986). Whereas potency was maintained with the deletion

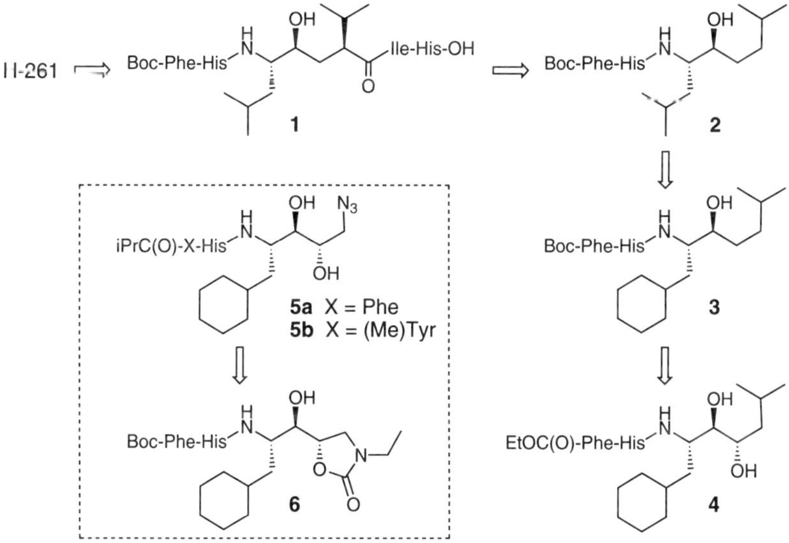

H-261 ⇒ Boc-Phe-His-N(H)-CH(CH2CH(CH3)2)-CH(OH)-CH2-CH(iPr)-C(O)-Ile-His-OH (1) ⇒ Boc-Phe-His-N(H)-CH(CH2CH(CH3)2)-CH(OH)-CH2CH2CH(CH3)2 (2)

⇓

Boc-Phe-His-N(H)-CH(CH2-cyclohexyl)-CH(OH)-CH2CH2CH(CH3)2 (3)

⇓

EtOC(O)-Phe-His-N(H)-CH(CH2-cyclohexyl)-CH(OH)-CH(OH)-CH2CH(CH3)2 (4)

iPrC(O)-X-His-N(H)-CH(CH2-cyclohexyl)-CH(OH)-CH(OH)-CH2N3
5a X = Phe
5b X = (Me)Tyr

⇓

Boc-Phe-His-N(H)-CH(CH2-cyclohexyl)-CH(OH)-(3-ethyl-2-oxo-oxazolidin-5-yl) (6)

Figure 3. Renin inhibitors incorporating transition-state mimics with enhanced affinity for the active site.

of His-Pro from H-261 (compound **1,** Table I), inhibitor **2,** which lacked all C-terminal residues except for the valine side chain, was 750-fold less active. Replacement of the isobutyl side chain of statine with cyclohexylmethyl had been shown to enhance potency (Boger *et al.,* 1985) and this modification (inhibitor **3**) restored significant activity. The putative tetrahedral intermediate for amide bond hydrolysis is a dihydroxylated species, yet statine and the hydroxyethylene isostere each bear only a single hydroxyl. Addition of a second hydroxyl afforded *erythro*-glycol **4** with renin inhibitory activity equivalent to that of H-261. Inhibitors **5** and **6** incorporate related transition-state analogues that proved useful in subsequent inhibitor optimization studies (Rosenberg *et al.,* 1989, 1990a). The speed with which the medicinal chemistry group was able to successfully develop novel, tightly binding transition-state mimics was the direct result of a close working relationship between the chemistry and biochemistry groups, coupled with the simplicity and low material requirements of the renin-inhibition assays that the latter had established.

3.2. Models to Evaluate Pharmacological Responses

The next challenge was to establish appropriate *in vivo* animal models. When the target enzyme is conserved from species to species and, therefore, the drug is expected to be effective in multiple species, the whole animal pharmacology is limited only by reproducing the pathophysiological model. However, when compounds are primate-specific, both the sensitivity of the animal tested and the disease model must be considered. As expected, the testing of primate-specific compounds for efficacy was most predictive of the human response when nonhuman primates were employed as the experimental model. This was especially crucial in the early stages of discovering compounds and establishing structure–activity relationships when most compounds were not very potent. Early renin inhibitors could be screened for efficacy in cynomolgus monkeys (Kleinert *et al.,* 1988c), marmosets (Wood *et al.,* 1985), or human renin-infused rats (Pals *et al.,* 1990). We selected the cynomolgus monkey as our efficacy species of choice.

It was anticipated that animals and humans with normal blood pressure and normal baseline plasma renin activity (PRA) would not respond to RAS blockade. Salt depletion activates the RAS, elevates the baseline PRA, and renders the normal experimental subject sensitive to renin inhibition. Either a low-salt diet and/or diuretic therapy can successfully establish this salt-depleted, high-renin state. Our early compounds were tested intravenously for hypotensive activity in the salt-depleted monkey, a normotensive, high-plasma-renin model. Because of the high level of sensitivity to renin inhibition, this model was susceptible to relatively weak inhibitors and served as a screening guide for the discovery of increasingly potent agents.

Compounds **4** and **5a** represented the most advanced renin inhibitors that the project had then prepared. Both inhibitors caused dose-related (0.01–1 mg/kg) reductions of blood pressure when administered via the intravenous route to anesthetized, salt-depleted monkeys (Kleinert *et al.*, 1988a; Luly *et al.*, 1988), thereby confirming that this model was indeed appropriate for compound evaluation. The discovery of specific, potent, intravenously active renin inhibitors that produced the desired cardiovascular effects had been achieved. Neither these compounds, nor other structurally related inhibitors, however, elicited significant hypotensive responses following oral or intraduodenal (i.d.) dosing. It was clear that further optimization would be required to achieve oral activity.

3.3. Molecular Weight, Proteolytic Stability, and Aqueous Solubility

We had successfully developed several strategies to reduce molecular weight while maintaining *in vitro* potency. Thus, compounds **4** (mol.wt. = 600), **5a** (mol.wt. = 599), and **6** (mol.wt. = 655) represented a significant improvement compared with polypeptide renin inhibitors such as SCRIP (mol.wt. = 1037). These renin inhibitors were substantially larger, however, than peptidic ACE inhibitors with demonstrated oral bioavailability in humans, such as captopril (mol.wt. = 217) and enalapril (mol.wt. = 376). It has been generally accepted that compounds with molecular weights above a threshold limit of 500 are excreted in appreciable quantities into the bile (Klassen and Watkins, 1984; Plattner and Norbeck, 1990). In fact, the primary route of elimination of all renin inhibitors reported to date is the liver (Kleinert *et al.*, 1990; Adedoyin *et al.*, 1993). The hepatic extraction of renin inhibitors may be based not only on their molecular weights, but also on the highly lipophilic nature of these molecules or the possibility that they are eliminated bound to renin, which is also cleared by the liver. As luck would have it, the hepatic route of elimination would be preferred to renal excretion for renin inhibitors, because the primary indications for these agents are hypertension and heart failure, cardiovascular conditions often associated with compromised kidney function.

It was unclear whether molecular weight had been reduced to an extent sufficient to achieve reasonable plasma drug levels following oral administration. Unfortunately, our structure–activity studies had not provided us with an obvious path for further size reductions. Additional factors, however, might also have been limiting oral absorption, including susceptibility to proteolytic enzymes, insufficient aqueous solubility, or other parameters that we had yet to identify.

3.3.1. PROTEOLYTIC STABILITY AND EVALUATION OF *IN VIVO* ABSORPTION

Several *in vitro* assays provided a rapid assessment of the stability of these renin inhibitors in various biological settings (Bolis *et al.*, 1987). Importantly, no enzymatic degradation was observed on incubation in plasma. Prototype inhibitors were also stable to liver, intestinal, and kidney homogenates. However, specific cleavage at the Phe-His peptide bond was effected by both purified chymotrypsin and crude pancreatic protease. Examination of the published specificity requirements for chymotrypsin revealed several interactions that were critical for efficient substrate binding. Replacing phenylalanine with either (*O*-methyl)tyrosine, which is too large for the chymotrypsin hydrophobic pocket, or a benzyl succinate residue, which cannot make a critical hydrogen bond, maintains potency against renin while stabilizing the inhibitor toward chymotrypsin-mediated degradation (Rosenberg *et al.*, 1987; Plattner *et al.*, 1988).

Now our laboratory had to concentrate its efforts on models of oral bioavailability. Test compounds can be evaluated for oral activity by oral dosing in conscious animals or by direct i.d. administration. The i.d. route allows the animals to be studied under anesthesia and also allows drug to be deposited directly at the site of intestinal absorption. Although i.d. and oral administration are not equivalent because deposition of drug right into the intestinal lumen avoids the acid pH of the stomach and ensures a high concentration of intact drug at the site of absorption, the i.d. absorption model was a good first screen for oral activity.

Oral administration of compound **4** to conscious rats confirmed that only trace amounts reached the systemic circulation (Table II; Luly *et al.*, 1988). These experiments, however, could not discern between poor intestinal absorption and extensive hepatic extraction. After considering numerous *in vitro, in situ,* and *in vivo* models, we settled on a simple, straightforward rat model. To better evaluate absorption, future compounds were administered via the i.d. route to anesthetized rats. Plasma drug levels were determined by HPLC or a renin inhibition assay (Rosenberg *et al.*, 1989) from samples taken at 10 and 30 min from both the peripheral systemic and portal circulation in the same animals. Although this model was insufficient for the determination of bioavailability, the data provided a high-throughput estimate of both absorption from the intestine and extraction by the liv-

Table II

Comparison of the Aqueous Solubility and Absorption Following Intraduodenal Administration (10 mg/kg) to Rats of Early Renin Inhibitors

		Plasma drug level (C_{max}, ng/ml)	
Compound	Solubility (mg/ml)	Portal	Systemic
4	0.025[a]	—	31 ± 11[b]
5b	0.029[c]	53 ± 29	15 ± 13
Enalkiren	>10[a]	1100 ± 500	170 ± 70
A-65317	4.0[d]	420 ± 260	2.5 ± 0.1

[a] Water, 37°C.
[b] Oral administration.
[c] pH 6.5 phosphate buffer, 37°C.
[d] pH 7.4 phosphate buffer, 37°C.

er. Inhibitor **5b** (Fig. 3, Table I) is a representative inhibitor incorporating (*O*-methyl)tyrosine (Rosenberg *et al.,* 1989). Intraduodenal administration to rats demonstrated that it was poorly absorbed from the intestine, leading to systemic plasma drug levels similar to those obtained with compound **4**. Thus, stabilizing these renin inhibitors to proteolytic degradation was not sufficient to impart oral bioavailability.

3.3.2. INHIBITOR SOLUBILITY

Following these disappointing results, we focused our attention on physicochemical properties that could be quickly and easily measured and that might potentially correlate with either intestinal absorption or biliary excretion. Our primary analysis was aqueous solubility, although octanol–water partition coefficient data were also obtained for selected inhibitors. We also briefly used an *in vitro* permeability assay that employed isolated perfused rat intestinal segments (Rosenberg *et al.,* 1989). Ultimately, we found that this last protocol was not as rapid as our i.d. rat model and that it was more efficient to proceed directly to the *in vivo* system.

Compounds **4** (Kleinert *et al.,* 1988b) and **5b** (Rosenberg *et al.,* 1989) were in fact quite insoluble (Table II), leading us to speculate that this was the underlying factor behind the lack of intestinal absorption. We employed several strategies to enhance aqueous solubility, which culminated in the design of enalkiren (A-64662, Fig. 4; Kleinert *et al.,* 1988b, 1990) and A-65317 (Rosenberg *et al.,* 1990b). The solubility of the former is enhanced by a basic nitrogen at the N-terminus whereas the latter employs neutral but polar residues at both termini. As outlined in Tables I and II, both inhibitors are highly potent and possess aqueous solubilities some 1000-fold greater than was observed for compounds **4** and **5b**.

Enalkiren and A-65317 represented a new generation of renin inhibitors. On i.v. administration to anesthetized, salt-depleted monkeys, both compounds elicited hypotensive responses that were greater and of longer duration than we had seen with earlier structures (Kleinert *et al.,* 1990; Rosenberg *et al.,* 1990b). Nanomolar compounds, like enalkiren, showed a brisk onset of action typically reducing blood

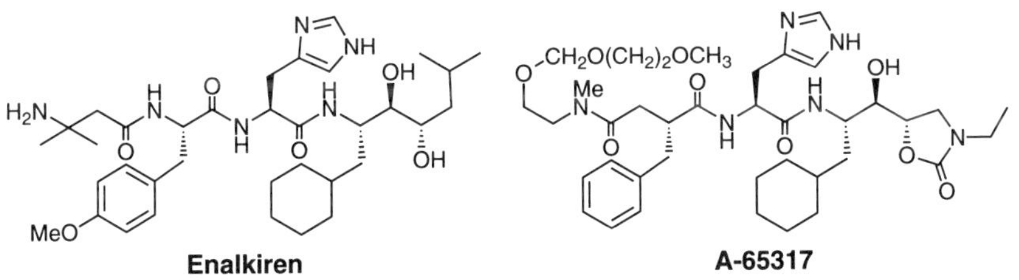

Figure 4. Renin inhibitors possessing both aqueous solubility and stability toward proteolysis.

pressure within 5 min of administration of an i.v. bolus. The nadir of the hypotensive response occurred at approximately 30 min following dosing (Kleinert *et al.*, 1988c). The duration and recovery of the biological activity were dose-related. Additionally, a modest blood pressure response was observed following i.d. administration. This activity was sufficient for enalkiren to be chosen for early clinical experiments.

3.3.3. CLINICAL EXPERIENCE WITH THE SOLUBLE AND STABLE INHIBITOR ENALKIREN

Studies in experimental animals predicted the human i.v. response to renin inhibition. Enalkiren (A-64662) was one of the first and most extensively tested renin inhibitors to be studied in humans. Clinical pharmacology was investigated in normal, healthy volunteers, essential hypertensive patients, and patients with congestive heart failure.

As described above in the animal studies, normal subjects were sensitized to the effects of renin inhibition by pretreatment with the diuretic furosemide and/or a fixed sodium intake diet. The first clinical study with an Abbott renin inhibitor compared vehicle with progressively increasing i.v. doses (0.001–0.1 mg/kg) of enalkiren in eight normal, healthy men on a 100 meq/day sodium diet (Delabays *et al.*, 1989). As noted when enalkiren was given to salt-depleted monkeys, dose-related reductions in PRA and plasma AII were observed in these normals. Peak inhibition of PRA occurred 5 min postdosing and the magnitude and duration of the effect were dose-related. Interestingly, despite the observed biochemical responses to enalkiren, no significant reductions in blood pressure or heart rate were seen in these normal subjects. Nevertheless, this study proved that enalkiren was well tolerated in humans, pharmacologically active, and that inhibition of PRA alone was not adequate to lower blood pressure in a normotensive human. Later, enalkiren successfully lowered blood pressure in hypertensive patients after single i.v. doses with an exaggerated response elicited by pretreatment with a diuretic (Weber *et al.*, 1990), as well as after multiple dosing which led to dose-related antihypertensive activity of surprisingly significant duration (Boger *et al.*, 1990). Further, i.v. enalkiren was safe and effective in improving the hemodynamic profile of patients with congestive heart failure (Neuberg *et al.*, 1991). Unfortunately, enalkiren was established to be only approximately 2% orally bioavailable in humans (Cavanaugh *et al.*, 1989).

Although increased aqueous solubility appeared to confer improved *in vivo* efficacy, results from the i.d. rat model were definitive and discouraging. Whereas portal drug levels for A-65317 were higher than those observed with **4** and **5b,** essentially no drug reached the systemic circulation, indicating almost complete hepatic extraction. Similarly, the plasma drug levels for enalkiren appeared somewhat improved (Luly *et al.*, unpublished results), but subsequent oral dosing experiments demonstrated that enalkiren was less than 2% bioavailable in dogs and monkeys as was confirmed in humans (Kleinert *et al.*, 1990, 1992a). Despite the lack of oral activity, the discovery and characterization of enalkiren and A-65317 was a significant milestone for the project, one that required 3-1/3 years of research and the synthesis and biological evaluation of 1400 inhibitors to achieve.

3.4. Renin Inhibitors with Oral Bioavailability

Clearly neither solubility nor proteolytic stability was sufficient to achieve good bioavailability. Therefore, in an attempt to identify those factors that would affect oral absorption and hepatic extraction, we began a systematic evaluation of the relationship between physicochemical properties, structure, and plasma drug levels. For these studies, the i.d. rat model gained increased importance as a biological screen and it was the high-throughput nature of this model that permitted the ultimate success of this approach.

By varying structural parameters at all readily accessible sites in our renin inhibitors, substitutions for histidine proved to have the most profound effects on absorption and biliary excretion (Tables III and IV; Rosenberg *et al.*, 1993a). Incorporating (thiazol-4-yl)alanine, which is less basic than histidine and lacks a

Table III

C-Terminal Oxazolidinone Renin Inhibitors Containing Heterocycle-Substituted Alanine Residues at the Histidine Site

No.	R	IC_{50} (nM)[b]	log P[c]	Solubility[d]	i.d. rat experiments[a]: Systemic	i.d. rat experiments[a]: Portal
7	Imidazol-4-yl (His)	1.3	2.20	2.8	34 ± 20	390 ± 140
8	Pyrazol-3-yl	21	2.49	2.5	7 ± 2	58 ± 23
9	(1-Methyl)imidazol-4-yl	58	2.16	2.1	43 ± 17	350 ± 230
10	Thiazol-4-yl	8.1	2.83	0.46	300 ± 50	1100 ± 500

[a] 10 mg/kg, 30 min plasma drug level, mean ± SEM, $n = 3$, determined by a renin inhibition assay.
[b] Human plasma renin, pH 7.4.
[c] Octanol/water, pH 7.4.
[d] mg/ml, pH 7.4 phosphate buffer, 37°C.

Table IV
C-Terminal Glycol Renin Inhibitors Containing Heterocycle-Substituted Alanine Residues at the Histidine Site

No.	R	IC_{50} (nM)[b]	log P[c]	Solubility[d]	i.d. rat experiments[a] Systemic	Portal
11	Imidazol-4-yl (His)	1.6	3.51	ND[e]	11 ± 1	170 ± 30
12	Pyrazol-3-yl	3.5	4.01	0.12	170 ± 80	1110 ± 500
13	(1-Methyl)imidazol-4-yl	110	3.71	0.095	320 ± 240	3300 ± 500
14	Thiazol-4-yl	8.3	4.27	0.016	210 ± 20	2500 ± 400

[a] 10 mg/kg, 30 min plasma drug level, mean ± SEM, $n = 3$, determined by a renin inhibition assay.
[b] Human plasma renin, pH 7.4.
[c] Octanol/water, pH 7.4.
[d] mg/ml, pH 7.4 phosphate buffer, 37°C.
[e] Not determined.
Reprinted with permission from Rosenberg *et al.*, 1993a, *J. Med. Chem.* **36:**449–459. Copyright 1993 American Chemical Society.

potential site for conjugation, into this position caused a remarkable enhancement of portal and systemic plasma drug levels for both oxazolidinone and glycol-derived structures (compounds **10** and **14**). Other heterocycle-substituted alanine derivatives improved absorption only for inhibitors incorporating the glycol transition-state mimic. An immediate conclusion from this study was that neither aqueous solubility nor octanol–water partition coefficient data were useful for predicting the pharmacokinetic profile of a given inhibitor. Both parameters were largely controlled by the nature of the C-terminal group and, for a particular transition-state mimic, varied little between the different histidine replacements. Instead, our structural studies led us to the empirical conclusion that optimum structures should contain a single, solubilizing substituent at the C- or N-terminus combined with a lipophilic histidine-site residue. These guidelines allowed us to design subsequent renin inhibitors with a reasonable degree of confidence that they would be well absorbed. It remains unknown whether other physicochemical measurements might have correlated with absorption. von Geldern *et al.* (1996) recently demonstrated a relationship between Δ log *P*, a parameter that we did not determine routinely, and absorption for a series of peptide-derived endothelin antagonists.

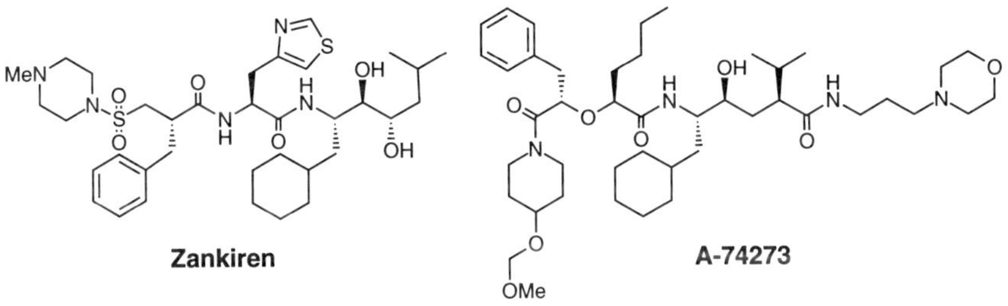

Zankiren A-74273

Figure 5. Orally active renin inhibitors zankiren (A-72517) and A-74273.

3.4.1. NONPEPTIDE RENIN INHIBITORS WITH ORAL ACTIVITY: A-74273

Several considerations went into our final inhibitor design. Inherent aqueous solubility, or the ability to be formulated as a salt, would be necessary to ensure that sufficient dissolution occurred for the compound to be absorbed following oral administration. Also, preclinical pharmacology and clinical data generated from enalkiren suggested that *in vitro* activity should be improved. We pursued multiple chemical series and experienced success on several fronts. Because we were unsure whether we would succeed with dipeptide core renin inhibitors, we continued to optimize a series of nonpeptide renin inhibitors that we had also discovered (Boyd *et al.,* 1992). This process led to the discovery of A-74273 (Fig. 5), which was more potent than enalkiren, possessed good inherent solubility, and incorporated a basic group at the C-terminus for salt formation. As outlined in Table V, this compound was well absorbed in the i.d. rat model despite a molecular weight approaching 800 (mol.wt. – 787).

Table V

Characteristics of the Orally Active Renin Inhibitors Zankiren (A-72517) and A-74273

				i.d. rat experiments[a]	
Compound[b]	IC_{50} (nM)[c]	log P[d]	Solubility[e]	Systemic	Portal
Zankiren	1.1	4.6	0.00081	440 ± 150	ND[f]
A-74273	3.1	4.3	0.21	330 ± 170	2200 ± 300

[a] 10 mg/kg, peak plasma drug level, mean ± SEM, n = 3, determined by a renin inhibition assay.
[b] See Fig. 5 for structures.
[c] Human plasma renin, pH 7.4.
[d] Octanol/water, pH 7.4.
[e] mg/ml, pH 7.4 phosphate buffer, 37°C.
[f] Not determined.

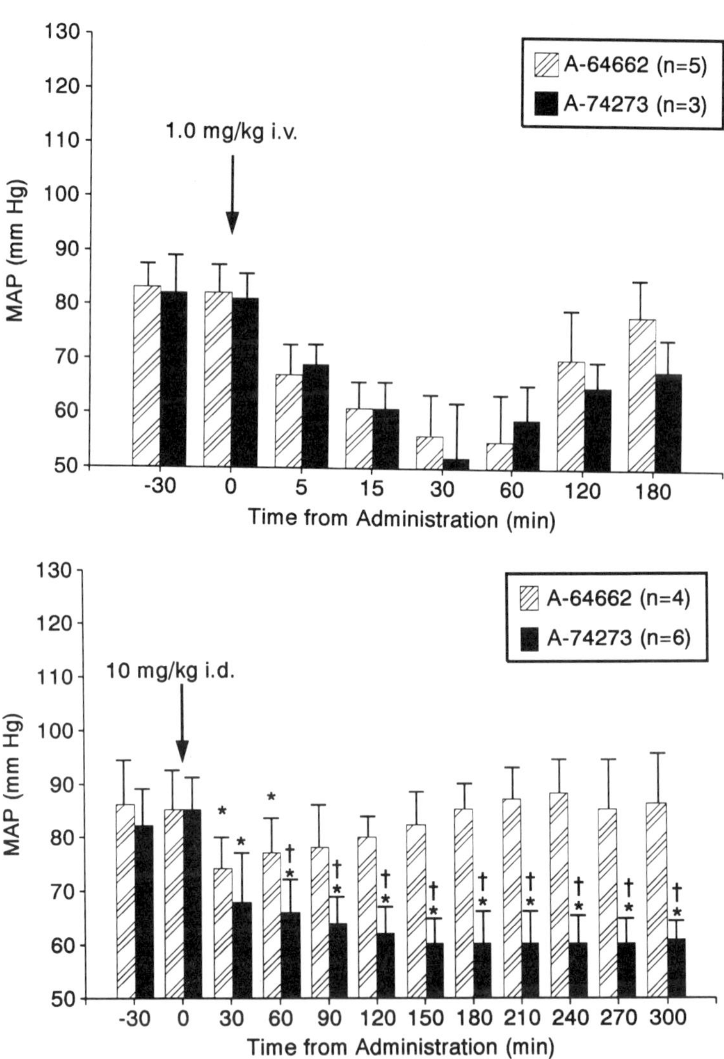

Figure 6. Comparison of mean arterial pressure (MAP) in response to enalkiren (A-64662) and A-74273 administered by intravenous route (top panel) or intraduodenally (bottom panel) to anesthetized, salt-depleted monkeys. Data are depicted as mean ± SEM. $*p < 0.05$ compared with time 0; $\dagger p < 0.05$ compared with enalkiren. Reprinted with permission from Boyd *et al.*, 1992, *J. Med. Chem.* **35:**1735–1746. Copyright 1992 American Chemical Society.

Figure 6 compares enalkiren and the nonpeptide renin inhibitor, A-74273, given intravenously and intraduodenally (Boyd *et al.,* 1992). Both compounds exhibited comparable hemodynamic profiles when injected intravenously in salt-depleted, anesthetized monkeys (top panel). However, the profiles clearly diverged to imply that only A-74273 was absorbed in sufficient quantities to significantly reduce mean arterial pressure in both magnitude and duration (bottom panel). Notice that the only detected i.d. activity of enalkiren occurred rapidly and briefly. Deposition of enalkiren into the duodenum provides a high concentration gradient across the intestinal lumen that drives a small fraction of the dose into the bloodstream. In contrast, A-74273 appeared to be absorbed slowly and consistently over the 3 hr of observation, as the mean arterial pressure was slowly reduced and sustained at a hypotensive nadir. This nonpeptide renin inhibitor was also given at 10 mg/kg to conscious, salt-depleted dogs either intravenously or orally and the hypotensive responses were recorded (Fig. 7; Kleinert *et al.,* 1992a). Statistically significant reductions in blood pressure were observed in all treated animals receiving drug by either route of delivery.

The classic definition of bioavailability is the dose-normalized ratio of the integrated plasma drug level–time curves from an agent given by both the i.v. and

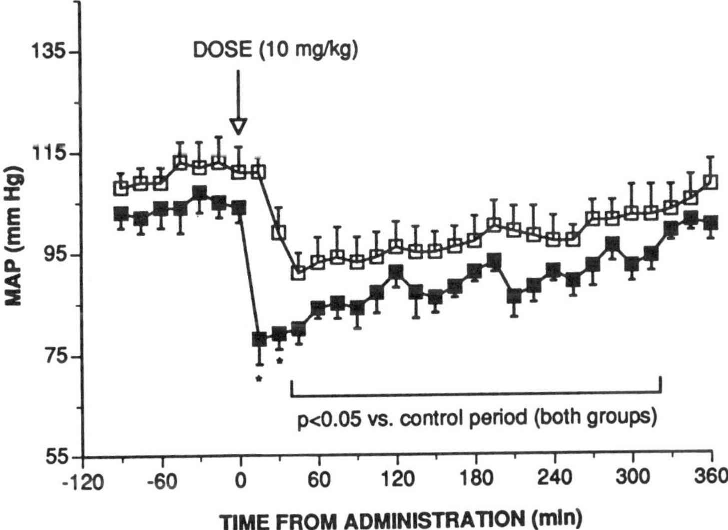

Figure 7. Mean arterial pressure (MAP) responses in conscious, salt-depleted dogs to A-74273 given at 10 mg/kg either intravenously (■, $n = 6$) or orally (□, $n = 8$). $p < 0.05$ values were significantly different from predosing control values. $^*p < 0.05$ versus predosing control values for intravenous group. Reproduced with permission from Kleinert *et al.*, 1992a, *Hypertension* **20**:768–775. Copyright 1992 American Heart Association.

oral (or i.d.) routes of administration. These calculations confirmed the superior absorption of A-74273. Bioavailability in the monkey (i.d. administration) and dog (oral administration) were determined to be 16 ± 4 and 54 ± 13%, respectively (Boyd *et al.,* 1992; Kleinert *et al.,* 1992a). The achievement of high oral bioavailability was remarkable for these complex structures. However, the synthesis of nonpeptides, such as A-74273, was complicated and would not be expected to be cost-effective on the manufacturing scale. Therefore, A-74273 did not become a clinical drug development candidate.

It is often tempting to assess bioavailability by comparing pharmacological activity in response to oral and i.v. routes of administration. This method can only approach accuracy when highly bioavailable compounds are tested. Notice in Fig. 7 that the line graph for the oral route is, for all intents and purposes, superimposable on the i.v. route blood pressure response. Because i.v. administration is considered to represent 100% bioavailability and the area under the curve for both responses are comparable, one could conclude that A-74273 is highly orally bioavailable as was indeed confirmed by direct plasma drug level analysis. However, a note of caution is warranted when trying to estimate oral bioavailability by biological activity of a highly potent compound, for only a small amount of absorbed drug may be required to show an exaggerated acute response, leading the observer to believe that the compound was well absorbed. The renin inhibitor ditekiren provides an example of this potential pitfall. By comparing the hypotensive responses following oral and i.v. administration to hog renin-infused ganglion-blocked rats, it was concluded that the bioavailability of this compound was greater than 10% (Pals *et al.,* 1986). Subsequently, definitive experiments in the rat showed that bioavailability was in fact only 1.3% (Rush *et al.,* 1991).

3.4.2. PEPTIDE-DERIVED RENIN INHIBITORS WITH ORAL ACTIVITY: ZANKIREN

We simultaneously pursued dipeptide core renin inhibitors, a series with which we were gaining enormous experience and expertise. In the dipeptide core series, we hoped to incorporate the C-terminal glycol as it tended to confer greater potency than other transition-state mimics and also because it was the most easily synthesized. This effort culminated with the discovery of zankiren (A-72517, Fig. 5). Each of the issues identified as necessary for oral activity was addressed in the design of zankiren: The sulfonamide linkage enhances potency, the *N*-methyl piperazine provides a site for salt formation, and the remaining structural features maintain good absorption (Rosenberg *et al.,* 1993b). Zankiren was well absorbed in the i.d. rat model (Table V), and despite the extremely low solubility of the free base, preclinical formulation studies revealed that the HCl salt was sufficiently soluble (10 mg/ml) for oral dosing. Most crucial to the question of oral bioavailability is consistency, reproducibility, and low variability within and be-

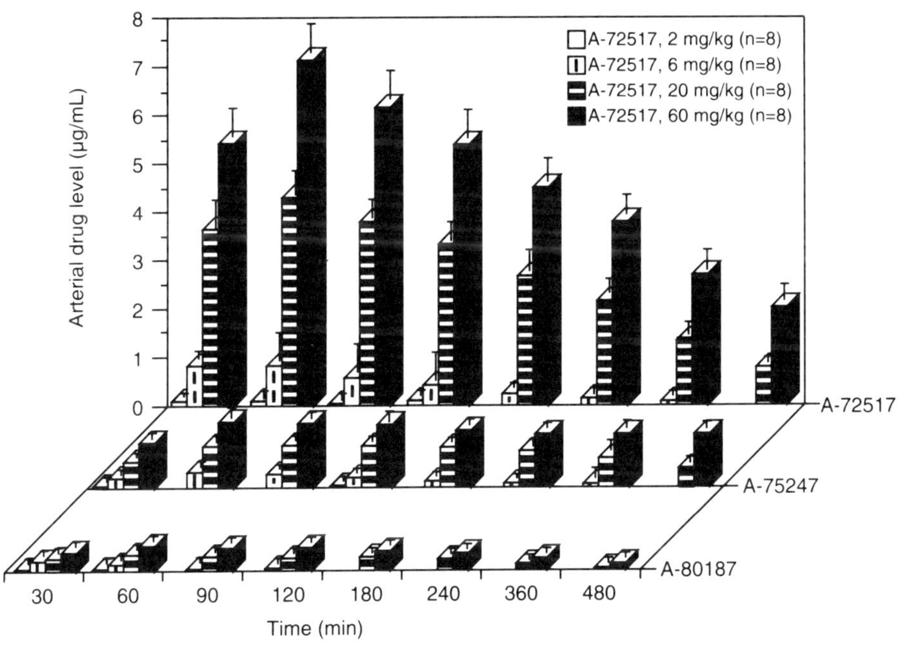

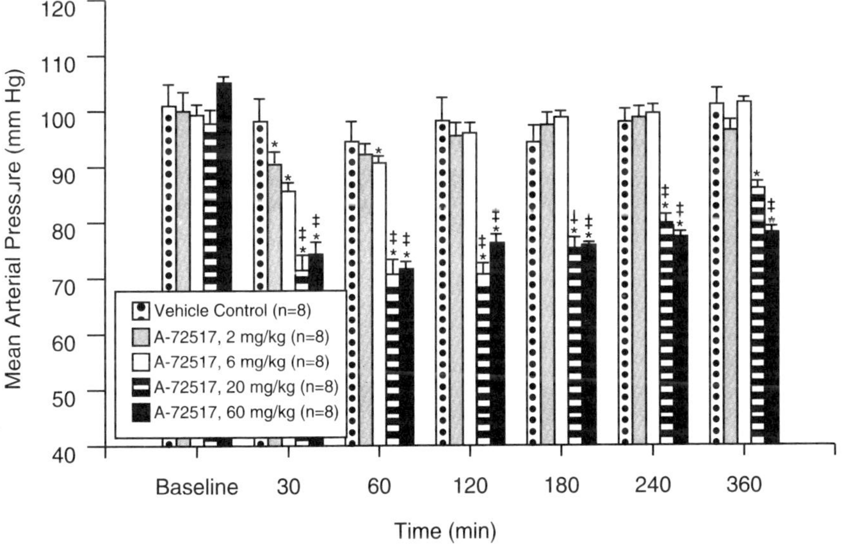

Figure 8. Mean arterial pressure (MAP) responses to zankiren (A-72517) given orally to conscious, salt-depleted dogs (top panel) with corresponding blood levels of parent drug (A-72517) and two metabolites (A-75247 and A-80187) measured by HPLC (bottom panel). Values are shown as mean ± SEM. Asterisk indicates a significant difference from baseline values. Dagger indicates a significant difference compared with the vehicle control group. Statistical significance accepted at $p < 0.05$. Reprinted from Kleinert *et al.*, 1992b, *Science* **257:**1940–1943. Copyright 1992 American Association for the Advancement of Science.

tween subjects. A general rule of thumb is that 10% or greater oral bioavailability will result in an acceptable to ideal variability around the mean. The higher the oral bioavailability is, the lower the variability. Oral bioavailability in the conscious monkey, dog, rat, and ferret was found to be 8.1 ± 1.0, 53 ± 8, 24 ± 9, and 32 ± 14%, respectively (Kleinert *et al.,* 1992b). Intraduodenal dosing experiments confirmed that the lower bioavailability in monkeys was the result of enhanced hepatic extraction in that species. Although zankiren was 100 times less potent against dog renin than against human plasma renin, the combination of pretreatment with salt depletion, the high bioavailability in the dog, and relatively high doses allowed for the conscious, orally dosed dog to show dose-related reductions in blood pressure. Figure 8 shows these hypotensive responses and corresponding blood levels as determined by HPLC of parent drug and two metabolites, A-75247 (desmethyl zankiren) and A-80187 (zankiren *N*-oxide), that are relatively inactive against dog renin, but are potent against human renin (Kleinert *et al.,* 1992b).

Zankiren was the first peptide-based renin inhibitor to demonstrate meaningful oral absorption in any species and the only renin inhibitor to give high circulating plasma drug levels after oral administration to human subjects. In a study in hypertensive patients, zankiren was safe, absorbed, and antihypertensive when given in tablet form (Boger *et al.,* 1993). Following the discovery of its predecessor enalkiren, the path to zankiren required two additional years of research and the synthesis and biological evaluation of over 1000 compounds, fully one quarter of which were tested for absorption in the i.d. rat as the primary, quick screening model. Which animal model best predicted oral bioavailability in humans? There is no animal model that consistently behaves like the human for all chemical agents. This is true for the monkey as well. There are examples where dogs, rats, or monkeys are predictive of human oral bioavailability for particular molecules. For the renin inhibitor zankiren, the dog, not the monkey, best predicted human bioavailability. Although zankiren was shown to be safe, effective, and well absorbed in humans, clinical development was not pursued beyond phase II clinical trials because of internal priority changes within the Pharmaceutical Development organization.

4. CONCLUSIONS

It is hoped that this chapter illustrates the decision-making processes that are associated with a drug discovery effort. The success of the renin inhibitor project was the direct result of the tight integration of the medicinal chemistry, biochemistry, pharmacology, and pharmacokinetic groups. As obstacles arose, a variety of approaches, often simultaneous, were taken to overcome them. Ultimately, although there were logical reasons for the choice of each of the compounds that the project synthesized, most of them were found not to lie on the critical path and were illustrative of the numerous dead ends that are to be expected in any scientific endeavor. Only the productivity of the medicinal chemists coupled with efficiency of our primary biological tests allowed sufficient avenues to be explored that we were able to achieve the project goals. Finally, our willingness to study enalkiren in the clinic early in our program provided valuable feedback and guidance for the discovery of the next generation of renin inhibitor.

REFERENCES

Adedoyin, A., Perry, P. R., and Wilkinson, G. R., 1993, Hepatic elimination in the rat of ditekiren (U-71038), a renin inhibitor pseudohexapeptide, *Drug Metab. Dispos.* **21:**184–188.

Atkinson, A. B., and Robertson, J. I. S., 1979, Captopril in the treatment of clinical hypertension and cardiac failure, *Lancet* **2:**836–839.

Boger, J., Lohr, N. S., Ulm, E. H., Poe, M., Blaine, E. H., Fanelli, G. M., Lin, T.-Y., Payne, L. S., Schorn, T. W., LaMont, B. I., Vassil, T. C., Stabilito, I. I., Veber, D. F., Rich, D. H., and Bopari, A. S., 1983, Novel renin inhibitors containing the amino acid statine, *Nature* **303:**81–84.

Boger, J., Payne, L. S., Perlow, D. S., Lohr, N. S., Poe, M., Blaine, E. H., Ulm, E. H., Schorn, T. W., LaMont, B. I., Lin, T.-Y., Kawai, M., Rich, D. H., and Veber, D. F., 1985, Renin inhibitors. Synthesis of subnanomolar, competitive transition-state analogue inhibitors containing a novel analogue of statine, *J. Med. Chem.* **28:**1779–1790.

Boger, R. S., Glassman, H. N., Cavanaugh, J. H., Schmitz, P. J., Lamm, J., Moyse, D., Cohen, A., Kleinert, H. D., and Luther, R. R., 1990, Prolonged duration of blood pressure response to enalkiren, the novel dipeptide renin inhibitor in essential hypertension, *Hypertension* **15:**835–840.

Boger, R. S., Glassman, H. N., Thys, R., Gupta, S. K., Hippensteel, R. L., and Kleinert, H. D., 1993, Absorption and blood pressure response to the new orally active renin inhibitor, A-72517, in hypertensive patients, *Am. J. Hypertens.* **6:**103A.

Bolis, G., Fung, A. K. L., Greer, J., Kleinert, H. D., Marcotte, P. A., Perun, T. J., Plattner, J. J., and Stein, H. H., 1987, Renin inhibitors. Dipeptide analogues of angiotensinogen incorporating transition-state, nonpeptidic replacements at the scissile bond, *J. Med. Chem.* **30:**1729–1737.

Bott, R., Subramanian, E., and Davies, D. R., 1982, Three-dimensional structures of the complex of the *Rhizopus chinensis* carboxyl proteinase and pepstatin at the 2.5 Å resolution, *Biochemistry* **21:**6956–6962.

Boyd, S. A., Fung, A. K. L., Baker, W. R., Mantei, R. A., Armiger, Y.-L., Stein, H. H., Cohen, J., Egan, D. A., Barlow, J. L., Klinghofer, V., Verburg, K. M., Martin, D. L., Young, G. A., Polakowski, J. S., Hoffman, D. J., Garren, K. W., Perun, T. J., and Kleinert, H. D., 1992, C-terminal modifications of nonpeptide renin inhibitors: Improved oral bioavailability via modification of physicochemical properties, *J. Med. Chem.* **35:**1735–1746.

Cavanaugh, J., Lamm, J., Moyse, D., Hoyos, P., Glassman, H., Dube, L., Boger, R., and Luther, R., 1989, Safety and pharmacologic profile following oral administration of the novel dipeptide renin inhibitor, A-64662, *J. Clin. Pharmacol.* **29:**861.

Delabays, A., Nussberger, J., Porchet, M., Waeber, B., Danekas, L., Boger, R., Glassman, H., Kleinert, H., Luther, R., and Brunner, H. R., 1989, Hemodynamics and humoral effects of a new renin inhibitor enalkiren in normal humans, *Hypertension* **13:**941–947.

Erdös, E. G., and Skidgel, R. A., 1986, The unusual substrate specificity and the distribution of human angiotensin I converting enzyme, *Hypertension* **8**(Suppl. I)**:**I-34–I-37.

Fox, A. J., Lalloo, U. G., Belvisi, M. G., Bernareggi, M., Chung, K. F., and Barnes, P. J., 1996, Bradykinin-evoked sensitization of airway sensory nerves: A mechanism for ACE-inhibitor induced cough, *Nature Med.* **2:**814–817.

Goa, K. L., and Wagstaff, A. J., 1996, Losartan potassium: A review of its pharmacology, clinical efficacy and tolerability in the management of hypertension, *Drugs* **51:**820–845.

Greenlee, W. J., 1990, Renin inhibitors, *Med. Res. Rev.* **10:**173–236.

Israili, Z. H., and Hall, W. D., 1992, Cough and angioneurotic edema associated with angiotensin-converting enzyme inhibitor therapy, *Ann. Intern. Med.* **117:**234–242.

James, M. N. G., Sielecki, A., Salituro, F., Rich, D. H., and Hofmann, T., 1982, Conformational flexibility in the active site of aspartyl proteinases revealed by a pepstatin fragment binding to penicillopepsin, *Proc. Natl. Acad. Sci. USA* **79:**6137–6141.

Karlberg, B. E., 1993, Cough and inhibition of the renin–angiotensin system, *J. Hypertens.* **11**(Suppl. 3):S49–S52.

Klassen, C. D., and Watkins, J. B., III, 1984, Mechanisms of bile formation, hepatic uptake, and biliary excretion, *Pharmacol. Rev.* **36:**1–67.

Kleinert, H. D., Martin, D., Chekal, M., Young, G., Rosenberg, S., Plattner, J. J., and Perun, T. J., 1988a, Cardiovascular actions of the primate-selective renin inhibitor A-62198, *J. Pharmacol. Exp. Ther.* **246:**975–979.

Kleinert, H. D., Luly, J. R., Marcotte, P. A., Perun, T. J., Plattner, J. J., and Stein, H., 1988b, Improvements in the stability and biological activity of small peptides containing novel Leu-Val replacements, *FEBS Lett.* **230:**38–42.

Kleinert, H. D., Martin, D., Chekal, M., Kadam, J., Luly, J. R., Plattner, J. J., Perun, T. J., and Luther, R. R., 1988c, Effects of renin inhibitor A-64662 in monkeys and rats with varying baseline plasma renin activity, *Hypertension* **11:**613–619.

Kleinert, H. D., Luly, J. R., Bopp, B. A., Verburg, K. M., Hoyos, P. A., Karol, M. D., Plattner, J. J., Luther, R. R., and Stein, H. H., 1990, Profile of the renin inhibitor, enalkiren (Abbott-64662), *Cardiovasc. Drug Rev.* **8:**203–219.

Kleinert, H. D., Stein, H. H., Boyd, S., Fung, A. K. L., Baker, W. R., Verburg, K. M., Polakowski, J. S., Kovar, P., Barlow, J., Cohen, J., Klinghofer, V., Mantei, R., Cepa, S., Rosenberg, S., and Denissen, J. F., 1992a, Discovery of a well-absorbed, efficacious renin inhibitor, A-74273, *Hypertension* **20:**768–775.

Kleinert, H. D., Rosenberg, S. H., Baker, W. R., Stein, H. H., Klinghofer, V., Barlow, J., Spina, K., Polakowski, J., Kovar, P., Cohen, J., and Denissen, J., 1992b, Discovery of a peptide-based renin inhibitor with oral bioavailability and efficacy, *Science* **257:**1940–1943.

Luly, J. R., BaMaung, N., Soderquist, J., Fung, A. K. L., Stein, H., Kleinert, H. D., Marcotte, P. A., Egan, D. A., Bopp, B., Merits, I., Bolis, G., Greer, J., Perun, T. J., and Plattner, J. J., 1988, Renin inhibitors. Dipeptide analogues of angiotensinogen utilizing a dihydroxyethylene transition-state mimic at the scissile bond to impart greater inhibitory potency, *J. Med. Chem.* **31:**2264–2276.

Neuberg, G. W., Kukin, M. L., Penn, J., Medina, N., Yushak, M., and Packer, M., 1991, Hemodynamic effects of renin inhibition by enalkiren in chronic congestive heart failure, *Am. J. Cardiol.* **67:**63–66.

Overlack, A., 1996, ACE inhibitor-induced cough and bronchospasm: Incidence, mechanisms and management, *Drug Safety* **15:**72–78.

Pals, D. T., Thaisrivongs, S., Lawson, J. A., Kati, W. M., Turner, S. R., DeGraaf, G. L., Harris, D. W., and Johnson, G. A., 1986, An orally active inhibitor of renin, *Hypertension* **8:**1105–1112.

Pals, D. T., Lawson, J. A., and Couch, S. J., 1990, Rat model for evaluating inhibitors of human renin, *J. Pharmacol. Methods* **23:**239–245.

Peach, M. J., 1977, Renin–angiotensin system: Biochemistry and mechanisms of action, *Physiol. Rev.* **57:**313–370.

Plattner, J. J., and Norbeck, D. W., 1990, Obstacles to drug development from peptide leads, in: *Drug Discovery Technologies* (C. R. Clark and W. H. Moos, eds.), pp. 92–126, Ellis Horwood, Chichester.

Plattner, J. J., Greer, J., Fung, A. K. L., Stein, H., Kleinert, H. D., Sham, H. L., Smital, J. R., and Perun, T. J., 1986, Peptide analogues of angiotensinogen. Effect of peptide chain length on renin inhibition, *Biochem. Biophys. Res. Commun.* **139:**982–990.

Plattner, J. J., Marcotte, P. A., Kleinert, H. D., Stein, H. H., Greer, J., Bolis, G., Fung, A. K. L., Bopp, B. A., Luly, J. R., Sham, H. L., Kempf, D. J., Rosenberg, S. H., Dellaria, J. F., De, B., Merits, I., and Perun, T. J., 1988, Renin inhibitors. Dipeptide analogues of angiotensinogen utilizing a structurally modified phenylalanine residue to impart proteolytic stability, *J. Med. Chem.* **31:**2277–2288.

Rosenberg, S. H., 1995, Renin inhibitors, in: *Progress in Medicinal Chemistry,* Volume 32 (G. P. Ellis and D. K. Luscombe, eds.), pp. 37–114, Elsevier, Amsterdam.

Rosenberg, S. H., Plattner, J. J., Woods, K. W., Stein, H. H., Marcotte, P. A., Cohen, J., and Perun, T. J., 1987, Novel renin inhibitors containing analogues of statine retro-inverted at the C-termini: Specificity at the P_2 histidine site, *J. Med. Chem.* **30:**1224–1228.

Rosenberg, S. H., Woods, K. W., Kleinert, H. D., Stein, H., Nellans, H. N., Hoffman, D. J., Spanton, S. G., Pyter, R. A., Cohen, J., Egan, D. A., Plattner, J. J., and Perun, T. J., 1989, Azido-glycols: Potent, low molecular weight renin inhibitors containing an unusual post scissile site residue, *J. Med. Chem.* **32:**1371–1378.

Rosenberg, S. H., Dellaria, J. F., Kempf, D. J., Hutchins, C. W., Woods, K. W., Maki, R. G., de Lara, E., Spina, K. P., Stein, H. H., Cohen, J., Baker, W. R., Plattner, J. J., Kleinert, H. D., and Perun, T. J., 1990a, Potent, low molecular weight renin inhibitors containing a C-terminal heterocycle: Hydrogen bonding at the active site, *J. Med. Chem.* **33:**1582–1590.

Rosenberg, S. H., Woods, K. W., Sham, H. L., Kleinert, H. D., Martin, D. L., Stein, H., Cohen, J., Egan, D. A., Bopp, B., Merits, I., Garren, K. W., Hoffman, D. J., and Plattner, J. J., 1990b, Water soluble renin inhibitors: Design of a subnanomolar inhibitor with a prolonged duration of action, *J. Med. Chem.* **33:**1962–1969.

Rosenberg, S. H., Spina, K. P., Woods, K. W., Polakowski, J., Martin, D. L., Yao, Z., Stein, H. H., Cohen, J., Barlow, J. L., Egan, D. A., Tricarico, K. A., Baker, W. R., and Kleinert, H. D., 1993a, Studies directed towards the design of orally active renin inhibitors 1. Some factors influencing the absorption of small peptides, *J. Med. Chem.* **36:**449–459.

Rosenberg, S. H., Spina, K. P., Condon, S., L., Polakowski, J., Yao, Z., Kovar, P., Stein, H. H., Cohen, J., Barlow, J. L., Klinghofer, V., Egan, D. A., Tricarico, K. A., Perun, T. J., Baker, W. R., and Kleinert, H. D., 1993b, Studies directed towards the design of orally active renin inhibitors 2. Development of the efficacious, bioavailable renin inhibitor (2*S*)-2-benzyl-3-(1-methylpiperidin-4-ylsulfonyl)propionyl-3-(thiazol-4-yl)-L-alanine amide of (2*S*,3*R*,4*S*)-2-amino-1-cyclohexyl-3,4-dihy droxy-6-methylheptane (A-72517), *J. Med Chem.* **36:**460–467.

Rush, B. D., Wilkinson, K. F., Zhong, W. Z., Closson, S. K., Lakings, D. B., and Ruwart, M. J., 1991, Absolute oral bioavailability of ditekiren, a renin inhibitor peptide, in conscious rats, *Int. J. Pharm.* **73:**231–237.

Steinberg, M. I., Wiest, S. A., and Palkowitz, A. D., 1993, Nonpeptide angiotensin II receptor antagonists, *Cardiovasc. Drug Rev.* **11:**312–358.

Sunman, W., and Sever, P. S., 1993, Non-angiotensin effects of angiotensin-converting enzyme inhibitors, *Clin. Sci.* **85:**661–670.

Szelke, M., Jones, D. M., Atrash, B., Hallett, A., and Leckie, B. J., 1983, Novel transition-state analogue inhibitors of renin, in: *Peptides, Structure and Function. Proceedings of the Eighth American Peptide Symposium* (V. J. Hruby and D. H. Rich, eds.), pp. 579–582, Pierce Chemical Co., Rockford, IL.

Tewksbury, D. A., Dart, R. A., and Travis, J., 1981, The amino terminal amino acid sequence of human angiotensinogen, *Biochem. Biophys. Res. Commun.* **99:**1311–1315.

von Geldern, T. W., Hoffman, D. J., Kester, J. A., Nellans, H. N., Dayton, B. D., Calzadilla, S. V., Marsh, K. C., Hernandez, L., Chiou, W., Dixon, D. B., Wu-Wong, J. R., and Opgenorth, T. J., 1996, Azole endothelin antagonists. 3. Using Δ log P as a tool to improve absorption, *J. Med. Chem.* **39:**982–991.

Weber, M. A., Neutel, J. M., Essinger, I., Glassman, H. N., Boger, R. S., and Luther, R. R., 1990, Assessment of renin dependency of hypertension with a dipeptide renin inhibitor, *Circulation* **81:**1768–1774.

Wood, J. M., Gulati, N., Forgiarini, P., Fuhrer, W., and Hofbauer, K. G., 1985, Effects of a specific and long-acting renin inhibitor in the marmoset, *Hypertension* **7:**797–803.
Wood, J. M., Cumin, F., and Maibaum, J., 1994, Pharmacology of renin inhibitors and their application to the treatment of hypertension, *Pharmacol. Ther.* **61:**325–344.
Workman, R. J., and Burkitt, D. W., 1979, Pepsin inhibition by a high specific activity radioiodinated derivative of pepstatin, *Arch. Biochem. Biophys.* **194:**157–164.

Chapter 3

The Discovery and Development of Angiotensin II Antagonists

David J. Carini, David D. Christ, John V. Duncia, and Michael E. Pierce

1. INTRODUCTION

In 1982, work was begun at DuPont on a series of nonpeptide angiotensin II antagonists. These compounds lower blood pressure in animals by blocking the renin–angiotensin system at the level of the angiotensin II (Ang II) receptor. This work eventually led to the discovery of losartan (Fig. 1) in 1986 (Carini *et al.*, 1991; Duncia *et al.*, 1992). In 1990, DuPont entered into a joint agreement with Merck for the development of the angiotensin II antagonists. This collaboration significantly accelerated the advancement of losartan, which was first marketed in Europe in 1994 and in the United States in early 1995. The introduction of losartan represented the first antihypertensive drug with a novel mechanism of action to become available in over a decade. An indirect consequence of the codevelopment of the Ang II antagonists by DuPont and Merck was the creation in 1991 of the DuPont Merck Pharmaceutical Co. as a joint venture between the two parent companies.

Losartan, the primary candidate in this program, made it through development without any problems arising that would have forced the dropping of this compound. However, a variety of significant issues were encountered during the preclinical and early clinical development of losartan that required the efforts of both Discovery and Development to address. Some of these issues were problems

---

David J. Carini, David D. Christ, John V. Duncia, and Michael E. Pierce • DuPont Merck Pharmaceutical Company, Experimental Station, Wilmington, Delaware 19880-0500.

Integration of Pharmaceutical Discovery and Development: Case Studies, edited by Borchardt *et al.*, Plenum Press, New York, 1998.

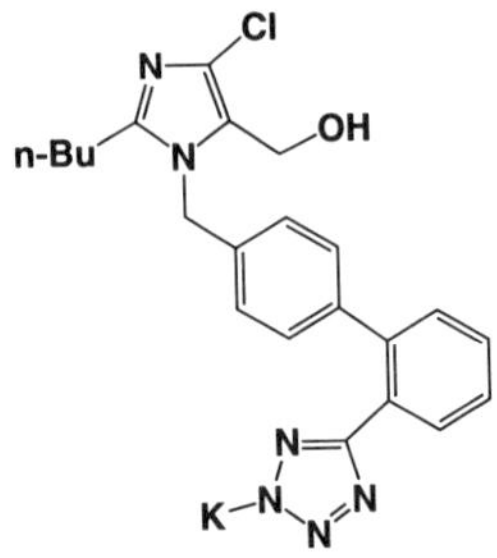

Figure 1. Losartan.

that needed to be solved if losartan was to advance. Others were anticipated as potential problems that might arise during losartan's development. Among these issues were the following:

- The successful development of an acidic tetrazole derivative in humans was unprecedented prior to losartan. The potential metabolism of the tetrazole ring was therefore uncertain. Synthetically, the preparation of large quantities of the 2-(tetrazol-5-yl)biphenyl side chain of losartan was considered a significant challenge. Finally, the chemical stability of the tetrazole ring was uncertain.
- In some animal species, losartan forms a major, active metabolite that contributes to the antihypertensive activity and duration of losartan. The ability of humans to produce this metabolite was a concern. The question of whether the metabolite might be developed was considered.
- As losartan was the first of a new class of compounds, it was considered very important to establish as early as possible that an Ang II receptor antagonist would be effective at blocking the hypertensive properties of Ang II in humans.
- The existence of multiple subtypes of the angiotensin II receptor had been established, and it was found that losartan is a selective antagonist of one of these receptor subtypes. The possible clinical advantages and disadvantages of a selective Ang II antagonist was an open question.

In this chapter we will discuss how these concerns were answered by the combined efforts of the development and discovery groups at DuPont, Merck, and DuPont Merck.

2. DEVELOPMENT OF A TETRAZOLE DERIVATIVE

2.1. Chemical Stability and Potential Toxicity of Tetrazoles

At the time we discovered losartan, there were no drugs on the market or in development that contained an acidic tetrazole residue. Thus, there were few or no toxicity data available on such compounds, especially in humans. Of concern was the possibility that the tetrazole ring might decompose and that the by-products might be toxic. The literature on tetrazoles indicated that 5-phenyltetrazole, low-molecular-weight tetrazoles, and a number of metal derivatives were explosive above their melting points (Benson, 1947). Decomposition products include hydrazoic acid, ammonia, nitrogen, reactive nitrenes, as well as NO_x. However, 5-substituted tetrazoles are very stable to base, forming salts, and are moderately stable to acids, as well as to oxidizing and reducing agents (Benson, 1967). Two possible mechanisms for the decomposition of tetrazoles are: (1) a spontaneous retro-3+2 cycloaddition mechanism or (2) tautomerization to the iminoyl azide followed by elimination of hydrazoic acid or its salt (Fig. 2). The latter mechanism is not unreasonable considering that a popular tetrazole synthesis involves the formation of an iminoyl azide using sodium azide or hydrazoic acid followed by the rapid tautomerization to the tetrazole (Fig. 3) (Duncia *et al.*, 1991), and these two steps might be reversible. In addition, some iminoyl azides are known to fail to tautomerize fully to their respective tetrazoles (Butler, 1977a).

Another potential liability of tetrazoles is their photolytic susceptibility. Tetrazolide anions, for example, liberate 2 moles of nitrogen and a carbene which undergoes insertion and addition reactions (Butler, 1977b). However, no nitrile products or N_2-elimination products were detected either during storage of losartan or in its metabolism products. The tetrazole turned out to be a very stable entity. Losartan's successful development thus represents the incorporation of the

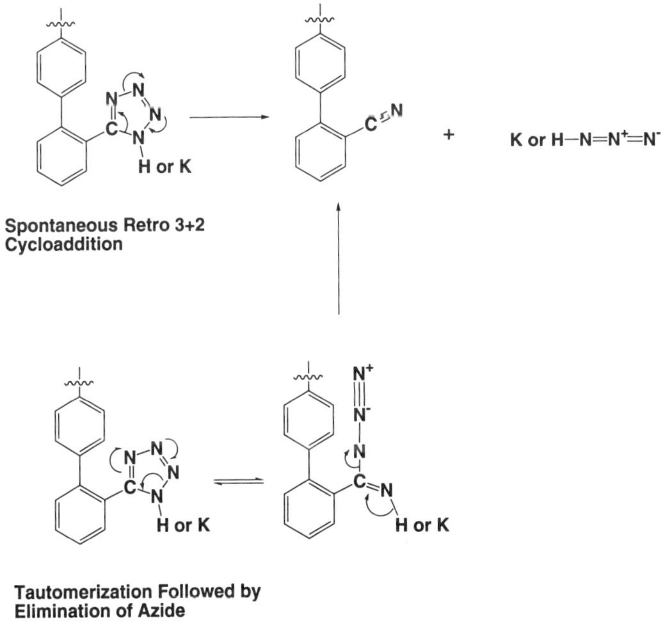

Figure 2. Two possible mechanisms for spontaneous tetrazole decomposition.

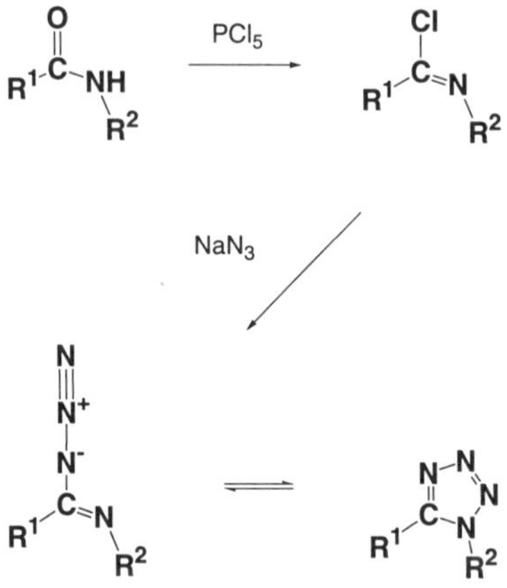

Figure 3. Synthesis of tetrazoles via an iminoyl azide intermediate.

tetrazole group into the medicinal chemical arsenal of stable and nontoxic functional groups.

2.2. Metabolism of Tetrazoles

At the time of losartan's discovery, little was known in the literature about the metabolism of tetrazoles. Once losartan went into development, it was found that the tetrazole moiety of Ang II antagonists becomes glucuronidated as shown in Fig. 4 (Stearns *et al.*, 1992; Colletti and Krieter, 1994). This results in a shorter duration of action in rhesus monkeys and dogs after i.v. administration. It was well known from the structure–activity relationships, developed during the discovery

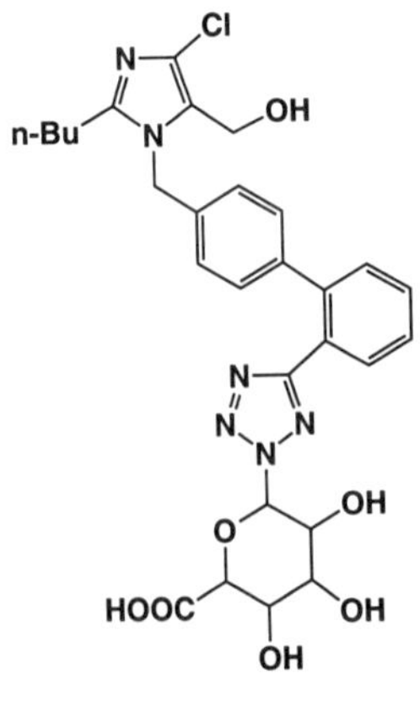

Figure 4. Glucuronidated metabolite of losartan.

of losartan, that the removal of the acidic group from the biphenyl reduces the binding affinity of the Ang II antagonists by about two or three orders of magnitude (Duncia *et al.*, 1992). The glucuronidated metabolites have their acidic tetrazole group masked by a sugar moiety making it no longer acidic (the glucuronic acid metabolite contains a carboxylic acid residue, but apparently it is located in the wrong place and thus the metabolite binds poorly to the angiotensin II receptor). Thus, there was concern that rapid glucuronidation might cause losartan to have a short half-life in humans. As it turned out, losartan can be dosed once a day and glucuronidation is not a practical problem.

2.3. The Search for Tetrazole Replacements

Because little was known about the stability, toxicity, and metabolism of tetrazoles, other acidic isosteres were investigated for use in potential backup development candidates. In addition, difficulties were initially encountered in scaling up the synthesis of losartan using the original laboratory synthesis. All of these reasons propelled the search for other acidic isosteres that could effectively replace the tetrazole ring and might be easier to synthesize on a large scale. Table I summarizes all of the isosteres investigated at DuPont and at Merck, as well as isosteres employed in Ang II antagonists from other companies.

The 1,2,3-triazole **2** appears to mimic the tetrazole **1,** but it was inactive. It was hypothesized that there is a positive charge in the receptor site that binds to the negatively charged tetrazole group (Duncia *et al.*, 1990). It has been shown recently through site-directed mutagenesis experiments that the charged site is a Lys^{199} residue acting in concert with a His^{256} residue in the Ang II receptor (Noda *et al.*, 1995). Therefore, compounds employing acidic isosteres that are ionized appreciably at physiological pH should have higher affinity for the receptor. Increasing the acidity of the triazole rings with electron-withdrawing groups, such as in compounds **3** to **6,** improved the binding somewhat. The steric hindrance caused by the protruding electron-withdrawing groups could be lowering the affinity, although **6** is probably just not acidic enough. The sulfonic acid **7** is very potent, but the trifluoroacetamide **8** is not acidic enough and therefore it binds poorly. The trifluoromethanesulfonamide group of **9,** although acidic, does not impart good binding affinity in the biphenyl series, but it does so for nonbiphenyls such as Glaxo's GR138950 and GR159763 (Middlemiss and Watson, 1994). Amide **10** is not acidic and therefore binds poorly to the Ang II receptor. Substituted amides, such as hydroxamic acids **11** to **13,** are also not acidic enough and thus do not bind well. Sulfonated carboxamide **14** and hydrazide **15** are most likely acidic enough, but must fail to meet some other criteria.

A series of very effective isosteres were discovered initially at Merck and at Hoechst and were later employed at DuPont Merck. These isosteres are represented by the acylsulfonamide **17,** acyl sulfamide **18,** sulfonylcarbamate **20,** and

Table I
Tetrazole Isosteres Found among Angiotensin II Antagonists

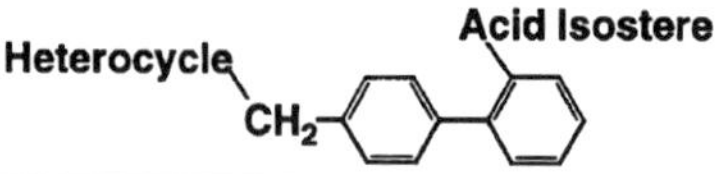

Heterocycle—CH_2—biphenyl—Acid Isostere

No	Carboxylic acid isostere	Heterocycle	AT_1 (nM)[a]	Reference
1	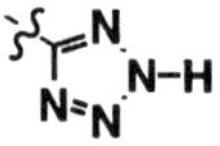 N, N–H, N=N	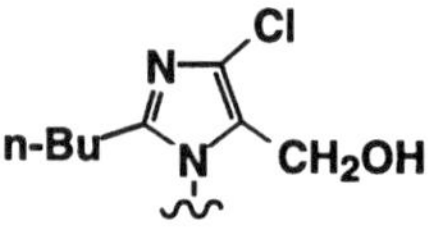 Cl, N, n-Bu, N, CH_2OH	19	Carini *et al.* (1991)
2	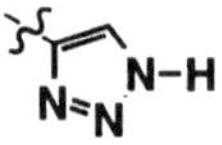 N–H, N=N	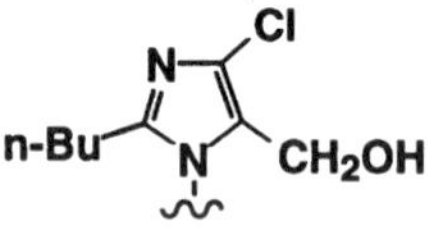 Cl, N, n-Bu, N, CH_2OH	9,600	Duncia *et al.* (1992)
3	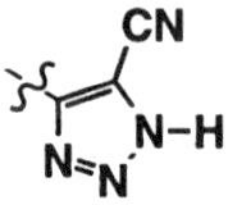 CN, N–H, N=N	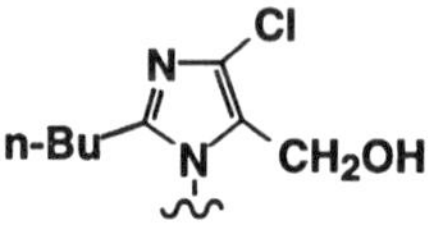 Cl, N, n-Bu, N, CH_2OH	280	Carini *et al.* (1991)
4	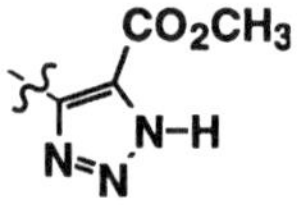 CO_2CH_3, N–H, N=N	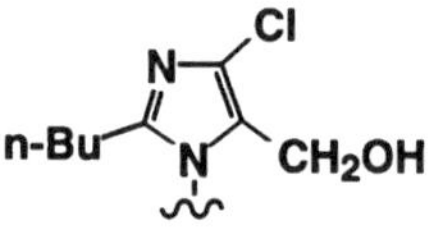 Cl, N, n-Bu, N, CH_2OH	260	Carini *et al.* (1991)
5	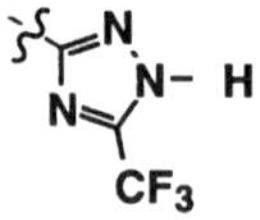 N, N– H, N, CF_3	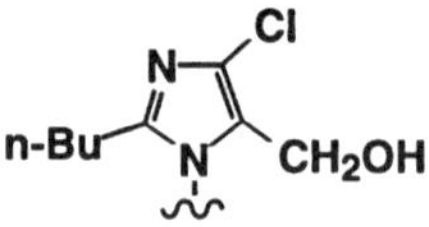 Cl, N, n-Bu, N, CH_2OH	370	Carini *et al.* (1991)
6	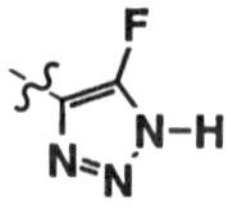 F, N–H, N=N	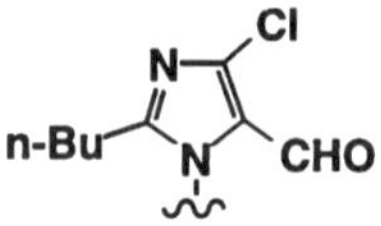 Cl, N, n-Bu, N, CHO	1,300	Carini *et al.* (1991)
7	$-SO_3H$	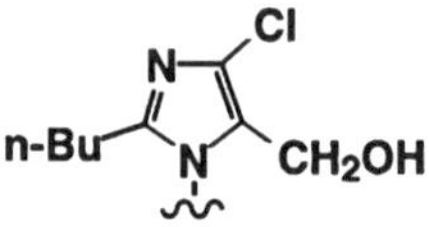 Cl, N, n-Bu, N, CH_2OH	79	Duncia (unpublished results)
8	$-NH-CO-CF_3$	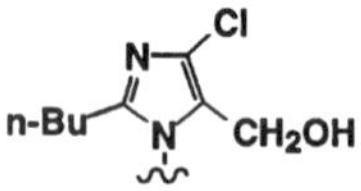 Cl, N, n-Bu, N, CH_2OH	6,300	Carini *et al.* (1991)
9	$-NHSO_2CF_3$	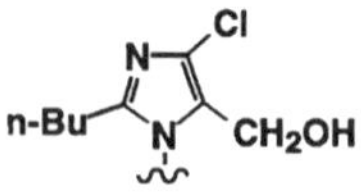 Cl, N, n-Bu, N, CH_2OH	190	Carini *et al.* (1991)
10	$-CO-NH_2$	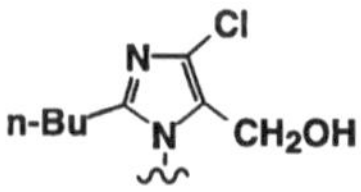 Cl, N, n-Bu, N, CH_2OH	35,000	Carini *et al.* (1991)

Table I (*Continued*)

No	Carboxylic acid isostere	Heterocycle	AT_1 (nM)[a]	Reference
11	**-CO-NH-OH**	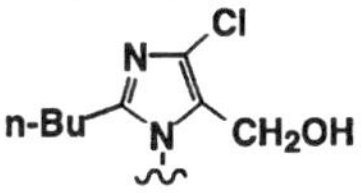 Cl, N, n-Bu, N, CH_2OH	4,100	Carini *et al.* (1991)
12	**·CO-NH-OCH$_3$**	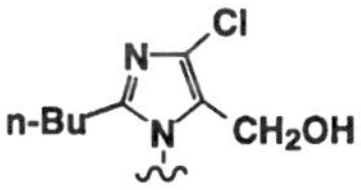 Cl, N, n-Bu, N, CH_2OH	2,900	Carini *et al.* (1991)
13	**-CO-NH-OCH$_2$Ph**	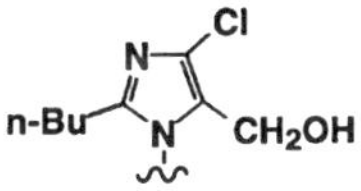 Cl, N, n-Bu, N, CH_2OH	4,900	Carini *et al.* (1991)
14	**-CO-NH-SO$_2$Ph**	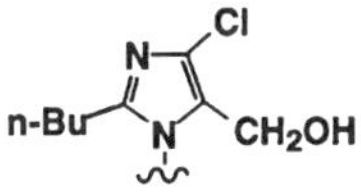 Cl, N, n-Bu, N, CH_2OH	140	Carini *et al.* (1991)
15	**-CO-NH-NH-SO$_2$-Ph**	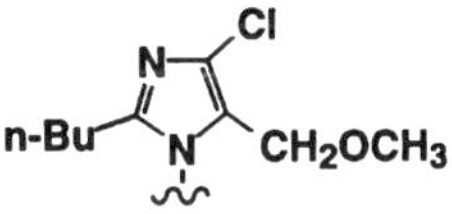 Cl, N, n-Bu, N, CH_2OCH_3	200	Carini *et al.* (1991)
16	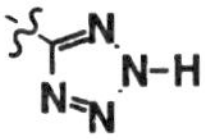 N, N-H, N=N	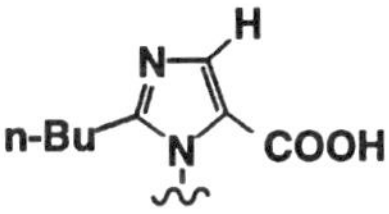 H, N, n-Bu, N, COOH	2.9	Naylor *et al.* (1994)
17	**-SO$_2$-NH-CO-Ph**	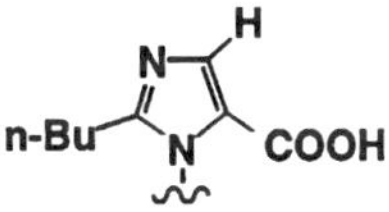 H, N, n-Bu, N, COOH	6.2	Naylor *et al.* (1994)
18	**-NH-SO$_2$-NH-CO-Ph**	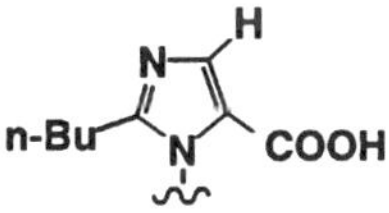 H, N, n-Bu, N, COOH	6.0	Naylor *et al.* (1994)
19	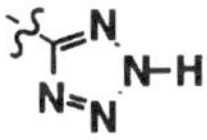 N, N-H, N=N	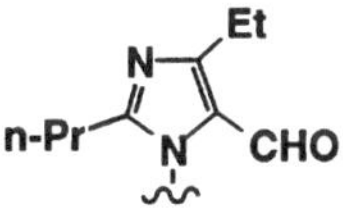 Et, N, n-Pr, N, CHO	2.7	Carini *et al.* (1994)
20	**-SO$_2$-NH-CO-O-n-Bu**	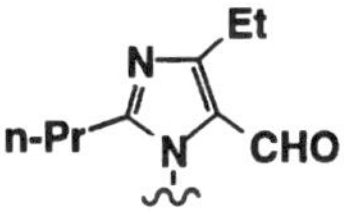 Et, N, n-Pr, N, CHO	2	Quan *et al.* (1994)
21	**SO$_2$-NH-CO-NH-n-Pr**	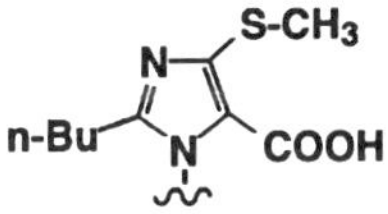 S-CH_3, N, n-Bu, N, COOH	0.48	Deprez *et al.* (1995)
22	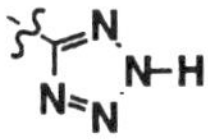 N, N-H, N=N	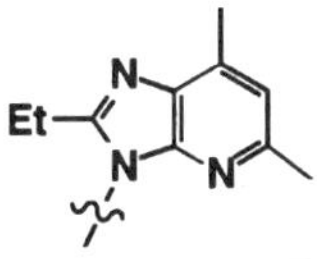 N, Et, N, N	0.3	Mantlo et al. (1991)

(continued)

Table I (*Continued*)

No	Carboxylic acid isostere	Heterocycle	AT_1 (nM)[a]	Reference
23	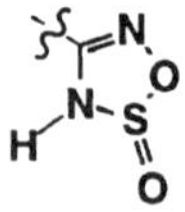	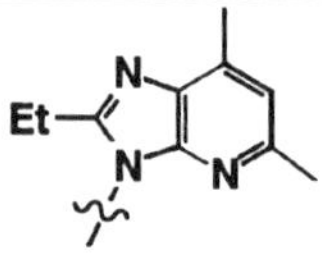	0.7	Kim *et al.* (1994)
24	Same as No. 23	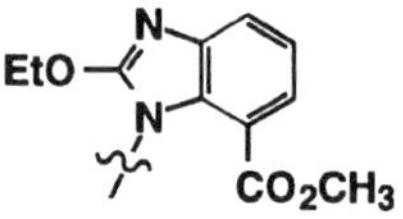	460	Kohara *et al.* (1994)
25	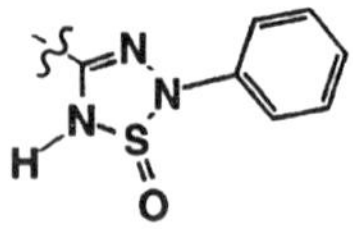	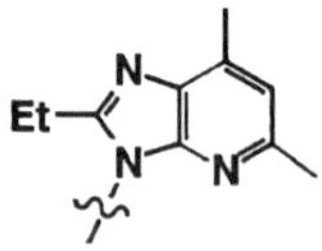	6	Kim *et al.* (1994)
26	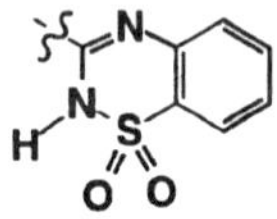	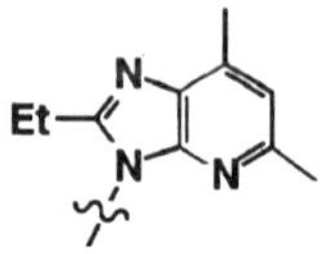	120	Kim *et al.* (1994)
27	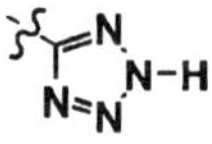	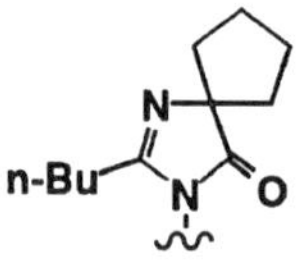	1.3	Ferrari *et al.* (1994)
28	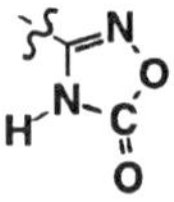	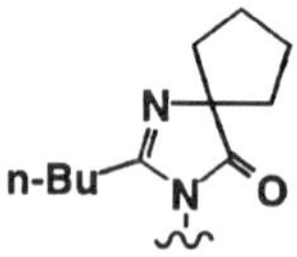	1.7	Ferrari *et al.* (1994)
29	Same as No. 28	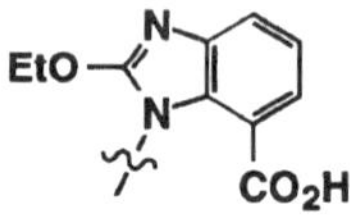	420	Kohara *et al.* (1996)
30	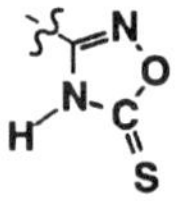	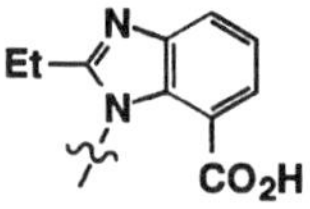	34	Kohara *et al.* (1996)
31	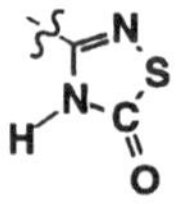	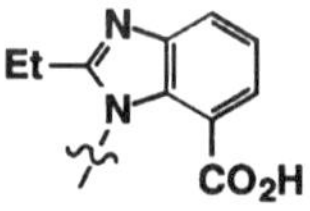	69	Kohara *et al.* (1996)
32	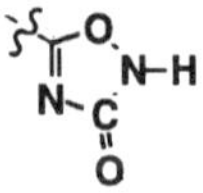	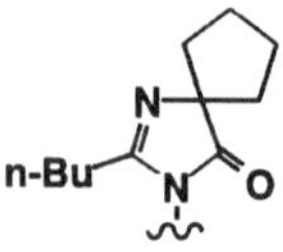	230	Ferrari *et al.* (1994)

Table I (*Continued*)

No	Carboxylic acid isostere	Heterocycle	AT_1 (nM)[a]	Reference
33	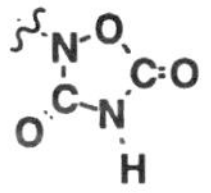	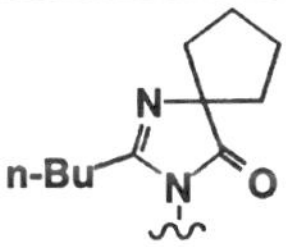	1.4	Ferrari *et al.* (1994)
34	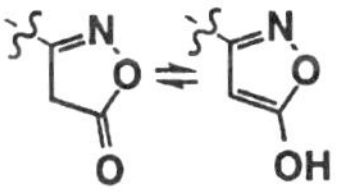	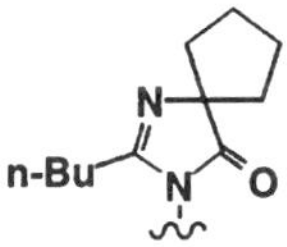	25	Ferrari *et al.* (1994)
35	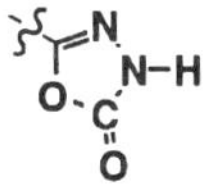	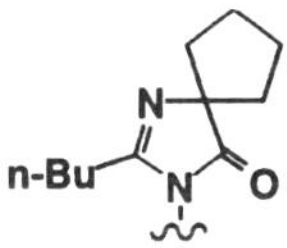	60	Ferrari *et al.* (1994)
36	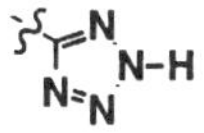	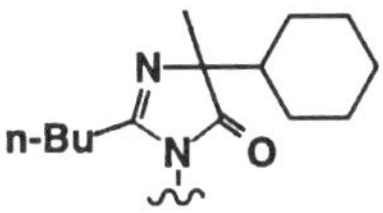	5.2	Ferrari *et al.* (1994)
37	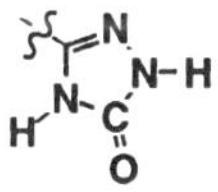	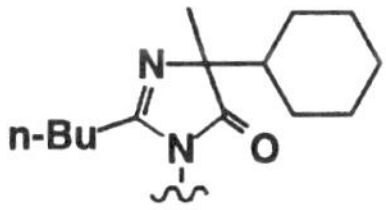	100	Ferrari *et al.* (1994)
38	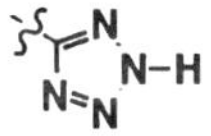	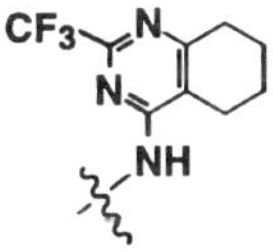	3	Soll *et al.* (1993)
39	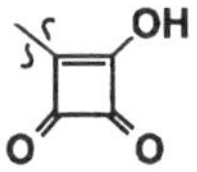	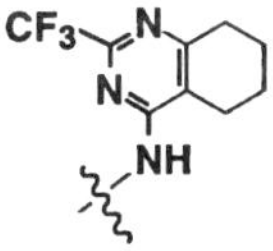	25	Soll *et al.* (1993)

[a] IC_{50} value for the angiotensin-type 1 receptor subtype. Different assay conditions were employed throughout the various references.

sulfonylurea **21** (tetrazoles such as **16** and **19** have been included in Table I as standards for comparison of binding affinities). These latter compounds are better able to locate the negative charge at the appropriate position relative to the biphenyl. For example, Fig. 5 reveals that the distance from the carbon on the biphenyl containing the tetrazole to the acidic nitrogen atoms is 2.6 Å to N-1 and 3.7 Å to N-2. For the carboxylic acid group of EXP7711, the distance is only 2.2 Å , and for the sulfonated carboxamide group of **14** it is 2.3 Å. Both of these distances are most

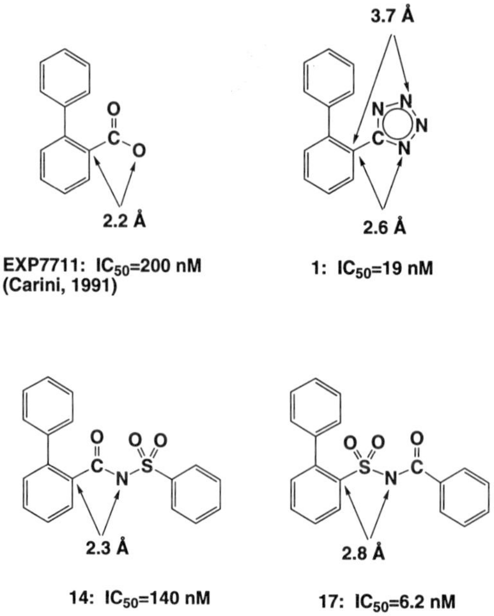

3.7 Å

2.2 Å

2.6 Å

EXP7711: IC_{50}=200 nM
(Carini, 1991)

1: IC_{50}=19 nM

2.3 Å

2.8 Å

14: IC_{50}=140 nM

17: IC_{50}=6.2 nM

Figure 5. Distances between the acid functionality-substituted carbon atom and the atom with the localized negative charge.

likely too short to reach the positive charge in the receptor site. For the acylsulfonamide group of **17,** the distance from the corresponding biphenyl carbon atom to the acidic nitrogen atom is 2.8 Å and greater than 3Å to the carbonyl oxygen which can also bear the negative charge.

MK-996 (L-159,282, Fig. 6) bears a benzoylsulfonamide group as a tetrazole replacement. This compound maintains the potency (IC_{50} = 0.2 nM), duration of action, and bioavailability of related biphenyltetrazoles (Chakravarty *et al.,* 1994; Chang *et al.,* 1994). Also, the sulfonamide group of MK-996 does not undergo the glucuronidation seen with the tetrazoles. For these reasons, MK-996 was selected for development.

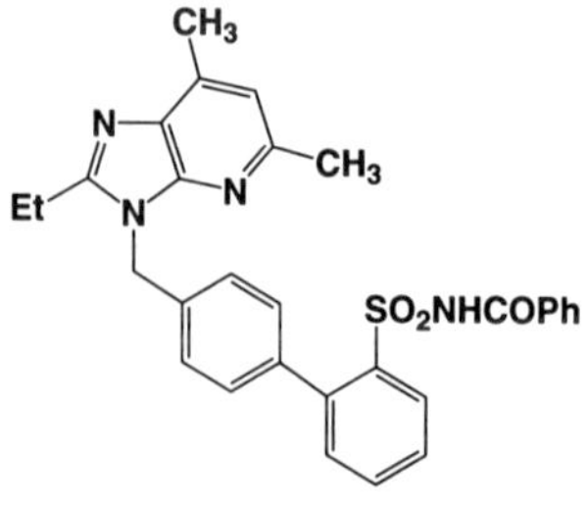

CH_3

CH_3

Et

$SO_2NHCOPh$

Figure 6. MK-996 (L-159,282).

Table II
Acidic Heterocyclic Tetrazole Isosteres Not Found among the Angiotensin II Antagonists

No.	Heterocycle	Ref.	No.	Heterocycle	Ref.
40	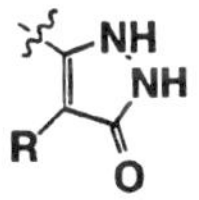	Kees *et al.* (1995)	**45**	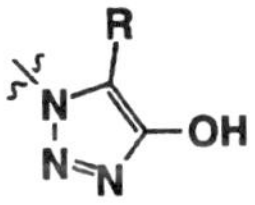	Kees *et al.* (1995)
41	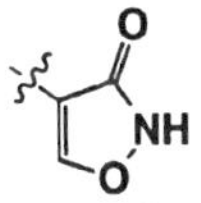	Kees *et al.* (1995)	**46**	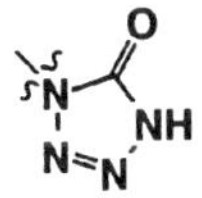	Kees *et al.* (1995)
42	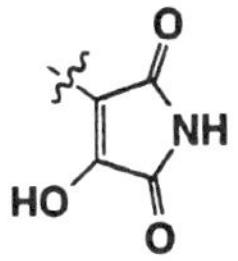	Kees *et al.* (1995)	**47**	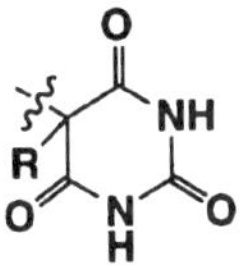	Villemin and Labiad (1990)
43	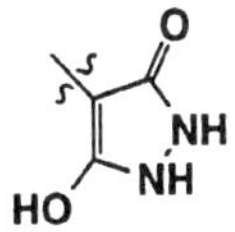	Kees *et al.* (1995)	**48**	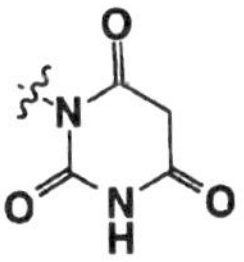	Villemin and Labiad (1990)
44	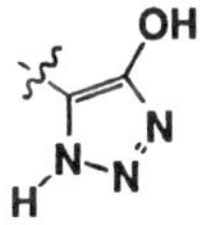	Kees *et al.* (1995)	**49**	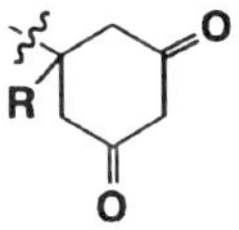	Villemin and Labiad (1990)

In addition to the acidic heterocycles already mentioned, there are others that have emanated from Merck (compounds **22, 23, 25, 26**) (Kim *et al.*, 1994), Sanofi (compounds **27, 28, 32–37**) (Ferrari *et al.*, 1994), Wyeth-Ayerst (compounds **38, 39**) (Soll *et al.*, 1993), and Takeda (compounds **24, 29–31**) (Kohara *et al.*, 1996). The heterocycles that give rise to stronger binding affinities have their negative charge localized at a distance greater than 2.3 Å as per the above discussion. It is not clear, however, why certain heterocycles are better than others with respect to binding affinity. Some heterocyclic acid isosteres that have not appeared in Ang II antagonists and that might be suitable are summarized in Table II.

2.4. Synthetic Availability of Biphenyltetrazoles

Another concern in the development of losartan was our ability to prepare it on a commercial scale. DuPont's first Medicinal Chemistry synthesis of losartan, **1,** is outlined in Fig. 7. Imidazole **50** was prepared by reacting the methyl imidate of valeronitrile with dihydroxyacetone in the presence of ammonia followed by

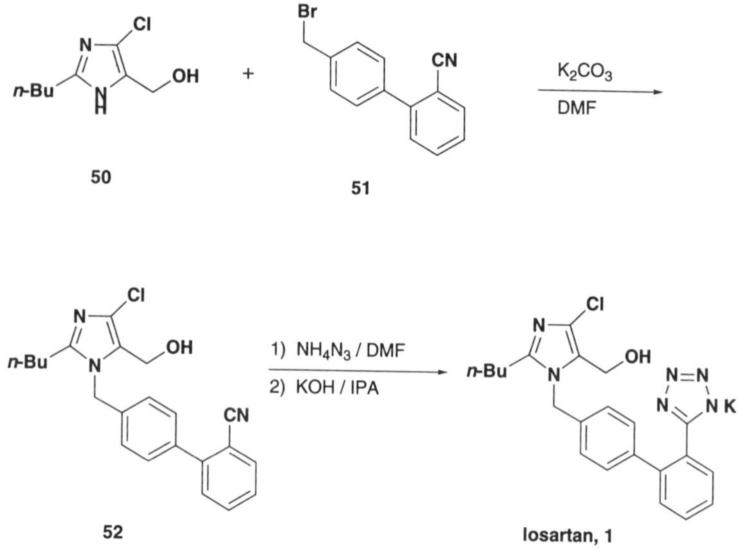

Figure 7. Original synthetic route to losartan.

chlorination of the resulting imidazole with *N*-chlorosuccinimide. The biphenylnitrile fragment **51** was prepared from *o*-anisic acid (Meyers and Mihelich, 1975). Conversion of the acid to the oxazoline followed by displacement of the methoxy group with *p*-tolylmagnesium bromide gave the biphenyl moiety. The oxazoline was then converted to the nitrile by treatment with phosphorous oxychloride. Finally the material was brominated with *N*-bromosuccinimide and a radical initiator. The two major problems with this route were the nonregioselective alkylation of the hydroxymethylimidazole **50** with the (bromomethyl)biphenylnitrile **51** and the subsequent conversion of the nitrile to the tetrazole. The alkylation of **50** with **51** under a variety of conditions tended to give 45 to 50% of the desired regioisomer, which then required column chromatography to isolate in 25 to 35% yields. The regioselectivity problem was readily solved based on the observation that imidazole-4-carboxaldehyde is alkylated by dimethylsulfate under neutral conditions to give largely the desired 1,5-substituted product (Hubball and Pyman, 1928). Hydroxymethylimidazole **50** was oxidized with manganese dioxide to the aldehyde **53,** followed by alkylation with **51** in DMF giving 90 to 93% of the desired regioisomer, which was reduced *in situ* to **52** with sodium borohydride (Fig. 8). Isolated yields of **52** ranged from 70 to 75% without the requirement for chromatography. A subsequent study of the factors that influence the alkylation regioselectivity indicated that the alkylation of **53** was via its potassium salt; the improved regioselectivity was not related to alkylation of the neutral aldehyde (Pierce *et al.*, 1993).

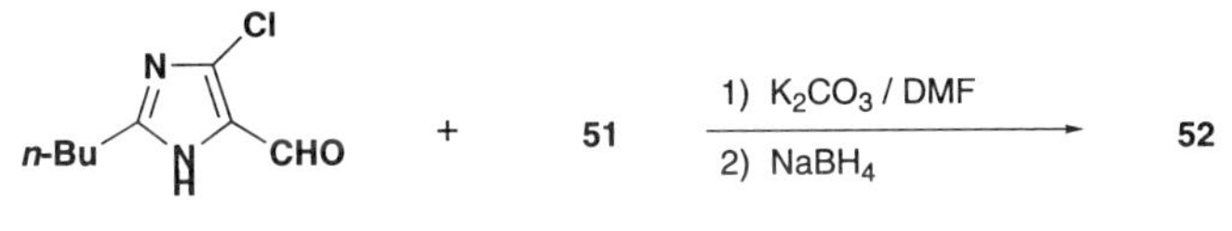

Cl
N
n-Bu
N
H
CHO
53
+ 51
1) K_2CO_3 / DMF
2) $NaBH_4$
52

Figure 8. Regioselective alkylation of an imidazole precursor to losartan.

The second obstacle for making large quantities of losartan was the tetrazole-forming reaction. The initial approach was to convert nitrile **52** to losartan by the classical ammonium azide reaction in dimethylformamide (Finnegan *et al.*, 1958). This reaction with the highly hindered nitrile was extremely sluggish, requiring a large excess of ammonium chloride/sodium azide and 4–5 days at 100 to 110 °C. Under these conditions, significant product decomposition was observed, with a concomitantly large heat of reaction [−66 kcal/mole determined by accelerated rate calorimetry (ARC)] and variable, low yields (0–40 %) of product after chromatographic purification. Additionally, there was a major concern over the safety of this procedure as ammonium azide tends to sublime and is shock sensitive (Bretherick, 1990). A synthetic study for preparing *o*-biphenyl tetrazoles was initiated using routes and reagents deemed to minimize the hazard of this reaction. The substrates considered were nitriles, imidates, amidines, thioimidates, and amidrazones. The trimethylsilylazide reaction with the biphenylnitrile seemed promising because of the relative stability (decomposition above 250 °C) of the reagent and its lack of shock sensitivity (Birkoffer and Ritter, 1965; Birkoffer and Wegner, 1988). This reagent, however, was too unreactive, although partial conversion occurred with BF_3 catalysis in CCl_4. Of the reagents investigated, the trialkyltin azides performed the best. These non-shock-sensitive azides had previously been shown to convert electronically deactivated nitriles to tetrazoles (Reichle, 1964; Thayer and West, 1964; Thayer, 1966; Sisido *et al.*, 1971). Trimethyltin azide, readily prepared from trimethyltin chloride and sodium azide (Luitjen *et al.*, 1962), could be used to convert the hindered biphenylnitrile **52** to the trimethylstannyl derivative of losartan by refluxing in xylenes for 24 to 30 hr. Unlike the reaction with ammonium azide, product decomposition was not observed by ARC or HPLC. The use of nonpolar solvents such as toluene or xylenes was required because solvents such as DMF tended to complex with the tin reagents and decrease the reactivity. Removal of the tin residue was considered essential as trimethyltin azide is a known mutagen and trialkyltin compounds in general are quite toxic. Complete trimethyltin removal was problematic. The trimethyltin residue could largely be removed by treating the stannyl tetrazole derivative with anhydrous HCl, thereby crystallizing the free tetrazole and preparing trimethyltin chloride which could be recycled. The best procedure was to hydrolyze the stannyl tetrazole, and then to trap the tetrazole as its triphenylmethyl

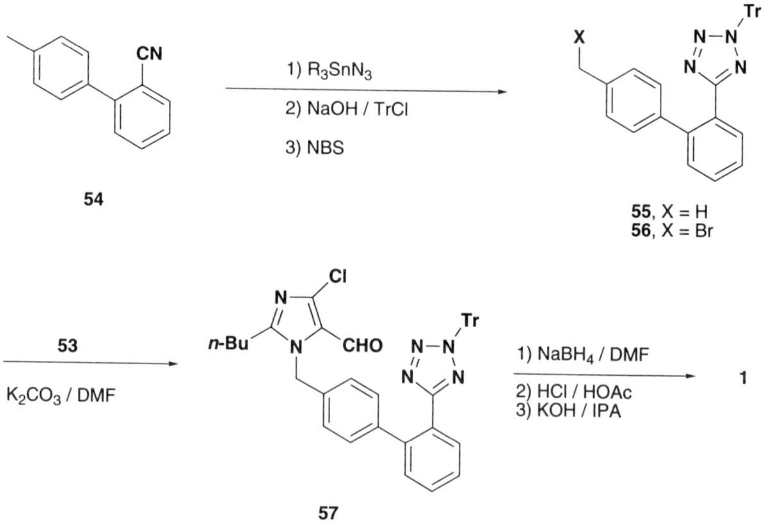

Figure 9. Intermediate synthetic route to losartan.

derivative, which could be crystallized essentially free of tin residues [less than 10 ppm as determined by inductively coupled plasma spectroscopy (ICP)]. This suggested the possibility of preparing the (triphenylmethyl)tetrazole at an earlier stage of synthesis. The benefits of this modified route (Fig. 9) are increased economics for the imidazole portion, easier introduction of the tetrazole moiety, and more operational steps prior to the isolation of the final product so that low levels of organotin compounds could be shed (Duncia *et al.*, 1991; Aldrich *et al.*, 1989). Also, the triphenylmethyl moiety serves as a tetrazole-protecting group for the subsequent benzylic bromination and alkylation steps. The reaction of *o*-tolylbenzonitrile, **54,** with tributyltin azide in refluxing toluene or xylenes proceeded to give a somewhat viscous mixture of the stannylated tetrazoles, which exist as oligomers in solution. Treatment with base followed by triphenylmethyl chloride allowed the triphenylmethyl tetrazole **55** to crystallize away from the resulting tributyltin oxides. Although somewhat less reactive than trimethyltin azide, the tributyl analogue was chosen because of lower cost and decreased potential toxicity. After bromination of **55** to give **56,** the regioselective alkylation/reduction steps were performed in a similar fashion as shown in Fig. 7. Deprotection with HCl/THF followed by titration with potassium hydroxide gave losartan (Carini *et al.*, 1991). This route was employed to prepare hundreds of kilograms of losartan, which was used in phase III of the clinical development.

Ultimately, the preparation of the tetrazole and the potential tin residues was eliminated as an issue by using commercially available 5-phenyltetrazole as a starting material. This required development of suitable methods for the coupling of the biphenyl in the presence of a tetrazole. Merck's Process Research labs discovered that the protected 5-phenyltetrazole **58** could be *o*-metallated and coupled with 4-iodotoluene under Negishi conditions to give **55** as shown in Fig. 10 (Mantlo *et al.,* 1991; Shuman *et al.,* 1991). The major drawbacks of this synthesis are the relatively high cost of 4-iodotoluene and the potential for nickel residues in the product. However, at DuPont Merck it was found that the *o*-metallated protected tetrazole could be converted to the boronic acid and, in turn, employed in a Suzuki coupling with the more readily accessible 4-bromotoluene (Lo and Rossano, 1992).

The collaboration between DuPont Merck and Merck led to development of the current losartan process, which is shown in Fig. 11. In this route, imidazole **53** is first alkylated with commercially available 4-bromobenzyl bromide, followed by reduction of the intermediate aldehyde, to give the (4-bromobenzyl)imidazole **61**. This route avoids the inherent mixture of nonbrominated, mono-, and dibromination products that are formed in the conversion of **54** to either **51** or **55** (Larsen *et al.,* 1994). Coupling of **61** with the boronic acid **62** under Suzuki conditions gives the protected precursor to losartan. Acid-catalyzed deprotection, followed by pH adjustment with NaOH, allows for selective precipitation of triphenylmethanol (which may be recycled into the process). Neutralization with aqueous H_2SO_4 gives losartan (free acid) in 80% overall yield from the imidazole carboxaldehyde **53**.

A program was conducted to identify a suitable salt for losartan. A number of salts were evaluated including potassium, sodium, lithium, calcium, magnesium, zinc, copper, iron, meglumine, choline, ethylenediamine, and ammonium. Based on stability, bioavailability, and water solubility, the potassium salt was chosen for product development. The process involved preparation of the salt with potassium hydroxide in isopropanol/water followed by azeotropically removing most of the water. Dilution with heptane then gave a slurry of product that was easily isolated and dried.

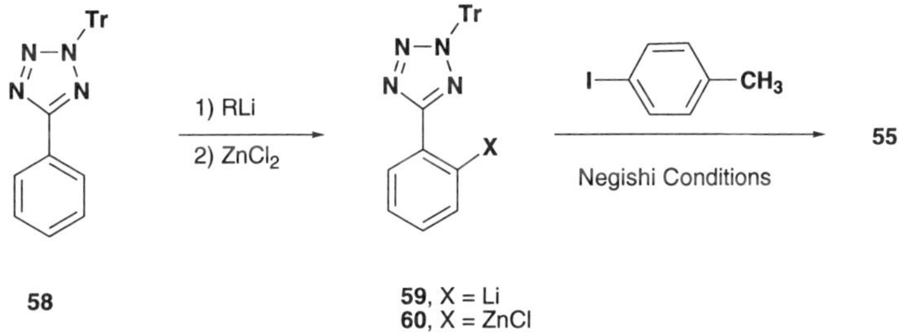

Figure 10. Alternate synthesis of the biphenyltetrazole intermediate.

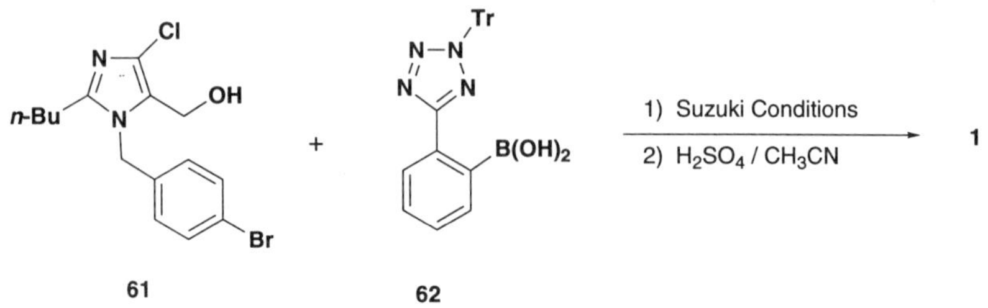

Figure 11. Commercial synthetic route to losartan.

3. AN ACTIVE METABOLITE OF LOSARTAN

3.1. Identification of EXP3174

The role of active metabolites in the therapeutic activity of drugs has long been recognized, and the species-selective formation of active metabolites can be a concern for the rapid development of new drugs. Early work with losartan revealed that its antihypertensive effect in rats was biphasic and lasted longer in rats than in dogs (Wong *et al.,* 1990a, 1991a), suggesting the species-selective formation of an active metabolite. Incubation of losartan with hepatic microsomes demonstrated the formation of a major, more polar metabolite by rats but not dogs and, more importantly, significant formation by human liver (Wong, unpublished results). This product was subsequently identified as the imidazole-5-carboxylic acid metabolite of the primary alcohol, and designated EXP3174 (Fig. 12). The species-selective formation of this metabolite was confirmed in later studies (Stearns *et al.,* 1992; Christ *et al.,* 1994). Further studies with the synthetic metabolite confirmed that it was a more potent Ang II antagonist than losartan (Wong *et al.,* 1990b). Early metabolism studies thus confirmed the species-selective formation of EXP3174 and demonstrated that it was likely that EXP3174 would be produced in humans dosed with losartan. These observations affected the design of

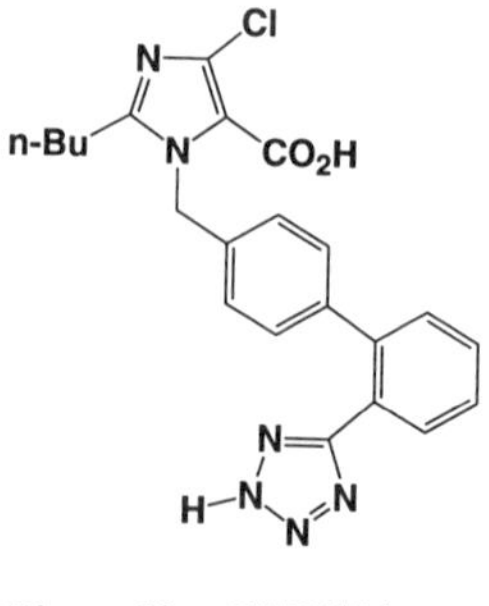

Figure 12. EXP3174.

the initial clinical studies, directing the development of a sensitive and specific analytical assay capable of measuring both losartan and EXP3174 in plasma. In clinical studies, it was confirmed that humans readily metabolize losartan to produce EXP3174. Losartan's long duration of action is related partly to the formation of EXP3174, which has a longer half-life than losartan itself ($t_{1/2}$ = 2 hr for losartan versus 6 hr for EXP3174) (Lo *et al.*, 1995).

3.2. Should We Develop EXP3174?

The discovery of EXP3174 as an active metabolite of losartan almost immediately raised the question of whether DuPont could develop this compound. As discussed above, the formation of EXP3174 is believed to contribute significantly to the antihypertensive effect and duration of action of losartan. Rats produce EXP3174 whereas dogs do so poorly or not at all. Prior to the clinical trials, there was still concern about the ability of humans to form the metabolite and therefore about the effectiveness of losartan in humans. The development of EXP3174 would have had the advantage of avoiding this issue entirely. A second consideration in favor of EXP3174's development was that it is significantly more potent, *in vitro* and *in vivo,* than losartan as both an angiotensin II antagonist and an antihypertensive (Carini and Duncia, 1993). Losartan inhibits the binding of [^{125}I]-Ang II to rat adrenal cortical microsomes with an IC_{50} of 19 nM, whereas EXP3174 is 10-fold more potent with an IC_{50} of 1.3 nM. When administered intravenously to a renal hypertensive rat, EXP3174 [ED_{30}(i.v.) = 0.04 mg/kg] is 20-fold more potent than losartan [ED_{30}(i.v.) = 0.80 mg/kg]. However, despite the superior intrinsic potency of EXP3174, its oral antihypertensive potency [ED_{30}(p.o.) = 0.66 mg/kg] is actually less than that of losartan [ED_{30}(p.o.) = 0.59 mg/kg]. The oral bioavailability of EXP3174 was determined to be only 12% in rats (Christ, unpublished results), whereas the bioavailability of losartan is 33% (Wong *et al.,* 1990b). Therefore, based on this initial experience in rats, the decision was made not to consider EXP3174 for development.

3.3. The Search for a Superior EXP3174 Analogue

Despite the decision not to develop EXP3174, it was clear that the discovery of an analogue of this compound possessing greater bioavailability and oral antihypertensive potency might be very desirable. Our efforts subsequently demonstrated that diacidic angiotensin II antagonists are often very potent but seldom very bioavailable. However, two compounds, DuP 532 (Fig. 13) (Carini *et al.,* 1993, 1994) and DMP 811 (Fig. 14) (Carini *et al.,* 1994), came sufficiently close to our goal to rate further interest. These two derivatives are both direct analogues of EXP3174 and are very closely related to each other structurally. DuP 532 has

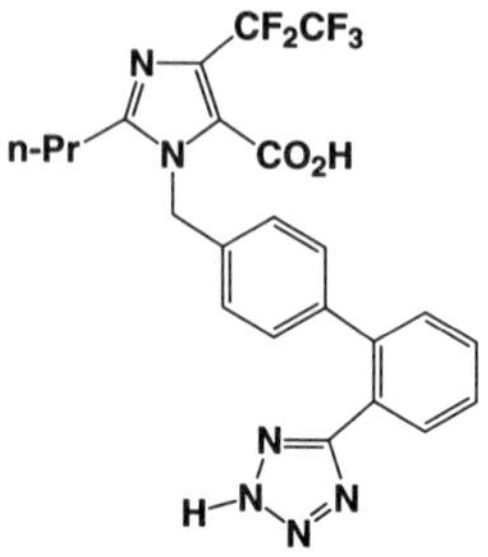

Figure 13. DuP 532.

3-fold greater oral antihypertensive activity than EXP3174 (see Table III) and a longer duration of action, whereas the oral antihypertensive activity of DMP 811 is 20-fold greater than that of EXP3174. However, the bioavailabilities of DuP 532 (Wong *et al.,* 1994) and DMP 811 do not differ significantly from that of EXP3174. Despite their modest bioavailabilities, the oral antihypertensive potency of DMP 811 and the increased duration of action shown by DuP 532 were sufficiently interesting that these compounds were placed into development as backup candidates to losartan.

Another way to avoid the need for the metabolic activation that is required with losartan would be to prepare a monoacidic Ang II antagonist with an intrinsic potency equal to or greater than that of EXP3174 and with good oral bioavailability. One such compound is Merck's L-158,809, an imidazo[4,5-*b*]pyridine derivative. L-158,809 (Fig. 15) is an exceedingly potent inhibitor of Ang II binding to its receptor [IC_{50} = 0.3 nM] (Mantlo *et al.,* 1991; Chang *et al.,* 1992) and a potent antagonist of the Ang II pressor response in conscious rats [ED_{50}(i.v.) = 0.029 mg/kg; ED_{50}(p.o.) = 0.023 mg/kg] (Siegl *et al.,* 1992). Finally, the oral bioavailability of L-158,809 was found to be approximately 100% in rats (Colletti and Krieter, 1994).

DuPont's initial reports on the discovery of nonpeptide angiotensin II receptor antagonists launched major efforts by many other pharmaceutical companies to discover their own Ang II antagonists, and a large number of competitive com-

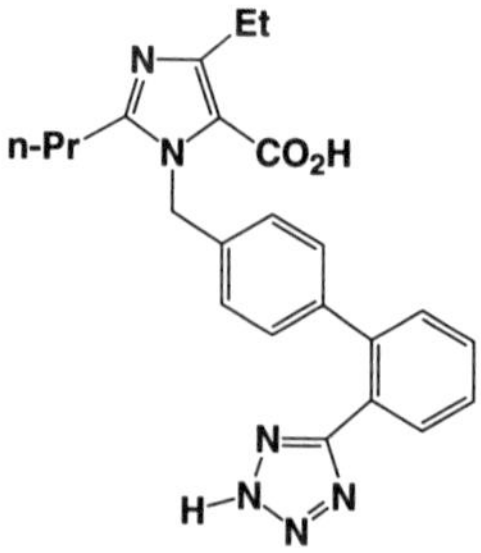

Figure 14. DMP 811.

Table III
Comparison of the Properties of Three Diacidic Ang II Antagonists

No.	i.v. ED_{30} (mg/kg)	Oral ED_{30} (mg/kg)	Oral bioavailability (%F)
EXP3174	0.038	0.66	12% (rats)
DuP 532	0.042	0.21	8% (rats), 12% (dogs)
DMP 811	0.005	0.03	11% (rats), 13–15% (dogs)

pounds have now been disclosed (Wexler *et al.,* 1996). These compounds include both monoacidic and diacidic derivatives, and many of them are claimed to be more potent, orally bioavailable, and long acting.

In hindsight the metabolism of losartan to EXP3174 is arguably an advantage. Humans do produce the metabolite very well, and losartan is a very effective antihypertensive in clinical use. The production of EXP3174 results in the relatively slow onset of full antihypertensive activity. This property may be responsible for the low incidence of dizziness in patients dosed with losartan.

4. EARLY EVALUATION OF LOSARTAN'S ACTIVITY IN HUMANS

One hallmark of the discovery and development of losartan was the commitment to proving activity in humans as soon as possible, in the United States or Europe. This commitment was facilitated by the excellent safety profile demonstrated by losartan in rodents and dogs, by the favorable solubility and stability profiles of the drug substance, and by the availability of a relevant surrogate endpoint for hypertension, the blockade of exogenous angiotensin I or angiotensin II vasopressor responses. The availability of relevant animal models and surrogate clinical markers is an important advantage for drugs targeted for cardiorenal diseases, an advantage that does not exist for the rapid discovery and development of drugs for other important therapeutic areas such as the dementias or AIDS. Although an

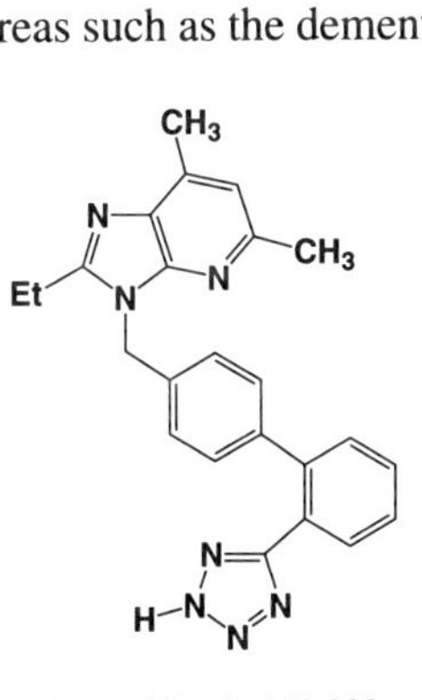

Figure 15. L-158,809.

integral consideration for drug development today, evaluating the activity of new chemical entities in humans in Europe before filing a formal IND application with the FDA was not as widespread in the mid-1980s. Measuring the plasma concentrations of losartan and EXP3174 in early studies was also an important objective.

The first clinical study was designed in collaboration with Professor H. R. Brunner of the Centre Hospitalier Universitaire Vaudois, Lausanne, Switzerland. Dr. Brunner and colleagues had long been leaders in characterizing the role of the renin–angiotensin axis in hypertension. They had developed a protocol for measuring the exogenously administered Ang I- or Ang II-mediated vasopressor responses in healthy volunteers and the effects of various agents on these responses.

Losartan was given orally to the first volunteer in January 1989, less than 3 years after it was first synthesized. Healthy, young volunteers were given single oral doses of losartan ranging from 2.5 to 40 mg, and the systolic blood pressure responses to i.v. doses of Ang I were recorded (Christen *et al.*, 1991). Losartan produced a dose-dependent decrease in systolic blood pressure after Ang I or Ang II challenge without clinically significant side effects or evidence of agonist activity. Moreover, antagonism was present 24 hr after the eighth oral dose of 40 mg. Subsequent analysis revealed that EXP3174 was present in plasma at greater concentrations than losartan and was eliminated more slowly (Munafo *et al.*, 1992), observations confirmed in later pharmacokinetic studies (Lo *et al.*, 1995). These studies illustrate the power of early drug evaluation in humans and were important in helping to define the future development program for losartan. These studies framed the likely starting doses for efficacy trials in hypertensive patients, demonstrated that single or multiple daily oral doses would be safe, well tolerated, and efficacious, and confirmed the importance of EXP3174.

5. SELECTIVE VERSUS BALANCED ANGIOTENSIN II RECEPTOR ANTAGONISTS

It has been established that there are two distinct subtypes of the angiotensin II receptor, designated AT_1 and AT_2 (Whitebread *et al.*, 1989; Chiu *et al.*, 1989; Chang and Lotti, 1991). This observation was made possible by the discovery of selective nonpeptide antagonists for each of these subtypes. Losartan is a highly selective AT_1 antagonist (Chiu *et al.*, 1990; Wong *et al.*, 1991b), whereas PD123177 (Fig. 16) and related compounds are AT_2-selective (Blankley *et al.*, 1991). The AT_1 receptor mediates virtually all of the known Ang II physiological functions, such as vasoconstriction and aldosterone release, and the utility of AT_1-selective antagonists, such as losartan, is now well established. On the other hand, the physiological role of the AT_2 has still not been clearly defined.

Early in the development of losartan, concern was expressed about the use of an AT_1-selective agent. It has been reported that blockade of the AT_1 receptor in

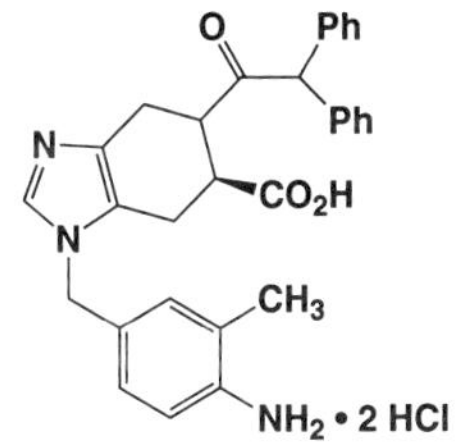

Figure 16. PD123177.

animals and humans causes an increase in plasma levels of angiotensin II (Goldberg *et al.*, 1993; Wong *et al.*, 1990c), and the consequences of exposing the unprotected AT_2 receptors to these increased Ang II levels were unknown. Fortunately, losartan has proven to be very safe in clinical use (Nelson *et al.*, 1995), and no effects attributable to AT_2 stimulation have been reported (Timmermans *et al.*, 1993) in humans or animals. However, because of this initial concern, the discovery of balanced AT_1/AT_2 antagonists became the goal of a collaborative effort between Merck and DuPont Merck. A second reason for pursuing balanced antagonists was the hope that such compounds would demonstrate clinical effects superior or complementary to the AT_1-selective agents.

The work on balanced antagonists was extensive, involving the efforts of a large number of research scientists for approximately 2 years, and will not be reviewed in detail here (for a review of balanced AT_1/AT_2 antagonists including our work, see Wexler *et al.*, 1996). Instead, a variety of compounds representing several series of balanced antagonists will be presented. Because of the lack of a known AT_2-mediated pharmacological effect, work directed toward balanced antagonists focused on producing compounds with equal affinity at the two receptor subtypes.

The most successful approach to balanced AT_1/AT_2 antagonists has been to modify AT_1-selective compounds to enhance their AT_2 affinities. Excellent

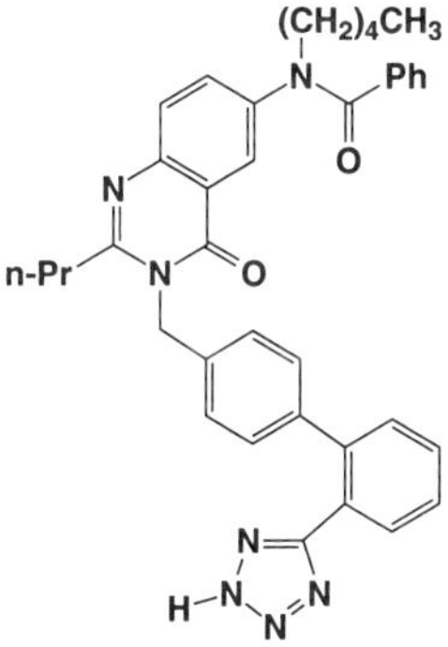

Figure 17. L-159,689.

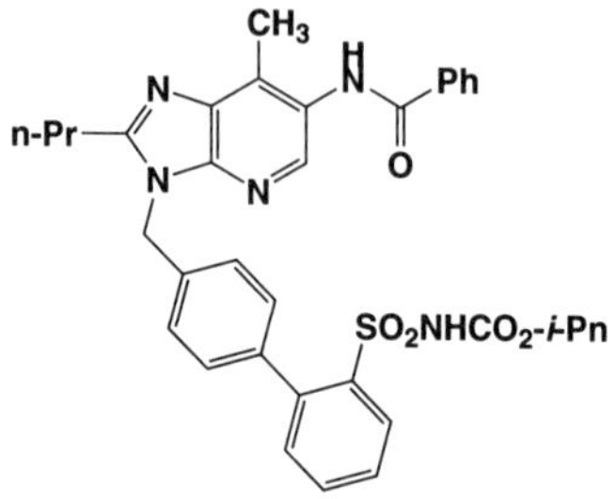

Figure 18. L-163,017.

AT_1/AT_2 balance has been achieved in several heterocyclic series. For example, Merck's quinazolinone L-159,689 (AT_1 IC_{50} = 1.7 nM; AT_2 IC_{50} = 0.7 nM) (Fig. 17) (de Laszlo *et al.,* 1993) possesses excellent AT_1/AT_2 balance with high affinities for both receptors.

In many series, AT_2 affinity and AT_1/AT_2 balance were achieved more readily when an acylsulfonamide group was employed as an isoteric replacement for the tetrazole ring (see above). Several examples from Merck of the use of acylsulfonamide substituents are the imidazopyridine L-163,017 (AT_1 IC_{50} = 0.24 nM; AT_2 IC_{50} = 0.29 nM) (Fig. 18) (Chang *et al.,* 1995), the quinazolinone L-163,579 (AT_1 IC_{50} = 0.57 nM; AT_2 IC_{50} = 0.39 nM) (Fig. 19) (Glinka *et al.,* 1994), and the triazolinone L-163,958 (AT_1 IC_{50} = 0.20 nM; AT_2 IC_{50} = 0.12 nM) (Fig. 20) (Chang and Greenlee, 1995), whereas from DuPont Merck there is the imidazole XR510 (AT_1 IC_{50} = 0.26 nM; AT_2 IC_{50} = 0.28 nM) (Fig. 21) (Quan *et al.,* 1995). All of these compounds have subnanomolar affinities for both the AT_1 and AT_2 receptors with excellent AT_1/AT_2 balance.

The concerns over the development of an AT_1-selective antagonist eventually proved unwarranted, and the development of a balanced AT_1/AT_2 antagonist as a backup was not necessary. The clinical utility of balanced AT_1/AT_2 antagonists, as well as AT_2-selective antagonists, is still uncertain. The answer to these questions awaits the determination of the physiological role of the AT_2 receptor.

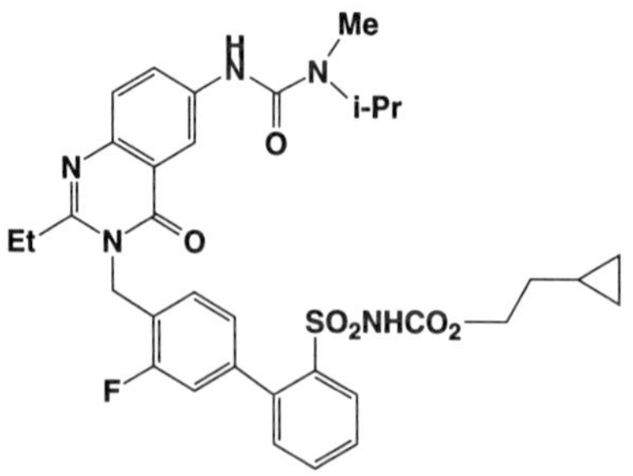

Figure 19. L-163,579.

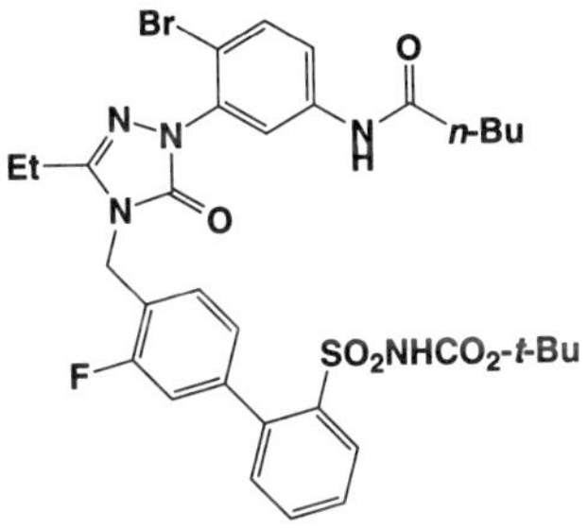

Figure 20. L-163,958.

6. CONCLUSION

During the development of losartan, various issues arose that could have stopped losartan and hampered the successful development of the angiotensin II antagonists in general. Some of these issues were real problems that had to be solved if losartan was to advance, such as the need for a commercially viable synthesis. Other issues, such as whether EXP3174 would form in humans, were anticipated as potential problems that should be addressed. Because of the combined efforts of discovery and development groups at DuPont, and later at Merck and DuPont Merck, the development of losartan was rapid. While some people were working to answer questions such as the antihypertensive efficacy of losartan in humans as quickly as possible, other people were searching for potential backups to losartan. Without an efficient integration of Discovery and Development efforts, the commercially successful development of any drug would be threatened, and it helped to give losartan a critical 2-year lead on the most advanced competitive Ang II antagonists.

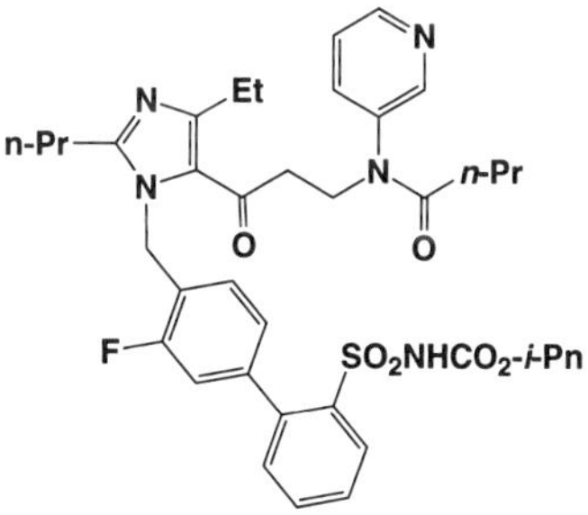

Figure 21. XR510.

REFERENCES

Aldrich, P. E., Duncia, J. V., and Pierce, M. E., 1989, Tetrazole intermediates to antihypertensive compounds, U.S. Patent 4,874,867.

Benson, F. R., 1947, The chemistry of the tetrazoles, *Chem. Rev.* **41:**5.

Benson, F. R., 1967, The tetrazoles, in: *Heterocyclic Compounds,* Volume 8 (R. Elderfield, ed.), pp. 1–104, Wiley, New York.

Birkoffer, L., and Ritter, A., 1965, New methods of preparative organic chemistry IV, the use of silylation in organic synthesis, *Angew. Chem. Int. Ed. Engl.* **4:**417.

Birkoffer, L., and Wegner, P., 1988, Trimethylsilyl azide, in: *Organic Syntheses,* Coll. Vol. 6, p. 1030, Wiley, New York.

Blankley, C. J., Hodges, J. C., Klutchko, S. R., Himmelsbach, R. J., Chocholowski, A., Connolly, C. J., Neergaard, S. J., Van Nieuwenhze, M. S., and Sebastian, A., 1991, A synthesis and structure–activity relationships of a novel series of nonpeptide angiotensin II receptor binding inhibitors specific for the AT_2 subtype, *J. Med. Chem.* **34:**3248–3260.

Bretherick, L., 1990, *Handbook of Reactive Chemical Hazards,* p. 1254, Butterworths, London.

Butler, R. N., 1977a, Recent advances in tetrazole chemistry, *Adv. Heterocycl. Chem.* **21:**379.

Butler, R. N., 1977b, Recent advances in tetrazole chemistry, *Adv. Heterocycl. Chem.* **21:**348.

Carini, D. J., and Duncia, J. V., 1993, The discovery and development of the nonpeptide angiotensin II receptor antagonists, in: *Advances in Medicinal Chemistry,* Volume 2 (B. E. Maryanoff and C. A. Maryanoff, eds.), pp. 153–195, JAI Press, London, and references cited therein.

Carini, D. J., Duncia, J. V., Aldrich, P. E., Chiu, A. T., Johnson, A. L., Pierce, M. E., Price, W. A., Santella, J. B., III, Wells, G. J., Wexler, R. R., Wong, P. C., Yoo, S.-E., and Timmermans, P. B. M. W. M., 1991, Nonpeptide angiotensin II antagonists: The discovery of a series of N-(biphenylylmethyl)imidazoles as potent, orally active antihypertensives, *J. Med. Chem.* **34:**2525–2547.

Carini, D. J., Chiu, A. T., Wong, P. C., Johnson, A. L., Wexler, R. R., and Timmermans, P. B. M. W. M., 1993, The preparation of (perfluoroalkyl)imidazoles as nonpeptide angiotensin II receptor antagonists, *Bioorg. Med. Chem. Lett.* **3:**895–898.

Carini, D. J., Ardecky, R. J., Ensinger, C. L., Pruitt, J. R., Wexler, R. R., Wong, P. C., Huang, S.-M., Aungst, B. J., and Timmermans, P. B. M. W. M., 1994, Nonpeptide angiotensin II receptor antagonists: The discovery of DMP 581 and DMP 811, *Bioorg. Med. Chem. Lett.* **4:**63–68.

Chakravarty, P. K., Naylor, E. M., Chen, A., Chang, R. S. L., Chen, T.-B., Faust, K. A., Lotti, V. J., Kivlighn, S. D., Gable, R. A., Zingaro, G. J., Schorn, T. W., Schaffer, L. W., Broten, T. P., Siegl, P. K. S., Patchett, A. A., and Greenlee, W. J., 1994, A highly potent orally active imidazo(4,5-b)pyridine biphenylacylsulfonamide (MK-996, L-159,282): A new AT_1 selective angiotensin II receptor antagonist, *J. Med. Chem.* **37:**4068–4072.

Chang, L. L., and Greenlee, W. J., 1995, Angiotensin II receptor antagonists: Nonpeptides with equivalent high affinity for both the AT_1 and AT_2 subtypes, *Curr. Pharm. Des.* **1:**407–424.

Chang, R. S. L., and Lotti, V. J., 1991, Angiotensin receptor subtypes in rat, rabbit, and monkey tissues: Relative distribution and species dependency, *Life Sci.* **49:**1485–1490.

Chang, R. S. L., Siegl, P. K. S., Clineschmidt, B. V., Mantlo, N. B., Chakravarty, P. K., Greenlee, W. J., Patchett, A. A., and Lotti, V. J., 1992, In vitro pharmacology of L-158,809, a new highly potent and selective angiotensin II receptor antagonist, *J. Pharmacol. Exp. Ther.* **262:** 133–138.

Chang, R. S. L., Bendesky, R. J., Chen, T.-B., Faust, K. A., Kling, P. J., O'Malley, S. A., Naylor, E. M., Chakravarty, P. K., Patchett, A. A., Greenlee, W. J., Clineschmidt, B. V., and Lotti, V. J., 1994, In vitro pharmacology of MK-996, a new potent and selective angiotensin II (AT_1) receptor antagonist, *Drug Dev. Res.* **32:**161–171.

Chang, R. S. L., Lotti, V. J., Chen, T.-B., O'Malley, S. S., Bendesky, R. J., Kling, P. J., Kivlighn, S. D., Siegl, P. K. S., Ondeyka, D., Greenlee, W. J., and Mantlo, N. B., 1995, In vitro pharmacology of

an angiotensin AT_1 receptor antagonist with balanced affinity for AT_2 receptors, *Eur. J. Pharmacol.* **294:**429–437.
Chiu, A. T., Herblin, W. F., McCall, D. E., Ardecky, R. J., Carini, D. J., Duncia, J. V., Pease, L. J., Wong, P. C., Wexler, R. R., Johnson, A. L., and Timmermans, P. B. M. W. M., 1989, Identification of angiotensin II receptor subtypes, *Biochem. Biophys. Res. Commun.* **165:**196–203.
Chiu, A. T., McCall, D. E., Price, W. A., Wong, P. C., Carini, D. J., Duncia, J. V., Wexler, R. R., Yoo, S.-E., Johnson, A. L., and Timmermans, P. B. M. W. M., 1990, Nonpeptide angiotensin II receptor antagonists. VII. Cellular and biochemical pharmacology of DuP 753, an orally active antihypertensive agent, *J. Pharmacol. Exp. Ther.* **252:**711–718.
Christ, D. D., Wong, P. C., Wong, Y. N., Hart, S. D., Quon, C. Y., and Lam, G. N., 1994, The pharmacokinetics and pharmacodynamics of the angiotensin receptor antagonist losartan potassium (DuP 753/MK 954) in the dog, *J. Pharmacol. Exp. Ther.* **268:**1199–1205.
Christen, Y., Waeber, B., Nussberger, J., Porchet, M., Borland, R. M., Lee, R. J., Maggon, K., Shum, L., Timmermans, P. B. M. W. M., and Brunner, H. R., 1991, Oral administration of DuP 753, a specific angiotensin II receptor antagonist, to normal male volunteers. Inhibition of pressor responses to exogenous angiotensin I and II, *Circulation* **83:**1333–1342.
Colletti, A. E., and Krieter, P. A., 1994, Disposition of the angiotensin II antagonist L-158,809 in rats and rhesus monkeys, *Drug Metab. Dispos.* **22:**183–188.
de Laszlo, S. E., Quagliato, C. S., Greenlee, W. J., Patchett, A. A., Chang, R. S. L., Lotti, V. J., Chen, T.-B., Scheck, S. A., Faust, K. A., Kivlighn, S. D., Schorn, T. S., Zingaro, G. J., and Siegl, P. K. S., 1993, L-159,689, a potent, orally-active, balanced affinity antagonist of the angiotensin II AT_1 and AT_2 receptors, *J. Med. Chem.* **36:**3207–3210.
Deprez, P., Guillaume, J., Becker, R., Corbier, A., Didierlaurent, S., Fortin, M., Frechet, D., Hamon, G., Heckmann, B., Heitsch, H., Kleemann, H.-W., Vevert, J.-P., Vincent, J.-C., Wagner, A., and Zhang, J., 1995, Sulfonylureas and sulfonylcarbamates as new non-tetrazole angiotensin II receptor antagonists. Discovery of a highly potent orally active (imidazolylbiphenylyl)sulfonylurea (HR 720), *J. Med. Chem.* **38:**2357–2377.
Duncia, J. V., Chiu, A. T., Carini, D. J., Gregory, G. B., Johnson, A. L., Price, W. A., Wells, G. J., Wong, P. C., Calabrese, J. C., and Timmermans, P. B. M. W. M., 1990, The discovery of potent nonpeptide angiotensin II receptor antagonists: A new class of potent antihypertensives, *J. Med. Chem.* **33:**1312–1329.
Duncia, J. V., Pierce, M. E., and Santella, J. B., III, 1991, Three synthetic routes to a sterically hindered tetrazole. A new one-step mild conversion of an amide into a tetrazole, *J. Org. Chem.* **56:**2395–2400.
Duncia, J. V., Carini, D. J., Chiu, A. T., Johnson, A. L., Price, W. A., Wong, P. C., Wexler, R. R., and Timmermans, P. B. M. W. M., 1992, The discovery of DuP 753, a potent, orally active nonpeptide angiotensin II receptor antagonist, *Med. Res. Rev.* **12**(2)**:**149–191.
Ferrari, B., Taillades, J., Perreaut, P., Bernhart, C., Gougat, J., Guiraudou, P., Cazaubon, C., Roccon, A., Nisato, D., Le Fur, G., and Breliere, F. C., 1994, Development of tetrazole bioisosteres in angiotensin II antagonists, *Bioorg. Med. Chem. Lett.* **4:**45–50.
Finnegan, W. G., Henry, R. A., and Lofquist, R., 1958, An improved synthesis of 5-substituted tetrazoles, *J. Am. Chem. Soc.* **80:**3908.
Glinka, T. W., de Laszlo, S. E., Siegl, P. K. S., Chang, R. S., Kivilghn, S. D., Schorn, T. W., Faust, K. A., Chen, T.-B., Zingaro, G. J., Lotti, V. J., and Greenlee, W. J., 1994, Development of balanced angiotensin II antagonists equipotent towards human AT_1 and AT_2 receptor subtypes, *Bioorg. Med. Chem. Lett.* **4:**2337–2342.
Goldberg, M. R., Tanaka, W., Barchowsky, A., Bradstreet, T. E., McCrea, J., Lo, M. W., McWilliams, E. J., Jr., and Bjornsson, T. D., 1993, Effects of losartan on blood pressure, plasma renin activity, and angiotensin II in volunteers, *Hypertension* **21:**704–713.
Hubball, W., and Pyman, F L., 1928, Glyoxaline-4(5)-formaldehyde, *J. Chem. Soc.* **1928:**21.

Kees, K. L., Caggiano, T. J., Steiner, K. D., Fitzgerald, J. J., Kates, M. J., Christos, T. E., Kulishoff, J. M., Jr., Moore, R. D., and McCaleb, M. L., 1995, Studies on new acidic azoles as glucose-lowering agents in obese, diabetic db/db mice, *J. Med. Chem.* **38:**617–628.

Kim, D., Mantlo, N. B., Chang, R. S. L., Kivlighn, S. D., and Greenlee, W. J., 1994, Evaluation of heterocyclic acid equivalents as tetrazole replacements in imidazopyridine-based nonpeptide angiotensin II receptor antagonists, *Bioorg. Med. Chem. Lett.* **4:**41–44.

Kohara, Y., Kubo, K., Imamiya, E., Wada, T., Inada, Y., and Naka, T., 1996, Synthesis and angiotensin II receptor antagonistic activities of benzimidazole derivatives bearing acidic heterocycles as novel tetrazole bioisosteres, *J. Med. Chem.* **39:**5228–5235.

Larsen, R. D., King, A. O., Chen, C. Y., Corley, E. G., Foster, B. S., Roberts, F. E., Yang, C., Lieberman, D. R., Reamer, R. A., Tschaen, D. M., Verhoeven, T. R., Reider, P. J., Lo, Y. S., Rossano, L. T., Brookes, A. S., Meloni, D., Moore, J. R., and Arnett, J. F., 1994, Efficient synthesis of losartan, a nonpeptide angiotensin II receptor antagonist, *J. Org. Chem.* **59:**6391.

Lo, M.-W., Goldberg, M. R., McCrea, J. B., Lu, H. L., Furtek, C. I., and Bjornsson, T. D., 1995, Pharmacokinetics of losartan, an angiotensin II receptor antagonist, and its active metabolite EXP3174 in humans, *Clin. Pharmacol. Ther.* **58:**641–649.

Lo, Y. S., and Rossano, L. T., 1992, Tetrazolylphenylboronic acid intermediates for the synthesis of AII receptor antagonists, U.S. Patent 5,130,439.

Luitjen, J. G. A., Janssen, M. J., and van der Kerk, G. J. M., 1962, New organotin compounds containing a tin-nitrogen linkage, *Recl. Trav. Chim. Pays-Bas* **81:**202.

Mantlo, N. B., Chakravarty, P. K., Ondeyka, D., Siegl, P. K. S., Chang, R. S. L., Lotti, V. J., Faust, A. K., Chen, T. B., Schorn, T. W., Sweet, C. S., Emmert, S. E., Patchett, A. A., and Greenlee, W. J., 1991, Potent, orally active imidazo[4,5-b]pyridine-based angiotensin II receptor antagonists, *J. Med. Chem.* **34:**2919–2922.

Meyers, A. I., and Mihelich, E. D., 1975, Oxazolines XXII. Nucleophilic aromatic substitution on aryl oxazolines. An efficient approach to unsymmetrically substituted biphenyls and *o*-alkyl benzoic acids, *J. Am. Chem. Soc.* **97:**7383.

Middlemiss, D., and Watson, S. P., 1994, A medicinal chemistry case study: An account of an angiotensin II antagonist drug discovery program, *Tetrahedron,* **50:**13049–13080, and references therein.

Munafo, A., Christe, Y., Nussberger, J., Shum, L. Y., Borland, R. M., Lee, R. J., Waeber, B., Biollaz, J., and Brunner, H. R., 1992, Drug concentration response relationship in normal volunteers after oral administration of losartan, an angiotensin II receptor antagonist, *Clin. Pharmacol. Ther.* **51:**513–521.

Naylor, E. M., Chakravarty, P. K., Costello, C. A., Chang, R. S., Chen, T.-B., Faust, K. A., Lotti, V. J., Kivlighn, S. D., Zingaro, G. J., Siegl, P. K. S., Wong, P. C., Carini, D. J., Wexler, R. R., Patchett, A. A., and Greenlee, W. J., 1994, Potent imidazole angiotensin II antagonists: Acyl sulfonamides and acyl sulfamides as tetrazole replacements, *Bioorg. Med. Chem. Lett.* **4:**69–74.

Nelson, E. B., Harm, S. C., Goldberg, M., Shahinfar, S., Goldberg, A., and Sweet, C. S., 1995, Clinical profile of the first angiotensin II (AT-1 specific) receptor antagonists, in: *Hypertension: Pathophysiology, Diagnosis, and Management* (J. H. Laragh and B. M. Brenner, eds.), pp. 2895–2916, Raven Press, New York.

Noda, K., Saad, Y., Kinoshita, A., Boyle, T. P., Graham, R. M., Hussain, A., and Karnik, S. S., 1995, Tetrazole and carboxylate groups of angiotensin receptor antagonists bind to the same subsite by different mechanisms, *J. Biol. Chem.* **270**(5)**:**2284–2289.

Pierce, M. E., Carini, D. J., Huhn, G. F., Wells, G. J., and Arnett, J. F., 1993, Practical synthesis and regioselective alkylation of methyl 4(5)-(pentafluoroethyl)-2-propylimidazole-5(4)-carboxylate to give DuP 532, a potent angiotensin II antagonist, *J. Org. Chem.* **58:**4642.

Quan, M. L., Olson, R. E., Carini, D. J., Ellis, C. D., Hillyer, G. L., Lalka, G. K., Liu, J., VanAtten, M. K., Chiu, A. T., Wong, P. C., Wexler, R. R., and Timmermans, P. B. M. W. M., 1994, Balanced angiotensin II receptor antagonists. I. The effects of biphenyl "ortho"-substitution on AT_1/AT_2 affinities, *Bioorg. Med. Chem. Lett.* **4:**63–68.

Quan, M. L., Chiu, A. T., Ellis, C. D., Wong, P. C., Wexler, R. R., and Timmermans, P. B. M. W. M., 1995, Balanced AT_1/AT_2 receptor antagonists. 4. XR510 and related 5-(3-amidopropanoyl)-imidazoles possessing equal affinity for the AT_1 and AT_2 receptors, *J. Med. Chem.* **38:**2938–2945.

Reichle, W. T., 1964, Preparation, properties, and thermal decomposition products of organoazides of silicon, germanium, tin, lead, phosphorous, and sulfur, *Inorg. Chem.* **3:**402.

Shuman, R. F., King, A. O., and Anderson, R. K., 1991, *o*-Lithiation process for the synthesis of 2-substituted 1-(tetrazol-5-yl)benzenes, U.S. Patent 5,039,814.

Siegl, P. K. S., Chang, R. S. L., Mantlo, N. B., Chakravarty, P. K., Ondeyka, D. L., Greenlee, W. J., Patchett, A. A., and Lotti, V. J., 1992, In vitro pharmacology of L-158,809, a new highly potent and selective nonpeptide angiotensin II receptor antagonist, *J. Pharmacol. Exp. Ther.* **262:**139–144.

Sisido, K., Nabika, K., and Isida, T., 1971, Formation of organotin-nitrogen bonds III. *N*-Trialkyltin-5-substituted tetrazoles, *J. Organomet. Chem.* **33:**337.

Soll, R. M., Kinney, W. A., Primeau, J., Garrick, L., McCaully, R. J., Colatsky, T., Oshiro, G., Park, C. H., Hartupee, C., White, V., McCallum, J., Russo, A., Dinish, J., and Wojdan, A., 1993, 3-Hydroxy-3-cyclobutene-1, 2-dione: Application of a novel carboxylic acid bioisostere to an in-vivo active non-tetrazole angiotensin-II antagonist, *Bioorg. Med. Chem. Lett.* **3:**757–760.

Stearns, R. A., Miller, R. R., Doss, G. A., Chakravarty, P. K., Rosegay, A., Gatto, G. G., and Chiu, S.-H. L., 1992, The metabolism of DuP 753, a nonpeptide angiotensin II receptor antagonist, by rat, monkey, and human liver slices. *Drug Metab. Dispos.* **20:**281–287.

Thayer, J. S., 1966, Azide derivatives of organometallic compounds, *J. Organomet. Chem. Rev.* **1966:**157.

Thayer, J. S., and West, R., 1964, Trimethylazido compounds of group IVa elements, *Inorg. Chem.* **5:**889.

Timmermans, P. B. M. W. M., Wong, P. C., Chiu, A. T., Herblin, W. F., Benfield, P., Carini, D. J., Lee, R. J., Wexler, R. R., Saye, J. A., and Smith, R. D., 1993, Angiotensin II receptors and angiotensin II receptor antagonists, *Pharmacol. Rev.* **45:**205–251.

Villemin, D., and Labiad, B., 1990, Clay catalysis: Dry condensation of tetronic acid with aldehydes under microwave irradiation. Synthesis of 3-(arylmethylene)-2,4-(3H, 5H)-furandiones, *Synth. Commun.*, **20:**3207–3212.

Wexler, R. R., Greenlee, W. J., Irvin, J. D., Goldberg, M. R., Prendergast, K., Smith, R. D., and Timmermans, P. B. M. W. M., 1996, Nonpeptide angiotensin II receptor antagonists: The next generation in antihypertensive therapy, *J. Med. Chem.* **39:**625–656.

Whitebread, S., Mele, M., Kamber, B., and de Gasparo, M., 1989, Preliminary biochemical characterization of two angiotensin II receptor subtypes, *Biochem. Biophys. Res. Commun.* **163:**284–291.

Wong, P. C., Price, W. A., Chiu, A. T., Duncia, J. V., Carini, D. J., Wexler, R. R., Johnson, A. L., and Timmermans, P. B. M. W. M., 1990a, Nonpeptide angiotensin II receptor antagonists. VIII. Characterization of functional antagonism displayed by DuP 753, an orally active antihypertensive agent, *J. Pharmacol. Exp. Ther.* **252:**719–725.

Wong, P. C., Price, W. A., Chiu, A. T., Duncia, J. V., Carini, D. J., Wexler, R. R., Johnson, A. L., and Timmermans, P. B. M. W. M., 1990b, Nonpeptide angiotensin II receptor antagonists. XI. Pharmacology of EXP3174: An active metabolite of DuP 753, an orally active antihypertensive agent, *J. Pharmacol. Exp. Ther.* **255:**211–217.

Wong, P. C., Price, W. A., Chiu, A. T., Duncia, J. V., Carini, D. J., Wexler, R. R., Johnson, A. L., and Timmermans, P. B. M. W. M., 1990c, Hypotensive action of DuP 753, an angiotensin II antagonist, in spontaneously hypertensive rats. Nonpeptide angiotensin II receptor antagonists: X, *Hypertension* **15:**459–468.

Wong, P. C., Hart, S. D., Duncia, J. V., and Timmermans, P. B. M. W. M., 1991a, Nonpeptide angiotensin II receptor antagonists. Studies with DuP 753 and EXP3174 in dogs, *Eur. J. Pharmacol.* **202:**323–330.

Wong, P. C., Barnes, B., Chiu, A. T., Christ, D. D., Duncia, J. V., Herblin, W. F., and Timmermans, P. B. M. W. M., 1991b, Losartan (DuP 753), an orally active nonpeptide angiotensin II receptor antagonist, *Cardiovasc. Drug Rev.* **9:**317–339.

Wong, Y. N., Holm, K. A., Burcham, D. L., Huang, S.-M., and Quon, C. Y., 1994, The pharmacokinetics and metabolism of DuP 532, a non-peptide angiotensin II receptor antagonist, in rats and dogs, *Biopharm. Drug Dispos.* **15:**53–63.

Chapter 4

Development of an Orally Active Tripeptide Arginal Thrombin Inhibitor

Robert T. Shuman and Paul D. Gesellchen

1. INTRODUCTION

Blockage of diseased arteries resulting from thrombotic occlusions causes life-threatening heart attacks, strokes, and peripheral vascular disease. During normal hemostasis, blood components do not interact with intact endothelium. However, exposure of flowing blood to the subendothelial layers of a damaged vessel wall initiates a complex cascade that gives rise to the rapid deposition of platelets, insoluble fibrin, white blood cells, as well as many more blood components (Goldsmith and Turitto, 1986). This accumulating thrombus mass eventually will occlude the vessel and stop blood flow to downstream tissues. Morbidity and mortality from cardiovascular disorders, such as acute myocardial infarction produced by a blockage of a coronary artery, deep venous thrombosis, and thrombotic stroke could potentially be decreased with parenterally administered anticoagulants (Gold, 1990; Wagner and Hubbell, 1990). On subsequent release of the patient from the hospital, administered oral anticoagulants would be preferred to prevent future thrombotic episodes (Stein et al., 1989).

The "coagulation cascade" is a series of proteolytic enzymatic reactions in which inactive zymogens are converted to active enzymes whose biochemical relationship can be represented as two distinct pathways, termed the *extrinsic* and *intrinsic* pathways (Fig. 1). These reactions collectively lead to the formation of

Robert T. Shuman and Paul D. Gesellchen • Lilly Research Laboratories, Eli Lilly and Company, Indianapolis, Indiana 46285.

Integration of Pharmaceutical Discovery and Development: Case Studies, edited by Borchardt *et al.*, Plenum Press, New York, 1998.

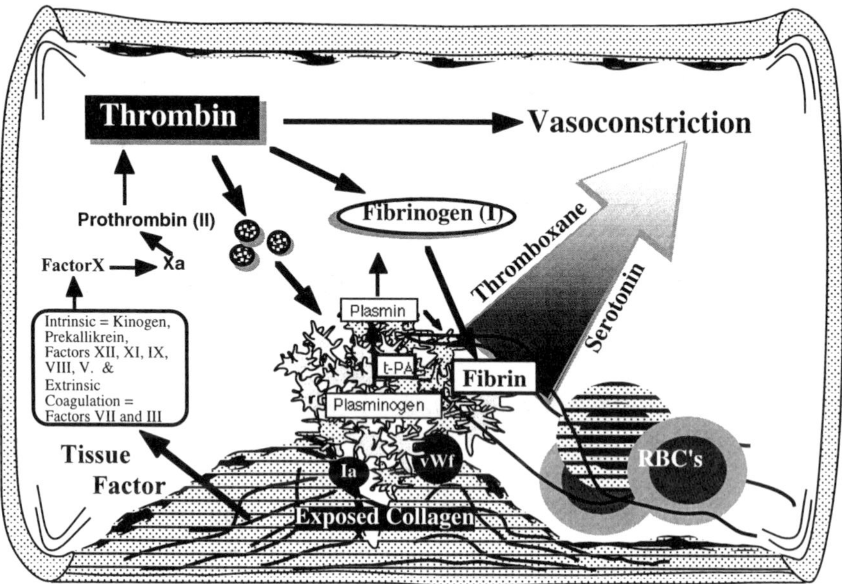

Figure 1. Representation of the interactive mechanisms giving rise to clot formation, regulation, and dissolution. Ruptured atherosclerotic plaque exposes tissue factor, which triggers the coagulation process. Simultaneously, the fibrinolytic cascade is activated, which converts plasminogen to plasmin in order to digest fibrin. Platelet aggregation causes thromboxane and serotonin secretion, resulting in smooth muscle-mediated vasoconstriction. Adapted from Jackson *et al.*, 1996, *Clin. Appl. Thromb./Hemost.* **2:**259–267, Lippincott–Raven Publishers.

activated factors V, X, and prothrombin, all bound to a lipid surface called the *prothrombinase complex* (Davies and Thomas, 1981). This complex converts prothrombin to thrombin, the terminal enzyme of the cascade. There are 12 principal coagulation factors in the reaction illustrated in Fig. 1. Seven of these coagulation factors (XII, prekallikrein, XI, IX, VII, X, and II) belong to a class of enzymes called *serine proteases.* These enzymes are so classified because they all have a serine residue as part of their charge transfer system in the active site. This charge transfer system consists of the amino acid residues histidine, serine, and aspartic acid.

Of all of the potential targets in the coagulation system with which to interfere, thrombin has emerged as the most attractive target. It exists as an inactive zymogen, prothrombin, and is only activated in blood after vascular injury, whereupon thrombin is rapidly generated at the site of vascular injury. Thrombin also plays a central role in platelet recruitment and aggregation (Smith, 1980). In addition, the thrombotic response is amplified, and modulated through feedback loops in the coagulation cascade. Fibrin formation is caused by thrombin proteolysis of four arginyl-glycine bonds in the plasma protein fibrinogen. The activated soluble fibrin molecules spontaneously polymerize into soluble oligomers, then to insoluble fibers that comprise, in part, the structural protein matrix of a blood clot. Small concentrations of thrombin (10^{-8} M) cause rapid clotting of blood which contains approximately 10^{-5} M fibrinogen (Blomback *et al.,* 1977). Therefore, the enzyme thrombin becomes a central mediator of thrombus formation and fibrin deposition in the pathogenesis of thromboembolic diseases (Goldsmith and Turitto, 1986).

To offset fibrin formation through the coagulation process, the fibrinolytic system is triggered. The fibrinolytic pathway dissolves fibrin (Blomback *et al.,* 1978). The key step in fibrinolysis is the conversion of plasminogen to the serine protease plasmin by tissue plasminogen activator (t-PA) (Fig. 1). Plasmin can digest either fibrinogen, fibrin monomers, or clot bound fibrin and all of these functions are a part of the normal process for maintaining blood flow. Thus, thrombolytic therapy is the pharmaceutical application of fibrinolysis that attempts to open a vessel occluded by a thrombus. Practically, this is accomplished by administering t-PA to induce plasmin formation, which dissolves the clot. Therefore, it is critical that any thrombin inhibitor that is going to be used clinically does not interfere with the fibrinolytic system serine proteases at pharmacologically relevant concentrations.

Presently, venous and arterial thrombotic conditions are treated with heparin or warfarin. Heparin administration is preferred for acute therapy. Unfortunately, heparin is not an optimal anticoagulant for several reasons. It acts indirectly on thrombin by accelerating the inhibitory effect of endogenous antithrombin III (the main physiological inhibitor of thrombin) (Amerena *et al.,* 1990). Because antithrombin III levels can vary in plasma and because surface-bound thrombin seems resistant to this indirect mechanism, heparin can be an ineffective treatment. Therefore, heparin is not effective in antagonizing the activity of clot bound thrombin.

Oral administration of an anticoagulant is preferred for chronic therapy with antithrombotic drugs and currently warfarin is the drug of choice. Warfarin inhibits multiple steps of the coagulation cascade by interfering with the vitamin K-dependent gamma carboxylation of prothrombin, as well as clotting factors VII, IX, and X (Amerena *et al.,* 1990). Warfarin therapy requires dose titration and anticoagulant activity must be monitored regularly. Harmful interactions between warfarin and many other drugs are common, and hemorrhage is the most common side effect (Smith *et al.,* 1988).

Parenteral and oral administration of a thrombin inhibitor may provide advantages over heparin and warfarin. Some advantages would include a rapid onset of activity and improved safety because only the target enzyme thrombin would be inhibited. During the past decade, there has been a virtual explosion of activity in the design, synthesis, and biological evaluation of peptide and peptide-mimetic thrombin inhibitors. This chapter will highlight some of the structure–activity relationships (SAR), biological evaluation, toxicology, and clinical data on a series of peptide arginal thrombin inhibitors.

2. IDENTIFICATION OF LEAD COMPOUNDS

Historically, antithrombotic compounds from early efforts were largely unsuccessful because of the difficulty in demonstrating antithrombotic activity in animal models (Okimoto *et al.,* 1975). The identification of a tripeptide sequence, D-phenylalanyl-L-prolyl-L-arginine (D-Phe-Pro-Arg), as a key sequence that mimicked the fibrinogen cleavage site led to a better understanding of the structural requirements for antithrombin activity (Bajusz *et al.,* 1978). Scientists from the Hungarian Institute of Drug Research (HIDR) in Budapest, Hungary, used the substrate analogue approach in the design of synthetic inhibitors of thrombin (Pozsyay *et al.,* 1981). Bajusz *et al.* (1981) showed that Boc-D-phenylalanyl-prolyl-arginine aldehyde (compound **12,** Boc-D-Phe-Pro-Arg-H, Table I) was a potent inhibitor of thrombin. Studies suggested that the manner of interaction of fibrinogen with thrombin involved binding with a specific sequence on the fibrinogen A-chain ($Phe^{8}...Val^{16}Arg^{17}$) where the residues between Phe^{8} and Val^{16} would be situated to allow the key amino acids to be in close proximity for binding to fibrinogen. It was the expectation that the D-Phe-Pro-Arg sequence would mimic the $Phe^{8}...Val^{16}Arg^{17}$ interaction (Bajusz *et al.,* 1981). The C-terminal aldehyde (arginal) group was added to the molecule to produce a transition-state inhibitor. As described by Bajusz *et al.* (1983), the aldehyde portion of the arginine carbonyl forms a covalent bond with the serine hydroxyl (Ser^{195}) in the active site of thrombin.

The discovery of the tripeptide arginals led to the observation of activity for Boc-D-Phe-Pro-Arg-H (**12**) in a rabbit model of thrombosis (Bagdy *et al.,* 1992). Compound **12** clearly demonstrated respectable activity *in vitro,* although it exhibited poor selectivity against the enzymes plasmin and t-PA (Table I). Compound **12** exhibited comparable activity to heparin in a rat model of arterial thrombosis (Shuman *et al.,* 1992). The next step was to improve the thrombin inhibitory potency and selectivity of **12.** However, because of the lack of an X-ray crystal structure of thrombin during this time period, the initial approach was to modify the P3 residue (D-Phe) of the tripeptide as a result of the ease of synthesis of aromatic amino acids and the ready availability of starting materials. The synthesis of the inhibitors generally followed procedures described by Shuman *et al.* (1995) and Bajusz *et al.* (1990).

Table I
In Vitro Inhibitory Activity[a] and Selectivity of Selected Tripeptide Arginals[b]

No.	Compound[c]	Thrombin	Plasmin	t-PA	P/Th[d]	t-PA/Th[e]
1	TFA-D-Phg(αEt)-Azt-Arg-H	0.0097	4.3	62	440	6,400
2	Boc-D-Chg-Pro-Arg-H	0.012	0.081	2.3	7	190
3	Boc-D-Phg-Azt-Arg-H	0.013	0.15	6.6	12	510
4	Boc-D-Phg-Pro-Arg-H	0.016	0.098	8.7	6	540
5	Boc-D-2-Nag-Pro-Arg-H	0.018	0.055	12	3	670
6	Boc-D-Phe-Azt-Arg-H	0.018	0.36	0.90	20	50
7	D-1-Tiq-Pro-Arg-H	0.019	1.5	430	80	23,000
8	Boc-D-Phg(αMe)-Azt-Arg-H	0.022	3.9	120	180	5,500
9	Boc-D-1-Nag-Pro-Arg-H	0.023	0.62	20	30	870
10	Boc-DL-Phg(3,4 Cl)-Pro-Arg-H	0.028	0.18	31	6	1,100
11	Boc-D-Phg(F)-Pro-Arg-H	0.030	0.085	18	4	600
12	Boc-D-Phe-Pro-Arg-H	0.045	0.19	0.95	4	20
13	Boc-D-Thg-Pro-Arg-H	0.073	0.15	13	2	180
14	Boc-D-Phe-Thz-Arg-H	0.088	0.34	23	4	260
15	*R*(+)-MeCH(Ph)CO-Pro-Arg-H	0.094	18	>10	190	>100
16	*R*(+)-EtCH(Ph)CO-Pro-Arg-H	0.13	21	>100	160	>800
17	*R*(+)-MeOCH(Ph)CO-Pro-Arg-H	0.25	15	860	60	3,400
18	Boc-D-Phe-DL-Pro(5,5Me)-Arg-H	0.29	11	>30,000	38	>50,000
19	D-MePhe-Pro-Arg(αMe)-H	1.0	850	>30,000	850	>30,000
20	Boc-D-Phe-Pip-Arg-H	1.7	8.1	48	5	30
21	D-1-Tiq-Pro-Arg(αMe)-H	4.9	740	>30,000	150	>6,000

[a]IC_{50} (μM), bovine thrombin, human plasmin, and human t-PA.
[b]Arranged in order of decreasing thrombin inhibitory potency.
[c]Symbols and abbreviations are in accordance with the recommendations of the IUPAC-IUB Commission on Biochemical Nomenclature (*Eur. J. Biochem.*, 1984, **138:**9); structures of unnatural amino acids are shown in Fig. 2.
[d]Plasmin–thrombin IC_{50} ratio.
[e]t-PA–thrombin IC_{50} ratio.

2.1. *In Vitro* Structure–Activity Relationships

A number of analogues were prepared and evaluated for their ability to inhibit thrombin, plasmin, and t-PA. For these compounds to be therapeutically useful, it was important that they not inhibit the fibrinolytic processes through inhibition of the enzymes plasmin and t-PA (Chandler *et al.*, 1974). A crude measure of the predicted therapeutic usefulness of these inhibitors was obtained by examination of the ratios of plasmin to thrombin, or t-PA to thrombin IC_{50}; thus, higher values denote greater selectivity (Tables I and II).

One approach in the SAR was to add conformational restriction to the P3 position by introduction of sterically demanding amino acids. The substitution of the phenylalanine residue in **12** with a phenylglycine residue gave **4,** which exhibited a

Table II
Thrombin Inhibitory Activity,[a] Selectivity, and Relative Oral Activity of R-Arg-H

No.	R[b]	Thrombin[c]	P/Th[d]	t-PA/Th[e]	% relative activity[f]
22	D-MePhe-Pro-	0.0089	80	2,400	12
23	TFA-D-Phg(αMe)-Azt-	0.012	510	11,000	6
24	D-MePhg-Pro-	0.018	30	2,300	18
25	Ac-D-Phg(αMe)-Azt-	0.019	400	13,000	16
26	*R*-1-Piq-Pro-	0.019	40	1,800	54
27	D-3-Tiq-Pro-	0.025	15	3,600	18
28	Boc-DL-Phg($3CF_3$)-Pro-	0.027	5	560	<1
29	D-Pip-Pro-	0.027	252	7,700	38
30	D-Pip-Azt-	0.027	114	3,200	32
31	Ac-D-Phg(3,4Cl)-Pro-	0.031	5	500	1
32	Ac-D-Chg-Pro-	0.041	43	740	3
33	*R*-3-Piq-Pro-	0.046	5	650	46
34	D-MePhg(OH)-Pro-	0.048	66	8,100	24
35	Ac-D-Phg(OMe)-Pro-	0.049	24	3,400	26
36	TFA-D-Phg(αMe)-Pro-	0.064	700	>50,000	4
37	D-Pro-Pro-	0.089	43	13,500	36
38	D-Thz-Pro-	0.25	7	230	16
39	D-Pyr-Pro-	2.7	2	430	40

[a]IC_{50} (μM).
[b]Symbols and abbreviations are in accordance with the recommendations of the IUPAC-IUB Commission on Biochemical Nomenclature (*Eur. J. Biochem.*, 1984, **138:**9); structures of unnatural amino acids are shown in Fig. 2.
[c]Bovine thrombin.
[d]Plasmin–thrombin IC_{50} ratio.
[e]t-PA–thrombin IC_{50} ratio.
[f]Index of relative oral activity expressed as a percent.

3-fold increase in potency with respect to its ability to inhibit thrombin. This change was unexpected, as the same modification in a similar series of thrombin inhibitors had been reported to produce a 10-fold decrease in potency (Bajusz *et al.*, 1983). This gave the first indication that the lipophilic binding pocket in thrombin (P3 position of the inhibitor) may have room to accept a diverse group of structures. An examination of molecular modeling based on the X-ray crystal structure of the enzyme trypsin (a serine protease with similar specificity) confirmed this hypothesis. Therefore, a probe of the structural diversity of the P3 residue was undertaken. Replacement of the phenylglycine in **4** with a variety of unnatural amino acids (Fig. 2) that varied in steric bulk parameters and conformational flexibility gave variations in antithrombotic potency and selectivity (**2, 5, 9–11,** and **13**). Replacement of the Boc-amino group of phenylglycine in **4** with various groups resulted in decreased antithrombotic potency but some analogues had improved selectivity (**15–17** and **24**). A similar finding was reported by Bajusz *et al.* (1984) on a series of phenylalanine modifications, which eventually led that group to a more potent thrombin inhibitor, compound **22** (Bajusz *et al.*, 1987). Replacement of phenylglycine with the constrained amino acid D-1-car-

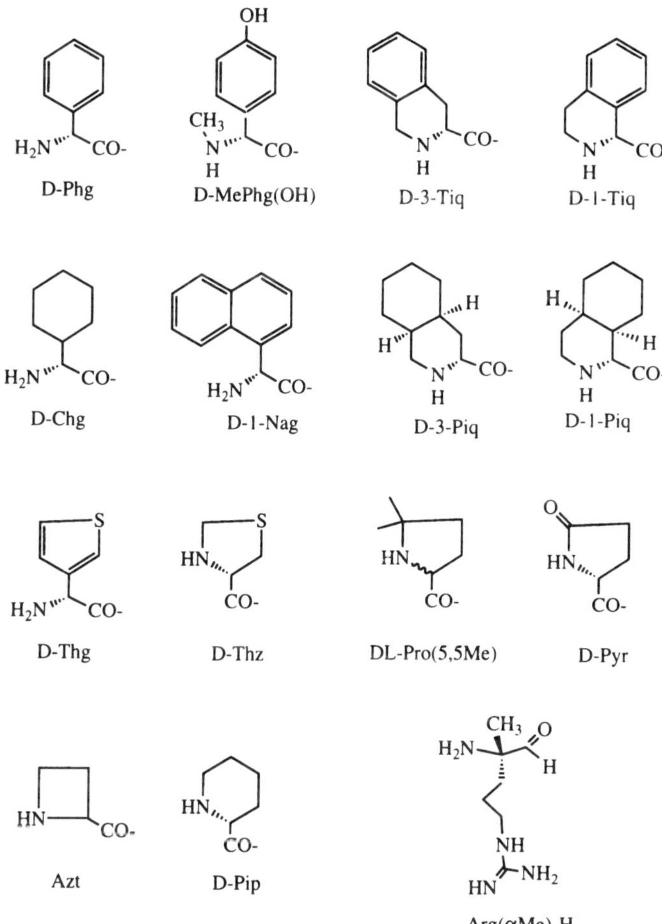

Figure 2. Structures of unnatural amino acids.

boxy-1,2,3,4-tetrahydroisoquinoline resulted in an analogue (**7**) that exhibited no significant loss in potency. However, it exhibited a high degree of selectivity for thrombin relative to t-PA as indicted by a t-PA/thrombin ratio of 23,000.

Another focus of the SAR was modification of the P1 and P2 residues. Bagdy had shown that C-terminal amino acid aldehydes undergo epimerization at the α-carbon (Bagdy *et al.,* 1992). Under certain conditions the arginal (P1) residue in the tripeptide thrombin inhibitors would epimerize (Tomori *et al.,* 1984). In an attempt to eliminate the potential for racemization at the P1 position of these arginals, C^{α}-methylarginine aldehyde was substituted for arginine aldehyde, which resulted in analogues **19** and **21**. These analogues exhibited >100-fold loss in thrombin inhibition; however, improved selectivity versus the other serine proteases was observed. Modification of the P2 position was explored in order to investigate the influence of proline on enzyme selectivity and potency. The compounds synthesized exhibited dramatic changes in potency and selectivity (**6, 14, 18,** and **20**) with the azetidine-2-carboxylic acid (Azt) substitution conferring improved potency.

These results prompted the incorporation of Azt at P2 and Phg at P3 in the same analogue. The resulting analogue (**3**) demonstrated a slight increase in thrombin inhibitory potency and little improvement in selectivity. The replacement of the phenylglycine residue in **3** with the conformationally constrained α-methylphenylglycine (**8**) resulted in a 2-fold loss in potency toward thrombin but a 10-fold increase in selectivity for thrombin versus both plasmin and t-PA. The replacement of the Boc group in **8** with the highly electronegative protecting group trifluoroacetyl (**23**) resulted in a 2-fold increase in potency and selectivity. Additional modifications of the amino protecting group, the α-alkyl group, and the P3 residue of **8** led to compounds **1, 25,** and **36,** which demonstrated improved potency for **1** and enhanced selectivity for **25** and **36**.

A systematic investigation of the SAR resulted in the development of more potent agents. However, wide variations in the specificity of these compounds were observed (Shuman *et al.*, 1993). The wide range of inhibitory effects toward plasmin and t-PA shown in Tables I and II suggests that certain of the arginals would not interfere with t-PA-mediated fibrinolysis, such as **1, 7, 8, 25,** and **36,** whereas other compounds like **6, 12,** and **20** could potentially interfere.

2.2. *In Vivo* Structure–Activity Relationships

Selected compounds were evaluated in animal models of thrombosis and the correlation between *in vitro* enzyme activity and *in vivo* anticoagulation was determined. The targets for the selection of a compound to be evaluated in humans were good selectivity *in vitro* with an acceptable antithrombotic potency and efficacy in animal models of thrombosis.

The rat was used as the primary animal model because of its small size and ease of study (Smith, 1980). The arterial–venous (AV) shunt thrombosis model in the rat was used because it is dependent primarily on fibrin deposition and it may mimic the clinical condition in which blood circulates through an artificial external device such as a cardiopulmonary bypass machine or a kidney dialysis machine. The rat $FeCl_3$-induced arterial injury model was also used because it is representative of arterial injury in which platelets are involved (Smith *et al.*, 1988; Kurz *et al.*, 1990). Further evaluation of the most promising candidates was performed in the dog. The canine antithrombotic model and the canine thrombolysis model were chosen because they represent models of coronary artery disease (Jackson *et al.,* 1992, 1996). In these models, the time to occlusion is measured following electrical injury of a coronary artery. A successful candidate for clinical evaluation should have little or no potential for bleeding liability. Thus an estimation of bleeding liability was obtained in the anesthetized dog by measurement of the bleeding time in the gingiva of the left jaw (Jackson *et al.,* 1993).

The antithrombotic effects of heparin and compounds **4** and **12** were compared in these models. The results of the rat $FeCl_3$-induced arterial injury model and the rat AV shunt model are summarized in Table III. Heparin and compounds **4** and **12** were found to be efficacious in the AV-shunt model. However, whereas compounds **4** and **12** exhibited equal potency in the arterial model versus the AV-shunt model, much larger doses of heparin were required in the AV-shunt model. This finding is consistent with the interpretation that heparin is less potent in platelet-dependent thrombosis than in fibrin-dependent thrombosis. The three compounds were then evaluated in a canine antithrombotic model. The standard clinically relevant dose of heparin (80 U/kg bolus plus 30 U/kg per hr infusion) was used (Smith and Sundboom, 1981). In the study reported by Jackson *et al.* (1992), heparin prolonged the time to occlusion. However, bleeding time was prolonged significantly (fourfold versus control) with this dose of heparin. Compounds **4** and **12** were effective antithrombotics in the canine study at a dose of 0.5 and 1.0 mg/kg per hr, respectively (Shuman *et al.,* 1992). The three doses of **4** studied (0.5, 1.0, and 2.0 mg/kg per hr) prolonged time to occlusion whereas only the highest two doses of **12** studied (1.0 and 2.0 mg/kg per hr) prolonged time to occlusion (Fig. 3).

Compounds **4** and **12** were compared with heparin as adjuncts to thrombolysis in the canine coronary thrombolysis model. Heparin was ineffective in pre-

Table III

Potency Comparison of Arginals in Rat Thrombosis Models[a]

Compound no.	AV shunt[b]	$FeCl_3$ arterial injury[c]
4	2.1	2.9
7	0.8	1.0
9	NT[d]	6.4
12	5.5	5.2
22	0.8	2.1
24	1.3	2.2
Heparin, bolus	26 U/kg	107 U/kg
Heparin, infusion	53 U/kg/hr	309 U/kg/hr

[a]Drug infused 15 min before allowing blood to flow through AV shunt or applying $FeCl_3$.
[b]ED_{50} = infusion dose mg/kg/hr required to reduce by 50% the control thrombus.
[c]$ED_{2\times}$ = infusion dose mg/kg/hr required to double the control time to occlusion (26 ± 1 min) after 35% $FeCl_3$ application.
[d]NT, not tested.

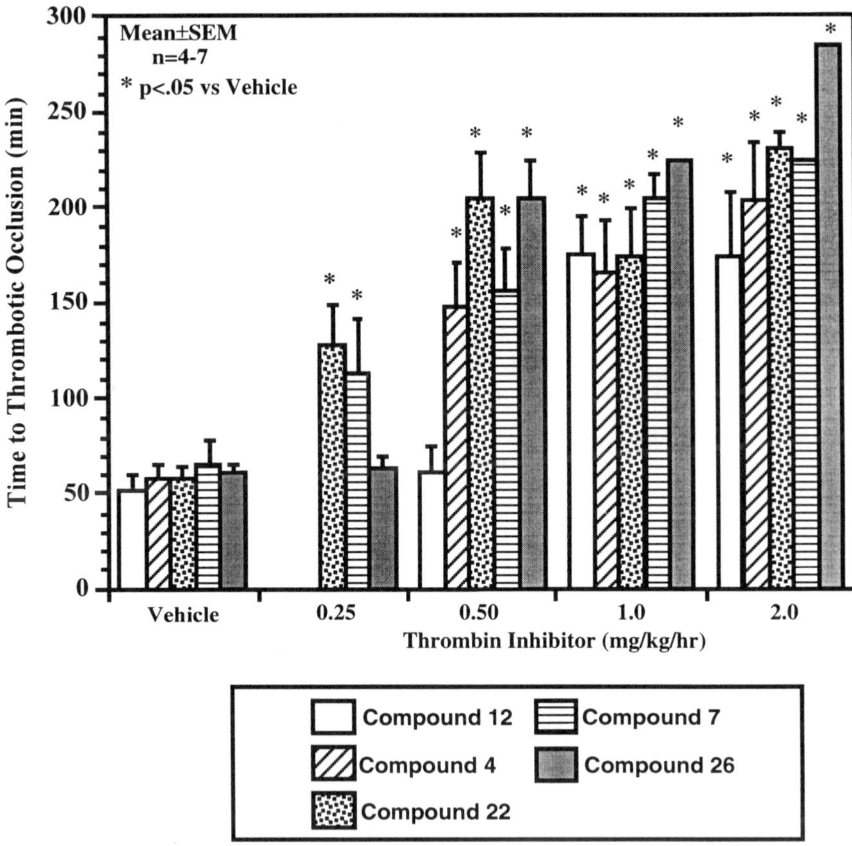

Figure 3. Dose response of thrombin inhibitors in canine thrombogenesis model.

venting or delaying the onset of reocclusion after thrombolysis (Jackson *et al.,* 1993). In fact, heparin showed a trend to adversely affect time to reocclusion. Compound **12** at 1.0 mg/kg per hr was also found to be ineffective in this model (Jackson *et al.,* 1993). In addition, compound **4** was ineffective at prevention of reocclusion at 0.5 and 1.0 mg/kg per hr (Fig. 4). At 1.0 mg/kg per hr, compound **4** prolonged the time required for t-PA to successfully lyse the coronary thrombus (time to reperfusion) (Fig. 4). A possible explanation of the delayed time to reperfusion could be inhibition of plasmin-mediated digestion of the fibrin thrombus. This rationale is supported by the *in vitro* data in Table I demonstrating that **4** inhibits plasmin. This inhibition of plasmin could theoretically be responsible for the delay in reperfusion at the dose of 1.0 mg/kg per hr.

Clearly, compounds **4** and **12** are very effective inhibitors of thrombin and

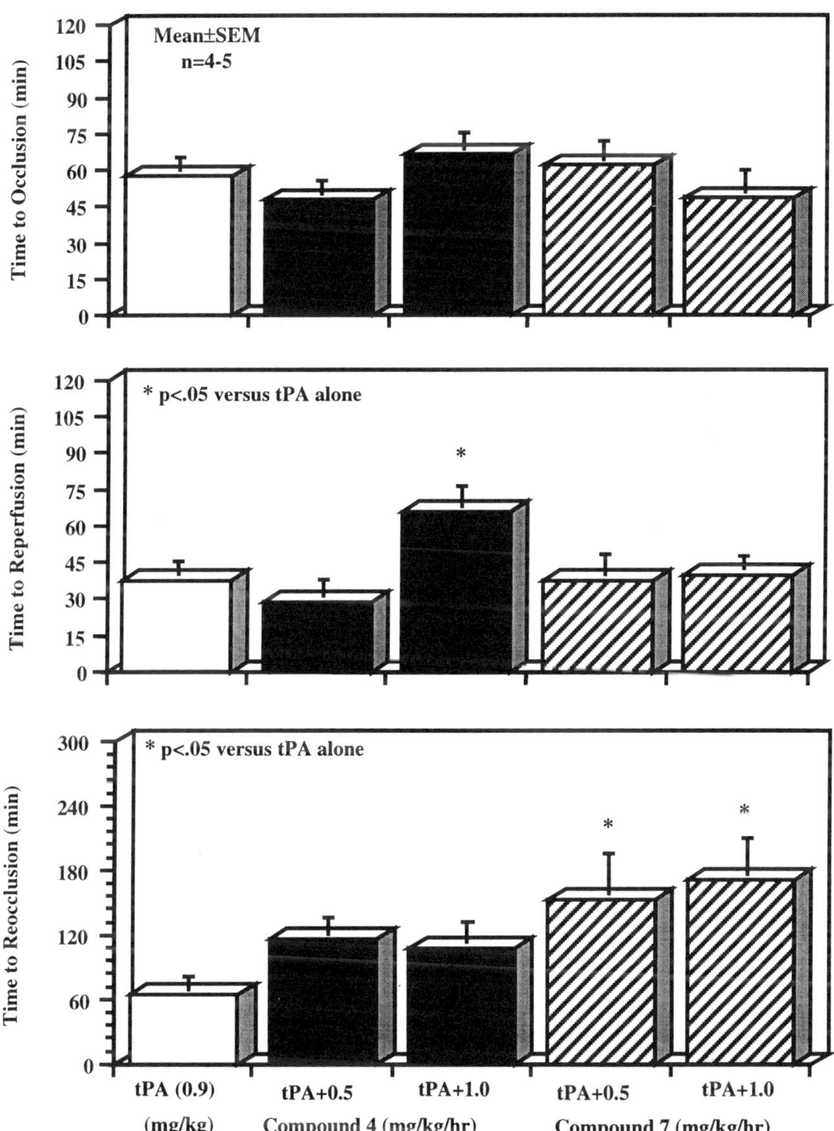

Figure 4. Effects of thrombin inhibitors on thrombolytic efficacy in canine model of coronary thrombosis. Animals received 1-hr infusion of t-PA or adjunctive therapy with simultaneous infusion of t-PA plus thrombin inhibitor (drugs were administered a total of 2 hr).

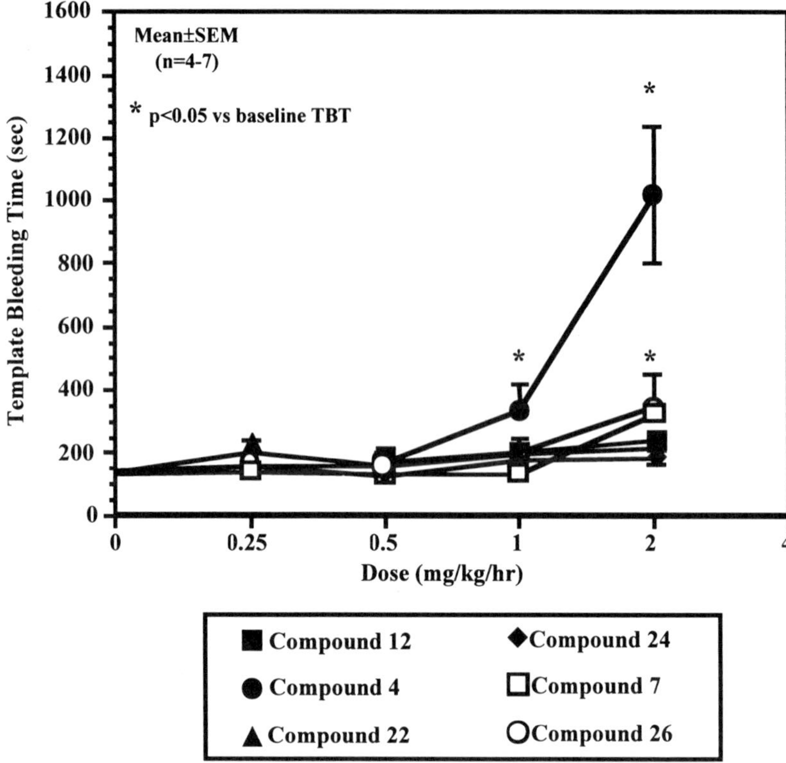

Figure 5. Effects of thrombin inhibitors on template bleeding time.

have an antithrombotic effect in these models. Thus, the demonstration of considerable therapeutic potential for this class of compounds led to the evaluation of other candidates from the *in vitro* SAR. The results of the *in vivo* SAR in the rat are summarized in Table III. All compounds studied caused dose-dependent antithrombotic responses. Compounds **7, 22,** and **24** exhibited the lowest antithrombotic doses in the rat AV-shunt model and the rat $FeCl_3$-induced arterial injury model. Compounds **4, 7, 12,** and **22** prolonged time to occlusion in the canine antithrombotic model at minimum effective doses of 0.5, 1.0, and 2.0 mg/kg per hr, respectively (Fig. 3). In addition, compounds **7** and **22** were examined in the canine coronary thrombolysis model as an adjunct to thrombolysis. The time to reocclusion was significantly prolonged for both compounds **7** (Fig. 4) and **22** (Jackson *et al.,* 1993) at doses of 0.5 and 1.0 mg/kg per hr. Compounds **4, 7, 12, 22,** and **24** were found to have very little effect on template bleeding time at the doses tested. Only compound **4** caused a significant increase in bleeding time at 1.0 and 2.0 mg/kg per hr (Fig. 5).

3. DEVELOPMENT OF PARENTERAL CLINICAL CANDIDATE

3.1. Development of Licensed Compound (Efegatran)

Compounds **22** (generic name efegatran) and **7** were shown to be efficacious antithrombotic agents in well-characterized rat and dog models. No other pharmacological effects could be detected even at very large doses (20–40 times the efficacious dose). Based on these and other results, compounds **7** and **22** were selected for further evaluation.

The intravenous bolus and continuous infusion plasma pharmacokinetics of compound **22** in animals were determined using a stereospecific HPLC method (Ruterbories *et al.,* 1992). Peptide arginals exist in aqueous media as an equilibrium of three principal physical forms (aldehyde hydrate and two epimeric cyclic hemiaminals) (Tomori *et al.,* 1984). This equilibrium causes pure peptide arginals to epimerize to a mixture of two diastereomeric peptide arginals in nonacidic aqueous solutions. This transformation process causes efegatran to epimerize to its inactive isomer (D-MePhe-Pro-D-Arg-H, i.e., the DLD-isomer). This represents an inactivation pathway and the rate of conversion affects the *in vivo* antithrombotic efficacy. The rate of conversion was examined *in vitro* by incubating efegatran in human plasma at 37°C, then determining the percentage of DLD-efegatran at selected intervals. Efegatran was extracted from plasma samples by solid-phase extraction and then derivatized with 2,4-dinitrophenylhydrazine (DNPH) to prevent hydrate and epimer formation thereby preventing racemization of efegatran to its inactive isomer (DLD-efegatran). The HPLC analysis at 360 nm was carried out after solid-phase extraction of the DNPH-treated samples. The DLD-efegatran concentration was 6% at the start of the experiment and increased to 30% after 2 hr and 44% after 4 hr. The transformation rate was linear during the first 2 hr of incubation with a rate of inversion of approximately 12%/hr, and thereafter decreased to about 7%/hr. The decrease in conversion with time suggests that at some time beyond 4 hr, the efegatran and DLD-efegatran concentrations would reach equilibrium (50–50 mixture). There was excellent agreement between efegatran plasma concentrations determined by HPLC and by thrombin time coagulation tests; thus, this inversion product appeared to have little or no thrombin inhibitory activity.

The toxicological evaluation of efegatran was performed in mice, rats, and dogs (Smith *et al.,* 1996). In mice, the median lethal bolus intravenous dose was 22.3 mg/kg for males and 26.4 mg/kg for females. Death occurred within 15 min after dosing. No deaths were seen at 15 mg/kg in males and at 20 mg/kg in females. Signs of toxicity included convulsions, exophthalmus, bradypnea, dyspnea, cyanosis, and respiratory arrest. In rats, the median lethal bolus intravenous dose of efegatran was 41.0 mg/kg for males and 38.0 mg/kg for females. Death occurred within 60 min after dosing. No deaths were seen at 15 mg/kg in males or females.

Signs of toxicity included excitation, tachypnea, tremors, and cyanosis. Bleeding at the injection site was observed in the highest two dose groups (40 and 60 mg/kg). In dogs, the estimated lethal bolus intravenous dose was greater than 40 mg/kg.

Although similar data were observed with compound **7**, compound **22** was selected for clinical evaluation.

3.2. Summary of Clinical Data on Efegatran

The anticoagulant activity and pharmacokinetics of efegatran were assessed in Phase 1 dose-escalation studies with intravenous infusion from 15 min duration to 48 hr. Efegatran was well tolerated at all doses administered. No serious life-threatening events were experienced by any subject during the studies. Mild events such as headache, dizziness, and phlebitis at the site of injection occurred. Headaches or dizziness occurred with similar frequency in the treatment and placebo group. Phlebitis was diminished by dilution of the efegatran infusion solution (Jackson *et al.,* 1996).

Efegatran then underwent further tests in Phase 2 protocols in patients with unstable angina or acute myocardial infarction. Intravenous infusion produced predictable and stable anticoagulant activities that reached steady state by about 2 hr with no accumulation of effect, up to doses of 0.84 mg/kg per hr. The elimination half-life in volunteers was 35 min, in excellent agreement with the preclinical disposition studies, with total plasma clearance of 0.4 liters/hr per kg. The pharmacokinetic/pharmacodynamic model developed from clinical trials conducted in healthy volunteers appears to be predictive of the effects observed in unstable angina patients. Approximately 85% of steady-state concentrations were achieved 2 hr after starting a constant-rate infusion. The 35-min half-life should represent a useful range for clinical evaluation of a parenteral anticoagulant compound. The short half-life provides the safety advantage of rapid reversal of effects gained on stopping the infusion.

4. DEVELOPMENT OF AN ORAL CANDIDATE

Subsequent to the successful progression of efegatran to clinical trials, a new focus was initiated to develop an oral antithrombotic agent. Because it was established that the arginal class of compounds had considerable therapeutic potential, the decision was made to use these compounds as the starting point for an oral SAR. The criteria established for an oral antithrombotic candidate were: an acceptable oral bioavailability in the rat and dog, antithrombotic efficacy in rat and dog models of thrombosis, little potential for bleeding liability as measured by canine template bleeding time, and an acceptable toxicological profile.

The rat served as the primary animal model for the estimation of relative oral

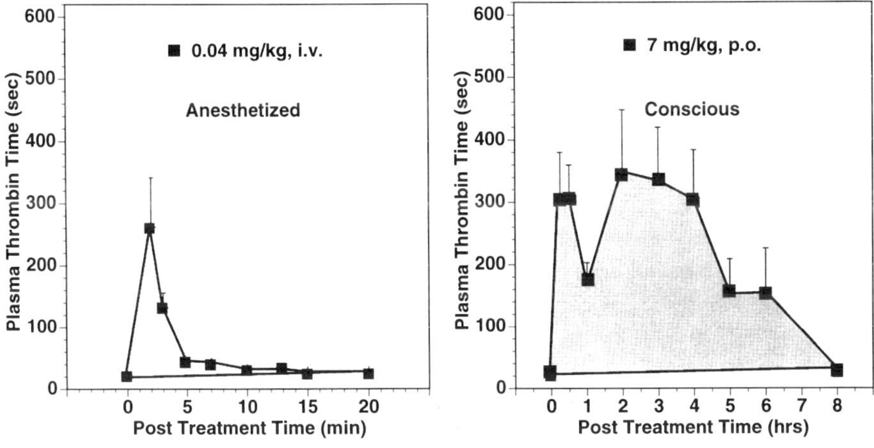

Figure 6. Determination of the time course of plasma thrombin time and area under the curve after administration of compound **26** to rats.

activity. Because of the extra effort required to develop a bioanalytical assay for each analogue prepared in the SAR, absolute bioavailability was not determined. Instead, bioactivity, as measured by changes in plasma thrombin time (TT), served as an index of plasma drug concentration to estimate relative bioavailability. TT represents the time required for a plasma sample to coagulate after addition of a standard amount of thrombin. The time course after intravenous administration was determined in anesthetized animals because the response was completed within 20 min. In contrast, fasted conscious rats were used to study the time course after oral treatment because the response usually persisted as long as 8 hr. The area under the curve (AUC) of the plasma TT time course was determined and adjusted for the different doses. The oral exposure was termed *relative oral activity* and was calculated for each compound using the equation at the bottom of Fig. 6. A typical time course after intravenous and oral dosing using this protocol is shown for compound **26** in Fig. 6 (Shuman *et al.*, 1995). The estimation of relative oral activity for the various analogues in the SAR is summarized in Table II along with their thrombin inhibitory potency and selectivity.

4.1. *In Vivo* Oral Bioavailability

The first compound evaluated orally in a rat was the lead structure compound **12,** which had a relative oral activiy of less than 1% (data not shown). Interestingly, oral administration of the parenteral candidates efegatran and compound **7** to rats gave relative oral activities of 12 and 24%, respectively. This twofold difference in the index of oral exposure found with compound **7** versus efegatran led to the hypothesis that modifications of the P3 residue could have a beneficial effect on oral absorption. Therefore, assuming that this might be predictive of oral activity in humans for this series of compounds, it became the goal to significantly improve on the relative oral activity in rats. In addition, the oral evaluation of efegatran in humans was addressed. The oral bioavailability of efegatran in humans was determined to be less than 6% by dosing a solution of efegatran to fasted human volunteers (unpublished data of Roberts and Lucas). Given the 12% relative oral activity in rats for efegatran, this species appeared to have approximately the same oral exposure profile as that found in humans.

A series of nitrogen modified aryl-substituted phenylglycine derivatives were prepared and evaluated (compounds **24, 28, 31, 34,** and **35**). The relative oral activity of compounds **28** and **31** was decreased compared with efegatran. In the case of compounds **24, 34,** and **35,** only slight improvements in relative oral activity were observed. Conformationally constrained α-methylphenylglycines or cyclohexylglycine in P3 with either trifluoroacetyl or acetyl as amino protecting groups and azetidine or proline in the P2 position (**23, 25, 32,** and **36**) resulted in improved selectivity. However, a fivefold loss in potency for compound **36** relative to **23** was observed with no substantial improvements in the relative oral activity. The conformationally constrained phenylalanine analogue **27** lost selectivity with no improvement in relative oral activity. Replacement of the P3 residue with a constrained amino acid that adds lipophilicity to the molecule (**29**) had no significant effect on potency; however, its relative oral activity improved twofold. This finding led to investigation of the optimal ring size in the P3 position to attain potent thrombin inhibition, good selectivity, and improved oral absorption. Compounds with decreasing ring size and increasing lipophilicity were prepared and evaluated (**37–39**). These modifications attenuated the selectivity and potency; however, the relative oral activity remained high in each case except for thiazolidine substitution (**38**). Replacement of the residue in the P2 position of compound **29** with azetidine-2-carboxylic acid (**30**) had no significant impact on potency, selectivity, or relative oral activity. This result allowed for the continued use of the less expensive proline residue in the P2 position. Replacement of the D-1-Tiq in **7** with a constrained amino acid of increased size and lipophilicity (*cis*-perhydroisoquinoline-3-carbonyl=R-3-Piq) resulted in analogue **33**. This modification reduced thrombin inhibition by twofold and substantially decreased selectivity, but a twofold increase in relative oral activity from 24% for compound **7** to 46% for compound **33** was observed. In contrast, substitution with *cis*-perhydroisoquinoline-1-carboxylic acid (R-1-Piq, **26**) caused no loss in potency. Compound **26** had improved selectivity, compared with compound **33,** and had no loss in relative oral activity. Changes in the P3 moiety were useful not only for increased absorption but also for increased selectivity. Insertion of saturated rings in P3 (i.e., Pip, Pro,

1- or 3-Piq) dramatically improved relative oral activity with compound **26** yielding the highest relative oral activity (54%).

4.2. Oral Dosing in Efficacy Models

Efegatran and compound **26** were studied at various times after oral administration in the rat AV-shunt thrombosis model. To minimize the effect of anesthesia in the oral study, groups of conscious rats were treated at various time intervals and were anesthetized 15 min before thrombus determination. Both compounds caused dose-dependent reductions in the weight of the formed thrombus. The results demonstrate that both efegatran and compound **26** reduced thrombus weight in a time-dependent manner with an ED_{50} of 18.4 and 8.1 mg/kg, respectively, after a single oral dose of 20 mg/kg. Compound **26** exhibited a significantly greater reduction in thrombus weight 1 hr after oral dosing and the antithrombotic effect persisted significantly longer than efegatran. In addition, compound **26** had an $ED_{2\times}$ of 6.6 mg/kg 1 hr after oral administration in the rat $FeCl_3$ model [oral dose required to double the control time to occlusion (22 ± 2 min) after 35% $FeCl_3$ application]. This improved oral potency of compound **26** can be explained by its greater relative oral activity and improved half-life in the rat (Shuman *et al.,* 1996).

An estimation of bleeding liability was performed in a dog by measurement of template bleeding times. Compound **26** was found to have very little effect on template bleeding time at the doses tested (Fig. 5). The only dose that caused a significant increase in bleeding time was 2.0 mg/kg per hr after intravenous infusion, which is four times the intravenous antithrombotic dose in the dog (Fig. 3).

4.3. Pharmacokinetics of Oral Candidate

The antithrombotic efficacy of compound **26** was evaluated in dogs given a single 5 mg/kg oral dose. Plasma concentrations of compound **26** were assayed by HPLC and the data are summarized in Fig. 7. The approximate occlusion time is indicated (X) after vessel injury initiated 240 min postdose. The data clearly demonstrate a relationship between whole blood thrombin time and parent drug concentration with efficacy up to 375 ±SEM min postdose. The pharmacokinetic profile of compound **26** was evaluated in four conscious dogs given a single 1 mg/kg intravenous dose and a 2 mg/kg oral dose in a crossover design. The plasma elimination half-life after oral administration was approximately 2 hr. The volume of distribution (0.3 liter/kg) was similar to the plasma compartment volume. The route of administration had no appreciable effect on total plasma clearance (0.1 liter/hr·kg), elimination rate constant (0.006 min^{-1}), or volume of distribu-

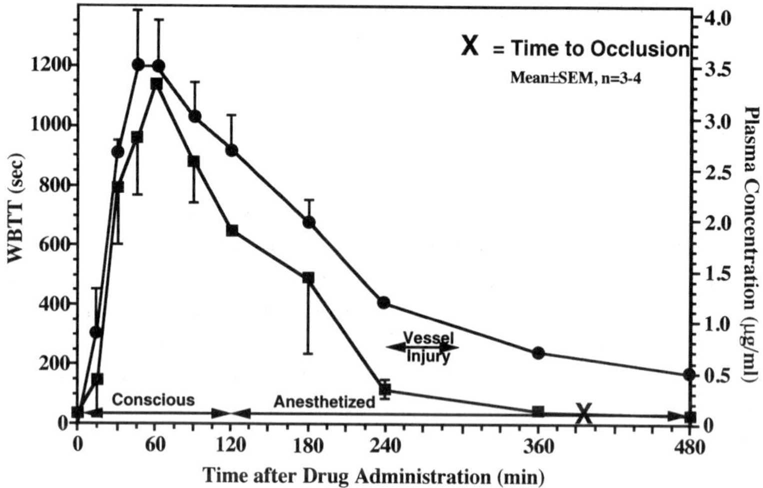

Figure 7. Effect of **26** (5 mg/kg, oral) on whole blood thrombin time (■) and parent drug concentrations (●) plotted vs time in a canine model of coronary artery thrombosis. Dogs were anesthetized 2 hr after dosing. X, approximate occlusion time. The control group occluded about 60 min following electrical injury.

tion. The absolute oral bioavailability was approximately 30% in dogs (unpublished data of Jackson *et al.* and Lindstrom *et al.*).

The relative oral activity of compound **26** in rats was approximately four- to fivefold greater than for efegatran. Both compounds demonstrated oral efficacy in rat thrombosis models. Compound **26** was the more potent antithrombotic agent and the effect persisted longer after oral administration in the rat models. Based on these and other results, compound **26** was chosen for further evaluation as an oral antithrombotic agent.

Oral administration of compound **26** to rats (10, 30, and 100 mg/kg bid) resulted in a dose-dependent increase in maximal plasma concentrations (C_{max}) on days 14 and 28 in both male and female rats. The C_{max} values associated with the 100 mg/kg bid dose group were slightly higher than projected from the C_{max} values observed for the 10 and 30 mg/kg bid dose groups on days 1, 14, and 28 in both male and female rats. However, the C_{max} versus dose relationship was very nearly linear. Areas under the plasma concentration versus time curves (AUC) in male rats were dose-dependent and were similar on days 1 and 28 indicating no accumulation of compound **26** in plasma. The times of maximal plasma **26** concentration (T_{max}) ranged from 0.5 to 2.0 hr. Most T_{max} values were observed after 1.0 hr and were independent of dose. The elimination half-life of compound **26** also was independent of dose and ranged from 1.1 to 2.0 hr with a mean value of 1.4 hr (unpublished data of Sandusky *et al.*).

After either single or multiple dosing, oral administration of compound **26** to dogs resulted in little or no increase in exposure of male dogs to compound **26,** as indicated by C_{max} and AUC values between the 10 and 20 mg/kg per day dose levels. Increased exposure was demonstrated at the 40 mg/kg per day dose level, especially after multiple dosing. Mean C_{max} values in male dogs on day 1 in the 10, 20, and 40 mg/kg per day dose groups were 3.6, 3.7, and 4.1 μg/mL, respectively. On day 25, the mean C_{max} values were 3.3, 4.2, and 15.8 μg/mL, respectively. Compound **26** T_{max} values ranged from 0.5 to 2 hr throughout the study and no sex, dose, or dose duration effects on T_{max} were observed. There was a slight prolongation of the elimination $T_{1/2}$ with increasing dose after a single dose in male dogs. No such trend was observed in female dogs after a single dose, or in male or female dogs after multiple dosing. Mean $T_{1/2}$ values ranged from 2.2 to 4.0 hr in male and female dogs in all dose groups after a single dose, and 3.8 to 4.8 hr after multiple dosing.

Elimination of radioactivity by rats after a single intravenous dose of ^{14}C-compound **26** was 49% of the dose in the urine within 6 hr of dosing, which increased to 63% after 120 hr. Elimination of radioactivity by rats after a single oral dose of ^{14}C-compound **26** was 89% of the dose in the feces and 7% in the urine within 24 hr. In rats after an oral dose of 2 mg/kg of ^{14}C-compound **26,** radioactivity concentrations in plasma and blood remained almost constant from 0.25 to 2 hr postdose. Thereafter, plasma and blood radioactivity concentrations declined with biphasic kinetics having initial half-lives of 76 and 84 min and terminal half-lives of 67 and 214 hr for plasma and blood radioactivity, respectively. The ratio of plasma to blood radioactivity show that drug-related material was contained mainly in the plasma compartment.

The binding of ^{14}C-compound **26** to rat, dog, and human plasma proteins *in vitro* was evaluated at drug concentrations of 2.0, 1.0, 0.5, and 0.1 μg/mL of plasma. The extent of binding was inversely related to plasma concentration in all three spieces. Binding was significantly lower in human plasma than in rat or dog plasma. In rat plasma, 49, 55, 59, and 60% of the radioactivity was bound at 2.0, 1.0, 0.5, and 0.1 μg/mL, respectively. In dog plasma, 51, 58, and 62% of the radioactivity was bound at 1.0, 0.5, and 0.1 μg/mL, respectively. In human plasma, 36.0, 39.8, 43.5, and 44.6% of the radioactivity was bound at 2.0, 1.0, 0.5, and 0.1 μg ^{14}C-**26**/mL, respectively (unpublished data of Lindstrom *et al.*).

4.4. Clinical Data for Oral Candidate

The first human study on compound **26** was conducted in the United Kingdom, to evaluate the safety and tolerance of single oral doses administered in the fasted and fed states. The pharmacokinetics and pharmacodynamics of single oral doses of compound **26** were evaluated in a single-blind, placebo-controlled, ran-

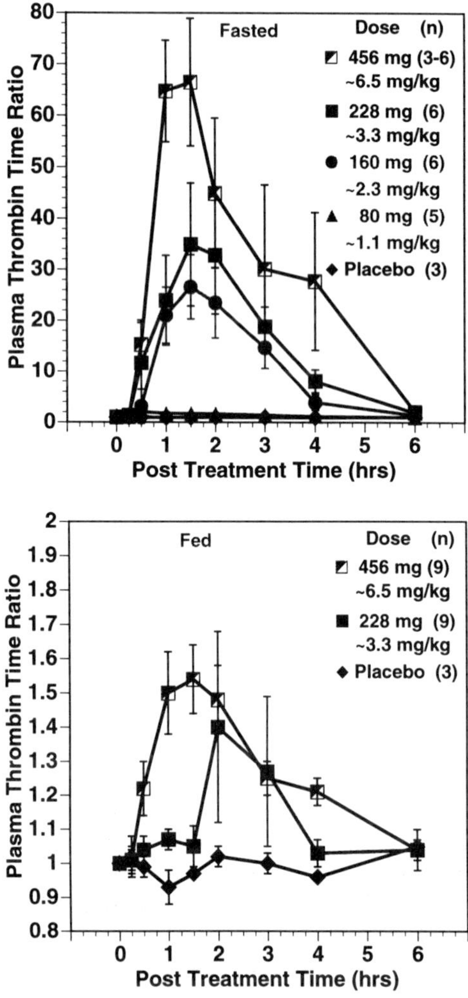

Figure 8. (Top) Time course of plasma thrombin time ratio of compound **26** in fasted volunteers. Subjects were fed at 7:00 PM the night before the experiment and no food was allowed after midnight. Upper limit of thrombin time assay is 1200 sec. Some values for the two highest doses exceeded that limit. Thrombin time ratio = experimental/control. Values represent the mean ± SEM. (Bottom) Time course of plasma thrombin time ratio of compound **26** in fed volunteers. Subjects were fed two eggs, bacon, two slices of toast, and orange juice 0.5–1.5 hr before drinking the drug aqueous solution (note the change in scale).

domized study. Eighteen healthy male volunteers participated in the study with each volunteer attending for four dose periods. Dosing periods one to three consisted of two doses of **26** and one dose of placebo, administered after an overnight fast. Escalating doses of 80, 160, 228, and 456 mg were administered as a solution (Fig. 8, top panel) (Shuman *et al.,* 1996). Because studies in rats demonstrated reduced oral activity of compound **26** in the fed state, all subjects received a dose of either 228 or 456 mg, administered 1 hr after breakfast during dose period four (Fig. 8, bottom panel). Compound **26** significantly prolonged TT when administered to fasted subjects. There was a direct correlation between plasma drug concentration and prolongation of TT. Concentrations of approximately 150 ng/mL were required to double TT. Thrombin times returned to baseline by 6 hr after administration of the 160-, 228-, and 456-mg doses in fasted subjects. Blood samples also were assayed by HPLC for concentrations of compound **26**. Maximum plasma concentrations were achieved approximately 2 hr after administration of the dosing solution. The elimination half-life was 1.8 hr when administered after fasting and was slightly prolonged to 3 hr when given with food. No adverse events were observed at any of the doses tested. Food had a negative effect on the bioavailability, and therefore, the oral activity, of compound **26**. When administered 1 hr after a meal, the relative bioavailability of compound **26** was reduced by 70–80%. The criteria established for evaluation of an oral antithrombotic in Phase 2 efficacy studies were acceptable pharmacokinetic and pharmacodynamic profiles in humans. In addition, the drug should have minimal variability of systemic exposure when taken with food. As these critical success factors were not achieved, the drug was withdrawn from further clinical evaluation.

5. CONCLUSION

The parenteral agent efegatran was chosen for clinical evaluation after extensive SAR studies and subsequent collaborations with the HIDR. It was studied extensively in Phase 1 and Phase 2 clinical trials to determine if it could provide superior benefits to heparin for cardiovascular patients with unstable angina or thrombolysis during acute myocardial infarction. Analysis of data from Phase 2 clinical trials demonstrated that efegatran exhibited equivalent efficacy to heparin. As a consequence further development with this parenteral agent was discontinued.

Compound **26** was discovered from the continuation of the SAR and was evaluated in Phase 1 trials as a potential oral antithrombotic agent. Although plasma anticoagulant activity of compound **26** was prolonged, in a dose- and time-dependent manner after oral administration of an aqueous solution, the half-life of the anticoagulant activity approximated 2 hr and practical utility may be limited. In addition, compound **26** exhibited a considerable reduction in oral exposure when administered immediately after eating. The development of a compound with a longer half-life and minimal food effects would present an opportunity for the development of a novel oral anticoagulant agent.

Recent reviews on thrombin inhibitors may provide additional insights for the reader (Scarborough, 1995; Edmunds and Rapundalo, 1996).

Acknowledgments

The authors thank Dr. Gerry Smith and Ms. Donetta S. Gifford-Moore for the *in vitro* analysis; Dr. Kennth Kurz, Mr. Alex Wilson, Mr. Dick Moore, and Mr. Tommy Smith for *in vivo* small animal pharmacology; Dr. Charles V. Jackson, Ms. Gail Crowe, and Mr. Harve Wilson for *in vivo* large animal pharmacology; Dr. Terry Lindstrom and Mr. Kenneth Ruterbories for parent drug evaluation and pharmacokinetic/pharmacodynamic analysis; Dr. Eiry W. Roberts, Blanche Singer, and Dr. Richard A. Lucas for clinical evaluation; Dr. Julie Satterwhite for human pharmacokinetic analysis; Dr. George Sandusky for toxicological evaluations; Mr. Robert Rothenberger and Mr. Charles Campbell for synthetic technical assistance and contributions to the development process.

REFERENCES

Amerena, J., Mashford, M. L., and Wallace, S., 1990, Adverse effects of anticoagulants, *Adverse Drug React. Acute Poisoning Rev.* **9**(1)**:**1.

Bagdy, D., Szabo, G., Bararas, E., and Bajusz, S., 1992, Inhibition by D-MePhe-Pro-Arg-H (GYKI-14766) of thrombus growth in experimental models of thrombosis, *Thromb. Haemost.* **68:**125–129.

Bajusz, S., Barabas, E., Tolnay, P., Szell, E., and Bagdy, D., 1978, Inhibition of thrombin and trypsin by tripeptide aldehydes, *Int. J. Pept. Protein Res.* **12:**217–221.

Bajusz, S., Szell, E., Barabas, E., and Bagdy, D., 1981, Structure–activity relationships among the tripeptide aldehyde inhibitors of plasmin and thrombin, in: *Peptides: Synthesis–Structure–Function, Proceedings of the Seventh American Peptide Symposium* (D. H. Rich and E. Gross, eds.), pp. 417–420, Pierce Chemical Co., Rockford, IL.

Bajusz, S., Bagdy, D., Barabas, E., Szell, E., and Dioszegi, M., 1983, Peptides acting upon haemostasis, in *Biomed. Signif. Pept. Res., Sect. Med. Hung. Acad. Sci. Annu. Gen. Meet.* (F. A. Ldszlo and F. Antoni, eds.), p. 227, Akad. Kiado, Budapest.

Bajusz, S., Szell Hasenohrl nee, E., Barabas, E., and Bagdy, D., 1984, U.S. Patent 4,478,745.

Bajusz, S., Szell Hasenohrl nee, E., Bagdy, D., Barabas, E., Dioszegi, M., Fittler, Z., Jozsa, F., Horvath, C., and Tomori nee Jozst, E., 1987, U.S. Patent 4,703,036.

Bajusz, S., Szell, E., Bagdy, D., Barabas, E., Horvath, G., Dioszegi, M., Fittler, Z., Szabo, G., Juhasz, A., Tomori, E., and Szilagyi, G., 1990, Highly active and selective anticoagulants: D-Phe-Pro-Arg-H, a free tripeptide aldehyde prone to spontaneous inactivation, and its stable N-methyl derivative, D-MePhe-Pro-Arg-H, *J. Med. Chem.* **33:**1729–1735.

Blomback, B., Hessel, B., Hogg, D., and Claesson, G., 1977, Substrate specificity of thrombin on protein and synthetic substrates, in: *Chemistry and Biology of Thrombin* (R. L. Lundbald, ed.), pp. 275–285, Ann Arbor Science, Ann Arbor, MI.

Blomback, B., Hessel, B., Hogg, D., and Therkildsen, L., 1978, A two-step fibrinogen–fibrin transition in blood coagulation, *Nature* **275:**501–505.

Chandler, A. B., Chapman, I., Erhardt, L. R., Roberts, W. C., Schwartz, C. J., Sinapius, D., Spain, D. M., Sherry, S., Ness, P. M., and Simon, T. L., 1974, Coronary thrombosis in myocardial infarction, *Am. J. Cardiol.* **34:**823–833.

Davies, M. J., and Thomas, T., 1981, The pathological basis and micro-anatomy of occlusive coronary thrombus formation in human coronary arteries, *Philos. Trans. R. Soc. London* **294:**225–229.

Edmunds, J. J., and Rapundalo, S. T., 1996, Thrombin and factor Xa inhibition, *Annu. Rep. Med. Chem.* **31:**51–60.

Gold, H. K., 1990, Conjunctive antithrombotic and thrombolytic therapy for coronary occlusion, *N. Engl. J. Med.* **323:**1483–1485.

Goldsmith, H. L., and Turitto, V. T., 1986, Rheological aspects of thrombosis and haemostasis: Basic principles and applications, *Thromb. Haemost.* **55:**415–435.

Jackson, C. V., Crowe, V. G., Frank, J. D., Wilson, H. C., Coffman, W., Utterback, B. G., Jakubowski, J. A., and Smith, G. F., 1992, Pharmacological assessment of the antithrombotic activity of the peptide thrombin inhibitor, D-methyl-phenylalanyl-prolyl-arginal (GYKI-14766), in a canine model of coronary artery thrombosis, *J. Pharmacol. Exp. Ther.* **261:**546–552.

Jackson, C. V., Wilson, H. C., Crowe, V. G., Shuman, R. T., and Gesellchen, P. G., 1993, Reversible tripeptide thrombin inhibitors as adjunctive agents to coronary thrombolysis: A comparison to heparin in a canine model of coronary artery thrombosis, *J. Cardiovasc. Pharmacol.* **21:**587–594.

Jackson, C. V., Satterwhite, J., and Roberts, E., 1996, Preclinical and clinical pharmacology of efegatran (LY294468): A novel antithrombin for the treatment of acute coronary syndromes, *Clin. Appl. Thromb./Hemost.* **22:**258–267.

Kurz, K. D., Main, B. W., and Sandusky, G. E., 1990, Rat model of arterial thrombosis induced by ferric chloride, *Thromb. Res.* **60:**269–280.

Okimoto, S., Hijikata, A., Kinjio, K., Kikumoto, R., Ohkuba, K., Tonomura, S., and Tamao, Y., 1975, Novel series of synthetic thrombin inhibitors having extremely potent and highly selective action, *Kobe J. Med. Sci.* **21:**43–51.

Pozsyay, M., Szabo, G. C. S., Bajusz, S., Sinonsson, R., Gaspar, R., and Elodi, P., 1981, Study of the specificity of thrombin with tripeptidyl-p-nitroanilide substrates, *Eur. J. Biochem.* **115:**491–495.

Ruterbories, K. J., Hanssen, B. R., and Lindstrom, T. D., 1992, *ISSX Proceedings, Fourth North American ISSX Meeting,* Bal Harbour, FL.

Scarborough, R. M., 1995, Anticoagulant strategies targeting thrombin and factor Xa, *Annu. Rep. Med. Chem.* **30:** 71–80.

Shuman, R. T., Rothenberger, R. B., Campbell, C. S., Smith, G. F., Jackson, C. V., Kurz, K. D., and Gesellchen, P. D., 1992, Prevention of reocclusion by a thrombin inhibitor (LY282056), in: *Peptides: Chemistry and Biology. Proceedings of the Twelth American Peptide Symposium* (J. A. Smith and J. E. Rivier, eds.), pp. 799–800, ESCOM Science Publishers, Leiden, The Netherlands.

Shuman, R., Rothenberger, R., Campbell, C., Smith, G., Gifford-Moore, D., and Gesellchen, P., 1993, Highly selective tripeptide thrombin inhibitors, *J. Med. Chem.* **36:**314–319.

Shuman, R. T., Rothenberger, R. B., Campbell, C. S., Smith, G. F., Gifford-Moore, D. S., Paschal, J. W., and Gesellchen, P. D., 1995, Structure–activity study of tripeptide thrombin inhibitors using alpha-alkyl amino acids and other conformationally constrained amino acid substitutions, *J. Med. Chem.* **38:**4446–4453.

Shuman, R. T., Rothenberger, R. B., Jackson, C. V., Roberts, E. W., Singer, B., Lucas, R. A., and Kurz, K. D., 1996, Oral activity of tripeptide aldehyde thrombin inhibitors, in: *Peptides: Chemistry and Biology. Proceedings of the Fourteenth American Peptide Symposium,* (J. A. Smith and J. E. Rivier, eds.), pp. 215–216, Mayflower Scientific Ltd. Publishers, West Midlands, England.

Smith, G. F., 1980, The mechanism of fibrin-polymer formation in solution, *Biochem. J.* **185:**1–11.

Smith, G. F., and Sundboom, J. L., 1981, Heparin and protease inhibition II: The role of heparin in the α_2-antithrombin inactivation of thrombin, plasmin and trypsin, *Thromb. Res.* **22:**115–133.

Smith, G. F., Neubauer, B. L., Sundboom, J. L., Best, K. L., Goode, R. L., Tanzer, L. R., Merriman, R. L., Frank, J. D., and Hermann, R. G., 1988, Correlation of the *in vivo* anticoagulant, antithrombotic, and antimetastatic efficacy of warfarin in rat, *Thromb. Res.* **150:**163–174.

Smith, G. F., Shuman, R. T., Craft, T. J., Gifford, D. S., Kurz, K. D., Jones, N. D., Chirgadze, N., Hermann, R. B., Coffman, W. J., Sandusky, G. E., Roberts, E., and Jackson, C. V., 1996, A family of arginal thrombin inhibitors related to efegatran, *Semin. Thromb. Hemost.* **22:**173–183.

Stein, B., Fuster, V., Halpering, J. L., and Chesebro, J. H., 1989, Antithrombotic therapy in cardiac disease. An emerging approach based on pathogenesis and risk, *Circulation* **80:**1501–1513.

Tomori, F., Szell, E., and Barabas, E., 1984, High-performance liquid chromatography of a new tripeptide aldehyde (GYKI-14166). Correlation between the structure and activity, *Chromatographia* **19:**437–442.

Wagner, W. R., and Hubbell, J. A., 1990, Local thrombin synthesis and fibrin formation in an *in vitro* thrombosis model result in platelet recruitment and thrombosis stabilization on collagen in heparinized blood, *J. Lab. Clin. Med.* **116:**636–650.

Chapter 5

Discovery and Development of an Endothelin A Receptor-Selective Antagonist PD 156707

Annette M. Doherty and
Andrew C. G. Uprichard

1. INTRODUCTION

The potent vasoconstrictor endothelin (ET) is implicated in several cardiovascular, pulmonary, renal, and cerebrovascular human diseases (Miller *et al.*, 1989; Giaid *et al.*, 1993; Takahashi *et al.*, 1994; Ferro and Webb, 1996; Patel, 1996) . Since the discovery of the ET family of peptides in 1988 (Yanagisawa *et al.*, 1988; Inoue *et al.*, 1989), there has been intensive interest in development of ET receptor antagonists in an attempt to define the physiological and pathophysiological role(s) of the ETs (Doherty, 1992; Peishoff *et al.*, 1995).

It is now known that endothelins (ET-1, ET-2, ET-3) are a family of 21-residue containing peptides that are derived by a two-step proteolytic cleavage from a protein precursor known as preproendothelin (Yanagisawa *et al.*, 1988; Inoue *et al.*, 1989). In the late 1980s, receptor subtypes were not known and there was relatively little known about the ET system and the biological effects mediated by this interesting family of peptides.

Several approaches to the discovery of ET mediators have been utilized (Fig. 1).

Annette M. Doherty • Department of Chemistry, Parke-Davis Pharmaceutical Research Division, Warner-Lambert Company, Ann Arbor, Michigan 48105. *Andrew C. G. Uprichard* • Department of Cardiac and Vascular Diseases, Parke-Davis Pharmaceutical Research Division, Warner-Lambert Company, Ann Arbor, Michigan 48105.

Integration of Pharmaceutical Discovery and Development: Case Studies, edited by Borchardt *et al.*, Plenum Press, New York, 1998.

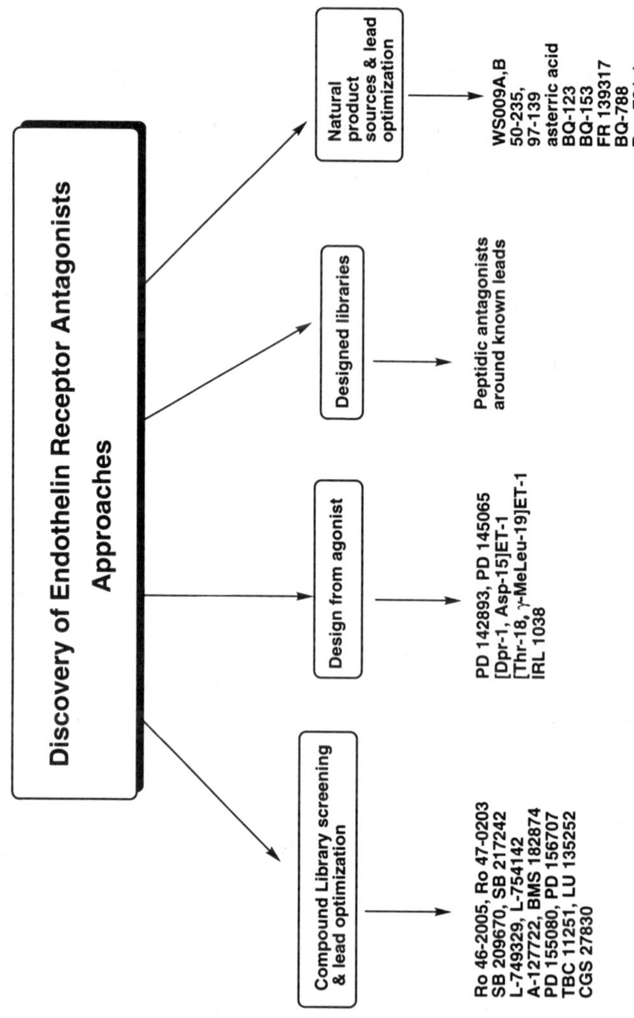

Figure 1. Approaches to the discovery of ET antagonists.

In the early days, many groups tried to develop antagonists with various degrees of selectivity from the structure of the agonist itself and this led to a number of potent peptide inhibitors (Doherty, 1992; Peishoff *et al.*, 1995). Later approaches have successfully led to several clinical candidates primarily via compound library screening and subsequent lead optimization (Clozel *et al.*, 1993; Peishoff *et al.*, 1995).

In 1990, two subtypes of ET receptors known as ET_A and ET_B were cloned and characterized from animal systems and subsequently the two ET subtypes (both of which are seven-transmembrane G-protein-coupled receptors) were also cloned from mammalian cells (Arai *et al.*, 1990; Sakurai *et al.*, 1990; Sakamoto *et al.*, 1991; Hosoda *et al.*, 1991). A third ET receptor subtype has been cloned from *Xenopus* dermal melanophores (Karne *et al.*, 1993) and heart (Kumar *et al.*, 1994), although this subtype has never been described in mammalian tissues.

The distribution of both ET_A and ET_B receptors has been studied in animal and human tissues and found to be widely localized throughout the body (Davenport *et al.*, 1993; Davenport and Maguire, 1994; Godfraind, 1994; Davenport *et al.*, 1995). In a wide variety of animal tissues, vasoconstriction occurs via activation of ET_A and/or ET_B receptors depending on the species and vascular bed under study (Clozel *et al.*, 1992; Sumner *et al.*, 1992; Moreland *et al.*, 1992; Tschudi and Luscher, 1994; Seo *et al.*, 1994; Sudjarwo *et al.*, 1994). The ET_B receptor has been shown to mediate nitric oxide release from endothelial cells and a vasodilator response *in vivo* (Clozel *et al.*, 1992). However, there continues to be some controversy as to the importance of ET_B receptors in mediating vasoconstrictor responses in mammalian tissues (Bax *et al.*, 1993; White *et al.*, 1994; Seo *et al.*, 1994; Davenport *et al.*, 1995). Davenport *et al.* (1995) reported that ET_A-mediated vasoconstriction plays a major role in some human vessels, such as coronary artery, but were unable to demonstrate ET_B-receptor-mediated contractions in human tissues using ET_B-selective agonists such as [Ala 1,3,11,15]ET-1 and BQ 3020 (Davenport and Maguire, 1994). Several groups have shown that the ET_B receptor agonist SRTX-6c can elicit vasoconstriction in human vessels although the magnitude of the response has been found to be considerably less than that observed for ET-1 itself (Bax *et al.*, 1993; White *et al*, 1994; Sudjarwo *et al.*, 1994). It is possible that downregulation of ET_B receptors in isolated tissues is responsible for these observations.

It has been demonstrated that the ET peptides and their receptors are important in fetal development and cardiovascular regulation through gene knock-out mouse experiments (Kurihara *et al.*, 1994). In addition, targeted disruption of the endothelin B-type receptor gene or the gene encoding ET-3 in a mouse has been shown to produce an autosomal recessive phenotype of white spotting and megacolon (Baynash *et al.*, 1994). Mutations in the ET_B gene can be demonstrated in familial and isolated cases of Hirschsprung disease (Puffenberger *et al.*, 1994).

At Parke-Davis we initially began our program searching for ET antagonists by study of structure–activity relationships of the agonist peptide itself and discovered several highly potent hexapeptide antagonists including PD 142893 (Cody *et al.*, 1992; Doherty *et al.*, 1993a,b) and PD 145065 (Cody *et al.*, 1993). A number of other peptide ET antagonists were also reported including BQ-123 (Ishikawa *et al.*, 1992), FR 139317 (Sogabe *et al.*, 1993), and Tak-044 (Masuda *et al.*, 1996), all derived from natural product library screening and lead optimization. BQ-788, a recently reported ET_B-selective antagonist derived from the BQ-123 pentapeptide, has also been described (Ishikawa *et al.*, 1994).

Because our own peptidic antagonists and those from other programs were not orally active and had a short duration of action, several companies including our group chose to screen chemical libraries against ET_A and/or ET_B receptors to uncover suitable leads with which to carry out structure–activity studies and lead optimization. We believed that an ET_A receptor-selective antagonist would be useful therapeutically to inhibit vasoconstriction and mitogenesis, because it would selectively block ET_A-mediated vasoconstriction without affecting ET_B-mediated vasodilation. Other groups have developed ET_A/ET_B balanced agents believing that the ET_B receptor may be important in mediating some of the biological effects of ET in specific disease states and indeed there is evidence for an upregulation of ET_B receptors in certain diseases such as congestive heart failure (Love *et al.*, 1995; Love and McMurray, 1996). The selectivity of various clinical candidates and subsequent studies in humans will eventually elucidate the relative importance of these two receptor subtypes in different diseases.

From the Parke-Davis program we have developed several series of nonpeptide endothelin antagonists and the discovery of our clinical candidate PD 156707 will be the focus of this review (Doherty *et al.*, 1995; Reynolds *et al.*, 1995a). A number of other nonpeptide ET antagonists from various pharmaceutical companies have also been reported. These include the Shionogi steroid analog 97–139 (Mihara *et al.*, 1994), and several balanced ET_A/ET_B nonpeptide antagonists, including Ro 46–2005 (Clozel *et al.*, 1993), Ro 47–0203 (bosentan) (Roux *et al.*, 1993), SB 209670 (Elliott *et al.*, 1994), SB 217242 (Ohlstein *et al.*, 1996), CGS 27830 (Mugrage *et al.*, 1993), and L-749,329 (Walsh *et al.*, 1994), have been described, in addition to the more ET_A-selective antagonists L-744,453 (Williams *et al.*, 1996), SB 209834 (Peishoff *et al.*, 1995), BMS 182874 (Stein *et al.*, 1994), TB 11251 (*et al.*, 1996), and A-127722 (Opgenorth *et al.*, 1996).

2. DISCOVERY OF PD 156707: MEDICINAL CHEMISTRY, PHARMACOLOGY, AND PHARMACOKINETICS

2.1. Identification of Lead Structures

We screened our chemical library of about 170,000 compounds for their capacity to inhibit specific [^{125}I]-ET-1 binding in rabbit renal artery vascular smooth muscle cells (VSMC), known to express only the ET_A receptor, using an assay system previously described (Doherty *et al.*, 1993a,b). Using this approach we discovered several series of nonpeptide antagonists from which we selected the butenolide class as one series to follow up with medicinal chemistry. We subsequently screened all compounds from this series against human cloned receptors (Reynolds *et al.*, 1995a).

We optimized the potency of an initial lead structure, PD 012527 (compound **1**; Table I), to discover potent orally active ET_A-selective antagonists and ET_A/ET_B balanced agents (Reynolds *et al.*, 1995a; Doherty *et al.*, 1995, 1996). Compound **1** showed micromolar binding affinity for the rabbit ET_A receptor (IC_{50} = 2 μM) and also inhibited [^{125}I]-ET-3-specific binding to rat cerebellum (ET_B) with an IC_{50} of 5 μM. The compound also inhibited ET-1-induced arachidonic acid release in rabbit renal artery VSMC with an IC_{50} of 5 μM showing that it was an ET_A functional antagonist (Reynolds and Mok, 1989). Compound **1** exhibited very weak inhibitory activity against ET-1-induced vasoconstriction in the ET_A-specific rabbit femoral artery (pA_2 ~5) and no activity was observed in inhibiting SRTX-6c-induced vasoconstriction in rabbit pulmonary artery, an assay that evaluates ET_B functional activity (Panek *et al.*, 1992; Doherty *et al.*, 1993b).

Table I
Structure–Activity Relationships from Screening Lead PD 012527 (**1**)

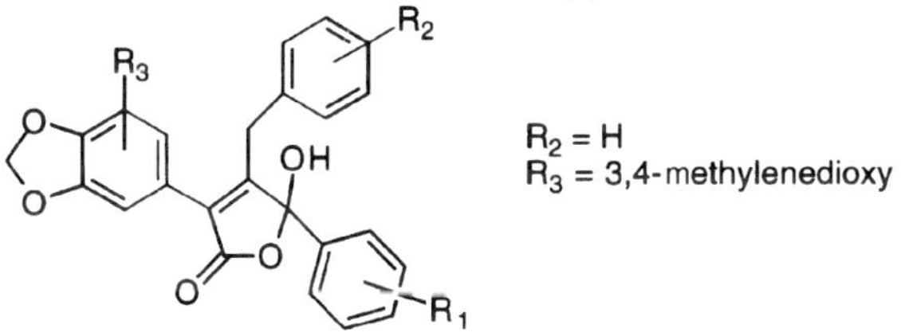

R_2 = H
R_3 = 3,4-methylenedioxy

		IC_{50} (nM)	
Compound	R_1	ET_A	ET_B
1	4-Cl	2200[a]	5,200[c]
		430[b]	27,000[d]
2	4-H	1300[a]	22,000[c]
		600[b]	30,000[d]
3	4-OMe	50[a]	500[c]
		7.4[b]	4,550[d]
4	4-Me	1200[a]	4,500[c]
		4900[b]	>25,000[d]
5	3,4-Cl_2	560[a]	—
		400[b]	>25,000[d]
6	4-OMe, 3-Me	12[a]	—
		12[b]	1,750[d]

[a]Rabbit renal artery vascular smooth muscle cells.
[b]Human ET_A (Ltk-expressed).
[c]Rat cerebellum.
[d]Human ET_B (CHO expressed).

2.2. Structure–Activity Relationships

Preliminary enhancement of the receptor binding affinity of compound **1** was achieved via application of the Topliss "Decision Tree" approach for lead optimization based on QSAR principles (Topliss, 1972, 1977). Topliss developed a nonmathematical, nonstatistical, and noncomputerized guide to the use of basic Hansch principles, including electronic, lipophilic, and steric considerations, for the optimization of activity of a lead structure containing benzene rings. We first applied this approach for optimization of the substituents R_1, R_2, and R_3 on each of the phenyl rings in the butenolide structure.

Compound **1** with 4-Cl substitution at the R_1 position was slightly less active than the unsubstituted R_1 phenyl ring analogue **2** in binding to both ET_A (rabbit) and ET_B (rat) receptors. The subsequent course of action called for synthesis of the 4-OMe analogue, compound **3**. The 4-Me (**4**) and 3,4-Cl_2 (**5**) analogues were also synthesized to check the validity of the approach. As can be seen from Table I, the 4-OMe analogue (**3**) was considerably more potent than the unsubstituted, 4-Cl, 4-Me, and 3,4-Cl_2 analogues (compounds **2, 1, 4, 5**) as expected for favorable substitution with a small $-\pi$ value and reasonably large $-\sigma$ value.

In order to enhance the $-\sigma$ effect still further, the 3-Me, 4-OMe analogue (**6**) was synthesized and, as expected, was found to be slightly more potent than the 4-OMe analogue (**3**) (Table I). We found that, in general, compounds were more potent against cloned human ET_A receptors compared with rabbit ET_A receptors. In contrast, compounds tended to be less potent against the human ET_B receptor compared with the rat ET_B receptor. The net result of these species differences was increased selectivity for ET_A versus ET_B in the human receptor systems. We applied the same approach to optimize the substituents on the remaining two phenyl rings utilizing 4-OMe at the R_1 position. At both the R_1 and R_2 positions, optimization of potency was achieved by increasing the lipophilicity and electron-donating power of the substituents on the aromatic ring. The same trends in species differences were observed. At the R_3 position, electron rich aromatic rings were also preferred, although the specific substitution pattern was important and the 3,4-methylenedioxy moiety afforded the most active compounds.

Having elucidated the importance of electron-donating substituents at R_2, we explored further ring substitutions to discover the potent trimethoxy analogue **7,** PD 156707, with subnanomolar affinity for the ET_A receptor (Fig. 2). PD 156707 2-benzo[1,3]dioxol-5-yl-4- (4-methoxyphenyl) -4-oxo-3-(3,4,5-trimethoxybenzyl)-but-2-enoate (Z) sodium salt was selected as a clinical candidate on the basis of its pharmacological and pharmacokinetic properties (*vide infra*).

We have investigated the structure–activity relationships and pharmacology of the nonpeptide orally active PD 156707 series of ET antagonists where the selectivity ratios for human ET_A and ET_B receptors have been varied from greater than 2000- to 20-fold. Compounds with increased lipophilicity at R_2 showed increased ET_B affinity and a more balanced ET_A/ET_B profile.

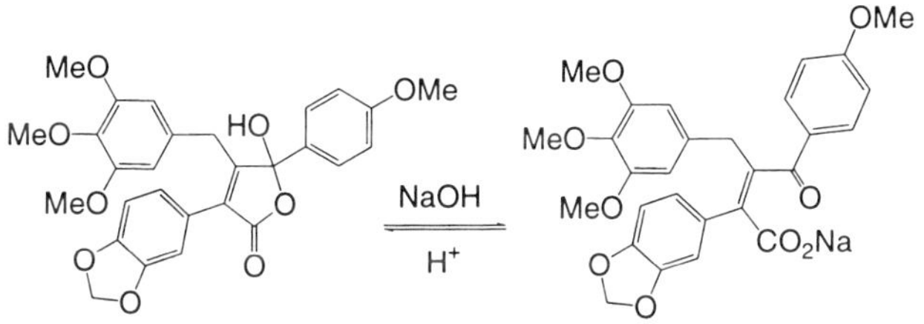

PD 156707 (7)

Figure 2. Structure of PD 156707, an ET_A-selective antagonist.

2.3. Pharmacokinetics/Selection

Our SAR investigation led to several promising analogues and selection of a suitable clinical candidate was based primarily on the potency, ET_A/ET_B receptor selectivity, oral bioavailability, and pharmacokinetic profile of the compounds. We selected a series of highly potent analogues and evaluated them more intensively (Table II, Fig. 3).

The best affinities for the human ET_A receptor were seen with PD 156707, 158312, and 158372. Regarding human ET_A/ET_B selectivities, the best compounds were PD 156707, 158372, 156453, 157781, and 158040 (Table II). Compounds with the best functional activity in inhibiting ET-1-induced vasoconstriction included PD 156707, 158312, 158372, and 158040. Of the series of compounds shown in Table II, the most potent (dosed at 10 mg/kg orally) in inhibiting thc pressor response to i.v. ET-1 in rat were PD 156707, 158312, 158372, and 156453.

Table II
Compound Selection Process

	IC_{50}(nM)			Pressor response to ET-1 (0.3 nM/kg i.v.), p.o. administration (10 mg/kg)
PD No.	ET_A	ET_B/ET_A	pA_2	
155080	7.8	442	6.3	35%
156707	0.3	1600	7.5	22%
158189	3.9	746	7.2	30%
158312	0.18	589	7.4	22%
158372	0.28	882	7.6	22%
156453	2.1	>1000	7.0	20%
155719	2.1	429	7.0	30%
157781	1.06	4651	7.1	40%
158040	2.90	1690	7.6	37%
158370	0.24	600	7.4	37%

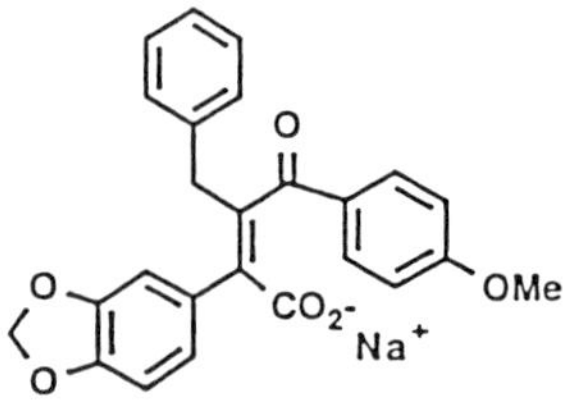

PD 155080-0015

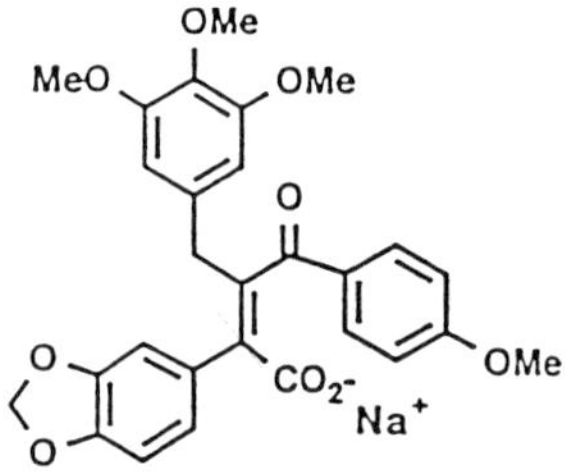

PD 156707-0015

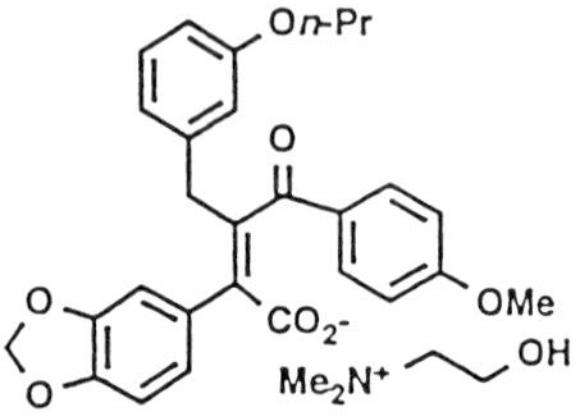

PD 155719-8239

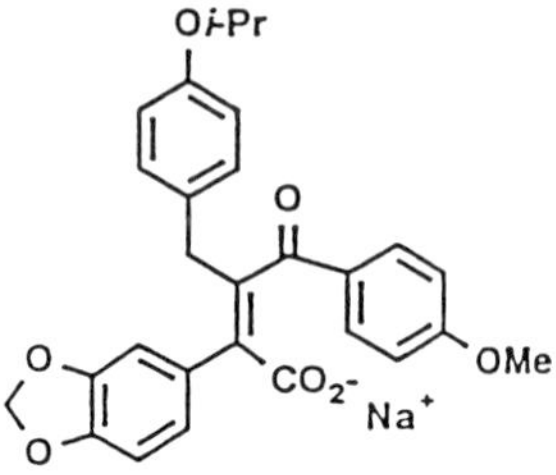

PD 156453-0015

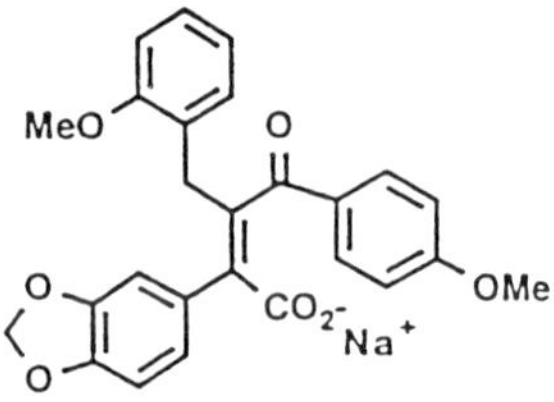

PD 158189-0015

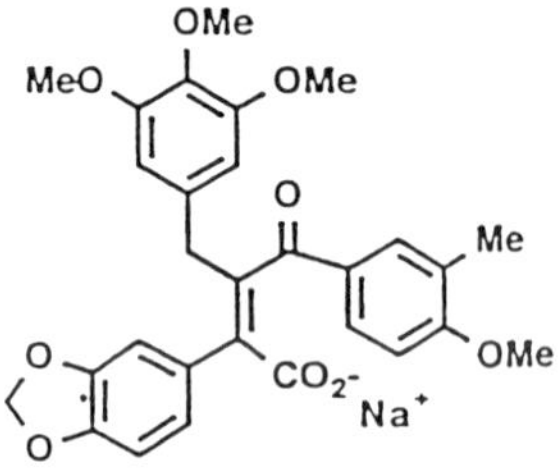

PD 158312-0015

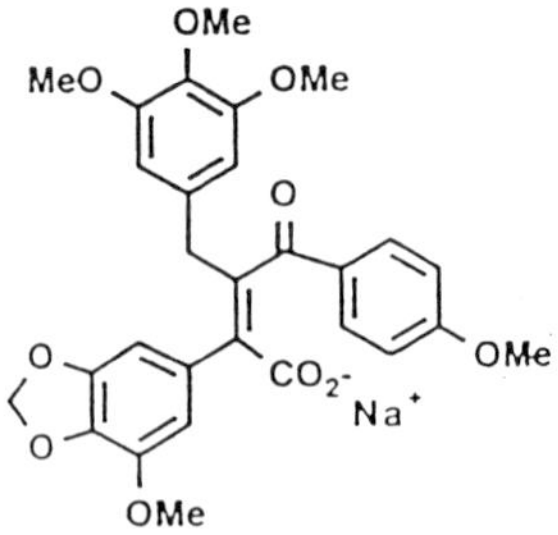

PD 158372-0015

Figure 3. Highly potent ET_A-selective antagonists from butenolide series.

Table III
Pharmacokinetics of Optimal ET_A Antagonists from Butenolide Series

Compound	Half-life (hr)	Bioavailability (%)
PD 155080	5	87
PD 156707	0.5–1	41
PD 155719	<0.3	<5
PD 158312	0.5–1	15
PD 148189	0.5–1	78
PD 158372	0.4–1	44
PD 156453	0.2–1.5	33

Pharmacokinetic properties were significantly influenced by structural modifications at R_2. The pharmacokinetics of three compounds in particular are considered herein, to highlight these differences within the series. ET_A antagonists PD 155719, 155080, and 156707 were studied in male Wistar rats following a 15 mg/kg i.v. or oral gavage dose (three animals per dose). Plasma concentrations were determined by a specific HPLC assay. The terminal elimination $t_{1/2}$ was 5 hr for PD 155080, 1 hr for PD 156707, and less than 5 min for PD 155719. After oral dosing, PD 155080 and 156707 were rapidly absorbed; oral bioavailabilities ranged from less than 5% for PD 155719 to 41% for PD 156707 and 87% for PD 155080 (Doherty *et al.*, 1995). Pharmacokinetic analysis of some other analogues from the series is summarized in Table III; from this analysis, PD 158312 and 156453 were also eliminated as they exhibited lower oral bioavailability than either PD 156707 or 158372. PD 156707 and 158372 consistently had the best overall profiles from our selection criteria and we chose the compound that offered the least potential for metabolism (one less methoxy group) and the one that was the easiest to access synthetically, namely, PD 156707. We found that the pharmacokinetic profiles of PD 156707 were similar across several species including monkey. In summary, PD 156707 was selected as the optimal overall candidate for clinical development, being highly potent and ET_A selective, easily prepared in large scale, rapidly absorbed and orally bioavailable in several species.

2.4. Chemistry/Chemical Development

The synthesis of PD 156707 has been elucidated (Doherty *et al.*, 1995). The four-step high-yielding synthesis has been scaled up to kilogram quantities for toxicology evaluation and clinical development. Conversion of the final butenolide to various salt forms was investigated as a part of the development analysis and the sodium salt (Fig. 2) was selected on the basis of its superior ease of preparation, lack of hygroscopicity, acceptable solid and solution stability, and high aqueous solubility (S. Babu *et al.*, unpublished results, Parke-Davis, 1996).

2.5. Biological Evaluation of PD 156707

PD 156707 has been shown to bind to human ET_A receptors with an approximately 800-fold higher affinity than to human ET_B receptors (Fig. 4). The K_i values of PD 156707 for human ET_A and ET_B receptors are 0.17 ± 0.04 and 133.8 ± 47.2 nM, respectively. PD 156707 is approximately 1300-fold more selective for ET_A versus ET_B receptors in rabbit tissues. The K_i values of PD 156707 for rabbit ET_A (rabbit renal artery VSMC) and ET_B (rabbit cerebellum) receptors are 0.28 ± 0.08 and 369.2 ± 105.7 nM, respectively (mean ± SEM; $n = 3$). In rat tissues, PD 156707 is about 100-fold more selective for ET_A versus ET_B receptors indicating the species difference noted previously between rat and human receptors for this series (Reynolds *et al.*, 1995b).

PD 156707 was found to be highly selective over a range of receptors and enzymes (Pan Labs screening). In addition, it was a highly ET_A-selective competitive antagonist in a variety of species including rabbit, dog, lamb, cat, monkey, and human tissues. The data suggest that the rat is the least sensitive species in terms of ET_A selectivity (Tables IV and V).

PD 156707 was found to inhibit functional responses to ET-1 including in-

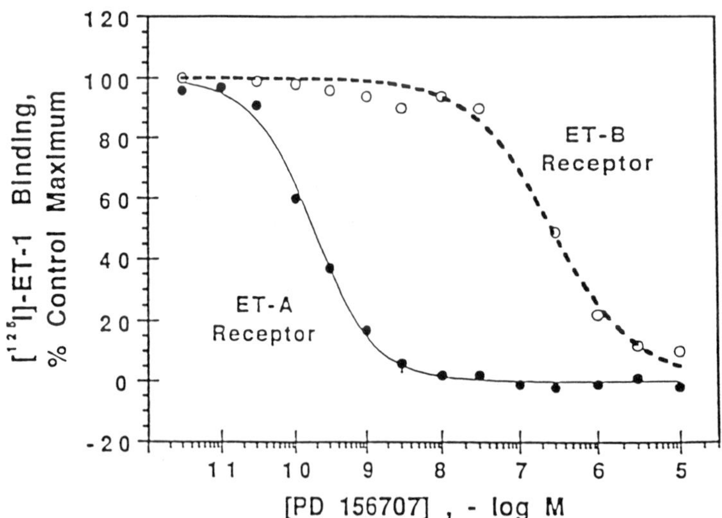

Figure 4. Inhibition of [^{125}I] ET-1 binding in membranes prepared from cells expressing either recombinant human ET_A or ET_B receptors by PD 156707. The analogue is highly selective for ET_A receptors.

Table IV
Species Comparisons in Binding Data for PD 156707

	IC_{50} (nM)		Ratio
	ET_A	ET_B	(B/A)
Human	0.35	443	1265
Rabbit	0.42	1010	2405
Pig	0.43	490	1139
Lamb	0.31	320	1032
Hamster	0.44	510	1159
Rat	0.42	110	261

ositol phosphate production in Ltk^- cells expressing recombinant human ET_A receptors with an IC_{50} of 2.4 nM (Reynolds *et al.*, 1995a) and arachidonic acid release in rabbit renal artery VSMC cells with an IC_{50} of 1.1 nM (Doherty *et al.*, 1995). PD 156707 was able to antagonize ET-stimulated vasoconstriction in rabbit femoral artery (RFA) and ET-3-stimulated contraction of rabbit pulmonary artery (RPA) which were used as models of ET_A- and ET_B-receptor-mediated vasoconstriction, respectively. Increasing concentrations of PD 156707 (0.1–10 μM) caused a rightward shift in the dose–response curve of ET-1-stimulated contraction of RFA, with a pA_2 value of 7.5. Much higher concentrations of PD 156707 (100 μM) were required to cause a rightward shift in the dose–response curve of ET-3-stimulated contraction of RPA, and PD 156707 had a pA_2 value of 4.7 in this tissue. PD 156707 was also able to reverse an established contraction of RFA rings induced by ET 1 (Reynolds *et al.*, 1995a).

The pA_2 values estimated for PD 156707 in human tissues (pA_2 of 7.6–8.1 at 300 nM) (Table VI) are better than the antagonism of ET-1 in RFA (Schild derived pA_2=7.5) (Davenport *et al.*, 1995). PD 156707 is 50 times more potent as an antagonist of ET-1 contractions in human vasculature *in vitro* (pA_2 ~ 9.2 at 3

Table V
Affinity (K_D) and Selectivity of PD 156707 for Human Endothelin Receptors[a]

	ET_A K_D (nM)	ET_B K_D (μM)	ET_A: ET_B	ET_A selectivity	*n*
Coronary artery	0.15 ± 0.06		100:0		3
Saphenous vein	0.50 ± 0.13	1.42 ± 0.02	83:17	2840-fold	3
Left ventricle	0.92 ± 0.38	13.3 ± 2.09	75: 25	144576-fold	3
Kidney	0.53 ± 0.49	0.57 ± 0.04	22:78	1076-fold	4

[a]Equilibrium dissociation constants (K_D) were derived using LIGAND from data pooled from the individual experiments. The ratio of ET_A to ET_B receptors identified in these tissues by PD 156707 was obtained by comparison of the B_{max} values. Receptor selectivity for PD 156707 was calculated by comparison of the ET_A K_D and ET_B K_D in each case. *n* values refer to the number of individuals from whom the tissues were obtained.

Table VI
Antagonism of Endothelin-1-Mediated Vasoconstriction by PD 156707 in Human Isolated Blood Vessels[a]

PD 156707 (nM)	Saphenous vein	Mammary artery	Coronary artery
3	9.23 ± 0.27 (n = 5)	9.17 ± 0.25 (n = 8)	
10	8.96 ± 0.20 (n = 7)	8.42 ± 0.18 (n = 9)	8.23 ± 0.26 (n = 6)
30	8.62 ± 0.22 (n = 8)	8.24 ± 0.11 (n = 8)	8.30 ± 0.14 (n = 8)
100	8.64 ± 0.35 (n = 5)	8.05 ± 0.14 (n = 6)	7.94 ± 0.13 (n = 8)
300	8.11 ± 0.30 (n = 6)	7.91 ± 0.31 (n = 3)	7.62 ± 0.11 (n = 5)
Mean	8.70 ± 0.13 (n = 31)	8.45 ± 0.11 (n = 34)	8.07 ± 0.09 (n = 27)

[a]For each concentration of PD 156707, data are the pA_2 values (mean ± SEM, from n individuals) estimated from the Gaddum–Schild equation assuming slope of one where pA_2 = log10{[concentration ratio 1]/concentration PD 156707 (M)}.

nM) than in the rabbit preparation (Table VI). This difference parallels the species differences in binding experiments, i.e., members of this series of butenolide analogueues have higher affinity for human ET_A receptors than rabbit ET_A receptors while the converse is true for ET_B receptors.

At an oral dose of 30 mg/kg, PD 156707 caused a 57% inhibition of the peak pressor response to ET-1, which represents full inhibition of the ET_A-mediated component of this response (Fig. 5). The IC_{50} value for PD 156707 inhibition of the ET-1-induced pressor response is approximately 1 mg/kg, p.o. whereas PD 156707 had no effect on the depressor response to ET-1 (30 mg/kg, p.o.) indicating ET_A selectivity *in vivo*. Furthermore, PD 156707 (30 mg/kg, p.o.) had no significant effect on basal blood pressure in normotensive rats.

2.6. Metabolism

We have evaluated the metabolism of PD 156707 in a variety of species (Iyer *et al.*, 1996). The methylenedioxyphenyl (1, 3-benzodioxole) moiety has been associated with metabolically derived metabolite-intermediate (MI) complex formation with the heme of cytochrome P450. Because PD 156707 contains such a moiety, the possibility of MI complex formation was explored in rat, dog, and human liver microsomes. Isosafrole, a methylene dioxyphenyl derivative known to form an MI complex, was employed as a positive control. PD 156707 incubated with liver microsomes and NADPH resulted in MI complex formation in all three species. Phenobarbital- and *b*-naphthoflavone-induced rat and dog liver microsomes showed a threefold increase in MI complex formation. PD 158881, an analogue of PD 156707 that contains a 3,5-dimethoxy group instead of the methylenedioxyphenyl moiety, did not demonstrate an MI complex in any of the three

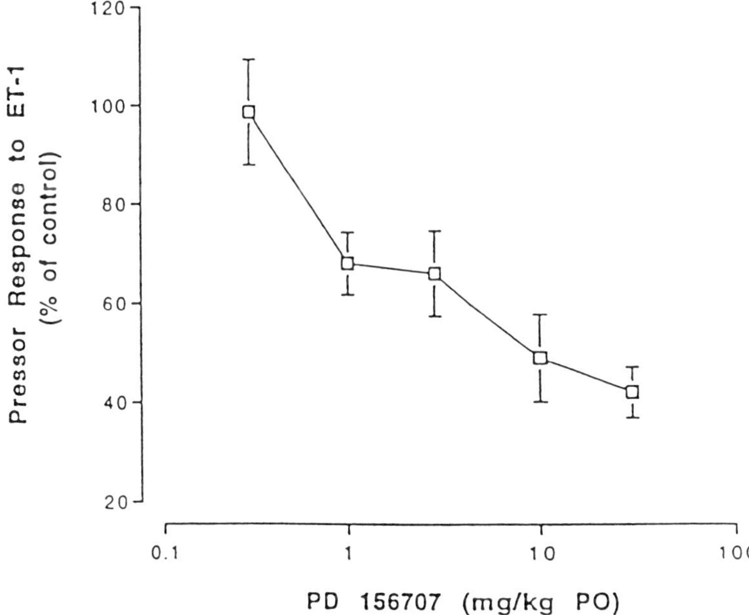

Figure 5. Oral activity of PD 156707; dose–response curve of PD 156707-induced inhibition of the pressor response to ET-1 in anesthetized, ganglionic blocked rats after oral administration. PD 156707 was administered by oral gavage (0.3–30 mg/kg, p.o.). The arterial blood pressure after ET-1 challenge (0.3 nmole/kg, i.v.) is plotted. Results are expressed as percent control pressor response (vehicle) to ET-1. Complete blockade of ET_A receptors results in a pressor response to ET-1 that is approximately 40% of control. Each point represents the computer-averaged response of six control rats and six treated with PD 156707.

microsomal systems. Subsequent metabolic studies of PD 156707 using human liver microsomes and cDNA-expressed human cytochrome P450 2B6 showed that both the *para*- and *meta*-methoxy groups of the trimethoxy phenyl were *O*-dealkylated, although the predominant metabolism was of the *para* derivative. In addition to demethylation, removal of the hydroxy group and *O*-glycuronidation are also major metabolic pathways after either i.v. or oral administration in a variety of species including dog and rat (H. Hallak *et al.*, unpublished results, 1996) (Fig. 6).

2.7. Assay Development

A carboxyl derivative of PD 156707, attached at the *meta* position of the trimethoxyphenyl, was synthesized and used to prepare the subsequent PD

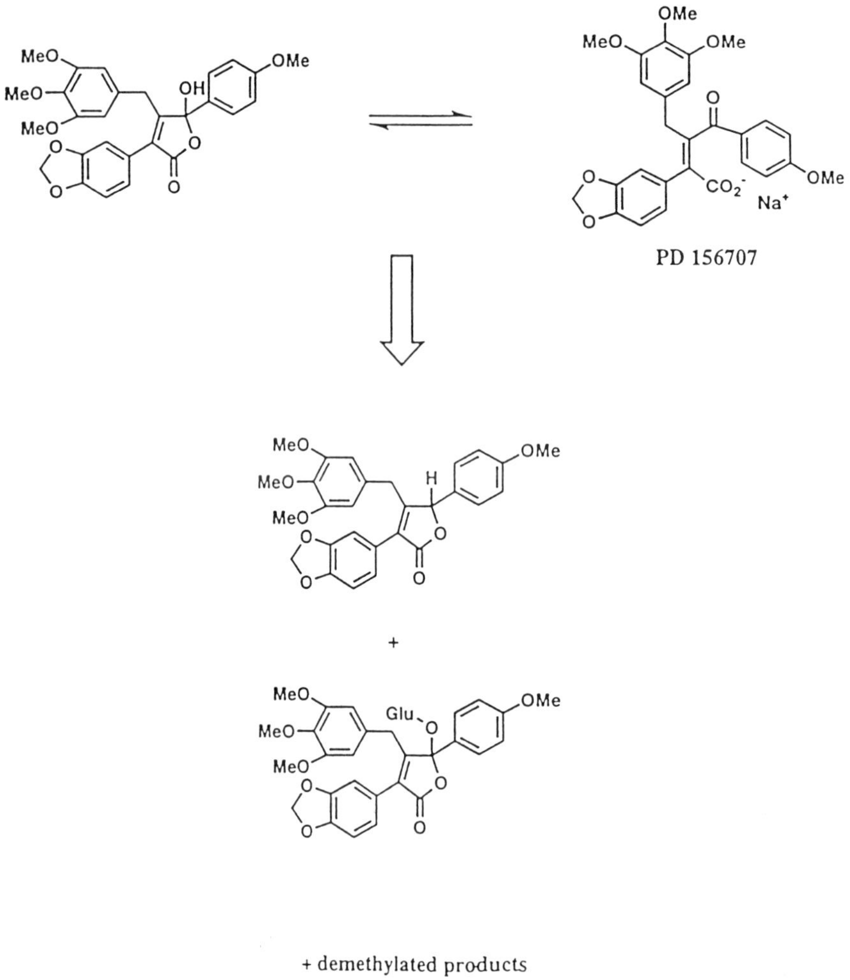

PD 156707

+

+ demethylated products

Figure 6. Metabolic pathways exhibited by PD 156707 *in vivo.*

156707-porcine thyroglobulin immunogen and the PD 156707-tyramine radiolabel and PD 156707-biotin label (Dudeck *et al.*, 1997) (Fig. 7). A nonisotopic enzyme immunoassay employing rabbit anti-PD 156707 antibody and the PD 156707-biotin analogue was eventually developed. Using a 1:2000 dilution of antiserum and 500 pg/ml of biotin, an assay resulted with an effective analytical range 25 to 4000 pg/ml of PD 156707.

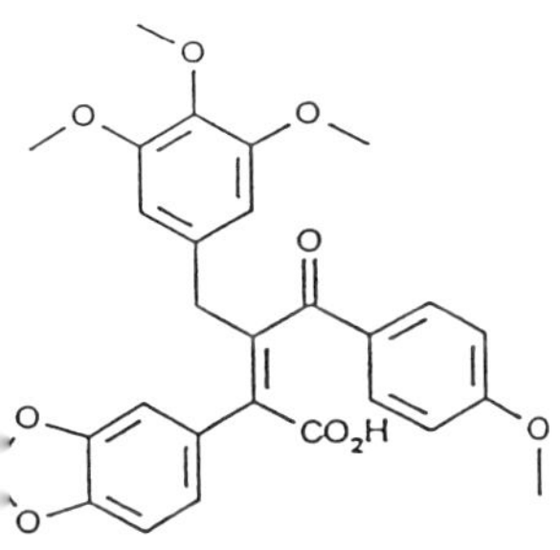

PD156707

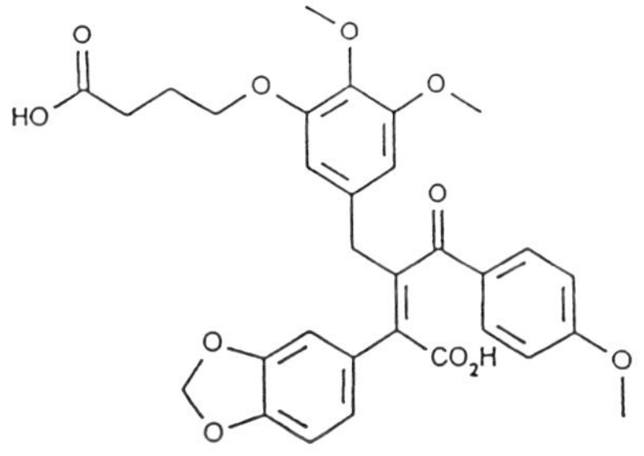

PD156707-CO_2H for Immunogen

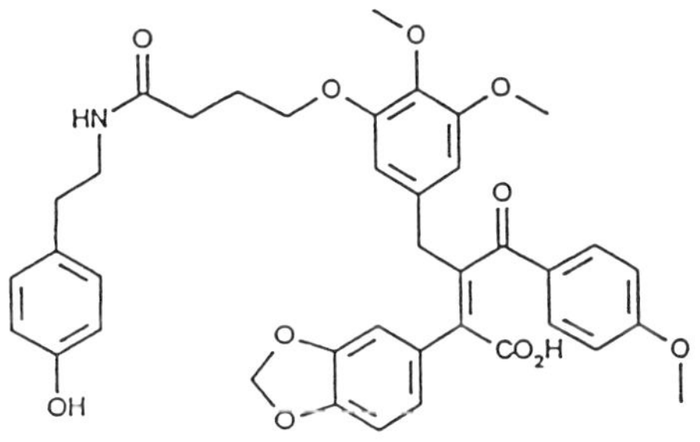

PD156707-CO_2NH-Tyramine for Radioiodination

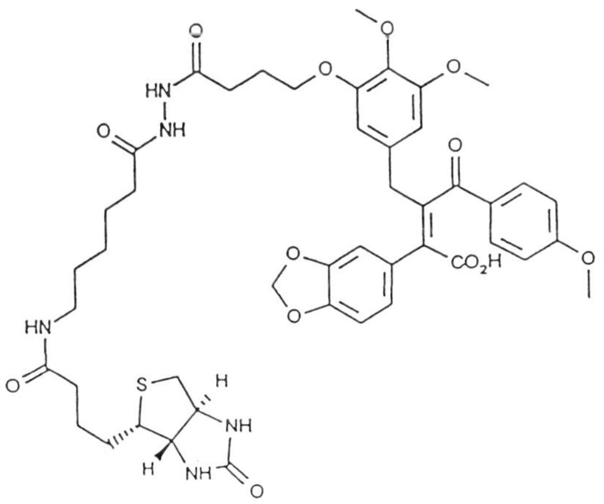

Figure 7. Development of tyramine- and biotin-labeled PD 156707 for radioimmunoassay use.

3. EFFICACY STUDIES: WHICH DISEASE STATES?

Historically, the pursuit of any pharmacological intervention has been based on the knowledge of a particular protein's involvement in a given disease state. Thorough research and understanding of, for example, the role of the adrenergic system in hypertension had resulted in the issuance of a patent for practolol in 1964, 16 years after Ahlquist's first description of the beta-receptor (Ahlquist, 1948). Time frames were shortened somewhat for ACE inhibitors, with the U.S. patent for captopril appearing just 7 years after the purification of ACE in 1970 (Dorer *et al.,* 1970). It was somewhat ironic, therefore, that advances in receptor cloning, mass screening, and structural biological techniques had arguably brought us to the point of identifying potent ET antagonists without any obvious therapeutic indication.

Broad pharmacological principles hold, however, and suggested areas to explore included those in which vasoconstriction (whether generalized or local) was thought to play an etiological role, as well as those in which ET levels were elevated, perhaps indicating a causative role (Allen *et al.,* 1993; Aoki *et al.,* 1994; Asbert *et al.,* 1993; Blazy *et al.,* 1994; Cassone *et al.,* 1996; Cody *et al.,* 1991; Estrada *et al.,* 1994; Ferri *et al.,* 1995; Heublein *et al.,* 1989; Isobe *et al.,* 1993; Kamoi *et al.,* 1990; Kaski *et al.,* 1995; Lerman *et al.,* 1992; Letizia *et al.,* 1995; McMurray *et al.,* 1992; Morelli *et al.,* 1995a,b; Nakamuta *et al.,* 1993; Perfetto *et al.,* 1995; Predel *et al.,* 1990; Rodeheffer *et al.,* 1992; Rosenberg *et al.,* 1993; Saito *et al.,* 1989, 1990; Shirakami *et al.,* 1994; Stewart *et al.,* 1991a,b; Stockenhuber *et al.,* 1992; Tomoda, 1993; Tsutamoto *et al.,* 1995; Uchida and Watanabe, 1993; Ziv *et al.,* 1992) (Fig. 8). For the purposes of the present review, we have chosen to describe those conditions where the greatest effort has been focused for the development of PD 156707 and related compounds, namely, hypertension, heart failure, pulmonary hypertension, and stroke.

3.1. Hypertension

It could be argued that the most obvious condition, and a very lucrative market, would be systemic ("essential") hypertension. Although mentioned as a potential etiological factor in the pathogenesis of hypertension in Yanagisawa's original description of ET (Yanagisawa *et al.,* 1988), plasma ET levels are not elevated in cases of uncomplicated hypertension in humans. One explanation for this may be that the abluminal secretion of the peptide generates increased local concentrations without any change in circulating levels. Experimental data have yielded conflicting results: Whereas some have demonstrated lowering of blood pressure in hypertensive models (Nishikibe *et al.,* 1993; Douglas *et al.,* 1994), early cardiovascular safety studies with PD 156707 in normal and hypertensive models in-

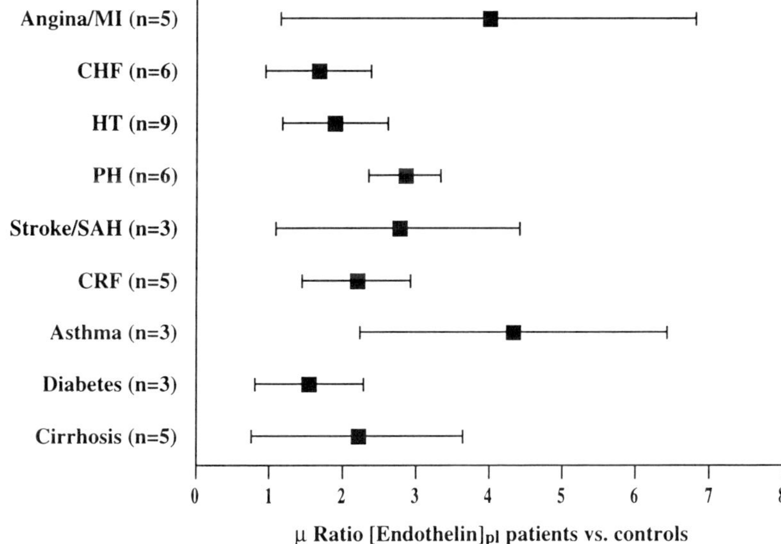

Figure 8. Pathological conditions associated with abnormalities of endothelin homeostasis in man. The *x* axis represents the mean plasma endothelin level relative to normal controls. Numbers in parentheses represent the number of studies from which the data are derived. MI, myocardial infarction; CHF, congestive heart failure; HT, hypertension; PH, pulmonary hypertension; SAH, subarachnoid hemorrhage; CRF, chronic renal failure.

dicated that the drug was not associated with significant falls in systemic blood pressure, whether given by the i.v. (Fig. 9) or oral (Fig. 10) route.

Recent attention has again been generated in this area with the observation that ET-1 +/− heterozygote mice demonstrate a sustained *elevation* in blood pressure (Kurihara *et al.,* 1994). While struggling with this concept, however, early clinical data would suggest that ET may in fact play a role in the maintenance of vascular tone in healthy volunteers (Haynes and Webb, 1994) and patients with hypertension (Warner *et al.,* 1996).

Although these data could suggest an indication for ET-receptor antagonists, our initial interest in hypertension was tempered by the difficult decision of whether there was an opportunity to commercialize a new antihypertensive agent in the cost-conscious era of managed care. Not only would any new agent have to compete against the $0.10 to $0.12 average daily cost for hydrochlorothiazide, but generic captopril, due in 1996, was expected to retail at a price of around $0.20 per day. Added to this, there was the realization that the use of blood pressure as a surrogate was coming under scrutiny both from a pharmacoeconomic point of view, as well as from a growing regulatory body that was asking whether "hard" clinical endpoints should be required for approval (FDA Cardiorenal Drugs Advisory Committee, October 20, 1995).

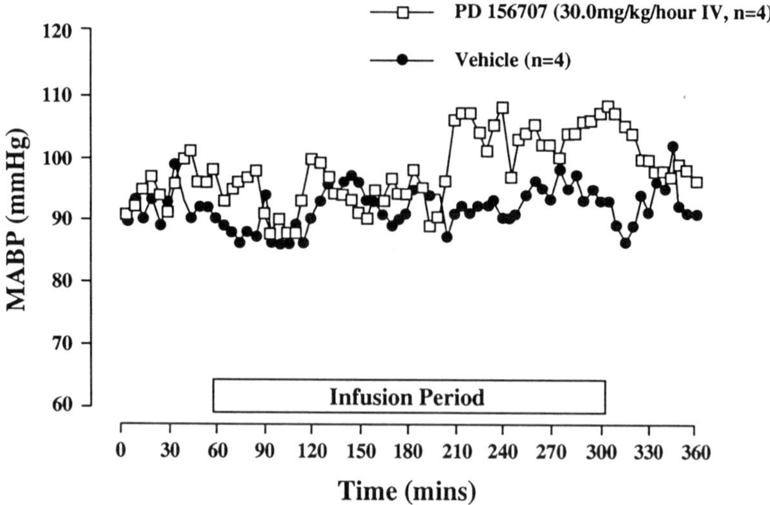

Figure 9. Effects of PD 156707 or vehicle on mean arterial blood pressure (MABP) in conscious dogs.

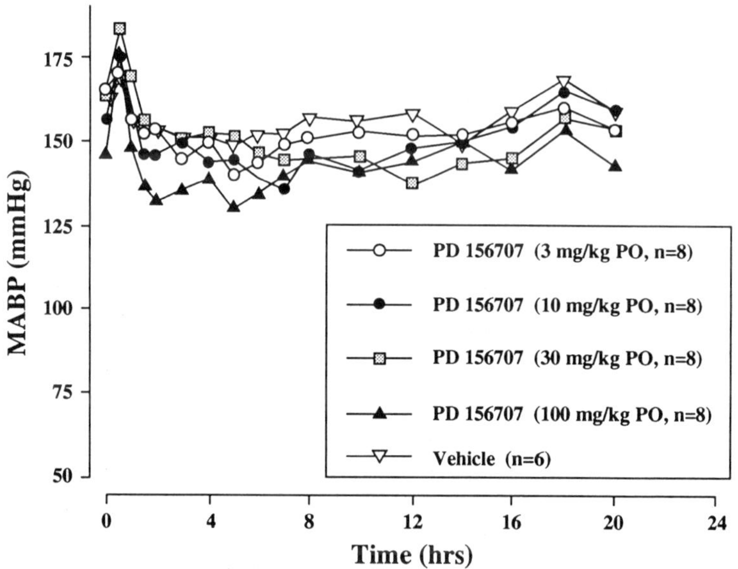

Figure 10. Effects of single oral doses of PD 156707 or vehicle on mean arterial blood pressure (MABP) in conscious, spontaneously hypertensive rats.

3.2. Heart Failure

Unlike uncomplicated hypertension, congestive heart failure (CHF) is a condition associated with marked elevations in circulating levels of big ET and ET-1 (Rodeheffer *et al.*, 1992; Wei *et al.*, 1994). Furthermore, it appears that higher levels may predict a worse outcome (Pacher *et al.*, 1993). Reasons for this may include the effects of ET on central and peripheral sympathetic activity (Wong-Dusting *et al.*, 1990; Matsumura *et al.*, 1994), antinatriuretic action (Miller *et al.*, 1989; Sorensen *et al.*, 1994) and/or effects on circulating levels of epinephrine (Boarder and Marriott, 1989), and aldosterone (Cozza *et al.*, 1989). The complex interplay between ET and the renin–angiotensin system has also been the subject of much attention (Rakugi *et al.*, 1990; Kawaguchi *et al.*, 1990; Scott-Burden *et al.*, 1991; Weber *et al.*, 1994).

The only clinical data to date appear to be those of Kiowski and co-workers who demonstrated favorable hemodynamic effects of infused bosentan in 24 patients with class III heart failure (Kiowski *et al.*, 1995). An interesting feature of this study was an observed doubling of circulating ET levels with bosentan; this phenomenon has been pursued in our own laboratories where we have concluded that it is a consequence of nonselective ET antagonism (Potoczak *et al.*, 1996). To date, there are very few preclinical data on the effects of ET antagonism in CHF. Teerlink described the hemodynamic effects of chronic oral bosentan in a rat coronary artery ligation model but avoided any mention of what might be regarded as more clinically meaningful endpoints (Teerlink *et al.*, 1994). A more recent study, however, suggested that the use of the ET_A-selective antagonist BQ 123 significantly improved the survival of rats with heart failure secondary to coronary artery ligation (Sakai *et al.*, 1996a). Our strategy in heart failure was to explore the potential of PD 156707 in a number of models utilizing several different endpoints (Table VII).

In a chronic rabbit model of heart failure resulting from rapid ventricular pacing, Spinale and co-workers demonstrated a significant improvement in LV geometry and pump function with s.c. pellets of PD 156707 versus a paced group treat-

Table VII
Studies with PD 156707 in Models of Heart Failure

Study	Model	Drug	Results
McConnell *et al.* (1996)	Paced dog	PD 156707	↓ mean arterial pressure ↓ total peripheral resistance
Spinale *et al.* (1996)	Paced rabbit	PD 156707	↑ LV pump function Normalization of myocyte contractile performance
Haleen *et al.* (Parke-Davis data)	Cardiomyopathic hamster	PD 156707	↑ LV pump function Fewer atrial thrombi

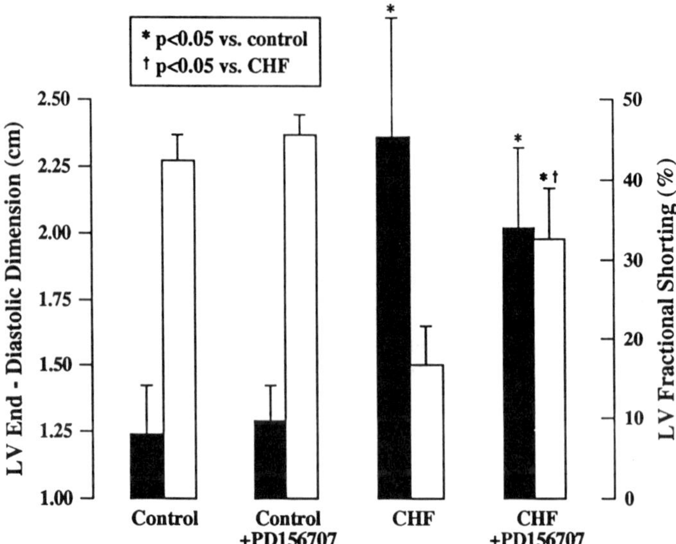

Figure 11. Effects of PD 156707 on left ventricular end-diastolic dimension and fractional shortening in a paced rabbit model of chronic heart failure. Long-term systemic drug exposure was achieved by means of subcutaneous pellets at a dose ($n = 8 \times 500$ mg) previously shown to inhibit ET-1 challenges after 3 days. Data presented as mean ± SD Adapted from Spinale *et al.* (1996).

ed with inert pellets (Spinale *et al.,* 1996). Treatment was associated also with reduced plasma norepinephrine levels and normalization of plasma renin activity. Finally, PD 156707 was associated with improved isolated myocyte contractile function and normalization of inotropic responsiveness (Fig. 11).

Zucker's group also demonstrated improvements in cardiac performance in their canine model; in this case, however, the effect was attributed to significant falls in systemic blood pressure (McConnell *et al.,* 1996). Our own experience with the cardiomyopathic hamster confirms the association of chronic ET_A receptor antagonism with oral PD 156707 and improved pump function at a 300-day endpoint; an additional observation in this model was the occurrence of fewer atrial thrombi in PD 156707-treated animals compared with controls.

3.3. Pulmonary Hypertension

An interesting observation from the evaluation of BQ 123 in rats with heart failure was the drug's effect on the pulmonary vasculature. In a follow-up publication, Sakai and colleagues described a significant reduction in right ventricular systolic pressure and central venous pressure without concomitant effects on the systemic vasculature (Sakai *et al.,* 1996b). This apparent predilection for the pulmonary bed was also seen in the hemodynamic study with bosentan in patients with heart failure: Bosentan reduced mean arterial pressure by 7.7% but pulmonary artery pressure by 13.7%; systemic vascular resistance by 16.5% but pulmonary vascular resistance by 33.2%. ET was shown to cause pulmonary vascular smooth muscle contraction and proliferation (Janakidevi *et al.,* 1992), and raised plasma ET-1 levels and immunoreactivity had been demonstrated previously in patients with pulmonary hypertension, but not other forms of lung disease (Stewart *et al.,* 1991b; Giaid *et al.,* 1993). In an earlier study, bosentan had attenuated ET-1-induced vasoconstriction in pulmonary arterial rings and isolated, perfused lungs (Eddahibi *et al.,* 1995) and the ET_A selective agent FR 139317 had produced greater effects in a canine model of pulmonary hypertension than in control dogs (Okada *et al.,* 1995). More recently, continued therapy with bosentan in a hypoxic rat model was associated with reversal of pulmonary hypertension, right heart hypertrophy, and pulmonary vascular remodeling despite continuing hypoxic exposure (Chen *et al.,* 1995).

Interest in this area led us to evaluate our receptor antagonists in a number of experimental models of pulmonary hypertension. In an ovine model of cardiopulmonary bypass (CPB), Fineman and colleagues studied the effects of the nonselective peptide antagonist, PD 145065 (Cody *et al.,* 1993). They found that preexisting increased pulmonary blood flow (achieved by means of an *in utero* placement of an aortopulmonary shunt) increased the response of the pulmonary circulation to CPB, and that this could be prevented by pretreatment with the antagonist. The authors concluded that these data suggested a role for ET-1 in post-CPB pulmonary hypertension, and that ET-1 receptor antagonists might decrease morbidity in children at risk for pulmonary hypertension after surgical repair of congenital heart lesions (Reddy *et al.,* 1996).

In a more recent rat study with s.c. pellets of PD 156707, McMurtry and coworkers were able to completely prevent the rise in pulmonary artery pressure seen with chronic exposure to hypoxia (personal communication). Interestingly, the phenomenon was most apparent with a low dose of drug (six pellets): A group of rats treated with eight pellets demonstrated less protection, suggesting perhaps some nonspecific ET_B receptor antagonism at these higher doses. Reference has already been made to the observation that PD 156707 is less ET_A-selective in the rat than any other species tested to date. If this were the case, it would add further credence to the suggestion that ET_A-selective antagonists may be more useful in the treatment of pulmonary hypertension.

Our own studies in-house have expanded McMurtry's observations in an attempt to identify a no-effect dose for PD 156707 in an acute hypoxic model. The data suggest an effect of the drug (i.v. infusion started 60 min prior to the onset of hypoxia) at doses as low as 0.3 μg/kg per hour. When given orally as a single dose 30 min prior to the onset of hypoxia, the drug was effective at doses as low as 30 μg/kg (Keiser *et al.,* 1997) (Fig. 12).

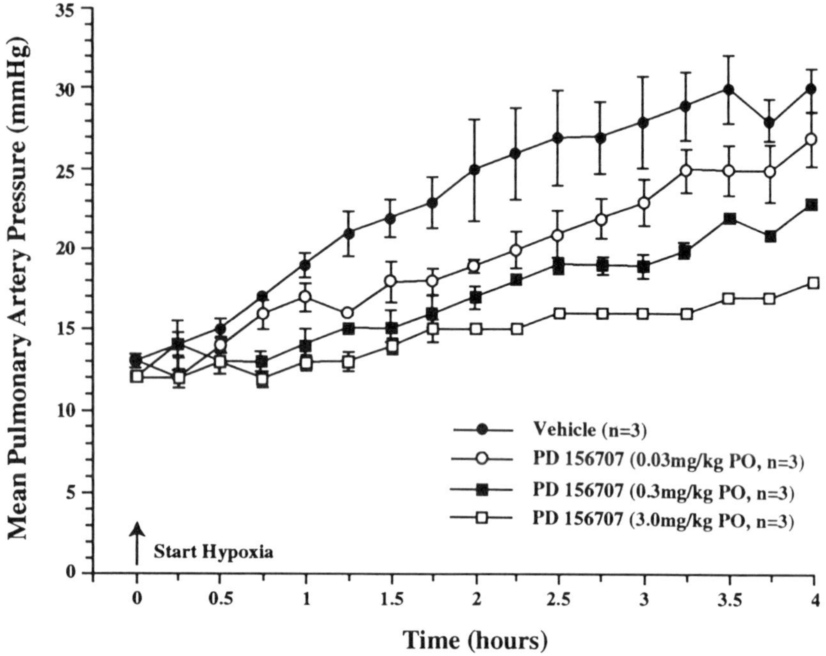
Mean Pulmonary Artery Pressure (mmHg)
35
30
25
20
15
10
5
0
Vehicle (n=3)
PD 156707 (0.03mg/kg PO, n=3)
PD 156707 (0.3mg/kg PO, n=3)
PD 156707 (3.0mg/kg PO, n=3)
Start Hypoxia
0
0.5
1
1.5
2
2.5
3
3.5
4
Time (hours)

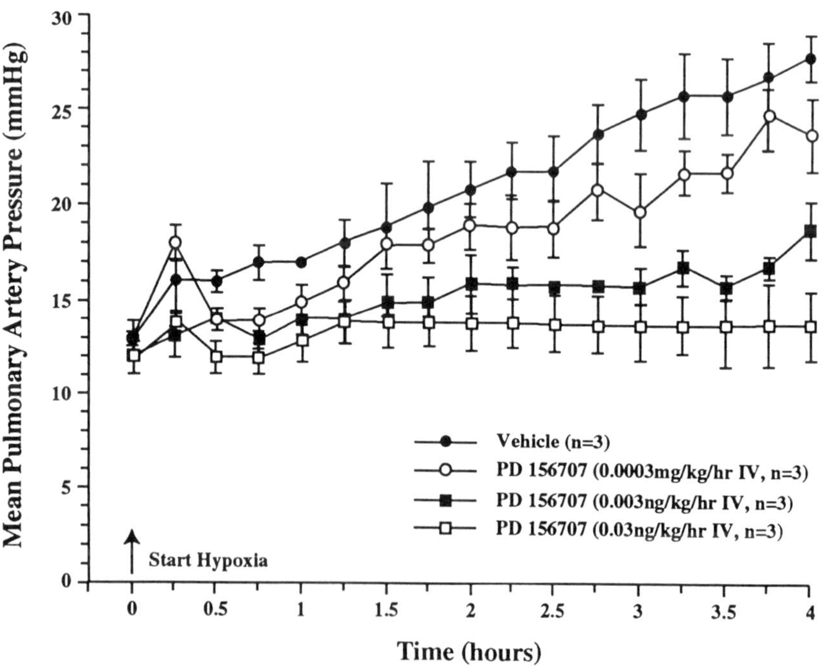
Mean Pulmonary Artery Pressure (mmHg)
30
25
20
15
10
5
0
Vehicle (n=3)
PD 156707 (0.0003mg/kg/hr IV, n=3)
PD 156707 (0.003ng/kg/hr IV, n=3)
PD 156707 (0.03ng/kg/hr IV, n=3)
Start Hypoxia
0
0.5
1
1.5
2
2.5
3
3.5
4
Time (hours)

3.4. Stroke

It had been known from the early days of the ET program that these drugs were profound cerebral vasoconstrictors *in vitro* and *in vivo* (Salom *et al.*, 1993). Raised ET levels had been detected in the plasma and cerebrospinal fluid of patients after stroke (Ziv *et al.*, 1992) and elevated ET immunoreactivity was a feature after focal ischemia in a rat model (Barone *et al.*, 1994). The role of ET, converting enzyme, and receptor antagonists has recently been reviewed by McCulloch's group from Glasgow, which has developed a series of elegant models ranging from direct measurement of pial arteriolar diameter to assessment of cerebral blood flow and determination of infarct size after focal and global ischemia (see Patel, 1996).

Using a feline model of focal cerebral ischemia, Patel and colleagues demonstrated a restoration of cerebral blood flow to normal within 6 hr of middle cerebral artery occlusion when PD 156707 (1.6 mg/kg i.v. bolus +2.7 mg/kg per hour) was infused 30 min after the insult (Patel *et al.*, 1995). In the same experiment, the volume of ischemic damage measured histologically was reduced by 45% in the PD 156707-treated group (Patel *et al.*, 1996) (Fig. 13). More recently, the same group demonstrated a 21% reduction in the volume of hemispheric infarction in the rat when PD 156707 was infused at 3 mg/kg per hour. A trend seen with a lower dose of drug (0.3 mg/kg per hour) did not reach statistical significance in this model (Takasago and McCulloch, 1997) (Fig. 13). It is of note, however, that earlier studies with bosentan had failed to show a similar protection in the rat, possibly because of antagonism of ET_B-mediated dilator responses. Similarly, although the ET_A-selective antagonist BQ 123 was effective in the spontaneously hypertensive rat (Patel and Wilson, 1995), it was without effect in the normotensive strain (Checkley *et al.*, 1995), raising speculation about blood–brain barrier penetration.

4. FUTURE PLANS

Demonstration of efficacy in a number of models of disease states has elevated the ET_A-selective antagonist PD 156707 to the status of lead compound and initiated formal toxicological testing. Being a new class of compounds, one does not have the benefit of experience in anticipating adverse effects of these drugs, so

Figure 12. Effects of oral (upper) and i.v. (lower) PD 156707 in preventing the hypoxia-associated rise in mean pulmonary artery pressure in conscious rats. Hypoxia was achieved by placing rats in individual 30-liter Plexiglas chambers and exposing them to 10% O_2:90% N_2 gas supplied at a rate of 3 liters/min per chamber. Drug was dosed orally 30 min before the onset of hypoxia or infused throughout the duration of the experiments, again starting 30 min before the onset of hypoxia. Calculated ID_{50}s were 0.14 mg/kg (oral) and 0.0016 mg/kg (i.v.).

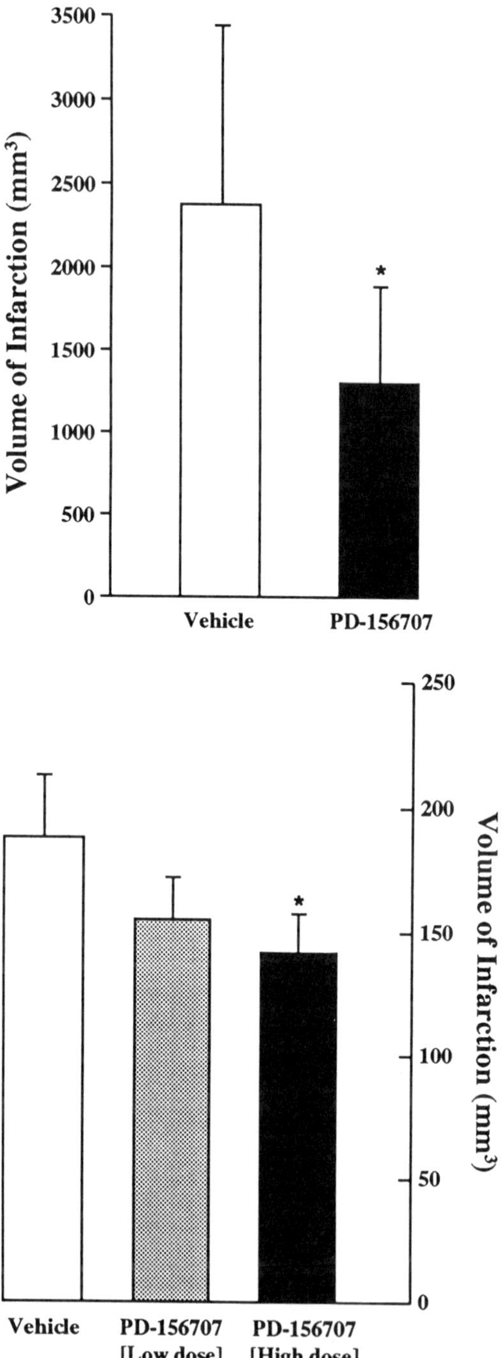

Figure 13. Effects of PD 156707 and vehicle on cerebral hemispheric infarction volume in the cat (upper panel) and rat (lower panel). The dose of PD 156707 in the cat was 1.6 mg/kg i.v. bolus + 2.7 mg/kg per hr; doses in the rat were 0.3 mg/kg per hr (low dose) and 3.0 mg/kg per hr (high dose). Data presented as mean ± SEM. Adapted from Patel *et al.*, (1996) and Takasago and McCulloch (1997).

it is possible that progress will be made more slowly than would be the case with "me-too" compounds. Nevertheless, at least two compounds that are ET_A/ET_B balanced antagonists (Ro 47-0203 and SB 209670) have reached clinical testing to date, so it is hoped that antagonism of ET receptors as a class effect will not be associated with an unacceptable toxicological profile. In addition, it is likely that the selectivity of the ET antagonist will influence the toxicological profile and therapeutic efficacy of this new class of pharmaceutical agents.

It can be seen that the single dose (2.7 mg/kg per hour) that produced a significant reduction in cerebral infarction in the cat was associated with a mean plasma level of 2.2 μg/ml. In heart failure, efficacy was seen in two models with levels as low as 41 and 50 ng/ml, but it is in pulmonary hypertension, however, that positive effects of the drug have been seen at the lowest doses: 0.54 and 0.56 ng/ml in parenteral and oral studies, respectively. As toxicology studies progress, it will be necessary to explore fully the therapeutic dose range of PD 156707 in each condition that we intend to pursue, such that a safety margin can be calculated for our clinical trials.

5. SUMMARY

PD 156707 is a highly potent, selective antagonist of the ET_A receptor that has demonstrated efficacy in a number of different disease models. The next few years will be exciting in the field of ET research as several compounds progress through clinical development. It is our hope that the efficacy data demonstrated to date with PD 156707 will some day be translated into real hope for the patients who are waiting beyond the confines of our research laboratories.

REFERENCES

Ahlquist, R. P., 1948, A study of the adrenotropic receptors, *Am. J. Physiol.* **153:**586–600.

Allen, S. W., Chatfield, B. A., Koppenhafer, S. A., Schaffer, M. S., Wolfe, R. R., and Abman, S. H., 1993, Circulating immunoreactive endothelin-1 in children with pulmonary hypertension, *Am. Rev. Respir. Dis.* **148:**519–522.

Aoki, T., Kojima, T., Ono, A., Unishi, G., Yoshijimia, S., Kameda-Hayashi, N., Yamamoto, C., Hirata, Y., and Kobayashi, Y., 1994, Circulating endothelin-1 levels in patients with bronchial asthma, *Ann. Allergy* **73:**365–369.

Arai, H., Hori, S., Aramori, I., Ohkubo, H., and Nakanishi, S., 1990, Cloning and expression of a cDNA encoding an endothelin receptor, *Nature* **348:**730–732.

Asbert, M., Gins, A., Gins, P., Jimnez, W., Claria, J., Salo, J., Arroyo, V., Rivera, F., and Rods, J., 1993, Circulating levels of endothelin in cirrhosis, *Gastroenterology* **104:**1485–1491.

Barone, F. C., Globus, M. Y.-T., Price, W. J., White, R. F., Storer, B. L., Feuerstein, G. Z., Busto, R., and Ohlstein, E. H., 1994, Endothelin levels increase in rat focal and global ischemia, *J. Cereb. Blood Flow Metab.* **14:**337–342.

Bax, W. A., Bos, E., and Saxena, P. R., 1993, Heterogeneity of endothelin/sarafotoxin receptors mediating contraction of the human saphenous vein, *Eur. J. Pharmacol.* **239:**267–268.

Baynash, A. G., Hosoda, K., Giaid, A., Richardson, J., Emoto, N., Hammer, R., and Yanagisawa, M., 1994, Interaction of endothelin-3 with endothelin-B receptor is essential for development of epidermal melanocytes and enteric neurons, *Cell* **79:**1277–1285.

Blazy, I., Dechaux, M., Charbit, M., Brocart, D., Souberbielle, J. C., Gagnadoux, M. F., Guillot, F., and Sachs, C., 1994, Endothelin-1 in children with chronic renal failure, *Pediatr. Nephrol.* **8:**40–44.

Boarder, M. R., and Marriott, D. B., 1989, Characterization of endothelin-1 stimulation of catecholamine release from adrenal chromaffin cells, *J. Cardiovasc. Pharmacol.* **13**(5)**:**S223–S224.

Cassone, R., Moroni, C., Parlapiano, C., Bondanini, F., Blefari, T., and Affricano, C., 1996, Endothelin-1 plasma levels in essential hypertension: Increased levels with coronary artery disease, *Am. Heart J.* **132**(5)**:**1048–1049.

Checkley, D., Pleeth, R., Breen, S., Waterton, J., and Wilson, C., 1995, Effect of BQ 123 on rat cerebral infarct size measured by diffusion- and T2-weighted MRI, *Proceedings of the 4th International Conference on Endothelin, London,* p. 174.

Chen, S. J., Chen, Y. F., Meng, Q. C., Durand, J., Dicarlo, V. S., and Oparil, S., 1995, Endothelin-receptor antagonist bosentan prevents and reverses hypoxic pulmonary hypertension, *J. Appl. Physiol.* **79:**2122–2131.

Clozel, M., Gray, G. A., Breu, V., Loffler, B.-M., and Osterwalder, R., 1992, The endothelin ET_B receptor mediates both vasodilatation and vasoconstriction in vivo, *Biochem. Biophys. Res. Commun.* **186:**867–873.

Clozel, M., Breu, V., Burri, K., Cassao, J.-M., Fischli, W., Gray, G. A., Hirth, G., Loffler, B.-M., Muller, M., Neidhart, W., and Ramuz, H., 1993, Pathophysiological role of endothelin revealed by the first orally active endothelin receptor antagonist, *Nature* **365:**759–761.

Cody, R. J., Haas, G. J., Binkley, P. F., Capers, Q., and Kelley, R., 1991, Plasma endothelin correlates with the extent of pulmonary hypertension in patients with chronic congestive heart failure, *Circulation* **85**(2)**:**504–509.

Cody, W. L., Doherty, A. M., He, J. X., DePue, P. L., Rapundalo, S. T., Hingorani, G. A., Major, T. C., Panek, R. L., Haleen, S., LaDouceur, D., Reynolds, E. E., Hill, K. E., and Flynn, M. A., 1992, Design of a functional antagonist of endothelin, *J. Med. Chem.* **35:**3301–3303.

Cody, W. L., Doherty, A. M., He, J. X., DePue, P. L., Waite, L. A., Topliss, J. G., Haleen, S. J., LaDouceur, D., Flynn, M. A., Hill, K. E., and Reynolds, E. E., 1993, The rational design of a highly potent combined ET_A and ET_B receptor antagonist, *Med. Chem. Res.* **3:**154–162.

Cozza, E. N., Gomez-Sanchez, C. E., Foecking, M. F., and Chiou, S., 1989, Endothelin binding to cultured calf adrenal zona glomerulosa cells and stimulation of aldosterone secretion, *J. Clin. Invest.* **84**(3)**:**1032–1035.

Davenport, A. P., and Maguire, J. J., 1994, Endothelin receptors in human coronary artery—Reply, *Trends Pharmacol. Sci.* **15:**136–137.

Davenport, A. P., O'Reilly, G., Molenaar, P., Maguire, J. J., Kuc, R. E., Sharkey, A., Bacon, C. R., and Ferro, A., 1993, Human endothelin receptors characterized using reverse transcriptase-polymerase chain reaction, in situ hybridisation and subtype selective ligands BQ-123 and BQ-3020: Evidence for expression of ET_B receptors in human vascular smooth muscle, *J. Cardiovasc. Pharmacol.* **22**(Suppl. 8)**:**S22–S25.

Davenport, A. P., O'Reilly, G., and Kuc, R. E., 1995, Endothelin ET_A and ET_B mRNA and receptors expressed by smooth muscle in the human vasculature: Majority of the ET_A sub-type, *Br. J. Pharmacol.* **114:**1110–1116.

Doherty, A. M., 1992, Endothelin: A new challenge, *J. Med. Chem.* **35:**1493–1508.

Doherty, A. M., Cody, W. L., DePue, P. L., He, J. X., Waite, L. A., Leonard, D. M., Leitz, N. L., Dudley, D. T., Rapundalo, S. T., Hingorani, G. P., Haleen, S. J., LaDouceur, D. M., Hill, K. E., Flynn, M. A., and Reynolds, E. E., 1993a, Structure–activity relationships of C-terminal endothelin hexapeptide antagonists, *J. Med. Chem.* **36:**2585–2594.

Doherty, A. M., Cody, W. L., He, J. X., DePue, P. L., Cheng, X.-M., Welch, K. M., Flynn, M. A., Reynolds, E. E., LaDouceur, D. M., Davis, L. S., Keiser, J. A., and Haleen, S. J., 1993b, In vitro and in vivo studies with a series of hexapepide endothelin antagonists, *J. Cardiovasc. Pharmacol.* **22**(Suppl. 8)**:**S98–S102.

Doherty, A. M., Patt, W. C., Edmunds, J. J., Berryman, K. A., Reisdorph, B. R., Plummer, M. S., Shahripour, A., Lee, C., Cheng, X.-M., Walker, D. M., Haleen, S. J., Keiser, J. A., Flynn, M. A., Welch, K. M., Hallak, H., Taylor, D. G., and Reynolds, E. E., 1995, Discovery of a novel series of orally active non-peptide endothelin-A (ET_A) receptor-selective antagonists, *J. Med. Chem.* **38:**1259–1263.

Doherty, A., Patt, W., Reisdorph, B., Repine, J., Walker, D., Flynn, M., Welch, K., Reynolds, E., and Haleen, S., 1996, Design and pharmacological evaluation of non-peptide ET_A selective and ET_A/ET_B receptor antagonists, in *Proceedings of the AFMC Symposium, Tokyo,* Sept. 1995 (M. Yamazaki, ed.), pp. 255–261, Blackwell, Oxford.

Dorer, F. E., Skeggs, L. T., Kahn, J. R., Lentz, K. E., and Levine, M., 1970, Angiotensin converting enzyme: Method of assay and partial purification, *Anal. Biochem.* **33:**102–113.

Douglas, S. A., Gellai, M., Ezekiel, M., and Ohlstein, E. H., 1994, BQ 123, a selective endothelin subtype A-receptor antagonist, lowers blood pressure in different rat models of hypertension, *J. Hypertens.* **12:**561–567.

Dudeck, R. C., Nordblum, G. D., Barkdale, C. M., Patt, W. C., Hallak, H., and Kindt, E., 1997, Development of an enzyme immunoassay for the ET_A selective endothelin antagonist PD 156707 in dog plasma employing a drug-biotin analog with enzyme-avidin conjugate, *Eighth International Symposium on Pharmaceutical and Biomedical Analysis, Orlando, Florida.*

Eddahibi, S., Raffestin, B., Clozel, M., Levame, M., and Adnot, S., 1995, Protection from pulmonary hypertension with an orally active endothelin receptor antagonist in hypoxic rats, *Am. J. Physiol.* **268:**H828–H835.

Elliott, J. D., Lago, M. A., Cousins, R. D., Gao, A., Leber, J. D., Erhard, K. F., Nambi, P., Elshourbagy, N. A., Kumar, C., Lee, J. A., Bean, J. W., DeBrosse, C. W., Eggleston, D. S., Brooks, D. P., Feuerstein, G., Ruffolo, R. R., Weinstock, J., Gleason, J. G., Peishoff, C. E., and Ohlstein, E. H., 1994, 1,3-Diarylindan-2-carboxylic acids, potent and selective non-peptide endothelin receptor antagonists, *J. Med. Chem.* **37:**1553–1557.

Estrada, V., Tellez, M. J., Moya, J., Fernandez-Durango, R., Egido, J., and Cruz, A. F., 1994, High plasma levels of endothelin-1 and atrial natriuretic peptide in patients with acute ischemic stroke, *Am. J. Hypertens.* **7:**1085–1089.

Ferri, C., Bellini, C., Desideri, G., Di Francesco, L., Baldoncini, R., Santucci, A., and De Mattia, G., 1995, Plasma endothelin-1 levels in obese hypertensive and normotensive men, *Diabetes* **44:**431–436.

Ferro, J. F., and Webb, D. J., 1996, The clinical potential of endothelin receptor antagonists in cardiovascular medicine, *Drugs* **51**(1)**:**12–27.

Giaid, A., Yanagisawa, M., Langleben, D., Michel, R. P., Levy, R., Shenneb, H., Kimura, S., Masaki, T., Duguid, W. P., and Stewart, D. J., 1993, Expression of endothelin-1 in the lungs of patients with pulmonary hypertension, *N. Engl. J. Med.* **328:**1732–1739.

Godfraind, T., 1994, Endothelin receptors in human coronary artery, *Trends Pharmacol. Sci.* **15:**136.

Haynes, W. G., and Webb, D. J., 1994, Contribution of endogenous generation of endothelin-1 to basal vascular tone, *Lancet* **344:**852–854.

Heublein, D. M., Rodeheffer, R. J., Cavero, P. G., Miller, W. L., Edwards, B. S., Redfield, M. M., and Burnett, J. C., Jr., 1989, Relationship between plasma endothelin and atrial natriuretic factor in humans with congestive heart failure, *Am. J. Hypertens.* **2:**37A.

Hosoda, K., Nakao, K., Arai, H., Suga, S., Ogawa, Y., Mukoyama, M., Shirakami, G., Saito, Y., Nakanishi, S., and Imura, H., 1991, Cloning and expression of human endothelin-1 receptor cDNA, *FEBS Lett.* **287:**23–26.

Inoue, A., Yanagisawa, M., Kimura, S., Kasuya, Y., Miyauchi, T., Goto, K., and Masaki, T., 1989, The human endothelin family: Three structurally and pharmacologically distinct isopeptides predicted by three separate genes, *Proc. Natl. Acad. Sci. USA* **86:**2863–2867.

Ishikawa, K., Fukami, T., Nagase, T., Fujita, K., Hayama, T., Niyama, K., Mase, T., Ihara, M., and Yano, M., 1992, Cyclic pentapeptide endothelin antagonists with ET_A selectivity. Potency and solubility enhancing modifications, *J. Med. Chem.* **35:**2139–2142.

Ishikawa, K., Ihara, M., Noguchi, K., Mase, T., Mino, N., Saeki, T., Fukuroda, T., Fukami, T., Ozaki, S., Nagase, T., Nishikibe, J., and Yano, M., 1994, Biochemical and pharmacological profile of a potent and selective endothelin B-receptor antagonist BQ-788, *Proc. Natl. Acad. Sci. USA* **91:**4892–4896.

Isobe, H., Satoh, M., Sakai, H., and Nawata, H., 1993, Increased plasma endothelin-1 levels in patients with cirrhosis and esophageal varices, *J. Clin. Gastroenterol.* **17**(3)**:**227–230.

Iyer, K., Sinz, M., Cheng, X., and Hallak, H., 1996, In vitro methylenedioxyphenyl metabolite-intermediate complex formation in rat, dog and human liver microsomes, *ISSX North American Meeting, San Diego, October.*

Janakidevi, K., Fisher, M. A., Del Vecchio, P. J., Tiruppathi, C., Figge, J., and Malik, A. B., 1992, Endothelin-1 stimulates DNA synthesis and proliferation of pulmonary artery smooth muscle cells, *Am. J. Physiol.* **263:**C1295–C1301.

Kamoi, K., Sudo, N., Ishibashi, M., and Yamaji, T., 1990, Plasma endothelin-1 levels in patients with pregnancy-induced hypertension [letter], *N. Engl. J. Med.* **322:**1486–1487.

Karne, S., Jayawickreme, C. K., and Lerner, M. R., 1993, Cloning and characterization of an endothelin-3 specific receptor (ET_C receptor) from *Xenopus laevis* dermal melanophores, *J. Biol. Chem.* **268:**19126–19133.

Kaski, J. C., Elliott, P. M., Salomones, O., Dickinson, K., Gordon, D., Hann, C., and Holt, D. W., 1995, Concentration of circulating plasma endothelin in patients with angina and normal coronary angiograms, *Br. Heart J.* **74:**620–624.

Kawaguchi, H., Sawa, H., and Yasuda, H., 1990, Endothelin stimulates angiotensin I to angiotensin II conversion in cultured pulmonary artery endothelial cells, *J. Mol. Cell. Cardiol.* **22:**839–842.

Keiser, J. A., Schroeder, R. L., Hallak, H., Uprichard, A. C. G., Doherty, A. M., and Haleen, S. J., 1997, Pharmacodynamics of PD 156707, a selective endothelin-A (ET_A) antagonist, in acute hypoxic pulmonary hypertension, *FASEB J.* **11**(3)**:**A36.

Kiowski, W., Sutsch, G., Hunziker, P., Muller, P., Kim, J., Oechslin, E., Schmitt, R., Jones, R., and Bertel, O., 1995, Evidence for endothelin-1-mediated vasoconstriction in severe chronic heart failure, *Lancet* **346:**732–736.

Kumar, C., Mwangi, V., Nuthulaganti, P., Wu, H.-L., Pullen, M., Brunb, K., Aiyar, H., Morris, R. A., Naughton, R., and Nambi, P., 1994, Cloning and characterization of a novel endothelin receptor from Xenopus heart, *J. Biol. Chem.* **269:**13414–13420.

Kurihara, Y., Kurihara, H., Suzuki, H., Kodama, T., Maemura, K., Nagal, R., Oda, H., Kuwaki, T., Cao, W.-H., Kamada, N., Jishage, K., Ouchi, Y., Azuma, S., Toyoda, Y., Ishikawa, T., Kumada, M., and Yazaki, Y., 1994, Elevated blood pressure and craniofacial abnormalities in mice deficient in endothelin-1, *Nature* **368:**703–710.

Lerman, A., Kubo, S. H., Tschumperlin, L. K., and Burnett, J. C., 1992, Plasma endothelin concentrations in humans with end-state heart failure and after heart transplantation, *J. Am. Coll. Cardiol.* **20**(4)**:**849–853.

Letizia, C., Cerci, S., De Ciocchis, A., D'Ambrosio, C., Scuro, L., and Scavo, D., 1995, Plasma endothelin-1 levels in normotensive and borderline hypertensive subjects during a standard cold pressor test, *J. Hum. Hypertens.* **9:**903–907.

Love, M. P., and McMurray, J. J. V., 1996, Endothelin in chronic heart failure; current position and future prospects, *Cardiovasc. Res.* **31**(5)**:**665–674.

Love, M. P., Haynes, W. G., Webb, D. J., and McMurray, J. J. V., 1995, Which endothelin receptors are functionally important in chronic heart failure? *Circulation* **92:**I-331.

Masuda, Y., Sugo, T., Kikuchi, T., Kawata, A., Satoh, M., Fujisawa, Y., Itoh, Y., Wakimasu, M., and Ohtaki, T., 1996, Receptor binding and antagonist properties of a novel endothelin receptor antagonist Tak-044 {cycle[D-α-aspartyl-3-[(4-phenylpiperazin-1-yl)carbonyl]-L-alanyl-L-a-aspartyl-D-2-(2-thienyl)glycyl-L-leucyl-D-tryptophyl]disodium salt}, in human endothelin A and endothelin B receptors, *J. Pharmacol. Exp. Ther.* **279**(2)**:**675–685.

Matsumura, K., Abe, I., Fukuhara, M., Tominaga, M., Tsuchihashi, T., Kobayashi, K., and Fujishima, M., 1994, Naloxone augments sympathetic outflow induced by centrally administered endothelin in conscious rabbits, *Am. J. Physiol.* **266**(4, Pt. 2)**:**R1403–R1410.

McConnell, P. I., Wang, W., Gallagher, K., and Zucker, I. H., 1996, The effects of a specific endothelin-1A (ETA) receptor antagonist on the development of chronic heart failure (HF) in the dog, *FASEB J.* **10:**A430.

McMurray, J. J., Ray, S. G., Abdullah, I., Dargie, H. J., and Morton, J. J., 1992, Plasma endothelin in chronic heart failure, *Circulation* **85**(4)**:**1374–1379.

Mihara, S., Nakajima, S., Matsumura, S., Kohnoike, T., and Fujimoto, M., 1994, Pharmacological characterization of a potent non-peptide endothelin receptor antagonist, 97–139, *J. Pharmacol. Exp. Ther.* **268:**1122–1128.

Miller, W. L., Redfield, M. M., and Burnett, J. C., 1989, Integrated cardiac, renal and endocrine actions of endothelin, *J. Clin. Invest.* **83:**317–320.

Moreland, S., McMullen, D. M., Delaney, C. L., Lee, V. G., and Hunt, J. T., 1992, Venous smooth muscle contains vasoconstrictor ET_B like receptors, *Biochem. Biophys. Res. Commun.* **184:**100–106.

Morelli, S., Ferri, C., Di Francesco, L., Baldoncini, R., Carlesimo, M., Bottoni, U., Properzi, G., Santucci, A., and Valesini, G., 1995a, Plasma endothelin-1 levels in patients with systemic sclerosis: Influence of pulmonary or systemic arterial hypertension, *Ann. Rheum. Dis.* **54:**730–734.

Morelli, S., Ferri, C., Polettini, E., Bellini, C., Gualdi, G. F., Pittoni, V., Valesini, G., and Santucci, A., 1995b, Plasma endothelin-1 levels, pulmonary hypertension, and lung fibrosis in patients with systemic sclerosis, *Am. J. Med.* **99:**255–260.

Mugrage, B., Moliterni, J., Robinson, L., Webb, R. L., Shetty, S. S., Lipson, K. E., Chin, M. H., Neale, R., and Cioffi, C., 1993, CGS 27830, a potent nonpeptide endothelin receptor antagonist, *Bioorg. Med. Chem. Lett.* **3:**2099–2104.

Nakamuta, M., Ohashi, M., Tabata, S., Tanabe, Y., Goto, K., Naruse, M., Naruse, K., Hiroshige, K., and Nawata, H., 1993, High plasma concentrations of endothelin-like immunoreactivities in patients with hepatocellular carcinoma, *Am. J. Gastroenterol.* **88**(2)**:**248–252.

Nishikibe, M., Tsuchida, S., Okada, M., Fukuroda, T., Shimamoto, K., Yano, M., Ishikawa, K., and Ikemoto, F., 1993, Antihypertensive effect of a newly synthesized endothelin antagonist, BQ-123, in a genetic hypertensive model, *Life Sci.* **52:**717–724.

Ohlstein, E. H., Nambi, P., Lago, A., Hay, D. W. P., Beck, G., Fong, K.-L., Eddy, E. P., Smith, P., Ellens, H., and Elliott, J. D., 1996, Nonpeptide endothelin antagonists. VI: Pharmacological characterization of SB 217242, a potent and highly bioavailable endothelin receptor antagonist, *J. Pharmacol. Exp. Ther.* **276:**609–615.

Okada, M., Yamashita, C., Okada, M., and Okada, K., 1995, Endothelin receptor antagonists in a beagle model of pulmonary hypertension: Contribution to possible potential therapy? *J. Am. Coll. Cardiol.* **25:**1213–1217.

Opgenorth, T. J., Adler, A. L., Calzadilla, V., Chiou, W. J., Dayton, B. D., Dixon, D. B., Gehrke, L. J., Hernandez, L., Magnuson, S. R., Marsh, K. C., Novosad, E. I., von Geldern, T. W., Wessale, J. L., Winn, M., and Wu-Wong, J. R., 1996, Pharmacological characterization of A-127722: An orally active and potent ET_A-selective receptor antagonist, *J. Pharmacol. Exp. Ther.* **276:**473–481.

Pacher, R., Bergler-Klein, J., Globits, S., Teufelsbauer, H., Schuller, M., Krauter, A., Ogris, E., Rodler, S., Wutte, M., and Hartter, E., 1993, Plasma big endothelin-1 concentrations in congestive heart failure patients with or without systemic hypertension, *Am. J. Cardiol.* **71:**1293–1299.

Panek, R. L., Major, T. C., Hingorani, G. P., Doherty, A. M., Taylor, D. G., and Rapundalo, S. T., 1992, Endothelin and structurally related analogs distinguish between endothelin receptor subtypes, *Biochem. Biophys. Res. Commun.* **183:**566–571.

Patel, J. B., and Wilson, C., 1995, Effect of BQ 123 in the SH rat focal ischaemia model, in: *Proceedings of the 4th International Conference on Endothelin, London,* p. 173.

Patel, T. R., 1996, Therapeutic potential of endothelin receptor antagonists in cerebrovascular disease, *CNS Drugs* **5**(4)**:**293–310.

Patel, T. R., Galbraith, S. L., McAuley, M. A., Doherty, A. M., Graham, D. I., and McCulloch, J., 1995, Therapeutic potential of endothelin receptor antagonists in experimental stroke, *J. Cardiovasc. Pharmacol.* **26**(3)**:**S412–S415.

Patel, T. R., Galbraith, S., Graham, D. I., Hallak, H., Doherty, A. M., and McCulloch, J., 1996, Endothelin receptor antagonist increases cerebral perfusion and reduces ischaemic damage in feline focal cerebral ischaemia, *J. Cereb. Blood Flow Metab.* **16:**950–958.

Peishoff, C. E., Lago, M. A., Ohlstein, E. H., and Elliott, J. D., 1995, Endothelin receptor antagonists, *Curr. Pharm. Des.* **1:**425–440.

Perfetto, F., Tarquini, R., Leonardis, V. de, Piluso, A., Lombardi, V., and Tarquini, B., 1995, Angiopathy affects circulating endothelin-1 levels in type 2 diabetic patients, *Acta Diabetol.* **32:**263–267.

Potoczak, R., Quenby-Brown, E., Haleen, S., Gallagher, K., Keiser, J., Doherty, A., and Uprichard, A., 1996, Circulating ET-1 levels are increased more by non-selective endothelin receptor blockade than ET_A or ET_B blockade in conscious dogs, *Circulation* **94**(4)**:**I-46.

Predel, H. G., Meyer-Lehnert, H., Backer, A., Stelkens, H., and Kramer, H. J., 1990, Plasma concentrations of endothelin in patients with abnormal vascular reactivity, *Life Sci.* **47:**1837–1843.

Puffenberger, E. G., Hosoda, K., Washington, S. S., Nakao, K., deWit, D., Yanagisawa, M., and Chakravarti, A., 1994, A missense mutation of the endothelin-B receptor gene in multigenic Hirschsprung disease, *Cell* **79:**1257–1266.

Rakugi, H., Tabuchi, Y., Nakamura, M., Nagano, M., Higashimori, K., Mikami, H., and Ogihara, T., 1990, Endothelin activates the vascular renin–angiotensin system in rat mesenteric arteries, *Biochem. Int.* **21**(5)**:**867–872.

Reddy, V. M., Hendricks-Munoz, K., Rajasinghe, H. A., Petrossian, E., Hanley, F. L., and Fineman, J. R., 1997, Post-cardiopulmonary bypass pulmonary hypertension in lambs with increased pulmonary blood flow: A role for endothelin-1, *Circulation* **95**(4)**:**1054–1061.

Reynolds, E. E., and Mok, L. L., 1989, Phorbol ester dissociates endothelin-stimulated phosphoinositide hydrolysis and arachidonic acid release in vascular smooth muscle cells, *Biochem. Biophys. Res. Commun.* **160:**868–873.

Reynolds, E. E., Keiser, J. A., Haleen, S. J., Walker, D. M., Davis, L. S., Olszewski, B., Taylor, D. G., Hwang, O., Welch, K. M., Flynn, M. A., Thompson, D. M., Edmunds, J. J., Berryman, K. A., Lee, C., Reisdorph, B. R., Cheng, X. M., Patt, W. C., and Doherty, A. M., 1995a, Pharmacological characterization of PD 156707, an orally active ET_A receptor antagonist, *J. Pharmacol. Exp. Ther.* **273:**1410–1417.

Reynolds, E. E., Hwang, O., Flynn, M. A., Welch, K. M., Cody, W. L., Steinbaugh, B., He, J. X., Chung, F.-Z., and Doherty, A. M., 1995b, Pharmacological differences between rat and human endothelin B receptors, *Biochem. Biophys. Res. Commun.* **209:**506–512.

Rodeheffer, R. J., Lerman, A., Heublein, D. M., and Burnett, J. C., 1992, Increased plasma concentrations of endothelin in congestive heart failure in humans, *Mayo Clin. Proc.* **67:**719–724.

Rosenberg, A. A., Kennaugh, J., Koppenhafer, S. L., Loomis, M., Chatfield, B. A., and Abman, S. H., 1993, Elevated immunoreactive endothelin-1 levels in newborn infants with persistent pulmonary hypertension, *J. Pediatr.* **123**(1)**:**109–114.

Roux, S. P., Clozel, M., Sprecher, U., Gray, G., and Clozel, J. P., 1993, Ro 47–0203, a new endothelin receptor antagonist reverses chronic vasospasm in experimental subarachnoid hemorrhage, *Circulation* **88:**I-170.

Sakai, S., Miyauchi, T., Kobayashi, M., Yamaguchi, I., Goto, K., and Sugishita, Y., 1996a, Inhibition of myocardial endothelin pathway improves long-term survival in heart failure, *Nature* **384:**353–355.

Saito, Y., Nakao, K., Mukoyama, M. Imura, H., 1990, Increased plasma endothelin levels in patients with essential hypertension. *N. Engl. J. Med.* **322**(3)**:**205.

Sakai, S., Miyauchi, T., Sakurai, T., Yamaguchi, I., Kobayashi, M., Goto, K., and Sugishita, Y., 1996b, Pulmonary hypertension caused by congestive heart failure is ameliorated by long-term application of an endothelin receptor antagonist, *J. Am. Coll. Cardiol.* **28**(6)**:**1580–1588.

Sakamoto, A., Yanagisawa, M., Sakurai, T., Takuwa, Y., Yanagisawa, H., and Masaki, T., 1991, Cloning and functional expression of human cDNA for the ET_B endothelin receptor, *Biochem. Biophys. Res. Commun.* **178:**656–663.

Sakurai, T., Yanagisawa, M., Takuwa, Y., Miyazaki, H., Kimura, S., Goto, K., and Masaki, T., 1990, Cloning of a cDNA encoding a non-isopeptide-selective subtype of the endothelin receptor, *Nature* **348:**732–735.

Salom, J. B., Torregrosa, G., Barbera, M. D., Jover, T., and Alborch, E., 1993, Endothelin receptors mediating contraction in goat cerebral arteries, *Br. J. Pharmacol.* **109:**829–830.

Scott-Burden, T., Resink, T. J., Hahn, A. W. A., and Vanhoutte, P. M., 1991, Induction of endothelin secretion by angiotensin II: Effects on growth and synthetic activity of vascular smooth muscle cells, *J. Cardiovasc. Pharmacol.* **17**(Suppl. 7)**:**S96–S100.

Seo, B., Oemar, B. S., Siebenmann, R., and Von Segesser, L., 1994, Both ET_A and ET_B receptors mediate contraction to endothelin-1 in human blood vessels, *Circulation* **89:**1203–1208.

Shirakami, G., Magaribuchi, T., Shingu, K., Kim, S., Saito, Y., Nakao, K., and Mori, K. 1994, Changes of endothelin concentration in cerebrospinal fluid and plasma of patients with aneurysmal subarachnoid hemorrhage, *Acta Anaesthesiol. Scand.* **38:**457–461.

Sogabe, K., Nirei, H., Shoubo, M., Nomoto, A., Ao, S., Notsu, Y., and Ono, T., 1993, Pharmacological profile of FR 139317, a novel, potent endothelin ET_A receptor antagonist, *J. Pharmacol. Exp. Ther.* **264:**1040–1046.

Sorensen, S. S., Madsen, J. K., and Pedersen, E. B., 1994, Systemic and renal effect of intravenous infusion of endothelin-1 in healthy human volunteers, *Am. J. Physiol.* **266:**(35)**:**F411–F418.

Spinale, F. G., Walker, J. D., Mukherjee, R., Iannini, J. P., Keever, A. T., and Gallagher, K. P., 1996, Concomitant endothelin receptor subtype-A blockade during the progression of congestive heart failure has direct and beneficial effects on left ventricular and myocyte function, *Circulation* **94**(8)**:**I-74.

Stein, P. D., Hunt, J. T., Floyd, D. M., Moreland, S., Dickinson, K. E. J., Mitchell, C., Liu, E. C.-K., Webb, M. L., Murugesan, N., Dickey, J., McMullen, D., Zhang, R., Lee, V. G., Serafino, R., Delaney, C., Schaeffer, T. R., and Kozlowski, M., 1994, The discovery of sulfonamide endothelin antagonists and the development of the orally active ET_A antagonist 5-(dimethylamino)-N-(3,4-dimethyl-5-isoxazolyl)-1-naphthalenesulfonamide, *J. Med. Chem.* **37:**329–331.

Stewart, D. J., Kubac, G., Costello, K. B., and Cernacek, P., 1991a, Increased plasma endothelin-1 in the early hours of acute myocardial infarction, *J. Am. Coll. Cardiol.* **18**(1)**:**38–43.

Stewart, D. J., Levy, R. D., Cernacek, P., and Langleben, D., 1991b, Increased plasma endothelin-1 in pulmonary hypertension: Marker or mediator of disease? *Ann. Intern. Med.* **114:**464–469.

Stockenhuber, F., Gottsauner-Wolf, M., Marosi, L., Liebisch, B., Kurz, R. W., and Balcke, P., 1992, Plasma levels of endothelin in chronic renal failure and after renal transplantation: Impact on hypertension and cyclosporin A-associated nephrotoxicity, *Clin. Sci.* **82:**255–258.

Sudjarwo, S. A., Hori, M., Tanaka, T., Matsuda, Y., Okada, T., and Karaki, H., 1994, Subtypes of endothelin ET_A and ET_B receptors mediating venous smooth muscle contraction, *Biochem. Biophys. Res. Commun.* **200:**627–633.

Sumner, M. J., Cannon, T. R., Mundin, J. W., White, D. G., and Watts, I. S., 1992, Endothelin ET_A and ET_B receptors mediate vascular smooth muscle contraction. *Br. J. Pharmacol.* **107:**858–860.

Takahashi, K., Totsune, K., and Mouri, T., 1994, Endothelin in chronic renal failure, *Nephron* **66:**373–379.

Takasago, T., and McCulloch, J., 1997, Endothelin-A receptor blockade reduces infarction after focal cerebral ischemia in the rat, *Cereb. Blood Flow Metab.* **17**(suppl. 1)**:**S-661.

Teerlink, J. R., Loffler, B.-M., Hess, P., Maire, J.-P., Clozel, M., and Clozel, J.-P., 1994, Role of endothelin in the maintenance of blood pressure in conscious rats with chronic heart failure, *Circulation* **90:**2510–2518.

Tomoda, H., 1993, Plasma endothelin-1 in acute myocardial infarction with heart failure, *Am. Heart J.* **125**(3)**:**667–672.

Topliss, J. G., 1972, Utilization of operational schemes for analog synthesis in drug design, *J. Med. Chem.* **15**(10)**:**1006–1011.

Topliss, J. G., 1977, A manual method for applying the Hansch approach to drug design, *J. Med. Chem.* **20**(4)**:**463–469.

Tschudi, M. R., and Luscher, T. F., 1994, Characterization of contractile endothelin and angiotensin receptors in human resistance arteries: Evidence for two endothelin and one angiotensin receptor, *Biochem. Biophys. Res. Commun.* **204:**685–690.

Tsutamoto, T., Hisanago, T., Fukai, D., Wada, A., Maeda, Y., Maeda, K., and Kinoshita, M., 1995, Prognostic value of plasma soluble intercellular adhesion molecule-1 and endothelin-1 concentration in patients with chronic congestive heart failure, *Am. J. Cardiol.* **76:**803–808.

Uchida, Y., and Watanabe, M., 1993, Plasma endothelin-1 concentrations are elevated in acute hepatitis and liver cirrhosis but not in chronic hepatitis, *Gastroenterol. Jpn.* **28**(5)**:**666–672.

Walsh, T. F., Fitch, K. J., Chakravarty, K., Williams, D. L., Murphy, K. A., Nolan, N. A., O'Brien, J. A., Lis, E. V., Pettibone, D. J., Kivlighn, S. D., Gabel, R. A., Zingaro, G. J., Krause, S. M., Siegl, P. K. S., Clineschmidt, B. V., and Greenlee, W. J., 1994, Discovery of L-749,329, a highly potent, orally active antagonist of endothelin receptors, *ACS National Meeting,* Washington, DC, MEDI 145.

Warner, T. D., Elliott, J. D., and Ohlstein, E. H., 1996, Meeting report. California dreamin' 'bout endothelin: Emerging new therapeutics, *Trends Pharmacol. Sci.* **17:**177–181.

Weber, H., Webb, M. L., Serafino, R., Taylor, D. S., Moreland, S., Norman, J., and Molloy, C. J., 1994, Endothelin-1 and angiotensin-II stimulate delayed mitogenesis in cultured rat aortic smooth muscle cells: Evidence for common signaling mechanisms, *Mol. Endocrinol.* **8:**148–158.

Wei, C.-M., Lerman, A., Rodeheffer, R. J., McGregor, C. G. A., Brandt, R. R., Wright, S., Heublein, D. M., Edwards, W. D., and Burnett, J. C., 1994, Endothelin in human congestive heart failure, *Circulation* **89:**1580–1586.

White, D. G., Garratt, H., Mundin, J. W., Sumner, M. J., Vallance, P. J., and Watt, I. S., 1994, Human saphenous vein contains both endothelin ET_A and ET_B contractile receptors, *Eur. J. Pharmacol.* **257:**307–310.

Williams, D. L., Murphy, K. L., Nolan, N. A., O'Brien, J. A., Lis, E. V., Pettibone, D. J., Clineschmidt, B. V., Krause, S. M., Veber, D. F., Naylor, E. M., Charkravarty, P. K., Walsh, T. F., Dhanoa, D. M., Chen, A., Bagley, S. W., Fitch, K. J., and Greenlee, W. J., 1996, Pharmacology of L-744,453, a novel nonpeptidyl endothelin antagonist, *Life Sci.* **58**(14)**:**1149–1157.

Wong-Dusting, H. K., La, M., and Rand, M. J., 1990, Mechanisms of the effects of endothelin on responses to noradrenaline and sympathetic nerve stimulation, *Clin. Exp. Pharmacol. Physiol.* **17:**269–273.

Yanagisawa, M., Kurihara, H., Kimura, S., Tomobe, Y., Kobayashi, M., Mitsui, Y., Yazaki, Y., Goto, K., and Masaki,T., 1988, A novel potent vasoconstrictor peptide produced by vascular endothelial cells, *Nature* **332:**411–415.

Ziv, I., Fleminger, G., Djaldetti, R., Achiron, A., Melamed, E., and Sokolovsky, M., 1992, Increased plasma endothelin-1 in acute ischemic stroke, *Stroke* **23:**1014–1016.

Chapter 6

Endothelin Receptor Antagonists

John D. Elliott, Eliot H. Ohlstein, Catherine E. Peishoff, Harma M. Ellens, and M. Amparo Lago

1. INTRODUCTION

The endothelins (ETs) are a family of three isopeptides, endothelin-1 (ET-1), ET-2, and ET-3 (Fig. 1), each of which is encoded in the human genome, and since their discovery in 1988 there have been many reports suggestive of a role for these extremely potent vasoconstrictor peptides in the etiology of disease (Ruffolo, 1995; Yanagisawa *et al.*, 1988). Much of the early evidence implicating the ETs in disease was indirect, linking elevations in endogenous ET levels with pathophysiology; however, more compelling data are now available based on animal model studies with receptor-specific antagonists (*vide infra*).

The ETs elicit their effects through binding to receptors of the G-protein-coupled seven-transmembrane-spanning superfamily, and two human receptor subtypes have been fully characterized through molecular cloning and expression (Arai *et al.*, 1990; Sakurai *et al.*, 1990). It is believed that the ET_A subtype, which is predominantly located on vascular smooth muscle, is the principal receptor subtype involved in ET-mediated vasoconstriction (Panek *et al.*, 1992). This subtype binds ET-1 and ET-2 with higher affinity than ET-3 and in addition to mediating vasoconstriction has also been implicated in stimulating cellular proliferation (Ohlstein *et al.*, 1992). The ET_B subtype, which binds all three ET peptide iso-

John D. Elliott, Eliot H. Ohlstein, Catherine E. Peishoff, Harma M. Ellens, and M. Amparo Lago • Department of Medicinal Chemistry, SmithKline Beecham Pharmaceuticals, King of Prussia, Pennsylvania 19406-0939.

Integration of Pharmaceutical Discovery and Development: Case Studies, edited by Borchardt *et al.*, Plenum Press, New York, 1998.

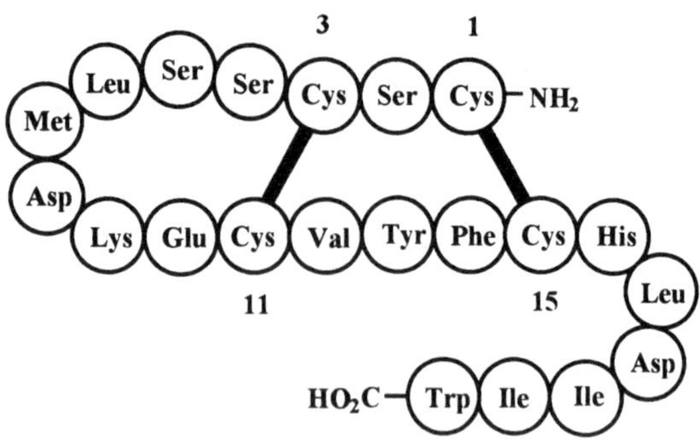

Endothelin-1

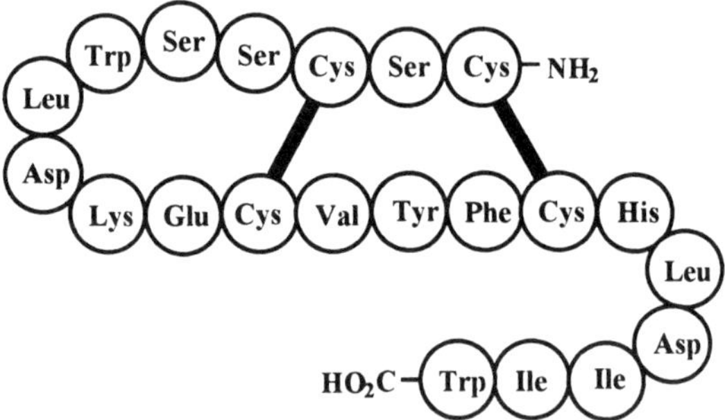

Endothelin-2

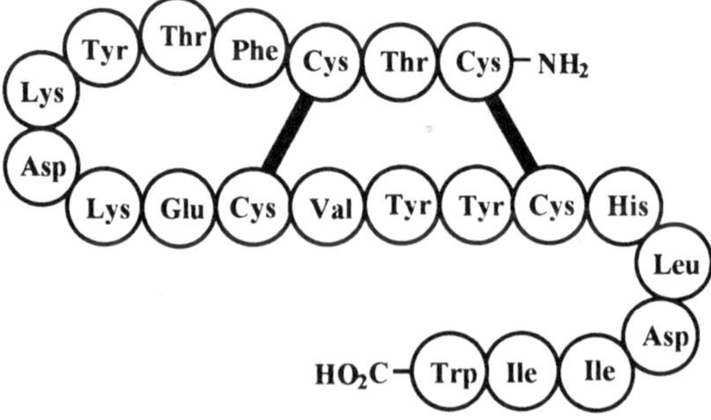

Endothelin-3

Figure 1. The endothelin peptides.

forms with equal and high affinity, mediates vasodilation through the release of endothelium-derived nitric oxide (DeNucci *et al.,* 1988), although this subtype also appears to be linked to vasoconstriction, in certain vascular beds (Warner *et al.,* 1993). At the present time, evidence for subtypes of the ET_B receptor exists (Webb and Meek, 1997), perhaps explaining its disparate roles; however, as yet, only one human subtype has been fully characterized. Despite the extensive work that has been performed in animal models of disease with antagonists of varying binding selectivities, it is not yet possible to assert the optimal binding profile of an ET antagonist for human therapy.

The biosynthesis of ET-1 involves as its final step proteolytic processing of an inactive precursor peptide termed *big ET-1* by the specific protease endothelin converting enzyme (ECE) (Fig. 2). Despite the fact that this scheme was proposed in the landmark publication identifying the ETs in 1988 (Yanagisawa *et al.,* 1988), relatively slow progress has been made toward specific ECE inhibitors. This is in part related to the difficulties experienced during attempts to purify and clone the enzyme, and it is only relatively recently that this has been achieved (Xu *et al.,* 1994). Available data suggest the existence of multiple ECE isoforms and efforts are ongoing in a number of laboratories to explore the therapeutic potential of agents that target this step in the ET cascade.

By way of contrast, efforts toward the identification of agents that impede the activation of cellular receptors specific for the ETs have met with more rapid success, and several pharmaceutical companies have such compounds in various stages of preclinical and clinical development (Lago *et al.,* 1996).

2. RATIONAL DESIGN OF SB 209670

Our involvement in the ET receptor antagonist area began with screening nonpeptide compounds from our G-protein coupled receptor (GPCR) ligand collection, the assembly of such a collection being inspired by the observation that certain common structural features are present in many of the known ligands of GPCRs. Thus, from a group of compounds selected for their known affinity to other GPCRs, or their structural similarity to such molecules, SK&F 66861 (**1,** R =

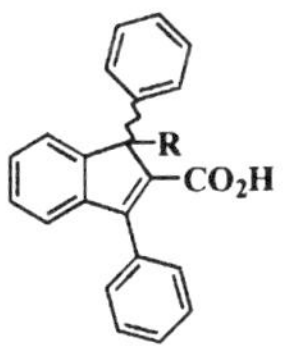

R
CO_2H

1

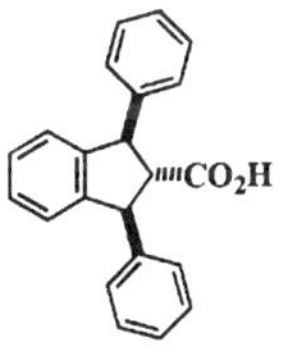

CO_2H

2

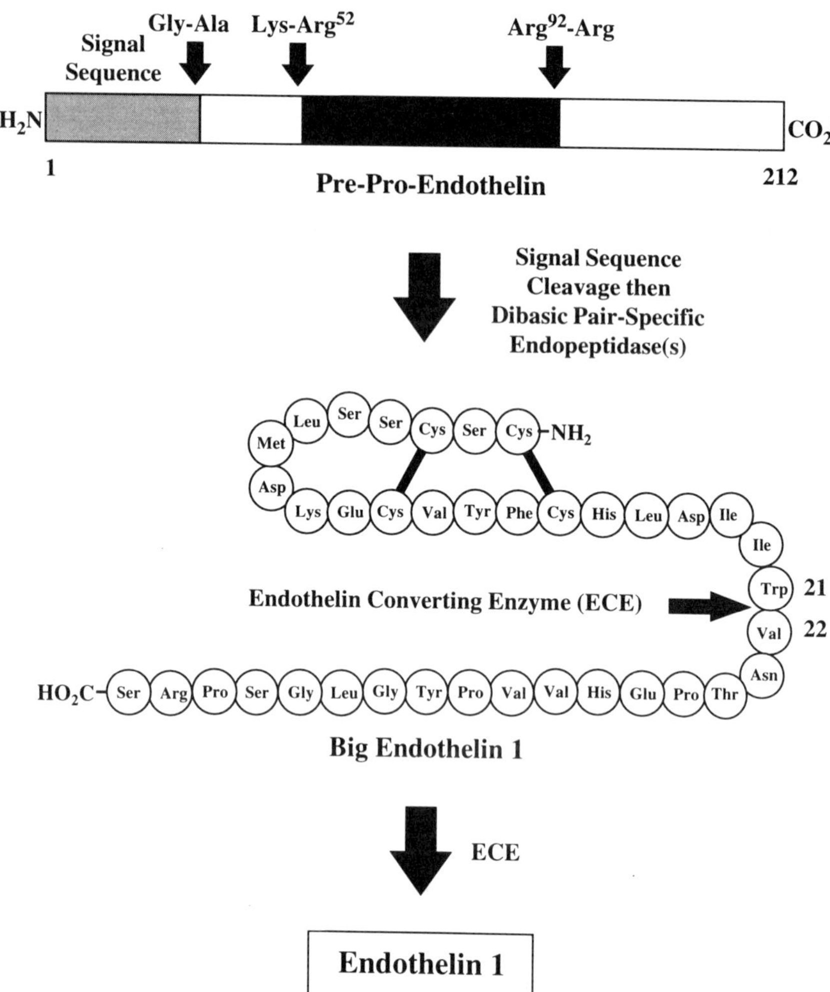

Figure 2. The biosynthetic pathway to endothelin 1. The pre-pro form of human ET-1 contains 212 amino acids, including a signal peptide sequence of 17 residues (lightly shaded section) which when cleaved gives pro-endothelin 1 (pro-ET-1). Dibasic pair-specific endoprotease cleavage of pro-ET-1 gives big ET-1 (darkly shaded section), which is cleaved by a specific metalloprotease, endothelin converting enzyme (ECE), to afford mature ET-1 (the ECE cleavage site is shown by the horizontal arrow between Trp21 and Val22 of big ET-1).

H) was identified as a weak antagonist of ET_A receptors (K_i = 7000 nM*), without measurable affinity toward the ET_B subtype (K_i > 30,000 nM). Furthermore, SK&F 66861 antagonizes the vasoconstrictor response to ET-1 in the rat aorta (K_b = 6600 nM) (Elliott *et al.*, 1994; Ohlstein *et al.*, 1994).

As part of our effort in the ET area, studies were initiated using ^{1}H-NMR spectroscopy to define the solution conformation of ET-1, and structural comparison of SK&F 66861 to low energy conformers of ET-1 thus obtained suggested several potential matches. Specifically, the 1- and 3-phenyl groups of SK&F 66861 can be overlaid on the aromatic rings of amino acid residues Tyr13 and Phe14 of ET-1 or on the aromatic rings of Trp21 and either Tyr13 or Phe14 (see Fig. 1). The vicinal location of residues Tyr13 and Phe14 on a helical region of the peptide, extending from around Asp8 to His16, allows orientation of the aromatic rings of these residues in a fashion resemblant of the pendant phenyl rings of SK&F 66861 and it is on this supposition that our peptidomimetic hypothesis is based. Overlays involving the C-terminal residue Trp21 and either Tyr13 or Phe14, however, cannot be discounted as none of the ^{1}H-NMR structures obtained for ET-1 contain conformational definition within the C-terminal six-residue "tail" of the peptide (Fig. 1) (Ruffolo, 1995, Chapter 5). Our early SAR investigations with SK&F 66861 demonstrated the critical contribution of the free carboxylic acid toward ET_A receptor affinity and based on overlays of the small molecule with residues Tyr13 and Phe14, outlined above, the acidic residues Glu10 and Asp18 became likely loci within the peptide for mimicry by this segment of the small molecule. Although alanine scanning of ET-1 (Hunt, 1992; Hunt *et al.*, 1991; Tam *et al.*, 1994) failed to support a receptor binding contribution of either Glu10 or Asp18, both residues are significantly linked to the functional activity of the peptide suggesting specific receptor interactions. Asp18 was ultimately chosen as the third overlay point for SK&F 66861 based on its more proximal location to Tyr13 and Phe14.

Given the electron-rich character of the aromatic ring in Tyr13 and the known tolerance of an electron-donating substituent on the aromatic ring of Phe14 (ET-3 possesses a tyrosine residue at position 14) (see Fig. 1), the incorporation of electron-donating substituents into the pendant phenyl rings at positions 1 and 3 is suggested by the peptidomimetic hypothesis. Attempts to effect such functionalization revealed the oxidative lability of the 1,3-diphenylinden-2-carboxylic acid nucleus and further SAR studies demonstrated an intolerance of receptor binding affinity to substitution at the 1-position, which would render such analogues oxidatively stable (e.g., **1,** R = Me, OH; K_is: ET_A > 30,000 nM). As an alternative to substitution at the 1-position, saturation of the indene five-membered ring was explored as a means of obtaining a stable framework on which to conduct further SAR studies, and in addition to the anticipated enhancement of stability it was dis-

*All binding data refer to affinities for the cloned human receptors.

covered that *trans,trans*-1,3-diphenylindan-2-carboxylic acid (**2**) possesses a similar ET receptor binding profile (K_is: ET_A = 11,000 nM, ET_B > 30,000 nM) to SK&F 66861. The reasoning outlined above inspired substitution of the pendant phenyl rings of **2,** and it was discovered that affinity is enhanced through placement of electron-donating substituents on both of these moieties. This led ultimately to compound **3,** which has significantly greater affinity for both ET_A and ET_B receptors (K_is: ET_A = 43 nM, ET_B = 760 nM) than the parent **2.** Further SAR studies with compound **3** showed that substitution of the benzo ring of the indane has much less impact than in the pendant phenyl rings and is tolerant of both electron-withdrawing and -donating substituents. These observations, taken together with the lack of measurable receptor affinity for the cyclopentane analogue of **2** (Bryan and Elliott, unpublished observations), suggest that the benzo phenyl ring fulfills more of a structural role rather than engaging in a direct receptor interaction. Although the relative insensitivity of the benzo ring toward substitution is disappointing from the perspective of defining a further area of the molecular framework with which to modulate affinity, this tolerance was used to advantage in that an electron-donating 5-substituent facilitates an early step in the synthesis of these compounds (Elliott *et al.,* 1994). Thus, compound **4,** which incorporates a 5-*n*-propoxyl substituent, has comparable receptor affinities (K_is: ET_A = 11 nM, ET_B = 1000 nM) to analog **3**.

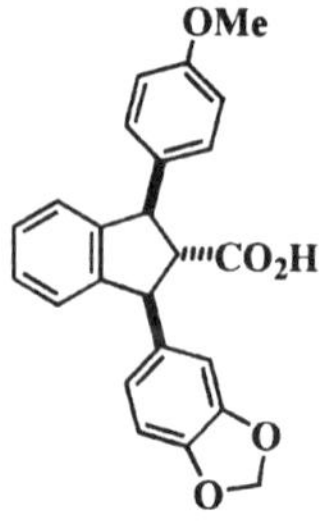

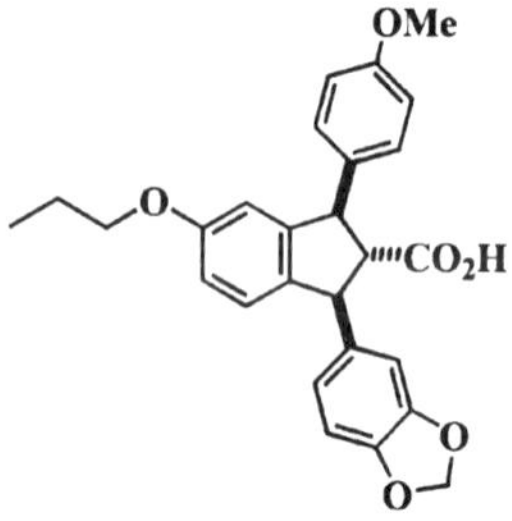

Although the ET_A receptor binding affinity of compound **4** shows marked enhancement over the lead structure SK&F 66861 (K_i = 11 versus 7000 nM), this was still viewed as modest by comparison with the natural agonist ET-1, which exhibits a K_d of 180 pM (Arai *et al.,* 1990). Early SAR studies conducted with the endothelin system revealed the critical importance of the C-terminal carboxylic acid for ET-1 receptor affinity (Nakajima *et al.,* 1989). Because our initial peptidomimetic hypothesis suggested that compound **4** is not taking advantage of the receptor interaction used by the C-terminus of the natural peptide, our attention turned to modification of **4** to engage this additional locus on the receptor. Although this hypothesis suggests the incorporation of an additional acidic residue into the indane structure, the ^{1}H-NMR structures of ET-1 fail to provide structural information concerning the C-terminal hexapeptide tail, a feature critical to the placement of this moiety. To overcome this problem, it was hypothesized that the conformationally well-defined cyclic pentapeptide antagonist BQ 123 (Ihara *et al.*, 1992) is a mimetic of the region of ET-1 from residues 18 through 21. Thus, a structure of BQ 123, determined by ^{1}H-NMR spectroscopy (Bean *et al.*, 1994), was used to generate a conformation of the tail of ET-1. In the resultant triple overlay of ET-1, BQ 123, and indane **2,** it appears that the additional acidic moiety should be appended to the *ortho* position of one of the pendant phenyl rings of **4** (Fig. 3). At the time of conception of this idea, a lack of knowledge of the absolute configuration of the most potent antipode of **4** prevented a prediction as to whether this acidic moiety should be placed on the 1- or the 3-phenyl substituent, but did indicate that the new side chain should comprise a carboxylic acid and a two- or three-atom linker. This proposal led to the preparation of SB 209670 (**5**), the first subnanomolar nonpeptide antagonist of the human ET_A receptor, which also possesses moderate affinity for the human ET_B subtype (K_is: ET_A = 0.43 nM, ET_B = 14.7 nM).

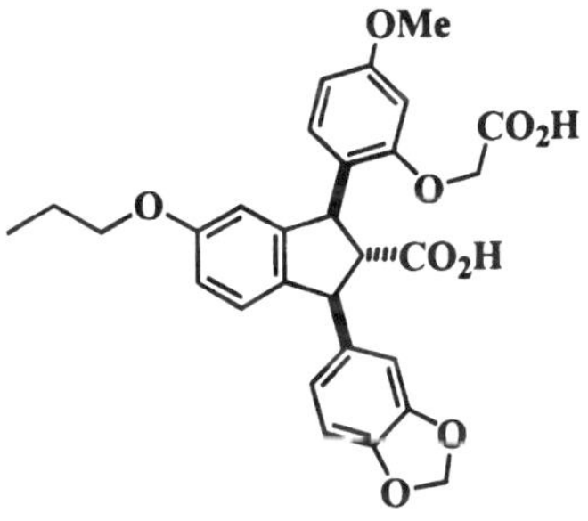

5

X-ray crystallographic characterization demonstrated the absolute configuration of SB 209670 to be as shown and, given that the absolute configuration of the most potent enantiomer of **4** is the same as that of SB 209670, our model would have correctly directed appendage of the additional carboxylic acid to the 3-phenyl ring. Furthermore, subsequent SAR studies have demonstrated that a 6-oxyacetic acid side chain on the 1-(3,4-methylenedioxy) phenyl substituent of **4** is actually deleterious to receptor binding affinity.

Although a peptidomimetic hypothesis has been successfully applied to the discovery of potent nonpeptide ET receptor antagonists, more direct evidence of an overlap of binding sites for peptide and nonpeptide ligands is emerging from site-directed mutagenesis studies (Krystek *et al.*, 1994; Lee *et al.*, 1994, 1995).

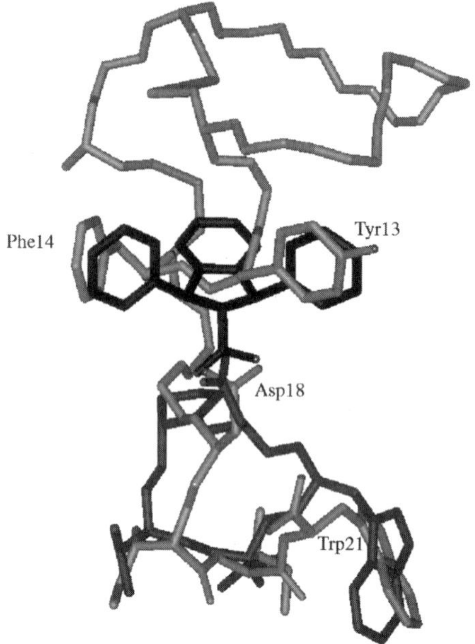

Figure 3. Modeling overlay of *trans,trans*-1,3-diphenylindan-2-carboxylic acid and an ^{1}H-NMR-derived conformational model of BQ 123 on an ^{1}H-NMR-derived conformational model of ET-1. A color representation of this figure appears on the facing page.

Mutagenesis of a lysine residue in transmembrane spanning domain 3 (TM3) of the ET_B receptor to alanine (K182A) totally abrogates the affinity of SB 209670 (Lee *et al.*, 1995). Although this mutation (K182A) does not affect the binding of ET-1, the affinities of the peptide agonists ET-3 and sarafotoxin 6c are markedly diminished. Although these data do not preclude a distinct binding site for ET-1, this seems unlikely, based on its homology with ET-3 and their similar solution structures (Ruffolo, 1995, Chapter 5). In any event, modulation of the affinities of both peptide agonists and small molecule antagonists by a single point mutation supports an overlap of their binding sites. Further binding studies to the ET_B, K182A mutant receptor with a variety of analogues of SB 209670 support a direct interaction of this basic residue with the indane-2-carboxyl of the antagonist (Lee *et al.*, 1995). Interestingly, the position of this residue is close to that occupied by the highly conserved aspartic acid residue in TM3 of the adrenergic receptors, which has been implicated as critical to the affinity of both agonists and antagonists (Strader *et al.*, 1987). These data coupled with mutagenesis studies on other receptors of this family (Strader *et al.*, 1988) are suggestive of a conserved ligand binding site region.

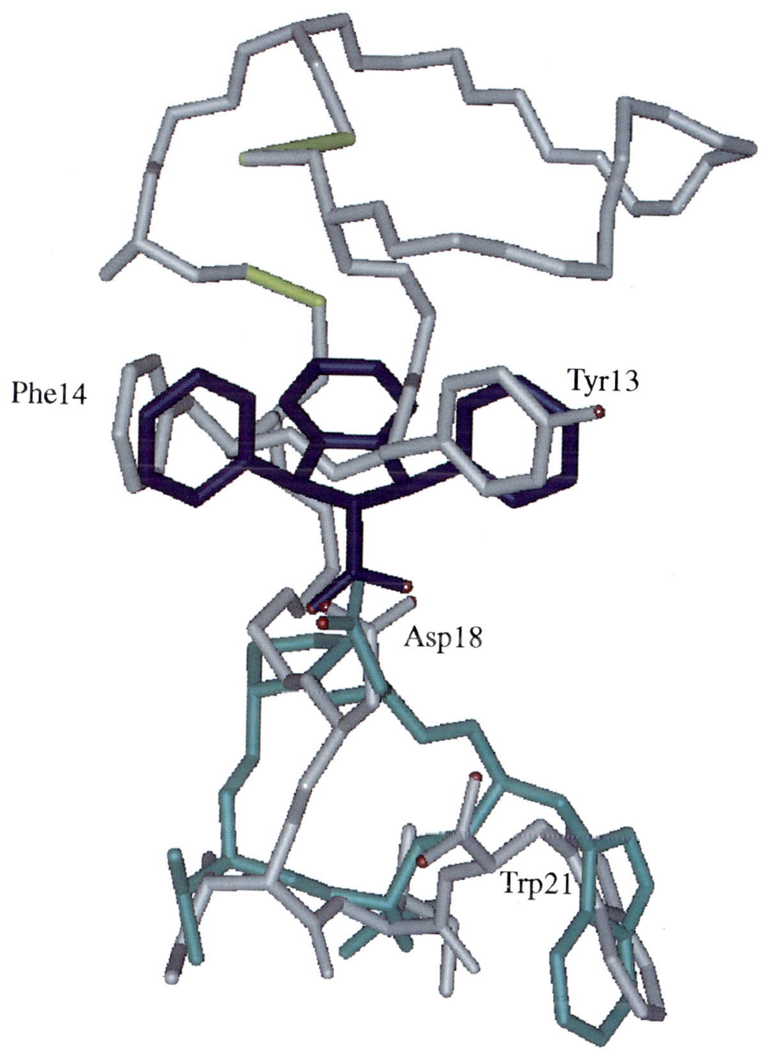

Figure 3. Modeling overlay of *trans,trans*-1,3-diphenylindan-2-carboxylic acid (dark blue) and an ^{1}H-NMR-derived conformational model of BQ 123 (cyan) on an ^{1}H-NMR-derived conformational model of ET-1 (gray).

3. PHARMACOLOGICAL, DRUG METABOLISM, AND PHARMACOKINETIC CHARACTERIZATION OF SB 209670

As would be anticipated from receptor binding studies, SB 209670 is a potent functional antagonist of ET-1 induced contraction mediated by the ET_A receptor subtype, and to a lesser extent antagonizes those effects occurring on ET_B receptor activation. Thus, SB 209670 antagonizes ET-1-induced contraction of the rat aorta (ET_A receptors) and sarafotoxin S6c-induced contraction of the rabbit pulmonary artery (ET_B receptors) with K_b values of 0.4 and 52 nM, respectively. In both tissues, Schild analysis of the concentration–response curves yields slopes of the regression lines not significantly different from unity, which is consistent with competitive antagonism (Ohlstein *et al.,* 1994a). Most significantly, SB 209670 is a potent antagonist of ET-1-induced contraction of human circumflex coronary arteries (K_b = 7 nM, determined using the racemate) (Ohlstein *et al.*, 1994a). SB 209670 is a selective antagonist of ET receptors in that it has no appreciable affinity (>10,000 nM) for a range of other G-protein-coupled receptors including the angiotensin II (AT-1) and vasopressin (V1) subtypes. When dosed intravenously, SB 209670 is efficacious in a number of animal models of disease thought to be mediated by the ETs. These models include: renal failure in the rat (Gellai *et al.*, 1994) and dog (Brooks *et al.*, 1994b), hypertension in the spontaneously hypertensive rat (SHR) (Ohlstein *et al.*, 1994), and ischemia-induced stroke in the gerbil (Ohlstein *et al.*, 1994b). The smooth muscle mitogenic effects of ET are known to be mediated via ET_A receptors in the rat (Ohlstein *et al.*, 1992) and the efficacy of SB 209670 in inhibiting neointimal proliferation following coronary artery balloon angioplasty in the rat supports a role for ET in this important clinical condition (Douglas *et al.*, 1994). Although SB 209670 is efficacious in reducing blood pressure, when dosed orally in the SHR, its oral bioavailability is only 4–5%.

The particularly significant data obtained with SB 209670 in models of renal failure supported the advancement of this compound as a clinical candidate. Thus, as shown in Fig. 4, in a uninephrectomized rat model of ischemia-induced renal failure, treatment with SB 209670 significantly reduced mortality.

4. SELECTION OF THE ORALLY BIOAVAILABLE CANDIDATE SB 217242

The low oral bioavailability of SB 209670 may limit its potential for chronic therapy and thus we sought to obtain an antagonist of similar high potency, but with markedly enhanced oral performance. Because SB 209670 is a dicarboxylic

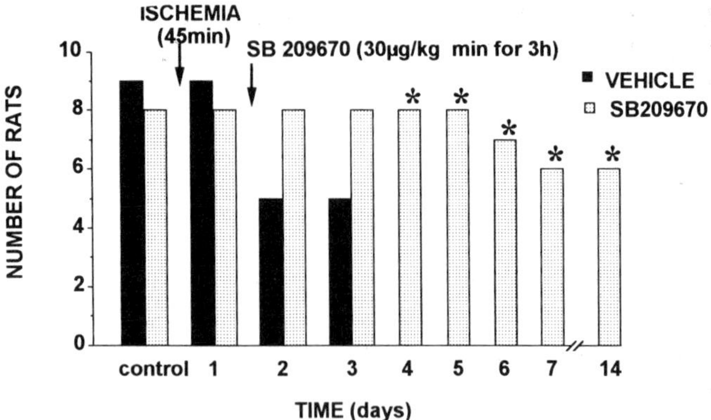

Figure 4. Reversal of severe acute renal failure in rats. Effects on survival of (±)-SB 209670 (stippled bars) or vehicle (solid bars) infused 24 hr after 45-min renal occlusion in the uninephrectomized, chronically instrumented, Sprague–Dawley rat. Basal values represent numbers before ischemia. The 24-hr control values serve as pretreatment controls. ★$p < 0.05$ versus vehicle.

acid, it seems likely that absorption could limit oral bioavailability and this is consistent with data generated for permeability across Caco-2 cell monolayers. Thus, SB 209670 proved to be even less permeant (0.0075 cm/hr) than the membrane-impermeant paracellular flux marker mannitol (0.011 cm/hr) (Ohlstein *et al.*, 1996). Extensive DMPK analysis in the rat supports the view that low oral bioavailability is related primarily to poor absorption rather than high first-pass elimination.

Transport across the intestinal epithelium can be divided into transcellular and paracellular processes. Transcellular transport includes the simple diffusion of lipophilic molecules across epithelial cells as well as carrier-mediated transport (e.g., in the transport of di- and tripeptides). Paracellular transport involves passive diffusion through the aqueous environment of the tight junctions between epithelial cells, this mode being generally restricted to small hydrophilic molecules such as mannitol (Ellens *et al.*, 1997). Inasmuch as the mucosal-to-serosal (0.0075 cm/hr) and serosal-to-mucosal (0.0055 cm/hr) fluxes for SB 209670 are essentially equivalent, and that transport does not correlate with changes in mannitol permeability, it appears that the indane dicarboxylic acid follows a passive transcellular mode of absorption. From these data, one would conclude that an enhancement of lipophilicity of SB 209670 could be beneficial to intestinal permeability and a program to screen a group of antagonists for permeability was initiated.

Permeability screening was conducted using rabbit large and small intestinal tissues to detect both active and passive transport, the expectation being that active transport would be detected in the small intestine and passive transport would be detected in either segment. Selected compounds were also examined using Caco-2 cell monolayers to ensure that a low flux in the tissues was not the result of an interaction with lamina propria components. This latter complication of *in vitro* screening using animal tissues would seemingly have little relevance to absorption in an *in vivo* setting, where a compound passing the epithelial barrier is carried away using the subepithelial capillary network. The permeability of SB 209670 is similar in small intestine (0.0034 cm/hr) and distal colon (0.0034 cm/hr) and these results correlate well with Caco-2 cell data (0.0075 cm/hr); however, in general, although the distal colonic measurements mirror Caco-2 cell results, this is not the case with data obtained using small intestinal tissue. These observations lead to speculation that the lamina propria may be imposing an additional barrier *in vitro*, as it is most extensive in the small intestine and almost nonexistent in the distal colon.

As anticipated, compound **3,** a monocarboxylic acid, showed markedly enhanced permeability over SB 209670 (0.1722 versus 0.0075 cm/hr in the rabbit distal colon); however, as alluded to earlier, the introduction of the second carboxylic acid side chain in SB 209670 is critical in that it provides a hundredfold increase in affinity for the ET_A receptor (K_i 43 versus 0.4 nM for compound **3** and SB 209670, respectively). The acylsulfonamide **6** (ET_A, K_i = 4 nM), although somewhat more potent than **3,** displayed permeability comparable to SB 209670 (0.0028 cm/hr), consistent with the acidity of the side chain acylsulfonamide being similar to that of a carboxylic acid. In the course of SAR studies with the indane series of antagonists, it was discovered that replacement of the oxyacetic acid moiety of SB 209670 with a hydroxyethoxy substituent provided an analogue SB 217242 (**7,** R = H) of only slightly diminished potency (ET_A, K_i = 1.1 nM, ET_B, K_i = 111 nM).

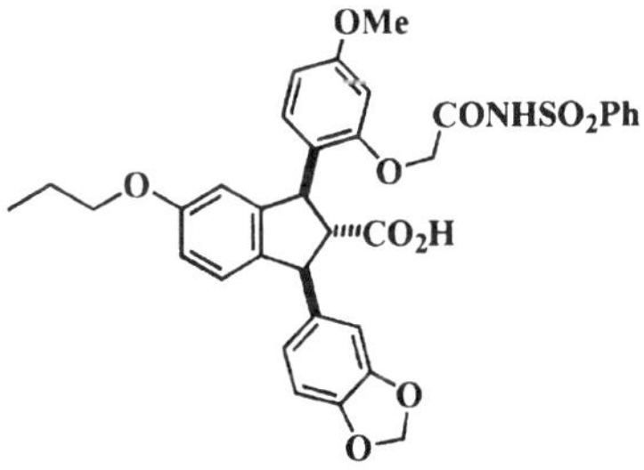

OMe CONHSO$_2$Ph O O CO$_2$H O O

6

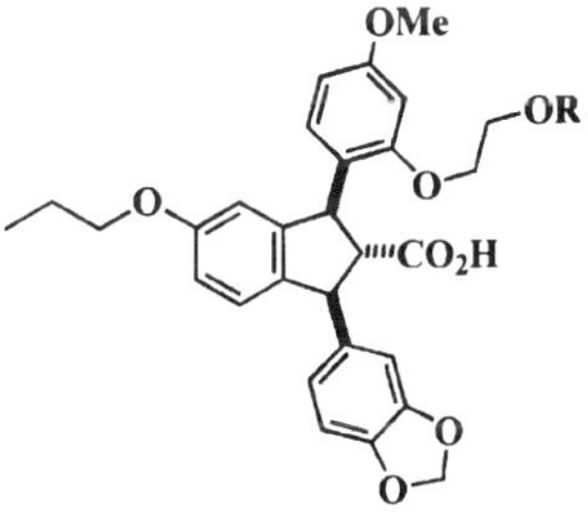

OMe OR O O CO$_2$H O O

7

The permeability of SB 217242 in rabbit distal colonic tissue is markedly greater than that of SB 209670 (0.0955 versus 0.0075 cm/hr) and this was confirmed using Caco-2 cell monolayers where the flux is 0.2045 cm/hr. Furthermore, as was the case with SB 209670, the mucosal-to-serosal and serosal-to-mucosal fluxes for SB 217242 are essentially equivalent, suggesting a similar passive transcellular mode of absorption. Further reduction of polarity of the side chain of SB 217242, through methylation of the terminal hydroxyl, yielded an analogue **7** (R

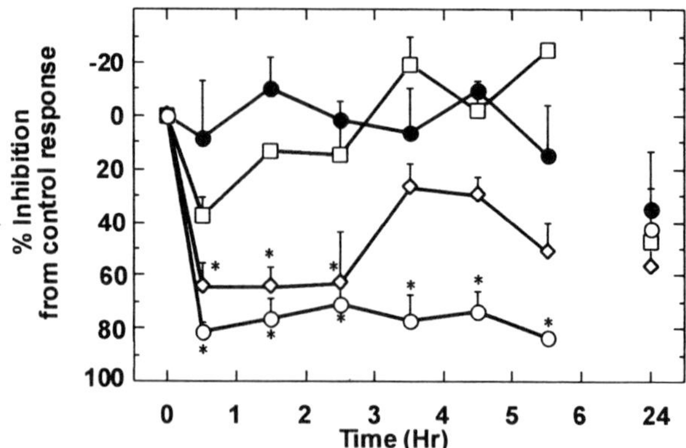

Figure 5. Effect of oral administration of SB 217242 [0.3 (□; n = 3), 3 (◇; n = 3), or 30 (○; n = 3) mg/kg] on the increases in mean arterial pressure observed in the conscious rat after repeated i.v. administration of 100 pmole/kg ET-1. ● (n = 5) represents the changes observed in control animals pretreated with an equal volume of vehicle. Values are means ± SEM. Asterisk indicates statistically different from time-matched vehicle-treated animals (P, 0.05, ANOVA).

= Me) of almost unchanged permeability (0.0911 cm/hr) but of slightly diminished potency (ET_A, K_i = 2.1 nM, ET_B, K_i = 400 nM). As a result of its identification via permeability screening, SB 217242 was submitted to extensive DMPK investigation, and as anticipated, oral bioavailability in the rat is dramatically enhanced over SB 209670 (66 versus 4%). Orally administered SB 217242 (0.3–30 mg/kg) produces a dose-dependent inhibition of the pressor response to exogenous ET-1 in conscious rats (Fig. 5); the effect of the 30-mg dose lasting for more than 5.5 hours. The plasma half-life of SB 217242 in rats following intraduodenal administration is 3.3 hr and systemic clearance is 27.3 ml/min per kg. Thus, in terms of both potency and DMPK profile, SB 217242 fulfills the requirements of an agent needed for chronic therapy.

As anticipated from earlier studies with SB 209670, SB 217242 is efficacious in animal models of stroke. ET has been implicated in the pathogenesis of both hemorrhagic and ischemic stroke (Ruffolo, 1995). Following ischemic stroke in humans, increased plasma levels of ET as well as ET receptors are observed and these changes have been correlated with infarct size and neurological deficits (Estrada *et al.,* 1994; Wei *et al.,* 1993; Ziv *et al.,* 1992). SB 217242 has been evaluated in a middle cerebral artery occlusion model of stroke in rats. Treatment with SB 217242 (3–15 mg/kg p.o.) significantly reduces the degree of cerebral hemispheric infarction and infarct volume (Fig. 6) (Barone *et al.,* 1995). The oral efficacy of SB 217242 demonstrates its ability to cross the blood–brain barrier, unlike its dicarboxylic acid counterpart, SB 209670, where activity was limited to intracerebroventricular administration. Studies with both compounds support a potential role for ET receptor antagonism as a therapeutic strategy for ischemic stroke.

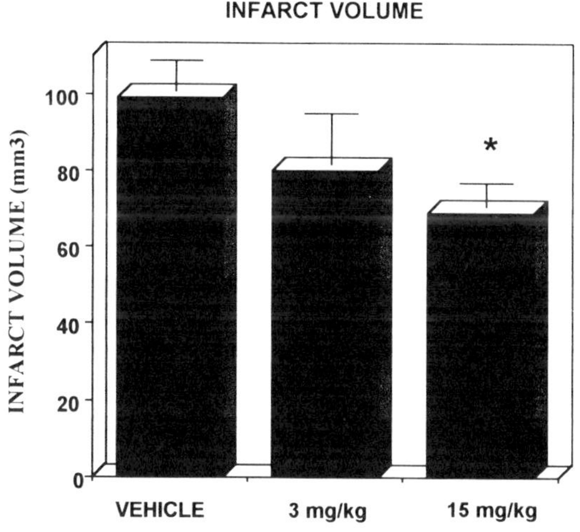

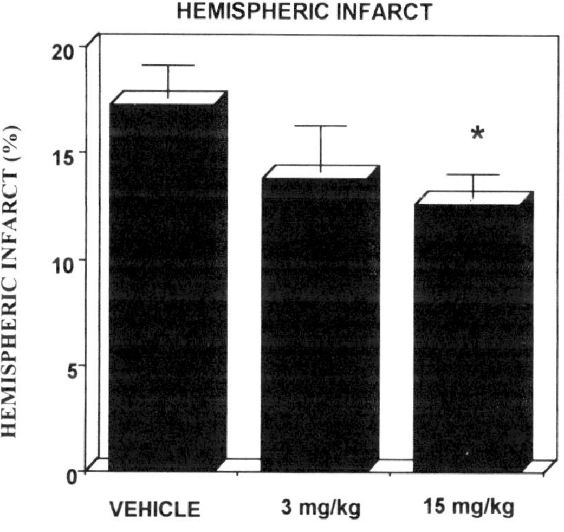

Figure 6. Hemispheric infarct and infarct volume for vehicle- and SB 217242-treated rats. SB 217242 at the 15 mg/kg treatment regimen significantly reduced infarction ($\star P < 0.05$, one-way ANOVA with Dunnett's test follow-up).

As a potential oral agent for the treatment of pulmonary hypertension associated with hypoxia, SB 217242 has been evaluated in a relevant animal model. Patients with pulmonary hypertension associated with a number of different diagnoses (e.g., congenital heart disease, collagen vascular disease, pulmonary thromboembolism, valvular heart disease, and congestive heart failure) all show an elevation of plasma or urinary ET-1 levels (Michael and Markewitz, 1996). It has

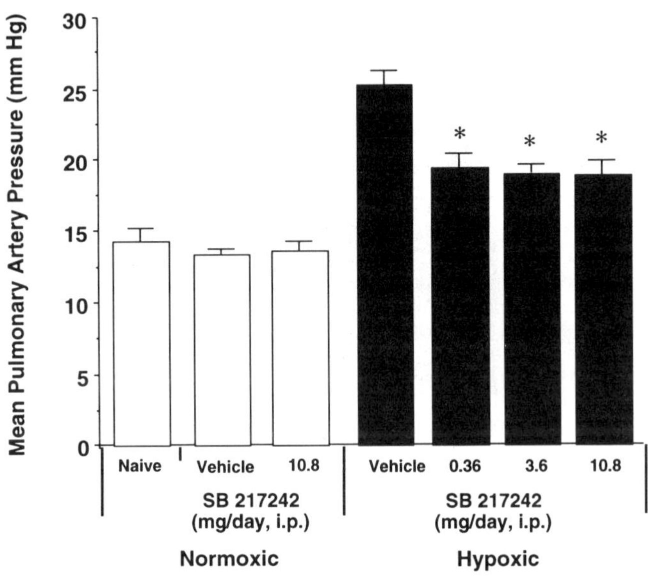

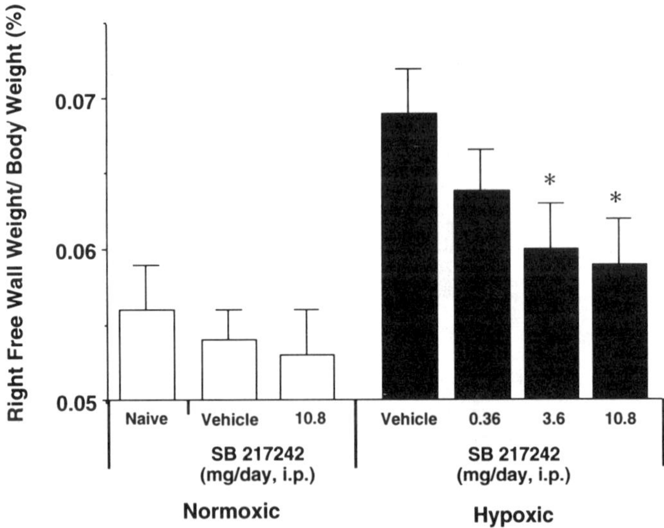

Figure 7. SB 217242 significantly inhibits the effects of hypoxia on mean pulmonary artery pressure and heart right free wall weight in the guinea pig; a model of pulmonary hypertension associated with hypoxic conditions.

been demonstrated in guinea pigs that chronic exposure to hypoxia results in a significant increase in pulmonary artery pressure and hypertrophy of the right ventricle (Underwood *et al.,* 1997). Administration of the ET receptor antagonist SB 217242 significantly inhibits these hypoxia-induced changes (Fig. 7) supporting clinical studies, currently under way, to evaluate its potential in the treatment of pulmonary hypertension associated with hypoxia.

5. CONCLUSION

In our effort to obtain antagonists of the ET receptors, rational design based on agonist structure played a crucial part in the strategy that led to SB 209670. *In vivo* pharmacological studies with SB 209670, both in our own laboratories and through the work of others, have done much to elucidate potential roles for ET in the etiology of disease and to establish the therapeutic potential of ET receptor antagonists. Hampered in our search for an orally effective agent by the poor bioavailability of SB 209670, intestinal permeability screening was used to discover SB 217242, an antagonist with excellent oral characteristics. The key contributions outlined above toward the design of SB 217242 emphasize the multidisciplinary approach used in today's drug discovery efforts.

REFERENCES

Arai, H., Hori, S., Aramori, I., Ohkubo, H., and Nakanishi, S., 1990, Cloning and expression of a cDNA encoding an endothelin receptor, *Nature* **348:**730–732.

Barone, F. C., White, R. F., Elliott, J. D., Feuerstein, G. Z., and Ohlstein, E. H., 1995, The endothelin receptor antagonist SB 217242 reduces cerebral focal ischemic brain injury, *J. Cardiovasc. Pharmacol.* **26**(Suppl. 3)**:**S404–S407.

Bean, J. W., Peishoff, C. E., and Kopple, K. D., 1994, Conformations of cyclic pentapeptide endothelin receptor antagonists, *Int. J. Peptide Protein Res.* **44:**223–232.

Brooks, D. P., DePalma, P. D., Gellai, M., Nambi, P., Ohlstein, E. H., Elliott, J. D., Gleason, J. G., and Ruffolo, R. R., Jr., 1994, Non-peptide endothelin receptor antagonists. III. Effect of SB 209670 and BQ 123 on acute renal failure in anesthetized dogs, *J. Pharmacol. Exp. Ther.* **271:**769.

DeNucci, G., Thomas, R., D'Orleans-Juste, P., Antunes, E., Walder, C., Warner, T. D., and Vane, J. R., 1988, Pressor effects of circulating endothelin are limited by its removal from the pulmonary circulation and by the release of prostacyclin and endothelium-derived relaxing factor, *Proc. Natl. Acad. Sci. USA* **85:**9797–9800.

Douglas, S. A., Louden, C., Vickery-Clark, L. M., Storer, B. L., Hart, T., Feuerstein, G. Z., Elliott, J. D., and Ohlstein, E. H., 1994, A role for endogenous endothelin-1 in neointimal formation after rat carotid artery balloon angioplasty, *Circ. Res.* **75:**190–197.

Ellens, H., Eddy, E. P., Lee, C.-P., Dougherty, P., Lago, A., Xiang, J.-N., Elliott, J. D., Cheng, H.-Y., Ohlstein, E., and Smith, P. L., 1997, In vitro permeability screening for identification of orally bioavailable endothelin receptor antagonists, *Adv. Drug Delivery Rev.* **23:**99–109.

Elliott, J. D., Lago, M. A., Cousins, R. D., Gao, A., Leber, J. D., Erhard, K. F., Nambi, P., Elshourbagy, N. A., Kumar, C., Lee, J. A., Bean, J. W., DeBrosse, C. W., Eggleston, D. S., Brooks, D. P., Feuerstein, G., Ruffolo, R. R., Weinstock, J., Gleason, J. G., Peishoff, C. E., and Ohlstein, E. H., 1994, 1,3-Diarylindan-2-carboxylic acids, potent and selective non-peptide endothelin receptor antagonists, *J. Med. Chem.* **37:**1553–1557.

Elliott, J. D., Bryan, D. L., Nambi, P., and Ohlstein, E. H., 1996, A novel series of non-peptide endothelin receptor antagonists, in: *Peptides: Chemistry, Structure and Biology* (P. T. P. Kaumaya and R. S. Hodges, eds.), pp. 673–675, Mayflower Scientific, Kingswinford.

Estrada, V., Tellez, M. J., Moya, J., Fernandez-Durango, R., Egido, J., and Cruz, A. F., 1994, High plasma levels of endothelin-1 and atrial natriuretic peptide in patients with acute ischemic stroke, *Am. J. Hypertens.* **7:**1085–1089.

Gellai, M., Jugus, M., Fletcher, T., DeWolf, R., and Nambi, P., 1994, Reversal of postischemic ARF with a selective ET_A receptor antagonist in the rat, *J. Clin. Invest.* **93:**900–906.

Hunt, J. T., 1992, SAR of endothelin deduced from monocyclic analogs, *Drug News Perspect.* **5:**78–82.

Hunt, J. T., Lee, V. G., Stein, P. D., Hedberg, A., Liu, E. C.-K., McMullen, D., and Moreland, S., 1991, Structure–activity relationships of monocyclic endothelin analogs, *Bio-Org. Med. Chem. Lett.* **1:**33–38.

Ihara, M., Noguchi, K., Saeki, T., Fukuroda, T., Tsuchida, S., Kimura, S., Fukami, T., Ishikawa, K., Nishikibe, M., and Yano, M., 1992, Biological profiles of highly potent novel endothelin antagonists selective for the ET_A receptor, *Life Sci.* **50:**247–255.

Krystek, S. R., Jr., Patel, P. S., Rose, P. M., Fisher, S. M., Kienzle, B. K., Lach, D. A., Liu, E. C., Lynch, J. S., Novotny, J., and Webb, M. L., 1994, Mutation of peptide binding site in transmembrane region of a G protein-coupled receptor accounts for endothelin receptor subtype selectivity, *J. Biol. Chem.* **269:**12383–12386.

Lago, M. A., Luengo, J. I., Peishoff, C. E., and Elliott, J. D., 1996, Endothelin antagonists, *Annu. Rep. Med. Chem.* **31:**81–90.

Lee, J. A., Elliott, J. D., Sutiphong, J. A., Friesen, W. J., Ohlstein, E. H., Stadel, J. M., Gleason, J. G., and Peishoff, C. E., 1994, Tyr-129 is important to the peptide ligand affinity and selectivity of human endothelin type A receptor, *Proc. Natl. Acad. Sci. USA* **91:**7164–7168.

Lee, J. A., Sutiphong, J. A., Longton, E. D., Peishoff, C. E., Stadel, J. M., Kumar, C., Ohlstein, E. H., Gleason, J. G., and Elliott, J. D., 1995, Lysine 182 of endothelin B receptor modulates agonist selectivity and antagonist affinity: Evidence for the overlap of peptide and non-peptide ligand binding sites, *Biochemistry* **33:**14543–14549.

Michael, J. R., and Markewitz, B. A., 1996, Endothelins and the lung, *Am. J. Respir. Crit. Care Med.* **154:**555–581.

Nakajima, K., Kubo, S., Kumagaye, S., Nishio, H., Tsunemi, M., Inui, T., Kuroda, H., Chino, N., Watanabe, T. X., Kimura, T., and Sakakibara, S., 1989, Structure–activity relationship of endothelin: Importance of charged groups, *Biochem. Biophys. Res. Commun.* **163:**424–429.

Ohlstein, E. H., Arleth, A., Bryan, H., Elliott, J. D., and Sung, C. P., 1992, The selective endothelin-A receptor antagonist BQ-123 antagonizes ET-1 mediated mitogenesis in vascular smooth muscle, *Eur. J. Pharmacol.* **225:**347–350.

Ohlstein, E. H., Beck, G. R., Jr., Douglas, S. A., Nambi, P., Lago, A., Gleason, J. G., Ruffolo, R. R., Jr., Feuerstein, G., and Elliott, J. D., 1994a, Nonpeptide endothelin receptor antagonists. II. Pharmacological characterization of SB 209670, *J. Pharmacol. Exp. Ther.* **271:**762–768.

Ohlstein, E. H., Nambi, P., Douglas, S. A., Edwards, R. M., Gellai, M., Lago, A., Leber, J. D., Cousins, R. D., Gao, A., Frazee, J. S., Peishoff, C. E., Bean, J. W., Eggleston, D. S., Elshourbagy, N. A., Kumar, C., Lee, J. A., Yue, T.-L., Brooks, D. P., Weinstock, J., Feuerstein, G., Poste, G., Ruffolo, R. R. Jr.; Gleason, J. G. and Elliott, J. D., 1994b, SB 209670, a rationally designed potent nonpeptide endothelin receptor antagonist, *Proc. Natl. Acad. Sci. USA* **91:**8052–8056.

Ohlstein, E. H., Nambi, P., Lago, A., Hay, D. W. P., Beck, G., Fong, K.-L., Eddy, E. P., Smith, P., Ellens, H., and Elliott, J. D., 1996, Nonpeptide endothelin receptor antagonists. VI: Pharmacological characterization of SB 217242, a potent and highly bioavailable endothelin receptor antagonist, *J. Pharmacol. Exp. Ther.* **276:**609–615.

Panek, R. L., Major, T. C., Hingorani, G. P., Doherty, A. M., Taylor, D. G., and Rapundalo, S. T., 1992, Endothelin and structurally related analogs distinguish between endothelin receptor subtypes, *Biochem. Biophys. Res. Commun.* **183:**566–571.

Ruffolo, R. R., Jr., 1995, *Endothelin Receptors from the Gene to the Human*, CRC Press, Boca Raton.

Sakurai, T., Yanagisawa, M., Takuwa, Y., Miyazaki, H., Kimura, S., Goto, K., and Masaki, T., 1990, Cloning of a cDNA encoding a non-isopeptide-selective subtype of the endothelin receptor, *Nature* **348:**732–735.

Strader, C. D., Sigal, I. S., Register, R. B., Candelore, M. R., Rands, E., and Dixon, R. A., 1987, Identification of residues required for ligand binding to the beta-adrenergic receptor, *Proc. Natl. Acad. Sci. USA* **84:**4384–4388.

Strader, C. D., Sigal, I. S., Candelore, M. R., Rands, E., Hill, W. S., and Dixon, R. A., 1988, Conserved aspartic acid residues 79 and 113 of the beta-adrenergic receptor have different roles in receptor function, *J. Biol. Chem.* **263:**10267–10271.

Tam, J. P., Liu, W., Zhang, J.-W., Galatino, M., Bertolero, F., Cristiani, C., Vaghi, F., and Castiglione, R. D., 1994, Alanine scan of endothelin: Importance of aromatic residues, *Peptides* **15:**703–708.

Underwood, D. C., Bochnowicz, S., Osborn, R. R., Luttman, M. A., and Hay, D. W. P., 1997, Nonpeptide endothelin receptor antagonists. X. Inhibition of endothelin-1 and hypoxia-induced pulmonary pressor responses in the guinea pig by the endothelin receptor antagonist, SB 217242, *J. Pharmacol. Exp. Ther.* **283:**1130–1137.

Warner, T. D., Allcock, G. H., Corder, R., and Vane, J. R., 1993, Use of the endothelin antagonists BQ 123 and PD 142893 to reveal three endothelin receptors mediating smooth muscle contraction and release of EDRF, *Br. J. Pharmacol.* **110:**777–782.

Webb, M. L., and Meek, T. D., 1997, Inhibitors of endothelin, *Med. Res. Rev.* **17:**17–67.

Wei, G. Z., Zhang, J., Sheng, S. L., Ai, H. X., Ma, J. C., and Lui, H. B., 1993, Increased plasma ET-1 in patients with acute cerebral infarction and actions on pial arterioles of rat, *Chin. Med. J.* **106:**917–921.

Xu, D., Emoto, N., Giaid, A., Slaughter, C., Kaw, S., Wit, D. D., and Yanigasawa, M., 1994, ECE-1: A membrane-bound metalloprotease that catalyzes the proteolytic activation of big endothelin-1, *Cell* **78:**473.

Yanagisawa, M., Kurihara, H., Kimura, S., Tomobe, Y., Kobayashi, M., Mitsui, Y., Yazaki, Y., Goto, K., and Masaki, T. A., 1988, A novel potent vasoconstrictor peptide produced by vascular endothelial cells, *Nature* **332:**411–415.

Ziv, I., Fleminger, G., Dyaldetti, R., Achiron, A., Melamed, E., and Sokolovsky, M., 1992, Increased plasma endothelin-1 in acute ischemic stroke, *Stroke* **23:**1014–1016.

Chapter 7

LHRH Antagonists

Fortuna Haviv, Eugene N. Bush, Judith Knittle, and Jonathan Greer

1. MECHANISM OF ACTION OF LHRH AGONISTS AND ANTAGONISTS

Luteinizing hormone-releasing hormone (LHRH), also called gonadotropin releasing hormone (GnRH), is a decapeptide hormone, pGlu-His-Trp-Ser-Tyr-Gly-Leu-Arg-Pro-GlyNH_2, which is released from the hypothalamus in a pulsatile fashion and binds to a specific receptor on the pituitary gland, thereby inducing the release of LH and FSH (Dutta, 1988; Filicori and Flamigni, 1988; Karten and Rivier, 1986). Subsequently, LH acts on the gonads to cause the release of reproductive hormones, in particular testosterone (T) in males and estradiol and progesterone in females. The present therapeutic use of LHRH agonists (Table I) is related to their ability to suppress sex hormones during chronic administration. In contrast, acute administration increases the levels of the reproductive hormones. This paradoxical effect is the result of downregulation of the LHRH receptor caused by high levels of LHRH agonist (Conn and Crowley, 1994). When LHRH agonists are administered to humans, they increase T over the first 4 to 7 days of administration, then slowly within 10 days the hormone levels drop to castrate. This initial hormonal surge may sometimes cause a temporary exacerbation of disease symptoms. Presently, the LHRH agonists are therapeutically utilized in various sex hormone-dependent diseases. The most common is prostate cancer (Garnick *et al.,* 1984). Suppression of T either by orchiectomy or by LHRH agonist

Fortuna Haviv, Eugene N. Bush, Judith Knittle, and Jonathan Greer • Abbott Laboratories, North Chicago, Illinois 60064-3500.

Integration of Pharmaceutical Discovery and Development: Case Studies, edited by Borchardt *et al.,* Plenum Press, New York, 1998.

Table I
Structures of LHRH Analogues
pGlu-His-Trp-Ser-Tyr-Gly-Leu-Arg-Pro-GlyNH$_2$

	Substitution	Name (source)
	Agonists	
1	As above	LHRH
2	DLeu^{6},Pro^{9}NHEt	leuprolide (TAP)
3	DSer(OtBu)^{6},Pro^{9}NHEt	buserelin (Hoechst)
4	D2Nal^{6}	nafarelin (Syntex)
5	DSer(OtBu)^{6},Azagly^{10}	goserelin (ICI)
6	DHis(Bzl)^{6},Azagly^{10}	histrelin (Roberts)
7	DTrp^{6},Pro^{9}NHEt	deslorelin
	Antagonists	
8	NAcΔ^{3}Pro^{1},D4FPhe^{2},DTrp^{3,6}	4F-Ant (Salk Inst.)
9	NAcD2Nal^{1},D4FPhe^{2},DTrp^{3},DArg^{6}	NalArg (Salk Inst.)
10	NAcD2Nal^{1},D4ClPhe^{2},DTrp^{3},DhArg(Et$_2$)^{6},DAla^{10}	detirelix (Syntex)
11	NAcD2Nal^{1},D4ClPhe^{2},D3Pal^{3},Arg^{5},DGlu(AA)^{6},DAla^{10}	NalGlu (Salk Inst.)
12	NAcD2Nal^{1},D4ClPhe^{2},D3Pal^{3},Lys(Nic)^{5},DLys(Nic)^{6},Lys(Isp)^{8},DAla^{10}	antide (Serono)
13	NAcD4ClPhe^{1,2},DBal^{3},DLys^{6},DAla^{10}	ORG-30850 (Organon)
14	NAcD2Nal^{1},D4ClPhe^{2},D3Pal^{3},DCit^{6},DAla^{10}	cetrorelix (Asta)
15	NAcD2Nal^{1},D4ClPhe^{2},D3Pal^{3},DhArg(Et$_2$)^{6},hArg(Et$_2$)^{8},DAla^{10}	ganirelix (Syntex)
16	NAcD2Nal^{1},D4ClPhe^{2},D3Pal^{3},NMeTyr^{5},DLys(Nic)^{6},Lys(Isp)^{8},DAla^{10}	A-75998 (TAP/Abbott)
17	NAcD2Nal^{1},D4ClPhe^{2},D3Pal^{3},Aph(atz)^{5},DAph(atz)^{6},Lys(Isp)^{8},DAla^{10}	azaline B (Salk Inst.)
18	NAcD2Nal^{1},D4ClPhe^{2},D3Pal^{3},DhCit^{6},Lys(Isp)^{8},DAla^{10}	antarelix (Europeptides)
19	[N(1-adamantyl)acetyl)Ser^{4},DTrp^{6},Pro^{9}NHEt](4–9)	(4–9) hexapeptide
20	[N(4FPP)D1Nal^{3},NMeTyr^{5},DLys(Nic)^{6},Lys(Isp)^{8},DAla^{10}](3–10)	A-76154

administration can induce clinical remission. To maintain chemical castration the drug is generally administered either daily by s.c. injection or by a more preferred route of a 1-month depot injection (Dlugi *et al.*, 1990). Currently, LHRH agonists are also used for endometriosis, uterine fibroids, *in vitro* fertilization, and precocious puberty (Filicori and Flamigni, 1988; Simon *et al.*, 1990).

Once the LHRH agonists (Table I, **2–6**) were shown to be therapeutically useful, the next goal was to develop LHRH antagonists that would suppress sex hormones from the onset of treatment. The development of LHRH antagonists has been much slower than that of agonists. It has continued for over 15 years (Karten, 1992). The research progress was hampered first by low potency and then by safety issues, which related to the propensity of the clinical candidates of the second generation to release histamine (Karten, 1992). Only the third generation of LHRH antagonists (Table I, **11–18**), which were discovered near the end of the 1980s, has reached the stage of advanced clinical studies (see Section 5.2).

2. STRUCTURAL DIFFERENCES OF LHRH AGONISTS AND ANTAGONISTS

Soon after the elucidation of the structure of LHRH, it became apparent that this decapeptide hormone has a very short half-life *in vivo,* mainly because of enzymatic degradation (Koch *et al.*, 1974; Redding *et al.*, 1973). This finding prompted an intensive synthetic effort by many research groups to increase the peptide's metabolic stability. The first enhancement in biological activity was achieved by substitution of the $Gly^{10}NH_2$ in LHRH with *N*-ethyl amide (Fujino *et al.*, 1973). The second major boost in potency was obtained on substitution of D-amino acid for the Gly at position 6 (Coy *et al.*, 1976; Monahan *et al.*, 1973). These two modifications led to the so-called "superagonists" of LHRH. Five of these agonists (**2–6**) (Karten and Rivier, 1986) are now available on the market in the United States as approved drugs.

Whereas the structures of LHRH agonists differ from the natural hormone in just one or two residues, those of the antagonists contain only three to five native amino acids. Almost all antagonists contain the same D-amino acids at positions 1, 2, 3, and 10; the structural differences are mainly at positions 5, 6, and 8 (Table I). A representative of the first generation of antagonists is 4F-Ant (**8**), which contains $NAc\Delta^3Pro$ at position 1 and D-amino acids at 2, 3, and 6 (Rivier *et al.*, 1981). Antagonist **8** was tested in humans and was found to be insufficiently potent, although no side effects were observed. Two representatives of the second generation are NalArg (**9**) (Rivier *et al.*, 1984) and detirelix (**10**) (Nestor *et al.*, 1988). Both antagonists were very potent *in vivo* with long durations of action. Unfortunately, when tested in humans they caused histamine-mediated systemic side effects (Karten, 1992). It was rationalized that this property of mast cell degranulation was caused by the proximity of the two basic amino acids at positions 6 and 8 along with the highly hydrophobic residues at positions 1, 2, and 3 (Karten and Rivier, 1986; Karten, 1992; Karten *et al.,* 1987). To reduce the peptide's hydrophobicity, Rivier and co-workers (Rivier *et al.,* 1986) substituted D3Pal* at position 3, and to increase the distance between the two basic residues one Arg was moved from position 6 to 5 and DGlu(AA) was substituted at 6 resulting in the antagonist NalGlu (**11**). This antagonist (**11**), which represents the first of the third generation, did not show any systemic side effects in humans other than some local skin reactions. Nevertheless, NalGlu was shown to be effective in suppressing T levels in man (Bagatell *et al.,* 1989; Pavlou *et al.,* 1989).

Additional structural modifications at positions 5, 6, and 8 have led to the present generation of antagonists (**12–18**) that are currently in clinical studies. Antide (**12**), which contains $Lys(Nic)^5$, $DLys(Nic)^6$ and $Lys(Isp)^8$ residues, was discovered by Folkers's group (Ljungqvist *et al.,* 1988). This decapeptide was the first antagonist with no tendency to release histamine (Ljungqvist *et al.,* 1987). A s.c. dose of 1.0 mg/kg of antide administered to ovariectomized cynomolgus monkeys suppressed LH for 5 days (Edelstein *et al.,* 1990; Leal *et al.,* 1988). The major drawback of antide is very low water solubility, which limited its efficacy in humans (Bagatell *et al.,* 1993). A-75998 (**16**), shown in Fig. 1, differs from antide only in position 5. It contains NMeTyr instead of Lys(Nic) (Haviv *et al.,* 1993a). This compound, as described in Sections 3.2 and 5.1, was very efficacious both in animal models and in humans. Organon's antagonist ORG-30850 (**13**) contains DBal at position 3 and DLys at 6 (Deckers *et al.,* 1989). This compound was very effective in suppressing LH in monkeys (Scott *et al.,* 1989). Its shortcoming is its low ED_{50} for histamine release (HR) (Karten, 1992). Ganirelix (**15**) is another antagonist, discovered by the Syntex group (Nestor *et al.,* 1992), that was designed to minimize the HR property of detirelix by substituting D3Pal at position 3 and $hArg(Et_2)$ at 8. Ganirelix was shown to be efficacious and safe in animals (Lee *et al.,* 1989; Vickery *et al.,* 1990) and was developed for clinical studies (see Section 5.2). The most reported antagonist is cetrorelix or SB-75 (**14**), which was discovered by Schally's group (Bajusz *et al.,* 1988). It contains D-citrulline at position 6, a residue that is hydrophilic but neutral. The compound was shown to be safe and effective in animals and proceeded to clinical studies (Reissmann *et al.,* 1994, 1996). Antarelix or EP-24332 (**18**) is an antagonist, discovered by the Europeptides group (Deghenghi *et al.,* 1993). Its structure resembles both cetrorelix and antide: It contains DhCit at position 6 and Lys(Isp) at 8. The compound is being developed for clinical studies. Another antagonist is azaline B (**17**), which was discovered by the group at the Salk Institute (Rivier *et al.,* 1992a). It contains Aph(atz)

*Abbreviations used: D2Nal, D-3-(2-naphthyl)alanine; D3Bal, D-3-(3-benzthienyl)alanine; D4ClPhe, D-3-(4-Cl-phenyl)alanine; D3Pal, D-3-(3-pyridyl)alanine; NMeTyr, *N*-α-methyl-tyrosine; DLys(Nic), D-lysine(*N*-ϵ-nicotinyl); Lys(Isp), lysine(*N*-ϵ-isopropyl); DGlu(AA), 4-(*p*-methoxybenzoyl)-D-2-aminobutyric acid; Aph(atz), 3-[4-(3′-amino)-1*H*-1′,2′,4′-triazolyl]-phenylalanine; $hArg(Et_2)$, $N^G,N^{G'}$-diethyl-homoarginine; Dpr, 2,3-diaminopropionic acid; 4FPP, 4-F-phenylpropionic acid.

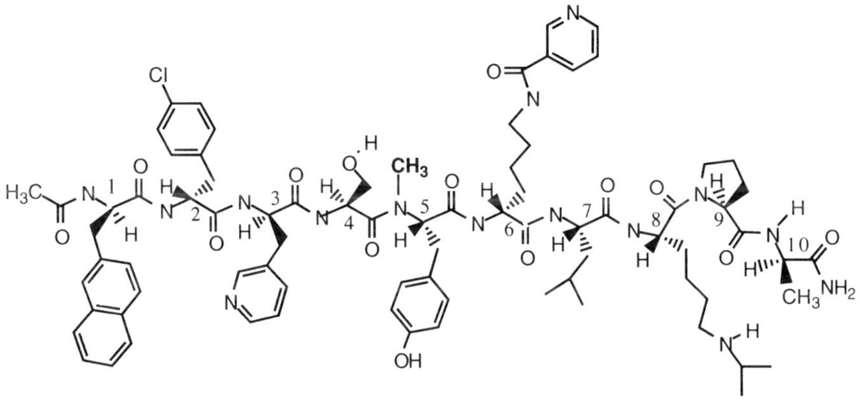

Figure 1. Chemical structure of A-75998.

at position 5 and DAph(atz) at 6. The compound was very efficacious and safe in animals and is currently in advanced development.

2.1. Reduction of Size of LHRH Analogues

All of the LHRH antagonists, which have reached the stage of clinical studies, are decapeptides (Table I). As a part of our interest in developing an orally active peptidomimetic or nonpeptidic antagonists of LHRH, we attempted to reduce the size of these peptides (Haviv *et al.*, 1989). Our early work in this area describes a series of hexapeptide LHRH analogues that contain the (4–9) fragment from several agonist structures, wherein the N-terminus was coupled to a carboxylic acid such as 3-(3-indolyl)propionic or 3-(1-naphthyl)propionic, to mimic the amino acid at position 3 (Haviv *et al.*, 1989). Interestingly, it was found that by varying the substituent at position 3 of these (4–9) reduced-size LHRH analogues, the compound could be easily transformed from agonist to antagonist. Also, the structure–activity relationship of the substituent at position 3 showed that there was an optimal size, length, and shape for receptor affinity, biological potency, and type of response. Additionally, the substituent at position 6 somehow feeds back to the residue at position 3 to change the compound from agonist to antagonist as the side chain gets larger. The most active antagonist in this series, the [*N*(1-adamantyl)acetyl)Ser^4,DTrp^6,Pro^9NHEt]LHRH (4–9) (**19**), bound to rat LHRH receptor with a 9.34 pK_I, equal to that of the endogenous LHRH hormone, and inhibited LH release *in vitro* with pA_2 of 8.5. *In vivo* this compound produced 70% suppression of LH when 1.2 mg/kg was administered by i.v. infusion to castrate rat over 120 min (Haviv *et al.*, 1989).

To improve both the *in vitro* potency and the *in vivo* duration of action, another reduced-size series was developed. This new series was designed based on the (3–9) fragment of the agonist [Phe^2,DTp^6,Pro^{9}NHEt]LHRH. Elimination of pGlu^1 from the N-terminus and substitution of Phe^2 with 3-(4-Cl-phenyl)propionic acid produced the antagonist [*N*-(3-(4Cl-phenylpropionyl))DTrp^{3,6},Pro^9-NHEt]LHRH, which had receptor binding affinity equal to the parent agonist and inhibited LH release with a pA_2 of 9.90 (Haviv *et al.,* 1994). However, this antagonist was inactive *in vivo*. On further systematic substitutions of D1Nal^3, NMeTyr^5, and DAla^{10}, it was possible to improve the pharmacokinetics and increase the *in vivo* potency up to the range of NalGlu. Nevertheless, the compound still had a too low ED_{50} value for histamine release. Finally, to reduce the HR property, Lys(Isp)^8 was substituted for Arg^8 resulting in A-76154 (**20**). This octapeptide antagonist has an ED_{50} for HR 10-fold higher than NalGlu. A-76154 suppressed LH levels by 90% in castrate rats at a dose of 30 μg/kg s.c. (Haviv *et al.,* 1994).

2.2. Enzymatic Stability of LHRH Analogues and Effect of *N*-methyl Substitution on Enzymatic Stability of LHRH Agonists

As indicated above, the substitutions of D-amino acid at position 6 and *N*-ethylamide at position 10 of LHRH increased metabolic stability of agonists (Coy *et al.,* 1975; Koch *et al.,* 1974). For example, in humans leuprolide's (**2**) half-life is 174 min versus 57 min for LHRH. Because chymotrypsin cleaves the Trp^3-Ser^4 bond in leuprolide (Haviv *et al.,* 1993b), to stabilize this bond NMeSer^4 was substituted in the peptide rendering it completely stable against enzymatic degradation, although it was 10-fold less active. The stability of leuprolide was further probed by separately substituting *N*-methyl at each peptidic bond. *N*-methylation of residue 2 increased stability by 7-fold, whereas that at positions 6, 7, 8, or 10 did not have any beneficial effect on stability. On the other hand, *N*-methylation of residues 3, 4, and 5 very effectively blocked chymotrypsin cleavage. These last results were rationalized by examining the three-dimensional structure of leuprolide substrate bound to chymotrypsin's active site (Haviv *et al.,* 1993b). The model showed that both main-chain NH of residues 3 and 5 of the substrate are involved in hydrogen bond interaction with the enzyme. Substitution of one of the amide hydrogens with methyl disrupts the hydrogen bond, and steric hindrance of the methyl group forces a distortion in the conformation of the substrate on the enzyme. The net result is the inability of chymotrypsin to cleave leuprolide. As a spin-off of this study on enzymatic stabilization, it was found that substitutions of the following agonists converted the parent agonists into antagonists: leuprolide with NMePhe^2 or NMe1Nal^3; deslorelin (**7**) with NMeHis^2; and nafarelin (**4**) with NMePhe^2 or NMeArg^8 (Haviv *et al.,* 1993b). The agonist/antagonist switch was influenced by a combination of the site of *N*-methylation and the substituents at positions 6 and 10. These findings substantially modify the known structure–activity relationship picture of LHRH antagonists.

2.3. Effect of *N*-methyl Substitution on Water Solubility of LHRH Antagonists. Discovery of A-75998

Because most antagonists contain five D-amino acids, they are stable against *in vitro* enzymatic degradation. Nevertheless, metabolic studies with ganirelix in the rat indicated the presence of the (5–10) fragment as one of the metabolites (Chan *et al.,* 1991). To eliminate this possibility, $NMeTyr^5$ was substituted in the structures of three antagonists: ORG-30850, cetrorelix, and antide. The goal was to study the effect of this peptide backbone substitution on *in vitro* and *in vivo* activities and safety. No major change in activity, either *in vitro* or *in vivo,* was observed with ORG-30850 following the *N*-methyl substitution at position 5. Introduction of $NMeTyr^5$ in cetrorelix showed a 5-fold increase in receptor binding and 2-fold increase in LH inhibition *in vitro. In vivo* in the castrate rat, 30 μg/kg of cetrorelix, administered subcutaneously, suppressed LH for 8 hr, somewhat longer than the $NMeTyr^5$ analogue. Substitution of $NMeTyr^5$ in antide produced A-75998, which exhibited 2- to 4-fold improvement in *in vitro* activity, although in castrate rats both antagonists showed similar LH suppression (Haviv *et al.,* 1993a). In advanced pharmacology testing and in humans, A-75998 was more potent than antide (see Section 3.2). Most interestingly, during the HPLC purification, it was observed that all of the antagonists containing $NMeTyr^5$ were more water soluble than their parent analogues. Comparative solubility studies showed that for the three antagonists containing NMeTyr, the water solubility was increased by 12- to 25-fold (Haviv *et al.,* 1993a). This effect was attributed to the better exposure of the peptide side chains to interaction with the aqueous solvent as a result of the peptide backbone distortion caused by the *N*-methyl substitution. This finding demonstrates that it is possible to increase water solubility of peptides without adding any hydrophilic groups. It also has solved one of the major hurdles in the drug development of LHRH antagonists, namely, poor water solubility. A-75998 was subjected to the battery of tests described in Sections 3.1 and 3.2. This peptide (**16**) was efficacious and safe in all *in vitro* and *in vivo* tests and was selected for clinical studies.

3. BIOLOGICAL TESTING STRATEGY

Critical to the discovery of an LHRH antagonist clinical candidate is the design of a biological testing strategy that is based on the anticipated therapeutic applications and that will select the most potent, effective, and safest compound for them. The testing strategy used in the discovery of A-75998 differs from the conventional methods previously adopted for LHRH antagonists.

There were several overall considerations that shaped the compound testing strategy. (1) Intrinsic potency was assessed first in an *in vitro* assay and not in an *in vivo* test, such as antiovulatory activity, where potency is dependent on pharmacokinetics which may be even further confused by problems of solubility and formulation. (2) In all *in vivo* testing, two routes of administration were used to mimic the expected clinical testing paradigm. Daily s.c. injections were employed, as this was the expected clinical route of administration in early phase I safety testing. Eventually, the clinical compound would be formulated for a monthly depot, so s.c. infusion using Alzet® minipumps was also tested to determine if the compound would be suitable for this route of administration. Indeed, the need to be able to formulate for a 30-days depot required the discovery of a very potent compound. (3) Another major consideration was safety. As previously indicated, the more potent LHRH antagonists of the second generation caused systemic HR responses (Karten, 1992). It was important, therefore, to exclude compounds with high HR potential early in testing.

Intrinsic potency was assessed by ligand binding affinity to the LHRH receptor on rat pituitary membranes (reported as pK_I) and by *in vitro* LH release from primary rat pituitary cell cultures (reported as pA_2) (Haviv *et al.,* 1989). Compounds that displayed sufficiently high pA_2 continued to *in vivo* testing in rats, dogs, and monkeys as described in Section 3.2. After passing all functional tests, the compounds were tested for safety considerations. Further *in vivo* HR tests were performed by monitoring hypotension and edema in rats. Eventually, a full hemodynamic study was carried out. A-75998 was the first compound to pass all of the tests described below and deemed suitable for clinical development.

3.1. *In Vitro* Testing of A-75998: Receptor Binding, Inhibition of LH Release, and Histamine Release

A-75998 possesses high affinity for rat pituitary LHRH receptor with a pK_I of 10.50. In cultured rat pituitaricytes, A-75998 inhibited leuprolide-induced LH release with a pA_2 of 11.20. A-75998 released histamine from rat peritoneal mast cells with an ED_{50} of 10 μg/ml (Haviv *et al.,* 1993a). This ED_{50} is approximately one order of magnitude higher than that observed with Nal-Glu (**11**), which in humans produces local skin reactions but no systemic reactions.

3.2. *In Vivo* Studies of A-75998 in Rat, Dog, and Monkey

When 30 μg/kg of A-75998 was injected subcutaneously into castrate male rats, LH levels were suppressed within 1 hr; maximum effects were observed with-

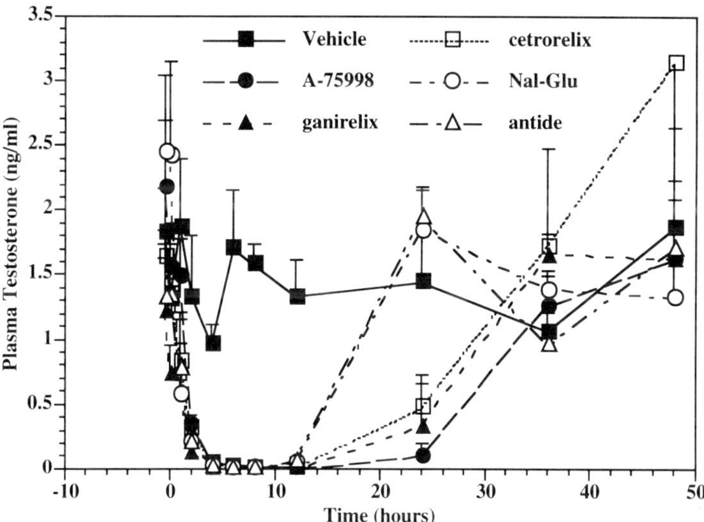

Figure 2. Suppression of plasma testosterone in male dogs by A-75998 and a number of third-generation LHRH antagonists after a single dose of 10 μg/kg administered at time 0.

in 4–6 hr. Plasma LH rose gradually thereafter and returned to pretreatment levels by 24 hr (Bush *et al.,* 1993; Haviv *et al.,* 1993a).

In a dose-ranging study of single s.c. injections of A-75998 in intact male dogs, a clear dose-dependent suppression of plasma T concentration was observed (Leal *et al.,* 1994). A dose of 10 μg/kg of A-75998 produced a significant 24-hr suppression of T, with return to pretreatment levels by 36 hr postinjection (Fig. 2). Ganirelix and cetrorelix tested at 10 μg/kg produced good T suppression for 12 hr, but only partial T suppression at 24 hr. At the same dose, NalGlu and antide suppressed T for only 12 hr. A-75998 is more potent than NalGlu and antide in dogs, as a longer-lasting T suppression was observed at equivalent doses (Fig. 2). When dogs received five daily injections of A-75998, a threefold higher dose was required to maintain suppression of T levels over the dosing period, compared with a single injection (Fig. 3). In the same test, ganirelix also decreased plasma T at 30 μg/kg per day. When A-75998 was infused subcutaneously for 3 days in dogs via Alzet® osmotic minipump at 7.5 and 15 μg/kg per day, T was suppressed to undetectable levels during the 3 days of treatment (Leal *et al.,* 1994).

Subcutaneous administration of 300 μg/kg of A-75998 to ovariectomized cynomolgus monkeys suppressed serum LH levels to undetectable by 24 hr after injection and kept them suppressed for 5–6 days (Gordon *et al.,* 1994b). Daily s.c. injections of A-75998 to intact adult male cynomolgus monkeys for 30 days at doses of 100 and 300 μg/kg per day were fully effective in producing and maintain-

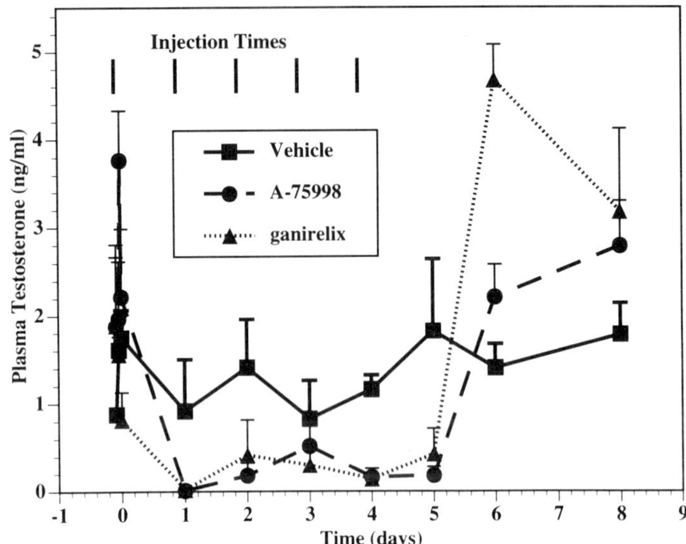

Figure 3. Effect of multiple-dose administration of A-75998 and ganirelix on plasma testosterone in male dogs. Compounds were injected s.c. daily for 5 days, at a dose of 30 μg/kg per day.

ing full suppression of serum T levels for the duration of treatment (Fig. 4), whereas a dose of 30 μg/kg per day was not. Intact male cynomolgus monkeys receiving an i.v. infusion of A-75998 from an Alzet® osmotic minipump for 1 week, had sustained suppression of T at doses of 100 and 200 μg/kg per day (Gordon *et al.*, 1994a).

3.3. Pharmacokinetics of A-75998 in Rat, Dog, and Monkey

A-75998 was injected into rats by i.v. bolus and subcutaneously at a dose of 100 μg/kg (Table II) and plasma A-75998 concentrations were measured by a specific radioimmunoassay. The areas under the drug's plasma level versus time curve (AUC) were similar, indicating that A-75998 was well absorbed from an s.c. injection site. Pharmacokinetic parameters for A-75998 were also determined in dog and monkey at a 30 μg/kg i.v. dose (Table II). Clearance was highest in rat, intermediate in dog, and lowest in monkey with a ratio of 40:2.5:1. Both half-life and volume of distribution improved in the two larger species. The slowest clearance in the monkey is related to the very low volume of distribution in this species.

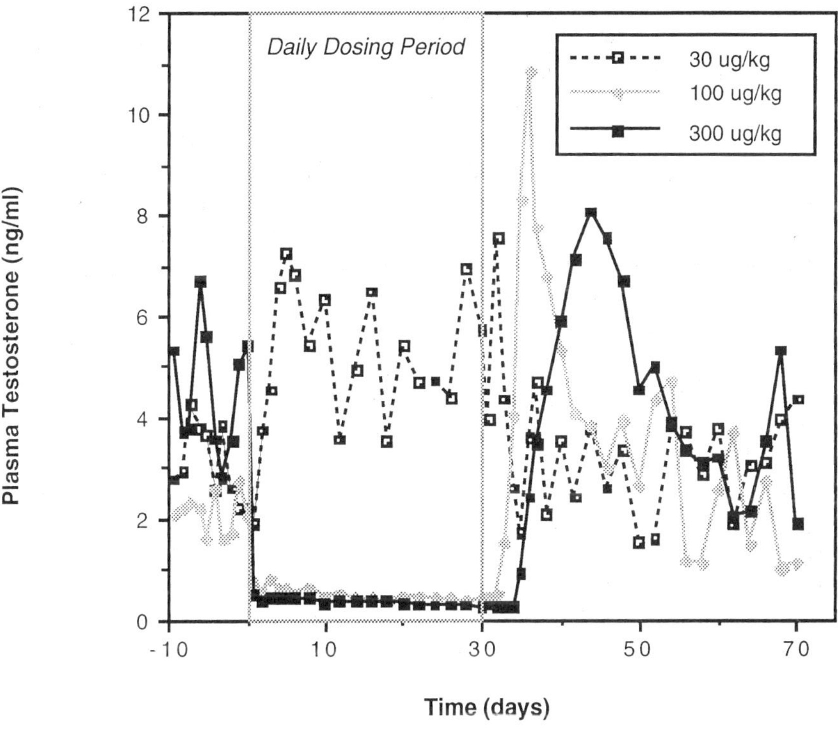

Figure 4. Dose-dependent serum testosterone suppression in response to daily s.c. injections of A-75998 in male cynomolgus monkeys for 30 days.

4. AGGREGATION AND FORMULATION OF A-75998

Although A-75998 is considerably more water soluble than many of the other peptide antagonists (Haviv *et al.*, 1993a), the compound presented significant problems in developing a suitable formulation for clinical administration. The

Table II
Pharmacokinetic Parameters for A-75998

Species	Route	Dose (μg/kg)	AUC (ng/ml per hr)	Clearance (ml/hr per kg)	$t_{1/2}$ (hr)	V_D (ml/kg)
Rat	i.v.	100	127.2	837.6	1.14	1374
	s.c.	100	148.0	685.2		
Dog	i.v.	30	671.4	54.19	7.99	648
Monkey	i.v.	30	1393.8	21.55	5.75	179

original clinical formulation consisted of 5% dextrose (D5W) in acetate buffer pH 4.5. In this vehicle, drug could not be prepared at high concentrations because the solutions formed gels on standing.

Accordingly, a study of the solubility properties of A-75998, using the method of dynamic light scattering (DLS), was initiated (Bush *et al.,* 1996; Cannon *et al.,* 1995). This method enables the analysis of particle size distributions in solutions of particles ranging in size from several micrometers down to about 1 nm, corresponding to compounds with a molecular mass as low as 1000 Da (Matayoshi and Krill, 1998). A sample of the clinical lot of drug that was visually clear, with no precipitate or haze apparent to the naked eye, was examined by DLS (Fig. 5). It appeared that all of the compound exists in some form of aggregate ranging in size from 22 nm (about 1500 molecules, assuming an approximately spherical shape) up to 2.1 μm (3.0×10^8 molecules). The size and distribution of aggregates varied with time and preparation. A wide variety of clinically acceptable solvent systems, pH, and temperatures were examined in an attempt to produce a stable and monomeric form of A-75998. The best formulation was the 2-hydroxypropyl-β-cyclodextrin (HPCD; also called Encapsin®). Thirty percent HPCD solutions at pH 4.5 in acetate buffer preserve A-75998 completely as monomers giving only 1.9-nm particles by DLS (Fig. 5).

To determine whether there is any pharmacological relevance to this aggregation phenomenon, the effect of the degree of aggregation of A-75998 was tested *in vivo* (Bush *et al.,* 1996). A well-characterized aggregated preparation of A-75998 in D5W acetate buffer pH 4.5 was compared with a nonaggregated

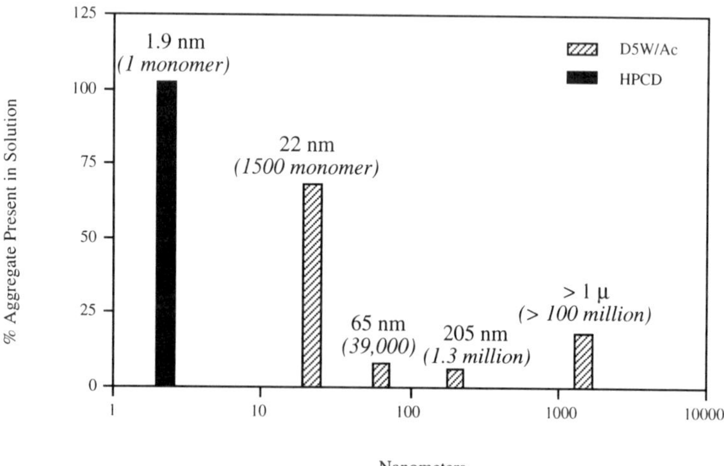

Figure 5. Distribution of aggregate sizes for A-75998 formulated in D5W acetate buffer pH 4.5 as measured by dynamic light scattering. Also shown is a monomeric solution of A-75998 formulated in Encapsin® at pH 7.4.

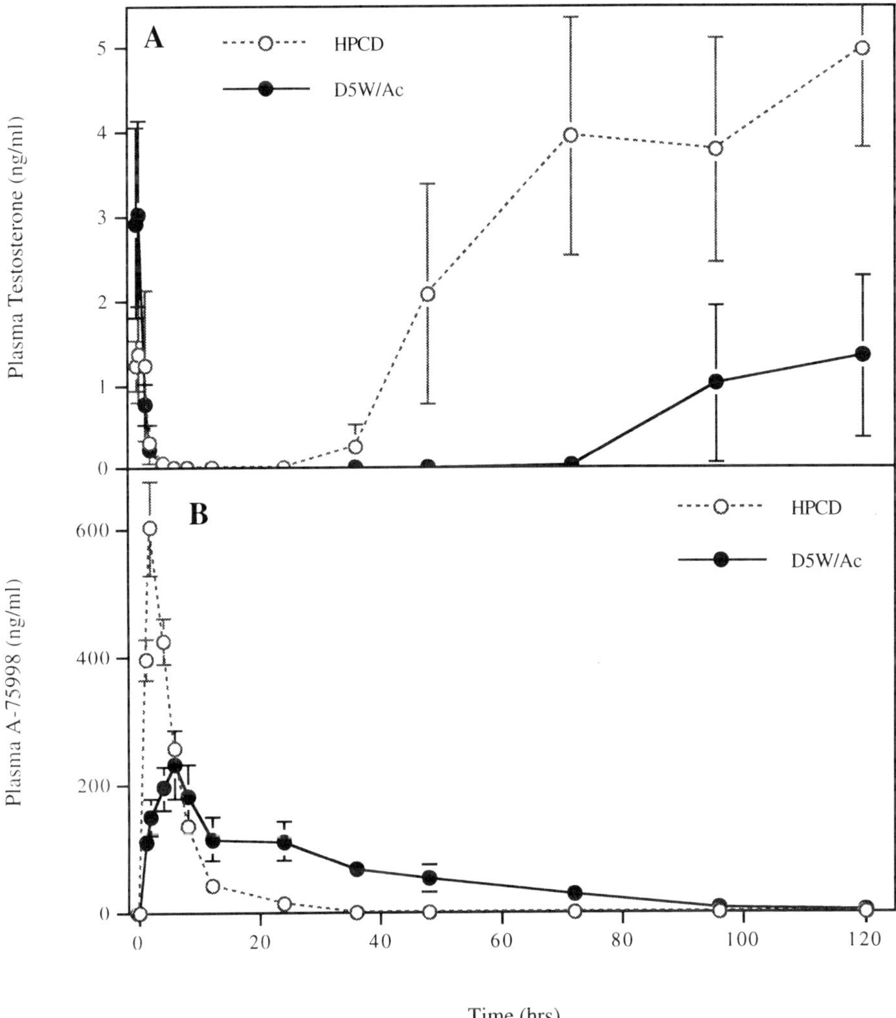

Figure 6. (A) Testosterone suppression produced by HPCD and D5W/Ac formulations given to male dogs at a concentration of 3 mg/ml s.c. (B) A-75998 plasma levels reflect the above testosterone suppression results.

monodispersed one in 30% HPCD. A 100 μg/kg dose of the drug was administered to dogs at a concentration of 3 mg/ml s.c., serial blood samples were collected, and plasma A-75998 as well as T levels were measured (Fig. 6). In the HPCD formulation group, T was suppressed to undetectable levels for up to 36 hr and then gradually returned to baseline (Fig. 6A). Blood levels of A-75998 were detectable for up to 36 hr (Fig. 6B). However, in the D5W formulation, there was a dramatic increase in duration of T suppression between 36 and 72 hr and a parallel increase reflected in the A-75998 plasma levels with significant levels detectable up to 96 hr. It appears that the aggregated solution in D5W deposits at the s.c. injection site resulting in prolonged drug blood levels and T suppression. In contrast, the HPCD formulation (Fig. 5) stabilizes the compound in the monomer form, even at pH 7.4, a pH compatible with tissue fluid, minimizing the possibility of precipitation at the site of injection and delivering compound into the bloodstream effectively.

The above experiments show clearly that the aggregation state of compounds like A-75998 is a crucial property that must be carefully studied both for purposes of proper formulation of a drug for the clinic as well as for potential effects on the physiological and pharmacological performance of the drug. For A-75998, the use of Encapsin® provided a single monomer species to produce a unique, reproducible, and stable formulation of this compound.

5. LHRH ANTAGONISTS IN CLINICAL EVALUATION

5.1. Clinical Study of A-75998

A-75998 was administered to healthy adult men in a prospectively randomized study. Subjects ranged in age from 19 to 45 (mean 32 years). A-75998 was dosed subcutaneously as a single bolus, in separate groups, at 0.01, 0.03, 0.1, 0.3, 1.0, 2.0, 3.0, 5.0, and 10.0 mg. Each group consisted of six subjects receiving a set dose of A-75998 and two placebo controls. T levels were measured before dosing and afterwards for 72 hr. The results are shown in Fig. 7. Diurnal cycling of T levels in adult males could be seen for the placebo control and the lower doses. Transient partial suppression was first observed at the 0.3-mg dose. T suppression was at or near castrate levels (<50 ng/dl) for the higher doses, 2.0 mg and above, and suppression was sustained for up to 36 hr at the highest two doses. All subjects returned or were proceeding to baseline by 72 hr after dosing. No unusual safety concerns were noted in this study. Mild, transient injection site reactions were observed at doses of 1.0 mg and above. A-75998 was found to be efficacious and safe at doses up to 10.0 mg. Further clinical studies with this compound are ongoing.

5.2. Current LHRH Antagonists in Clinical Studies

NalGlu was the first representative of the third generation of antagonists to be tested in humans. It did not cause any systemic side effects but did produce some skin reactions. In healthy men a daily dose of 5 mg over 7 days was effec-

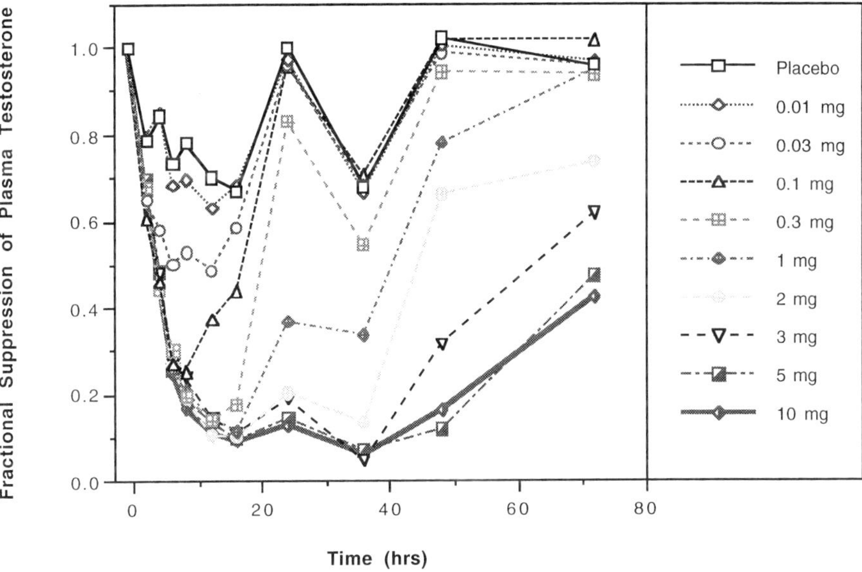

Figure 7. Results of single-dose clinical study of A-75998 in normal male subjects.

tive in reducing T levels by 80% (Pavlou *et al.,* 1989). As a clinical probe, NalGlu was also tested in humans for male contraception (Pavlou *et al.,* 1989). The compound was not further pursued for drug development (Rivier *et al.,* 1992b). Other antagonists, which are either in clinical studies or being so considered, include: ORG-30850, antide, ganirelix, cetrorelix, antarelix, A-75998, and azaline B (Table I). Antide was administered intravenously to healthy men at doses ranging from 10 to 50 μg/kg. At the highest dose, T levels were suppressed to 50–70% of the baseline (Bagatell *et al.,* 1993); however, the low water solubility of the compound precluded further development of the drug. Phase I clinical studies with cetrorelix in men and women have been reported (Behre *et al.,* 1994; Sommer *et al.,* 1994). A 3-mg dose of cetrorelix administered subcutaneously to normal men was effective in suppressing T by 75% within 8 hr of administration (Klingmuller *et al.,* 1993). Additional clinical studies with cetrorelix in prostate cancer and benign prostatic hyperplasia patients have been reported (Gonzalez-Barcena *et al.,* 1994). The compound is also being evaluated for treatment of endometriosis and controlled induction of ovulation (Diedrich *et al.,* 1994). Phase I clinical studies of ganirelix in postmenopausal women have examined pharmacokinetics and endocrine effects (Rabinovici *et al.,* 1992). In this group a 6-mg dose of ganirelix suppressed LH, FSH, and free alpha-subunit by 70%. In two separate groups of premenopausal women, 1- and 2-mg doses of ganirelix administered daily over 8 days suppressed estradiol levels by over 50% (Nelson *et al.,* 1995).

6. SUMMARY

After almost two decades, the research on LHRH antagonists has produced a number of decapeptides that are currently in clinical studies. The structures of these antagonists, unlike the agonists, differ substantially from that of LHRH. Five of the ten amino acids are unnatural and of D configuration. The structural combination of a hydrophobic N-terminus (residues 1, 2, and 3) and a basic/hydrophilic C-terminus (residues 6 and 8) was thought to be responsible for some HR reactions encountered with the second generation of LHRH antagonists. This side effect was greatly reduced by substituting the appropriate combination of amino acids at positions 5, 6, and 8. The next hurdle in the drug development of LHRH antagonists was solubility and aggregation. In the case of A-75998, water solubility was improved by 12- to 25-fold via substitution of NMeTyr at position 5. However, based on DLS analysis, the aqueous solutions still contained some large aggregates that were not visible to the naked eye. This formation of aggregates was eliminated on formulating A-75998 in Encapsin®. In men, a single s.c. dose of 2 mg of A-75998 suppressed T to the castrate levels for over 30 hr. Other LHRH antagonists including ganirelix and cetrorelix are also in phase I/II clinical studies. Clinical studies with cetrorelix in prostate cancer, *in vitro* fertilization, and benign prostate hypotrophy have been reported.

REFERENCES

Bagatell, C. J., McLachlan, R. I., de Krester, D. M., Burger, H. G., Vale, W. W., Rivier, J. E., and Bremner, W. J., 1989, A comparison of the suppressive effects of testosterone and a potent new gonadotropin-releasing hormone antagonist on gonadotropin and inhibin levels in normal men, *J. Clin. Endocrinol. Metab.* **69:**43–48.

Bagatell, C. J., Conn, P. M., and Bremner, W. J., 1993, Single dose administration of the gonadotropin-releasing hormone antagonist, Nal-Lys (antide) to healthy men, *Fertil. Steril.* **60:**680–685.

Bajusz, S., Kovacs, M., Gazdag, M., Bokser, L., Karashima, T., Csernus, V. J., Janaky, T., Gouth, J., and Schally, A. V., 1988, Highly potent antagonists of luteinizing hormone-releasing hormone free of edematogenic effects, *Proc. Natl. Acad. Sci. USA* **85:**1637–1641.

Behre, H. M., Bockers, A., Schlingheider, A., and Nieschlag, E., 1994, Sustained suppression of serum LH, FSH and testosterone and increase of high-density lipoprotein cholesterol by daily injection of GnRH antagonist cetrorelix over 8 days in normal men, *Clin. Endocrinol.* **40:**241–248.

Bush, E. N., Nguyen, A. T., Diaz, G. J., Love, S. K., Mikusa, J. P., Cybulski, V. A., Carlson, R. P., Haviv, F., Fitzpatrick, T. D., Nichols, C. J., Swenson, R. E., Mort, N. A., Johnson, E. S., Dodge, P. W., Knittle, J., and Greer, J., 1993, Effects of A-75998 and other antagonists of gonadotropin-releasing hormone (GnRH) in castrate male rats, *Endocr. J.* **1:**291–297.

Bush, E. N., Leal, J. A., Cybulski, V. A., Rhutasel, N., Diaz, G. J., Bammert, G., Haviv, F., Nichols, C., Swenson, R. E., Mort, N., Matayoshi, E., Krill, S., McChesney-Harris, L., Ruiz, L., Johnson, E. S., Knittle, J., Dodge, P. W., and Greer, J., 1996, Effect of formulation, concentration, and route of administration of the GnRH antagonist A-75998 on pharmacokinetics and testosterone suppression in male dogs, in: *10th International Congress of Endocrinology,* San Francisco, Abstr. P-3-240.

Cannon, J. B., Krill, S. L., and Porter, W. R., 1995, Physicochemical properties of A-75998, an antagonist of luteinizing hormone releasing hormone, *J. Pharm. Sci.* **84:**953–958.
Chan, R. L., Hsieh, S. C., Haroldsen, P. E., Ho, W., and Nestor, J. J., 1991, Disposition of RS-26306, a potent luteinizing hormone antagonist, in monkeys and rats after single intravenous and subcutaneous administration, *Drug Metab. Dispos.* **19:**858–864.
Conn, P. M., and Crowley, W. F., 1994, Gonadotropin-releasing hormone and its analogs, *Annu. Rev. Med.* **45:**391–405.
Coy, D. H., Labrie, F., Savary, M., Coy, E. J., and Schally, A. V., 1975, LH-releasing activity of potent LH-RH analogs *in vitro, Biochem. Biophys. Res. Commun.* **67:**576–582.
Coy, D. H., Vilchez-Martines, J. A., Coy, E. J., and Schally, A. V., 1976, Analogs of luteinizing hormone releasing hormone (LHRH) with increased biological activity produced by D-amino acid substitutions in position six, *J. Med. Chem.* **19:**423–425.
Deckers, G. H. J., Kloosterboer, H. J., and Loozen, H. J. J., 1989, Properties of a new LHRH antagonist (Org 30850), in: *71st Annual Meeting of the Endocrine Society,* Seattle, Abstr. 923.
Deghenghi, R., Boutignon, F., Wuthrich, P., and Lenaerts, V., 1993, Antarelix (EP 24332) a novel water soluble LHRH antagonist, *Biomed. Pharmacother.* **47:**107–110.
Diedrich, K., Diedrich, C., Santos, E., Zoll, C., Al-Hasani, S., Reissmann, T., Krebs, T., and Klingmuller, D., 1994, Suppression of the endogenous luteinizing hormone surge by the gonadotrophin-releasing hormone antagonist cetrorelix during ovarian stimulation, *Hum. Reprod.* **9:**788–791
Dlugi, M. A., Miller, J. D., Knittle, J., and Lupron Study Group, 1990, Lupron Depot (luprolide acetate for depot suspension) in the treatment of endometriosis: A randomized, placebo-controlled double-blind study, *Fertil. Steril.* **54:**419–427.
Dutta, A. S., 1988, Luteinizing hormone-releasing hormone (LHRH) agonists, *Drugs Future* **13:**43–57.
Edelstein, M. C., Gordon, K., Williams, R. F., Danforth, D. R., and Hodgen, G. D., 1990, Single dose long-term suppression of testosterone secretion by a gonadotropin-releasing hormone antagonist (antide) in male monkeys, *Contraception* **42:**209–214.
Filicori, M., and Flamigni, C., 1988, GnRH agonists and antagonists: Current clinical status, *Drugs* **35:**63–82.
Fujino, M., Shinagawa, S., Yamazaki, I., Kobayashi, S., Obayashi, M., Fukuda, T., Nakayama, R., White, W. F., and Rippel, R. H., 1973, [DesGlyNH_2^{10}, Pro-ethylamide^9]LHRH. A highly potent analog of luteinizing hormone releasing hormone, *Arch. Biochem. Biophys.* **154:**488–489.
Garnick, M. B., Glode, M., and Lupron Study Group, 1984, Leuprolide versus diethylstilbestrol for metastatic prostate cancer, *N. Engl. J. Med.* **311:**1281–1286.
Gonzalez-Barcena, D., Vadillo-Buenfil, M., Gomez-Orta, F., Fuentes-Garcia, M., Cardenas-Cornejo, I., Graef-Sanchez, A., Comaru-Schally, A. M., and Schally, A. V., 1994, Responses to the antagonist analog of LH-RH (SB-75, cetrorelix) in patients with benign prostatic hyperplasia and prostatic cancer, *Prostate* **24:**84–92.
Gordon, K., Williams, R. F., Greer, J., Bush, E. N., Haviv, F., Herrin, M., and Hodgen, G. D., 1994a, A-75998: A fourth generation GnRH antagonist: I. Preclinical studies in male primates, *Endocrine* **2:**1133–1139.
Gordon, K., Williams, R. F., Greer, J., Bush, E. N., Haviv, F., Herrin, M., and Hodgen, G. D., 1994b, A-75998: A fourth generation GnRH antagonist: II. Preclinical studies in female primates, *Endocrine* **2:**1141–1144.
Haviv, F., Palabrica, C. A., Bush, E. N., Diaz, G., Johnson, E. S., Love, S., and Greer, J., 1989, Active reduced-size hexapeptide analogues of luteinizing hormone-releasing hormone, *J. Med. Chem.* **32:**2340–2344.
Haviv, F., Fitzpatrick, T. D., Nichols, C. J., Swenson, R. E., Mort, N. A., Bush, E. N., Diaz, G., Nguyen, A., Holst, M. R., Cybulski, V. A., Leal, J. A., Bammert, G., Rhutasel, N. S., Dodge, P. W., Johnson, E. S., Cannon, J. B., Knittle, J., and Greer, J., 1993a, The effect of NMeTyr^5 substitution in luteinizing hormone-releasing hormone antagonists, *J. Med. Chem.* **36:**928–933.
Haviv, F., Fitzpatrick, T. D., Swenson, R. E., Nichols, C. J., Mort, N. A., Bush, E. N., Diaz, G., Bam-

mert, G., Nguyen, A., Rhutasel, N. S., Nellans, H. N., Hoffman, D. J., Johnson, E. S., and Greer, J., 1993b, Effect of N-methyl substitution of the peptide bonds in luteinizing hormone-releasing hormone agonists, *J. Med. Chem.* **36:**363–369.

Haviv, F., Fitzpatrick, T. D., Nichols, C. J., Bush, E. N., Diaz, G., Bammert, G., Nguyen, A., Johnson, E. S., Knittle, J., and Greer, J., 1994, *In vitro* and *in vivo* activities of reduced-size antagonists of luteinizing hormone-releasing hormone, *J. Med. Chem.* **37:**701–705.

Karten, M. J., 1992, An overview of GnRH antagonist development: Two decades of progress, in: *Modes of Actions of GnRH and GnRH Analogs* (W. F. Crowley and P. M. Conn, eds.), pp. 277–297, Elsevier, Amsterdam.

Karten, M., and Rivier, J. E., 1986, Gonadotropin-releasing hormone analog design. Structure–function studies toward the development of agonists and antagonists: Rationale and perspective, *Endocr. Rev.* **7:**44–66.

Karten, M. J., Hook, W. A., Siraganian, R. P., Coy, D. H., Folkers, K., Rivier, J. E., and Roeske, R. W., 1987, In vitro histamine release with LHRH analogs, in: *LHRH and its Analogs: Contraception and Therapeutic Applications, Part 2* (B. H. Vickery and J. J. J. Nestor, eds.), pp. 179–190, MTP Press, Lancaster.

Klingmuller, D., Schepke, M., Enzweiler, C., and Bidlingmaier, F., 1993, Hormonal responses to the new potent GnRH antagonist cetrorelix, *Acta Endocrinol.* **128:**15–18.

Koch, Y., Baram, T., Chobsieng, P., and Fridkin, M., 1974, Enzymatic degradation of luteinizing hormone-releasing hormone (LHRH) by hypothalamic tissue, *Biochem. Biophys. Res. Commun.* **61:**95–103.

Leal, J. A., Williams, R. F., Danforth, D. R., Gordon, K., and Hodgen, G. D., 1988, Prolonged duration of gonadotropin inhibition by a third generation GnRH antagonist, *J. Clin. Endocrinol. Metab.* **67:**1325–1327.

Leal, J. A., Bush, E. N., Holst, M. R., Cybulski, V. A., Nguyen, A. T., Rhutasel, N. S., Diaz, G. J., Haviv, F., Fitzpatrick, T. D., Nichols, C. J., Swenson, R. E., Mort, N. A., Carlson, R. P., Dodge, P. W., Knittle, J., and Greer, J., 1994, A-75998 and other GnRH antagonists suppress testosterone in male beagle dogs. A comparison of single injection, multiple injections and infusion administration, *Endocrine* **2:**921–927.

Lee, C. H., VanAntwerp, D., Hedley, L., Nestor, J. J. J., and Vickery, B. H., 1989, Comparative studies on the hypotensive effect of LHRH antagonists in anesthetized rats, *Life Sci.* **45:**697–702.

Ljungqvist, A., Feng, D. M., Tang, P. F., Kubota, M., Okamoto, T., Zhang, Y. W., Bowers, C. Y., Hook, W. A., and Folkers, K., 1987, Design, synthesis and bioassays of antagonists of LHRH which have high antiovulatory activity and release negligible histamine, *Biochem. Biophys. Res. Commun.* **148:**849–856.

Ljungqvist, A., Feng, D. M., Hook, W., Shen, Z. X., Bowers, C., and Folkers, K., 1988, Antide and related antagonists of luteinizing hormone release with long action and oral activity, *Proc. Natl. Acad. Sci. USA* **85:**8236–8240.

Matayoshi, E., and Krill, S., 1998, Manuscript in preparation.

Monahan, M. W., Amoss, M. S., Anderson, H. A., and Vale, W., 1973, Synthetic analogs of the hypothalamic lutenizing hormone releasing factor with increased agonist or antagonist properties, *Biochemistry* **12:**4616–4620.

Nelson, L. R., Fujimoto, V. Y., Jaffe, R. B., and Monroe, S. E., 1995, Suppression of follicular phase pituitary–gonadal function by a potent new gonadotropin-releasing hormone antagonist with reduced histamine-releasing properties (ganirelix), *Fertil. Steril.* **63:**963–969.

Nestor, J. J., Tahilramani, R., Ho, T. L., McRae, G. I., and Vickery, B. H., 1988, Potent, long-acting luteinizing hormone releasing hormone antagonists containing new synthetic amino acids: N,N′-dialkyl-D-homoarginines, *J. Med. Chem.* **31:**65–72.

Nestor, J. J., Tahilramani, R., Ho, T. L., Goodpasture, J. C., Vickery, B. H., and Ferrandon, P., 1992, Potent gonadotropin releasing hormone antagonists with low histamine-releasing activity, *J. Med. Chem.* **35:**3942–3948.

Pavlou, S. N., Wakefield, G. B., Schlechter, N. L., Lindner, J., Souza, K. H., Kamilaris, T. C., Konidaris, S., Rivier, J. E., Vale, W. W., and Toglia, M., 1989, Mode of suppression of pituitary and gonadal function after acute or prolonged administration of a luteinizing hormone-releasing hormone antagonist in normal men, *J. Clin. Endocrinol. Metab.* **68:**446–454.

Rabinovici, J., Rothman, P., Monroe, S. E., Nerenberg, C., and Jaffe, R. B., 1992, Endocrine effects and pharmacokinetic characteristics of a potent new gonadotropin-releasing hormone antagonist (ganirelix) with minimal histamine-releasing properties: Studies in postmenopausal women, *J. Clin. Endocrinol. Metab.* **75:**1220–1225.

Redding, T. W., Kastin, A. J., Gonzales-Barcena, D., Coy, D. H., Coy, E. J., Schalch, D. S., and Schally, A. V., 1973, The half-life, metabolism and excretion of tritiated luteinizing hormone-releasing hormone (LH-RH) in man, *J. Clin. Endocrinol. Metab.* **37:**626–631.

Reissmann, T., Engel, J., Kutscher, B., Bernd, M., Hilgard, P., Peukert, M., Szelenyi, I., Reichert, S., Gonzales-Barcena, D., Nieschlag, E., Comaru-Schally, A. M., and Schally, A. V., 1994, Cetrorelix, *Drugs Future* **19:**228–237.

Reissmann, T., Klenner, T., Deger, W., Hilgard, P., McGregor, G. P., and Voigt, K., 1996, Pharmacological studies with cetrorelix (SB-75), a potent antagonist of luteinizing hormone-releasing hormone, *Eur. J. Cancer* **32A:**1574–1579.

Rivier, J., Rivier, C., Perrin, M., Porter, J., and Vale, W. W., 1981, GnRH analogs: Structure activity relationships, in: *LHRH Peptides as Female and Male Contraceptives* (G. I. Zatuchni, J. D. Shelton, and J. J. Sciarra, eds.), pp. 13–23, Harper & Row, New York.

Rivier, J., Rivier, C., Perrin, M., Porter, J., and Vale, W., 1984, LHRH analogs as antiovulatory agents, in: *LHRH and its Analogs* (B. H. Vickery, J. J. J. Nestor, and E. S. E. Hafez, eds.), pp. 11–22, MTP Press, Lancaster.

Rivier, J. E., Porter, J., Rivier, C. L., Perrin, M., Corrigan, A., Hook, W. A., Siraganian, R. P., and Vale, W. W., 1986, New effective gonadotropin releasing hormone antagonists with minimal potency for histamine release in vitro, *J. Med. Chem.* **29:**1846–1851.

Rivier, J., Porter, J., Hoeger, C., Theobald, P., Craig, A. G., Dykert, J., Corrigan, A., Perrin, M., Hook, W. A., Siraganian, R. P., Vale, W., and Rivier, C., 1992a, Gonadotropin-releasing hormone antagonists with N omega-triazolylornithine, -lysine, or -p-aminophenylalanine residues at positions 5 and 6, *J. Med. Chem.* **35:**4270–4278.

Rivier, J. E., Theobald, P., Hoeger, C., Craig, A. G., Perrin, M., Porter, J., Corrigan, A., Koerber, S., Hagler, A., Vale, W., and Rivier, C., 1992b, GnRH antagonists: A synopsis, *Contraception* **46:**109–112.

Scott, R. T. J., Gordon, K., Williams, R. F., and Hodgen, G. D., 1989, New long-acting GnRH antagonist: Accelerated GnRH test response in primates, in: *71st Annual Meeting of the Endocrine Society,* Seattle, Abstr. 216.

Simon, A., Birkenfeld, A., and Schenker, J. G., 1990, Gonadotropin releasing hormone (GnRH): Mode of action and clinical applications. A review, *Int. J. Fertil.* **35:**350–362.

Sommer, L., Zanger, K., Dyong, T., Dorn, C., Luckhaus, T., Diedrich, K., and Klingmuller, D., 1994, Seven-day administration of the gonadotropin-releasing hormone antagonist cetrorelix in normal cycling women, *Eur. J. Endocrinol.* **131:**280–285.

Vickery, B. H., McRae, G., Lee, C. H., Nerenberg, C. A., Ferrandon, P., and Nestor, J. J. A., 1990, A new highly potent LHRH antagonist with low histamine releasing activity has unusually high oral bioavailability, in: *72nd Annual Meeting of the Endocrine Society,* Atlanta, Abstr. 1375.

Chapter 8

LHRH Agonists

Kenneth W. Funk, Jonathan Greer, and Akwete L. Adjei

1. INTRODUCTION

Recent advances in genetic engineering have increased our knowledge as to how biochemical species are manufactured and released in the body to protect homeostasis in humans. Tools utilized in this field of science have also fostered significant understanding of disease, etiology of various disease states, and biochemical mechanisms invoked by the body to combat disease. The end result is that new therapeutic paradigms with significantly different drug chemistries have been created. Examples of these drug systems include hormones, enzymes, genes, immunomodulators, and neurotransmitters of all sorts. Of the hormones, insulins and luteinizing hormone analogues are probably the most pervasive in clinical therapeutics today. The majority of these drugs are either sourced from biological origins and therefore are considered natural entities, or are partially chemically synthesized especially in those instances where a completely synthetic process is not feasible. Regardless of source, it is imperative that manufacturing conditions, specifications and controls, formulation modalities, as well as clinical and preclinical requirements be succinctly described in order to reliably control effectiveness of these drug systems in people who use them. This chapter focuses on one class of hormonal drugs, luteinizing hormone-releasing hormones (LHRH). Various aspects of pharmaceutical development, from drug candidate selection

Kenneth W. Funk, Jonathan Greer, and Akwete L. Adjei • Abbott Laboratories, North Chicago, Illinois 60064

Integration of Pharmaceutical Discovery and Development: Case Studies, edited by Borchardt *et al.*, Plenum Press, New York, 1998.

through pivotal clinical studies, are chronicled and should thus make it a useful text for scientists in industry, academia, and regulatory agencies worldwide.

1.1. Background

LHRH is the primary factor controlling reproductive function in vertebrates. It acts as a messenger between the hypothalamus and the anterior pituitary, thus regulating the release of gonadotropins, which control sexuality, ovulation, and spermatogenesis. LHRH is synthesized and stored in the hypothalamus in neurons, which project to the median eminence, and is released in periodic bursts into the hypophyseal portal circulation. the 1971 isolation, structure elucidation, and synthesis of porcine LHRH by Schally and co-workers (Schally *et al.*, 1971a,b; Matsuo *et al.*, 1971a,b; Baba *et al.*, 1971) prompted intense activities in synthesis of analogues by various laboratories around the world. Porcine LHRH, a decapeptide (Matsuo *et al.*, 1971a; Baba *et al.*, 1971), is structurally identical to ovine, bovine, human, and rat LHRH (Amoss *et al.*, 1971, Burgus *et al.*, 1972; Schally *et al.*, 1973, 1978, 1980) and has the following amino acid sequence:

pGlu-His-Trp-Ser-Tyr-Gly-Leu-Arg-Pro-Gly-NH_2

Structurally, LHRH is schematically represented as follows:

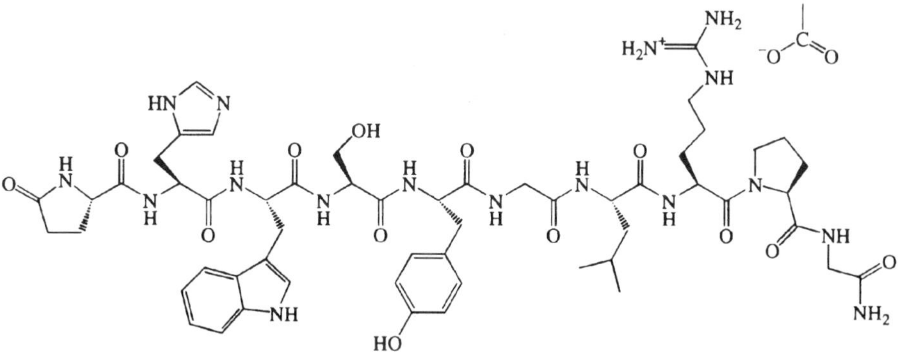

Molecular formula: $C_{55}H_{75}N_{17}O_{13} \cdot xCH_3COOH$

1.1.1. PRECLINICAL FINDINGS

A number of studies demonstrated that injection of LHRH into animals causes secretion of the gonadotropins luteinizing hormone (LH) and follicle-stimulating hormone (FSH) by the anterior pituitary (Vale *et al.*, 1977; Schally, 1978; Schally *et al.*, 1971c,d; Kastin *et al.*, 1972) subsequently resulting in trophic and steroidogenic effects on gonadal tissues. These data also demonstrate that responsiveness of rats to LHRH varies during the rat estrous cycle, being greatest in proestrus and estrus (Martin *et al.*, 1974; Gordon and Reichlin, 1974). This variation in sensitivity may be related to circulating estrogen levels (Kanematsu *et al.*, 1974) and the capacity of cycling animals to either produce antibodies or prevent the preovulatory surge of LH and FSH following administration of various doses of LHRH (Arimura *et al.*, 1973, 1974a, 1976; Koch *et al.*, 1973; Makino *et al.*, 1973; Fraser *et al.*, 1975). These findings suggest that LHRH may be effective in preventing ovulation and thus act as a potent contraceptive, or that chronic administration may be useful in modulating diseases mediated by the gonadotropins.

1.1.2. CLINICAL REQUIREMENTS

Clinical usefulness of any drug depends on a number of factors including safety, efficacy, pharmacokinetic profile, and *in vivo* stability. For LHRH and its analogues, survival against metabolizing enzymes in the body may be the single most important factor underscoring their potential as clinical candidates. Pharmacokinetic studies with tritiated and synthetically derived LHRH revealed that this decapeptide rapidly degrades in blood by enzymatic cleavage of the pGlu-His moiety, which is excreted along with some amount of the parent compound by the kidney (Redding *et al.*, 1973). Other studies demonstrated that the plasma half-life of LHRH is about 57 min in humans but only 7 min in the rat (Redding and Schally, 1973), suggesting that the degree of enzymatic deactivation of this compound may vary among species. For example, *in vitro* data using homogenates of rat and pig hypothalami demonstrated that this decapeptide might be easily cleaved (Griffith *et al.*, 1974; Koch *et al.*, 1974) between amino acid residues 6 and 7. Thus, therapeutic usefulness of LHRH would be limited by its short biological half-life and also by rapid inactivation by the liver, kidney, hypothalamus, and anterior pituitary gland (Sandow *et al.*, 1974). Stabilization of this peptide by chemical modification and possibly potentiation of its biological effectiveness would thus be imperative for it to become relevant in palliative treatment of diseases mediated by the gonadotropins.

1.2. Drug Candidate Selection

The rather weak potency of LHRH, coupled with its instability to metabolizing enzymes led in the early 1970s to intense research to synthesize analogues that would be stable while being significantly more effective than the natural hormone. Early studies showed that pGlu, His, and Trp played a functional role in the biological potency of LHRH and that simple substitutions or deletions in these positions decreased or abolished LHRH potency (Schally *et al.,* 1973; WHO, 1982). The data also demonstrated that considerable potency of the peptide is retained by substitution of these amino acids with other moieties possessing similar acid–base and hydrogen bonding capacity, or suitably oriented aromatic nuclei capable of generating similar electronic interactions. Structure–activity relationship (SAR) studies were thus initiated. These studies guided chemists to synthesize a series of compounds whose biological potencies were determined by selective rat pituitary assays. In this regard, biochemists, biologists, and pharmacologists considerably enhanced effectiveness of the drug discovery effort by providing rapid and timely bioassay support, an activity that was paramount in guiding the selection of drug candidates from thousands of analogues obtained via SAR studies. Table I lists some of the key agonists of LHRH deemed successful from a drug design and bioactivity standpoint. Results from these studies demonstrated that positions 2 and 3 were the preferred sites for substitution or deletion to generate inhibitory activity (Arnold *et al.,* 1974; Prasad *et al.,* 1976; Coy *et al.,* 1974a,b, 1975a, 1976; Geiger *et al.,* 1972; Fujino *et al.,* 1972a,b; Monohan *et al.,* 1973). Unfortunately, studies with the tripeptide pGlu-His-Trp or its amide yielded inactive compounds (Schally *et al.,* 1973) compared with LHRH, indicating that other residues on the molecule might be essential for biological activity of the peptide (Geiger *et al.,* 1972). For example, substitution of D-pGlu, Trp, β-pyrazolyl-3-Ala, and D-His at either position 1 or 2 yielded weakly active compounds with biological potencies ranging from about 8 to 50% relative to LHRH (Geiger *et al.,* 1972; Fujino *et al.,* 1972a; Monohan *et al.,* 1973). Moreover, substitution of various groups at position 3, e.g., 2-naphthylalanine, also yielded very weakly active compounds in bioassays for LH release compared with native LHRH (Geiger *et al.,* 1972; Fujino *et al.,* 1972a).

The failure to obtain very active compounds by substitutions at positions 1 and 2 led chemists to examine modification of LHRH in other positions on the molecule as alternatives to those earlier described. Substitutions at positions 4–10 yielded very active analogues in both LH release and ovulation studies. For example, substitution of D-Ala, D-Phe, and D-Trp at position 6 yielded compounds that were about 6 to 9 times more potent than native LHRH whereas substitution of des-Gly at position 10 yielded an agonist that was only 3 to 5 times as active as native LHRH (Arimura *et al.,* 1974b,c; Rippel *et al.,* 1975a; Vilchez-Martinez *et al.,* 1974; Fujino *et al.,* 1973a,b, 1974a; Coy *et al.,* 1975b; The Leuprolide Study Group, 1984; Meldrum *et al.,* 1982). These data thus prompted further elaboration of the molecule simultaneously at multiple sites. *In vitro* bioassay data as well as *in vivo* steroidogenic effects with a number of compounds having substitutions in positions 6 and 10 yielded superactive analogues of LHRH some of which are included in Table I. For example, D-Ser(Bu^t)^{6}desGly^{10} and D-Leu^6-desGly^{10} demonstrated biological potencies about 19 to 35 times greater than native LHRH. Phar-

Table I
Biological Potency of LHRH Analogues as a Function of Chemical Substitution

LHRH analogue	Biological potency relative to LHRH
D-pGlu^{1}-LHRH	8%
Trp^{2}-LHRH [a]	40%
(β-Pyrazolyl-3-Ala)^{2}-LHRH[a,b]	19%
D-His^{2}-LHRH[c,d]	10%
2-Naphthylala^{3}-LHRH [d]	52%
Pentamethylphenylalanine^{3}-LHRH[d]	34–70%
desGly^{10}-LHRH ethylamide[e–n]	300–500%
desGly-2,2,2-trifluoroethylamide-LHRH[m,n]	900%
D-Ala^{6}-LHRH[k,p]	600–700%
D-Phe^{6}-LHRH[k,p]	1000–1300%
D-Trp^{6}-LHRH[k,p]	1000–1300%
[D-Ser(Bu^{t})^{6}desGly^{10}]LHRH ethylamide (buserelin)	1900–3500%
[D-Leu^{6}-desGly^{10}]LHRH ethylamide (leuprolide)	3000–6000%
N-Me-Ser^{4}-D-Trp^{6}-leuprolide (Abbott-67062)	1500–3600%
p-Cl-Phe^{3}-leuprolide (Abbott -64926)	1500–3600%
Ac-Sar^{1}-N-Me-Ser^{4}-D-Trp^{6}-leuprolide (Abbott-67610)	1500–3600%
[D-Ser(Bu^{t})-^{6}AzGly^{10}]LHRH (Zoladex)	1900–3500%
[D-Trp^{6}-NMeLeu^{7}-desGly^{10}]LHRH ethylamide (Lutrelin)	1900–3500%

[a]Geiger *et al.* (1972).
[b]Fujino *et al.* (1972a).
[c]Coy *et al.* (1974a).
[d]Monohan *et al.* (1973).
[e]Arimura *et al.* (1974b).
[f]Rippel *et al.* (1975a).
[g]Vilchez-Martinez *et al.* (1974).
[h]Arimura *et al.* (1974c).
[i]Fujino *et al.* (1974a).
[j]Fujino *et al.* (1973a).
[k]Fujino *et al.* (1973b).
[l]Coy *et al.* (1975b).
[m]The Leuprolide Study Group (1984).
[n]Meldrum *et al.* (1982).
[p]Coy *et al.* (1974b).

macodynamic studies with [D-Leu^{6}-desGly^{10}]LHRH ethylamide in female cycling rats indicated that, like the d-Ser(Bu^{t})^{6}desGly^{10} analogue, it has a much prolonged *in vivo* activity (Vilchez-Martinez *et al.,* 1974; Arimura *et al.,* 1974c) compared with LHRH. This prompted further preclinical testing to promote [D-Leu^{6}-desGly^{10}]LHRH ethylamide, also called leuprolide, to clinical candidate status. These studies established that leuprolide consistently demonstrated 30- to 60-fold increased biological potency relative to LHRH. Also, doses as low as 0.0001 mg/kg sustained LH release (Coy *et al.,* 1974c, 1976; Karten and Rivier, 1986; Fujino *et al.,* 1974b; Rippel *et al.,* 1975b) in immature male rats over a 6-hr period. Similarly, FSH release increased by about 15-fold while its ovulation inducing activity rose to about 50- to 80-fold that of the parent hormone (Coy *et al.*, 1974b; Arimura *et al.*, 1974b,c; Rippel *et al.*, 1975a; Vilchez-Martinez *et al.*, 1974; Fujino *et al.*, 1974a). *In vitro* studies with anterior pituitary cells in monolayer cultures consistently confirmed the noted increases in LH- and FSH-releasing activity of leuprolide to the extent that these biochemical endpoints have now become standards for determining the potency of drug formulations for *in vivo* studies (Karten and Rivier, 1986). Similar studies utilizing subcutaneous injections of other superagonists such as [D-Ser(Bu^t)^{6}desGly^{10}]LHRH ethylamide (Hoechst's analogue buserelin) confirmed the position 6 and 10 criterion as demonstrated by leuprolide by releasing about 19 times as much LH and about 17 times as much FSH compared with a similar dose of LHRH (Lemay *et al.*, 1984).

The increase in biological potency of these peptides was clearly a primary objective in the SAR studies, but latency and *in vivo* stability was a close second (Karten and Rivier, 1986; Coy *et al.*, 1974c, 1975c; Fujino *et al.*, 1974b; Rippel *et al.*, 1975b; Sandow *et al.*, 1978; Dutta *et al.*, 1978; Corbin *et al.*, 1984). Dutta *et al.* (1978) demonstrated that LHRH analogues containing azaglycine in position 10 and a D-amino acid in position 6 are comparable to buserelin regarding ovulation induction in androgen-sterilized constant-estrus female rats. The increase in biological activity of the superactive LHRH analogues with substitutions in positions 6 and 10 was attributed either to enhanced binding to pituitary receptors or to a slower inactivation, or a combination of both factors (Monohan *et al.*, 1973; Corbin *et al.*, 1984; Besser, 1974; Marks and Stern, 1974). Monohan *et al.* (1973) also suggested that the greater activity of D-Ala^6-LHRH may be related to changes in conformation; the stabilized type II′ β-bend of this analogue apparently creates a greater affinity to pituitary receptors than is seen with LHRH. These may be the reasons why both [D-Ala^6-desGly^{10}]LHRH ethylamide and [D-Leu^6-desGly^{10}]LHRH ethylamide (Marks and Stern, 1974) are less readily degraded by brain enzymes than is LHRH.

Acute dosing of leuprolide by injection in the systemic circulation induced the release of LH and FSH from the anterior pituitary. It demonstrated a longer biological half-life in plasma so that chronic and long-term administration paradoxically desensitized the pituitary resulting in a reversible biochemical castration via downregulation of LHRH receptors. For this reason, leuprolide is effective therapy for many hormonally sensitive diseases such as prostatic carcinoma (The Leuprolide Study Group, 1984), endometriosis (Meldrum *et al.*, 1982; Lemay *et al.*, 1984), and precocious puberty as well as uterine fibroids. Like most LHRH analogues, leuprolide acetate causes regression of dimethylbenzanthracene (DMBA)-induced mammary tumors, reduced size of sex organs, and reduced gonadotropin and sex steroid levels in both males and females (Lemay *et al.*, 1984). It is noteworthy that leuprolide is blocked at both ends so that not only is its stability to metabolizing enzymes enhanced, but also it is readily stable in aqueous media ranging from pH ~4 to 7. The compound's chemical structure is shown below.

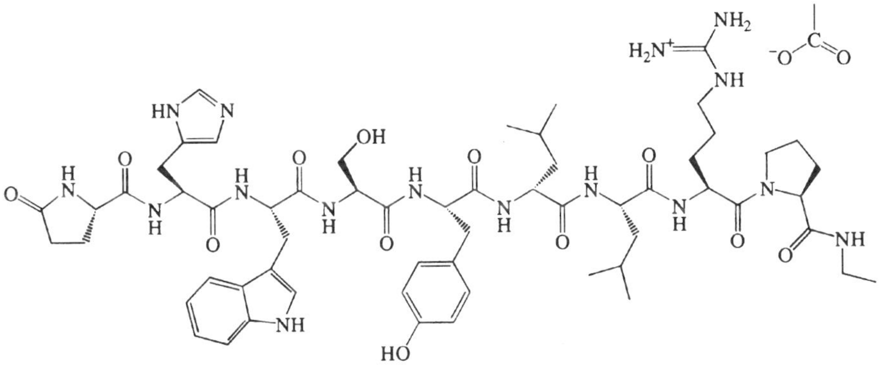

Molecular formula: $C_{59}H_{84}N_{16}O_{12} \cdot xCH_3COOH$

2. PHYSICAL CHEMISTRY AND CHEMICAL CHARACTERIZATION

Leuprolide has three ionization sites, namely, the imidazolyl nitrogen of histidine ($pK_a \approx 6.0$), the phenolic hydroxyl of tyrosine ($pK_a \approx 10.0$), and the guanidine nitrogen of arginine ($pK_a \approx 13.0$). The tryptophan moiety does not ionize in water and therefore does not participate in salt formation at typical formulation pH. Because the guanidine nitrogen is extremely basic, this peptide as synthesized exists in the protonated form and is generally associated with at least 1 mole of acetic acid. The compound therefore exists as an acetate salt, a hydrophilic ion pair ($-NH_3^+ \cdot -OOCCH_3$) that exists in ionized form across a wide pH range of physiologic interest.

The impact of salt form on partitioning behavior of leuprolide was investigated using alkyl sulfonic acids (C = 1, 4, 6, 8, 10), salicylic acid, acetic acid, and dehydrocholic acid (Adjei *et al.,* 1993). The distribution behavior was studied as a function of pH and counterion concentration. Results showed that methane and butane sulfonate did not help partitioning of leuprolide into octanol as a model lipid system for biomembranes although there is a slight improvement in lipophilicity of the drug with increasing pH. For the C_6–C_{10} alkyl sulfonates the partitioning increases significantly in the following order: hexane < octane < decane sulfonate (Adjei *et al.,* 1993). Data for salicylate and acetate indicated a marginal effect on partitioning of leuprolide. Similarly, results obtained for dehydrocholate showed no improvement in lipophilicity of the drug suggesting that the acid might be too weak (high pK_a) and may be sterically hindered from forming an effective ion pair. It was observed that increase in lipophilicity of various leuprolide ion pairs was proportional to the extent of ionization of the imidazolyl nitrogen of histidine, the type of counterion, and the number of lipophilic counterions per molecule. The data further demonstrated that lipophilicity of the ion pairs was proportional to pK_a

of the acid from which the anion was derived, i.e., sulfonic acid ($pK_a < 2.0$) > salicylic acid ($pK_a \approx 3.0$) > dehydrocholic acid ($pK_a > 6.0$), suggesting that alkyl sulfonate salts would favor nonaqueous dispersions of leuprolide in contrast to either acetate or cholate salts of the drug. For the alkyl sulfonate series a plot of log K (where K represents ion pair equilibrium constant) versus number of carbon atoms in the alkyl chain yielded a straight line with a slope of 0.5 per methylene group. This value is in good agreement with literature values of the Hansch π constant for a methylene group.

2.1. Bulk Drug Synthesis

From the mid-1970s the strategy of choice for preparing research quantities of LHRH agonists has been stepwise elongation by solid phase peptide synthesis (SPPS). This rapid assembly method served the discovery effort well as thousands of LHRH analogues were prepared and tested. Parallel to the prolific SAR effort was the increasing availability of new condensing agents, protecting groups, racemization suppressants, and solid-phase supports. A wide variety of strategies and orthogonal tactics emerged for assembly, deprotection, cleavage, and purification of LHRH analogues.

Compared with many of today's agonists or antagonists, leuprolide has a relatively simple sequence in that it contains all natural amino acids with only one residue in the D-configuration. The substitution of D-amino acids and C-terminal derivatization combined with various global protection schemes at the time provided synthetic challenges. To the uninitiated, these challenges might be considered synthetically trivial by today's standards, which now encompass the inclusion of unnatural amino acids, *N*-methylated amino acids, retrosequences, and so forth. However, the need to prepare clinical and commercial quantities of this drug constitutes a daunting task. Like most early LHRH agonists, leuprolide was first prepared by the standard SPPS technique utilizing dicyclohexylcarbodiimide-mediated coupling of t-Boc amino acids. These were assembled on a Merrifield resin but were later replaced by a benzhydrilamine resin. Acidolytic cleavage with HF was the method of choice for deprotection and cleavage. Purification by ion-exchange and gel-filtration chromatography subsequently completed the synthetic procedure.

Regrettably, yields from the SPPS technique revealed that this synthetic procedure was not viable for commercialization compared with standard solution chemistry techniques. For this reason, combined efforts by discovery and development personnel at Abbott Laboratories and the Takeda Chemical Company in Japan were initiated to evaluate several solution-phase routes including stepwise chain elongation and various segment condensation schemes. Early attempts to synthesize this nonapeptide were met with varying and sometimes disappointing degrees of success. Segment condensation schemes (e.g., attachment of pGlu-His, pGlu-His-Trp) encountered solubility problems, low yields, and racemization of the activated carboxyls, in some cases above 30%. Couplings at other locations along the peptide chain were met with more success, but with varying degrees of optical compromise and purification difficulties. Since that time, segments of the peptide have been successfully synthesized with various combinations of N-terminal, C-terminal, and side-chain protection schemes. Benzyloxycarbonyl and *t*-butyloxycarbonyl amino acids were the two commercially viable candidates for this process. The method included incorporation of benzyloxycarbonyl amino acids with dicyclohexylcarbodiimide-mediated couplings and minimal side-chain protection. The absence of reducible moieties or potential catalyst poisons in the sequence allowed the use of catalytic hydrogenation for deprotection during the synthetic process. This led to the development of a rapid, scalable, clean, and cost-effective process to efficiently manufacture bulk leuprolide acetate. As a result, leuprolide is now commercially synthesized from three segments, which are assembled stepwise from CBZ- and Boc-protected amino acids with minimal side-chain protection. Two segments (I and II) are combined to form an intermediate sequence, which, after saponification, is condensed with segment III to form the complete peptide sequence in a [(I + II) + III] scheme as illustrated below. The preparation of the three segments is carried out in 100- to 300-gallon glass-lined and stainless-steel reactors. Removal of the CBZ group at each step is achieved by palladium-catalyzed hydrogenolysis in 50-gallon reactors at about 45 psig. All but two of the intermediate peptides crystallize as solids and are isolated and fully characterized against appropriate standards. The scale ranges up to 40 kg for many of the intermediates.

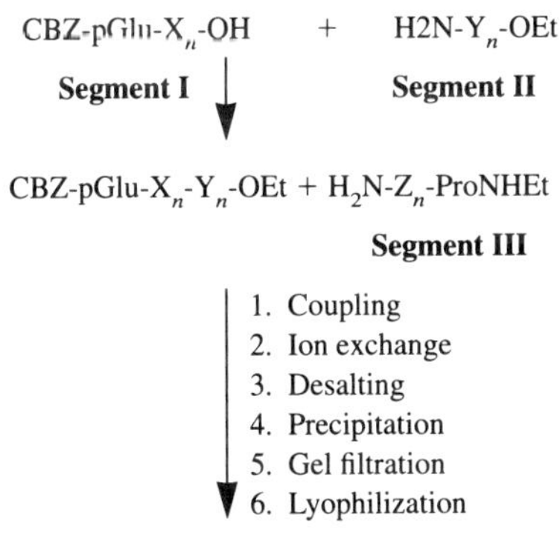

pGlu-His-Trp-Ser-Tyr-D-Leu-Leu-Arg-ProNHEt

Purification of bulk peptides usually requires some form of column chromatography. LHRH agonists or antagonists prepared by SPPS most often require reverse-phase chromatography to ensure adequate separation of optical isomers and other drug-related impurities. This is usually accomplished with various eluents or salt buffer systems. At a minimum it is necessary to ensure that the peptide is in the correct salt form, i.e., acetate for leuprolide, which is achieved with ion chromatography. In our laboratories, after ion-exchange chromatography, leuprolide is purified usually in 2-kg batches by desalting, precipitation, and gel-filtration chromatography. The product pool, now in dilute acetic acid, is passed through in-line cartridge filters to remove particulates and pyrogens. The solution is then immediately lyophilized in trays over about 3 days. The entire assembly, purification, and isolation process, which is comprised of 15 individual steps, including preparation of selected amino acid derivatives and intermediates, takes about 4 to 6 months if run in a stepwise sequence. It is noteworthy that the convergent segment strategy used for leuprolide acetate allows overall process time per batch to be minimized. This technique should thus be applicable to the synthesis of other LHRH analogues.

2.2. Manufacturing Controls

The traditional concept of a hormone is a substance, secreted by a ductless endocrine gland into the bloodstream, that has a stimulating, and in some instances a trophic effect on a target organ. However, the locality and diversity of actions by a hormone drug may be numerous (Roth *et al.,* 1982). Many neuroactive hormones and peptide growth factors, examples being the pituitary glycoproteins TSH, LH, and FSH, besides being naturally and intrinsically heterogeneous, are produced in cells distributed far more widely than the sites from which they were originally thought to be secreted. For this reason, some of these are believed to act as modulators of, e.g., neurological transmitters rather than as prime agonists. Furthermore, various precursor and metabolized forms of traditional peptide hormones could conceivably have their own characteristic biological effects (Roth *et al.,* 1982; Storring *et al.,* 1981; Robertson and Diczfalusy, 1977). For this reason, in considering procedures and controls for synthesis of LHRH analogues, two types of heterogeneity in preparations must be considered: that of artifacts (i.e., impurities) caused by extraction and that caused by handling (i.e., degradates) during manufacture and or storage. The identity of these undesirable compounds as well as their biological properties may be unknown, yet it is conceivable that small differences related to addition, substitution, deletion, or altered conformation in structure of the peptide could lead to a profound difference in pharmacological function. It is good scholarship then to question what toxicity tests may be applied to a synthetic peptide product, each batch of which contains an unknown mixture of unknown peptide analogues, in unknown quantities, and with unknown biological effects. All such heterogeneity gives rise to two questions. First, as the purity of a product can profoundly affect tests of identity, what are the extent and nature of impurities in a production batch? Second, what particular molecular mixture is to be called the "hormone": (1) in formal endocrinology, (2) for selection of material to use as a standard, and (3) for control and compendial specifications?

Because hormones are inherently heterogeneous and each isohormone often has a slightly different biological action, it is necessary to agree on a specific characterization technique, either a chemical method or a bioassay procedure that appropriately defines the mixture that is called the "hormone." Further, analogues of peptide hormones and other artifactual forms derived from biological extraction procedures or from synthesis need identification by characterization of their biological actions. Thus, theoretical and pragmatic differences between bioassays and ligand assays must be compared in order to have sufficient justification as to which method may be used for purposes of identification in pharmacopeias and other control specifications. All such tests rely heavily on the availability of attested reference materials such as those established by the International Conference on Harmonization (ICH, 1996). At this point, we note that peptide products that require definitions for compendia include those made from natural sources, those totally synthesized, those made by recombinant DNA procedures, and those made by combinations of such methods. For all such peptide hormone products, the nature and rigor of identification and the type and timing of control procedures should thus be related reasonably to the nature and use of the product. A manufacturer does (at least) those tests on his product that are required to show that it complies with specifications that were agreed to by the licensing authority when he received his product license; these include specifications for identity and purity of intermediates during production and in-process control. For LHRH, we now discuss tests developed by Abbott Laboratories to support bulk drug manufacture of the agonist leuprolide acetate. Pertinent impurities of the drug substance, standard control procedures, and multiple lot data are presented that demonstrate control of the synthetic process used to manufacture this drug.

2.3. Physical Characteristics and Methods

Although the current ICH Guidelines on impurities in new drug substances (ICH, 1996) do not apply to peptides, FDA has supplied Guidance for Industry as an informal communication under 21 CFR 10.90(b)(9) regarding submission of CMC (chemistry, manufacturing, and controls) information for the synthesis of peptide substances by solid or classical solution-phase methodology (Chiu, 1994). Like other LHRH agonists, standards for controlling the manufacture of leuprolide acetate must deal with the problems of activation, coupling, protection, removal of protecting groups, and methods of isolation and purification as these procedures will in some measure leave undesired by-products that must be eliminated or minimized. Therefore, to ensure the purity of leuprolide the following physical characteristics are examined.

2.3.1. PHYSICAL APPEARANCE

Lyophilized powders of leuprolide and other LHRH agonists are white. The occurrence of color in the bulk leuprolide acetate or dilute solutions of this peptide normally indicates the presence of one or more degradants or impurities introduced during processing. This drug contains an indole nucleus (i.e., tryptophan), which, like other readily oxidizable or UV-sensitive residues, degrades (Bodanszky, 1993) to yield colored quinone-like residues. These degradation products as well as monoxides, dioxides, kynurenines, and Schiff-base by-products may be identified by LC-MS and can be removed by treatment with activated charcoal just prior to final lyophilization. It is noteworthy that fines from decolorization carbons, filter-aids, chromatographic resins, or other air- or solvent-borne particulates may be introduced during bulk drug manufacture. These need to be eliminated or at least controlled by filtration through a 0.22 to 0.45-μm Millipore filter or equivalent. Clarity of leuprolide acetate for example is usually determined by inspecting a solution of 1% of the peptide in 1% acetic acid for the presence of color and insoluble particulates.

2.3.2. SPECIFIC ROTATION $[\alpha]_D$

Specific rotation is a measure of the combined optical integrity of all chiral centers in the molecule and hence may be related directly to the purity of the bulk drug substance. Low levels of epimerization at one or more chiral centers may not have a significant impact on the optical rotation to make it fall outside a specified range of ± 2–3° or standard deviations. The specific rotation of leuprolide acetate is, however, directly affected by pH or the presence of residual solvent, and as such care must be taken to evaluate this drug or its isolated intermediates under consistent conditions.

2.3.3. PARTICLE SIZE

Lyophilization of leuprolide acetate yields heterogeneous powders thus requiring milling as a final step in the manufacturing process. Although not producing a state of complete homogeneity, milling to certain particle size requirements helps ensure that samples for analytical testing are representative of the batch. This drug, like most LHRH agonists, is hygroscopic and as such requires much care to protect against moisture pickup during milling. Parenteral dosage forms of the drug utilize aqueous media and thus particle size requirements may be of little concern. However, formulation presentations such as inhalation aerosols requiring particle sizes between 1 and 10 μm for deep lung penetration may call for rigid controls of the bulk drug particle size to ensure dose uniformity.

2.4. Chemical Characterization and Methods

Synthetic procedures used for leuprolide acetate require complex chemistries as well as time- and resource-intensive purification methods. Initial development processes for this drug produced yields in the 20 to 50% range. Separation of degradants and other-drug related impurities proved to be a formidable effort. In particular, chiral separation of optical and isomeric forms as well as purification of the bulk drug from its fragmented forms, required sophisticated analytical tools many of which needed to be installed on line as integral components of the sequential assembly of the peptide. Of these, chromatography proved to be most beneficial in helping to resolve the final bulk drug substance from its impurities. Optical isomers, notably impurities introduced from raw materials or formed during activation, coupling, deprotection, cleavage, or on isolation and workup, were identified by thin-layer and high-performance liquid chromatography (TLC/HPTLC) as well as mass spectral techniques such as fast atom bombardment (FAB-MS), electrospray (ES-MS), plasma or laser desorption, and nuclear magnetic resonance spectroscopy (^{1}H-NMR and ^{13}C-NMR). It is noteworthy that not all of these techniques will be employed to characterize other LHRH agonists, but the selected tests should at least provide adequate information about the entire covalent structure of the molecule. Several of these techniques have been reviewed elsewhere (Adjei and Hsu, 1993). Although racemization is often assumed to be minimal on a small scale and controllable on a commercial scale, it has recently been demonstrated that there is a need for multiple analytical methods for the assessment of optical purity (Malspeis *et al.,* 1984) as levels of just a few tenths of a percent may be considered unacceptable from a biological activity or toxicity standpoint. The discussion as to the level of impurities required for peptides, synthetic or isolated, is ongoing. Unknowns for general organic pharmaceuticals are required to be controlled at or below the 0.1% level. Whether or not this specification limit must be applied to peptides remains to be clarified. It is clear, however, that good science must be used to keep impurities at their lowest possible and practical levels in order to simplify toxicological requirements for the bulk drug substance.

2.5. Moisture and Acetic Acid

Leuprolide acetate, like most peptides, is hygroscopic, and this presented serious challenges during bulk drug manufacture from a stability and drug potency standpoint. Analysis of variable drug potencies as a function of age, batch size, and impurity profile demonstrated that the peptide backbone retained bound and unbound molecules of water by hydrogen bonding. The results also demonstrated that counterions paired to acidic or basic side chains and terminal functional groups were often present to various degrees often as high as 10 to 20% and that this was based on environmental conditions (i.e., humidity and temperature) during assembly of the peptide. For this reason, a clear delineation of the relationship between total weight and grams of activity was needed for accurate assessment of drug potency. The impact of humidity was reduced significantly by installing controlled environmental conditions in both manufacturing as well as analytical test areas. Sample weighing chambers, namely, glove boxes furnished with nitrogen purge lines, were utilized. Drug assay methods were also developed that took into account a correction for moisture and acetic acid. Moisture levels were determined coulometrically such as the type provided by the Aquatest™ IV Karl Fisher titrimetric procedure. A method for acetic acid comprising ion chromatography equipped with FID and automatic temperature programming was also included in specifications. The instrumentation included a Waters 6000A pump with Schoeffel 770 UV spectrophotometer, Spectra-Physics SP4100 integrator/recorder, and either a Superox-FA (30M $\times$ 0.53 mm) or a 1.2-μm film polyethylene glycol ester column or equivalent. A typical injection volume for the assay was about 1.0 μl with oven, injector, and detector temperatures of 100, 200, and 250°C, respectively. Integration of these methods enhanced the purity of leuprolide to about 96%, the balance largely being excess free acetic acid from the salting step in the manufacturing process.

2.6. Amino Acid Analysis

Utilization of the solution-based synthesis procedure for leuprolide unleashed a number of purification problems during scaleup from pilot- to production-size quantities of this peptide. In particular, a variety of peaks were found in samples of the drug. Concentrations of these materials varied from batch to batch implying that controls of the process, as well as sources of starting and or intermediate materials may be implicated. To resolve the process, many of the bulk substance contaminants were first separated by HPLC and LC-MS. Results revealed the presence of several optical isomers with similar molecular weight thus requiring identification by alternate methods. This was accomplished by preparing samples of the unknown isomers using solid-phase synthesis followed by spiking of the synthetic materials into the respective HPLC chromatograms and LS-MS spectra. The characterization of the final bulk drug substance was accomplished by amino acid analysis to yield sequence information and enantiomorphic identities of each contaminant. It is noteworthy that during the hydrolysis of peptide sequences, natural and unnatural amino acid residues could degrade to varying extents impacting on recovery efficiency. For leuprolide, the tryptophan and serine residues were found to easily degrade by oxidation while the pyrrolidone ring of pyroglutamic acid would often open to yield glutamic acid. These problems were resolved with appropriate standards and hydrolysis conditions that enhanced the reliability of the calculated molar ratios. Other hydrolysis media utilized in some of these studies included alkali and alkyl or aryl sulfonic acids but these were found to be reactive with other amino acids such as Arg and Ser and were therefore abandoned.

3. FORMULATION CHEMISTRY OF LEUPROLIDE ACETATE

As discussed earlier, systemic concentrations of leuprolide induce the release of LH and FSH from the anterior pituitary. Like most LHRH analogues, this drug possesses a long biological half-life in plasma and chronic administration paradoxically desensitizes pituitary receptors resulting in what is often described as reversible biochemical castration. This biological effect was used as a pharmacological marker and pharmacodynamic endpoint to investigate a number of formulation presentations for human and veterinary use. The studies utilized various animal models in an effort to uncover any species-specific or immunogenic differences that might be present. The studies also evaluated the impact of a number of formulation variables on both stability as well as bioavailability of leuprolide. Some selected aspects of these earlier studies are summarized below.

3.1. *In Vitro* Studies

Like most peptides, stability of leuprolide in liquid formulations was a critical product development issue shortly after this peptide was elevated to clinical candidate status. Preformulation and formulation studies were therefore initiated. These evaluated survival of leuprolide in typical formulation media and its compatibility with packaging components as well as device systems. The studies included interfacial phenomena and surface energetics of the drug in relation to aggregation and fragmentation of the peptide when formulated with typical pharmaceutical excipients. Because leuprolide is not orally active, most of these early presentations included aqueous and nonaqueous liquids as well as semisolids for use as injection products. Several of these formulations contained stabilizers and dispersants, examples being lipophilic ion pairs, polylactic/polyglycolide, gelatin, D-mannitol, and semipolar emulsified surfactant systems. Emulsions containing components were chosen as to their impact on drug lipophilicity at varying pH and ion-pair concentration. One such formulation consisted of 5 mg/ml leuprolide acetate, 2 mg/ml decane sulfonic acid, up to 10% water, and about 1 to 5% Emulphor EL-719 as surfactant. This formulation used a mixture of ethyl alcohol and safflower oil as the nonaqueous fraction. Gelatin-based microspheres dispersed in a water-in-oil carrier system were also explored. These formulation presentations used safflower oil as a vehicle of choice. *In vitro* stability studies were conducted as a function of time and temperature with the bulk lyophilized drug powder as a control. Results demonstrated satisfactory stability of the drug at least through the duration of *in vivo* studies. Both real-time stability data as well as Arrhenius kinetic projections yielded product T_{90}s (time taken for leuprolide potencies to fall to 90% of initial values) ranging from about 3 to 48 months, suggesting that unlike most peptides, solid or semisolid and liquid formulations of leuprolide acetate would be commercially viable.

3.2. *In Vivo* Studies

In vivo studies utilized various animal models (e.g., dogs, rats, New Zealand rabbits, monkeys, and pigs) to assess safety as well as pharmacokinetics (i.e., distribution, metabolism, and elimination) of leuprolide following acute and chronic administration of formulations. For example, several groups of crossbred finishing pigs consisting of boars and barrows approximately 120 days old with a weight range of about 175–200 lb were selected for bioavailability and pharmacodynamic studies to support veterinary applications. Blood sampling regimens were on the order of a few hours to several days after drug administration. Concentrations of leuprolide in these samples were determined bioanalytically. The tests utilized receptor-binding activity to monitor leuprolide concentrations, but in some cases LH, FSH, estradiol, and testosterone release were measured as surrogates of pharmacological activity of leuprolide. As many of these tests involve receptor-binding assays, rat pituitary plasmas were prepared as a source of LHRH receptors. Because the tests require tracers, [^{125}I]leuprolide was prepared by the cloramine-T method and purified by ion-exchange chromatography over carboxymethylcellulose. ^{125}I-labeled [Tyr^5]leuprolide and an antibody capable of recognizing the tripeptide antigenic determinant X-Leu-Arg-Pro-NHEt were then coincubated at 4°C and after equilibration, the fraction of bound tracer was separated by centrifugation. The resulting pK_i values, being the negative log of the equilibrium dissociation constant, were then determined and used as estimates of *in vivo* activity of the parent drug.

In one such study, a single i.m. dose of 50 μg/kg leuprolide acetate was administered to pigs from sustained-release formulations comprised of an emulsion and a microsphere oil dispersion. Maximal serum drug concentrations determined by RIA of the oil suspension were about 15–23 ng/ml compared with an average of about 31 ng/ml for the microsphere system. Results with both formulation types indicated a rapid burst of serum leuprolide during the first hour of drug administration. Peak serum drug concentrations, C_{max}, occurred at approximately 10–30 min postdosing after which rapid elimination may have contributed to a lowering of systemic concentrations beyond the limit of detection by about days 5–7 of the

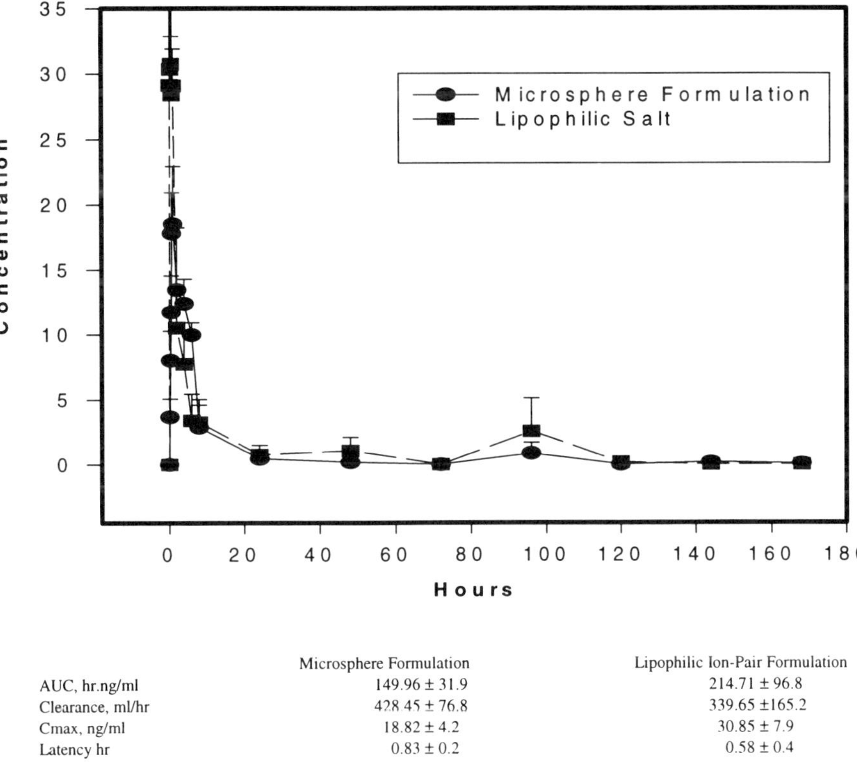

	Microsphere Formulation	Lipophilic Ion-Pair Formulation
AUC, hr.ng/ml	149.96 ± 31.9	214.71 ± 96.8
Clearance, ml/hr	428.45 ± 76.8	339.65 ±165.2
Cmax, ng/ml	18.82 ± 4.2	30.85 ± 7.9
Latency hr	0.83 ± 0.2	0.58 ± 0.4

Figure 1. Plasma concentrations of leuprolide following administration of oil dispersion.

study. Mathematical analysis of the serum data yielded half-lives for the drug in the range of 2–4 hr with a serum latency of 0.50 ± 0.17 hr for the oil suspension compared with 0.58 ± 0.42 hr for the microsphere dispersion. Serum durations of about 120–170 hr and clearance rates of approximately 430 ml/kg per hr were predicted for both formulations. The RIA for leuprolide concentrations in the serum samples are summarized in Fig. 1. Findings from this study suggested that the subject formulations sustained serum levels above 1 ng/ml for at least 8 hr except in group 1–3 pigs where drug concentrations fell below the detectable limit after the first day of drug administration. These results are generally lower than data typically observed in humans after i.m. administration of Lupron Depot®, but it is noteworthy that systemic concentrations of leuprolide remained above the limit of quantitation for at least up to 7 days postdosing, suggesting either presentation may need significant refinement to be clinically viable. The rapid distribution of leuprolide from porcine serum about 24 hr postdosing for essentially all of the pigs used in the study is unexpected because the formulation is designed to provide constant release of drug from the site of administration. Furthermore, the results do not correlate with data obtained after administration of the second dose on day 21; serum leuprolide concentrations on day 21 were somewhat higher for group 2 (50 μg/kg) than for group 3 (200 μg/kg). These observations suggest that there may be a faster distribution and systemic clearance of leuprolide in pigs compared with humans, or that quasi-dose-dependent pharmacokinetics may be involved in the *in vivo* release of leuprolide from these two formulations.

Absorption and distribution of peptide drugs in the body may vary based on the port of drug entry to the body. This could have a significant impact on clinical usefulness of these drugs especially in those cases where absorption is largely limited by deactivation at the delivery site. In considering veterinary applications for leuprolide, a number of injection sites were explored in pigs. In one such study, multiple doses of a slow-release formulation were administered by i.m. injection into the neck, rear leg, and gonads. Serum AUC data with 0, 50, and 200 μg/kg leuprolide from this study are summarized in Fig. 2. Of the three injection sites investigated, the mean AUC data estimated at all dosages following drug administration to the neck were the lowest. There was some aberration in the serum data for the two remaining injections sites, namely, rear leg and gonads, with plasma durations and latencies favoring the gonads (Fig. 2). This apparent site-dependent pharmacokinetic phenomenon for leuprolide in the pig implied that serum latency may be several days by i.m. injection into the gonads compared with approxi-

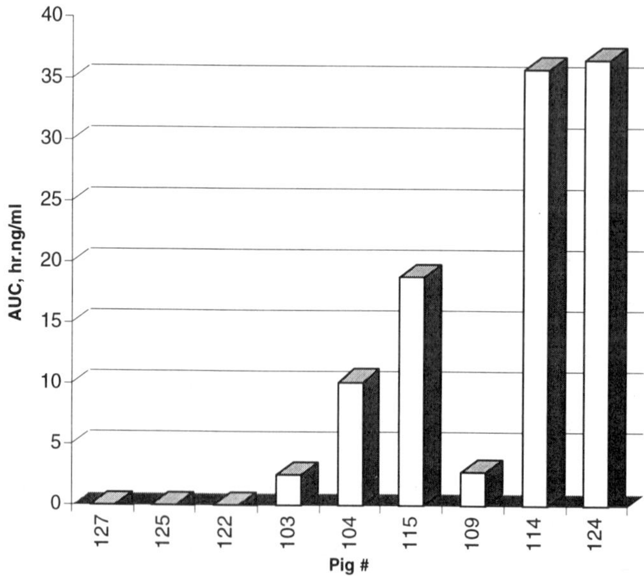

Figure 2. Impact of dose on AUC through 42 days after injection of leuprolide in pigs.

mately 2 days via the neck or rear leg. Pathology and organoleptic data generated at the conclusion of this study indicated the absence of odor-producing compounds during steroidogenesis particularly in animals dosed with leuprolide via the gonads. Results of this analysis implied that injection site as well as the dose of leuprolide are critical for efficacy. Pharmacodynamic studies were therefore initiated. Data from these studies indicated significant reduction in plasma testosterone and a sufficient delay in testicular development in the pig at dosages ranging from 25 to 100 μg/kg. These data suggest that i.m. administration of Lupron Depot® in the neonatal boar may provide a useful preventive measure in certain veterinary applications requiring either elimination of odors in the meat or downregulation of pituitary receptors so as to control sexual function and weight gain. In humans, a number of studies were conducted to investigate the effectiveness of injection, nasal, and inhalation presentations of leuprolide to regulate endogenous LHRH. Tests with injectable dosage forms were very successful and have since been used to secure worldwide claims for leuprolide in the palliative management of prostatic cancer, endometriosis, and uterine fibroids. Other applications involving precocious puberty and *in vitro* fertilization may not be far from being granted, at least in the United States.

4. CLINICAL DEVELOPMENT

Having demonstrated satisfactory pharmacological and toxicological profiles of a drug in various formulation presentations, one is faced with the many challenges of the clinical development process, i.e., safety, efficacy, and control of the drug product. Of these, the most daunting activity involves scaleup and reproducibility of the manufacturing process used for the bulk drug substance. The following sections summarize primary manufacturing controls established for leuprolide acetate as a model LHRH agonist.

4.1. Standards and Controls

The advent of supersensitive analytical techniques such as GC-MS and FAB-MS has enhanced the chemist's capabilities to detect impurities in bulk drug substances at levels that were unimaginable only 20 or so years ago. Rigorous testing and tighter specifications are now required for registration of new clinical candidates as the quality and depth of standards should be commensurate with the state of technology. Requirements for standards and controls on leuprolide acetate have increased steadily dating back to April 1985 when it was first approved for commercial use in the United States. This time-dependent tightening of specifications, as expected, should be typical for any bulk drug as processes are refined and controls associated with manufacture become more predictable. Release tests for identity of bulk leuprolide acetate mirror USP requirements and these include (1) comparison of the HPLC retention time to that obtained for a reference standard, (2) amino acid analysis by ion chromatography, and (3) comparison of the sample's IR spectra to that of a reference standard. Methods for bulk drug purity testing include optical rotation and HPLC assays for potency and total impurities that may be present. As discussed earlier, levels of moisture and acetic acid are monitored by coulometric Karl Fisher titration and gas chromatography, respectively. Inorganic impurities and endotoxin levels are minimized by controls in the manufacturing process. Analysis of these impurities is done by sulfated ash and the limulus amebocyte lysate methods. A summary of essentially all of the current compendial tests and specifications for leuprolide acetate is presented in Table II.

Table II

Comparison of Release Specifications and USP Specifications for Leuprolide Acetate

Specification	Release	USP
HPLC assay[a]	97.0–103.0%	97.0–103.0%
Appearance	Fine powder or fluffy material	
Color	White to off-white	
Solution (1% acetic acid)	Practically clear; practically colorless	
Identification (IR)	Qualitatively identical to standard	Qualitatively identical to standard
Identification (AAA)[b]	See Table V	See Table V
Moisture by Karl Fisher	NMT[c] 5%	NMT 5%
$[\alpha]_D$ (1% acetic acid)	−42.0 to −38.0°	−42.0 to −38.0°
Residue on ignition	NMT 0.3%	
Identification (HPLC)	Retention time similar	Retention time similar
Total impurities (HPLC)	NMT 2.5%	NMT 2.5%
Unknowns	No single impurity greater than 0.5%	No single impurity greater than 0.5%
Acetyl-leuprolide	NMT 1.0%	NMT 1.0%
[D-His^2]	NMT 0.5%	NMT 0.5%
[L-Leu^6]	NMT 0.5%	NMT 0.5%
[D-Ser^4]	NMT 0.5%	NMT 0.5%
Acetic acid (GC)	4.7–9.0%	4.7–9.0%
Bacterial endotoxin	NMT 16.67 IU/ml	NMT 16.67 IU/ml
pH	5.5–7.5	
Particle size (mean)	NMT 285 μm	
Particle size (90% v/v)	NMT 525 μm	
Turbidity	NLT[d] 90%	
Microbial limit test	LT[e] 50 microorganisms/g	

[a]Calculated relative to the anhydrous free base.
[b]Amino acid analysis.
[c]NMT, not more than.
[d]NLT, not less than.
[e]LT, less than.

4.2. Physical and Chemical Characterization

Validated analytical methods combined with adequate manufacturing process controls ensure run-to-run consistency and reproducibility of leuprolide acetate bulk drug. Although each lot must pass all testing before release for dosage form preparation, certain critical tests are needed for early assessment of the overall quality of a batch. Therefore, HPLC analysis for potency and impurity levels as well as moisture and acetic acid content are generally run first.

4.2.1. ANALYTICAL METHODS AND RESULTS

An improved HPLC method able to discriminate between most of the optical isomers has recently been submitted in monographs to the U.S. and European Pharmacopoeia for approval utilizing a pH 3.0 triethylammonium phosphate mobile-phase buffer. The instrumentation and test procedures include the following: Spectra-Physics model SP-8800 quaternary pump and SP-8880 autosampler equipped with a model SP-200 detector. The instrument uses a YMC-Pack, 3μ ODS-A (4.5 mm × 100 mm) column, and the mobile phase is a buffer consisting of a 2:3 mixture of 150 mM triethylammonium phosphate and 85:15 *n*-propanol-acetonitrile at a pH of about 3.0. The flow rate is approximately 1.0–1.5 ml/min and the detector wavelength is 220 nm, 0.2 AUFS. A typical integrator setting is as follows: attenuation 4, 20 μl injection volume, and 0.5 cm/min chart speed. The sample and the standard solutions are usually tested at concentrations of approximately 1.0 and 0.01 mg/ml in the mobile phase, respectively. Under these conditions the limit of detection of the assay is approximately 1.0 μg at impurity levels of about 0.1%. A typical chromatogram obtained for leuprolide acetate using this analytical procedure is shown in Fig. 3. Representative HPLC data that demonstrate lot-to-lot uniformity of leuprolide acetate are presented in Table III. Key optical isomers and other drug-related impurities have been isolated and/or synthesized such that their identity can now be determined by their relative retention times. Table IV compares real and relative retention times for several stereoisomers of leuprolide acetate utilizing the improved HPLC method described earlier. A typical chromatogram showing resolution of these point isomers is presented in Fig. 4. Although this HPLC method sufficiently demonstrates control of the bulk drug manufacturing process, further method refinement may be necessary to resolve [D-pGlu^{1}]leuprolide from [D-His^{2}]leuprolide, and [D-Ser^{4}]leuprolide from [D-Leu^{7}]leuprolide, which coelute with this method.

4.2.2. SPECIFIC ROTATION $[\alpha]_D$

The specific rotation for leuprolide acetate is also corrected for and reported on the anhydrous acetic acid free basis. A clear, colorless solution is made by dissolving 50 mg of the peptide in 5.0 ml of 1.0% acetic acid at 25°C. The solution is placed in

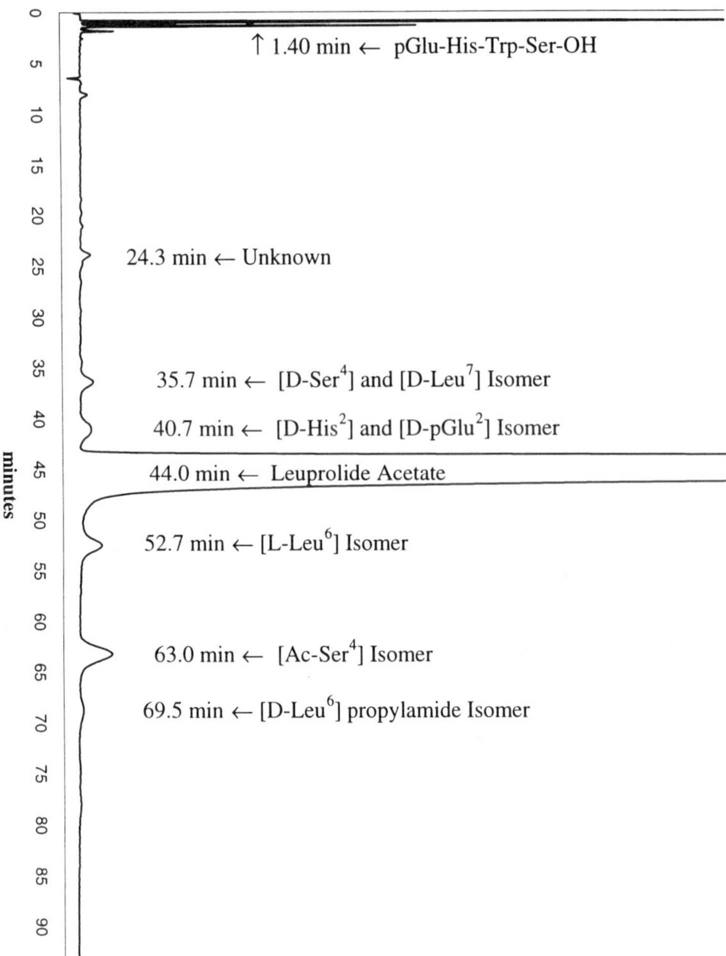

Figure 3. Typical chromatogram of leuprolide acetate.

a 1-dm micropolarimeter cell and the angular rotation measured using sodium light. The resulting angular rotation is then compared with a 1% acetic acid blank. Results for the specific rotation of leuprolide acetate are generally between −42.0 and −38.0°. Typical data for representative production batches are included in Table III.

4.2.3. AMINO ACID ANALYSIS

The HPLC method for i.d. testing of leuprolide acetate (see USP) is complemented by amino acid analysis. Data for the functional amino acids may be ob-

Table III
Typical Analytical Release Data for Lots of Leuproı ıate

Lot run no.	Moisture (%)	Acetic acid	Total impurities	[Ac-Ser[4]]	[D-His[2]]	[L-Leu[6]]	[D-Ser[4]]	Largest single unknown impurity	[α][a]	pH	ROI
A	1.2	7.6	1.6	0.5	0.2	0.3	0.1	0.2	−39.0	5.7	0.0
B	1.3	7.5	1.5	0.5	0.2	0.3	0.1	0.3	−40.6	5.9	0.0
C	1.0	7.6	1.5	0.5	0.2	0.3	0.1	0.2	−40.4	5.8	0.0
D	2.2	7.4	1.7	0.5	0.2	0.3	0.1	0.3	−41.1	5.8	0.0
	1.7	7.6	1.8	0.5	0.2	0.3	0.1	0.3	−40.1	5.8	0.0
E	1.0	8.0	1.8	0.6	0.2	0.3	0.1	0.3	−40.0	5.7	0.0
F	1.6	7.4	1.7	0.6	0.2	0.3	0.1	0.3	−40.1	5.8	0.0
G	1.0	7.8	1.5	0.5	0.1	0.2	0.1	0.2	−40.1	5.7	0.0

[a]Calculated on the anhydrous acetic acid free basis.

Table IV
Stereoisomers of Leuprolide Acetate in Real Time (RT) and Relative Retention Times (RRT)

	Optical isomer	RT	RRT
1	[D-Trp^3] leuprolide acetate [a]	29.7	0.68
2	[D-His^2, D-Ser^4] leuprolide acetate [b]	32.0	0.73
3	[D-Ser^4] leuprolide acetate [b,c]	35.7	0.81
4	[D-Leu^7] leuprolide acetate [c]	35.7	0.81
5	[D-pGlu^1] leuprolide acetate [d]	40.7	0.93
6	[D-His^2] leuprolide acetate[b,d]	40.7	0.93
7	Leuprolide acetate	44.0	1.00
8	[Leu^6] leuprolide acetate [b]	52.7	1.20
9	[D-Tyr^5] leuprolide acetate [a]	59.5	1.35
10	[Ac-Ser^4] leuprolide acetate[b]	63.0	1.43
11	[D-Pro^9] leuprolide acetate[a]	70.9	1.61
12	[D-Arg^8] leuprolide acetate[a]	94.8	2.15

[a]An unobserved isomer.
[b]An observed impurity.
[c][D-Ser^4]- and [D-Leu^7] leuprolide acetate coelute in this HPLC method.
[d][D-pGlu^1]- and [D-His^2] leuprolide acetate coelute in this HPLC method.

tained by ion chromatography. Typical results from this test for a production batch of the drug are shown in Table V. These results confirm that the respective amino acids are present in their correct molar ratios. The method description is as follows: Leuprolide acetate, approximately 65 mg, is hydrolyzed in 2.0 ml of 6 N HCl for 16 hr at 120°C in an evacuated hydrolysis tube. An aliquot of the amino acid hydrolysate is taken to dryness by lyophilization or equivalent technique and is subsequently reconstituted into 10 ml pH 2.2 citrate buffer. The mole ratio of each amino acid and ethylamine is obtained by comparison of the respective peak responses against those of a standard mixture of pure amino acids. The analysis may be performed on a Durrum (Dionex) D-500 amino acid analyzer or its equivalent. The system includes a DEC PDP-8/M computer, an ASR-33 teletype, and a Honeywell "Electronik" Model 196 recorder.

4.3. Pathology and Toxicology

Toxicologically, most LHRH agonists are considered to be very safe pharmaceuticals. These peptidic drugs are potent inhibitors of gonadotropin secretion following chronic administration of therapeutic doses of about 0.005–0.010 mg/kg per day. This leads to suppression of ovarian and testicular steroidogenesis but the effect is reversible on discontinuation of drug therapy. For example, in humans, s.c. administration of leuprolide acetate has been shown to result in an initial in-

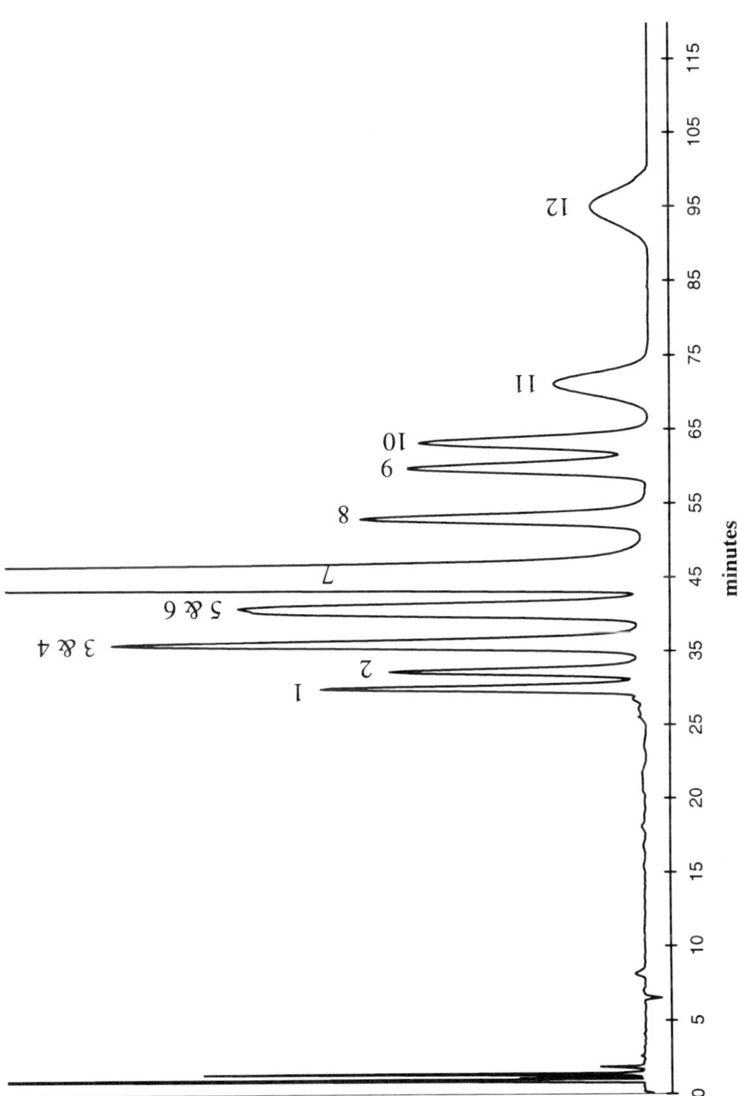

Figure 4. Resolution of point isomers of leuprolide acetate.

Table V
Amino Acid Analysis of the Acid Hydrolysate of Leuprolide Acetate

	Mole ratio		
Amino acid	Found	Theory	Specification
Ser	0.91	1.0	Present
Glu	1.02	1.0	0.85–1.1
Pro	1.00	1.0	0.85–1.1
Leu	2.01	2.0	1.8–2.2
Tyr	1.02	1.0	0.85–1.1
Trp	0.77	1.0	Present
His	1.02	1.0	0.85–1.1
Arg	1.04	1.0	0.85–1.1
Ethylamine[a]	0.99	1.0	

[a]Not an amino acid, but derived from the C-terminal proline ethylamide residue.

crease in circulating levels of LH and FSH subsequently leading to a transient increase in levels of gonadal steroids, namely, testosterone and dihydrotestosterone in males, and estrone and estradiol in females. Chronic administration, however, results in decreased LH and FSH causing biochemical castration in males and reduction of estrogen to postmenopausal levels in premenopausal females. These decreases are seen after about 2–4 weeks of therapy with durations of treatment up to about 5 years.

Besides their impact on the gonads, very few significant clinical events have been reported in the literature for LHRH analogues. A rare evidence of anaphylactic reaction to synthetic GnRH (Factrel®) has been reported (Malspeis *et al.*, 1984). Pharmacologically, these drugs may be contraindicated in pregnancy or in females expecting to become pregnant, as doses of 0.00024 to 0.024 mg/kg leuprolide acetate (i.e., about 1/600 to 1/6 of the human dose) administered to rabbits on day 6 of pregnancy produced dose-related increases in fetal abnormalities (Package Insert, 1996). These fetal malformations were not demonstrable in rats although the data showed increased fetal mortality and decreased fetal weights at high doses in both rats and rabbits. Toxicological events such as these are not surprising as they are logical consequences of the alterations in hormonal levels brought about by LHRH agonists.

A number of mild adverse events have been reported for leuprolide acetate following chronic administration to humans (Package Insert, 1996). Transient increases in testosterone after about a week or so of therapy produced bone pain and headache in a small number of patients. In a few cases, temporary worsening of existing hematuria and urinary tract obstruction occurred as well as temporary weakness and paresthesia of the lower limbs. Incidents relating to decreased libido and hot flashes have also been reported. Overdose of up to 500 times the human dose resulted in dyspnea, decreased activity, and local irritation at injection sites, but dosages of up to 20 mg/day for 2 years caused no adverse effects differing from those observed with a 1 mg/day dose regimen in rats. Data from clinical studies revealed that in general, drug-related reactions following prolonged use of leuprolide injections were comparable to those reported for diethylstilbestrol (DES) (Package Insert, 1996). In all cases, however, these clinical events were related to physiological effects of the drug, i.e., decreased testosterone. In less than 5% of patients, adverse reactions were reported involving the cardiovascular system (angina, cardiac arrhythmias, myocardial infarction, pulmonary emboli), gastrointestinal system (diarrhea, dysphagia, gastrointestinal bleeding, rectal polyps), endocrine (decreased libido, thyroid enlargement), skeletal system (joint pain), central nervous system (anxiety, blurred vision, lethargy, memory disorder, mood swings, nervousness, paresthesia, neuropathy, syncope/blackouts, taste disorder), and the integument (skin carcinoma, dry skin, ecchymosis, hair loss, itching). Again, the reported clinical events were somewhat comparable to systemic reactivity to chronic DES therapy.

4.4. Clinical Pharmacokinetics and Pharmacodynamics

As discussed earlier, oral bioavailability of LHRH agonists is poor with results generally on the order of about 0.05% relative to i.v. administration. Subcutaneous presentations of these compounds, exemplified by leuprolide acetate, are equivalent to i.v. with both routes having a plasma half-life of about 3 hr. Like most peptides, liver metabolism is extensive although catabolism, distribution, and excretion pathways have not been convincingly demonstrated in humans. In a number of clinical studies, single doses of 7.5 mg Lupron Depot® by i.m. injection to healthy male volunteers yielded a characteristic initial increase in plasma concentrations of the drug. C_{max} values ranged from 4.6 to 10.2 ng/ml at 4 hr postdosing. Unfortunately, intact leuprolide and inactive metabolites could not be quantified by the assay. Following the initial rise, leuprolide concentrations started to plateau within 2 days after dosing and remained relatively stable in the range of 0.3 to 0.9 ng/ml for about 4 to 5 weeks. Studies have also shown that the mean steady-state volume of distribution of leuprolide following i.v. administration to male volunteers (Package Insert, 1996) is about 27 liters. *In vitro* binding to plasma proteins is high and ranges between 43 and 49% in healthy male subjects. Intravenous administration of 1-mg bolus dosages revealed a mean systemic clearance of about 7.6 liters/hr, with a terminal elimination half-life of about 3 hr based on two-compartment pharmacokinetic modeling.

In rats and dogs, administration of [^{14}C]leuprolide acetate yielded smaller inactive fragments, namely, a pentapeptide (metabolite I), tripeptides (metabolites II and III), and a dipeptide (metabolite IV), these metabolites being further catabolized to still other smaller inactive fragments. In prostate cancer patients, only metabolite I was quantifiable by the RIA method. Peak plasma concentrations, approximately 6% of the C_{max} of the parent drug, occurred at about 2–6 hr postdosing with the levels falling to about 20% that of leuprolide 1 week later (Package Insert, 1996). Less than 5% total of the parent and metabolite I was recovered in the urine after administration of the 3.75-mg i.m. product. However, this excretion study employed only five prostate cancer patients and the results may thus be equivocal.

5. CONCLUSIONS

Our discussion thus far indicates that by integrating the physical, biochemical, and clinical sciences, a number of potent analogues of LHRH, for example leuprolide, were identified as clinically relevant, safe, and effective therapeutic agents for ameliorating diseases mediated by the gonadotropins. Purity standards for this class of peptide drugs, at least as demonstrated by leuprolide, suggest that these compounds can be manufactured under controls generally applied to conventional drugs. Clearly then, one must answer three crucial questions. First, are these drugs safe? As discussed in this chapter, toxicological data in various animals administered with drug at several times the human dose revealed no meaningful responses attributable to these analogues other than their expected pharmacodynamic effect on the gonads. Thus, these drugs appear safe even for clinical applications that require chronic administration. It is important, however, that dose requirements relative to therapeutic benefit be carefully balanced against the impact of these drugs on hormonal activity. Second, are these LHRH agonists efficacious? Again, as indicated, these drugs effectively downregulate pituitary receptors thus causing biochemical castration, a condition on which their *in vivo* activity is based. Certainly, the literature makes it abundantly clear that injectable products of leuprolide acetate provide effective treatment modalities for prostate cancer, endometriosis, and uterine fibroids. As such, therapeutic efficacy of these drugs is rather high subsequently prompting their continued use in clinical therapeutics today. Third and finally, are these drug products manufacturable under appropriate and relevant controls? Specifications and analytical methods used for leuprolide acetate were discussed at length earlier. Data summaries provided in the discussion suggest that those methods for bulk substance synthesis as well as procedures used for finished goods are robust and rugged enough to enable systematic and consistent manufacture of various formulation presentations of leuprolide. This is definitely true for other LHRH analogues as well.

ACKNOWLEDGMENTS

The authors wish to express their sincere thanks to our Drug Metabolism Department and Dr. Eugene Bush for performing the bioassays. Animal studies to support this work were conducted by Dr. Billy Day, D.V.M., and his research team at the University of Missouri, Department of Animal Science, Columbia, Missouri. Their technical assistance is also very much appreciated.

REFERENCES

Adjei, A. L., and Hsu, L., 1993, Stability and Characterization of Protein and Peptide Drugs: Case Histories (Y. J. Wang and R. Pearlman, eds.), pp. 159–199, Plenum Press, New York.

Adjei, A. L., Garren, J., Menon, G., Rao, S., and Vadnere, M., 1993, Effect of ion-pairing on octanol-water partitioning of peptide drugs. I: The nonapeptide leuprolide acetate, *Int. J. Pharm.* **90:**141.

Amoss, M., Burgus, R., Blackwell, R., Vale, W., Fellows, R., and Guillemin, R., 1971, Purification, amino acid composition and N-terminus of the hypothalamic luteinizing hormone releasing factor (LRF) of ovine origin, *Biochem. Biophys. Res. Commun.* **44:**205.

Arimura, A., Sato, H., Kumasaka, T., Worobec, R. B., Debeljuk, L., Dunn, J. D., and Schally, A. V., 1973, Production of antiserum to LH-RH associated with marked atrophy of gonads in rabbits; characterization of the antibody and development of a radioimmunoassay for LH-RH, *Endocrinology* **93:**1092.

Arimura, A., Debeljuk, L., and Schally, A. V., 1974a, Blockade of preovulatory surge of gonadotropins LH and FSH and of ovulation by anti-LH-RH serum in rats, *Endocrinology* **95:**323.

Arimura, A.. Vilchez-Martinez, J. A., and Schally, A. V., 1974b, In vivo comparison of LH-RH and FSH-RH activities of [des-Gly^{10}][Pro^{9}ethylamide]-LH-RH, [desGly^{10}][Pro^{9}ethylamide]-LH-RH, and LH-RH using immature male rats, *Proc. Soc. Exp. Biol. Med.* **146:**17.

Arimura, A., Vilchez-Martinez, J. A., Coy, D. H., Coy, E. J., Hirotsu, Y., and Schally, A. V., 1974c, [D-Ala^{6}, desGly-NH$_{2}$^{10}]-LH-RH-ethylamide: A new analogueue with unusually high LH-RH/FSH-RH activity, *Endocrinology* **95:**1174.

Arimura, A., Shiino, M., de la Cruz, K. G., Rennels, E. G., and Schally, A. V., 1976, Effect of active and passive immunization with LH-RH on serum LH, FSH levels and the ultrastructure of the pituitary gonadotrophs in castrated male rats, *Endocrinology* **99:**291.

Arnold, W., Flouret, G., Morgan, R., Rippel, R., and White, W., 1974, Synthesis and biological activity of some analogues of the gonadotropin releasing hormone, *J. Med. Chem.* **17:**314.

Baba, Y., Matsuo, H., and Schally, A. V., 1971, Structure of porcine LH and FSH-releasing hormone. II: Confirmation of the proposed structure by conventional sequential analysis, *Biochem. Biophys. Res. Commun.* **44:**459.

Besser, G. M., 1974, Hypothalamus as an endocrine organ, I, *Br. Med. J.* **3:**560.

Bodanszky, M., 1993, *Principles of Peptide Synthesis,* 2nd ed., p. 159, Springer-Verlag, Berlin.

Burgus, R., Butcher, M., Amoss, M., Ling, N., Monohan, M., Rivier, J., Fellows, R., Backwell, R., Vale, W., and Guillemin, R., 1972, Primary structure of the ovine hypothalamic luteinizing hormone-releasing factor (LRF), *Proc. Natl. Acad. Sci. USA* **69:**278.

Chiu, Y., 1994, *Guidance for Industry for the Submission of Chemistry, Manufacturing, and Controls Information for Synthetic Peptide Substances.* Center for Drug Evaluation and Research, Center for Biologics Evaluation and Research, Food and Drug Administration.

Corbin, A., Bex, F. J., and Jones, R. C., 1984, Comparison of LH-RH agonist (AG) and antagonist (ANT): Antifertility and therapeutic developments, *J. Steroid Biochem.* **20:**1369.

Coy, D. H., Coy, E. J., Hirotsu, Y., Vilchez-Martinez, J. A., Schally, A. V., Van Nispen, J. W., and Tesser, G. I., 1974a, Investigation of the role of tryptophan in the luteinizing hormone releasing hormone, *Biochemistry* **13:**3550.

Coy, D. H., Coy, E. J., Schally, A. V., Vilchez-Martinez, J. A., Debeljuk, L., Carter, W. H., and Arimura, A., 1974b, Stimulatory and inhibitory analogues of luteinizing hormone releasing hormone, *Biochemistry* **13:**323.

Coy, D. H., Coy, E. J., Schally, A. V., Vilchez-Martinez, J., Hirotsu, Y., and Arimura, A., 1974c, Synthesis and biological properties of [D-Ala^6-, des-GlyNH$_2$^{10}]-LH-RH ethylamide, a peptide with greatly enhanced LH and FSH-releasing activity, *Biochem. Biophys. Res. Commun.* **57:**335.

Coy, D. H., Coy, E. J., and Schally, A. V., 1975a, Structure activity relationships of the LH and the FSH releasing hormone, *Res. Methods Neurochem.* **3:**393.

Coy, D. H., Vilchez-Martinez, J. A., Coy, E. J., Nishi, N., Arimura, A., and Schally, A. V., 1975b, Polyfluoroalkylamine derivatives of luteinizing hormone-releasing hormone, *Biochemistry* **14:**1848.

Coy, D. H., Labrie, F., Savary, M., Coy, E. J., and Schally, A. V., 1975c, LH-releasing activity of potent LH-RH analogues in vitro, *Biochem. Biophys. Res. Commun.* **67:**576.

Coy, D. H., Vilchez-Martinez, J. A., Coy, E. J., and Schally, A. V., 1976, Analogues of luteinizing hormone releasing hormone (LH-RH) with increased biological activity produced by D-amino acid substitutions in position six, *J. Med. Chem.* **19:**423.

Dutta, A. S., Furr, B. J. A., Giles, M. B., and Valcaccia, B., 1978, Synthesis and biological activity of highly active alpha-aza analogueues of luliberin, *J. Med. Chem.* **21:**1018.

Fraser, H. M., Jeffcoate, S. L., Gunn, A., and Holland, D. T., 1975, Effect of active immunization to luteinizing hormone releasing hormone on gonadotropin levels in ovariectomized rats, *J. Endocrinol.* **64:**191.

Fujino, M., Kobayashi, S,, Obayashi, M., Fukuda, T., Shinagawa, S., Yamazaki, I., Nakayama, R., White, W. F., and Rippel, R. H., 1972a, Syntheses and biological activities of analogues of luteinizing hormone releasing hormone (LH-RH), *Biochem. Biophys. Res. Commun.* **49:**698.

Fujino, M., Kobayashi, S., Obayashi, M., Shinagawa, S., Fukuda, T., Kitada, C., Nakayama, R., Yamazaki, I., White, W. F., and Rippel, R. H., 1972b, Structure–activity relationships in the C-terminal part of luteinizing hormone releasing hormone (LH-RH), *Biochem. Biophys. Res. Commun.* **49:**863.

Fujino, M., Shinagawa, S., Yamazaki, I., Kobayashi, S., Obayashi, M., Fukuda, T., Takayama, R., White, W. F., and Rippel, R. H., 1973a, [DesGly-NH$_2$^{10}, Pro-ethylamide^9]-LH-RH. A highly potent analogue of luteinizing hormone releasing hormone, *Arch. Biochem. Biophys.* **154:**488.

Fujino, M., Shinagawa, S., Obayashi, M., Kobayashi, S., Fukuda, T., Yamazaki, I., Nakayama, R., White, W. F., and Rippel, R. H., 1973b, Further studies on the structure–activity relationships in the C-terminal part of luteinizing hormone-releasing hormone, *J. Med. Chem.* **16:**1144.

Fujino, M., Yamazaki, I., Kobayashi, S., Fukuda, T., Shinagawa, S., Nakayama, R., White, W. F., and Rippel, R. H., 1974a, Some analogues of luteinizing hormone releasing hormone (LH-RH) having intense ovulation-inducing activity, *Biochem. Biophys. Res. Commun.* **57:**1248.

Fujino, M., Fukuda, T., Shinagawa, S., Kobayashi, S., Yamazaki, I., Nakayama, R., Seely, J. H., White, W. F., and Rippel, R. H., 1974b, Synthetic analogues of luteinizing hormone releasing hormone (LH-RH) substituted in position 6 and 10, *Biochem. Biophys. Res. Commun.* **60:**406.

Geiger, R., Wissmann, H., Konig, W., Sandow, J., Schally, A. V., Redding, T. W., Debeljuk, L., and Arimura, A., 1972, Synthesis and biological evaluation of 4-alanine-luteinizing hormone-releasing hormone ([Ala-4l-LH-RH), *Biochem. Biophys. Res. Commun.* **49:**1467.

Gordon, J. H., and Reichlin, S., 1974, Changes in pituitary responsiveness to luteinizing hormone-releasing factor during the rat estrous cycle, *Endocrinology* **94:**974.

Griffith, E. C., Hooper, K. C., Jeffcoate, S. L., and Holland, D. T., 1974, The presence of peptidases in the rat hypothalamus inactivating luteinizing hormone-releasing hormone (LH-RH), *Acta Endocrinol. (Copenhagen)* **77:**435.

International Conference on Harmonization, 1996, Guidelines Availability: Impurities in New Drug Substances: Notice, *Fed. Regis.* **61**(3), January 4.

Kanematsu, S., Scaramuzzi, R. J., Hilliard, J., and Sawyer, C. H., 1974, Patterns of ovulation-inducing LH release following coitus, electrical stimulation and exogenous LH-RH in the rabbit, *Endocrinology* **95:**247.

Karten, M. J., and Rivier, J. E., 1986, Gonadotropin-releasing hormone analogue design. Structure–function studies toward the development of agonists and antagonists: Rationale and perspective, *Endocr. Rev.* **7**(1):44.

Kastin, A. J., Schally, A. V., Gual, C., and Arimura, A., 1972, Release of LH and FSH after administration of synthetic LH-releasing hormone, *J. Clin. Endocrinol. Metab.* **34:**753.

Koch, Y., Chobsieng, P., Zor, V., Fridkin, M., and Lindner, H. R., 1973, Suppression of gonadotropin secretion and prevention of ovulation in the rat by antiserum to synthetic gonadotropin-releasing hormone, *Biochem. Biophys. Res. Commun.* **55:**623.

Koch, Y., Baram, T., Chobsieng, P., and Fridkin, M., 1974, Enzymic degradation of luteinizing hormone-releasing hormone (LH-RH) by hypothalamic tissue, *Biochem. Biophys. Res. Commun.* **61:**95.

Lemay, A., Maleux, R., Faure, N., Jean, C., and Fazekas, A. T. A., 1984, Efficacy and safety of LH-RH agonist treatment in 10 patients with endometriosis, *J. Steroid Biochem.* **20:**1379.

Makino, T., Takahashi, M., Yoshinaga, K., and Greep, R. O., 1973, Ovulation blockade in rats by rabbit anti-luteinizing hormone releasing factor serum, *Contraception* **8:**133.

Malspeis, L., Weinrib, A. B., Staubus, A. E., Arever, M. R., Balcerzak, S. P., and Niedhark, J. A., 1984, Clinical pharmacokinetics of 2′-deoxycoformycin, *Cancer Treat. Symp.* **2:**7.

Marks, N., and Stern, F., 1974, Enzymatic mechanisms for the inactivation of luteinizing hormone-releasing hormone (LH-RH), *Biochem. Biophys. Res. Commun.* **61:**1458.

Martin, J. E., Turey, L., Everett, J. W., and Fellows, R. E., 1974, Variations in responsiveness to synthetic LH releasing factor (LRF) in proestrous and diestrous-3 rats, *Endocrinology* **94:**556.

Matsuo, H., Baba, Y., Nair, R. M. G., Arimura, A., and Schally, A. V., 1971a, Structure of the porcine LH and FSH-releasing hormone. I: The proposed amino acid sequence, *Biochem. Biophys. Res. Commun.* **43:**1334.

Matsuo, H., Arimura, A., Nair, R. M. G., and Schally, A. V., 1971b, Synthesis of the porcine LH and FSH-releasing hormone by the solid phase method, *Biochem. Biophys. Res. Commun.* **45:**822.

Meldrum, D. R., Chang, R. J., Lu, J., Vale, W., Rivier, J., and Judd, H. L., 1982, Medical oophorectomy using a long-acting GnRH agonist—A possible new approach to the treatment of endometriosis, *J. Clin. Endocrinol. Metab.* **54**(5)**:**1081.

Micsbauer, L. J., 1995, Identification of Impurities in ABT-43818 by ESI LC/MS, Structural Chemistry Report No. 95:410:226.

Monohan, M. W., Amoss, M. S,, Anderson, H. A., and Vale, W., 1973, Synthetic analogues of the hypothalamic luteinizing hormone releasing factor with increased agonist or antagonist properties, *Biochemistry* **12:**4616.

Package Insert, 1996, Lupron® Injection and Lupron® Depot, *Physicians' Desk Reference* (50th ed.).

Prasad, K. U. M., Roeske, R. W., Weitl, F. L., Vilchez-Martinez, J., and Schally, A. V., 1976, Structure–activity relationships in luteinizing hormone-releasing hormone, *J. Med. Chem.* **19:**492.

Redding, T. W., and Schally, A. V., 1973, Synthesis of luteinizing hormone releasing hormone containing tritium-labeled pyroglutamic acid, *Life Sci.* **12:**23.

Redding, T. W., Kastin, A. J., Gonzalez-Barcena, D., Coy, D. H., Coy, E. J., Schalch, D. S., and Schally, A. V., 1973, The half-life, metabolism and excretion of tritiated luteinizing hormone-releasing hormone (LH-RH) in man, *J. Clin. Endocrinol. Metab.* **37:**626.

Rippel, R. H., Johnson, E. S., White, W. F., Fujino, M., Fukuda, T., and Kogayashi S., 1975a, Ovulation and gonadotropin-releasing activity of [D-Leu^6, desGlyNH$_2$^{10}, Pro-ethylamide^9]-GnRH, *Proc. Soc. Exp. Biol. Med.* **148:**1193.

Rippel, R. H., Johnson, E. S., White, W. F., Fujino, M., Fukuda, T., and Kogayashi, S., 1975b, Ovulation and gonadotropin-releasing activity of [D-Leu^6, des-GlyNH$_2$, Pro-ethylamide^9]-GnRH (38715), *Proc. Soc. Exp. Biol. Med.* **148:**1193.

Robertson, D. M., and Diczfalusy, E., 1977, Biological and immunological characterization of human luteinizing hormone: II. A comparison of the immunological and biological activities of pituitary extracts after electrofocusing using different standard preparations, *Mol. Cell. Endocrinol.* **9:**57.

Roth, J., LeRoith, D., Shiloach, J., Rosenzweig, J. L., Lesniak, M. A., and Havrankova, J., 1982, The evolutionary origins of hormones, neurotransmitters, and other intracellular chemical messengers, *N. Engl. J. Med.* **306:**523.

Sandow, J., Heptner, W., and Vogel, H. G., 1974, Studies on *in vivo* inactivation of synthetic LH-RH, in: *Hypothalamic Hypophysiotropic Hormones* (C. Gual and E. Rosenberg, eds.), p. 64, Excerpta Medica, Amsterdam.

Sandow, J., Rechenberg, W. V., Konig, W., Hahn, M., Jerzabek, G., and Fraser, H., 1978, Physiological studies with highly active analogueues of LH-RH, in: *Hypothalamic Hormones: Chemistry, Physiology, and Clinical Applications* (D. Gupta and W. Voelters, eds.), Weinheim: b Verlag Chemie, Tübingen, Germany, p. 307.

Schally, A. V., 1978, Aspects of hypothalamic regulation of the pituitary gland, *Science* **202:**18.

Schally, A. V., Arimura, A., Baba, Y., Nair, R. M. G., Matsuo, H., Redding, T. W., Debeljuk, L., and White, W. F., 1971a, Isolation and properties of the FSH and LH-releasing hormone, *Biochem. Biophys. Res. Commun.* **43:**393.

Schally, A. V., Nair, R. M. G., Redding, T. W., and Arimura, A., 1971b, Isolation of the LH and FSH-releasing hormone from porcine hypothalami, *J. Biol. Chem.* **246:**7230.

Schally, A. V., Arimura, A., Kastin, A. J., Matsuo, H., Baba, Y., Redding, T. W., Nair, R. M. G., Debeljuk, L., and White, W. F., 1971c, Gonadotropin-releasing hormone: One polypeptide regulates secretion of luteinizing and follicle stimulating hormones, *Science* **173:**1036.

Schally, A. V., Kastin, A. J., and Arimura, A., 1971d, Hypothalamic follicle-stimulating hormone (FSH) and luteinizing hormone (LH) regulating hormone: Structure, physiology, and clinical studies, *Fertil. Steril.* **22:**703.

Schally, A. V., Arimura, A., and Kastin, A. J., 1973, Hypothalamic regulatory hormones, *Science* **179:**341.

Schally, A. V., Coy, D. H., and Meyers, C. A., 1978, Hypothalamic regulatory hormones, *Annu. Rev. Biochem.* **48:**89.

Schally, A. V., Arimura, A., and Coy, D. H., 1980, Recent approaches to fertility control based on derivatives of LH-RH, *Vitam. Horm. (N.Y.)* **38:**257.

Storring, P. L., Zaidi, A. A., Mistry, Y. G., Fröysa, B., Stenning, B. E., and Diczfalusy, E., 1981, A comparison of preparations of highly purified human pituitary FSH: Differences in the FSH potencies as determined by *in vivo* bioassay, *in vitro* bioassay and immunoassay, *J. Endocrinol.* **91:**352.

The Leuprolide Study Group, Garnick, M. M., and thirty other participants including Max, D. T., from Abbott, 1984, Leuprolide versus diethylstilbestrol for metastatic prostate cancer, *N. Engl. J. Med.* **311:**1281.

Vale, W., Rivier, C., and Brown, M., 1977, Regulatory peptides of the hypothalamus, *Annu. Rev. Physiol.* **39:**473.

Vilchez-Martinez, J. A., Coy, D. H., Arimura, A., Coy, E. J., Hirotsu, Y., and Schally, A. V., 1974, Synthesis and biological properties of [D-Leu^6]-LH-RH and [D-Leu^6, desGly-NH_2^{10}]-LH-RH ethylamide, *Biochem. Biophys. Res. Commun.* **59:**1226.

World Health Organization, 1982, 32nd Report, WHO Expert Committee on Biological Standardization, *WHO Tech. Rep. Ser.* **673.**

Chapter 9

Discovery and Development of Somatostatin Agonists

Peter Marbach, Wilfried Bauer, David Bodmer, Ulrich Briner, Christian Bruns, Andrea Kay, Ioana Lancranjan, Janos Pless, Friedrich Raulf, Rodney Robison, John Sharkey, Thomas Soranno, Barbara Stolz, Peter Vit, and Gisbert Weckbecker

1. INTRODUCTION

Somatostatin was discovered in the laboratories of Professor R. Guillemin at the Salk Institute in La Jolla, California (Brazeau *et al.*, 1973; Guillemin, 1992), and was first described as hypothalamic growth hormone (GH)-release inhibiting factor. Within a few years, more and more information accumulated about its ubiquitous distribution in different regions of the body, including the pancreas and gastrointestinal tract, and on its more general inhibitory functions on hormones such as insulin, glucagon, gastrin, and other gastrointestinal hormones, as well as on enzymes such as those from the exocrine pancreas. These characteristics suggested that somatostatin had enormous therapeutic potential, and early clinical investigations substantiated hopes for applications in the treatment of hypersecretory states

Peter Marbach, Wilfried Bauer, David Bodmer, Ulrich Briner, Christian Bruns, Ioana Lancranjan, Janos Pless, Friedrich Raulf, Barbara Stolz, Peter Vit, and Gisbert Weckbecker • Novartis Pharma AG, Basel, Switzerland CH-4002. *Andrea Kay, Rodney Robison, John Sharkey, and Thomas Soranno* • Novartis Pharma Ltd., East Hanover, New Jersey 07936.

Integration of Pharmaceutical Discovery and Development: Case Studies, edited by Borchardt *et al.*, Plenum Press, New York, 1998.

of GH and in diabetes, as well as in therapy for peptide-secreting, gastroentero-pancreatic (GEP) endocrine tumors including carcinoid ones.

The very short biological half-life of somatostatin represented a challenge to peptide chemists all over the world to design and synthesize more stable analogues, in order to achieve an easy route of administration and a longer duration of action. Analogues with more specific activity profiles were needed to overcome the multiplicity of biological actions associated with the natural molecule.

It is interesting to note that somatostatin was discovered mainly through the application of two innovative technologies, namely, investigation of the direct effect of hypothalamic extracts on the release of GH in monolayer cultures of rat pituitary cells, and the measurement of GH levels using a specific radioimmunoassay. These methodologies were used to characterize the analogues, leading to the selection of those having a high potency for GH inhibition, or as it is now understood, a high selectivity for somatostatin receptor subtype 2, which is the dominant receptor in the pituitary gland. Somatostatin receptor research developed comparatively recently, and the different receptor subtypes present in different tissues were not characterized until the 1980s. Nevertheless, this chapter opens with a brief summary of somatostatin receptors, as they are fundamental to an understanding of the development of new delivery forms of somatostatin analogues for specific applications in oncology, and for the design of analogues for tumor imaging and tumor therapy.

2. SOMATOSTATIN RECEPTORS

2.1. Heterogeneity of Somatostatin Receptors

On the basis of radioligand-binding studies, it has been suggested that there are at least two different somatostatin receptor subtypes that exhibit somatostatin-14- and -28-selective binding properties, termed SS-1/SRIF-1 (Reubi, 1984) and SS-2/SRIF-2 (Martin *et al.,* 1991). Functional studies supported the concept of somatostatin receptor heterogeneity as somatostatin-14 and -28 and various short synthetic somatostatin analogues were shown to differ in their abilities to inhibit the release of neurotransmitters and/or hormones. Photoaffinity labeling and purification studies provided further evidence for the existence of somatostatin receptor subtypes. The final proof was provided by the cloning of five somatostatin receptor subtype genes.

2.2. The Somatostatin Receptor Gene Family

In 1992–1993, molecular cloning revealed the existence of a whole new gene family of somatostatin receptors comprising at least five different genes for structurally related receptor proteins (Bruns *et al.,* 1994; Patel *et al.,* 1995; Reisine and Bell, 1995). They belong to the superfamily of G-protein-coupled receptors with seven transmembrane domains whose ligands include neurotransmitters, peptide hormones, and olfactory molecules.

The first receptors to be identified, human sst_1 and sst_2, were cloned from genomic DNA using a polymerase chain reaction (Yamada *et al.,* 1992a). At the same time, another approach, that of expression library cloning of cDNA, was successfully applied to rat sst_2 in our laboratories (Kluxen *et al.,* 1992). The discovery of these first genes paved the way to the identification of the other receptors by homology cloning, using sst_1 and sst_2 probes to screen genomic or cDNA libraries. There ensued a race to clone the remaining three human receptor subtypes, sst_3 (Yamada *et al.,* 1992b), sst_4 (Rohrer *et al.,* 1993), and sst_5 (Panetta *et al.,* 1993).

The availability of five cloned receptors provided the unique possibility of building molecular receptor models to investigate the ligand–receptor interaction. Mutagenesis experiments proved that the specificity of octreotide for sst_2 is defined primarily by only two amino acids in transmembrane domains VI and VII, a phenylalanine and an asparagine (Kaupmann *et al.,* 1995).

Besides these differences, all sst subtypes associate with heterotrimeric G-proteins and are able to mediate the inhibition of adenylyl cyclase activity when transfected into Chinese hamster ovary or COS cells. Coupling to G-proteins is needed for the high affinity of the receptor in regard to ligand binding. Coupling of individual ssts to protein tyrosine phosphatases, Na^+/H^+ exchangers, cGMP-dependent protein kinases, phospholipase A_2, as well as K^+ and Ca^{2+} channels, have been described (Bruns *et al.,* 1995; Patel *et al.,* 1995; Reisine and Bell, 1995). It remains to be elucidated whether these different signaling mechanisms are cell- or subtype-specific parts of parallel or independent signal transduction pathways.

2.3. Tissue Distribution

The expression of ssts was determined at the mRNA level by various methods, and showed a distinct but overlapping pattern of expression. All five receptor subtypes were found to be expressed in the brain and pituitary gland. Remarkable levels of expression were found in peripheral tissues, e.g., in the adrenal glands (high levels of sst_2) and pancreas (sst_2, sst_3, and sst_5), in addition to low levels of mRNA in many other tissues (Raulf *et al.,* 1994). Many tumors, especially those of neuroendocrine origin, express sst_2. The simplistic correlation of tissue expression of one specific subtype with a particular physiological response to somatostatin, e.g., GH inhibition, is hampered by the simultaneous expression of two or more sst subtypes in a single tissue, even in a given cell type.

Table I
The Binding Properties of Human sst_1–sst_5 Receptors Expressed in CHO Cells (sst_1–sst_4) or COS-1 Cells (sst_5)

	pK$_i$ [−log M]				
Peptide	sst_1	sst_2	sst_3	sst_4	sst_5
Somatostatin-14	9.2	9.7	9.6	8.9	9.4
Somatostatin-28	9.2	9.7	9.5	9.1	9.4
CGP 23996	8.4	9.1	9.4	8.7	8.6
Octreotide	6.6	9.5	8.3	<6.0	8.3
BIM 23014	6.3	9.3	7.2	<6.0	8.4
MK 678	<6.0	10.1	7.5	<6.0	7.9
RC 160	6.8	10.1	7.7	6.8	8.3

2.4. Pharmacology

The establishing of cell lines that express stably one of the five somatostatin receptors provided the invaluable opportunity of studying the subtype-specific receptor pharmacology *in vitro,* and to screen for new somatostatin analogues and mimetics. A variety of somatostatin analogues have been used to characterize the different binding properties of the five cloned receptors (Bruns *et al.,* 1995, 1996; Patel *et al.,* 1995; Reisine and Bell, 1995). In general, the natural peptide hormones somatostatin-14 and -28 show very minor differences in their high-affinity binding ($pK_i \leq 9$) toward sst_1–sst_5 (Table I). However, the short synthetic analogues such as octreotide, BIM 23014, MK 678, and RC 160 display different binding profiles that are very similar for all of the mentioned analogues: High-affinity binding is observed only for sst_2 and rat sst_5, whereas sst_3 and human sst_5 display intermediate affinities. Only very low affinities could be demonstrated for sst_1 and sst_4. Great effort is currently being invested in the development of subtype-specific compounds. The identification of these will help in the analysis of certain subtype-specific regulatory effects. The development of specific antagonists will be a prerequisite for differentiating signal transduction pathways in cells or tissues where two or more sst receptor subtypes are expressed simultaneously.

3. DISCOVERY AND DEVELOPMENT OF SANDOSTATIN®

The development of somatostatin analogues by our group commenced in 1974. There was a strong foundation for the project, which was part of the company's general research program on the regulation of anterior pituitary hormones. On the one hand, we had outstanding expertise in peptide chemistry, as demonstrated by pioneering work on neurohypophyseal hormones, adrenocorticotropic hormone analogues, and salmon calcitonin. On the other hand, there was already considerable emphasis on prolactin research, and the development of dopaminergic drugs such as bromocriptine (a peptide-ergot derivative) was in its early stages. The indications that were envisaged for an inhibitor of GH secretion were hypersecretory states of somatotropin, and in particular, GH-secreting adenomas of the pituitary gland and late-stage vascular complications in diabetes.

The research program started with a large series of noncyclical analogues, i.e., with a reduced cystine bridge. However, the poor chemical stability of these peptides limited their use and compelled the chemists to design cyclic structures. Analogues were protected against proteolytic enzyme attack by introducing unnatural D-amino acids, mainly at the amino-terminal end, and by protecting the carboxy-terminal against carboxypeptidases using amidation, esterification, reduction, and the like.

Our group was not the only team worldwide involved in the design of somatostatin analogues. In particular, the group under Guillemin investigated the structure–activity relationships of "their" tetradecapeptide (Rivier *et al.,* 1975; Vale *et al.,* 1978). They achieved a marked improvement in biological activity by replacing the tryptophan residue in position 8 by its D-isomer and confirmed that omission of the first two amino acids outside the disulfide bridge did not compromise the biological activity associated with somatostatin.

The systematic work by our group on analogues with an intact, 12-amino-acid ring structure culminated in 1978 with the synthesis of a somatostatin analogue, SDZ 36-465 [D-Phe-Cys-Nle-Asn-Phe-Phe-D-Trp-Lys-Thr-Phe-Thr-Ser-Cys-Asp(diol)], which was developed as far as the early clinical stages. In rats this analogue had been shown to be 20 times more potent than somatostatin in inhibiting GH after intravenous injection, but only twice as potent when injected intramuscularly. It was 15 and more than 40 times more specific than somatostatin with regard to the inhibition of GH relative to insulin, after intravenous and intramuscular application, respectively. However, in further investigations using rhesus monkeys, SDZ 36-465 did not have a higher potency than somatostatin and the specificity profile demonstrated in rats was barely apparent in monkeys. Experiments in humans should give more information on the predictability of results obtained with one or the other species, as well as on the most predictive mode of application.

Preliminary investigations in humans clearly showed no significant advantage over natural somatostatin in terms of potency or specificity. Similar findings were published by Adrian *et al.* (1979), who infused different analogues into patients with metastatic endocrine tumors. No analogue had a significantly more potent effect on basal hormone secretion than somatostatin, and none showed consistently different specificity in terms of suppression.

The conclusion was that a dramatic change in analogue design was needed. One reason for the disappointing performance of somatostatin analogues in humans could have been the very low metabolic stability of the peptides, which necessitated their continuous infusion to achieve a sufficient duration of action. Methods of shortening the peptide sequence, in order to achieve more rigid and hopefully more stable analogues, became of considerable interest.

It is to the credit of J. Rivier and W. Vale, at the time members of the research group of Guillemin, that they consistently tried to ascertain the contribution of each individual amino acid to the biological activity, and to find the minimal essential sequence (Vale *et al.,* 1978). Their systematic work on structure–activity relationships resulted in the design of smaller analogues containing the sequence Phe^{7}-Lys^{8}-Trp^{9}-Thr^{10}, as occurs in natural somatostatin. The most active of these, Des-$AA^{1,2,4,5,12,13}$,[D-Trp^{8}]somatostatin, was also found to have a prolonged duration of action in the rat. This analogue had been included in the study of Adrian *et al.* (1979) on tumor patients, where it was shown to be less potent than, and of comparable duration of action to, somatostatin.

Important contributions to the understanding of structure–activity relationships came from the laboratories of Merck, Sharp and Dohme. Veber *et al.* (1978, 1984), using computer-assisted modeling techniques and proton NMR studies, first achieved conformationally restricted bicyclic analogues and, subsequently, high-potency cyclic hexapeptides.

Despite the synthesis of many peptide analogues by different research groups, compounds with improved duration of action or specificity were yet to be developed. Nevertheless, there was now a sound fundamental knowledge of the structure–activity relationships of the somatostatin molecule in our group, entertaining hopes that the design of small yet highly potent and metabolically stable analogues should be possible.

During development of SDZ 36-465, chemists in our group used a new approach for the synthesis of analogues with smaller molecular structures, starting from the minimal essential sequence Phe^{7}–Thr^{10}, enclosed within a Cys-Cys bridge (Bauer *et al.,* 1982b). Exocyclic N- and C-terminal addition of the missing essential residues resulted in analogues with high biological activities (Table II). Inhibitory effects of these structures on GH secretion were tested *in vitro* and *in vivo.*

Readdition of the Phe-6 (compound **3**) at the N-terminal A brought the highest increase in activity, when the residue was attached in the reversed D-configuration. Only then could the phenyl side chain point down below the β-sheet structure and occupy part of the conformational space of the essential Phe-6 currently occupied by the Cys bridge in the new reduced-size analogues. Furthermore, this protected the N-terminus against metabolic degradation by enzymes.

Further optimization of the C-terminal residue B with corresponding structural elements of somatostatin finally resulted in biological activity considerably surpassing that of the parent peptide. With the introduction of amino alcohols, where the carboxyl group is reduced to alcohol, a combination of both high activity and metabolic stability was reached. SMS 201-995 (octreotide) with its terminal carboxyl group of L-Thr reduced to a -CH_2OH alcohol (Thr-ol), represented

Table II
Structures and Biological Activities of Somatostatin Analogues Having the Generic Structure A-Cys-Phe-D-Trp-Lys-Thr-Cys-B[a]

	Terminal residues		Potency (%)	
	A	B	*In vitro*	*In vivo*
1	H	NH_2	0.1	1.4
2	H	D-Ser(NH_3)	0.1	3.3
3	D-Phe	NH_2	4	165
4	D-Phe	D-Ser(NH_3)	12	680
5	D-Phe	D-Thr(NH_2)	47	1160
6	D-Phe	D-Thr(ol)	54	1100
Octreotide	D-Phe	L-Thr(ol)	300	7000

[a]Results are expressed as relative potencies of GH inhibition (somatostatin = 100%). *In vitro* activities were measured in rat pituitary primary cell cultures; *in vivo* potencies in male rats were determined 15 min after intramuscular administration under pentobarbital anesthesia.

the culmination of the search for potent, small-molecule analogues. The increase in potency of GH inhibition was 5000-fold that of the weak cyclic hexapeptide lead (compound **1**) in *in vivo* experiments. Thus, octreotide was selected for further development (Bauer *et al.,* 1982a).

Octreotide crystals suitable for X-ray studies were recently obtained, and two different conformations were found among the three molecules in the asymmetric unit (Pohl *et al.,* 1995). All three have the expected type II′ β-turn around D-Trp-Lys but differ in the terminal regions. The crystal structure is stabilized by a network of inter- and intramolecular H-bonds in addition to the solvent (water) network. This may also explain the presence of different conformations. The molecules that differ from the regular, flat, antiparallel β-sheet structure have less H-bonds with neighboring octreotide molecules but more with those of the surrounding water; these conformations may represent those favored in aqueous solutions. The crystal structures illustrate the conformational flexibility even of reduced-size cyclic analogues, which makes it difficult to draw conclusions about active-site geometries.

3.1. Synthesis of Octreotide

Octreotide was first synthesized using classical fragment condensation strategy. A modern solid-phase process was developed in parallel, but the introduction of the terminal amino alcohol Thr-ol caused problems. Synthetic methods of liberating the Thr-ol peptide at the end of a solid-phase synthesis required severe conditions to achieve cleavage from the resin, resulting in product mixtures that were difficult to purify. Thus, new acid-labile linkers were developed for the solid phase

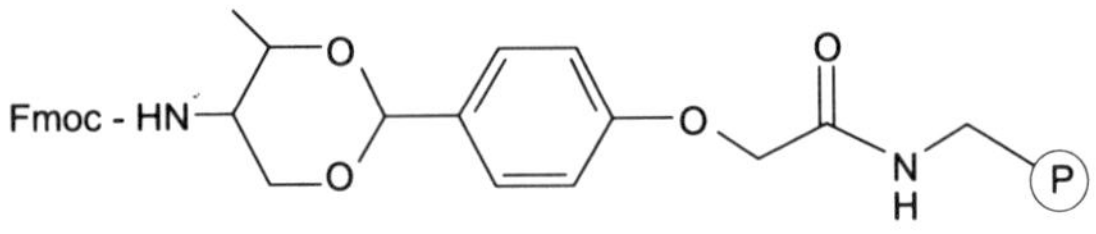

Figure 1. Fmoc-Thr-ol-acetal anchor for the solid-phase synthesis of octreotide.

synthesis of peptides with C-terminal Thr-ol. A new anchoring principle was achieved, based on cyclic acetal formation between the two hydroxy groups of Thr-ol and *p*-formyl-phenoxyacetic acid. This new Fmoc-Thr-ol-acetal anchor (Fig. 1) can be attached to amino-functionalized polymers, and is fully compatible with the base-labile Fmoc solid-phase strategy. Finally, it allows cleavage under very mildly acidic conditions (Mergler *et al.*, 1991) and purification of the product in a single process.

3.2. Pharmacodynamic Tests

The outstanding *in vivo* results obtained using rat models were confirmed in conscious male rhesus monkey (Fig. 2). Compared with the natural hormone somatostatin, octreotide showed a long duration of action following subcutaneous application and a favorable GH–insulin selectivity profile. A dose of 1 μg/kg caused a less pronounced effect on insulin release than 316 μg/kg of native somatostatin-14 although its effect on GH release was more potent and longer lasting. The pharmacological effects of octreotide have been described in more detail elsewhere (Bauer *et al.*, 1982a; Marbach *et al.*, 1988, 1992).

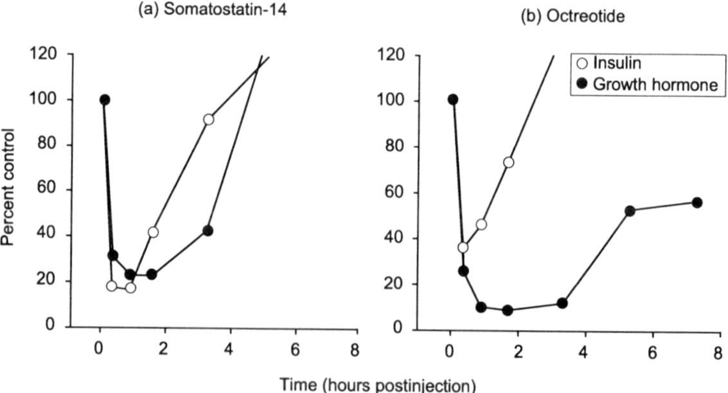

Figure 2. Specific effects of somatostatin-14 (316 μg/kg) and octreotide (1 μg/kg) in monkeys after subcutaneous administration. GH and insulin concentrations were determined by radioimmunoassay. Reprinted by permission of Springer-Verlag.

3.3. Pharmacokinetic Studies

The availability of a specific octreotide radioimmunoassay, which was already early in its development phase (Marbach *et al.,* 1985), allowed investigation of octreotide pharmacokinetics in different animal species, and after various routes of administration. The radioimmunoassay was an important tool for the definition of pharmacokinetic–pharmacodynamic relationships (Marbach *et al.,* 1992), and it facilitated the optimization and characterization of new delivery forms of octreotide, such as the long-acting release (LAR) formulations (Grass *et al.,*1996; see Sections 4 and 5), during both preclinical and clinical development.

The distribution, excretion, and metabolism of octreotide have been studied in the rat (Lemaire *et al.,* 1989). It was radiolabeled either with ^{3}H in the gamma and delta positions of the Lys residue, or with ^{14}C in the methylene group of D-Trp.

Following intravenous administration, octreotide has a short distribution phase. Its tissue concentrations were similar when determined either by radioimmunoassay or by whole-body autoradiography, which suggests that the distribution of ^{3}H or ^{14}C radioactivity observed 0.5 hr after intravenous administration mostly represents unmetabolized octreotide. High levels of the compound were found in the blood, kidney, liver, and blood vessel walls, whereas concentrations in the brain were insignificant. Clear differences in the distribution pattern in assumed target and nontarget organs were observed. After 4 hr, plasma concentration had fallen to less than 2% of the initial dose, whereas in the pancreas, an organ with high receptor density, the levels of octreotide were still one-third of the 0.5-hr value.

Unmetabolized drug accounted for most of the radioactivity detectable in plasma, urine, and bile, but only traces of intact octreotide were detectable in the feces. However, in the intestinal tract, extensive degradation could be shown. These results demonstrated the metabolic stability of octreotide *in vivo:* About 50 and 20% of the applied dose was excreted unchanged in bile and urine, respectively.

The *in vivo* fate of octreotide in the rat may be characterized as having a small volume of distribution, low hepatic metabolism, high and rapid biliary excretion, and showing degradation as it continues through the intestinal tract.

3.4. Toxicology

A full range of preclinical safety studies have been performed with octreotide. These included acute studies in mice and rats and repeat dose studies in mice, rats, dogs, and monkeys. *In vitro* mammalian and nonmammalian and *in vivo* mammalian genetic toxicity studies were conducted to assess the mutagenic potential of octreotide. Chronic studies included 6- and 12-month studies in both rats and dogs. A full reproductive toxicity program was performed in rats and rabbits (segments I, II, and III), and carcinogenicity studies (using the subcutaneous route) were carried out in both mice and rats.

Octreotide showed low toxicity in all species tested. No evidence of mutagenic or genotoxic potential was noted. Minor pharmacological effects seen in some studies included reduced body weight gain in all species, diarrhea in dogs, a reduced growth rate of rat pups born to treated dams, and elevated plasma glucose concentrations in some rat studies. The reduced growth of treated rat offspring was considered to be a consequence of GH inhibition. Reproductive studies of octreotide in animals have demonstrated no adverse effects on fertility or general reproductive performance and no evidence of teratogenic potential.

Two effects observed in these studies required further evaluation. First, injection site sarcomas were found in the 52-, 104-, and 116-week rat chronic toxicity/carcinogenicity studies, varying from well-differentiated fibrosarcomas to polymorphocellular or giant-cell sarcomas. Follow-up investigations concluded that these were most probably caused by tissue damage resulting from the low pH of the injected material. Injection site sarcomas after repeated subcutaneous administration have been seen in chronic rat studies, not only with pharmaceutical compounds but also with numerous other materials (Grasso, 1976; Theiss, 1982).

Second, there was a slightly increased incidence of uterine (endometrial) adenocarcinoma in the 104-week rat carcinogenicity study [vehicle control, 0/60 rats, saline control, 4/60 rats; 1.25 mg/kg octreotide (high dose), 9/60 rats]. This increased incidence of malignant tumors was not associated with an increase in benign tumors or proliferative lesions of the endometrium. The incidence at the high dose was statistically significant when the control groups were combined. However, the evidence of estrogen dominance with endometritis, coupled with the absence of corpora lutea and uterine glandular and luminal dilatation, suggests that the tumors may be associated with a hormonal imbalance. Estrogen dominance with endometritis is known to be associated with uterine tumors in aged rats (Flueckiger *et al.,* 1983). It was therefore concluded that both of these effects are specific to the rat and do not present a hazard to humans.

3.5. Clinical Development

Following preclinical pharmacological and toxicological investigations, octreotide was developed further and entered clinical trials. In humans, GH levels are characterized by diurnal profiles and most of the time are undetectable, i.e., below the limits of quantification of bioanalytical methods. In the early 1980s, standard radioimmunoassays had detection limits of around 2 ng/ml, including one developed by our group (Marbach *et al.,* 1978). Single, rising-dose tolerability studies in healthy male volunteers were followed by investigations of the pharmacodynamic effects on stimulated GH secretion (Marbach *et al.,* 1985). These studies rapidly showed that the advantageous pharmacodynamic profile for octreotide seen in animals also prevailed in humans: high activity, prolonged half-life, and a favorable specificity profile. In particular, the biological stability of octreotide allowed administration of the drug using multiple subcutaneous injections rather than continuous infusion.

Not only could octreotide be tested in different clinical situations such as acromegaly and in GEP tumor patients, but the discovery of a stable analogue opened a new era in the field of somatostatin research. Worldwide collaborative research projects and clinical studies between our group and leading endocrinologists and gastroenterologists quickly revealed an extraordinarily broad therapeutic potential for a somatostatin analogue. An open-minded, research-oriented development strategy was therefore a prerequisite for optimal exploitation of all of the possible opportunities (O'Dorisio, 1986; Krejs, 1987; Bloom, 1990; Farthing, 1990).

Octreotide was consequently introduced to the marketplace in 1987 under the trade name Sandostatin® for the pharmacological therapy of acromegaly and for the symptomatic treatment of GEP endocrine tumors, including carcinoid tumors. Later, registrations followed for the control of refractory diarrhea associated with AIDS, and for the prevention of complications following pancreatic surgery.

However, full clinical development for some promising indications was not possible because of various limitations, such as the need for multiple subcutaneous injections, and the extremely long-term nature of the clinical trials. Therefore, conditions such as diabetes, sleep apnea, gastric ulcers, and tumors, which had been identified in experimental and preliminary clinical studies as likely candidates for treatment with octreotide, could not be followed up beyond clinical investigational status.

4. DEVELOPMENT OF SANDOSTATIN® LAR®

Sandostatin® is generally administered by subcutaneous injection, two or three times daily. However, several studies have indicated that it would be more effective if given by constant subcutaneous infusion (Harris *et al.,* 1995), although this is impractical in routine use and is therefore restricted to specific cases. The search for alternative routes of application and formulations of the drug has therefore been a challenge since early in its development. The nasal and transdermal routes were also considered, as both could be optimized to achieve high relative bioavailabilities. The nasal formulation, in particular, underwent successful clinical testing. Another strategy exploited the extensive technical and clinical expertise within our group in biodegradable polymers for LAR preparations. The dopaminergic, prolactin release-inhibiting ergot compound bromocriptine had been developed in such a formulation, using microencapsulation in a polymer formed from poly-DL-lactide-coglycolide glucose, and was marketed as Parlodel LAR®. The same principle was adopted and optimized for octreotide.

4.1. Manufacture

Sandostatin® LAR® is produced as microcapsules using an organic phase separation process. These contain the active material uniformly distributed throughout a polymer matrix. The manufacturing procedure involves (1) dispersion of the drug in a polymer-containing solution, (2) formation of the microcapsules, (3) hardening of the microcapsules to facilitate collection, (4) collection and washing of the microcapsules, and (5) drying of the microcapsules to remove residual solvents. All of these processes were thoroughly investigated and optimized, as were the selection of the best polymer with respect to molecular weight, structural polymerization (star polymerization with glucose induction), carbohydrate ratios, and the percentage loading of the drug into the polymer (5% could be achieved). Thus, the formulation could be optimized aiming at low dissolution and long-acting profiles accompanied by smooth, *in vivo* biodegradation at the site of injection. As well as *in vitro* testing, this involved careful characterization of *in vivo* release profiles in rabbits.

4.2. Preclinical Studies

A preclinical safety assessment was conducted in which Sandostatin® LAR® was administered intramuscularly to assess its local and systemic toxicity. Two single-dose studies in rats and rabbits and a repeat-dose study in rats were also carried out to evaluate the response of muscle tissue to Sandostatin® LAR® suspended in 0.5% sodium carboxymethylcellulose. Additionally, the systemic toxicity was evaluated in rats.

In the single-dose intramuscular studies, 1.0 and 25.0 mg Sandostatin® in 20- and 500-mg microcapsules was administered to rats and rabbits, respectively. The animals were sacrificed at nine different intervals between 2 and 92 days postinjection. Scanning electron microscopy evaluation of the injection sites indicated that the microcapsules were completely biodegraded by day 75 postinjection. Figure 3b illustrates this degradation (in this case, showing octreotide pamoate LAR microcapsules 60 days after intramuscular injection into a rabbit). Localized subacute granulomatous myositis was observed by light microscopy: this minimal foreign body reaction was considered to be insignificant.

In the repeat-dose systemic toxicity study, rats were given 2.5 mg Sandostatin® in 50-mg microcapsules, intramuscularly once every 4 weeks for 6 months, with a 3-month recovery period. Using light microscopy, at 60 days after the final injection, only remnants of microcapsules could be found at the injection sites and

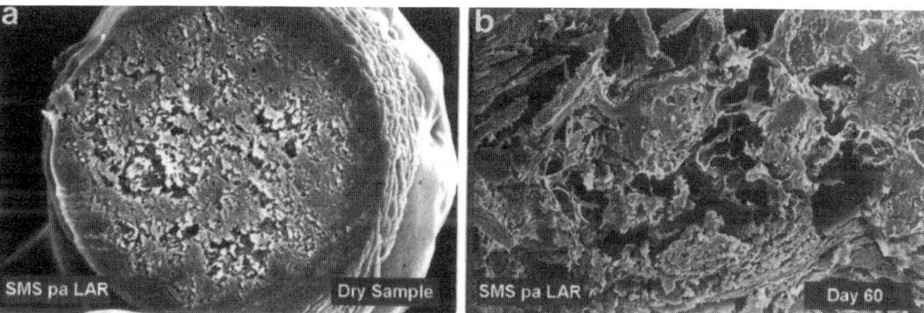

Figure 3. Scanning electron micrographs of octreotide pamoate LAR microcapsules, showing (a) a dry sample microcapsule and (b) a microcapsule dissected from the intramuscular injection site 60 days after administration. Bioerosion is approximately 85% complete with only a small remnant of solid polymer remaining in the left center of the micrograph.

biodegradation was complete by 90 days. A minimal decrease in body weight gain was seen in males but not in females. No other significant changes were observed during the *in vivo* phase or at gross necropsy. Histopathological changes were limited to the muscle injection sites and consisted of localized granulomatous myositis surrounding pockets of microcapsules.

4.3. Clinical Studies

Sandostatin® LAR® subsequently underwent clinical development (Kaal *et al.*, 1995; Lancranjan *et al.*, 1996). It could be demonstrated, both in healthy subjects and in acromegalic patients, that the LAR formulation provides drug profiles in humans characterized by an immediate release after the intramuscular injection. Thereafter, octreotide concentrations (1) decrease and remain negligible for 2–7 days, (2) increase to a dose-dependent plateau lasting 3–4 weeks, and finally (3) decrease steadily until the octreotide has completely disappeared, within 10–12 weeks after the injection. This release profile allows the compound to be injected at 4-week intervals, achieving constant plasma concentrations under steady-state conditions after the third injection. In fact, the plasma concentration profiles also mimic those observed using continuous subcutaneous pump infusions with respect to fluctuation indices.

Efficacy and safety profiles for Sandostatin® LAR® in acromegalic patients and in patients with GEP tumors are at least as good as those observed with standard Sandostatin® therapy (Lancranjan *et al.*, 1996). The replacement of subcutaneous injections two or three times daily, however, by only one intramuscular depot injection per month makes this LAR formulation attractive to the patients. Sandostatin® LAR® has so far been registered in several European countries.

5. ONCOLAR™: TECHNICAL DEVELOPMENT OF A NEW LAR FORMULATION OF OCTREOTIDE

OncoLAR™ (octreotide pamoate LAR) is a new microcapsule depot formulation of octreotide pamoate salt intended for repeated deep intramuscular administration of larger doses of octreotide. The rationale for this new formulation has been to achieve high doses of 60–160 mg or more, in single-use vials, considering the maximum tolerable administration volume after reconstitution with the vehicle. A minimum 10–20% drug loading has been calculated to be necessary. An additional goal has been to use an industrial manufacturing process with low (organic) solvent volumes and terminal sterilization if possible.

5.1. Manufacture

Development of OncoLAR™ started by investigating microcapsules with various loading levels of octreotide acetate and manufactured using a special emulsion process (Bodmer *et al.,* 1992). This very simple process requires only an organic phase for dissolution of the polymer and dispersion of the drug substance and a viscous aqueous gelatin solution for generation of the microcapsules. No further phase inducers, hardening or drug retaining substances, as required for similar systems, are necessary (see Section 4.1). Nevertheless, there were limitations in achieving the high dosages required, including the need for a totally aseptic process because of instability of the polymer during terminal sterilization and inappropriate drug release profiles when the drug was loaded at above 6%. Both of these hurdles could be overcome by switching from octreotide acetate to the pamoate salt (Bodmer *et al.,* 1994), and by several modifications to the manufacturing procedure, which still remains an organic phase separation process. The major features of OncoLAR™ can be summarized as follows: (1) 20% octreotide (29% octreotide pamoate)-loaded microcapsules, (2) sustained *in vivo* drug release for at least 1 month with an early onset, (3) a relatively simple manufacturing process, and (4) a stable product with an adequate shelf life, of at least 2 years.

5.2. Preclinical Studies

Five additional safety studies were performed in animals to demonstrate the safety of the new pamoate salt form of microencapsulated octreotide. These included an acute study in mice, two single-dose 90-day studies in rats and rabbits comparing the original acetate form and the new pamoate formulation of Sandostatin® LAR®, a repeat-dose rat study comparing the two salt forms, and an Ames test. *In vivo* biodegradation studies of the two salt forms of Sandostatin® LAR® using scanning electron microscopy indicated that microcapsule erosion commenced at around day 30 postinjection and was complete by day 75. In all cases, whether for systemic or local toxicity, the results for these two salt forms were comparable. The mutagenicity test for octreotide pamoate showed no effects of the drug treatment.

6. ANTIPROLIFERATIVE EFFECTS OF SINGLE-AGENT OCTREOTIDE

As well as having antisecretory effects, somatostatin inhibits cell proliferation in a number of *in vitro* tumor cell models (Weckbecker *et al.,* 1993). Accordingly, somatostatin analogues cause potent growth inhibition in various types of cancer cells, cultured *in vitro* or grown as tumors in rodents. The antiproliferative effect of octreotide *in vitro* has been demonstrated, e.g., in ZR-75-1 human breast tumor cells (Weckbecker *et al.,* 1992a). At nanomolar concentrations, octreotide inhibited both serum- and growth factor-driven cell proliferation. Similar potent effects were observed in the AR42J rat pancreatic tumor cell model. Tumors derived from both ZR-75-1 and AR42J cells were highly responsive to continuous treatment with octreotide, delivered either by subcutaneously implanted osmotic minipumps or in the form of OncoLAR™ (Fig. 4).

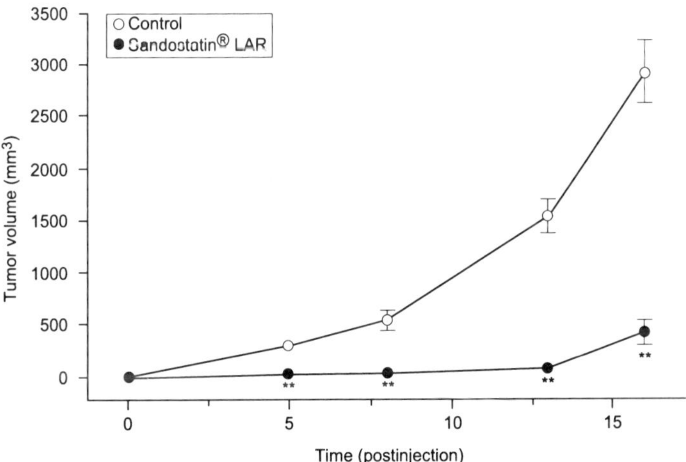

Figure 4. Inhibition of the growth of AR42J tumors in nude mice following a single injection of octreotide pamoate LAR (50 mg/kg). Because most of the LAR pellets are degraded by 8 weeks after the injection, the daily dose amounts to approximately 0.9 mg/kg. Data are means ± SE. $^{**}p < 0.01$.

6.1. Mechanism of Antiproliferative Action

The described antiproliferative effect is correlated with the presence of somatostatin receptor subtype sst_2, as detected by reverse transcription polymerase chain reaction and receptor binding assays on cultured cells and on tumors derived from the ZR-75-1 and the AR42J cell line. Buscail *et al.* (1994) also studied the role of sst_2 in mediating inhibitory effects on cell growth: In sst_2-expressing COS-7 and NIH 3T3 cells they found that octreotide and RC-160 inhibited very potently serum-driven cell proliferation with IC_{50} values in the picomolar range. These results suggest that a direct antiproliferative effect of octreotide through sst_2 could be effective in a wide range of tumors (Weckbecker *et al.*, 1993).

Only limited information is available on the postreceptor events that are essential for the antiproliferative response to somatostatin analogues, including octreotide. The interaction of somatostatin with its receptor elicits various signaling responses, of which the activation and/or translocation of phosphotyrosine phosphatases is apparently the prerequisite for inhibition of cell growth. Buscail *et al.* (1994) demonstrated a positive correlation between the activation of a phosphotyrosine phosphatase and antiproliferative effects of RC-160 or octreotide in COS-7 and NIH 3T3 cells expressing sst_2. Recruitment of phosphotyrosine phosphatase 1C to the cell surface by octreotide (Srikant and Shen, 1996) may be an early event in the signaling cascade leading to cell growth inhibition. Membrane-associated phosphotyrosine phosphatase 1C can promote association through the SH2 domains to phosphorylated epidermal growth factor (EGF) receptor, thereby blocking EGF receptor-induced mitogenic signaling.

The antiproliferative action of octreotide might also be mediated indirectly by the downregulation of a number of tumor growth stimuli including insulinlike growth factor 1 (IGF-1), EGF, or GH (Serri *et al.*, 1992; Huynh and Pollak, 1994). New capillaries originate from preexisting blood vessels by a process termed *angiogenesis.* Tumor-induced angiogenesis is a prerequisite for tumor growth, and it has been suggested in a number of independent studies that indirect effects of octreotide on tumor growth may also include antiangiogenic properties. Barrie *et al.* (1993) studied the effects of various somatostatin analogues related to octreotide in an *in vivo* model of angiogenesis (chick embryo chorioallantoic membrane). Both octreotide and RC-160 potently inhibited angiogenesis, whereas derivatives that were devoid of GH-suppressing properties failed to interfere with the formation of new vessels. Reubi *et al.* (1994) reported that veins surrounding human cancer tissue (colonic adenocarcinomas, carcinoids, renal cell carcinomas, and lymphomas) show high-levels of somatostatin receptor expression. As Tyr^3-octreotide also showed high affinity binding, it is likely that the tumor-associated veins express sst_2 and/or sst_5. Danesi and Del Tacca (1996) confirmed the antiangiogenic properties of octreotide, using various *in vivo* angiogenesis systems such as the rat cornea and mesentery models.

6.2. Route of Administration and Plasma Levels

It is noteworthy that the anticancer effects of octreotide were dose-dependent, and optimal at plasma concentrations between 5 and 15 ng/ml as measured by radioimmunoassay. These levels are higher than those used in endocrine or antisecretory treatments. Importantly, the administration route has a significant impact on the therapeutic efficacy in that octreotide administered continuously (via a subcutaneously implanted minipump) was at least 10-fold more active than a twice-daily intraperitoneal injection regimen (Weckbecker *et al.,* 1992a,b). Thus, there is a solid basis for using the recently developed LAR formulation for clinical studies in breast and pancreatic cancer patients.

6.3. Octreotide as a Potentiator of Standard Anticancer Regimens

Because the antiproliferative mechanism of somatostatin analogues is unique and these agents have a wide therapeutic window, the potential benefit of combining octreotide with standard antitumor treatment strategies for cancer, particularly of the breast and pancreas, has been investigated.

6.3.1. COMBINATIONS WITH ENDOCRINE THERAPIES

Tamoxifen administration and ovariectomy are well-established therapies for breast cancer although their effectiveness is limited. Their use is based on the simple concept of antagonizing estrogen-stimulated growth. Tamoxifen has also been shown to modulate the expression of growth factors and their binding proteins, which contributes to its antineoplastic activity. In an effort to study the interactions between octreotide and either tamoxifen or ovariectomy with respect to antineoplastic activity, the 7,12-dimethylbenz(*a*)anthracene (DMBA)-induced mammary carcinoma rat model was used (Weckbecker *et al.,* 1994). Ovariectomy usually leads to a marked regression of preexisting tumors, which regrow after a few weeks. However, when 6 weeks of high-dose octreotide infusion was initiated shortly after the operation, tumor regrowth was suppressed in most of the rats. Treatment of DMBA tumor-bearing rats for 6 weeks with single-agent tamoxifen or octreotide induced a significant but incomplete inhibition of tumor development (volume and number of tumors per animal). However, combined treatment with both agents induced markedly greater tumor growth suppression, which persisted for weeks after termination of drug administration. Based on these preclinical studies, the combination of tamoxifen and octreotide is currently being tested in clinical trials in metastatic and adjuvant breast cancer patients (see Section 7.2).

6.3.2. COMBINATIONS WITH CYTOTOXIC AGENTS

The treatment of cancer with cytotoxic drugs is routinely carried out as combination therapy in order to achieve additive or synergistic antitumor effects while reducing the incidence and severity of side effects. We have explored the modulatory role of octreotide in combination with the cytotoxic agents taxol, 5-fluorouracil, doxorubicin, and mitomycin, which are used in the treatment of various malignancies such as pancreatic, breast, and colon cancer (Weckbecker, 1991; Pratt *et al.*, 1994). AR42J cells, which express sst_2, were exposed to the drug combinations *in vitro* and with selected combination therapies in AR42J tumor-bearing nude mice. The dose-dependent antiproliferative effects of taxol, doxorubicin, and mitomycin were synergistically enhanced by octreotide. Combinations of octreotide and 5-fluorouracil resulted in either additive or, at high concentrations of the cytotoxic agent, synergistic interactions (Weckbecker *et al.*, 1996). The potential of the combination of octreotide and 5-fluorouracil for the treatment of pancreatic cancer is consequently being studied in clinical trials (see Section 7.2).

In AR42J tumor-bearing nude mice, the combination of doxorubicin and octreotide was well tolerated. Tumor growth inhibition was clearly more pronounced with this drug combination relative to treatment with either single agent (Fig. 5). There was a tendency for octreotide to protect animals from doxorubicin-induced side effects and death.

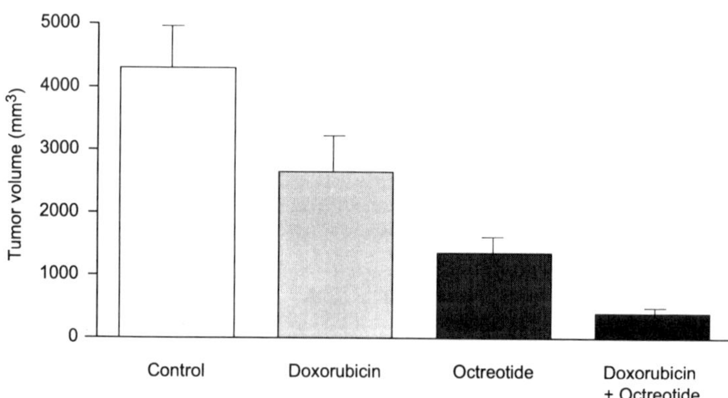

Figure 5. Effect of treatment with octreotide and doxorubicin, alone and in combination, on the size of AR42J tumors in nude mice. Mice received either 0.9% NaCl infusions (control), octreotide infusions at a rate of 0.72 mg/kg per day, doxorubicin (3 mg/kg intraperitoneally, twice weekly), or a combination of both drugs at the same dosages as in the single-agent arms. Tumor volumes were measured on day 22 of treatment. Data are means ± SE.

Taken together, these data clearly suggest that octreotide can potentiate the anticancer effects of both endocrine treatments and cytotoxic regimens without causing increased overt toxicity.

7. DEVELOPMENT OF OCTREOTIDE FOR ONCOLOGICAL USES BEYOND THE CONTROL OF DISEASE-RELATED SYMPTOMS IN GEP TUMORS

In addition to the control of disease-related symptoms of carcinoid tumors, VIPomas (pancreatic tumors characterized by release of vasoactive intestinal peptide), and other GEP tumors, objective tumor regression has been observed in some cases. The reported frequency of objective tumor shrinkage varies between 0 and 31% depending on the publication (Anthony *et al.,* 1993; Arnold *et al.,* 1993; DiBartolomeo *et al.,* 1996). Objective response rates of 10–20% (De Vries *et al.,* 1993) and disease stabilization in approximately one-third of patients (Arnold *et al.,* 1996) have been quoted. A dose-dependent effect was suggested by Anthony *et al.* (1993) in a phase I trial in which doses ranged from 1500 to 6000 μg/day. No maximum tolerated dose was found; the dose administered was limited by the volume of injection and not the safety profile of octreotide.

7.1. Somatostatin Receptor Binding and Growth Factor Suppression

Human tumors that express somatostatin receptors include adenocarcinomas and neuroendocrine tumors (Lamberts *et al.,* 1991; Weckbecker *et al.,* 1993). Specifically, there is a high incidence of somatostatin receptors in pituitary tumors, endocrine pancreatic tumors, carcinoid and other apudomas, neuroblastomas, meningiomas, Merkel cell tumors, small-cell lung carcinomas, and lymphomas (Reubi *et al.,* 1992). An intermediate level of expression is seen in adenocarcinomas of breast cancer and these is a low incidence of receptors in prostate, ovarian, cervical, endometrial, renal, gastric, and colorectal cancers (Reubi *et al.,* 1992).

The suppression of tumor growth factors expands the potential targets for octreotide therapy beyond tumors that express the sst_2 and sst_5 receptors. Gastrin, secretin, and cholecystokinin are growth factors for both normal and neoplastic tissues in the gastrointestinal tract (Johnson, 1981; Reichlin, 1983). IGF-1 is another potential tumor growth factor that is suppressed by octreotide. Tamoxifen also suppresses IGF-1 (Colletti *et al.,* 1989). The combination of tamoxifen and octreotide has been shown to suppress IGF-1 *in vivo* more quickly and profoundly than tamoxifen alone (Pollak *et al.,* 1996). Tumor types in which the IGFs have been implicated as potential growth factors are those of the human breast and pancreatic carcinomas (Myai *et al.,* 1984; Peyrat *et al.,* 1990; Bergmann *et al.,* 1996).

7.2. Clinical Trials

A phase I trial of the immediate release formulation of octreotide was performed in patients with metastatic breast cancer who had failed treatment with hormonal or chemotherapy (Somlo *et al.,* 1993).

Five cancer indications were selected by our group to be investigated in phase II clinical trials based on either the presumed presence of somatostatin receptors that would bind octreotide, or the ability of octreotide to suppress potential tumor growth factors. These tumor types were: metastatic breast cancer (estrogen receptor-positive or -negative), small-cell lung cancer, gastric cancer, colorectal cancer, and adenocarcinoma of the pancreas.

The pancreatic cancer trial was the only phase II trial of this group that appeared positive, utilizing matched historical controls to evaluate survival. The trial involved 49 patients with unresectable stage II, III, or IV adenocarcinoma of the pancreas, and 44 of these were evaluable. The median survival of patients treated with high-dose octreotide was 5.9 months, compared with 4.0 months in patients receiving only best supportive care (Buechler *et al.,* 1994). These results echoed those of an independent trial performed by Ebert *et al.* (1994) in which low-dose (200 μg) versus high-dose (2000 μg) octreotide, three times daily, was evaluated in unresectable pancreatic cancer patients: The two groups had median survivals of 4.0 and 6.0 months, respectively. The investigators in both the high-dose-only study and the low-versus-high-dose study also noted that some patients had improved appetites and weight gain, as well as a better quality of life.

Currently, clinical trials are ongoing to determine the efficacy of the long-acting preparation of octreotide, OncoLAR™, in combination with antiestrogens (tamoxifen) in the adjuvant treatment of breast cancer and in the treatment of metastatic breast cancer. The primary efficacy endpoint for the metastatic breast trial is progression-free survival. The metastatic breast cancer trial will evaluate the patients' quality of life using a European Organization for Research on the Treatment of Cancer quality of life questionnaire (QLQ-30, version 2.0). Two clinical trials in unresectable pancreatic cancer are also ongoing: In one the efficacy of OncoLAR™ in combination with 5-fluorouracil is being evaluated and in the other, OncoLAR™ is being studied as monotherapy.

8. RADIOLABELED OCTREOTIDE ANALOGUES

A fascinating project with respect to oncological applications is the *in vivo* targeting of sst_2-positive tumors and their metastases in patients, using intravenous injection of radiolabeled octreotide analogues (Krenning *et al.,* 1995).

8.1. Imaging of Tumors with OctreoScan®

The concept of targeting sst_2-expressing tumors for imaging purposes was first realized using ^{125}I- and ^{123}I-labeled Tyr^3-octreotide (Krenning *et al.*, 1989; Schirmer *et al.*, 1993). However, the hepatobiliary clearance of this analogue results in relatively high nonspecific accumulation of radioactivity in both the liver and intestine, rendering the interpretation of scintigrams of the abdominal region more difficult. In addition, the labeling procedure is cumbersome and time-consuming. To overcome these major drawbacks, the I label was replaced by ^{111}In, which also improves the scintigraphy data obtained at later time points by virtue of its longer half-life.

To facilitate the labeling of octreotide with ^{111}In, a new analogue was developed by our group in collaboration with Erasmus University (Rotterdam, The Netherlands) and Mallinckrodt Medical (St. Louis, MO). It contains a diethylamine tetramine penta-acetic acid (DTPA) group coupled to the α-NH_2 group of the N-terminal D-Phe^1 residue (DTPA-D-Phe^1-octreotide) (Bakker *et al.*, 1991). *In vitro* receptor binding studies using ^{111}In-DTPA-D-Phe^1-octreotide showed nanomolar affinities to sst_2; the unlabeled analogue displayed somatostatinlike inhibitory effects on GH release *in vitro* and *in vivo*. After injection of ^{111}In-DTPA-D-Phe^1-octreotide into tumor-bearing rats, the radioligand rapidly binds to sst_2 on the tumor cells (Bruns *et al.*, 1993). The tumor may be visualized by gamma camera scintigraphy 5 min after injection of the radioligand. The rapid appearance of radioactivity in the urine clearly indicates the effective renal clearance of the peptide radioligand. There is little uptake of radioactivity in the intestines and liver, which facilitates the localization of small tumors and their metastases in the abdominal region.

The metabolic properties of ^{111}In-DTPA-D-Phe^1-octreotide were found to be similar in rats and humans. After intravenous administration, ^{111}In-DTPA-D-Phe^1-octreotide is rapidly cleared from the circulation by the kidneys (Krenning *et al.*, 1993). ^{111}In-DTPA-D-Phe^1-octreotide was introduced to the market as OctreoScan®111 in 1994. Since then, the demonstrable incidence of somatostatin receptors in many different tumor types is increasing, from 60% in neuroendocrine tumors (insulinomas) for example, to 96 or 100% (carcinoid and small-cell lung cancer, respectively) (Krenning *et al.*, 1993). The use of somatostatin receptor scintigraphy for the localization and staging of tumors should allow optimal therapy to be selected for each patient (Krenning *et al.*, 1995).

8.2. Tumor Radiotherapy with SMT 487

Radiotherapy with radiolabeled somatostatin analogues would be an extension of the sst_2 imaging approach. One of the preferred radionuclides suggested for radiotherapy is ^{90}Y, which is a hard beta-particle emitter with a mean range of about 5 mm in tissue. It is commercially available in sufficient amounts and in a

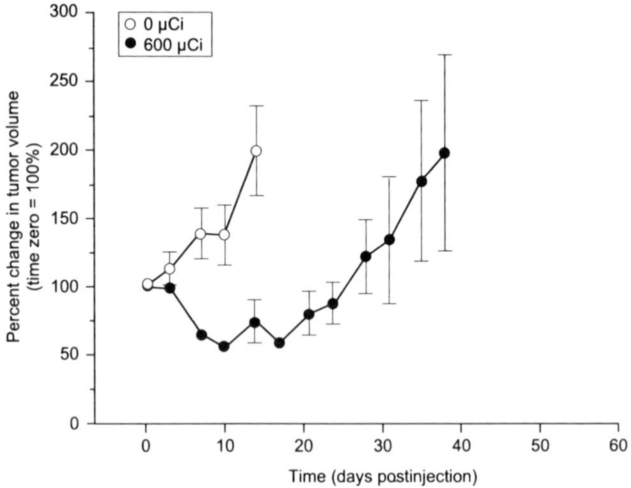

Figure 6. SMT 487: radiotherapeutic effect in human small-cell lung cancer (NCI-H69)-bearing nude mice after a single injection of 600 μCi. Unlabeled analogue was used as a control.

no-carrier-added form that allows the preparation of a radioligand with high specific activity. ^{90}Y has a short half-life (64.1 hr), which is compatible with the fast clearance rate usually expected for peptide–ligand conjugates.

A newly synthesized somatostatin analogue, SMT 487, consists of the targeting vehicle Tyr^3-octreotide and the chelating moiety 1,4,7,10-tetraazacyclododecane-1,4,7,10-tetraacetic acid (DOTA). Y–DOTA complexes are extremely stable ($\log K = 24.9$). The radioligand exhibits nanomolar binding affinity to sst$_2$ receptors and accumulates specifically in sst$_2$-expressing tumors *in vivo.* In an experimental mouse tumor model, an sst$_2$-expressing human small-cell lung tumor, a dose-dependent regression down to 50% of the initial tumor volume was observed in response to a single injection of SMT 487 (Fig. 6).

Both single and repeated treatment with SMT 487 resulted in a significant increase in survival rate as a consequence of tumor shrinkage.

Receptor-targeted radiotherapy with ^{90}Y-labeled octreotide analogues represents a new strategy for the treatment of sst$_2$-expressing tumors, opening up entirely new possibilities for tumor-targeted treatment of patients suffering sst$_2$-expressing cancer. At present, SMT 487 is in early clinical trials, which include patients with carcinoid and GEP tumors, small-cell lung cancer, and lymphomas.

9. SUMMARY AND OUTLOOK

The history of Sandostatin® started in 1973 with the discovery of somatostatin, which was first described as a natural hypothalamic, GH release-inhibiting factor.

The drug discovery phase involved biochemical approaches using *in vitro* pituitary cell culture models and *in vivo* animal experiments. The entire development process was a collaborative approach, both between preclinical and clinical development teams within our company as well as between researchers within and outwith it. This strategy made it possible to integrate into the development process all of the findings that emerged as the somatostatin research progressed. These included discoveries such as the ubiquitous distribution of somatostatin and its prominent role as an endogenous regulator of hormones and growth factors, and its antiproliferative activity against neoplastic cells.

The wide range of potential therapeutic applications for Sandostatin® demanded the development of alternative formulations to circumvent cumbersome multiple daily subcutaneous injections. The LAR formulations designed specifically for low-dose endocrinological and gastrointestinal applications (Sandostatin® LAR®) and for high-dose oncological applications (OncoLAR™) are attractive alternatives for the patient.

An important finding that emerged during the development of Sandostatin® was the discovery of somatostatin receptor expression in many tumors. A profound understanding of structure–activity relationships permitted design of analogues for tumor imaging (OctreoScan®) and target-directed radiotherapy (SMT 487).

The discovery and development of Sandostatin® was based mainly on classical chemical, biochemical, and pharmacological approaches. However, our view of the physiology and pathophysiology of somatostatin was broadened considerably by the influence of molecular biology. In particular, the cloning of five somatostatin receptor subtypes was a great stimulus to current research on somatostatin analogues with subtype profiles differing from that of octreotide (mainly having a high affinity toward sst_2). New subtype-selective somatostatin analogues, as well as universal ones, might provide new methods of treatment. Furthermore, rational drug design and molecular modeling using experimentally refined receptor models should help in the successful development of new somatostatin peptidomimetics and antagonists with clinical potential, especially in indications related to the CNS. It still remains to be shown whether all of these fundamentally new approaches will give rise to the discovery and development of new, therapeutically useful somatostatin analogues.

REFERENCES

Adrian, T. E., Barnes, A. J., Long, R. G., O'Shaughnessy, D. J., Brown, M. R., Rivier, J., Vale, W., Blackburn, A. M., and Bloom, S. R., 1979, The effect of somatostatin analogs on secretion of growth, pancreatic, and gastrointestinal hormones in man, *J. Clin. Endocrinol. Metab.* **53:**675–681.

Anthony, L., Johnson, D., Hande, K., Shaff, M., Winn, S., Krozely, M., and Oates, J., 1993, Somatostatin analogueue phase I trials in neuroendocrine neoplasms, *Acta Oncol.* **32:**217–223.

Arnold, R., Neuhaus, C., Benning, R., Schwerk, W. B., Trautmann, M. E., Joseph, K., and Bruns, C., 1993, Somatostatin analogue sandostatin and inhibition of tumor growth in patients with metastatic endocrine gastroenteropancreatic tumors, *World J. Surg.* **17:**511–519.

Arnold, R., Trautmann, M. E., Creutzfeldt, W., Benning, R., Benning, M., Neuhaus, C., Juergensen, R., Stein, K., Schaefer, H., Bruns, C., and Dennler, H. J., 1996, Somatostatin analogueue octreotide and inhibition of tumour growth in metastatic endocrine gastroenteropancreatic tumours, *Gut* **38:**430–438.

Bakker, W. H., Alberts, R., Bruns, C., Breeman, W. A. P., Hofland, L. J., Marbach, P., Pless, J., Pralet, D., Stolz, B., Koper, J. W., Lamberts, S. W. J., Visser, T. J., and Krenning, E. P., 1991, ^{111}In-DTPA-DPhe^{1}-octreotide, a potential radiopharmaceutical for imaging of somatostatin receptor-positive tumors: Synthesis, radiolabelling and in vitro validation, *Life Sci.* **49:**1583–1591.

Barrie, R., Woltering, E. A., Hajarizadeh, H., Mueller, C., Ure, T., and Fletcher, W. S., 1993, Inhibition of angiogenesis by somatostatin and somatostatin-like compounds is structurally dependent, *J. Surg. Res.* **55:**446–450.

Bauer, W., Briner, U., Doepfner, W., Haller, R., Huguenin, R., Marbach, P., Petcher, T. J., and Pless, J., 1982a, SMS 201–995: A very potent and selective octapeptide analogue of somatostatin with prolonged action, *Life Sci.* **31:**1133–1140.

Bauer, W., Briner, U., Doepfner, W., Haller, R., Huguenin, R., Marbach, P., Petcher, T. J., and Pless, J., 1982b, Structure–activity relationships of highly potent and specific octapeptide analogues of somatostatin, in: *Proceedings of the European Peptide Symposium, Prague* (K. Blaha and P. Malon, eds.), pp. 583–588, de Gruyter, Berlin.

Bergmann, U., Yokoyama, M., Berger, H. G., and Korc, M., 1996, Insulin-like growth factor-II is expressed in human islet cells and is mitogenic in pancreatic cancer cells, *Proc. Am. Assoc. Cancer Res.* **37:**1504A.

Bloom, S. R. (ed.), 1990, Somatostatin analogueues–peptide therapy in action: Gastro-enteropancreatic tumors, *Digestion* **45**(Suppl.1)**:**1–43.

Bodmer, D., Kissel, T., and Traechslin, E., 1992, Factors influencing the release of peptides and proteins from biodegradable parenteral depot systems, *J. Control. Rel.* **21:**129–138.

Bodmer, D., Gengenbacher, T., and Esparza, I., 1994, Stabilization of pharmacologically active compounds in sustained release compositions, European Patent EP 0 626 170A2.

Brazeau, P., Vale, W., Burgus, R., Ling, N., Butcher, M., Rivier, J., and Guillemin, R., 1973, Hypothalamic polypeptide that inhibits the secretion of immunoreactive pituitary growth hormone, *Science* **179:**77–79.

Bruns, C., Stolz, B., Albert, R., Marbach, P., and Pless, J., 1993, OctreoScan®111 for imaging of a somatostatin receptor-positive islet cell tumor in rat, *Horm. Metab. Res.* **27**(Suppl.)**:**5–11.

Bruns, C., Weckbecker, G., Raulf, F., Kaupmann, K., Schoeffter, P., Hoyer, D., and Lübbert, H., 1994, Molecular pharmacology of somatostatin-receptor subtypes, *Ann. N.Y. Acad. Sci.* **733:**138–146.

Bruns, C., Weckbecker, G., Raulf, F., Lübbert, H., and Hoyer, D., 1995, Characterization of somatostatin receptor subtypes, *Ciba Found. Symp.* **190:**89–101.

Bruns, C., Raulf, F., Hoyer, D., Schloos, J., Lübbert, H., and Weckbecker, G., 1996, Binding properties of somatostatin receptor subtypes, *Metabolism* **45**(Suppl. 1)**:**17–20.

Buechler, M., Sulkowski, U., Pederzoli, P., Arnold, R., Dinse, P., Mietlowski, W., and Israel, R., 1994, A phase II study of octreotide (SMS) in advanced pancreatic cancer, *Proc. Am. Soc. Clin. Oncol.* **13:**A642.

Buscail, L., Delesque, N., Estève, J.-P., Saint Laurent, N., Prats, H., Clerc, P., Robberecht, P., Bell, G. I., Liebow, C., Schally, A. V., Vaysse, N., and Susini, C., 1994, Stimulation of tyrosine phosphatase and inhibition of cell proliferation by somatostatin analogueues: Mediation by human somatostatin receptor subtypes SSTR1 and SSTR2, *Proc. Natl. Acad. Sci. USA* **91:**2315–2319.

Colletti, R. B., Roberts, J. D., Devlin, J. T., and Copeland, K. C., 1989, Effect of tamoxifen on plasma insulin-like growth factor I in patients with breast cancer, *Cancer Res.* **49:**1882–1884.

Danesi, R., and Del Tacca, M., 1996, Effects of octreotide on angiogenesis, in: *Octreotide: From Basic Science to Clinical Medicine* (C. Scarpignato, ed.), pp. 234–245, Karger, Basel.

De Vries, E. G., Kema, I. P., Slooff, M. J., Verschueren, R. C., Kleibeuker, J. H., Mulder, N. H., Sleijfer, D. T., and Willemse, P. H., 1993, Recent developments in diagnosis and treatment of metastatic carcinoid tumours, *Scand. J. Gastroenterol. Suppl.* **200:**87–93.

DiBartolomeo, M., Bajetta, E., Buzzoni, R., Mariani, L., Carnaghi, C., Somma, L., Zilembo, N., and diLeo, A., 1996, Clinical efficacy of octreotide in the treatment of metastatic neuroendocrine tumors, *Cancer* **77**:402–408.

Ebert, M., Friess, H., Beger, H. G., and Buchler, M. W., 1994, Role of octreotide in the treatment of pancreatic cancer, *Digestion* **55**(Suppl. 1):48–51.

Farthing, M. J. G. (ed.), 1990, Somatostatin analogueues–peptide therapy in action: Perspectives in gastroenterology, *Digestion* **45**(Suppl. 1):45–87.

Flueckiger, E., Richardson, B., Donatsch, P., and Enz, A., 1983, Neuroendocrine aspects of the aging rat, in: *Aging Brain and Ergot Alkaloids* (A. Agnoli, ed.), pp. 61–71, Raven Press, New York.

Grass, P., Marbach, P., Bruns, C., and Lancranjan, I., 1996, Sandostatin LAR (microencapsulated octreotide acetate) in acromegaly: Pharmacokinetic and pharmacodynamic relationships, *Metabolism* **45**(Suppl. 2):27–30.

Grasso, P., 1976, Review of tests for carcinogenicity and their significance to man, *Clin. Toxicol.* **9**:745–760.

Guillemin, R.,1992, Somatostatin: The early days, *Metabolism* **41**(Suppl.2):2–4.

Harris, A. G., Kokoris, S. P., and Ezzat, S., 1995, Continuous versus intermittent subcutaneous infusion of octreotide in the treatment of acromegaly, *J. Clin. Pharmacol.* **35**:59–71.

Huynh, H. T., and Pollak, M. N., 1994, Enhancement of tamoxifen-induced suppression of insulin-like growth factor I gene expression and serum level by somatostatin analogue, *Biochem. Biophys. Res. Commun.* **203**:253–259.

Johnson, L. R., 1981, Effects of gastrointestinal hormones on pancreatic growth, *Cancer* **47**:1640–1645.

Kaal, A., Frystyk, J., Skjaerbaek, C., Nielsen, S., Jorgensen, J. O. L., Bruns, C., Marbach, P., Lancranjan, I., Weeke, J., and Orskov, H., 1995, Effects of intramuscular microsphere-encapsulated octreotide on serum growth hormone, insulin-like growth factors (IGFs), free IGFs, and IGF-binding proteins in acromegalic patients, *Metabolism* **44**(Suppl. 1):6–14.

Kaupmann, K., Bruns, C., Raulf, F., Weber, H. P., Mattes, H., and Lübbert, H., 1995, Two amino acids, located in transmembrane domains VI and VII, determine the selectivity of the peptide agonist SMS 201–995 for the SSTR2 somatostatin receptor, *EMBO J.* **14**:727–735.

Kluxen, F. W., Bruns, C., and Lübbert, H., 1992, Expression cloning of a rat brain somatostatin receptor cDNA, *Proc. Natl. Acad. Sci. USA* **89**:4618–4622.

Krejs, G. J. (ed.), 1987, Gastrointestinal endocrine tumors: Diagnosis and management; role of somatostatin analogue SMS 201–995, *Am. J. Med.* **83**(5B):1–99.

Krenning, E. P., Bakker, W. H., Breeman, W. A. P., Koper, J. W., Kooij, P. P. M., Ausema, L., Lameris, J. S., Reubi, J. C., and Lamberts, S. W. J., 1989, Localization of endocrine related tumors with radioiodinated analogueue of somatostatin, *Lancet* **1**:242–245.

Krenning, E. P., Kwekkeboom, D. J., Bakker, W. H., Breeman, W. A. P., Kooij, P. P. M., Oei, H. Y., Van Hagen, P. M., Postema, P. T. E., De Jong, M., Reubi, J. C., Visser, T. J., Reijs, A. E. M., Hofland, L. J., Koper, J. W., and Lamberts, S. W. J., 1993, Somatostatin receptor scintigraphy with [^{111}In-DTPA-D-Phe^{1}]- and [^{123}I-Tyr^{3}]-octreotide: The Rotterdam experience with more than 1000 patients, *Eur. J. Nucl. Med.* **20**:283–292.

Krenning, E. P., Kwekkeboom, D. J., Pauwels, S., Kvols, L. K., and Reubi, J. C., 1995, Somatostatin receptor scintigraphy, in: *Nuclear Medicine Annual* (L. M. Freeman, ed.), pp. 1–50, Raven Press, New York.

Lamberts, S. W. J., Krenning, E. P., and Reubi, J. C., 1991, The role of somatostatin and its analogues in the diagnosis and treatment of tumors, *Endocr. Rev.* **12**:450–482.

Lancranjan, I., Bruns, C. H., Grass, P., Jaquet, P., Jervell, J., Kendall-Taylor, P., Lamberts, S. W. J., Marbach, P., Orskov, H., Pagani, G., Sheppard, M., and Simionescu, L., 1996, Sandostatin LAR: A promising therapeutic tool in the management of acromegalic patients, *Metabolism* **45**(Suppl. 1):67–71.

Lemaire, M., Azria, M., Dannecker, R., Marbach, P., Schweitzer, A., and Maurer, G., 1989, Disposition of Sandostatin, a new synthetic somatostatin analogueue, in rats, *Drug Metab. Dispos.* **17**:699–705.

Marbach, P., Goetz, U., Veteau, J. P., and Wagner, H., 1978, RIA analysis by means of non-linearized response functions, in: *Radioimmunoassay and Related Procedures in Medicine,* pp. 383–397, IAEA, Vienna.

Marbach, P., Neufeld, M., and Pless, J., 1985, Clinical applications of somatostatin analogues, *Adv. Exp. Med. Biol.* **188:**339–353.

Marbach, P., Andres, H., Azria, M., Bauer, W., Briner, U., Buchheit, K. H., Doepfner, W., Lemaire, M., Petcher, T. J., Pless, J., and Reubi, J. C., 1988, Chemical structure, pharmacodynamic profile and pharmacokinetics of SMS 201–995 (Sandostatin), in: *Sandostatin in the Treatment of Acromegaly* (S. W. J. Lamberts, ed.), pp. 53–60, Springer-Verlag, Berlin.

Marbach, P., Briner, U., Lemaire, M., Schweitzer, A., and Terasaki, T., 1992, From somatostatin to Sandostatin: Pharmacodynamics and pharmacokinetics, *Metabolism* **41**(Suppl. 2)**:**1–10.

Martin, J. L., Chesselet, M. F., Raynor, K., Gonzales, C., and Reisine, T., 1991, Differential distribution of somatostatin receptor subtypes in rat brain revealed by newly developed somatostatin analogues, *Neuroscience* **41:**581–593.

Mergler, M., Hellstern, H., Wirth, W., Langer, W., Gysi, P., and Prikoszowich, W., 1991, A new acid-labile linker for the solid-phase synthesis of peptides with C-terminal threoninol, Poster presented at the Twelfth American Peptide Symposium, Boston.

Myai, Y., Shiu, R. P. C., Bhaumick, B., and Bala, M., 1984, Receptor binding and growth-promoting activity of insulin-like growth factors in human breast cancer cells (T-47D) in culture, *Cancer Res.* **44:**5486–5490.

O'Dorisio, T. M. (ed.), 1986, Neuroendocrine disorders of the gastroenteropancreatic system; clinical applications of the somatostatin analogue SMS 201–995, *Am. J. Med.* **81**(6B)**:**1–101.

Panetta, R., Greenwood, M. T., Warszynska, A., Demchyshyn, L. L., Day, R., Niznik, H. B., Srikant, C. B., and Patel, Y. C., 1994, Molecular cloning, functional characterization, and chromosomal localization of a human somatostatin receptor (somatostatin receptor type 5) with preferential affinity for somatostatin-28, *Mol. Pharmacol.* **45:**417–427.

Patel, Y. C., Greenwood, M. T., Panetta, R., Demchyshyn, L., Niznik, H., and Srikant, C. B., 1995, The somatostatin receptor family, *Life Sci.* **5:**1249–1265.

Peyrat, J. P., Bonneterre, J., Vennin, P. H., Jammes, H., Beuscart, R., Hecquet, B., Djiane, J., Lefebvre, J., and Demaille, A., 1990, Insulin-like growth factor 1 receptors (IGF1-R) and IGF1 in human breast tumors, *J. Steroid Biochem. Mol. Biol.* **37:**823–827.

Pohl, E., Heine, A., Sheldrick, G., Dauter, Z., Wilson, K., Kallen, J., Huber, W., and Pfaeffli, P., 1995, Structure of octreotide, *Acta Crystallogr.* **D51:**48–59.

Pollak, M. N., Ingle, J. N., Susman, V. J., Kugler, J. W., Nickerson, T., and Deroo, B., 1996, Enhancement of tamoxifen-induced suppression of serum IGF-1 levels in metastatic breast cancer patients by co-administration of somatostatin analogueue octreotide, *Proc. Am. Assoc. Cancer Res.* **37:**1167A.

Pratt, W. B., Ruddon, R. W., Ensminger, W. D., and Maybaum, J., 1994, *The Anticancer Drugs,* 2nd ed., Oxford University Press, London.

Raulf, F., Pérez, J., Hoyer, D., and Bruns, C., 1994, Differential expression of five somatostatin receptor subtypes, SSTR1–5, in the CNS and peripheral tissue, *Digestion* **55**(Suppl. 3)**:**46–53.

Reichlin, S., 1983, Somatostatin, *N. Engl. J. Med.* **309:**1495–1501.

Reisine, T., and Bell, G. I., 1995, Molecular biology of somatostatin receptors, *Endocr. Rev.* **16:**427–442.

Reubi, J. C., 1984, Evidence for two somatostatin-14 receptor subtypes in rat brain cortex, *Neurosci. Lett.* **49:**259–263.

Reubi, J. C., Krenning, E., Lamberts, S. W. J., and Kvols, L., 1992, In vitro detection of somatostatin receptors in human tumors, *Metabolism* **41:**104–110.

Reubi, J. C., Horisberger, U., and Laissue, J., 1994, High density of somatostatin receptors in veins surrounding human cancer tissue: Role in tumor–host interaction? *Int. J. Cancer* **56:**681–688.

Rivier, J., Brown, M., and Vale, W., 1975, D-Trp^8-somatostatin: An analogue of somatostatin more potent than the native molecule, *Biochem. Biophys. Res. Commun.* **65:**746–751.

Rohrer, L., Raulf, F., Bruns, C., Buettner, R., Hofstaedter, F., and Schüle, R., 1993, Cloning and characterization of a fourth human somatostatin receptor, *Proc. Natl. Acad. Sci. USA* **90:**4196–4200.

Schirmer, W. J., O'Dorisio, T. M., Schirmer, T. P., Mojzisik, C. M., Hinkle, G. H., and Martin, E. W., 1993, Intraoperative localization of neuroendocrine tumors with ^{125}I-Tyr^3-octreotide and a hand held gamma detecting probe, *Surgery* **114:**745–752.

Serri, O., Brazeau, P., Kachra, Z., and Posner, B., 1992, Octreotide inhibits insulin-like growth factor-I hepatic gene expression in the hypophysectomized rat: Evidence for a direct and indirect mechanism of action, *Endocrinology* **130:**1816–1821.

Somlo, G., Vogel, C., Flamm Honig, S., Linnartz, R., and Israel, R., 1993, A phase I study of octreotide (SMS) in patients with advanced breast cancer, *Proc. Am. Soc. Clin. Oncol.* **12:**A172.

Srikant, C. B., and Shen, S. H., 1996, Octapeptide somatostatin analogue SMS 201–995 induces translocation of intracellular PTP1C to membranes in MCF-7 human breast adenocarcinoma cells, *Endocrinology* **137:**3461–3468.

Theiss, J. C., 1982, Utility of injection site tumorigenicity in assessing the carcinogenicity risk of chemicals to man, *Regul. Toxicol. Pharmacol.* **2:**213–222.

Vale, W., Rivier, J., Ling, N., and Brown, M., 1978, Biological and immunologic activities and application of SRIF analogues, *Metabolism* **27**(Suppl. 1)**:**1391–1401.

Veber, D., Holly, F. W., Paleweda, W. J., Nutt, R. F., Bergstrand, S. J., Torchiana, M., Glitzer, M. S., Saperstein, R., and Hirschmann, R., 1978, Conformationally restricted bicyclic analogues of somatostatin, *Proc. Natl. Acad. Sci. USA* **75:**2636–2640.

Veber, D., Saperstein, R., Nutt, R. F., Freidinger, R. M., Brady, S. F., Curley, P., Perlow, D. S., Paleweda, W. J., Dylion Colton, C., Zacchei, A. G., Tocco, D. J., Hoff, D. R., Vandlen, R. L., Gerich, J. E., Hall, L., Mandarino, L., Cordes, E. H., Anderson, P. S., and Hirschmann, R., 1984, A super active cyclic hexapeptide analogue of somatostatin, *Life Sci.* **34:**1371–1378.

Weckbecker, G., 1991, Biochemical pharmacology and analysis of fluoropyrimidines alone and in combination with modulators, *Pharmacol. Ther.* **50:**367–424.

Weckbecker, G., Liu, R., Tolcsvai, L., and Bruns, C., 1992a, Antiproliferative effects of the somatostatin analogueue octreotide (SMS 201–995) on ZR-75–1 human breast cancer cells in vivo and in vitro, *Cancer Res.* **52:**4973–4978.

Weckbecker, G., Tolcsvai, L., Liu, R., and Bruns, C., 1992b, Preclinical studies on the anticancer activity of the somatostatin analogueue octreotide, SMS 201–995, *Metabolism* **41:**99–103.

Weckbecker, G., Raulf, F., Stolz, B., and Bruns, C., 1993, Somatostatin analogues for diagnosis and treatment of cancer, *Pharmacol. Ther.* **60:**245–264.

Weckbecker, G., Tolcsvai, L., Stolz, B., Pollak, M., and Bruns, C., 1994, Somatostatin analogue octreotide enhances the antineoplastic effects of tamoxifen and ovariectomy on 7,12-dimethylbenz(*a*)anthracene-induced rat mammary carcinomas, *Cancer Res.* **54:**6334–6337.

Weckbecker, G., Raulf, F., Tolcsvai, L., and Bruns, C., 1996, Potentiation of the antiproliferative effects of anticancer drugs by octreotide in vitro and in vivo, *Digestion* **57:**22–28.

Yamada, Y., Post, S. R., Wang, K., Tager, H. S., Bell, G. I., and Seino, S., 1992a, Cloning and functional characterization of a family of human and mouse somatostatin receptors expressed in brain, gastrointestinal tract, and kidney, *Proc. Natl. Acad. Sci.* USA **89:**251–255.

Yamada, Y., Reisine, T., Law, S. F., Ihara, Y., Kubota, A., Kagimoto, S., Seino, M., Seino, Y., Bell, G. I., and Seino, S., 1992b, Somatostatin receptors, an expanding gene family: Cloning and functional characterization of human SSTR3, a protein coupled to adenylyl cyclase, *Mol. Endocrinol.* **6:**2136–2142.

Chapter 10

Factors Impacting the Delivery of Therapeutic Levels of Pyrone-Based HIV Protease Inhibitors

Guy E. Padbury, Gail L. Zipp, Francis J. Schwende, Zhiyang Zhao, Kenneth A. Koeplinger, Kong Teck Chong, Thomas J. Raub, and Suvit Thaisrivongs

1. INTRODUCTION

1.1. HIV Protease as a Therapeutic Target

The rapid and continuing spread of acquired immune deficiency syndrome (AIDS) has resulted in an intensive worldwide effort to identify and develop therapeutic agents to arrest the replication of the causative virus of this disease, the human immunodeficiency virus (HIV-1). The fact that four chapters of this book are devoted to the development of anti-HIV therapeutics is reflective of the intensity of the research activity currently being invested in this disease area. Retroviruses, including HIV, possess a number of unique enzyme activities that are not present in humans and serve as rational targets for therapeutic intervention. Foremost of these targets are the virus-encoded reverse transcriptase and protease enzymes for which

---

Guy E. Padbury, Gail L. Zipp, Francis J. Schwende, Zhiyang Zhao, Kenneth A. Koeplinger, Kong Teck Chong, Thomas J. Raub, and Suvit Thaisrivongs • Pharmacia & Upjohn, Inc., Kalamazoo, Michigan 49007.

Integration of Pharmaceutical Discovery and Development: Case Studies, edited by Borchardt *et al.*, Plenum Press, New York, 1998.

a number of small-molecule therapeutics have received FDA approval and several others are under preclinical or clinical testing (Johnston and Hoth, 1993). Aspects of the Pharmacia & Upjohn (PNU) HIV reverse transcriptase inhibitor program are discussed in a separate chapter of this book. The virus-encoded aspartyl protease mediates the posttranslational cleavage of two HIV mRNA-derived polyproteins, gag (p55) and gag-pol (p160), into individual peptides during the final stages of viral maturation (Peng *et al.,* 1989; Oroszlian and Luftig, 1990; Debouck and Metcalf, 1990). It has been well established that when the retroviral protease is catalytically defective, viral maturation in HIV-infected cell culture is blocked, and consequently, infection is arrested (Kohl *et al.,* 1988; Louis *et al.,* 1989). The viral protease is a paradigm for rational drug design. It is a member of a well-characterized mechanistic set of aspartyl proteases, and as such, much is known by inference regarding its structure and mechanism. However, the HIV-1 enzyme is unique among all known aspartyl proteases in having a C2 symmetrical dimer structure, which, accordingly, offers a virus-selective therapeutic target. The enzyme has been cloned, expressed, purified in large quantities, and has been manipulated genetically to probe structure–activity relationships (SAR). It is amenable to crystallization for high-resolution structural characterization of the inhibitor binding site and enzyme–inhibitor complexes. Each monomer contributes a catalytic β-carbonyl group from the side chain of an aspartyl residue brought in close proximity by the protein fold. Mechanistic studies indicate that the two aspartyl residues facilitate general acid–base peptide cleavage where the nucleophile is, most likely, an activated water molecule (Hyland *et al.,* 1991a,b; Jaskolski *et al.,* 1991).

From a drug development standpoint, there are no readily available *in vivo* models of AIDS. As a consequence, the potency of potential HIV protease inhibitors was assessed using *in vitro* methods. Two *in vitro* approaches were used for the primary assessment for HIV-1 protease inhibitory activity: (1) inhibition of recombinantly expressed HIV-1 protease enzyme and (2) the inhibition of viral replication in cell culture. In the former, compounds were assessed as competitive inhibitors (K_i) of the cleavage of a peptide substrate by the purified enzyme. In the latter, human white blood cells, either primary peripheral blood mononuclear cells (PBMC), or established transformed T-cell lines such as H-9 or MT-4 in cell culture, were infected with HIV. The replication of virus in either the presence or absence of drug was measured. Activity of a compound was expressed in terms of the drug concentration required to inhibit viral replication by 50% (IC_{50}) or 90% (IC_{90}) relative to drug-free controls. Secondary activity assessments included further *in vitro* determinations of activity against other infected cell lines, clinical isolates, strains resistant to other anti-HIV drugs, and the extent of cross-reactivity with normal mammalian aspartyl proteases including pepsin, renin, and the cathepsins. The HIV-1 protease enzyme and its inhibitors have been reviewed extensively over recent years and the reader is referred to these sources for further detail (Huff, 1991; Meek, 1992; Tomasselli *et al.,* 1996; Chong, 1996; Darke and Huff, 1994).

1.2. Pyrone-Based Inhibitors

PNU's HIV protease inhibitor program was an outgrowth of an earlier effort to identify peptidic inhibitors of the human aspartyl protease, renin (Greenlee, 1990; Tomasselli *et al.*, 1990). Initial efforts to develop inhibitors of HIV protease were focused on peptidomimetic compounds containing transition-state inserts in place of the dipeptide cleavage site of the normal enzyme substrate (PNU-75875, Fig. 1) (Ashorn *et al.*, 1990; Thaisrivongs *et al.*, 1991). The low oral bioavailability, rapid biliary excretion, and complicated syntheses of the peptide-derived compounds limited their utility as potential therapeutic agents (Thaisrivongs, 1994; Redshaw, 1994; West and Fairlie, 1995; Plattner and Norbeck, 1990). As a consequence, research was subsequently directed toward the identification of inhibitors with reduced peptidic nature and improved pharmacokinetic properties using a combination of *de novo* high-volume template screening followed by three-dimensional crystal structure-based template optimization. Employing a high-volume broad screening effort for HIV-1 protease inhibitory activity of a set of 5000 dissimilar compounds from the PNU collection, the 4-hydroxycoumarin, warfarin (Fig. 1), was identified as a weak inhibitor (IC_{50} = 30 μM). Warfarin was also subsequently reported as having an antiviral effect on HIV replication and spread (Bourinbaiar *et al.*, 1993) and a number of warfarin derivatives have been reported as competitive inhibitors of HIV protease by other workers (Tummino *et al.*, 1994a,b; Vara Prasad *et al.*,1994, 1995; Lunney *et al.*, 1994). On the basis of the 4-hydroxycoumarin structure, additional compounds from a similarity search of the PNU collection were tested as potential inhibitors, and another 4-hydroxycoumarin, phenprocoumon (Fig. 1), was found with significantly improved inhibitory activity (K_i = 1 μM) (Thaisrivongs *et al.*, 1994). Very importantly, both warfarin and phenprocoumon had been used as oral therapeutic agents in humans, suggesting a promising lead nonpeptide structure for an orally bioavailable therapeutic (Hirsh *et al.*, 1992). To facilitate the lead optimization and refinement process, crystal structures of the HIV-1 protease inhibitor complex formed the basis of iterative cycles of structure-based design of more active analogues. Iterative cycles identified the pyrone PNU-96988 (K_i = 38 nM, IC_{90} = 10 μM, Fig. 1) as the first clinical candidate (Thaisrivongs *et al.*, 1994, 1995). Further structure-based drug design led to the discovery of the sulfonamide-containing PNU-103017 (K_i < 1 nM, IC_{90} = 5 μM, Fig. 1) as the second-generation clinical candidate (Romines *et al.*, 1995a,b; Skulnick *et al.*, 1995). Most recently, compounds in the 5,6-dihydro-4-hydroxy-2-pyrone templates, such as PNU-140690 (K_i < 0.007 nM, IC_{90} = 0.1 μM, Fig. 1), were shown to have a greater than 100 fold increase in potency and by the time this book is published this compound will be under clinical testing (Thaisrivongs *et al.*, 1996a,b). For the purposes of this chapter, selected examples from each "generation" of the pyrone-based inhibitors will be used to illustrate the integration of drug delivery concepts into the drug discovery and development process.

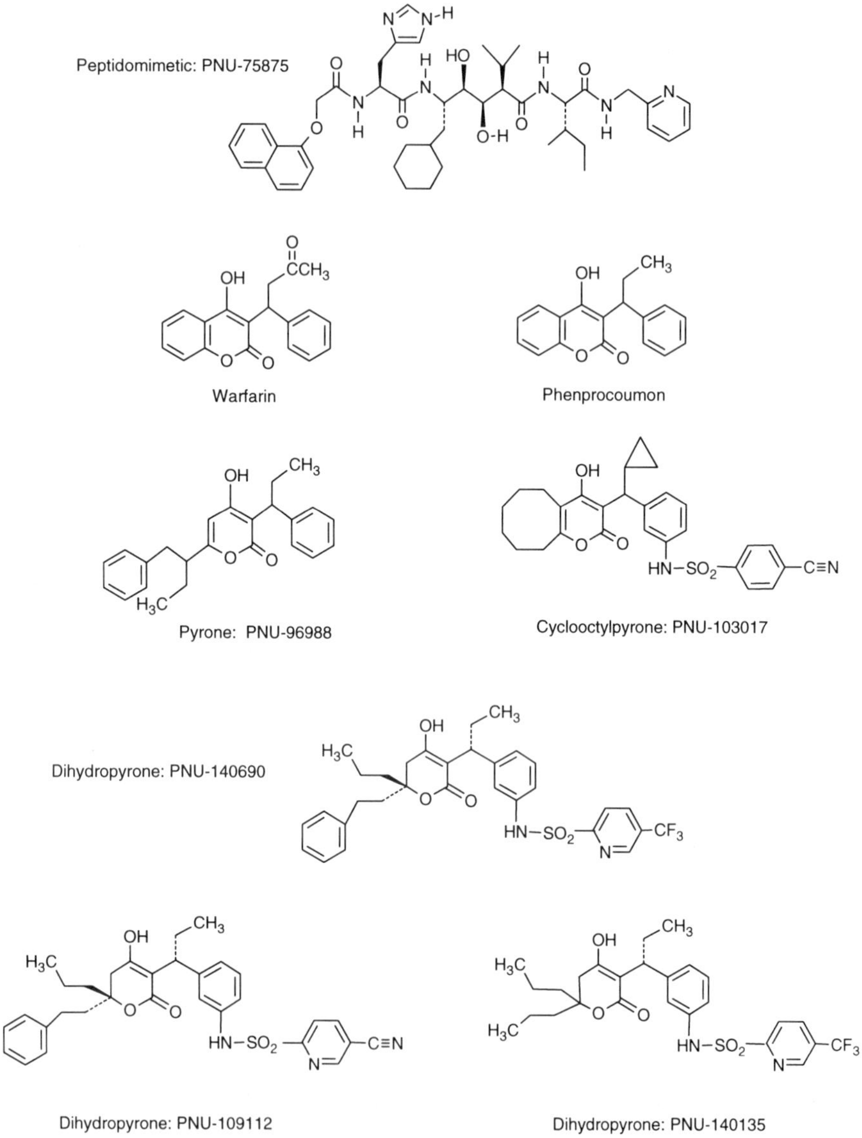

Peptidomimetic: PNU-75875

Warfarin

Phenprocoumon

Pyrone: PNU-96988

Cyclooctylpyrone: PNU-103017

Dihydropyrone: PNU-140690

Dihydropyrone: PNU-109112

Dihydropyrone: PNU-140135

Figure 1. Chemical structures of selected HIV protease inhibitors.

1.3. Factors that Affect Drug Delivery

Intuitively, the goal of any drug discovery program is to identify a novel medicinal compound that can be delivered to the biologically relevant target site at pharmacological concentrations in a predictable and reproducible manner. It is undoubtedly safe to state that most new therapeutic agents fail to possess all of the attributes of a "perfect drug" (*vide infra*), but more often reflect a balance of multiple properties such that they reliably achieve their intended pharmacological effect. Key factors that can play a role in determining the viability of a potential new pharmaceutic intended for oral administration include: (1) intrinsic potency; the primary determinant of the amount of drug that needs to be delivered to achieve the desired therapeutic effect; (2) physicochemical properties including solubility, ionization constants (pK_a), lipophilicity ($\log P$), crystal and/or salt forms; reflecting those factors that control the ability of the drug to be presented in an absorbable form at the site of absorption; (3) absolute oral bioavailability of the compound; encompassing the extent of absorption and the fraction of drug that makes it through the first pass metabolism effect of the gut and liver into the systemic circulation; (4) protein binding and/or red blood cell partitioning; factors that decrease the effective circulating drug concentration available for pharmacological activity; and (5) pharmacokinetics; including the mechanism and rate of clearance of the compound and the extent and duration of systemic drug exposure. Although these parameters can be listed separately for point of discussion, it must be borne in mind that these are not independent variables and that they are intimately linked to each other. It is not uncommon in the drug discovery process that structural or property changes introduced into a given chemical template to selectively modulate a specific characteristic have secondary effects on other pivotal properties such that the gain achieved is more than offset by a deleterious effect on one or more of the other drug attributes.

1.4. Life in a Perfect World

A simplistic view of "life in a perfect world," from a drug delivery standpoint, is graphically illustrated in Fig. 2. The optimal drug demonstrates: (1) a high absolute oral bioavailability reflecting minimal first-pass metabolism and/or absorption limitations in order to minimize the potential for large inter- and intrasubject variability, thereby facilitating effective patient management; (2) a half-life of 12–16 hr, long enough to be amenable to a single or twice daily dosing regimen increasing the probability of patient compliance, but short enough to minimize the time needed to achieve steady-state drug plasma concentrations and ensure a rapid decrease in systemic drug exposure after cessation of treatment; (3) drug plasma levels that exceed some minimal therapeutic concentration for an extended period

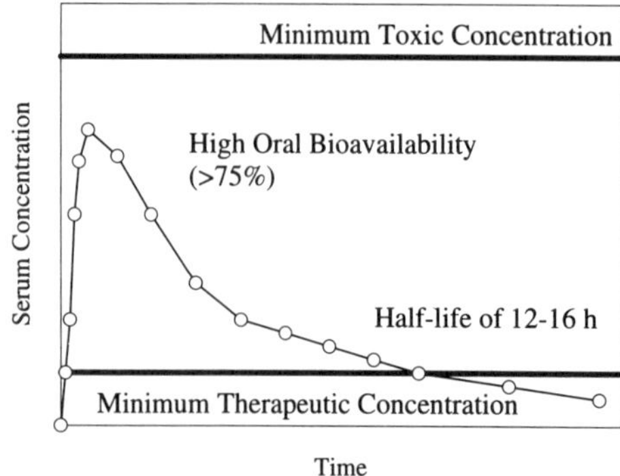

Figure 2. Hypothetical drug serum concentration versus time profile for an ideal drug.

of time after dosing to ensure maximum therapeutic effect; and (4) a large therapeutic safety margin such that the minimum toxicological drug concentration is several orders of magnitude higher than the therapeutic drug concentrations thereby ensuring patient safety.

2. EFFICACY

2.1. Effect/Importance of Protein Binding

The warfarin and coumarin class of compounds are known to be extensively bound to plasma protein and this binding has been well documented to modulate both the activity and kinetics of these compounds (Yacobi and Levi, 1975). Given that the pyrone-based protease inhibitors were derived from this template, it was a logical extension to assess the extent and impact of protein binding on the pharmacological properties of the most promising pyrone analogues. Representative compounds of each pyrone inhibitor subclass were found to be extensively protein bound under conditions of the viral replication inhibition cell culture assay containing 10% (v/v) fetal bovine serum (FBS) (Fig. 3). Mechanistic studies demonstrated that the pyrone inhibitors bound specifically, and with high affinity ($K_d < 6$ μM), to the warfarin site IIA of albumin (He and Carter, 1992). It is generally accepted that only unbound (free) drug is available for bioactivity (Fig. 4). Experiments conducted with human PBMC demonstrated that the drug albumin binding isotherm for the pyrone inhibitors was extremely steep (Fig. 5) and that cellular drug uptake was a linear function of the unbound drug concentration in the incubation media (Fig. 5 inset). Furthermore, reduced apparent *in vitro* antiviral potency was observed with the pyrones in the viral replication assay with increasing amounts of serum protein. This effect is well illustrated by the data for

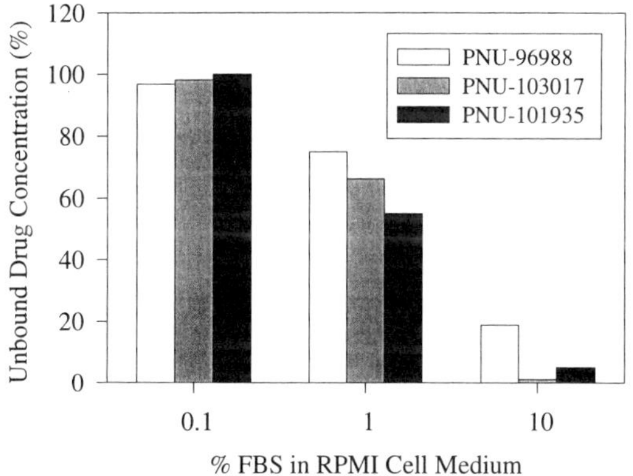

Figure 3. Unbound drug concentration as a function of fetal bovine serum (FBS) in antiviral replication assay cell culture medium for representative pyrone (PNU-96988), cyclooctylpyrone (PNU-103017), and dihydropyrone (PNU-101935) HIV-1 protease inhibitors.

the cyclooctylpyrone PNU-103017: Increasing the protein concentration in cell culture media from 2% to 10% FBS (v/v) supplemented with 13 mg/ml human serum albumin (HSA) yielded a three log increase in the total drug concentration required to inhibit viral replication by 50% (IC_{50}) (Fig. 6a). However, when these same data were considered in the context of the unbound drug concentration, the spread in IC_{50} values decreased to less than twofold (Fig. 6b), well within the experimental error of a cell culture-based assay. Similar effects have been reported with several other antiviral drugs including both reverse transcriptase and protease inhibitors (Baba *et al.,* 1993; DeCamp *et al.,* 1992; Kageyama *et al.,* 1994). In contrast, plasma protein binding has been shown to have relatively little effect on the *in vitro* activity of drugs that have a weak affinity or low extent of serum protein binding such as delavirdine or zidovudine (azidothymidine, AZT) (Baba *et al.,*

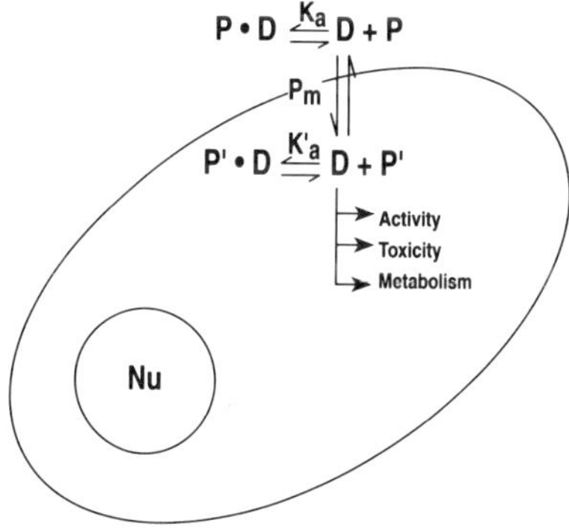

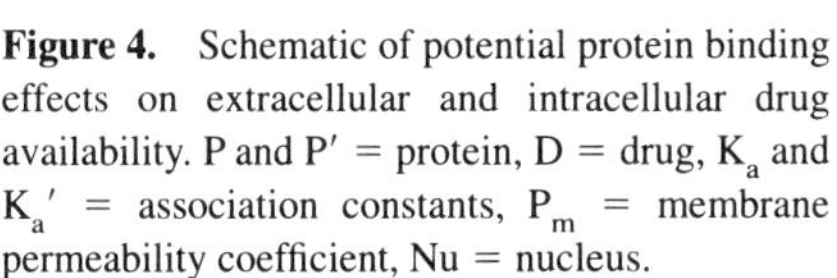
Figure 4. Schematic of potential protein binding effects on extracellular and intracellular drug availability. P and P′ = protein, D = drug, K_a and K_a' = association constants, P_m = membrane permeability coefficient, Nu = nucleus.

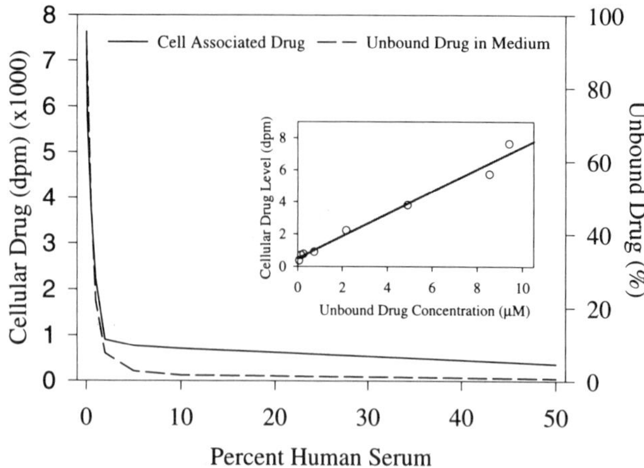

Figure 5. *In vitro* cellular uptake by human peripheral blood lymphocytes of cyclooctylpyrone HIV-1 protease inhibitor (PNU-103017) as a function of human serum protein binding.

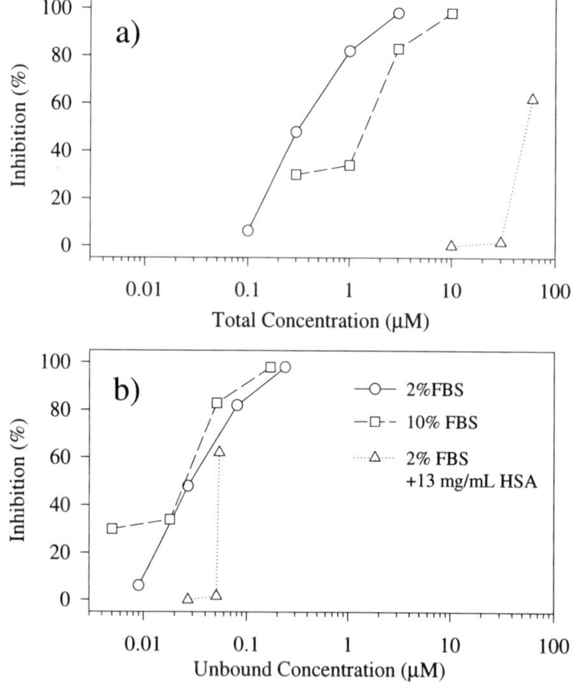

Figure 6. *In vitro* HIV viral replication inhibiton by cyclooctylpyrone HIV-1 protease inhibitor (PNU-103017) as a function of total (panel a) and unbound (panel b) drug concentration in the presence of increasing protein concentration. FBS, fetal bovine serum; HSA, human serum albumin.

1993; K. T. Chong, unpublished data, 1996), respectively, confirming that both extent and affinity of the drug–protein interaction contribute to the impact on observed biological activity. The results from these experiments raised the possibility that decreases in apparent IC_{50} or IC_{90} observed during the SAR development of the pyrone inhibitors might have been related to changes in protein binding rather than an inherent improvement in the intrinsic potency of the chemical template. Ideally, an *in vitro* screening assay for antiviral activity should rank the compounds only on their intrinsic ability to inhibit the viral replication and intrinsic cellular toxicity. As the data for PNU-103017 highlighted, in the case of the pyrone protease inhibitors, it was unequivocally established that the unbound IC_{90} best estimated the intrinsic potency in cell culture. To incorporate this rationale into the pyrone inhibitor testing paradigm, a high-volume fluorescence-based warfarin displacement albumin binding assay was developed and validated versus classical methods (e.g., equilibrium dialysis) (Epps *et al.,* 1995; Koeplinger and Zhao, 1996). Binding constants (K_d) of new analogues to serum albumin were determined and used to predict the unbound drug concentration under conditions of the viral replication assay (10% FBS) such that analogue potency could be compared on the basis of the unbound IC_{90}. Binding data for the most promising leads were confirmed via equilibrium dialysis.

2.2. Clinical Targets

The limitations protein binding imparted on the *in vitro* antiviral activity of the pyrone inhibitors were also relevant to the *in vivo* situation. Therefore, target clinical drug plasma concentrations had to be based on free drug concentrations. In other words, the unbound *in vitro* activity measurements (IC_{90}) had to be translated to unbound *in vivo* plasma drug concentrations. This approach is illustrated in Table I. For a series of the pyrone-based compounds that exhibited only a 10-

Table I
Target Clinical Serum Concentrations for Selected Pyrone-Based HIV Protease Inhibitors Corrected for Protein Binding *in Vitro* and *in Vivo*[a]

Compound	Series	Total IC_{90} (μM)	fu CCM	Unbound IC_{90} (μM)	fu human serum	Target clinical serum concentration (μM)
PNU-96988	Pyrone	10	0.17	1.7	0.005	340
PNU-103017	Cyclooctylpyrone	5	0.011	0.055	0.0002	275
PNU-104904	Dihydropyrone	2	0.2	0.4	0.0049	81
PNU-140232	Dihydropyrone	1	0.016	0.016	0.0036	4.4

[a]Abbreviations: IC_{90}, drug concentration required to inhibit *in vitro* HIV-1 viral replication by 90%; fu, fraction unbound; CCM, cell culture media containing 10% (v/v) fetal bovine serum. Calculations: unbound IC_{90} = total IC_{90} × fu in CCM; clinical target concentration = unbound IC_{90} × fu in human serum.

fold range in IC_{90} values based on total drug concentration, when *in vitro* and *in vivo* protein binding was taken into account, the range between analogues translated into a 100-fold difference in clinical target concentrations. In addition, the clinical target concentrations increased from 4- to 55-fold from the apparent IC_{90} measured directly in cell culture. As a consequence, the extent of binding to animal and human plasma proteins was determined by equilibrium dialysis on a routine basis for the most promising analogues and viability of new leads judged on the compound's ability to achieve therapeutic systemic drug levels, based on the free drug concentrations, in rat and dog after a modest oral dose ($\leq$10 mg/kg).

3. PHARMACOKINETICS

3.1. Total versus Unbound Intrinsic Clearance

As noted earlier, extensive protein binding not only has the potential for reducing effective drug concentrations for pharmacological endpoints but also has the potential for impacting other biological processes including the clearance and elimination mechanisms for a given drug. To understand this, two fundamental pharmacokinetic parameters must be defined. For a compound cleared primarily via hepatic mechanisms (*vide infra*), total (bound and unbound) and unbound intrinsic clearances are defined as follows:

$$\text{Total (bound + unbound) clearance } (CL_{tot}) = \frac{\text{Dose}_{iv}}{\text{AUC}_{iv}}$$

$$\text{Unbound intrinsic clearance } (CL_i^u) = \frac{Q \times CL_{tot}}{fu \times (Q - CL_{tot})}$$

where AUC is the area under the drug plasma concentration versus time curve, Q is hepatic plasma flow, and fu is the fraction of unbound drug in plasma (Wilkinson, 1986).

Evaluation of representative compounds from each pyrone-based series in rat, dog, and monkey demonstrated these compounds to consistently have low total clearance and high intrinsic clearance for the unbound fraction in all three species using literature values for warfarin as a comparator (Table II) (Yacobi and Levi, 1975; Williams *et al.*, 1976; Nagashima and Levi, 1969). These data confirmed the pyrone-based inhibitors to be restricted-clearance compounds, that is, compounds for which protein binding is a limiting factor in clearance. As a consequence of the high hepatic elimination capacity, reflected in the high intrinsic

Table II
Total and Unbound Intrinsic Clearances in Rat, Dog, and Monkey for Warfarin and Representative Pyrone (PNU-96988), Cyclooctylpyrone (PNU-103017), and Dihydropyrone (PNU-106893) HIV-1 Protease Inhibitors[a]

Species	Compound	CL_{tot} (liter/hr/kg)	fu	CL_{int}^{u} (liter/hr/kg)
Rat	Warfarin	0.022	0.0153	1.4
	PNU-96988	0.59	0.005	165
	PNU-103017	0.075	0.0002	405
	PNU-106893	0.73	0.005	226
Dog	Warfarin	0.011	0.07	0.16
	PNU-96988	0.094	0.005	20
	PNU-103017	0.04	0.001	41
	PNU-106893	0.226	0.007	40
Monkey	Warfarin	0.013	0.001	13
	PNU-96988	1.61	0.0131	>Q
	PNU-103017	0.2	0.0002	1148
	PNU-106893	0.765	0.013	118

[a]Abbreviations: fu, fraction unbound; CL_{tot}, total (bound + unbound) clearance; CL_{int}^{u}, unbound intrinsic clearance.

clearance values, the key to future improvement of the pyrone-based inhibitors was not to be derived from reduced protein binding, but was dependent on the design of analogues with increased intrinsic potency and/or significantly reduced unbound intrinsic clearance.

3.2. Factors Affecting Clearance

In an effort to provide useful molecular design information to the synthetic effort, a dual approach to determine the structural elements that contributed to the high unbound intrinsic clearance of the pyrone inhibitors was initiated: (1) the elucidation of the clearance route and mechanism and (2) evaluation of selected series of analogues to explore the intrinsic clearance SAR. Because of potency considerations, these two efforts focused on the cyclooctylpyrone and dihydropyrone analogue series.

3.2.1. ROUTE AND MECHANISM OF CLEARANCE

Radiolabel studies demonstrated that hepatobiliary clearance was the major route of elimination for the pyrone inhibitors with more than 80% of the dose recovered in the feces after intravenous dose administration. Biliary/fecal profiling indicated negligible excretion of unchanged parent drug and the presence of sev-

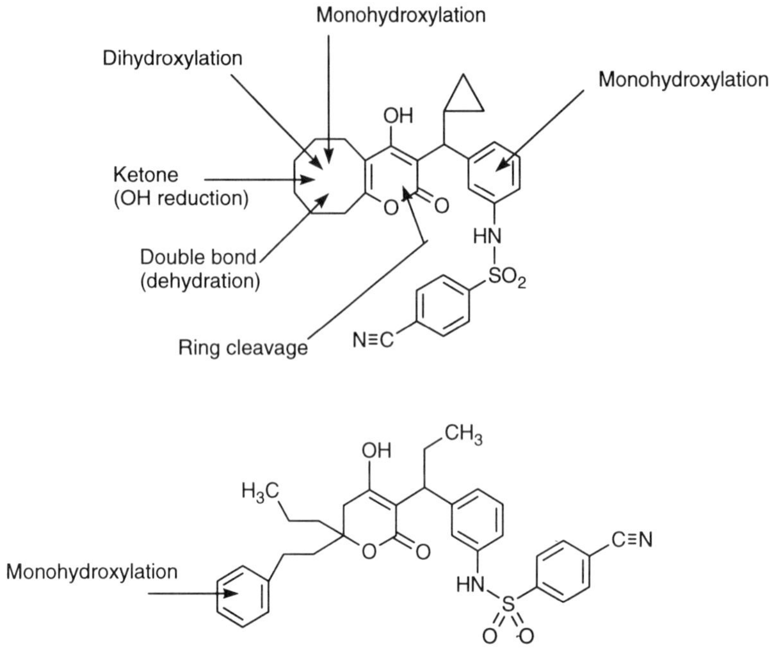

Figure 7. Primary oxidative biotransformation pathways identified for cyclooctylpyrone and dihydropyrone HIV-1 protease inhibitors.

eral major metabolites. Mass spectral characterization of the major fecal metabolites indicated extensive phase I oxidative metabolism, primarily hydroxylation, on the macrocycle of the cyclooctylpyrones and aromatic substituents on C-6 of the dihydropyrones (Fig. 7). *In vitro* studies using human and animal hepatic microsomes and purified enzyme preparations demonstrated that the oxidative metabolism of the pyrone inhibitors was primarily mediated by the cytochrome P450 3A isoform, and that dihydropyrone analogues with alkyl and aromatic substituents in *S*-configuration at C-6 were substrates for the cytochrome P450 2D6 isoform, a polymorphically expressed isoform in humans. Conjugation of the pyrone hydroxyl function was implicated to potentially play a more important role than oxidative metabolism in the clearance of the dihydropyrone analogues. Overall, however, the data suggested that the dihydropyrone analogues offered a reduced number of metabolic soft sites than the cyclooctylpyrone series of inhibitors.

3.2.2. CLEARANCE–STRUCTURE–ACTIVITY RELATIONSHIP

At the time the issue of the high unbound intrinsic clearance crystallized, a rich compound library was readily available enabling the evaluation of the effect of key structural features on the intrinsic clearance of dihydro- and cyclooctylpy-

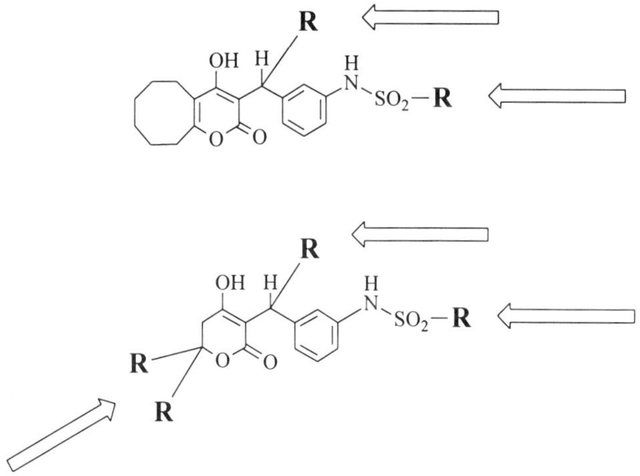

Figure 8. Sites of chemical modification evaluated by unbound intrinsic clearance–structure–activity relationship determination in rat for cyclooctylpyrone and dihydropyrone HIV-1 protease inhibitors.

rone inhibitors. Given that the trends in total and intrinsic clearance were consistent across the preclinical species (Table II), the rat was selected for this evaluation. Toward this end, the intrinsic clearance of a series of analogues with structural permutations at the bridge methylene, sulfonamide, and pyrone C-6 (Fig. 8) were evaluated in the rat and with *in vitro* hepatic microsomal incubations. Results from this evaluation (Fig. 9) identified analogues covering a three log range of in-

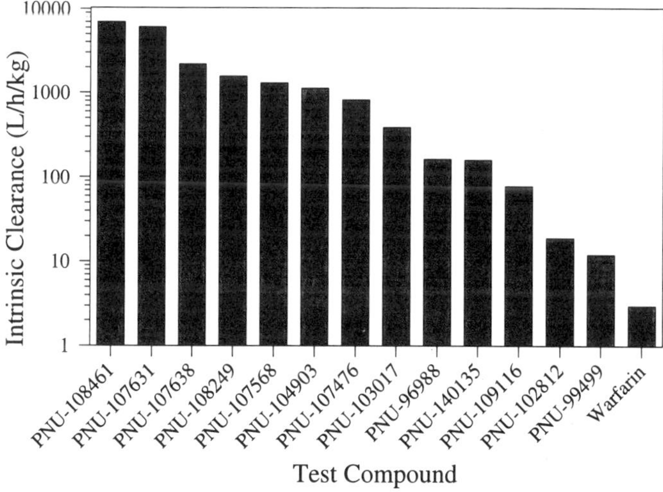

Figure 9. Summary of unbound intrinsic clearances in rat for selected cyclooctylpyrone and dihydropyrone HIV-1 protease inhibitors.

trinsic clearances facilitating conclusive identification of key structural elements influencing the magnitude of intrinsic clearance. In general, the intrinsic clearances for the cyclooctylpyrones were higher than those for the dihydropyrone series. Substitution at the bridge methylene of either series or at C-6 of the dihydropyrones did not significantly affect clearance. Analogues lacking a sulfonamide substituent had significantly lower clearance than analogues with the sulfonamide functional group. The substructure of the sulfonamide substituent had a measurable effect on clearance with lower clearances observed with polar ionizable groups. Analogues with intrinsic clearances lower than warfarin were not identified. Consistent with the mechanistic studies on clearance discussed previously, the clearance SAR effort demonstrated that the dihydropyrone series exhibited, although not optimal, the lowest intrinsic clearance of the pyrone-based inhibitors and that the sulfonamide group played a major role in determining the magnitude of intrinsic clearance. Unfortunately, however, the sulfonamide group was required to achieve submicromolar IC_{90} values in the viral replication assay. The structural information derived from the clearance evaluation was used to help focus template optimization during the final stages of this effort (*vide infra*).

3.3. Absolute Oral Bioavailability versus Systemic Exposure

The ability to achieve the unbound IC_{90} for the pyrone inhibitors was dependent not only on a reduced unbound intrinsic clearance but also on good drug absorption. As with any drug discovery program, an assessment of the pharmacokinetics and absolute oral bioavailability of promising compounds in preclinical species is required for successful drug development. Typically these evaluations begin in rodents, which require minimal quantities of drug, and progress to the larger nonrodent species, such as dog or monkey, as a compound continues to show promise and is scaled up synthetically. The pyrone-based inhibitors were no exception to this approach, and the pharmacokinetics of over 100 compounds were assessed in the rat and 25 compounds in monkey or dog over the duration of this research effort. As noted, a high oral bioavailability is a desired attribute for a potential new therapeutic, although it must be cautioned that high oral bioavailability is not always reflective of the ability of a compound to achieve pharmacologically relevant systemic drug concentrations. Consider the rat oral bioavailability data for three dihydropyrone inhibitors presented in Fig. 10. Clearly, the data for PNU-140135 represent an ideal biopharmaceutical situation where oral bioavailability and the resulting systemic drug exposure, expressed as the average systemic drug concentration over a 24-hr period (C_{avg}), were high and C_{avg} exceeded the target antiviral IC_{90} based on unbound drug concentrations. In contrast, PNU-106893 demonstrated good oral bioavailability but because of a higher total sys-

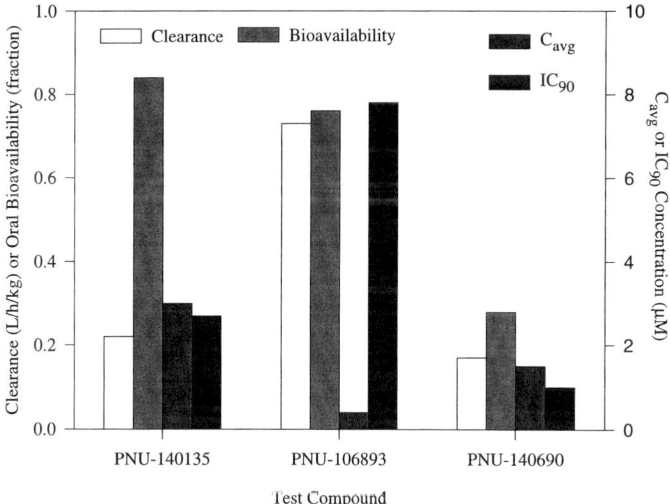

Figure 10. Total systemic clearance, absolute oral bioavailability, average systemic drug levels (C_{avg}), and *in vitro* antiviral potency (IC_{90}) for three representative dihydropyrone HIV-1 protease inhibitors in the male Sprague–Dawley rat. $n = 3$ per compound; intravenous dose 5 mg/kg, oral dose 10 mg/kg; C_{avg} defined as area under the drug plasma concentration versus time curve (AUC); dosing interval(I), IC_{90} = drug concentration required to inhibit *in vitro* HIV-1 viral replication by 90%.

temic clearance, the systemic drug exposure was markedly lower than PNU-140135 and the target IC_{90} for this compound. By further comparison, PNU-140690, which demonstrated only a modest bioavailability, emerges as a viable analogue as the C_{avg} exceeded the target IC_{90} value for this compound. As a consequence, viability of the pyrone inhibitors was assessed not only on absolute oral bioavailability but also on the ability to achieve therapeutic drug levels based on the unbound IC_{90} after a modest oral dose.

4. LIFE IN THE REAL WORLD

4.1. Selection of a Viable Chemical Template

The scientific literature is replete with examples of rational-based drug design in which the "Achilles heel" of a given molecule is mechanistically designed out of the pharmacophore. More often than not, however, viable new drug entities reflect an evolution of modest improvements of several compound attributes resulting in the subtle maturation of a pharmacologically active chemical to a viable therapeutic entity. The pyrone-based HIV-1 protease inhibitors are representative of this latter drug development scenario. The first-generation pyrone PNU-96988 and second-generation cyclooctylpyrone PNU-103017 analogues demonstrated target clinical drug plasma concentrations in excess of 100 μM (Table I) and, as a result, were, realistically, considered clinical concept assessment compounds. Both compounds were pursued through phase I single and multiple dose clinical trials in order to gain insight on the human tolerance, clinical pharmacokinetics, and biopharmaceutic properties of this structural class of inhibitors. The rapid testing of a series of sequentially improved drug candidates in humans to validate preclinical models for subsequent discovery iterations is a paradigm that is now being employed routinely by the pharmaceutical industry. Preclinical animal models cannot be applied to the drug discovery programs effectively if they can only be utilized retrospectively. For preclinical models to be used proactively in order to increase the chance of clinical success, experience in human early in the discovery process is requisite. Results from the clinical trials with PNU-96988 and PNU-103017 indicated both compounds to be reasonably well tolerated in human up through oral doses of ≥1000 mg, that the pharmacokinetics trends observed in human were consistent with those observed in rat and dog but not those in monkey and rabbit (Fig. 11), that the dog reliably predicted relative oral formulation performance in human, and that the disodium salt form of the drugs had suitable biopharmaceutic properties for solid dosage formulation development. Of the three

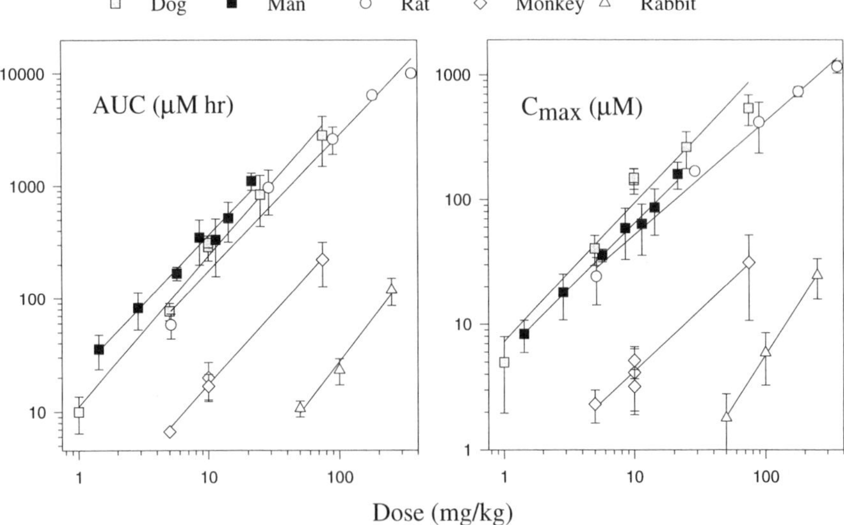

Figure 11. Systemic exposure (AUC and C_{max}) of PNU-103017 as a function of dose in rat, rabbit, dog, monkey, and human after a single oral administration.

pyrone-based templates, the dihydropyrone analogues emerged as the most viable series based on a number of factors: (1) this series demonstrated superior intrinsic (unbound) potency with nanomolar K_is against the HIV-1 protease and viral replication IC_{90}s below 10 μM, (2) the binding affinity to the warfarin IIA site of albumin was generally 10-fold less for the dihydropyrones than the cyclooctylpyrone analogues, (3) the cellular permeability and partitioning was significantly greater for the dihydropyrone series than the cyclooctylpyrones (data not discussed in this chapter), (4) the dihydropyrone core structure offered a reduced number of oxidative metabolic "soft sites," thus (5) minimizing the unbound intrinsic clearance for this series.

4.2. Identification of a Final Clinical Candidate

Once focused on the dihydropyrone series of inhibitors, the final structure–activity interactions centered on optimization of the sulfonamide structure and substituent groups on C-6 of the pyrone ring based on antiviral activity and intrinsic clearance considerations. From this effort, three compounds, PNU-109112, PNU-140135, and PNU-140690, emerged as potential clinical candidates (Fig. 1, Table III). PNU-109112 was subsequently eliminated from consideration after it was found that the sulfonamide substituent of this analogue was susceptible to chemical and enzymatic cleavage and that adequate systemic drug levels to ensure an adequate safety margin could not be achieved in the rat. Although the least soluble of the three candidates, PNU-140690 was selected for clinical development based on its greater *in vitro* potency and the fact that adequate systemic drug levels for toxicological evaluation could not be achieved for PNU-140135 in the dog because of a strong emetic response. The absolute oral bioavailability of PNU-140690 was limited (~30%) in both rat and dog (Table III) and mechanistic studies demonstrated that absorption limitations, presumably related to the finite solubility of the compound and not first-pass metabolism effects, were the major barrier to higher drug delivery after oral dosing. The modest bioavailability of PNU-140690 notwithstanding, after oral administration of 10 mg/kg, PNU-140690 plasma levels in excess of the target clinical drug plasma concentration (~1 μM) were maintained for greater than 4 hr after dosing in dog and rat (Fig. 12) and represented a significant advancement over the pyrone- and cyclooctylpyrone-based inhibitors (Fig. 12). Initial phase I clinical data with PNU-140690 appear to corroborate systemic drug delivery data obtained in animals. After a single 700-mg oral dose to normal healthy male volunteers, PNU-140690 plasma concentrations of greater than 1 μM were maintained for at least 6 to 8 hr after dosing and is viewed as promising as the compound proceeds to clinical trials in HIV-positive individuals.

Table III
Summary of Key Attributes for Three Dihydropyrone HIV-1 Protease Inhibitors, PNU-109112, PNU-140135, and PNU-140690

	109112	140135	140690
Molecular weight	559	540	602
$c \log P$	6.0	7	7.4
Intrinsic solubility (μg/mL)	0.7	0.9	0.1
pK_a	5.6, 8.2	5.4, 8.2	5.4, 8.1
Chiral centers	2	1	2
HIV-1 protease K_i (nM)	0.032	0.002–4	0.003–7
IC_{90} total (μM)	0.16–0.28	0.06–1.0	0.10–0.24
IC_{90} unbound (μM)	0.0020–36	0.0016–60	0.0019–46
Fraction unbound in human plasma	0.0017	0.0022	0.002
Target plasma concentration (μM)	1.2–2.1	7.5–11	0.96–2.3
Rat pharmacokinetics ($n = 3$)			
(5 mg/kg i.v., 10 mg/kg p.o.)			
Total clearance (liters/hr/kg)	0.59 ± 0.43	0.22 ± 0.057	0.17 ± 0.09
Intrinsic clearance (liters/hr/kg)	482	112	380
Volume of distribution (liters/kg)	0.47 ± 0.043	0.20 ± 0.011	0.51 ± 0.14
Terminal half-life (hr)	1.6 ± 0.65	1.68 ± 0.29	5.4 ± 0.30
C_{max} (μM)[a]	2.2 ± 0.47	13.3 ± 7.3	2.8–10.9
t_{max} (hr)[a]	0.67–4	1–4	1–8
Absolute oral bioavailability (%)	22 ± 15	84 ± 30	28–45
Dog pharmacokinetics ($n = 3$)			
(5 mg/kg i.v., 10 mg/kg p.o.)			
Total clearance (liters/hr/kg)	0.25 ± 0.069	0.13 ± 0.013	0.26 ± 0.053
Intrinsic clearance (liters/hr/kg)	185	67	>460
Volume of distribution (liters/kg)	0.089 ± 0.041	0.18 ± 0.006	0.14 ± 0.010
Terminal half-life (hr)	1.38 ± 0.39	2.5 ± 0.36	0.93 ± 0.31
C_{max} (μM)	26 ± 18	19 ± 10	3.9–9.8
t_{max} (hr)	0.67	1–2	2
Absolute oral bioavailability (%)	40 ± 17	55 ± 18	8–20

[a] C_{max}, maximal plasma concentration; t_{max}, time of C_{max}.

Acknowledgments

The discovery and development of the pyrone-based HIV protease inhibitors reflects the scientific efforts of a large multidisciplinary team within Pharmacia & Upjohn, Inc. The authors would like to acknowledge and thank the following key individuals for their contributions and selected use of their data: Harvey I. Skulnick, Steve R. Turner, Joseph W. Strohbach, Ruben A. Tommasi, Paul A. Aristoff, Thomas M. Judge, Ronald B. Gammill, Jeanette K. Morris, Karen R. Romines, Robert A Crusciel, Roger R. Hinshaw, W. Gary Tarpley, Janet C. Lynn, Miao-Miao Horng, Paul K. Tomich, Eric P. Seest, Lester A. Dolak, W. Jeffrey Howe, Gina M. Howard, Robert J. Dalga, Lisa N. Toth, Grace J. Wilson, Lihua Shiou, Karen F. Wilkinson, Bob D. Rush, Mary J. Ruwart, Serena Cole, Rennee M. Zaya, Thomas

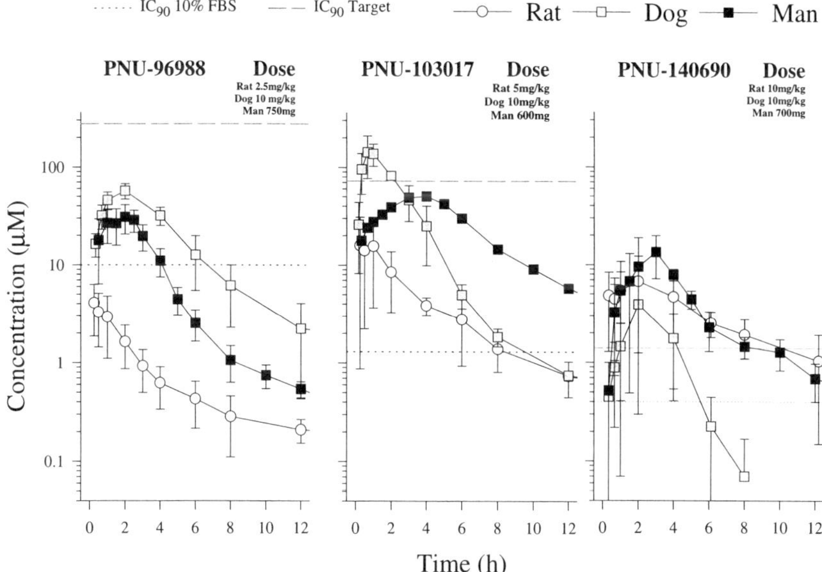

Figure 12. Plasma drug concentration versus time profiles in male Sprague–Dawley rat, male beagle dog, and male human following a single oral solution administration of PNU-96988, PNU-103017, or PNU-140690. $n = 3$–6 per species; FBS, fetal bovine serum; IC_{90} 10% FBS = total drug concentration required to inhibit *in vitro* HIV-1 viral replication by 90%; IC_{90} Target = IC_{90} corrected for protein binding *in vitro* and *in vivo*.

J. Kakuk, Musiri N. Janakiraman, Keith D. Watenpaugh, Dennis J. Epps, Phil Burton, Randy Wald, Rick C. Steenwyk, Philip M. Sanders, and Peggy Possert.

REFERENCES

Ashorn, P., McQuade, T. J., Thaisrivongs, S., Tomasselli, A. G., Tarpley, W. G., and Moss, B., 1990, An inhibitor of the protease blocks maturation of human and simian immunodeficiency viruses and spread of infection, *Proc. Natl. Acad. Sci. USA* **87:**7472–7476.

Baba, M., Yuasa, S., Niwa, T., Yamamoto, M., Yabuuchi, S., Takashima, H., Ubasawa, M., Tanaka, H., Miyasaka, T., Walker, R. T., Balzarini, J., DeClerco, G., and Shigeta, S., 1993, Effect of human serum on the in vitro anti-HIV-1 activity of 1-[(2-hydroxyethoxy)methyl]-6-(phenylthio)thymine (HEPT) derivatives as related to their lipophilicity and serum protein binding, *Biochem. Pharmacol.* **45:**2507–2512.

Bourinbaiar, A. S., Tan, X., and Nagorny, R., 1993, Effect of the oral anticoagulant, warfarin, on HIV-1 replication and spread, *AIDS* **7:**129–130.

Chong, K. T., 1996, Recent advances in HIV-1 protease inhibitors, *Exp. Opin. Invest. Drugs* **5:**115–124.

Darke, P. L., and Huff, J. R., 1994, HIV protease inhibitors target for the treatment of AIDS, in: *Advances in Pharmacology,* Volume 25 (T. J. August, M. W. Ander, and F. Murad, eds.), Academic Press, San Diego, pp. 399–454.

Debouck, C., and Metcalf, B. W., 1990, Human immunodeficiency virus protease: A target for AIDS therapy, *Drug Dev. Res.* **21:**1–17.

DeCamp, D. L., Babe, L. M., Salto, R., Lucich, J. L., Koo, M. S., Kahl, S. B., and Craik, C. S., 1992, Specific inhibition of HIV-1 protease by boronated porphyrins, *J. Med. Chem.* **35:**3426–3428.

Epps, D. E., Raub, T. J., and Kezdy, F. J., 1995, A general, wide-range spectrofluorometric method for measuring the site-specific affinities of drugs toward human serum albumin, *Anal. Biochem.* **227:**342–350.

Greenlee, W. J., 1990, Renin inhibitors, *Med. Res. Rev.* **10:**173–236.

He, X. M., and Carter, D. C., 1992, Atomic structure and chemistry of human serum albumin, *Nature* **358:**209–215.

Hirsh, J., Dalen, J. E., Deykin, D., Poller, L., and Bussey, H., 1992, Oral anticoagulants. Mechanism of action, clinical effectiveness, and optimal therapeutic range, *Chest* **108:**231S–246S.

Huff, J. R., 1991, HIV protease: A novel chemotherapeutic target for AIDS, *J. Med. Chem.* **34:**2305–2314.

Hyland, L. J., Tomaszek, T. A., Jr., Roberts, G. D., Carr, S. A., Magaard, V. W., Bryan, H. L., Fakhoury, S. A., Moore, M. L., Minnich, M. D., Culp, J. S., DesJarlais, R. L., and Meek, T. D., 1991a, Human immunodeficiency virus-1 protease. 1. Initial velocity studies and kinetic characterization of reaction intermediates by ^{18}O isotope exchange, *Biochemistry* **30:**8441–8453.

Hyland, L. J., Tomaszek, T. A., Jr., and Meek, T. D., 1991b, Human immunodeficiency virus-1 protease. 2. Use of pH rate studies and solvent kinetic isotope effects to elucidate details of chemical mechanism, *Biochemistry* **30:**8454–8463.

Jaskolski, M., Tomasselli, A. G., Sawyer, T. K., Staples, D. G., Heinrikson, R. L., Schneider, J., Kent, S. B., and Wlodawer, A., 1991, Structure at 2.5-Å resolution of chemically synthesized human immunodeficiency virus type 1 protease complexed with a hydroxyethylene-based inhibitor, *Biochemistry* **30:**1600–1609.

Johnston, M. I., and Hoth, D. F., 1993, Present status and future prospects for HIV therapies, *Science* **260:**1286–1293.

Kageyama, S., Anderson, B. D., Hoesterey, B. L., Hayashi, H., Kiso, Y., Flora, K. P., and Mitsuya, H., 1994, Protein binding of human immunodeficiency virus protease inhibitor KNI-272 and alteration of its in vitro antiretroviral activity in the presence of high concentrations of proteins, *Antimicrob. Agents Chemother.* **38:**1107–1111.

Koeplinger, K. A., and Zhao, Z., 1996, Chromatographic measurement of drug–protein interaction: Determination of HIV protease inhibitor–serum albumin association, *Anal. Biochem.* **243:**66–73.

Kohl, N. E., Emini, E. A., Schleif, W. A., Davis, L. J., Heimbach, J. C., Dixon, R. A. F., Scolnick, E. M., and Sigal, I. S., 1988, Active human immunodeficiency virus protease is required for viral infectivity, *Proc. Natl. Acad. Sci. USA,* **85:**4686–4690.

Louis, J. M., Smith, C. A., Wondrak, E. M., Mora, P. T., and Oroszlan, S., 1989, Substitution mutations of the highly conserved arginine 87 of HIV-1 protease result in loss of proteolytic activity, *Biochem. Biophys. Res. Commun.* **164:**30–38.

Lunney, E. A., Hagen, S. E., Domagala, J. M., Humblet, C., Kosinski, J., Tait, B. D., Warmus, J. S., Wilson, M., Ferguson, D., Hupe, D., Tummino, P. J., Baldwin, E. T., Bhat, T. N., Liu, B., and Erickson, J. W., 1994, A novel nonpeptide HIV-1 protease inhibitor: Elucidation of the binding mode and its application in the design of related analogues, *J. Med. Chem.* **37:**2664–2677.

Meek, T. D., 1992, Inhibitors of HIV-1 protease, *J. Enz. Inhib.* **6:**65–98.

Nagashima, R., and Levi, G., 1969, Comparative pharmacokinetics of coumarin anticoagulants V: Kinetics of warfarin elimination in rat, dog, and rhesus monkey compared to man, *J. Pharm. Sci.* **58:**845–849.

Oroszlan, S., and Luftig, R. B., 1990, Retroviral proteinases, *Curr. Top. Microb. Immunol.* **157:**153–185.

Peng, C., Ho, B. K., Chang, T. W., and Chang, N. T., 1989, Role of human immunodeficiency virus type 1-specific protease in core protein maturation and viral infectivity, *J. Virol.* **63:**2550–2556.

Plattner, J. J., and Norbeck, D. W., 1990, Obstacles to drug development from peptide leads, in: *Drug Discovery Technologies* (C. R. Clark and W. H. Moos, eds.), pp. 92–120, Ellis Horwood, Chichester.

Redshaw, S., 1994, Inhibitors of HIV proteinase, *Exp. Opin. Invest. Drugs* **3:**273–286.

Romines, K. R., Watenpaugh, K. D., Tomich, P. K., Howe, W. J., Morris, J. K., Lovasz, K. D., Mulichak, A. M., Finzel, B. C., Lynn, J. C., Horng, M.-M., Schwende, F. J., Ruwart, M. J., Zipp, G. L., Rush, B. D., Wilkinson, K. F., Possert, P. G., Dalga, R. J., and Hinshaw, R. R., 1995a, Use of medium-sized cycloalkyl rings to enhance secondary binding: Discovery of a new class of human immunodeficiency virus (HIV) protease inhibitors, *J. Med. Chem.* **38:**1884–1891.

Romines, K. R., Watenpaugh, K. D., Howe, W. J., Tomich, P. K., Lovasz, K. D., Morris, J. K., Janakiraman, M. N., Lynn, J. C., Horng, M.-M., Chong, K. T., Hinshaw, R. R., and Dolak, L. A., 1995b, Structure-based design of nonpeptidic HIV protease inhibitors from a cyclooctylpyranone lead structure, *J. Med. Chem.* **38:**4463–4473.

Skulnick, H. I., Johnson, P. D., Howe, W. J., Tomich, P. K., Chong, K. T., Watenpaugh, K. D., Janakiraman, M. N., Dolak, L. A., McGrath, J. P., Lynn, J. C., Horng, M.-M., Hinshaw, R. R., Zipp, G. L., Ruwart, M. J., Schwende, F. J., Zhong, W. Z., Padbury, G. E., Dalga, R. J., Shiou, L., Possert, P. G., Rush, B. D., Wilkinson, K. F., Howard, G. M., Toth, L. N., Williams, M. G., Kakuk, T. J., Cole, S. L., Zaya, R. M., Lovaz, K. D., Morris, J. K., Romines, K. R., Thaisrivongs, S., and Aristoff, P. A., 1995, Structure-based design of sulfonamide-substituted non-peptidic HIV protease inhibitors, *J. Med. Chem.* **38:**4968–4971.

Thaisrivongs, S., 1994, HIV protease inhibitors, *Annu. Rep. Med. Chem.* **17:**133–144.

Thaisrivongs, S., Tomasselli, A. G., Moon, J. B., Hui, J., McQuade, T. J., Turner, S. R., Strohbach, J. W., Howe, W. J., Tarpley, W. G., and Heinrikson, R. L., 1991, Inhibitors of the protease from human immunodeficiency virus: Design and modeling of a compound containing a dihydroxyethylene isostere insert with high binding affinity and effective antiviral activity, *J. Med. Chem.* **34:**2344–2356.

Thaisrivongs, S., Tomich, P. K., Watenpaugh, K. D., Chong, K. T., Howe, W. J., Yang, C. P., Strohbach, J. W., Turner, S. R., McGrath, J. P., Bohanon, M. J., Lynn, J. C., Mulichak, A. M., Spinelli, P. A., Hinshaw, R. R., Pagano, P. J., Moon, J. B., Ruwart, M. J., Wilkinson, K. F., Rush, B. D., Zipp, G. L., Dalga, R. J., Schwende, F. J., Howard, G. M., Padbury, G. E., Toth, L. N., Zhao, Z., Koeplinger, K. A., Kakuk, T. J., Cole, S. L., Zaya, R. M., Piper, R. C., and Jeffery, P., 1994, Structure-based design of HIV protease inhibitors: 4-Hydroxycoumarins and 4-hydroxy-2-pyrones as non-peptide inhibitors, *J. Med. Chem.* **37:**3200–3204.

Thaisrivongs, S., Watenpaugh, K. D., Howe, W. J., Tomich, P. K., Dolak, L. A., Chong, K. T., Turner, S. R., Strohbach, J. W., Mulichak, A. M., Janakiraman, M. N., Moon, J. B., Lynn, J. C., Horng, M.-M., Hinshaw, R. R., and Pagoan, P. J., 1995, Structure-based design of HIV protease inhibitors: Novel carboxamide-containing 4-hydroxycoumarins and 4-hydroxy-2-pyrones as potent non-peptidic inhibitors, *J. Med. Chem.* **38:**3624–3637.

Thaisrivongs, S., Skulnick, H. I., Turner, S. R., Strohbach, J. W., Tommasi, R. A., Johnson, P. D., Aristoff, P. A., Judge, T. M., Gammill, R. B., Morris, J. K., Romines, K. R., Crusciel, R. A., Hinshaw, R. R., Chong, K. T., Tarpley, W. G., Poppe, S. M., Slade, D. E., Lynn, J. C., Horng, M.-M., Tomich, P. K., Seest, E. P., Dolak, L. A., Howe, W. J., Howard, G. M., Schwende, F. J., Toth, L. N., Padbury, G. E., Wilson, G. J., Shiou, L., Zipp, G. L., Wilkinson, K. F., Rush, B. D., Ruwart, M. J., Koeplinger, K. F., Zhao, Z., Cole, S., Zaya, R. M., Kakuk, T. J., Janakiraman, M. N., and Watenpaugh, K. D., 1996a, Structure-based design of HIV protease inhibitors: Sulfonamide-containing 5,6-dihydro-4-hydroxy-2-pyrones as non-peptic inhibitors, *J. Med. Chem.* **39:**4349–4353.

Thaisrivongs, S., Romero, D. L., Tommasi, R. A., Janakiraman, M. N., Strohbach, J. W., Turner, S. R., Biles, C., Morge, R. R., Johnson, P. D., Aristoff, P. A., Tomich, P. K., Lynn, J. C., Horng, M.-M., Chong, K. T., Hinshaw, R. R., Howe, W. J., Finzel, B. D., and Watenpaugh, K. D., 1996b, Structure-based design of HIV protease inhibitors: 5,6-dihydro-4-hydroxy-2-pyrones as effective non-peptic inhibitors, *J. Med. Chem.* **39:**4630–4642.

Tomasselli, A. G., Hui, J. O., Sawyer, T. K., Staples, D. J., Bannow, C., Reardon, I. M., Howe, W. J., DeCamp, D. L., Craik, C. S., and Heinrikson, R. L., 1990, Specificity and inhibition of proteases from human immunodeficiency viruses 1 and 2, *J. Biol. Chem.* **265:**14675–14683.

Tomasselli, A. G., Thaisrivongs, S., and Heinrikson, R. L., 1996, Discovery and design of HIV protease inhibitors as drugs for treatment of AIDS, *Adv. Antiviral Drug Des.* **2:**173–228.

Tummino, P. J., Ferguson, D., and Hupe, D., 1994a, Competitive inhibition of HIV-1 protease by 4-hydroxy-benzopyran-2-ones and by 4-hydroxy-6-phenylpyran-2-ones, *Biochem. Biophys. Res. Commun.* **200:**1658–1664.

Tummino, P. J., Ferguson, D., and Hupe, D., 1994b, Competitive inhibition of HIV-1 protease by warfarin derivatives, Biochem. *Biophys. Res. Commun.* **201:**290–294.

Vara Prasad, J. V. N., Para, K. S., Lunney, E. A., Ortwine, D. F., Dunbar, J. B., Jr., Ferguson, D., Tummino, P. J., Hupe, D., Tait, B. D., Domagala, J. M., Humblet, C., Bhat, T. N., Liu, B., Guerin, D. M. A., Baldwin, E. T., Erickson, J. W., and Sawyer, T. K., 1994, Novel series of achiral, low molecular weight, and potent HIV-1 protease inhibitors, *J. Am. Chem. Soc.* **116:**6989–6990.

Vara Prasad, J. V. N., Para, K. S., Tummino, P. J., Ferguson, D., McQuade, T. J., Lunney, E. A., Rapundalo, S. T., Batley, B. L., Hingorani, G., Domagala, J. M., Gracheck, S. J., Bhat, T. N., Liu, B., Baldwin, E. T., Erickson, J. W., and Sawyer, T. K., 1995, Nonpeptidic potent HIV-1 protease inhibitors: (4-hydroxy-6-phenyl-2-oxo-2H-pyran-3-yl)thiomethanes that span P1-P2′ subsites in a unique mode of active site binding, *J. Med. Chem.* **38:**898–905.

West, M. L., and Fairlie, D. P., 1995, Targeting HIV-1 protease: A test of drug-design methodologies, *Trends Pharm. Sci.* **16:**67–75.

Wilkinson, G. R., 1986, Plasma binding and hepatic drug elimination, in: *Drug–Protein Binding* (M. M. Reidenberg and S. Erill, eds.), pp. 299–316, Praeger Scientific, New York.

Williams, R. L., Schary, W. L., Blaschke, M. D., Meffin, P. J., Melmon, K. L., and Rowland, M., 1976, Influence of acute viral hepatitis on disposition and pharmacologic effect of warfarin, *Clin. Pharmacol. Ther.* **20:**9097.

Yacobi, A., and Levi, G., 1975, Comparative pharmacokinetics of coumarin anticoagulants XIV: Relationship between protein binding, distribution, and elimination kinetics of warfarin in rats, *J. Pharm. Sci.* **64:**1660–1664.

Chapter 11

The Integration of Medicinal Chemistry, Drug Metabolism, and Pharmaceutical Research and Development in Drug Discovery and Development

The Story of Crixivan®, an HIV Protease Inhibitor

Jiunn H. Lin, Drazen Ostovic, and Joseph P. Vacca

1. INTRODUCTION

Drug research encompasses a number of diverse disciplines united by a common goal, the development of novel therapeutic agents. In short, the search for new drugs involves two steps: drug discovery and drug development. The former consists of setting up a working hypothesis of the target enzyme (or receptor) for a particular disease, establishing suitable models to test biological activities, and screening the *in vitro* and *in vivo* biological activities of new drug molecules. The latter is to gather data for toxicity and efficacy evaluation of the new drug candidates.

Once the working hypothesis is established, medicinal chemists use a variety of empirical and semiempirical structure–activity relationships to modify the

---

Jiunn H. Lin • Drug Metabolism, Merck Research Laboratories, West Point, Pennsylvania 19486. *Drazen Ostovic* • Pharmaceutical Research and Development, Merck Research Laboratories, West Point, Pennsylvania 19486. *Joseph P. Vacca* • Medicinal Chemistry, Merck Research Laboratories, West Point, Pennsylvania 19486.

Integration of Pharmaceutical Discovery and Development: Case Studies, edited by Borchardt *et al.*, Plenum Press, New York, 1998.

chemical structure of a compound to maximize its *in vitro* activity. However, good *in vitro* activity cannot be extrapolated to good *in vivo* activity, unless a drug has good bioavailability and a desirable duration of action. There is a growing awareness of the key roles that pharmacokinetics and metabolic processes play as determinants of *in vivo* drug action. Many drug companies now include pharmacokinetics and drug metabolism as part of their screening processes in the selection of drug candidates. Thus, industrial drug metabolism scientists have emerged from their traditional supportive role in drug development to assume important functions in the drug discovery efforts.

Because of ethical constraints, relevant pharmacological and toxicological assessments have to be studied extensively in laboratory animals prior to the first administration of drug in humans. Therefore, one of the fundamental challenges drug metabolism scientists face in drug discovery and development is the extrapolation of metabolic and pharmacokinetic assessment from animals to humans. Furthermore, because of the time constraints and the small quantities of each compound available in the early discovery stage, the studies are often limited to one or two animal species. Therefore, the selection of animal species and the experimental design of studies are crucial in providing a reliable prediction of drug absorption and elimination in humans. A good compound could be excluded on the basis of results from an inappropriate animal species or poor experimental design.

Recent surveys indicate that the entire process from the synthesis of a new chemical entity to its approval as a drug requires 10 to 15 years. Roughly, one-third of this time is needed for preclinical studies. Therefore, considerable investments have been made prior to clinical studies. It is very costly to begin again with a new compound after clinical studies show a drug to be metabolically and/or pharmacokinetically unsatisfactory. It is desirable to obtain information on the metabolic processes in human as early as possible. Fortunately, the availability of human liver tissues, together with the explosion of our knowledge of various drug-metabolizing enzymes at the molecular level, allows us to obtain early information on metabolic processes of a new drug candidate. In addition, the advance of commercial instrumentation for LC-MS/MS and the development of high-field NMR techniques have further strengthened our capability to study the metabolism of new drugs at the early stage of drug development.

The purpose of this chapter is to illustrate the integration of pharmacokinetics and drug metabolism in drug discovery and development, using the HIV protease inhibitor program as an example.

2. DISCOVERY OF L-735,524 (CRIXIVAN®)

HIV-1, the causative agent of acquired immunodeficiency disease (AIDS), is a member of the retrovirus subfamily and, like other retroviruses, contains three major genes (*gag, pol,* and *env*) (Ratner *et al.*, 1985; Toh *et al.*, 1985). The products of the *gag* gene include the major structural proteins of the virus nucleocapsid; the *env* gene encodes the membrane proteins of the mature virus; and the *pol* gene encodes three enzymes: a protease, the reverse transcriptase, and an endonuclease. The *gag* and *pol* gene products are expressed as polyproteins that are processed by the HIV protease as an essential step for virus maturation. The virally encoded protease is a member of the aspartyl proteinase family, and like other retroviral enzymes of this class, the catalytically competent form of the enzyme is a symmetrical homodimer in which each monomer contributes one-half of the active site. Mutations within the HIV protease coding region that inactivated the enzyme resulted in the expression of nonprocessed *gag* and *pol* gene products and noninfectious viral particles (Kohl *et al.,* 1988). Mutations within the *gag* gene that prevented proteolytic processing also rendered the virus noninfectious. Viruses containing such mutant *gag* genes, when coexpressed with wild-type virus, inhibited all viral replication, suggesting that even partial inhibition of processing leads to antiviral activity. Finally, potent inhibitors of the HIV protease completely block viral replication in cell culture (Huff, 1991; West and Fairlie, 1995). These observations suggested that this enzyme constituted an attractive target for antiviral therapy.

In late 1987, L-364,505 (compound **1,** Fig. 1) was identified as a potent HIV protease inhibitor *in vitro* (IC_{50} = 1 nM) and was weakly active in an assay that measured a compound's ability to completely stop the spread of infection in a cell culture ($CIC_{95} > 100$ μM). With this discovery, a chemistry effort at Merck was initiated to develop potent, orally bioavailable HIV protease inhibitors. L-364,505 was originally discovered in a renin inhibitor program and potent inhibitors of this enzyme were based on substrate analogues containing a transition-state mimic. Replacement of the scissile amide bond with a variety of nonhydrolyzable isosteres has yielded highly potent and specific inhibitors (Greenlee, 1990). However, at that time no renin inhibitor with adequate oral bioavailability in animals had been identified and brought forth into clinical trials. The discovery of a useful inhibitor based on this class of compounds was viewed as an insurmountable task. Nevertheless, work was initiated on this lead structure to identify smaller, less-peptide-like structures. Fortunately, minor modification of L-364,505 to give L-682,679 (compound **2,** Fig. 1) eliminated the renin activity of this series and maintained the HIV-1 protease activity. Further work led to L-687,908 (compound **3,** Fig. 1), which was the optimal compound in this series with regards to activity ($IC_{50} < 0.03$ nM; CIC_{95} = 12 nM) (Vacca *et al.,* 1991). Unfortunately, the compound was not orally bioavailable in animals and was too insoluble in acceptable vehicles for use as an intravenous agent.

An alternate, less-peptide-like series of compounds was also being developed in parallel and this effort resulted in the discovery of L-685,434 (compound **4,** Fig. 2) (Lyle *et al.,* 1991), which was a potent inhibitor of the enzyme (IC_{50} = 0.3 nM) and in cell culture (CIC_{95} = 400 nM). Although this compound contained no amino acids, it still had no oral bioavailability in animals. One possible reason for

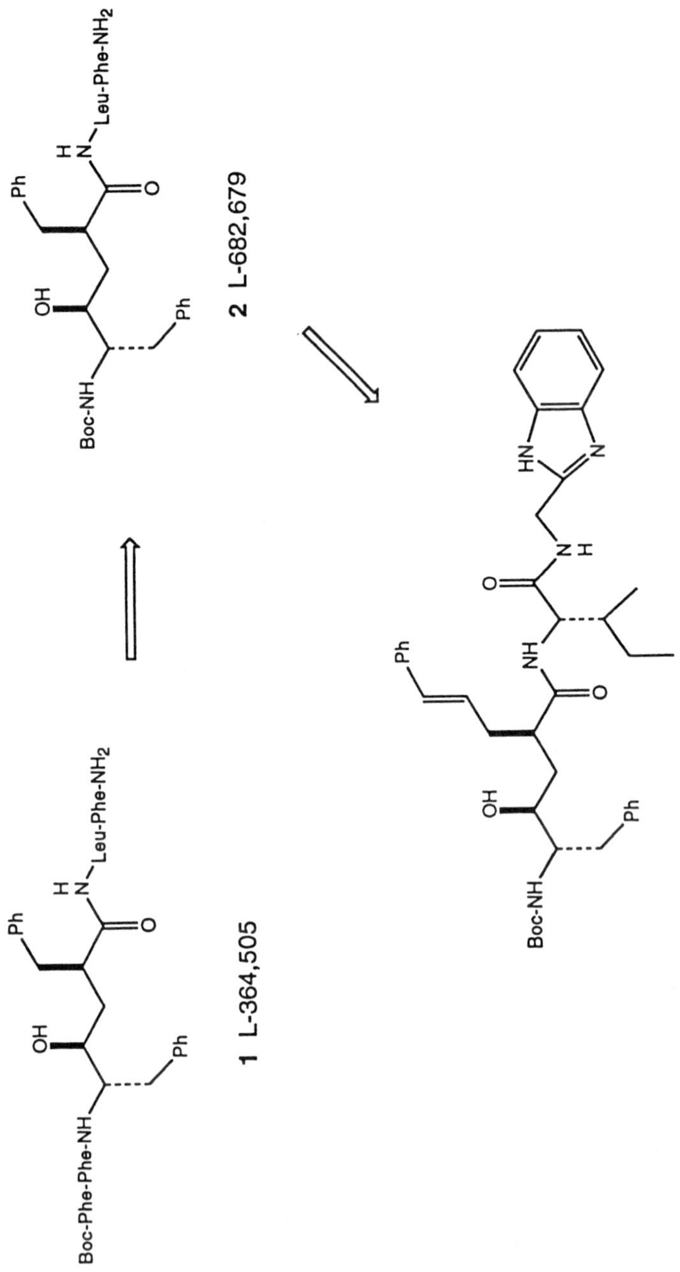

1 L-364,505

2 L-682,679

3 L-687,908

Figure 1. Structures of compounds 1–3.

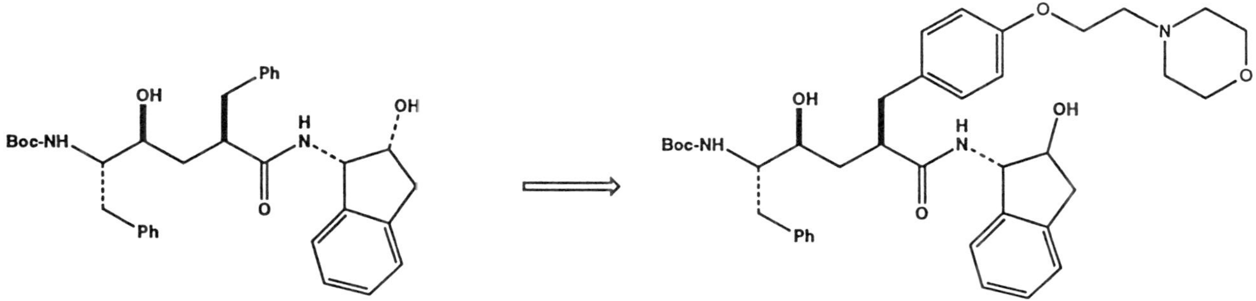

4 L-685,434

5 L-689,502
5% Orally Bioavailable

Figure 2. Structures of compounds **4** and **5**.

this deficiency in this compound and related analogues may have been their low aqueous solubility, which makes intestinal absorption nearly impossible. A possible solution would be to incorporate a weakly basic amine into these molecules. Molecular modeling studies of this series determined possible positions for adding solubilizing groups to these molecules without compromising antiviral activity. Several compounds were synthesized that were more soluble and were very active in cell culture. These were next administered to dogs as aqueous solutions and the plasma levels determined by HPLC after extraction of the plasma samples. The best compound from this series was L-689,502 (Thompson *et al.,* 1992) (compound **5,** Fig. 2), which was potent (IC_{50} = 0.45 nM, CIC_{95} = 12 nM) and was 5% orally bioavailable in dogs. Although this compound failed in subsequent toxicology studies, it represented an important program milestone because it demonstrated the type of physical properties that would be needed in an eventual development candidate.

Another compound series that was being explored is represented by L-687,630 (compound **6,** Fig. 3), which is a cyclized version of L-685,434 and was an attempt to reduce the number of secondary amide bonds in our inhibitors, thus increasing their absorption (Vacca *et al.,* 1994b). Further modification of this compound led to the more potent compound L-700,497 (compound **7,** Fig. 3). This compound was found to have good oral absorption in rats administered as a solution in 20% ethanol/water. Further modification of this series gave the highly potent lactam L-731,723 (compound **8,** Fig. 3). This compound had low absorption when given to dogs as a suspension in methocel but the plasma levels were improved after grinding the compound to a particle size of less than 10 μm (Hungate *et al.,* 1994). Unfortunately, further development of this compound was terminated because of unexpected activity in some ancillary assays.

3. IMPROVEMENT OF SOLUBILITY

Drug absorption is influenced by many factors. The two most important factors that affect both the extent and the rate of absorption are lipophilicity and solubility. In general, the higher the lipophilicity of a drug, the greater is its metabolic clearance and the shorter is its $t_{1/2}$. On the other hand, the lower the solubility of a drug, the poorer is its absorption.

It had been reported (Roberts *et al.,* 1990) that the Hoffmann-La Roche compound Ro 31-8959 (sanquinavir; Fig. 4) had modest oral bioavailability in rats, and we surmised that it was most likely related to the decahydroisoquinoline amine in its backbone, which enhances its water solubility. We were interested in combining this basic amine into our hydroxyethylene inhibitor series in order to increase solubility and oral absorption. Molecular modeling studies (Holloway *et al.,* 1994) with models of saquinavir and L-685,434 (compound **4,** Fig. 4) in the L-689,502

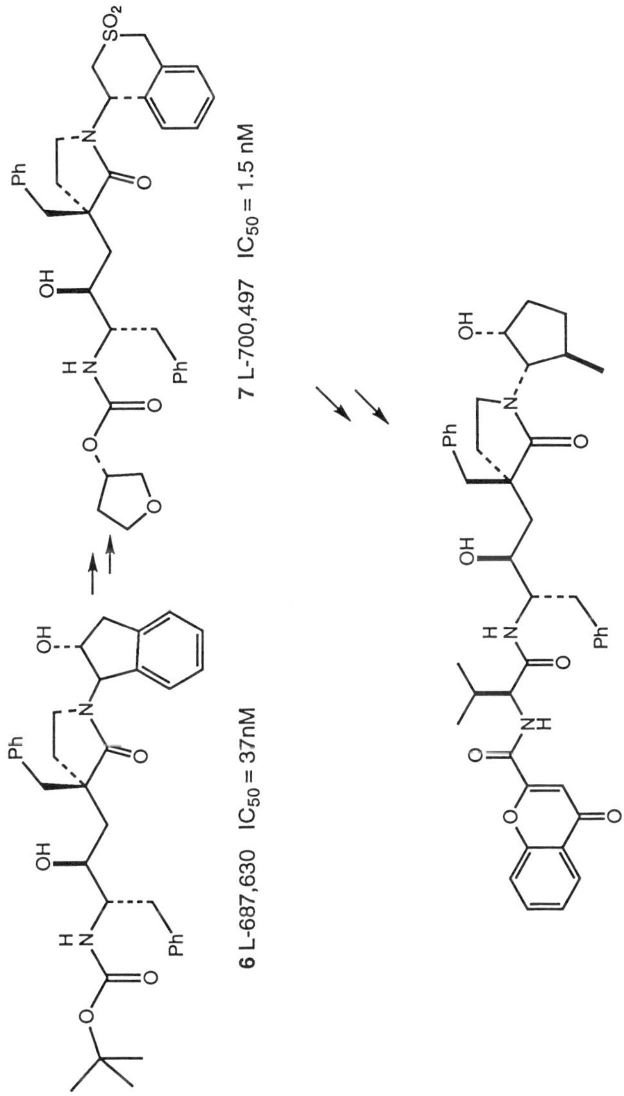

6 L-687,630 IC_{50} = 37nM

7 L-700,497 IC_{50} = 1.5 nM

8 L-731,723 IC_{50} = 0.1 nM

Figure 3. Structures of compounds **6–8**.

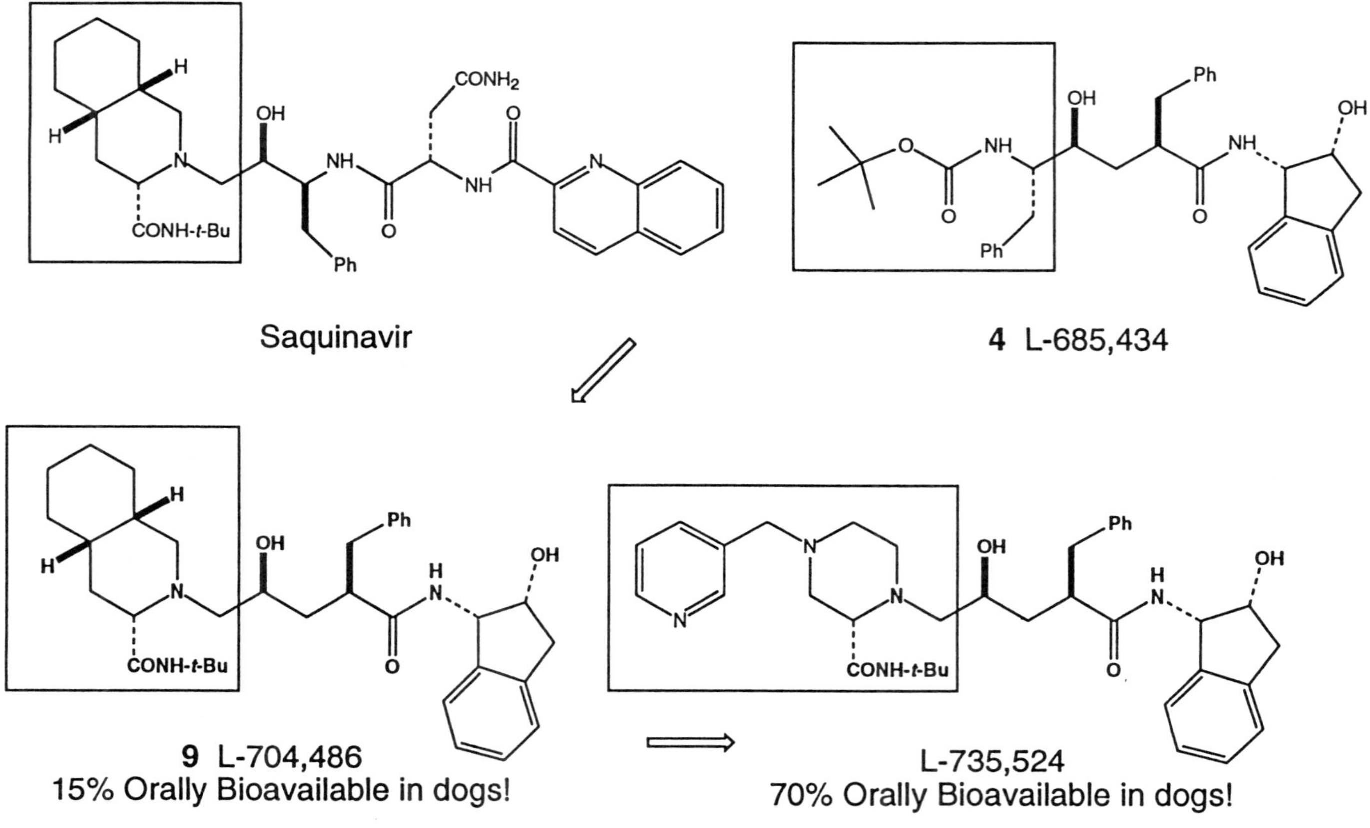

Figure 4. Improvement of oral bioavailability.

(compound **5,** Fig. 2) inhibited site indicated that the two compounds could be superimposed on each other with the decahydroisoquinoline overlaying the Boc-Phe position and the Phe-Asn-Qua superimposed on the indane portion of L-685,434. This suggested compound **9** (Fig. 4) as the target, which was synthesized and found to be a good HIV-1 protease inhibitor (IC_{50} = 7.8 nM) with modest oral absorption in dogs. Further enhancements in potency and solubility (Table I) were made by modifying the decahydroisoquinoline group of compound **9** (Fig. 4) and the optimal compound was found in L-735,524 (MK-639; indinavir; Crixivan®) (Fig. 5), which contains a 4-(3-pyridylmethyl)-2-*t*-butylcarboxamidino-piperazine in place of the Boc-Phe portion of L-685,434. L-735,524 was found to be a potent (IC_{50} = 0.3 nM, CIC_{95} = 50 nM), orally bioavailable protease inhibitor (Fig. 6) and was chosen for further development (Vacca *et al.*, 1994b; Dorsey *et al.*, 1994). The X-ray structure of L-735,524 complexed with HIV-1 and 2 protease has recently been reported (Chen *et al.*, 1994) and the compound occupies the active site in a manner as predicted in modeling studies.

4. PHYSICOCHEMICAL PROPERTIES OF MK-639 (INDINAVIR)

The first physical form of the drug that was evaluated in detail as a potential candidate for pharmaceutical development was the crystalline free base monohydrate. This drug form had aqueous solubility of less than 0.02 mg/ml at the native pH of 7–7.5 and its solubility was highly pH dependent (Fig. 7).

When the free base monohydrate was dosed in dogs as a suspension in 0.5%

Table I
Effect of Solubility of HIV Protease Inhibitors on Drug Absorption in Dogs (10 mg/kg p.o.)

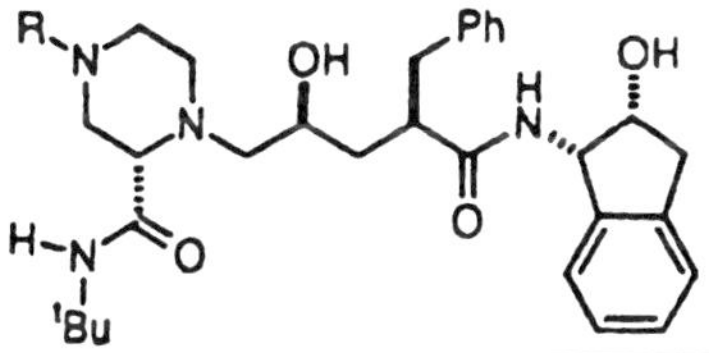

Compound[a]	R	C_{max} (μM)[b]	sol, pH 7.4 (5.2) (mg/ml)	log *P*
L-732,747	Benzyloxycarbonyl	<0.10	<0.001	4.67
L-735,482	8-Quinolinylsulfonyl	<0.10	<0.001	3.70
L-738,891	2,4-Difluorophenylmethyl	0.73 ± 0.15	0.0012 (0.03)	3.69
L-735,524	3-Pyridylmethyl	11.4 ± 2.3	0.07 (0.69)	2.92

[a]Each compound was delivered orally in 0.05 M citric acid; *n* = 2 except for L-735,524 (*n* = 4).
[b]C_{max}, maximum plasma concentration in dogs.

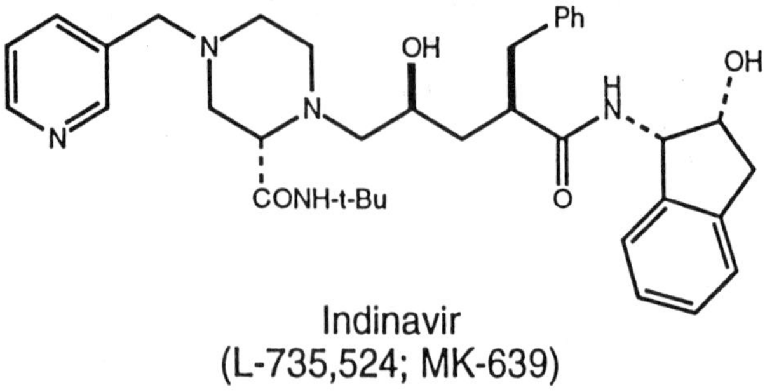

Indinavir
(L-735,524; MK-639)

Figure 5. Structure of indinavir (Crixivan®).

methocel, it showed relatively low and quite variable oral bioavailability. Oral bioavailability was higher and more reproducible from acidic solutions (Kwei *et al.*, 1995; Lin *et al.*, 1995a). A need for an acceptable soluble salt was recognized for development of a clinical dosage form. The pH solubility profile and pK_as of the drug, $pK_a1 = 3.68 \pm 0.14$, $pK_a2 = 6.00 \pm 0.01$, suggested that a fairly acidic salt would be needed in order to achieve a complete dissolution of relatively high anticipated clinical drug doses.

However, solution stability data showed that the drug was unstable in acidic solutions and that potential stability problems can be anticipated for an intrinsi-

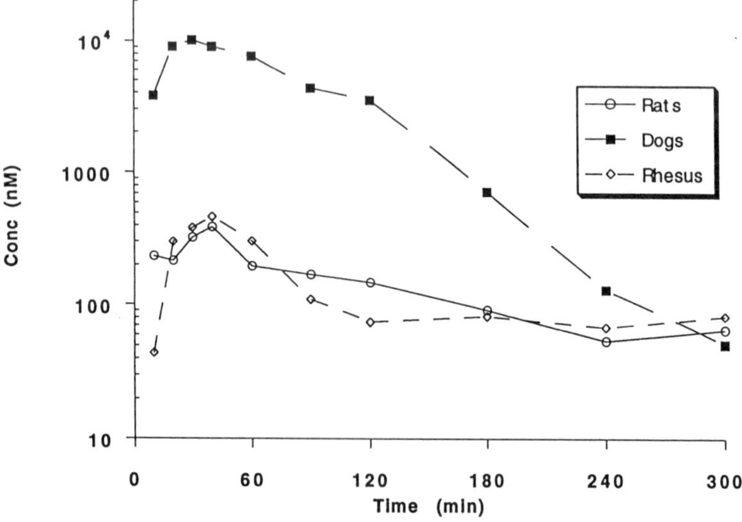

Figure 6. Plasma concentrations of L-735,524 (10 mg/kg in 0.05 M citric acid solution) in animals.

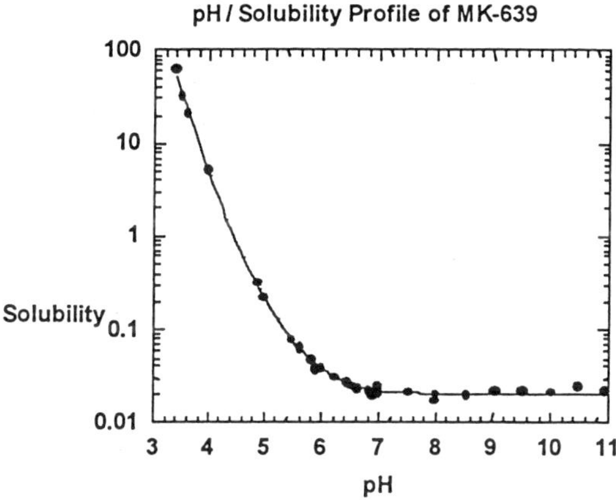

Figure 7. pH–solubility profile of MK-639.

cally acidic salt. pH–stability data are summarized in Table II. At all pH values, reactions followed the first-order rate law to completion. An example of degradation kinetics in solution is shown in Fig. 8.

The primary degradation pathway in solution and in the solid state is the formation of corresponding lactone and *cis*-aminoindanol degradates. The identification of major degradates allowed for easy quantitation of low levels of degradation (see Fig. 9).

Crystalline sulfate salt ethanolate was identified as the highly soluble drug form that showed excellent oral absorption in rats and dogs. The aqueous solubility of the sulfate salt was in excess of 500 mg/ml and the resulting pH of aqueous solutions was less than 3. The sulfate salt also proved excessively hygroscopic becoming deliquescent above 70% relative humidity (Fig. 10). X-ray powder diffraction studies showed that the drug undergoes physical form changes depending on relative humidity (see Fig. 11). Despite its potentially problematic

Table II
pH–Stability Data for MK-639

pH	Buffer	k_1 (hr^{-1}) at 40°C	$t_{1/2}$ (days) at 40°C
1	0.1 M HCl	2.16×10^{-3}	13
2	0.1 M maleate	1.14×10^{-3}	25
3	0.1 M citrate	7.12×10^{-4}	41
4	0.1 M citrate	3.36×10^{-4}	86
5	0.1 M citrate	1.10×10^{-4}	262
11	0.1 M carbonate (1/1 MeOH/H_2O)	1.23×10^{-3}	23

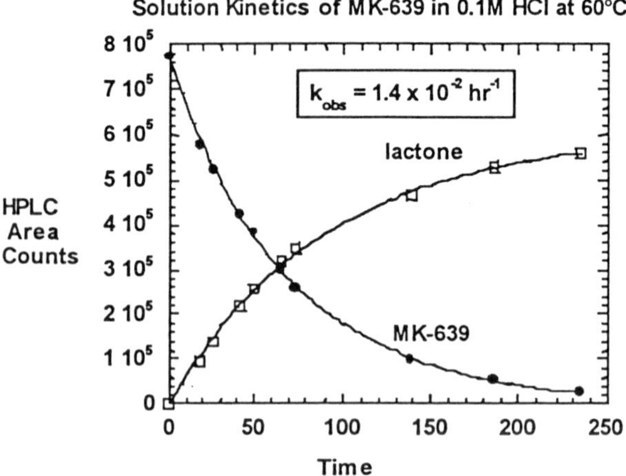

Figure 8. Solution kinetics of MK-639 in 0.1 M HCl at 60°C.

physicochemical properties, the sulfate salt ethanolate was selected as the form for development based on the superior pharmacokinetic profiles. This salt is extremely hygroscopic and can potentially convert to the amorphous material on standing, which may present a serious stability problem given the drug's intrinsic

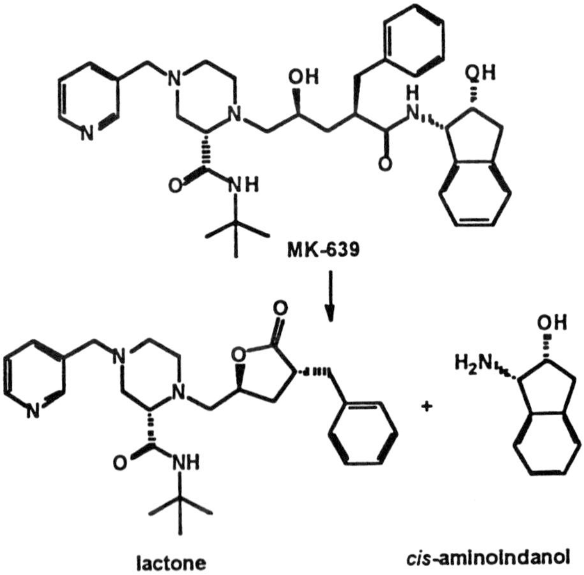

MK-639

lactone + *cis*-aminoindanol

Figure 9. Major degradates of MK-639.

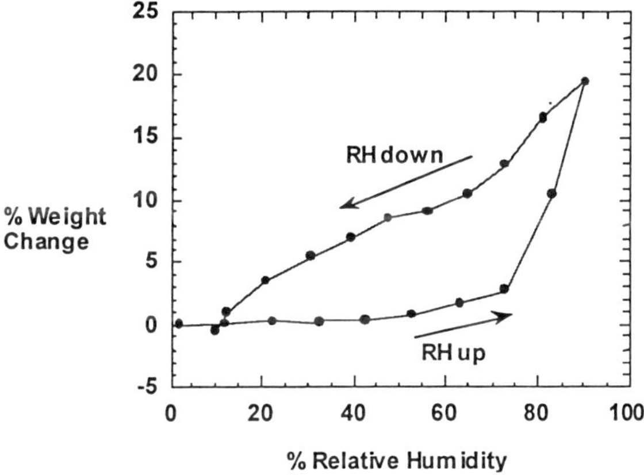

Figure 10. Hygroscopicity profile of MK-639 (sulfate salt) at 25°C.

acidity and acid sensitivity. For this reason, a series of solid-state stability and excipient compatibility studies was conducted under controlled humidity conditions. No significant difference was observed in the long-term stability of crystalline and amorphous neat drug stored at 33% RH. A shelf life greater than 2 years at 25°C and 33% RH was estimated from the data (Table III). The solid-state stability of neat and formulated drug deteriorated rapidly at 40°C and relative humidities above about 30%. Stability is acceptable at lower relative humidities (Table IV).

The above studies showed that as long as the drug is protected from atmospheric moisture during manufacture and storage, a satisfactory long-term stability can be expected. A probe formulation of MK-639 sulfate salt with standard anhydrous excipients in the hard gelatin capsule exhibited satisfactory chemical stability below about 30% RH. This formulation served as the basis for clinical and market formulation development. Based on the excessive hygroscopicity and stability–relative humidity dependence, handling of the bulk and formulated drug

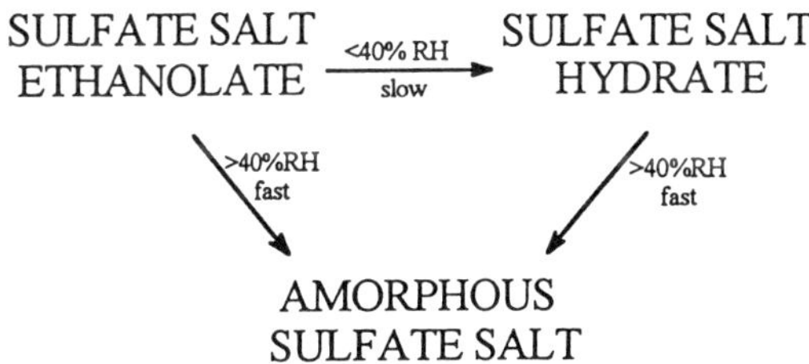

Figure 11. MK-639 physical form changes with relative humidity.

Table III
Long-Term Stability of Crystalline and Amorphous Neat Drug Stored at 33% RH

	% initial[a]			
	30°C, 33% RH		40°C, 33% RH	
Week	Crystalline	Amorphous	Crystalline	Amorphous
4	99.9	99.7	99.7	99.1
12	99.7	99.3	98.8	98.0
26	99.3	99.0	97.4	97.1

[a]Based on the appearance of lactone degradate.

below 30% RH and desiccation of the final product was recommended (Stelmach and Ostovic, 1996).

5. pH-DEPENDENT ORAL ABSORPTION

Although oral absorption is expected to be different among species because of their differences in gastrointestinal physiology and in activities of drug-metabolizing enzymes, the prediction of absorption in humans has been reasonably successful after appropriate application of pharmacokinetics and careful examination of the underlying mechanisms (Lin, 1995). In a survey, Clark and Smith (1984) showed that the fraction of absorption for a large variety of drugs is remarkably consistent between animals and humans; however, the bioavailability differs substantially among species, presumably a result of species differences in the magnitude of first-pass metabolism. This survey implies that the intrinsic absorption of

Table IV
Stability at Lower Relative Humidities

	% initial[a]				
	Neat drug		Powder blend[b]		Capsule[c]
	3 weeks	6 weeks	3 weeks	6 weeks	6 weeks
30°C, 11% RH	99.7	99.6	99.7	99.6	99.6
30°C, 33% RH	99.7	99.5	99.6	99.5	99.5
30°C, 54% RH	99.6	99.3	99.5	99.2	99.4
40°C, 11% RH	99.5	99.4	99.5	99.4	99.4
40°C, 33% RH	99.4	99.0	99.3	99.0	99.3
40°C, 54% RH	98.9	97.9	98.8	97.9	98.1

[a]Based on the appearance of lactone degradate.
[b]Drug, anhydrous lactose, and magnesium stearate.
[c]Powder blend encapsulated in hard gelatin capsules.

most drugs is similar in mammals and because the absorption process (passive diffusion) of a given drug is basically an interaction between the drug and the biomembrane.

Before MK-639 was selected for further development, detailed absorption kinetics of this drug were studied in rats and dogs to ensure that the drug would be well absorbed in humans. When MK-639 was given orally as a suspension in 0.5% methylcellulose (pH 6.5) at a dose of 10 mg/kg, the bioavailability was low in both rats and dogs, approximately 16%. However, when the same dose of MK-639 was given as a solution in citric acid (pH 2.5), the bioavailability increased four- to five-fold in dogs (72%; Fig. 12), but only slightly in rats (24%) (Lin *et al.*, 1995b). These results indicated that oral absorption of MK-639 is pH- and species-dependent.

The pH- and species-dependent differences in bioavailability observed in rats and dogs could be attributed to the species differences in the rate of gastric acid secretion and in the magnitude of hepatic first-pass metabolism. The aqueous solubility of MK-639 is pH-dependent, greater than 100 mg/ml at a pH below 3.5 and 0.03 mg/ml at pH 6. It is well known that gastric acid secretion is poor in dogs, but substantial in rats. When MK-639 was administered in 0.5% methylcellulose, a large portion of the drug in dogs, but not in rats, remained undissolved, resulting in poor absorption in dogs. On the other hand, when MK-639 was administered in citric acid, most of the drug would be in solution, allowing better absorption in dogs. The hypothesis of pH-dependent absorption was further supported by the finding that absorption of MK-639 was significantly increased in dogs after feeding. The increased absorption is believed to be caused by the stimulation of gas-

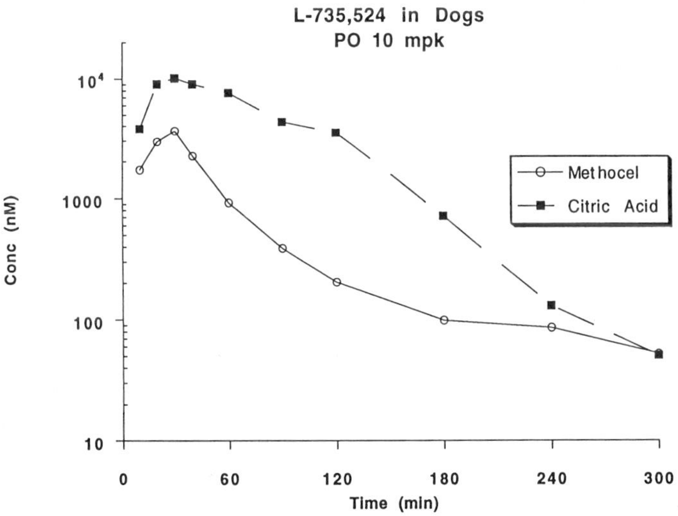

Figure 12. Formulations of L-735,524 (10 mg/kg p.o.) in dogs.

tric acid secretion in dogs by meal ingestion and the consequent lowering of gastric pH and enhancement of the solubility of the drug.

In contrast to the dog, the rat is a good gastric acid secretor. Thus, the low bioavailability observed in rats was mainly related to its high hepatic first-pass metabolism, rather than gastric acid secretion. In a separate study, the hepatic first-pass extraction of MK-639 was estimated to be about 70% by comparing the concentrations in the systemic circulation during portal or femoral vein infusion at steady state (Lin *et al.,* 1995a). When taking the hepatic first-pass metabolism into consideration, the extent of absorption of MK-639 in the rat was estimated to be about 53% for the methylcellulose suspension and 77% for the citric acid solution. The small difference in the extent of absorption (53 versus 77%) in rats after the administration of methylcellulose suspension and citric acid solution may reflect the small difference in the final pH of these two formulations after mixing with gastric juice. To test the hypothesis that the rate of gastric acid secretion is a primary determinant in the absorption of MK-639 in rats as well, a study was conducted in rats using famotidine, a potent H_2-receptor antagonist, and citric acid to modify the gastric–duodenal pH. As expected, pretreatment with famotidine resulted in substantial decreases in both C_{max} and AUC of MK-639 following the administration of the drug in methylcellulose suspension, but not in citric acid solution. These results confirmed the hypothesis of pH-dependent absorption.

After understanding the underlying mechanisms for the low bioavailability in rats and dogs, we predicted that MK-639 would be well absorbed in humans, because humans usually have a high gastric acid secretion and the first-pass effect of the drug would be less significant in humans based on *in vitro* metabolism studies. As expected, when MK-639 was given orally as capsules to AIDS patients (600 mg, ~10 mg/kg), the plasma profiles were similar to those in dogs receiving the same dose (10 mg/kg) in citric acid. The C_{max} and AUC in humans were 10 μM and 19 μM·hr, respectively, and the corresponding values for dogs were 11.5 μM and 12.5 μM·hr (Balani *et al.,* 1995) (data on file, Merck Research Laboratories).

6. *IN VITRO/IN VIVO* METABOLISM

Early information on human metabolism of a new drug is critical in predicting potential clinical drug–drug interactions and in selecting appropriate animal species for toxicity studies. It is required by the regulatory agencies that the animal species used in toxicity studies have metabolic patterns similar to humans. Well before MK-639 was administered to humans, metabolism of the drug was studied *in vitro* using precision-cut rat, dog, and human liver slices. Although limited to qualitative aspects, the metabolic profile of MK-639 obtained from human liver slices accurately reflects the metabolite pattern of urine samples collected from a clinical study (Chiba *et al.,* 1996). The major metabolic pathways in hu-

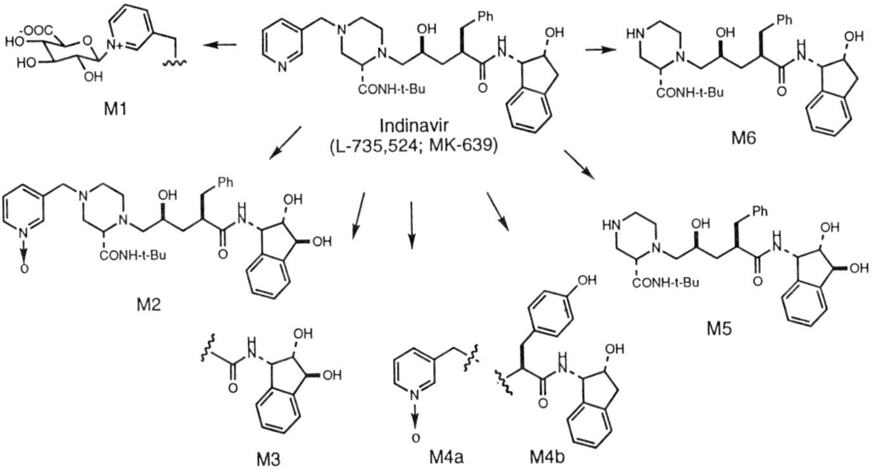

Figure 13. Proposed metabolic pathways of indinavir in humans.

man liver slices and urine were identified as: (1) glucuronidation at the pyridine nitrogen to yield a quaternized ammonium conjugation, (2) pyridine *N*-oxide, (3) *para*-hydroxylation of the phenylmethyl group, (4) 3′-hydroxylation of the indan, and (5) *N*-depyridomethylation (Fig. 13).

In vitro kinetic studies with human liver microsomes revealed that the oxidative metabolic reactions of MK-639 are all catalyzed by a single isozyme, CYP3A4 (Guengerich and Shimada, 1991). This conclusion is based on the results of the five *in vitro* approaches proposed by Guengerich and Shimada (Lin *et al.*, 1995b), namely, (1) chemical inhibition, (2) immunochemical inhibition, (3) metabolism by recombinant human P450 isoforms, (4) competitive effect on marker activities, and (5) a correlation analysis. Furthermore, the K_m value for each oxidative reaction was low, ranging from 0.8 to 3 μM. These results suggest that MK-639 may exhibit dose-dependent kinetics in humans when a high dose is employed. Indeed, both the C_{max} and AUC of MK-639 in AIDS patients increased in a greater than proportionate manner when the oral dose was increased from 100 to 1000 mg. The AUC increased from 0.5 μM·hr at 100 mg to 36 μM·hr at 1000 mg (data on file, Merck Research Laboratories).

7. BACKUP COMPOUNDS

Although MK-639 gives a reasonably good absorption profile, the drug has a comparatively short plasma $t_{1/2}$ (2–3 hr), resulting in a t.i.d. dosage regimen.

Thus, research was initiated for a backup drug with good oral bioavailability as well as a duration of action that allows a once-a-day dosage regimen.

Efforts to prolong the plasma $t_{1/2}$ included the replacement of the pyridine moiety with the furanopyridine moiety to yield L-754,394 (Fig. 14). Pharmacokinetic evaluation revealed that L-754,394 showed excellent absorption kinetics relative to MK-639 (Fig. 15). At the same oral dose (10 mg/kg), the C_{max} and AUC values of L-754,394 were 2.67 μM and 16.9 μM·hr, respectively, in rats, and 14.9 μM and 190 μM·hr, respectively, in dogs; the corresponding values for MK-639 were 0.44 μM and 0.85 μM·hr in rats, and 11.4 μM and 12.5 μM·hr in dogs (Chiba *et al.,* 1995).

However, detailed kinetic studies of L-754,395 showed that this drug exhibited time- and dose-dependent kinetics. In all animal species (rat, dog, and monkey) studied, the apparent clearance decreased when the dose was increased. The clearance decreased from 91 ml/min per kg at 0.5 mg/kg i.v. to 12 ml/min per kg at 10 mg/kg i.v. Apparently, the dose-dependency cannot be explained by Michaelis–Menten kinetics. L-754,394 in plasma declined log-linearly with time, but with an apparent $t_{1/2}$ that increased with dose. The apparent $t_{1/2}$ in rats increased from 20 min at 0.5 mg/kg i.v. to 120 min at 10 mg/kg i.v. Furthermore, L-754,394 exhibited time-dependent pharmacokinetics. After chronic i.v. doses for 7 days (1 mg/kg per day), the apparent clearance of L-754,394 in rats decreased

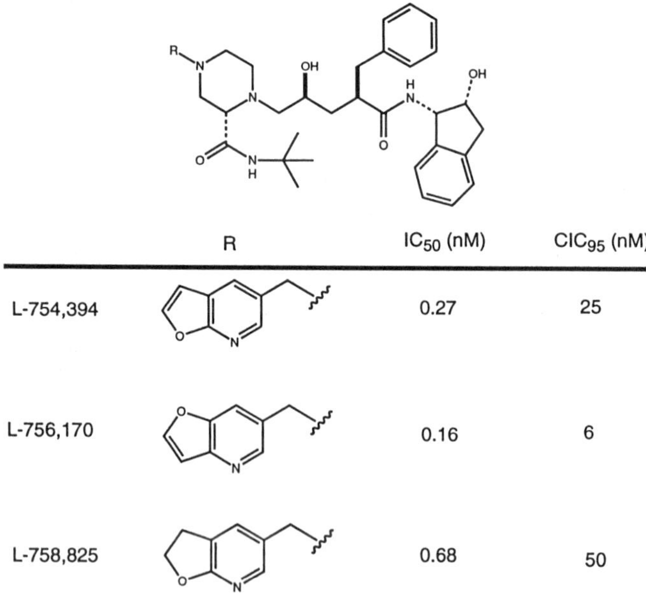

	R	IC_{50} (nM)	CIC_{95} (nM)
L-754,394		0.27	25
L-756,170		0.16	6
L-758,825		0.68	50

Figure 14. Indinavir analogues.

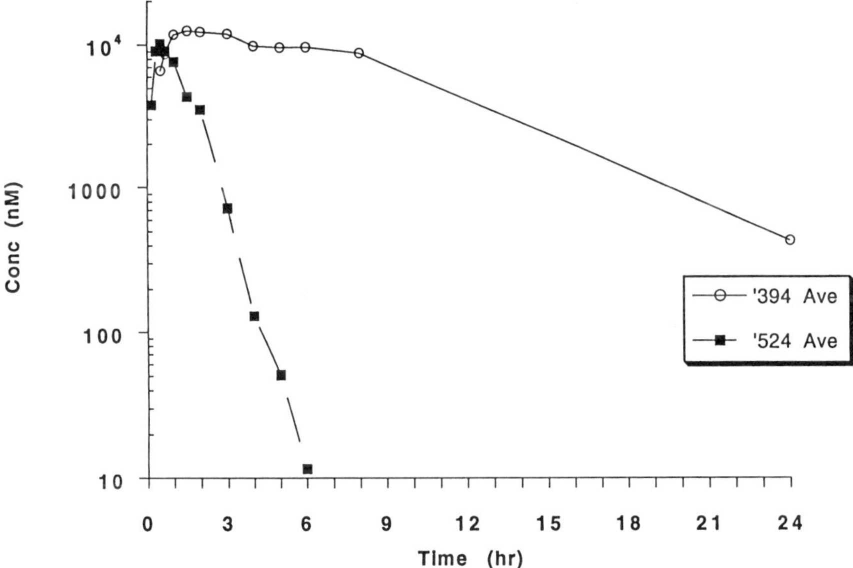

Figure 15. Plasma concentrations of L-735,524 and L-754,394 after oral administration to dogs of 10 mg/kg in 0.05 M citric acid.

from 87 ml/min per kg after the first dose to 25 ml/min per kg after the last dose. Similar results were observed in dogs and monkeys.

Later, *in vitro* microsomal studies revealed that L-754,394 is a potent mechanism-based inactivator (suicide enzyme inhibitor), and the time- and dose-dependent kinetics of the drug may be explained by the mechanism-based enzyme inactivation. *In vitro* spectral studies indicated that approximately 40 to 60% of the content of cytochrome P450 was inactivated when L-754,394 (10 μM) was incubated with rat, dog, and monkey liver microsomes in the presence of NADPH. Little or no inactivation of cytochrome P450 was observed when either NADPH or L-754,394 was omitted. In addition, L-754,394 selectively inhibited CYP2C11-dependent testosterone 2α- and 16α-hydroxylase activities and CYP3A1/2-dependent testosterone 6β-hydroxylase activity, but not CYP2D1/2-dependent bufuralol 1′-hydroxylase activity or CYP1A2-dependent phenacetin *O*-deethylase activity in rat liver microsomes.

Similarly, L-754,394 was found to be a very potent mechanism-based inactivator of human CYP3A4 and, to a lesser extent, of CYP2D6 (Sjoerdsma, 1981). The drug selectively inhibited human liver microsomal CYP3A4-dependent testosterone 6β-hydroxylase and CYP2D6-dependent bufuralol 1′-hydroxylase activities in a time- and concentration-dependent manner in the presence of NADPH. For testosterone 6β-hydroxylase, the inactivation kinetic constants, K_I

and k_{inact}, were 7.5M and 1.6 min^{-1}, respectively, while the partition ratio (moles product formed per moles enzyme inactivated) was approximately 1.35. Although there are many successful examples of suicide enzyme inhibitors that are currently used as drugs, it was decided not to develop L-754,394.

It is assumed that the furanopyridine moiety may play a role in inactivating cytochrome P450 enzymes and a possible route is outlined in Fig. 16. It was speculated that the furan olefin was epoxidized to afford a reactive intermediate which was then reacted in a covalent manner with the CYP-4503A enzyme in an irreversible fashion (Sahly *et al.,* 1996). To test this hypothesis, a series of structurally related compounds (Fig. 14) was examined. The addition of the fused furan ring to the pyridine moiety (L-754,394, L-756,170) resulted in a significant prolongation of the $t_{1/2}$ (>180 min) in dogs relative to that (~30 min) of L-735,524 when the same oral dose (10 mg/kg) was given. However, prolongation was not observed ($t_{1/2} \leq 60$ min) when the furan was replaced by dihydrofuran (L-758,825). Consistent with the *in vivo* observations in dogs, L-756,170, like L-754,394, showed mechanism-based inactivation on the activities of human hepatic CYP2D6 and 3A4, but not on CYP1A2 or CYP2C9. For those compounds with rapid elimination in dogs after p.o. dosing, there were no inhibitory effects on human P450 enzymes. These data strongly support the hypothesis that the furanopyridine moiety plays an important role in inactivating P450 enzymes. Based on this pharmacokinetic and metabolic information, medicinal chemists continue to search for a potent, longer-duration compound without the furanopyridine moiety.

8. CONCLUSION

History tells us that when given a reasonably active lead compound, medicinal chemists usually are able to increase the potency of the lead structure to a useful degree. The critical challenge for developing a clinically useful therapeutic agent is to improve oral bioavailability of the compounds to a practical level. Despite a wide variety of structural modifications to an early HIV-1 protease inhibitor lead, no general solution has emerged for the poor bioavailability that is characteristic of this molecular class. However, we have developed a series of potent inhibitors with increased polarity incorporated into the inhibitor backbone that led to an increase in aqueous solubility without compromising potency. Pharmacokinetic studies in dog showed an improvement in bioavailability from less than 5% for most previous inhibitors to greater than 20% for an initial analogue. Further development of this series led to the discovery of Crixivan®, which has now gained widespread use in the treatment of AIDS.

Although pharmacokinetics and molecular biochemistry have advanced greatly in recent years, it is not yet possible to predict all of the pharmacokinetic and metabolic parameters of a drug in human from animal studies or from *in vitro*

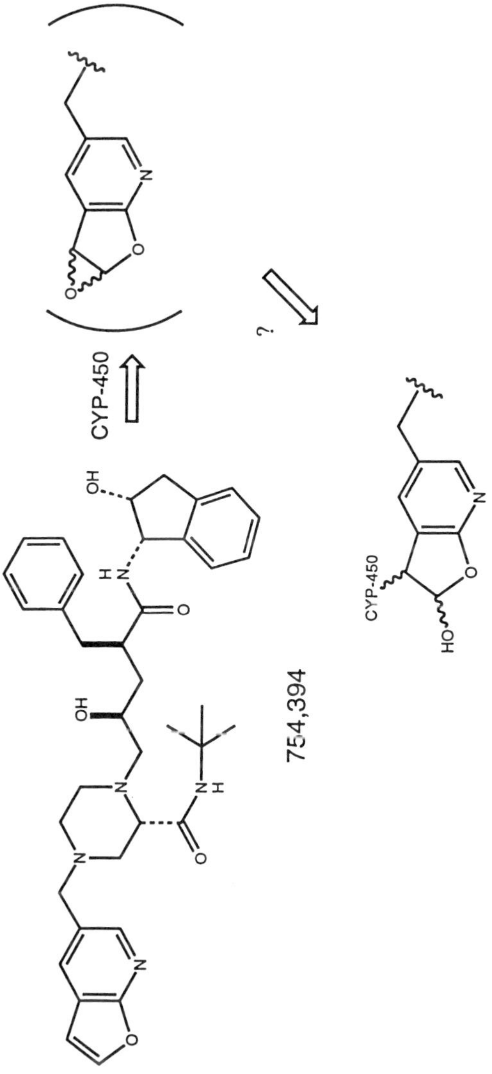

Figure 16. Possible mechanism for P450 inactivation by L-754,394.

studies. Nevertheless, under certain well-defined conditions, it may be possible to make reasonably good predictions as in the examples described in this chapter. Moreover, these examples illustrate the integration of drug metabolism in drug discovery and development.

REFERENCES

Balani, S. K., Arison, B. H., Mathai, L., Kauffman, L. R., Miller, R. R., Stearns, R. A., Chen, I.-W., and Lin, J. H., 1995, Metabolites of L-735,524, a potent HIV protease inhibitor, in human urine, *Drug Metab. Dispos.* **23:**266–270.

Chen, Z., Li, Y., Chen, E., Hall, D., Darke, P., Culberson, C., Shafer, J. A., and Kuo, L. C., 1994, Crystal structures of human immunodeficiency virus protease complexed with L-735,524—An orally bioavailable inhibitor of the HIV proteases, *J. Biol. Chem.* **269:**26344–26348.

Chiba, M., Nishime, J. A., and Lin, J. H., 1995, Potent and selective inactivation of human liver microsomal cytochrome P-450 isoforms by L-754,394, an investigational HIV protease inhibitor, *J. Pharmacol. Exp. Ther.* **275:**1527–1534.

Chiba, M., Hensleigh, M., Nishime, J. A., Balani, S. K., and Lin, J. H., 1996, Role of CYP3A4 in human metabolism of MK-639, a potent HIV protease inhibitor, *Drug Metab. Dispos.* **24:**307–314.

Clark, B., and Smith, D. A., 1984, Pharmacokinetics and toxicity testing, *Crit. Rev. Toxicol.* **12:**343–385.

Dorsey, B. D., Levin, R. B., McDaniel, S. L., Vacca, J. P., Guare, J. P., Darke, P. L., Zugay, J. A., Emini, E. A., Schleif, W. A., Quintero, J. C., Lin, J. H., Chen, I.-W., Holloway, M. K., Fitzgerald, P. M. D., Axel, M. G., Ostovic, D., Anderson, P. S., and Huff, J. R., 1994, L-735,524: The design of a potent and orally bioavailable HIV protease inhibitor, *J. Med. Chem.* **37:**3443–3451.

Greenlee, W. J., 1990, Renin inhibitors, *J. Med. Res. Rev.* **10:**173–236.

Guengerich, F. P., and Shimada, T., 1991, Oxidation of toxic and carcinogenic chemicals by human cytochrome P-450 enzymes, *Chem. Res. Toxicol.* **4:**391–407.

Holloway, M. K., Wai, J. M., Halgren, T. A., Fitzgerald, P. M. D., Vacca, J. P., Dorsey, B. D., Levin, R. B., Thompson, W. J., Chen, L. J., deSolms, S. J., Gaffin, N., Ghosh, A. K., Giuliani, E. A., Graham, S. L., Guare, J. P., Hungate, R. W., Lyle, T. A., Sanders, W. M., Tucker, T. J., Wiggins, M., Wiscount, C. M., Woltersdorf, O. W., Young, S. D., Darke, P. L., and Zugay, J. A., 1994, *A priori* prediction of activity for HIV-1 protease inhibitors employing energy minimization in the active site, *J. Med. Chem.* **38:**305–317.

Huff, J. R., 1991, HIV protease: A novel chemotherapeutic target for AIDS, *J. Med. Chem.* **34:**2305–2314.

Hungate, R. W., Chen, L. J., Starbuck, K. E., Vacca, J. P., McDaniel, S. L., Levin, R. B. Dorsey, B. D., Guare, J. P., Holloway, M. K., Whittier, W. L., Darke, P. L., Zugay, J. A., Schleif, W. A., Emini, E. A., Quintero, J. C., Lin, J. H., Chen, I.-W., Anderson, P. S., and Huff, J. R., 1994, Synthesis, antiviral activity, and bioavailability studies of *delta*-lactam derived HIV protease inhibitors, *Bioorg. Med. Chem.* **2**(9)**:**859–879.

Kohl, N. E., Emini, E. A., Schleif, W. A., Davis, L. A., Heimbach, J. C., Dixon, R. A. F., Scolnick, E. M., and Sigal, I. S., 1988, Active human immunodeficiency virus protease is required for viral infectivity, *Proc. Natl. Acad. Sci. USA* **85:**4686–4690.

Kwei, G. Y., Novak, L. B., Hettrick, L. A., Ostovic, D., Loper, A. E., Lui, C. Y., Higgins, R. J., Chen, I.-W., and Lin, J. H., 1995, Regiospecific intestinal absorption of the HIV protease inhibitor L-735,524 in beagle dogs, *Pharm. Res.* **12:**884.

Lin, J. H., 1995, Species similarities and differences in pharmacokinetics, *Drug Metab. Dispos.* **23:**1008–1021.

Lin, J. H., Chen, I.-W., Vastag, K. J., and Ostovic, D., 1995a, pH-dependent oral absorption of L-735,524, a potent HIV protease inhibitor, in rats and dogs, *Drug Metab. Dispos.* **23:**730–735.

Lin, J. H., Chiba, M., Chen, I.-W., Vastag, K. J., Nishime, J. A., Dorsey, B. D., Michelson, S. R., and McDaniel, S. L., 1995b, Time- and dose-dependent pharmacokinetics of L-754,394, an HIV protease inhibitor, in rats, dogs and monkeys, *J. Pharmacol. Exp. Ther.* **274:**264–269.

Lyle, T. A., Wiscount, C. M., Guare, J. P., Thompson, W. J., Anderson, P. S., Darke, P. L., Zugay, J. A., Emini, E. A., Schleif, W. A., Quintero, J. C., Dixon, R. A. F., Sigal, I. S., and Huff, J. R., 1991, Benzocycloalkyl amines as novel C-termini for HIV protease inhibitors, *J. Med. Chem.* **34:**1228.

Ratner, L., Haseltine, W., Patarca, R., Livak, K. J., Starcich, B., Josephs, S. F., Doran, E. R., Rafalski, J. A., Whitehorn, E. A., Baumeister, K., Ivanoff, L., Petteway, S. R., Jr., Pearson, M. L., Lautenberger, J. A., Papas, T. S., Ghrayeb, J., Chang, N. T., Gallo, R. C., and Wong-Staal, F., 1985, Complete nucleotide sequence of the AIDS virus, HTLV-III, *Nature* **313:**277–284.

Roberts, N. A., Martin, J. A., Kirchington, D., Broadhurst, A. V., Craig, J. C., Duncan, I. B., Galpin, S. A., Handa, B. K., Kay, J., Krohn, A., Lambert, R. W., Merrett, J. H., Mills, J. S., Parkes, K. E. B., Redshaw, S., Ritchie, A. J., Taylor, D. L., Thomas, G. J., and Machin, P. S., 1990, Rational design of peptide-based HIV proteinase inhibitors, *Science* **248:**358.

Sahly, Y., Balani, S. K., Lin, J. H., and Baillie, T. A., 1996, *In vitro* studies on the metabolic activation of the furanopyridine L-754,394, a highly potent and selective mechanism-based inhibitor of cytochrome P450 3A4, *Chem. Res. Toxicol.* **9:**1007–1012.

Sjoerdsma, A., 1981, Suicide enzyme inhibitors as potential drugs, *Clin. Pharmacol. Ther.* **30:**3–22.

Stelmach, C., and Ostovic, D., 1996, Physical and chemical characterization of the HIV protease inhibitor Crixivan®, *AAPS 10th Annual Meeting, Seattle.*

Thompson, W. J., Fitzgerald, P. M. D., Holloway, M. K., Emini, E. A., Darke, P. L., McKeever, B. M., Schleif, W. A., Quintero, J. C., Zugay, J. A., Tucker, T. J., Schwering, J. E., Homnick, C., Nunberg, J., Springer, J. P., and Huff, J. R., 1992, Synthesis and antiviral activity of a series of HIV-1 protease inhibitors with functionality tethered to the P1 or P1′ phenyl substituents: X-ray crystal structure assisted design, *J. Med. Chem.* **35:**1685–1701.

Toh, H., Ono, M., Saigo, K., and Miyata, T., 1985, Retroviral protease-like sequence in the yeast transposon Ty *1, Nature* **315:**691.

Vacca, J. P., Guare, J. P., deSolms, S. J., Sanders, W. M., Guiliani, E. A., Young, S. D., Darke, P. L., Zugay, J., Sigal, I. S., Schleif, W. A., Quintero, J. C., Emini, E. A., Anderson, P. S., and Huff, J. R., 1991, L-687,908, a potent hydroxyethylene-containing HIV protease inhibitor, *J. Med. Chem.* **34:**1225.

Vacca, J. P., Fitzgerald, P. M. D., Holloway, M. K., Hungate, R. W., Starbuck, K. E., Chen, L. J., Darke, P. L., and Huff, J. R., 1994a, Conformationally constrained HIV-1 protease inhibitors, *Bioorg. Med. Chem. Lett.* **4**(3):499–504.

Vacca, J. P., Dorsey, B. D., Schleif, W. A., Levin, R. B., McDaniel, S. L., Darke, P. L., Zugay, J., Quintero, J. C., Blahey, O. M., Roth, E., Sardana, V. V., Schlabach, A. J., Graham, P. I., Condra, J. H., Gotlib, L., Holloway, M. K., Lin, J. H., Chen, I.-W., Vastag, K., Ostovic, D., Anderson, P. S., Emini, E. A. and Huff, J. R., 1994b, L-735,524: An orally bioavailable human immunodeficiency virus type-1 protease inhibitor, *Proc. Natl. Acad. Sci. USA* **91:**4096–4100.

West, M. L., and Fairlie, D. P., 1995, Targeting HIV-1 protease: A test of drug-design methodologies, *Trends Pharmacol. Sci.* **16:**67–75.

Chapter 12

De Novo Design and Discovery of Cyclic HIV Protease Inhibitors Capable of Displacing the Active-Site Structural Water Molecule

George V. De Lucca, Prabhakar K. Jadhav, Robert E. Waltermire, Bruce J. Aungst, Susan Erickson-Viitanen, and Patrick Y. S. Lam

1. INTRODUCTION

Since the identification of HIV as the causative agent of AIDS, there has been a worldwide effort to find effective therapies for this disease. One of the most intense areas of research has been the effort to find effective inhibitors of the essential aspartic protease (PR) of HIV that processes the viral *gag* and *gag-pol* polyproteins into structural and functional proteins (Katz and Skalka, 1994). Inhibition of HIV-PR *in vitro* results in the production of progeny virions that are immature and noninfectious (Kohl *et al.*, 1988; Peng *et al.*, 1989). The abundance of structural information available on HIV-PR has made the enzyme an attractive target for computer-aided drug design strategies (Wlodawer and Erickson, 1993; Appelt, 1993; Ringe, 1994).

In clinical studies, several HIV-PR inhibitors have been shown to reduce the viral load and increase the number of $CD4^+$ lymphocytes in HIV-infected patients

---

George V. De Lucca, Prabhakar K. Jadhav, Robert E. Waltermire, Bruce J. Aungst, Susan Erickson-Viitanen, and Patrick Y. S. Lam • DuPont Merck Pharmaceutical Company, Experimental Station, Wilmington, Delaware 19880–0500.

Integration of Pharmaceutical Discovery and Development: Case Studies, edited by Borchardt *et al.*, Plenum Press, New York, 1998.

(Vella, 1994; Pollard, 1994; Vacca *et al.*, 1994; Wei *et al.*, 1995; Kempf *et al.*, 1995; Ho *et al.*, 1995; Kitchen *et al.*, 1995; Danner *et al.*, 1995). Saquinavir, ritonavir, indinavir, and nelfinavir have recently been approved by the FDA and are being used in AIDS therapy in combination with reverse transcriptase (RT) inhibitors. However, the daunting ability of the virus to rapidly generate resistant mutants (Jacobsen *et al.*, 1995; Markowitz *et al.*, 1995; Condra *et al.*, 1995; Ridky and Leis, 1995) suggests that there is an ongoing need for new HIV-PR inhibitors with superior pharmacokinetic and efficacy profiles.

The different approaches to the discovery of the various types of HIV-PR inhibitors have been extensively reviewed (Kempf, 1994; Vacca, 1994; Wlodawer, 1994; De Clercq, 1995; Darke and Huff, 1995; Kempf and Sham, 1996; De Lucca *et al.*, 1997). Leads have been identified through random screening and rational drug design. Regardless of how the leads were generated, a main feature of current work in HIV protease inhibitors is the extensive use of structural information and of computational/computer modeling techniques to optimize initial lead structures. This has been especially true of the HIV program at Dupont Merck, in which these techniques have been critical from lead generation to optimization to clinical candidates, as we will summarize in this chapter.

2. INITIATION OF PROGRAM AT DMPC

In 1988, during his lecture at the Du Pont Experimental Station on the structure of the Rous sarcoma virus (RSV) protease, Alex Wlodawer suggested that HIV-1 protease may also have a C_2 axis of symmetry (Miller *et al.*, 1989a; Wlodawer *et al.*, 1989). Based on this information, it was reasoned that a C_2 symmetric inhibitor would be more complementary to, and a potent inhibitor of, the C_2 symmetric enzyme. The initial lead compound, P9695, was synthesized using a pinacol coupling reaction of N-Boc-L-phenylalaninal with Caulton's reagent (Freudenberger *et al.*, 1989) and was found to be active against HIV-1 protease with IC_{50} = 500 nM (Jadhav *et al.*, 1994, 1995). The relative stereochemistry of P9695 was unequivocally established as *SRRS* by its alternative synthesis from D-mannitol (Jadhav and Woerner, 1992):

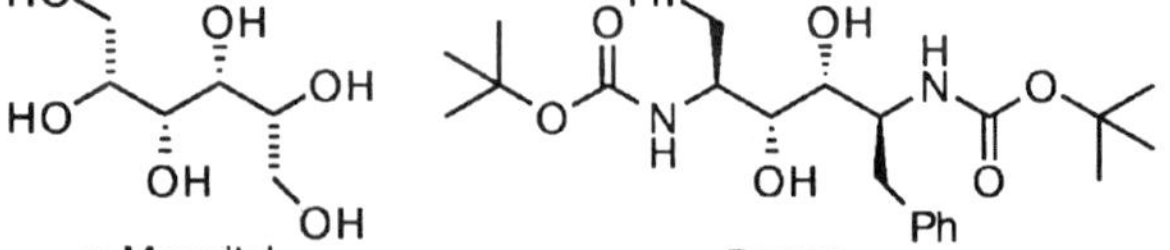

D-Mannitol P9695

Based on the analysis of the natural substrate of HIV-1 protease, P9695 was modified so as to interact with the S2/S3 subsites on the enzyme to give P9941, the first nanomolar inhibitor from our program.

Many analogues of P9941 were synthesized and Q8024 was found to be one of the most potent C_2 symmetric diols, both in the *in vitro* enzyme inhibition assay (K_i = 0.24 nM) and in the antiviral assay (IC_{90} = 50 nM). The analogue syn-

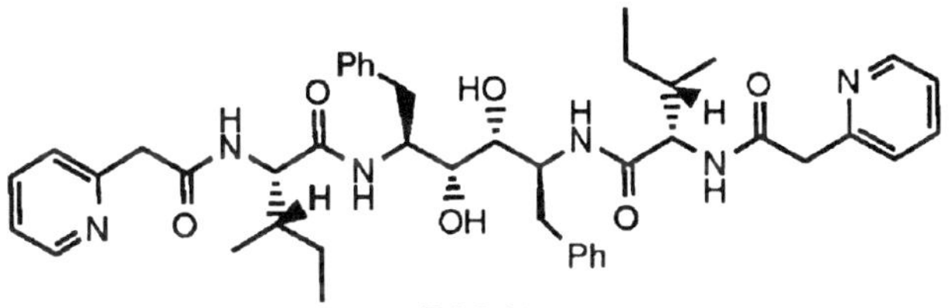

P9941

thesis effort also generated an excellent structure–activity relationship (SAR) that was valuable for subsequent design strategies.

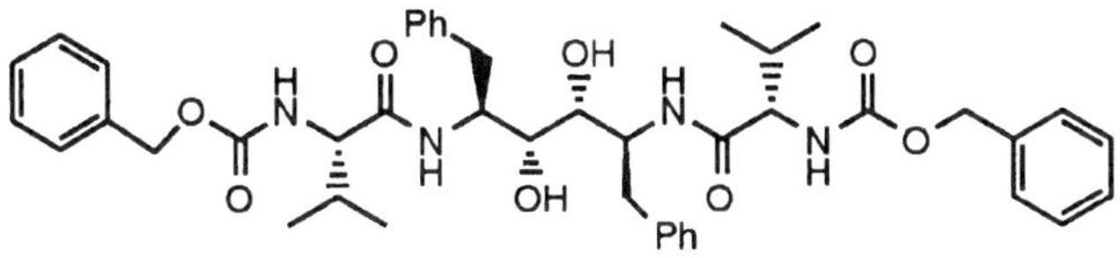

Q8024

Although we made rapid progress in producing more potent inhibitors, none showed significant oral bioavailability. All of these C_2 symmetric diols were very crystalline and highly insoluble in water as well as common organic solvents. These undesirable physical properties and the high molecular mass of C_2 symmetric diols (>700 Da) contributed to the difficulty of identifying an orally bioavailable, pharmaceutically useful compound from the linear C_2 symmetric diol series.

At about this time we discovered that other groups, and in particular the Abbott group (Kempf *et al.*, 1990; Erickson *et al.*, 1990), had independently discovered C_2 symmetric diols as HIV-PR inhibitors using essentially the same design strategy. In retrospect, it probably should not be surprising that in the current research climate, with the ease and speed of new structural information readily available, many similar ideas are concurrently conceived at different research organizations. This has been particularly true in the HIV-PR inhibitor area.

3. DESIGN OF CYCLIC UREAS

3.1. *De Novo* Design

The undesirable physical properties, the lack of oral bioavailability, and the ambiguous proprietary position prompted us to investigate alternative design strategies. A promising computational methodology that became available was the technique of searching data bases containing 3D molecular structures using a 3D pharmacophore model. This technique has been used to identify synthetic frameworks that can serve as the starting point for the design of nonpeptide inhibitors

(Martin, 1992). The use of this technique was incorporated as an important part of the design strategy to identify novel lead structures.

After we started our work in the C_2 symmetric diol series, high-resolution X-ray structures of linear inhibitors complexed with HIV-1PR became available (Miller *et al.,* 1989b; Swain *et al.,* 1990; Ringe, 1994). A common feature among these structures is the presence of a tetracoordinated structural water molecule linking the bound inhibitor to the flexible glycine-rich β-strands or "flaps" of the HIV-PR dimer (Fig. 1). This structural water molecule accepts two hydrogen bonds (H-bonds) from the backbone amide hydrogens of symmetry-related isoleucine residues Ile50 and Ile50′, and donates two H-bonds to the carbonyl oxygens flanking the transition-state mimetic of the inhibitor molecule. The incorporation of this structural water molecule into the inhibitor design to result in positive entropic and selectivity benefits became an important criterion in our design strategy, as summarized in Fig. 2 (Lam *et al.,* 1994).

Using the available structural information and SAR that we had established for the linear C_2 symmetric diols (Jadhav *et al.,* 1994, 1995), we were able to generate several pharmacophore models (Fig. 2A,B). The simplest model (Fig. 2C) was based on two key intramolecular distances: that between symmetric P1 and P1′ hydrophobic groups, and that from P1 and P1′ to H-bond donor/acceptor group(s) that bind to the catalytic aspartates. A 3D data base search with this pharmacophore model yielded the "hit" shown in Fig. 2D, which has the added benefit of incorporating a mimic for the structural water molecule.

Because a phenyl ring might not properly position all substituents of the inhibitor, a cyclohexanone ring (Fig. 2F) was chosen as the initial synthetic scaffold with the ketone oxygen as the structural water mimic. The cyclohexanone ring was enlarged to a seven-membered ring (Fig. 2G) to incorporate a diol functionality, as the SAR established for linear C_2 symmetric diols indicated that the diol imparts significant potency compared with corresponding mono-ol transition-state analogues (Jadhav *et al.,* 1994; Erickson *et al.,* 1990). This synthetic target was further modified to a cyclic urea (Fig. 2H) based on two considerations. First,

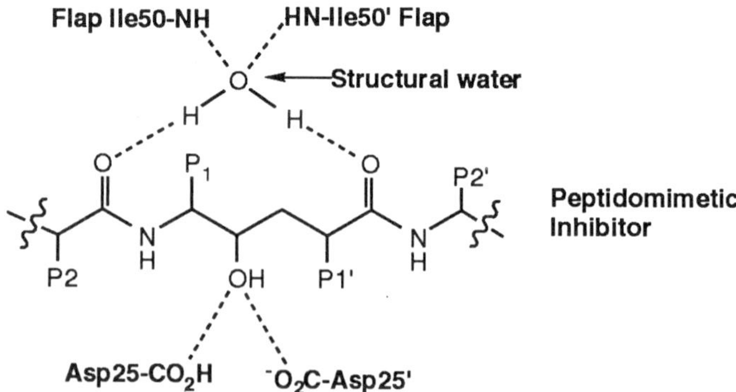

Figure 1. Peptidomimetic inhibitor binding at HIV-PR active site via a bridging structural water.

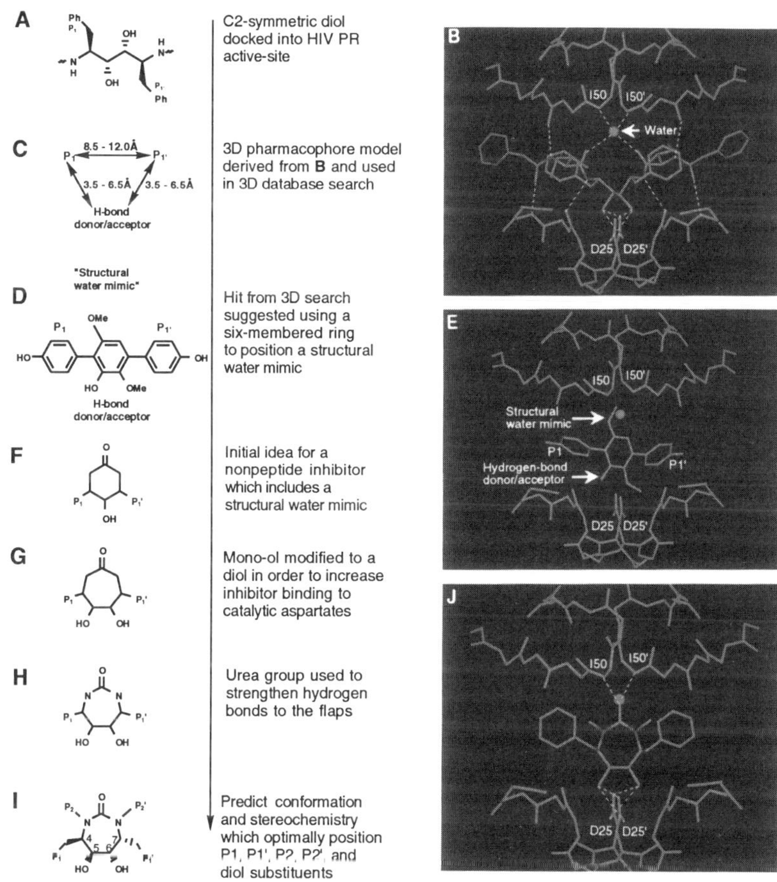

Figure 2. Strategy and steps involved in the design of cyclic urea inhibitors of HIV-PR. Reprinted with permission from Lam *et al.* (1994). Copyright 1994 American Association for the Advancement of Science.

cyclic ureas have precedent as excellent H-bond acceptors both in nature (Weber *et al.,* 1989) and in synthetic systems (Cram *et al.,* 1984; Cram and Lam, 1986). Second, it was realized that the seven-membered cyclic urea was synthetically accessible by cyclizing the precursor used in the linear C_2 symmetric diol series.

Critical to the design strategy is the qualitative prediction of the conformation of the cyclic ureas. The seven-membered ring cyclic ureas can exist in two pseudochair conformations (Fig. 3). When the nitrogens are unsubstituted, 1,3-diaxial strain dominates and conformer **2** with pseudodiequatorial benzyl groups is preferred. When the two nitrogens are substituted, the partial double bond character of the urea C–N bond introduces severe allylic 1,2-strain between the benzylic groups and the nitrogen substituents. This allylic 1,2-strain overcomes the 1,3-diaxial strain, and conformer **3** with pseudodiaxial benzyl groups is preferred. This

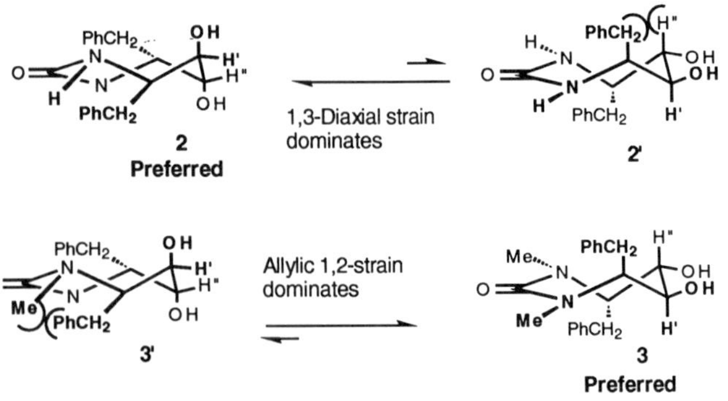

Figure 3. Conformational analysis of designed cyclic ureas predicting that **2** is preferred when the nitrogens are not substituted, whereas **3** is preferred when the nitrogens are substituted because of $A_{1,2}$ strain.

conformational prediction was subsequently confirmed by comparison of the single-crystal small-molecule X-ray analysis of the *N*-substituted and *N*-unsubstituted cyclic urea analogues (Lam *et al.*, 1996).

Using this type of conformational analysis, the predicted optimal stereochemistry for cyclic ureas with substituents on the nitrogens is 4*R*, 5*S*, 6*S*, 7*R* (Fig. 2I), which is derived from unnatural (D) phenylalanine. It is only with this stereochemistry that the substituents on N are directed toward the S2/S2′ sites of HIV-PR. This is in contrast to the linear C_2 symmetric diol inhibitors where natural (L) phenylalanine provides the optimal stereochemistry (Kempf *et al.*, 1993; Jadhav *et al.*, 1994).

3.2. Confirmation of Design

With the conformation and stereochemistry of the designed cyclic urea predicted, we proceeded to test our design ideas. Cyclic urea **2** (XK216), with allyl substituents, was the first (D)-phenylalanine-based cyclic urea synthesized, and we were gratified to find that it was a potent inhibitor with K_i = 5.2 nM (Table I). This was the *first indication of success.* Indeed, XK216 binds 1000-fold tighter than its enantiomer **3.** In addition to its high affinity, XK216 was also found to be orally bioavailable in rats (F = 49%). The high oral bioavailability of XK216 is probably attributable to its small size. Subsequently, after our disclosure of the cyclic ureas, we were informed that two other groups (personal communication) had also tried to cyclize their linear diaminodiol intermediates of their peptidal mimetics to make cyclic ureas. However, theirs were all inactive because they failed to recognize the conformational and stereochemical requirements of a seven-membered ring cyclic urea at the active site. This demonstrates the importance of careful modeling at the design stage.

Published X-ray structures of HIV-1PR revealed that the S2/S2′ pockets are essentially lipophilic except toward the edge of the pockets near the entrance to

Table I

P2/P2′ SAR of Symmetric Cyclic Urea Inhibitors of HIV Protease

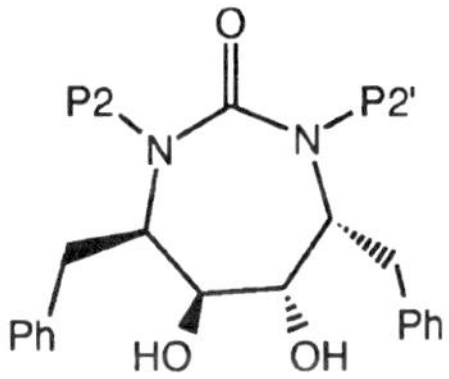

	P2/P2′	K_i[a] (nM)	IC_{90}[a] (μM)	Rat p.o. bioavailability[b]	
				C_{max} (μM)	F (%)
1	H	4500	>100	—	—
2	allyl	5.2	4.7	2.7	49
3	allyl (enantiomer of **2**)	4500	>100	—	—
4	methyl	5700	>141	—	—
5	*n*-ethyl	100	>132	—	—
6	*n*-propyl	8	54	—	—
7	*n*-butyl	1.4	0.68	—	—
8	*n*-pentyl	1.6	1.5	—	—
9	*n*-hexyl	4.6	>102	—	—
10	*n*-heptyl	260	>96	—	—
11	$CH_2CH_2OCH_3$	800	>114	—	—
12	2-dimethylaminoethyl	2660	106	—	—
13	2-trimethylaminoethyl	3150	69	—	—
14	cyclopropylmethyl	2.1	1.8	4.3	100
				9.2 (dog)	48 (dog)
15	cyclobutylmethyl	1.3	1.0	—	—
16	cyclopentylmethyl	4.3	1.7	0.2	—
17	cyclohexylmethyl	37	>96	—	—
18	3-methylisoxazolin-3-yl	3800	96	—	—
19	benzyl	3.0	0.83	1.3	—
20	α-naphthylmethyl	86	16	—	—
21	β-naphthylmethyl	0.31	3.9	0.38	—
22	*o*-fluorobenzyl	34	5.5	—	—
23	*m*-fluorobenzyl	3.0	0.71	—	—
24	*p*-fluorobenzyl	1.4	0.60	1.6	—
25	*m,m′*-difluorobenzyl	24	4.3	—	—
26	*m,m′*-dichlorobenzyl	625	7.8	—	—
27	*p*-hydroxymethyl benzyl	0.34	0.057	0.78	27
				2.8 (dog)	37 (dog)
28	*m*-hydroxymethylbenzyl	0.14	0.038	0.83	18
29	*p*-hydroxybenzyl	0.12	0.032	0.39	22
30	*m*-hydroxybenzyl	0.12	0.054	0.81	30
				2.0 (dog)	16 (dog)
31	*m*-aminomethylbenzyl	980	78	—	—
32	*m*-acetylaminobenzyl	0.42	0.14	0.32	—
33	*m*-aminobenzyl	0.28	0.13	2.25	71
	· $2CH_3SO_3H$			11.2 (dog)	79 (dog)
34	*m*-methylaminobenzyl	0.28	0.034	3.8	—
				1.3(dog)[c]	
35	*m*-*N,N*-dimethylaminobenzyl	0.06	0.038	0.07	

(continued)

Table I *(Continued)*

36	*m,p*-dihydroxybenzyl	0.038	0.74	—	—
37	*m*-carboxybenzyl	0.43	34	—	—
38	*m*-carbomethoxybenzyl	1.3	0.30	—	—
39	*m*-boronicbenzyl	0.011	0.04	<0.10	—
40	*m*-carboxamidobenzyl	0.060	0.71	—	—
41	*m-N*-methyl-carboxamidobenzyl	0.060	0.081	0.12	—
42	*m-N*-ethyl-carboxamidobenzyl	0.21	0.05	0.06	—
43	*m-O*-methyl-hydroxamicbenzyl	0.05	0.22	—	—
44	*m*-hydrazidobenzyl	0.018	10	—	—
45	*m*-acetylbenzyl	0.07	0.04	0.28	—
46	*m*-trifluoroacetylbenzyl	0.037	0.040	1.0	—
47	*m*-aldoximebenzyl	0.01	0.005	0.35	—
48	*m*-acetoximebenzyl	0.01	0.005	0.08	—
				0.51(gelucire)	
49	5-indazolyl-methyl	0.014	0.007	<0.10	—

[a]Values were measured as described previously (Lam *et al.*, 1996).
[b]Bioavailability was determined in groups of rats, unless otherwise indicated ($n = 4$ per group), dosed with compound in formulations containing propylene glycol, polyethylene glycol 400, water at 10 mg/kg. The maximum plasma concentration (C_{max}) is the observed peak plasma concentration after an oral dose. Oral bioavailability (F) was determined by the ratio AUC p.o./AUC/i.v., where AUC is the area under the plasma concentration–time curve from time zero to infinity and is normalized for dose.
[c]Dose was 5 mg/kg.

the active site. The SAR of the cyclic ureas is consistent with this observation. As the size of the *N*-substituent is increased incrementally from methyl to *n*-heptyl as in **4–10,** the potency increases. The optimal size is the *n*-butyl (**7**) with a K_i of 1.4 nM (Table I). In the cycloalkyl series, **14–17,** cyclobutylmethyl cyclic urea **15** was found to have the best K_i among the cycloalkylmethyl cyclic ureas. The hydrophobic nature of the S2/S2′ pockets was further demonstrated by the two to three order-of-magnitude decrease in binding when hydrophilic oxygen and nitrogen atoms are inserted into short alkyl side chains as in **11, 12,** and **13,** or into small cycloalkyls as in **18.**

The X-ray structures that were available and the docked models that we constructed incorporating the cyclic ureas showed that the S2/S2′ pockets are very large and should accommodate large substituents. The benzyl cyclic urea **19** was synthesized and had a K_i of 3.0 nM, and the β-naphthylmethyl cyclic urea **21** was found to be a subnanomolar inhibitor with a K_i of 0.31 nM. Modeling revealed that the β-naphthylmethyl could fit in only one orientation and that there is more space available at the *meta* versus the *ortho* or *para* positions of the P2/P2′ benzyl substituent. This information was useful in designing other analogues.

Because the benzyl cyclic urea **19** showed good potency, it became an attractive side chain for further analogue synthesis. A series of regioisomeric fluoro (**22–24**) substituents were introduced on P2/P2′ benzyl side chains. The *para* and *meta* positions are preferred over the *ortho* position. On the other hand, *m,m′* disubstitution as in **25** and **26** gave poorer binders. Models suggest that because one side of the P2 benzyl ring interacts with the wall of the S2 pocket, there is no room for an additional substituent at the m' position.

3.3. Molecular Recognition

The X-ray structures of HIV-1PR complexed with cyclic urea analogues (β-naphthylmethyl as well as some of the other early analogues) were determined (Lam *et al.,* 1994) soon after their discovery. All of the complexes show the seven-membered ring binding in the same conformation. Its axis of symmetry is nearly coincidental with that of the enzyme. The diols form multiple H-bonds with the catalytic Asp25/25′. The urea oxygen accepts two H-bonds from the backbone NH of Ile50/50′. Thus, the inhibitor links the protease catalytic aspartates to the flexible flaps via a H-bond network that does not include an intervening water molecule. The displacement of the structural water was further confirmed by NMR experiments (Grzesiek *et al.,* 1994). These were the first structural results that confirmed our original design predictions.

The single-crystal small-molecule X-ray structures of unbound *N*-substituted cyclic ureas have also been solved (J. C. Calabrese, unpublished results; Lam *et al.,* 1996). They all share the same ring conformation as the bound conformation. Fig. 4 shows an overlap of bound and "unbound" (small-molecule crystal structure) β-naphthylmethyl analogue. The similarity of the two structures suggests that the cyclic ureas are highly preorganized (Cram, 1986, 1988) for binding.

Variable temperature NMR studies from −70 to 90°C in methanol or DMSO indicate that cyclic ureas exist in a single conformation over this temperature range. Extensive NMR studies indicate that the ring conformation in water is similar to its solid-state X-ray conformation (Hodge *et al.,* 1998).

In general, preorganization includes, but is not limited to, conformational entropic penalty (Cram, 1986, 1988), hydrophobic collapse penalty (Rich, 1993), desolvation cost (Cram, 1986, 1988), and torsional strains (binding conformation not identical to the energetically global minimum conformation for the free drug in water). In our case, although it is not possible to dissect out these contributions, we estimate that the total is at least 4.8 kcal/mole (Lam *et al.,* 1996).

Three main factors are probably responsible for the potency of the *N*-substituted cyclic ureas:

1. The cyclic ureas are preorganized for high complementary binding to HIV-PR, with the conformational entropic penalties typically associated with binding a linear, flexible inhibitor being "prepaid" during synthesis rather than during binding.
2. Displacement of the water molecule is probably thermodynamically favorable (Dunitz, 1994).
3. Hydrophobic interactions between the cyclic urea and the S1/S1′ and the S2/S2′ subsites of HIV-PR are optimized with the preferred conformation and stereochemistry.

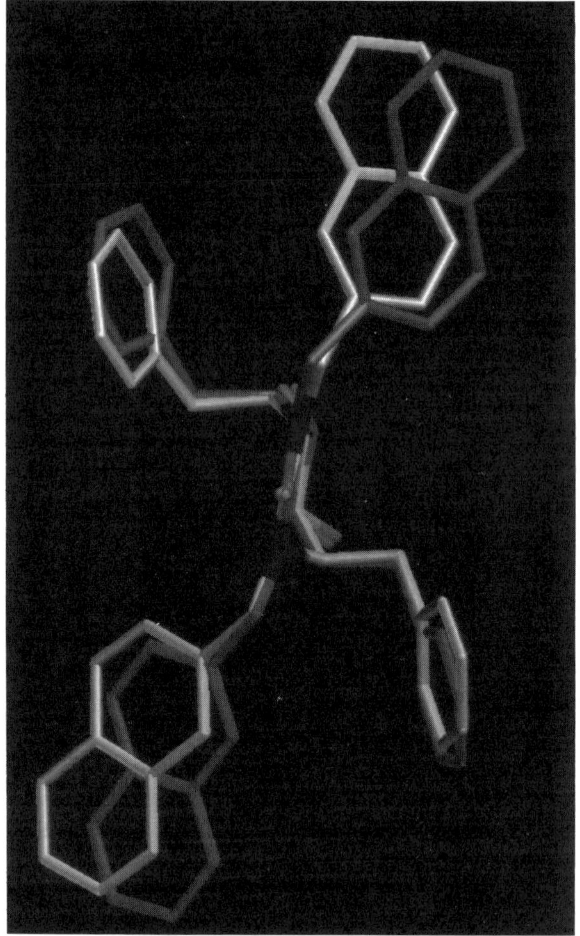

Figure 4. The bound conformation of β-naphthylmethyl analogue, obtained from X-ray analysis of the complex with HIV-PR, overlapped with the "unbound" (small-molecule crystal structure) β-naphthylmethyl analogue.

4. FIRST CLINICAL CANDIDATE DMP 323

4.1. Discovery and Optimization

The cyclic urea core structure is a symmetric, conformationally rigid scaffold designed to be complementary to the C_2 symmetric HIV-PR. The great conformational stability of the *N*-substituted cyclic ureas is invaluable in structure-based design because of the predictability that this stability provides as new substituents are added. This scaffold also provides a synthetic advantage in SAR studies, as cyclic ureas with symmetric P2/P2′ substituents can be prepared easily and optimal side chains rapidly identified. Although synthetically much more challenging, extensive analogue studies focusing on P1/P1′ have been carried out. However, because of lipophilicity, pharmacokinetic, and cost considerations, simple benzyl groups at P1/P1′ are generally preferred (Nugiel *et al.*, 1996).

Using the *N*-benzyl-substituted cyclic urea as a rigid scaffold, modeling (based on the complex of **21**/HIV-PR) revealed that the *N*-benzyl group serves two very important functions. First, it contributes an important hydrophobic interaction with the lipophilic S2 enzyme pocket. Second, it can serve as a scaffold for directing substituents from the *meta* and *para* position toward the S2/S3 subsites where there are several H-bond donors/acceptors, namely, the side chains and/or backbone amides of Asp29, Asp30, and Gly48.

To take advantage of these potential H-bond possibilities, hydroxy and hydroxymethyl groups were incorporated as in **27–30.** These compounds indeed have K_i values in the subnanomolar range. Moreover, because of the reduced lipophilicity, the translation from K_i to IC_{90} is greatly improved. For example, cyclic urea **27** (clog *P* and HPLC log*P* are 4.8 and 3.6, respectively) translates two orders of magnitude better than other subnanomolar inhibitors like **21** (clog*P* 9.2). The IC_{90}s of these cyclic ureas, **27–30,** are in the range of 0.032–0.057 μM.

Although increasing the potency of the cyclic ureas by an order of magnitude, these analogues maintained the oral bioavailability we had seen with our earlier compounds. The oral and i.v. pharmacokinetic profiles of these cyclic ureas were examined in the rat. At a dose of 10 mg/kg they exhibited good pharmacokinetics with a C_{max} of 0.39–0.83 μM (F = 18–30%) (Wong *et al.*, 1994).

We carried out additional studies in the dog with **27** and **30** based on their superior rat pharmacokinetic data (Table II). Both compounds showed higher blood levels and lower clearance values in the dog than in the rat, and in the case of the *p* hydroxymethyl cyclic urea, blood levels exceeded the IC_{90} for wild-type HIV (0.034 μg/ml) for 6 hr. Based on these data, **27** (DMP 323) was selected for preclinical development.

Because of the poor aqueous solubility of DMP 323, and lack of ionizable groups suitable for salt formation, several nonaqueous liquid formulations were examined in the dog. The formulation with the lowest interdog variability was selected for the subsequent human phase I study.

4.2. Chemistry and Process Development

The selection of DMP 323 for development created an immediate need for several kilograms of drug substance. Chemistry was given two critical short-term goals: Prepare 5 kg of DMP 323 over 6 months, and define a scalable process suitable for the preparation of at least 100 kg of DMP 323 within 1 year.

More than 5 kg of DMP 323 was prepared in the discovery group (Scheme 1) within 6 months of selection. The route differed from the original DMP 323 synthesis (Lam *et al.*, 1996) in the choice of P2 alkylating agent. Pinacol coupling of CBz-D-phenylalaninal **51** provided the diol **52** (>99% de after crystallization).

Table II
Pharmacokinetics of Substituted Benzylic Cyclic Ureas

	30	27	29	28
IC_{90} μM	0.054	0.057	0.032	0.038
Rat PK				
i.v. Cl (liters/hr per kg)	3.33	7.12	7.56	6.15
t half (hr)	1.2	0.95	2.25	1.08
p.o. C_{max} (μg/ml)	0.46	0.45	0.21	0.47
F (%)	30	27	22	18
t half (hr)	3.3	1.5	1.5	1.3
C 4 hr (μg/ml)	0.10	0.02	0.023	0.006
Dog PK				
i.v. Cl (liters/hr per kg)	1.33	1.48		
t half (hr)	2.53	1.80		
p.o. C_{max} (μg/ml)	1.07	1.48		
F (%)	16.0	38.0		
t half (hr)	1.05	1.75		
C 4 hr (μg/ml)	0.038	0.17		

The diol was protected as the bis-MEM ether **53.** Hydrogenolysis and cyclization provided cyclic urea **55,** which was then alkylated with the THP-protected benzyl chloride **56,** to give 98–99% of **57.** Deprotection of **57** provided DMP 323, which was isolated by chromatography to meet the specification of at least 98% purity.

A nonchromatographic synthesis of DMP 323 was clearly needed to produce

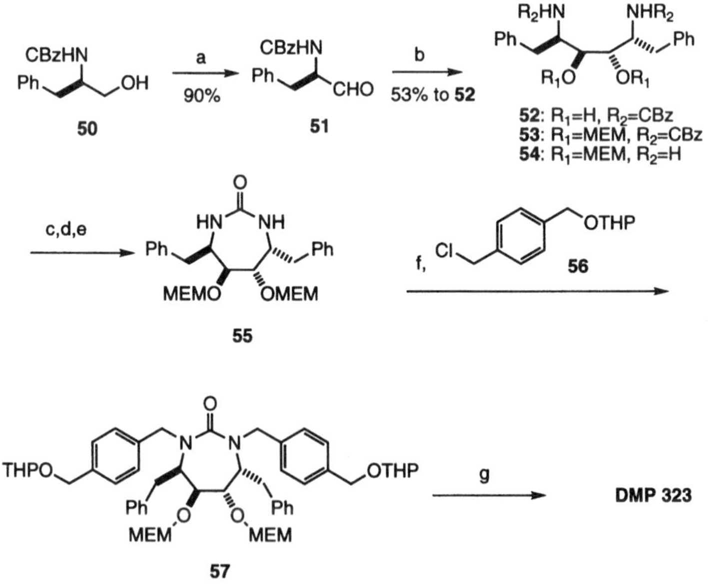

Scheme 1. (a) oxalyl choloide, DMSO, Et_3N; (b) Caulton's Reagent: (c) MEMCl, imidazole; (d) H_2, Pd/C; (e) CDI; (f) NaH, DMF; (g) HCl, H_2O.

over 100 kg (Scheme 2) (Pierce *et al.,* 1996). Initial efforts focused on identifying a diol protecting group that would provide a crystalline cyclic urea analogue of **55** that could be alkylated with 100% conversion. Conformational analysis indicated that in **55,** the MEM ethers were axial, but after alkylation they were required to be equatorial. The acetonide **62** was prepared with the assumption that bis-alkylation, not requiring a ring flip, would be more facile. Fortunately, this postulate proved to be true with alkylations of **62** routinely exceeding 99.5% conversion with a wide variety of electrophiles. Initially we sought to prepare **62** directly from **52** by acetonide formation, deprotection to the diamine, and cyclization. In practice, this was very difficult, as the cyclization required 200-fold dilution at 165°C to obtain 70% yield of **62.** To avoid such a volume-inefficient process, a surrogate protecting group, triethyl silyl (TES), was utilized to convert **52** in five steps to the highly crystalline intermediate **62** in 72% overall yield.

As alkylation of **62** with **56** did not provide a crystalline final intermediate, a series of alternative hydroxyl protecting groups were examined. The trityl protecting group was found to provide both a crystalline alkylating agent **63** and a crystalline final intermediate **64.** A final deprotection of **64** under acidic conditions completed the synthesis providing high-quality crystalline DMP 323.

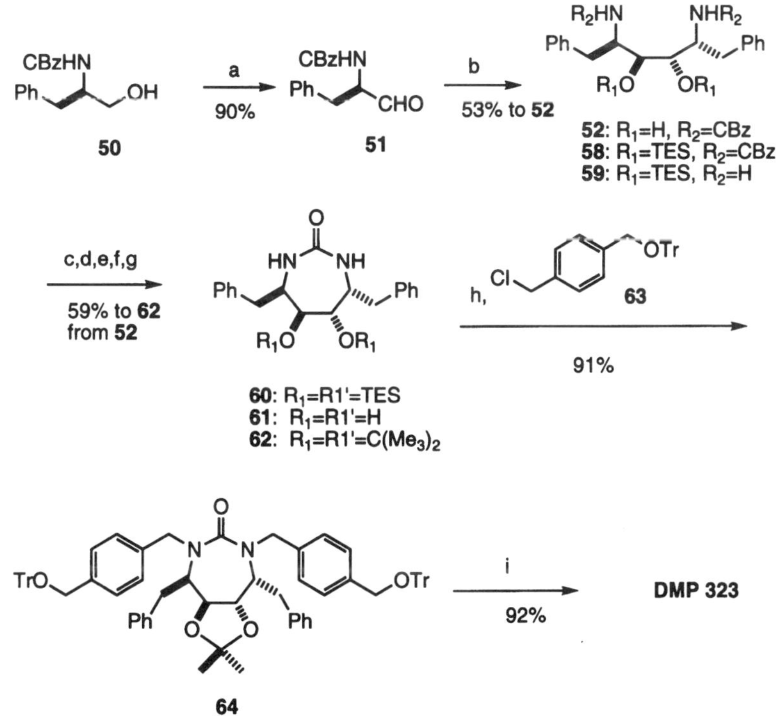

Scheme 2. (a) NaOCl, NaBr, TEMPO; (b) Caulton's Reagent; (c) TESCl, imidazole; (d) H_2, Pd/C; (e) CDI; (f) HCl, H_2O; (g) 2,2-dimethoxypropane, pTsOH; (h) KOtBu; (i) HCl, H_2O.

4.3. Clinical Study

Because the aqueous solubility (6 μg/ml) of DMP 323 is poor, several liquid formulations were examined in dogs. These liquid formulations were comprised of alcohol, propylene glycol, PEG 1450, water, and glycerin. Several of the individual components afforded considerable solubility: alcohol, 272 mg/ml; propylene glycol, 160 mg/ml; PEG 1450, 65 mg/ml. The final formulation is:

DMP 323	50.0 mg
Ethanol	0.263 ml
Glycerin	0.08 ml
Propylene glycol	0.30 ml
Polyethylene glycol, NF, 1450	200 mg
Purified water	to 1.0 ml

DMP 323 was examined in seronegative male volunteers with single doses ranging from 60 to 1200 mg. Disappointingly, blood levels at each dose in man showed a high degree of intersubject variation. For example, after a single dose of 750 mg, the values for C_{max} (in μg/ml) for five individual subjects were: 1.49, 0.165, 0.301, 0.341, and undetectable, resulting in a mean value of 0.46 ± 0.59 μg/ml. Further development of DMP 323 was discontinued.

At least two factors likely contributed to the variable and low plasma levels observed in man with DMP 323. First, the very poor solubility (6 μg/ml) of DMP 323 in aqueous media suggests that it may have precipitated on dosing, although this variability was not observed in other species dosed with cosolvent formulations. Second, metabolism of DMP 323 is both rapid and extensive (Christ *et al.*, 1993). As with several other HIV protease inhibitors (Chiba *et al.*, 1996; Kumar *et al.*, 1996), metabolism of DMP 323 is carried out by CYP 3A4. A major route of metabolism for DMP 323 is progressive oxidation of the hydroxymethylbenzyl side chain to the aldehyde and subsequently to the acid. Rats dosed with DMP 323 were found to contain significant metabolite in plasma corresponding to the monoacid form of DMP 323. In rat liver slices or human microsomes, a mixture of mono aldehyde, mono acid, aldehyde/acid, and bis acid was identified. Thus, rapid metabolism combined with poor absorption of compound resulting from precipitation in the stomach would result in low C_{max} and short apparent half-life.

5. SECOND CLINICAL CANDIDATE DMP 450

5.1. Discovery and Optimization

While the development of DMP 323 was proceeding, we intensified our analogue synthesis and design program. In designing the second generation of cyclic urea protease inhibitors, we sought to optimize physical properties and pharmacokinetics, while maintaining or improving potency. As pointed out earlier, the *N*-

benzyl substituent can serve as a scaffold for directing substituents toward the S2/S3 subsites where there are several H-bond donor/acceptor residues. Designing substituents that can better interact with these residues became an important way to increase the potency of our compounds.

Other functional groups in addition to hydroxyl groups, many with multiple H-bond donor and acceptor possibilities, were examined in order to increase potency and water solubility (Table I). Indeed, some cyclic ureas with P2/P2′ substituents capable of multiple H-bonding interactions that were examined, such as **36–44,** are an order of magnitude more potent enzyme inhibitors than DMP 323. However, many of them are too polar and the translation to antiviral potency is poor. Moreover, the oral bioavailability in rats of many of these compounds is poorer than DMP 323.

To address the poor physical property limitations observed with DMP 323, we examined a number of substituents with basic and acidic (**37**) functionalities. Attempts to introduce very highly basic groups (**31**) were not successful with regard to inhibitory potency, probably because of the introduction of a formal charge and the associated high desolvation penalty.

A number of symmetrical cyclic ureas containing aniline substituents at the P2/P2′ position were synthesized, including aminobenzyl (**33**), *N*-methylaminobenzyl (**34**), and *N,N*-dimethylaminobenzyl (**35**) substituted cyclic ureas (Table I). The aniline **33** (DMP 450) combined potency similar to DMP 323 with substantial water solubility (>130 mg/ml as the bis-mesylate salt) (Hodge *et al.*, 1996).

5.2. Safety and Pharmacokinetics

Anilines have been associated with potential carcinogenicity through the generation of reactive intermediates produced via oxidative metabolism of the aromatic amine. Early in the characterization of DMP 450, we defined a set of metabolism and genotoxicity assays designed to test the potential for formation of harmful metabolites of DMP 450.

First, the *in vitro* metabolism of DMP 450 was assessed in microsomes from various species. No evidence of hydroxylamine formation was observed.

Second, DMP 450 was tested for mutagenic activity in the *Salmonella–E. coli*–mammalian microsome reverse mutation screening assays (Ames test) in the presence and absence of metabolic activation by rat liver microsomes. Assays were conducted in various *Salmonella* and *E. coli* strains exposing the bacteria to levels of DMP 450 ranging from 10 to 5000 μg/bacterial lawn. DMP 450 did not induce apparent mutations in bacteria under the conditions of these assays.

Third, DMP 450 was tested in an *in vitro* assay for unscheduled DNA synthesis in rat liver primary cell cultures. DMP 450 did not induce significant changes in the nuclear labeling of rat primary hepatocytes over the concentration range examined.

Finally, DMP 450 was evaluated in an *in vitro* assay to determine the potential for the compound to induce chromosomal aberrations in Chinese hamster ovary (CHO) cells in the presence and absence of metabolic activation by rat liver microsomes. DMP 450 did not increase the incidence of chromosomal aberrations in this assay relative to controls.

Thus, DMP 450 was considered negative in this series of genotoxicity studies, and no evidence for the generation of undesirable reactive intermediates could be demonstrated *in vitro.*

DMP 450 was then studied extensively in rat, dog, rhesus monkey, and chimpanzee to define its pharmacokinetics (Hodge *et al.,* 1996). After i.v. administration, plasma concentrations declined in multiexponential fashion with terminal half-life (*t* half) ranging from 0.8 hr in the rhesus monkey to 3.6 hr in the dog. The systemic clearance ranged from a low of 0.21 liter/hr per kg in the dog to 4.7 liters/hr per kg in the rat. After oral administration, T_{max} varied among species ranging from 0.5 hr in the rat to 8 hr in the chimpanzee. C_{max} was highest in the dog (11.2 μM), and lowest in the chimpanzee (1.5 μM). Bioavailability was substantial in all species, with *F* ranging from 24% in the chimpanzee to 80% in the dog. Based on its potency, excellent pharmacokinetics, and acceptable safety profile, DMP 450 was selected for preclinical development.

5.3. Chemistry and Process Development

Selection of DMP 450 for development provided the challenges of rapidly preparing drug substance to support development and of identifying a commercial synthesis. The initial synthesis of DMP 450 used the same technology as was defined for DMP 323. Seven kilograms of DMP 450 was prepared by alkylation of **62** with 3-nitrobenzyl bromide, followed by deprotection and hydrogenation in the presence of methanesulfonic acid (Scheme 3).

Because this synthesis would not likely achieve the cost targets for commercialization, related to the cost of the D-amino acid starting material, a great deal of energy was expended on identifying alternative means to prepare the C_2 symmetric 1,4-diamine diol core (Jadhav and Woerner, 1992; Baker and Condon, 1993; Rossano *et al.,* 1995; Kang and Ryu, 1996; Nugiel *et al.,* 1996). The route finally selected for the commercial synthesis of DMP 450 is defined in Scheme 4.

The synthesis started with the commercially available protected form of C_2 symmetric L-tartaric acid **65.** Double reduction of **65** with DIBAL-H followed by reaction with dimethyl hydrazine provide the bis-hydrazone **66.** Chelation-controlled double addition of benzyl lithium proceeded in a highly diastereoselective fashion to provide **67** containing the desired four contiguous asymmetric centers (no detectable diastereomers were present in isolated salt **67**). Hydrogenation provided diamine **68.** Bis-reductive amination with 3-nitrobenzaldehyde provided the bis-secondary diamine **69.** An important discovery was that the acetonide-protected diamine **69** could be cyclized with phosgene at ~125°C in good yield to the desired cyclic urea product. Acid hydrolysis then gave a highly pure final intermediate **70.** The synthesis of DMP 450 was completed by hydrogenation to the desired aniline, methanesulfonic acid salt formation, and humidification to the trihydrate.

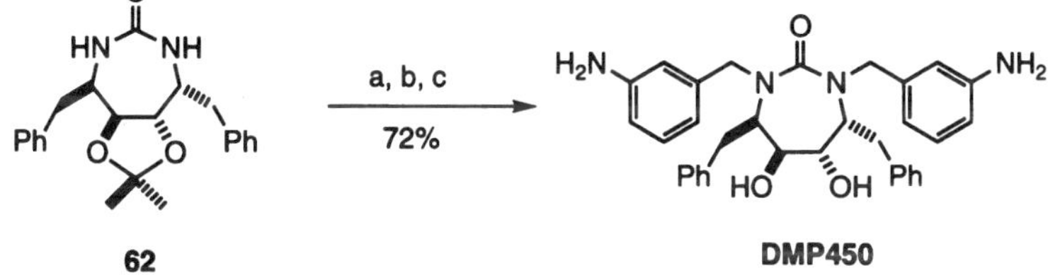

62 → (a, b, c; 72%) → DMP450

Scheme 3. (a) KOtBu, 3-nitrobenzyl bromide; (b) H_2SO_4, MeOH; (c) H_2, Pd/C; methanesulfonic acid, 2-propanol, H_2O

65 → (a,b) → [66] → (c,d; 72%) → 67 · 2 E-$HO_2CCHCHCO_2H$ → (e, f, g; 81%) → 68 · 2 p-Toluenesulfonic acid → (h; 96%) → 69 → (i,j; 84%) → 70 → (k; 81%) → DMP450

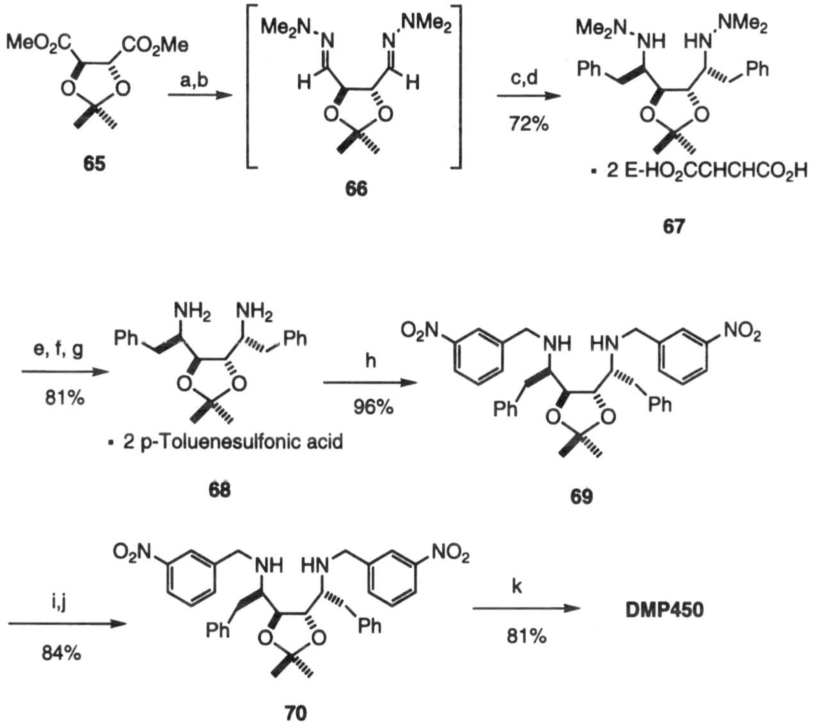

Scheme 4. (a) DIBAL-H; (b) H_2NNMe_2; (c) BnLi; (d) fumaric acid; (e) NaOH, MeOH; (f) Ra Ni, H_2; (g) p-toluenesulfonic acid; (h) 3-nitrobenzaldehyde, $NaBHOAc_3$; (i) phosgene, PhCl; (j) H_2SO_4, MeOH; (k) H_2, Pd/C, methanesulfonic acid, 2-propanol, H_2O.

5.4. Clinical Study

In phase I clinical studies in HIV seronegative male volunteers, DMP 450 showed substantial blood levels. With a single dose of 11 mg/kg the C_{max} was 6.5 μM and the level at 6 hr remained above 1 μM. The measured half-life in man (5.7

hr) is consistent with some degree of potential accumulation with multiple dosing every 6 to 8 hr. A multiple dose study using 1000 mg q.i.d. did, indeed, indicate an increase in trough level from 1.73 μM on day 2, to 3.2 μM by day 4. DMP 450 was well tolerated with no adverse effects noted in these studies (Hodge *et al.,* 1996).

At the same time, early clinical trial results with Indinavir had identified the potential for dramatic rebound in plasma RNA levels with concomitant emergence of HIV variants with multiple mutations in the protease coding regions (Condra *et al.,* 1995). In addition, the surprising clinical failure of SC-52151 (Bryant *et al.,* 1995) was ultimately ascribed to high plasma protein binding, and established that it was the relationship between the plasma level of free drug and the inherent drug potency that was the likely predictor of clinical efficacy.

To assess protein binding, the binding of ^{14}C-labeled DMP 450 to human plasma proteins was examined by equilibrium dialysis using undiluted human plasma and 5 μg/ml (9.4 μM) DMP 450. Binding to plasma proteins to the extent of 90–93% was observed. The effect of this plasma protein binding on the antiviral potency of DMP 450 was examined by conducting antiviral assays in the presence of the two major components of human plasma, namely, human serum albumin and α_1-acid glycoprotein at levels comparable to those found in the blood of AIDS patients. In the presence of 45 mg/ml serum albumin plus 1 mg/ml α_1-acid glycoprotein, the apparent antiviral potency measured as the concentration required to inhibit viral replication 90% (IC_{90}) was increased 4.5 to 8.4-fold depending on the methods utilized to monitor the extent of virus replication (Hodge *et al.,* 1996).

Figure 5 shows the plasma concentration versus time profiles for five species dosed with a single 10 mg/kg oral dose of DMP 450. The data are represented as the ratio of the plasma concentration at various times relative to the IC_{90} (144 nM) multiplied by the fold-increase in IC_{90} measured in the presence of human plasma proteins (average value of 6.45). It can be seen that plasma levels in man and in the dog exceed the level required for 90% inhibition of wild-type HIV for several hours, even accounting for losses of free drug caused by plasma protein binding. The aggressive dosing regimen of using 1000 mg q.i.d. would be sufficient to provide for 90% inhibition of wild-type HIV (929 nM, when adjusted for protein binding). These plasma levels, however, may not provide for adequate inhibition of mutant variants of HIV carrying amino acid substitutions within protease, which are likely to be present in the infected individual (Coffin, 1995). In order to focus internal resources on the identification of a third-generation cyclic urea with approximately 10-fold improvement over DMP 450, we outlicensed DMP 450 in 1996.

6. FUTURE CYCLIC UREAS

To discover superior inhibitors of HIV-PR, we have focused on simultaneous optimization of multiple properties. Our goal is to design an inhibitor that combines potency to wild-type and mutant strains of HIV, pharmacokinetic behavior, plasma protein binding propensity, and physical properties such that we can pro-

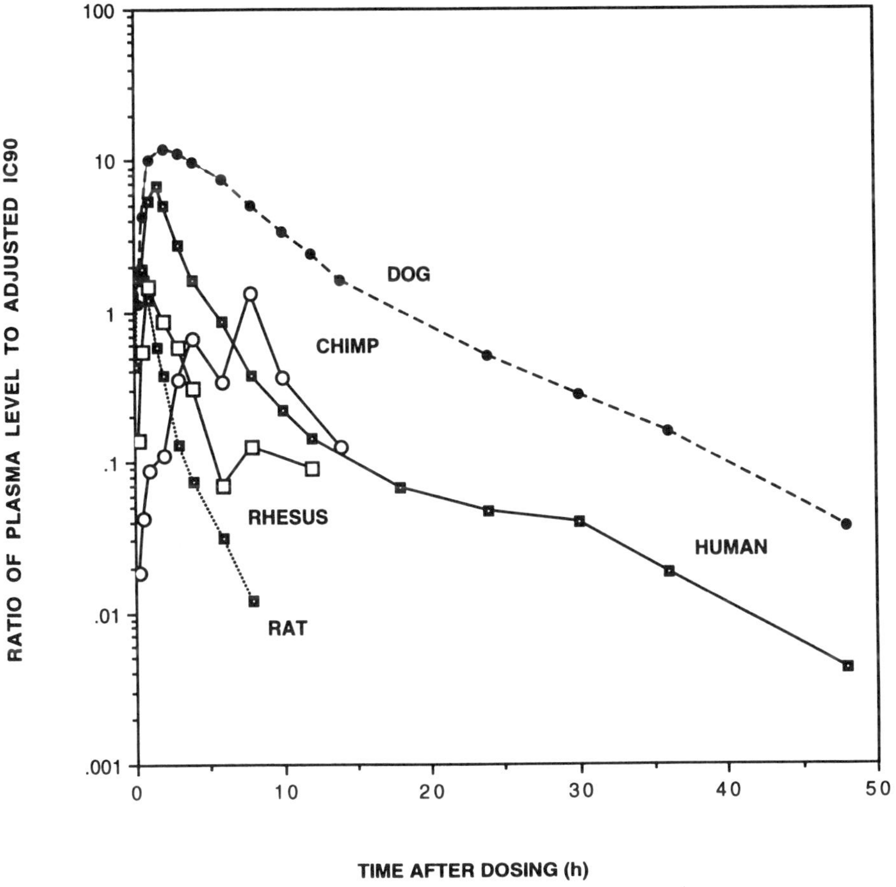

Figure 5. The plasma concentration versus time profiles for five species dosed with a single 10 mg/kg oral dose of DMP 450. The data are represented as the ratio of the plasma concentration at various times relative to the IC_{90} (144 nM) multiplied by the fold-increase in IC_{90} measured in the presence of human plasma proteins (average value of 6.45).

vide sufficient free drug at trough to inhibit both wild-type and mutant variants of HIV with b.i.d. or t.i.d. dosing.

6.1. Potency

Again using the *N*-benzyl analogue as a scaffold for directing substituents toward the S2/S3 sites with its array of H-bond donors and acceptors, inhibitors were designed to H-bond to the backbone of the wild-type enzyme. This may result not only in compounds with increased potency against wild-type virus, but also in ones that retain their effectiveness against mutant strains. This idea is based on the assumption that the protease of drug-resistant viruses would not display major alterations in the enzyme backbone (Jadhav *et al.*, 1997).

Indeed, functional groups having multiple H-bond donor and acceptor possibilities are an order of magnitude better enzyme inhibitors than DMP 323 or DMP 450. However, many are polar and the translation to antiviral potency is poor. Of particular note are the amides **40** and **41,** which show increasingly better translation as lipophilicity increases (Wilkerson *et al.*, 1996). Several more lipophilic heterocyclic amides were synthesized and evaluated and showed exceptionally potent antiviral activity. Besides heterocyclic amides, other compounds that translated enzyme potency into antiviral potency were the oxime analogues **47** and **48** (Han *et al.*, 1998) and the heterocyclic indazole analogue **49** (Rodgers *et al.*, 1996), with antiviral potency down to 5 nM.

6.2. Resistance Profile

Several cyclic urea amides were synthesized and evaluated against a panel of drug-resistant mutant viruses (Fig. 6, Table III). Cyclic urea amides XV652 and SD146 exhibited excellent profiles against the panel of drug-resistant mutants (Jadhav *et al.*, 1997).

The remarkable resistance profile of SD146 probably stems from its ability to form a large number of H-bonds with the backbone atoms and its extensive VDW contacts (Fig. 7). This relationship between the number of H-bonds to backbone atoms and resistance profile (as well as enzyme potency) has also been observed in other cyclic urea analogues (De Lucca *et al.*, 1998). However, in many cases substituents that are capable of forming multiple H-bonds are also very polar and are unable to translate their enzyme potency into antiviral potency. The heterocyclic amides are exceptional in their ability to form many H-bonds while being lipophilic enough to have excellent antiviral potency. SD146 is a very potent antiviral agent (IC_{90} = 5.1 nM) with an exceptional resistance profile. The knowledge gained from this study is useful for designing inhibitors with superior resistance profiles.

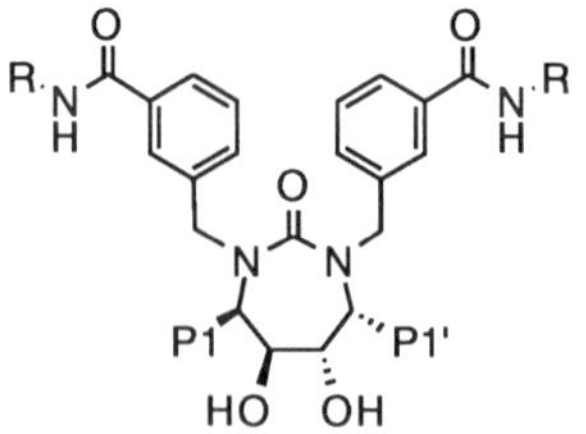

XV638 : P1 = P1' = benzyl; R = 2-thiazolyl
XV652 : P1 = P1' = benzyl; R = 2-imidazolyl
SD146 : P1 = P1' = benzyl; R = 2-benzimidazolyl

Figure 6. Structure of amides detailed in Table III.

Table III

Resistance of Reconstructed Mutant Viruses to Cyclic Urea Amides

				Resistance of constructed mutant viruses[a] (IC_{90} mutant / IC_{90} WT)						
Inhibitor	No. of H-bond Interactions	K_i (nM)[a]	IC_{90} (nM)[a]	82A	82F	84V	ABT 538 virus	48V/ 90M	MK639 virus	46I/47V/ 50V
DMP 323	10	0.34	57	2.8	7.1	22	93	0.3	18	11
DMP 450	10	0.28	130	2.5	5.3	9.9	49	1.5	27	7.5
XV638	12	0.027	4.2	2.2	0.9	1.2	24	0.2	8.7	23
XV652	14	0.014	19	2.7	1.6	0.7	0.2	0.1	0.4	0.4
SD146	14	0.024	5.1	0.8	0.3	0.6	1.2	0.3	0.7	1.0

[a]K_is, IC_{90}s for wild-type virus, and resistance of mutant viruses were measured as previously described (Jadhav *et al.*, 1997).

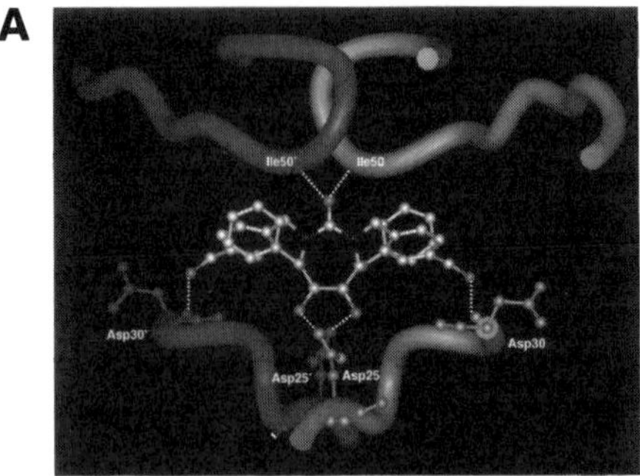

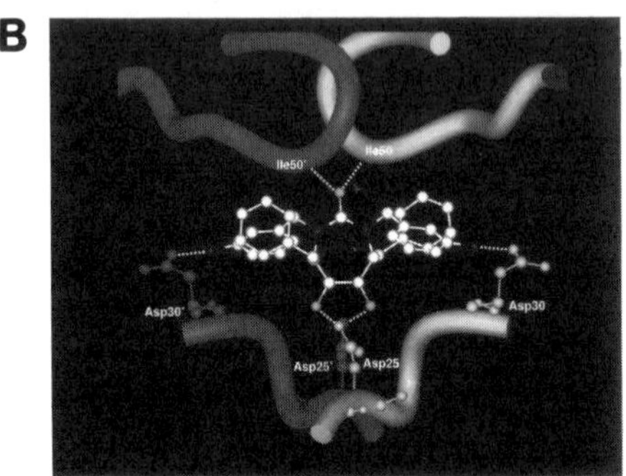

Figure 7. (A) X-ray structures of the complexes of DMP 323. (B) DMP 450 with HIV-1PR showing that the urea oxygen accepts hydrogen bonds from the protease flaps Ile50/50′ with the exclusion of the intervening structural water commonly found in linear peptidomimetic inhibitors. (C) Hydrogen bonding scheme for DMP323, DMP450, SD146, and XV638.

6.3. Pharmacokinetics

Unfortunately, because of its extreme insolubility in water and oils, to date, no formulation of the symmetrical SD146 could be developed for oral or i.v. administration to animals. Similarly, the potent, but symmetrical cyclic urea oxime and indazole analogues also showed low oral bioavailability.

Because pharmacokinetic behavior of these new, multiple H-bonding, analogues had become a key focus of our screening program, we have sought out reproducible, rapid methods to predict pharmacokinetic behavior utilizing high-throughput *in vitro* tests to reduce the number of compounds required for *in vivo* evaluation.

Factors reducing the extent of oral bioavailability include poor solubility or dissolution in the aqueous gastrointestinal fluids, poor diffusion through the intestinal membrane, and extraction or metabolism by the intestine or liver prior to reaching the systemic circulation. Chemical characteristics known to be associated with poor intestinal permeation include high molecular weight (Chadwick *et al.,* 1977) and the number of H-bonding functional groups (Conradi *et al.,* 1991).

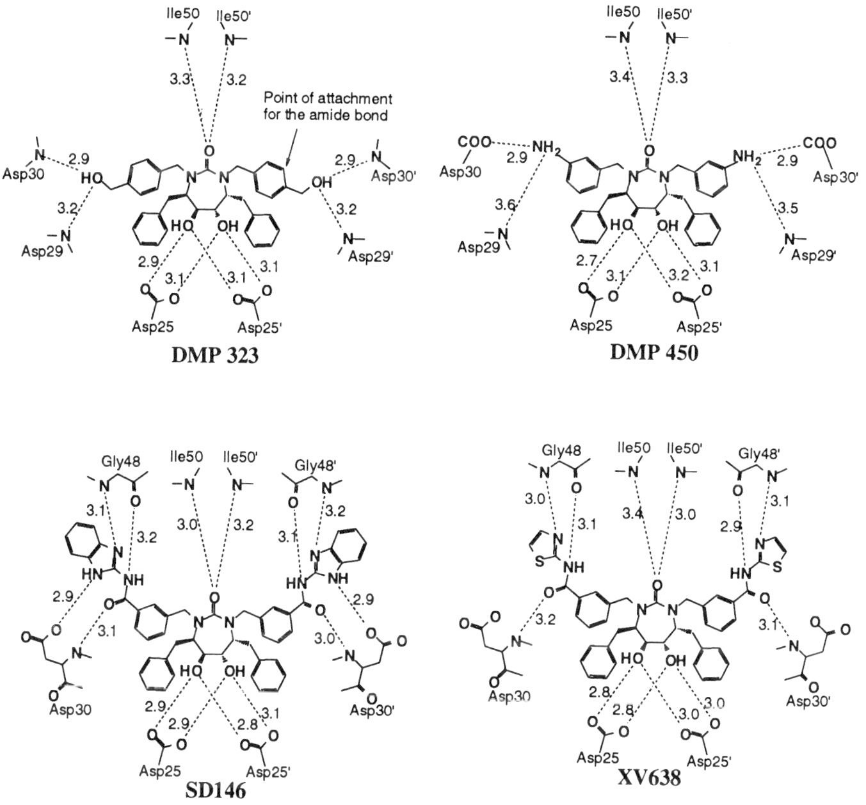

Figure 7. (*Continued*).

To evaluate intestinal permeability of cyclic urea HIV protease inhibitors, we measured and compared permeation rates through Caco-2 epithelial monolayers. Caco-2 cells, derived from a human colon adenocarcinoma, were cultured on microporous filter dishes to form monolayers that morphologically and functionally resemble the lower small intestine (Hidalgo *et al.*, 1989). For drugs whose absorption is not limited by slow dissolution, rates of permeation through Caco-2 monolayers have been shown to be well correlated with absorption percentages *in vivo* (Artursson and Karlsson, 1991; Ribadeneira *et al.*, 1996).

A number of reference compounds were examined in this model, to ensure that *in vitro* permeability under conditions that we used corresponds with *in vivo* absorption. We compared Caco-2 permeability coefficients with dog *in vivo* absorption properties. As shown in Fig. 8, there was a fairly good correlation between these parameters. Having established this *in vitro*/*in vivo* correlation, Caco-2 permeation studies were used to identify poorly permeable compounds that would not be expected to be absorbed *in vivo,* and to prioritize permeable compounds for *in vivo* testing.

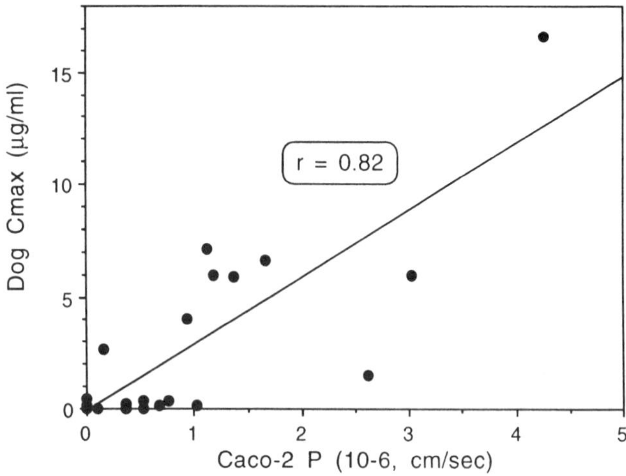

Figure 8. Comparison of Caco-2 permeability coefficients with dog *in vivo* absorption properties.

6.4. Design and Physicochemical Properties

Symmetric cyclic ureas are an extremely rigid scaffold complementary to HIV-PR that provides significant synthetic and cost advantages. However, these properties also proved to have significant limitations. On the other hand, nonsymmetric cyclic ureas offer the advantages of better solubility and greater flexibility in adjusting the physicochemical properties. They also offer greater flexibility, in designing enzyme interactions, than the symmetrical analogues. These potential benefits have prompted us to concentrate our analogue synthesis efforts on unsymmetrically *N*-substituted analogues (De Lucca *et al.*, 1998). In this way we can better address the often conflicting issues of solubility, potency, protein binding, oral bioavailability, and resistance profile.

7. CONCLUSION

Combining the Caco-2 cell assay, dog pharmacokinetic assessment on selected compounds, antiviral testing against wild-type and mutant variants, and antiviral testing in the presence of human plasma proteins, we can define the overall quality of a given compound. We can then select compounds for further preclinical evaluation.

Using this refined strategy we have been able to identify unsymmetrical analogues that have good potency, resistance profiles, and physicochemical properties, while maintaining excellent oral pharmacokinetics. The true test of our strategy and our assays' predictive power lies in phase I clinical trials.

We believe that many other opportunities are available (in the area of cyclic HIV-PR inhibitors capable of displacing the structural water) to find structurally different drug candidates with superior potency, mutation profiles, and pharmacokinetics. This area includes cyclic sulfamides, cyclic thioureas, cyclic sulfones, and azacyclic ureas. Extensive analogue SAR studies in these areas await further work.

REFERENCES

Appelt, K., 1993, Crystal structures of HIV-1 protease-inhibitor complexes, *Perspect. Drug Discovery Des.* **1**:23–48.

Artursson, P., and Karlsson, J., 1991, Correlation between oral drug absorption in humans and apparent drug permeability coefficients in human intestinal epithelial (Caco-2) cells, *Biochem. Biophys. Res. Commun.* **178**:880–885.

Baker, W. R., and Condon, S. L., 1993, Dipeptide isosteres. 1. Synthesis of dihydroxyethylene dipeptide isosteres via diastereoselective addition of alkyllithium reagents to N,N-dimethylhydrazones. Preparation of renin and HIV-1 protease inhibitor transition-state mimics, *J. Org. Chem.* **58**:3277.

Bryant, M., Getman, D., Smidt, M., Marr, J., Clare, M., Dillard, R., Lansky, D., DeCrecenzo, G., Heintz, R., Houseman, K., Reed, K., Stoszenbach, J., Talley, J., Vazquez, M., and Mueller, R., 1995, SC-52151, a novel inhibitor of the human immunodeficiency virus protease, *Antimicrob. Agents Chemother.* **39**:2229–2234.

Chadwick, V. S., Phillips, S. F., and Hofmann, A. F. 1977, Measurements of intestinal permeability using low molecular weight polyethylene glycols (PEG 400), *Gastroenterology* **73**:247–251.

Chiba, M., Hensleigh, M., Nishime, J. A., Balani, S. K., and Lin, J. H., 1996, Role of cytochrome P450 3A4 in human metabolism of MK-639, a potent human immunodeficiency virus protease inhibitor, *Drug Metab. Dispos.* **24**:307–314.

Christ, D. D., Meek, J. L., Farmer, A. R., and Larsen, B. S., 1993, Oxidative metabolism of the novel HIV protease inhibitor DMP 323 by rat, dog and human systems, *ISSX Proc.* **4**:230.

Coffin, J. M., 1995, HIV population dynamics in vivo: Implications for genetic variation, pathogenesis, and therapy, *Science* **267**:483–489.

Condra J. H., Schleif, W. A., Blahy, O. M., Gabryelski, L. J., Graham, D. J., Quintero, J. C., Rhodes, A., Robbins, H. L., Roth, E., Shivaprakash, M., Titus, D., Yang, T., Teppler, H., Squires, K. E., Deutsch, P. J., and Emini, E. A., 1995, in vivo emergence of HIV-1 variants resistant to multiple protease inhibitors, *Nature* **374**:569–571.

Conradi, R. A., Hilgers, A. R., Ho, N. F. H., and Burton, P. S., 1991, The influence of peptide structure on transport across Caco-2 cells, *Pharm. Res.* **8**:1453.

Cram, D. J., 1986, Preorganization—From solvents to spherands, *Angew. Chem. Int. Ed. Engl.* **25**:1039–1057.

Cram, D. J., 1988, The design of molecular hosts, guests and their complexes, *Science* **240**:760–767.

Cram, D. J., and Lam, P. Y. S.,1986, Host–guest complexation 37. Synthesis and binding properties of a transacylase partial mimic with imidazole and benzyl alcohol in place, *Tetrahedron* **42**:1607.

Cram, D. J., Dicker, I. B., Lauer, M., Knobler, C. B., and Trueblood, K. N., 1984, Host–guest complexation 32. Spherands composed of cyclic urea and anisyl units, *J. Am. Chem. Soc.* **106**:7150.

Danner, S. A., Carr, A., Leonard, J. M., Lehman, L. M., Gudiol, F., Gonzales, J., Raventos, A., Rubio, R., Bouza, E., Pintado, V., Aguado, A. G., de Lomas, J. G., Delgado, R., Borcees, J. C. C., Hsu, A., Valdes, J. M., Boucher, C. A. B., and Cooper, D. A., 1995, A short term study of the safety, pharmacokinetics, and efficacy of Ritonavir, an inhibitor of HIV-1 protease, *N. Engl. J. Med.* **333**:1528–1533.

Darke,P. L., and Huff, J. R., 1995, HIV protease as an inhibitor target for the treatment of AIDS, *Adv. Pharmacol.* **25**:399–454.

De Clercq, E., 1995, Toward improved anti-HIV chemotherapy: Therapeutic strategies for intervention with HIV infections, *J. Med Chem.* **38**:2491–2517.

De Lucca, G. V., Erickson-Viitanen, S., and Lam, P. Y. S., 1997, Cyclic HIV protease inhibitors capable of displacing the active-site structural water molecule. *Drug Discovery Today* **2**:6–18.

De Lucca, G. V., Kim, U. T., Liang, J., Cordova, B., Klabe, R. M., Garber, S., Bacheler, L. T., Lam, G. N., Wright, M. R., Logue, K. A., Erickson-Viitanen, S., Ko, S. S., and Trainor, G. L., 1998, Nonsymmetric P2/P2 cyclic urea HIV protease inhibitors. Structure-Activity relationship, bioavailability, and resistance profile of monoindazole–substituted P2 analogs, *J. Med. Chem.* **41:**2411–2423.

Dunitz, J. D., 1994, The entropic cost of bound water in crystals and biomolecules, *Science* **264:**670.

Erickson, J., and Kempf, D., 1994, Structure-based design of symmetric inhibitors of HIV-1 protease, *Arch Virol. Suppl.* **9:**19–29.

Erickson, J., Neidhart, D. J., VanDrie, J., Kempf, D. J., Wang, X. C., Norbeck, D. W., Plattner, J. J., Rittenhouse, J. W., Turon, M., Wideburg, N. E., Kohlbrenner, W. E., Simmer, R., Helfrich, R., Paul, D. A., and Knigge, M. F., 1990, Design, activity, and 2.8 Å crystal structure of a C2 symmetric inhibitor complexed to HIV-1 protease, *Science* **249:**527–532.

Freudenberger, J. H., Konradi, A. W., and Pedersen, S. F., 1989, Intermolecular pinacol cross coupling of electronically similar aldehydes. An efficient and stereoselective synthesis of 1,2-diols employing a practical vanadium(II) reagent, *J. Am. Chem. Soc.* **111:**8014–8016.

Grzesiek, S., Bax, A., Nicholson, L. K., Yamazaki, T., Wingfield, P., Stahl, S. J., Eyermann, C. J., Torchia, D. A., Hodge, C. N., Lam, P. Y. S., Jadhav, P. K., and Chang, C.-H., 1994, NMR evidence for the displacement of a conserved interior water molecule in HIV protease by a non-peptide cyclic urea-based inhibitor, *J. Am. Chem. Soc.* **116:**1581–1582.

Han, Q., Chang, C. H., Li, R., Ru, Y., Jadhav, P. K., and Lam, P. Y. S., 1998 Cyclic HIV protease inhibitors: Design and synthesis of orally bioavailable pyrazole P2/P2 cyclic ureas with improved potency. *J. Med. Chem.* **41:** 2019–2028.

Hidalgo, I. J., Raub, T. J., and Borchardt, R. T., 1989, Characterization of the human colon carcinoma cell line (Caco-2) as a model system for intestinal epithelial permeability, *Gastroenterology* **96:**736–749.

Ho, D. D., Neumann, A. U., Perelson, A. S., Chen, W., Leonard, J. M., and Markowitz, M., 1995, Rapid turnover of plasma virions and CD4 lymphocytes in HIV-1 infection, *Nature* **373:**123–126.

Hodge, C. N., Aldrich, P. E., Bacheler, L. T., Chang, C.-H., Eyermann, C. J., Garber, S., Grubb, M. F., Jackson, D. A., Jadhav, P. K., Korant, B., Lam, P. Y. S., Maurin, M. B., Meek, J. L., Otto, M. J., Rayner, M. M., Reid, C., Sharpe, T. R., Shum, L., Winslow, D. L., and Erickson-Viitanen, S., 1996, Improved cyclic urea inhibitors of the HIV-1 protease: Synthesis, potency, resistance profile, human pharmacokinetics and X-ray crystal structure of DMP 450, *Chem. Biol.* **3:**301–314.

Hodge, C. N., Lam, P. Y. S., Eyermann, C. J., Jadhav, P. K., Ru, Y., Fernandez, C. H., De Lucca, G. V., Chang, C. H., Kaltenbach III, R. F., Holler, E. E., Woerner, F. J., Daneker, W. K., Emmett, G. C., Calabrese, J. C., and Aldrich, P. E., 1998, Calculated and experimental low-energy conformations of cyclic urea HIV Protease Inhibitors, *J. Am. Chem. Soc.* **120:** 4570–4581.

Jacobsen, H., Yasargil, K., Winslow, D. L., Craig, J. C., Krohn, A., Duncan, I. B., and Mous, J., 1995, Characterization of human immunodeficiency virus type 1 mutants with decreased sensitivity to proteinase inhibitor Ro 31–8959, *Virology* **206:**527–534.

Jadhav, P. K., and Woerner, F. J., 1992, Synthesis of C2-symmetric HIV-1 protease inhibitors from D-mannitol, *Bioorg. Med. Chem. Lett.* **2:**353–356.

Jadhav, P. K., McGee, L. R., Shenvi, A., and Hodge, C. N., 1994, 1,4-Diamino-2,3-dihydroxybutanes, U.S. Patent 5,294,720.

Jadhav, P. K., McGee, L. R., Shenvi, A., and Hodge, C. N., 1995, 1,4-Diamino-2,3-dihydroxybutanes, U.S. Patent 5,430,155.

Jadhav, P. K., Ala, P., Woerner, F. J., Chang, C.-H., Garber, S. S., Anton, E. D., and Bacheler, L. T., 1997, Cyclic urea amides: HIV-1 protease inhibitors with low nanomolar potency against both wild type and protease inhibitor resistant mutants of HIV, *J. Med. Chem.* **40:**181–191.

Kang, S. H., and Ryu, D. H., 1996, Double asymmetric iodoamination: Synthesis of C_2 symmetric and meso-amino alcohols, *Chem. Commun.* **1996:**355.

Katz, R. A., and Skalka, A. M., 1994, The retroviral enzymes, *Annu. Rev. Biochem.* **63:**133–173.

Kempf, D. J., 1994, Design of symmetry-based, peptidomimetic inhibitors of human immunodeficiency virus protease, *Methods Enzymol.* **241:**334–354.

Kempf, D. J., and Sham, H. L., 1996, HIV protease inhibitors, *Curr. Pharm. Des.* **2:**225–246.

Kempf, D. J., Norbeck, D. W., Codacovi, L. M., Wang, X. C., Kohlbrenner, W. E., Wideburg, N. E., Paul, D. A., Knigge, M. F., Vasavanonda, S., Craig-Kennard, A., Saldivar, A., Rosenbrook, W., Jr., Clement, J. J., Plattner, J. J., and Erickson, J., 1990, Structure-based, C2 symmetric inhibitors of HIV protease, *J. Med. Chem.* **33:**2687–2689.

Kempf, D. J., Codacovi, L. M., Wang, X. C., Kohlbrenner, W. E., Wideburg, N. E., Saldivar, A., Vasavanonda, S., Marsh, K. C., Bryant, P., Sham, H. L., Green, B. G., Betebenner, D. A., Erickson, J., and Norbeck, D. W., 1993, Symmetry-based inhibitors of HIV protease. Structure–activity studies of acylated 2,4-diamino-1,5-diphenyl-3-hydroxypentane and 2,5-diamino-1,6-diphenylhexane-3,4-diol, *J. Med. Chem.* **36:**320.

Kempf, D. J., Marsh, K. C., Denissen, J., McDonald, E., Vasavanonda, S., Flentge, C., Green, B. G., Fino, L., Park, C., Kong, X., Wideburg, N. E., Saldivar, A., Ruiz, L., Kati, W. M., Sham, H. L., Robins, T., Stewart, K. D., Plattner, J. J., Leonard, J., and Norbeck, D., 1995, ABT-538 is a potent inhibitor of human immunodeficiency virus protease with high oral bioavailability in humans, *Proc. Natl. Acad. Sci. USA* **92:**2484–2488.

Kitchen, V. S., Skinner, C., Ariyoshi, K., Lane, E. A., Duncan, I. B., Burckhardt, J., Burger, H. U., Bragman, K., Pinching, A. J., and Weber, J. N., 1995, Safety and activity of Saquinavir in HIV infection, *Lancet* **345:**952–955.

Kohl, N. E., Emini, E. A., Schleif, W. A., Davis, L. J., Heimbach, J. C., Dixon, R. A. F., Scolnick, E. M., and Sigal, I. S., 1988, Active human immunodeficiency virus protease is required for viral infectivity, *Proc. Natl. Acad. Sci. USA* **85:**4686–4690.

Kumar, G. N., Rodrigues, D., Buko, A. M., and Denissen, J. F., 1996, Cytochrome P450-mediated metabolism of the HIV-1 protease inhibitor Ritonavir (ABT-538) in human liver microsomes, *J. Pharmacol. Exp. Ther.* **277:**412–431.

Lam, P. Y. S., Jadhav, P. K., Eyermann, C. J., Hodge, C. H., Ru, Y., Bacheler, L. T., Meek, J. L., Otto, M. J., Rayner, M. M., Wong, Y. N., Chang, C.-H., Weber, P. C., Jackson, D. A., Sharpe, T. R., and Erickson-Viitanen, S., 1994, Rational design of potent, bioavailable, nonpeptide cyclic ureas as HIV protease inhibitors, *Science* **263:**380–384.

Lam, P. Y. S., Ru, Y., Jadhav, P. K., Aldrich, P. E., De Lucca, G. V., Eyermann, C. J., Chang, C.-H., Emmett, G., Holler, E. R., Daneker, W. F., Li, L., Confalone, P. N., McHugh, R. J., Han, Q., Li, R., Markwalder, J. A., Seitz, S. P., Sharpe, T. R., Bacheler, L. T., Rayner, M. M., Klabe, R. M., Shum, L., Winslow, D. L., Kornhauser, D. M., Jackson, D. A., Erickson-Viitanen, S., and Hodge, C. N., 1996, Cyclic HIV protease inhibitors: Synthesis, conformational analysis, P2/P2′ structure–activity relationship, and molecular recognition of cyclic ureas, *J. Med. Chem.* **39:**3514.

Markowitz, M. M., Mo, H., Kempf, D. J., Norbeck, D. W., Bhat, T. N., Erickson, J. W., and Ho, D. D., 1995, Selection and analysis of human immunodeficiency virus type 1 variants with increased resistance to ABT-538, a novel protease inhibitor, *J. Virol.* **69:**701–706.

Martin, Y. C., 1992, 3D database searching in drug design, *J. Med. Chem.* **35:**2145–2154.

Miller, M., Jaskolski, M., Rao, J. K. M., Leis, J., and Wlodawer, A., 1989a, Crystal structure of a retroviral protease proves relationship to aspartic protease family, *Nature* **337:**576–579.

Miller, M., Schneider, J., Sathyanarayana, B. K., Toth, M. V., Marshall, G. R., Clawson, L., Selk, L., Kent, S. B. H., and Wlodawer, A., 1989b, Structure of complex of synthetic HIV-1 protease with a substrate-based inhibitor at 2.3 Å resolution, *Science* **246:**1149–1152.

Nugiel, D. A., Jacobs, K., Worley, T., Patel, M., Kaltenbach, R. F., III, Meyer, D. T., Jadhav, P. K., De Lucca, G. V., Smyser, T. S., Klabe, R. M., Bacheler, L. T., Rayner, M. M., and Seitz, S. P., 1996, Preparation and structure–activity relationship of novel P1/P1′ substituted cyclic urea-based human immunodeficiency virus type-1 protease inhibitors, *J. Med. Chem.* **39:**2156–2169.

Peng, C., Ho, B. K., Chang, T. W., Chang, N. T., 1989, Role of human immunodeficiency virus type 1-specific protease in core protein maturation and viral infectivity, *J. Virol.* **63:**2550–2556.

Pierce, M. E., Harris, G. D., Islam, Q., Radesca, L. A., Storace, L., Waltermire, R. E., Wat, E., Jadhav, P. K., and Emmett, G. C., 1996, Stereoselective synthesis of HIV-1 protease inhibitor, DMP 323, *J. Org. Chem.* **61:**444.

Pollard, R. B., 1994, Use of proteinase inhibitors in clinical practice, *Pharmacotherapy* **14:**21S–29S.

Ribadeneira, M. D., Aungst, B. J., Eyermann, C. J., and Huang, S.-M., 1996, Effects of structural modifications on the intestinal permeability of angiotensin II receptor antagonists and the correlation of in vitro, in situ and in vivo absorption, *Pharm. Res.* **13:**227.

Rich, D. H., 1993, Effect of hydrophobic collapse on enzyme–inhibitor interaction. Implication for the design of peptidal mimetics, in: *Perspectives in Medical Chemistry* (B. Testa, E. Kyburz, W. Fuhrer, and R. Giger, eds.) pp. 15–25, VCH, New York.

Ridky, T., and Leis, J., 1995, Development of drug resistance to HIV-1 protease inhibitors, *J. Biol. Chem.* **270:**29621–29623.

Ringe, D., 1994, X-Ray Structures of retroviral proteases and their inhibitor-bound complexes, *Methods Enzymol.* **241:**157–177.

Rodgers, J. D., Johnson, B. L., Wang, H., Greenburg, R. A., Erickson-Viitanen, S., Klabe, R. M., Cordova, B. C., Rayner, M. M., Lam, G. N., and Chang, C.-H., 1996, Potent cyclic urea HIV protease inhibitors with benzofused heterocycles as P2/P2′ groups, *Bioorg. Med. Chem. Lett.* **6:**2919–2924.

Rossano, L. T., Lo, Y. S., Anzalone, L., Lee, Y.-C., Meloni, D. J., Moore, J. R., Gale, T. M., and Arnett, J. F., 1995, A practical synthesis of nonpeptide cyclic ureas as potent HIV protease inhibitors, *Tetrahedron Lett.* **36:**4967.

Swain, A. L., Miller, M. M., Green, J., Rich, D. H., Schneider, J., Kent, S. B. H., and Wlodawer, A., 1990, X-ray crystallographic structure of a complex between a synthetic protease of human immunodeficiency virus 1 and a substrate-based hydroxyethylamine inhibitor, *Proc. Natl. Acad. Sci. USA* **87:**8805–8809.

Vacca, J. P., 1994, Design of tight-binding human immunodeficiency virus type 1 protease inhibitors, *Methods Enzymol.* **241:**311–334.

Vacca, J. P., Dorsey, B. D., Schleif, W. A., Levin, R. B., McDaniel, S. L., Drake, P. L., Zugay, J., Quintero, J. C., Blahy, O. M., Roth, E., Sardana, V. V., Schlabach, A. J., Graham, P. I., Condra, J. H., Gotlib, L., Holloway, M. K., Lin, J., Chen, I.-W., Vastag, K., Ostovic, D., Anderson, P. S., Emini, E. A., and Huff, J. R., 1994, L-735,524: An orally bioavailable human immunodeficiency virus type 1 protease inhibitor, *Proc. Natl. Acad. Sci. USA* **91:**4096–4100.

Vella, S., 1994, HIV therapy advances. Update on a proteinase inhibitor, *AIDS* **8**(Suppl 3)**:**25–29.

Weber, P. C., Ohlendorf, D. H., Wendoloski, J. J., and Salemme, F. R., 1989, Structural origins of high-affinity biotin binding to streptavidin, *Science* **243:**85.

Wei, X., Ghosh, S. K., Taylor, M. E., Johnson, V. A., Emini, E. A., Deutsch, P., Lifson, J. D., Bonhoeffer, S., Nowak, M. A., Hahn, B. H., Saag, M. S., and Shaw, G. M., 1995, Viral dynamics in human immunodeficiency virus type 1 infection, *Nature* **373:**117–122.

Wilkerson, W. W., Akamike, E., Cheatham, W. W., Hollis, A. Y., Collins, D., DeLucca, I., Lam, P. Y. S., and Ru, Y., 1996, HIV protease inhibitory bis-benzamide cyclic ureas: A quantitative structure–activity relationship analysis, *J. Med. Chem.* **39:**4299–4312.

Wlodawer, A., 1994, Rational drug design: The proteinase inhibitors, *Pharmacotherapy* **14:**9S–20S.

Wlodawer, A., and Erickson, J. W., 1993, Structure-based inhibitors of HIV-1 protease, *Annu. Rev. Biochem.* **62:**543–585.

Wlodawer, A., Miller, M., Jaskolski, M., Sathyanarayana, B. K., Baldwin, E., Weber, I. T., Selk, L. M., Clawson, L., Schneider, J., and Kent, S. B. H., 1989, Conserved folding in retroviral proteases: Crystal structure of a synthetic HIV-1 protease, *Science* **245:**616–621.

Wong, N. Y., Burcham, D. L., Saxton, P. L., Erickson-Viitanen, S., Grubb, M. F., Quon, C. Y., and Huang, S.-M., 1994, A pharmacokinetic evaluation of HIV protease inhibitors, cyclic ureas, in rats and dogs, *Biopharm. Drug Dispos.* **15:**535–544.

Chapter 13

Discovery and Development of the BHAP Nonnucleoside Reverse Transcriptase Inhibitor Delavirdine Mesylate

Wade J. Adams, Paul A. Aristoff, Richard K. Jensen, Walter Morozowich, Donna L. Romero, William C. Schinzer, W. Gary Tarpley, and Richard C. Thomas

1. INTRODUCTION, GOALS, AND STRATEGY

In 1981 came the first report of a newly identified and lethal condition, soon called *acquired immunodeficiency syndrome* (AIDS) (Gottlieb *et al.*, 1981). Fifteen years later, estimates of the number of persons infected with the human immunodeficiency virus (HIV), the causative agent of AIDS, exceeded 21 million individuals worldwide, and in many regions AIDS is the leading cause of death in men and women aged 15–49 years (Quinn, 1996). Thus, AIDS has rapidly become a global medical, economic, and social problem.

As we considered possible approaches toward therapeutic intervention in AIDS in the mid-1980s, several issues soon became apparent. First, there was no validated animal model that had been developed which could be readily utilized to evaluate potential drug candidates (a situation that continues more than a decade after HIV was identified and characterized). This necessitated an approach that in-

---

Wade J. Adams, Paul A. Aristoff, Richard K. Jensen, Walter Morozowich, Donna L. Romero, William C. Schinzer, W. Gary Tarpley, and Richard C. Thomas • Discovery Chemistry, Pharmacia & Upjohn, Inc., Kalamazoo, Michigan 49001-0199.

Integration of Pharmaceutical Discovery and Development: Case Studies, edited by Borchardt *et al.*, Plenum Press, New York, 1998.

volved selecting a compound that demonstrated activity as an antiviral in cell culture and taking it directly into humans without the benefit of an animal model to establish efficacy and therapeutic margin. As we were primarily interested in oral delivery because any useful agent was most likely to be administered chronically, this was particularly troubling. Thus, it was clear that antiviral evaluations in cell culture ultimately had to be very carefully coordinated with evaluations of pharmacokinetics and toxicology in appropriate animal models to ensure that meaningful progress toward promising drug candidates could be made and that the selection of the optimal clinical candidate could be realized. At the outset we wished to have only compounds whose properties were already well characterized enter the formal development system. Thus, in order to be successful, the drug discovery team needed representation from disciplines other than just virology and medicinal chemistry. The team also needed to have extensive participation from scientists representing drug delivery, drug metabolism, and drug safety disciplines.

A second issue was the inevitable development of drug resistance, always an issue in infectious diseases therapy and compounded in AIDS by the rapid mutation rate of the virus and, as was later discovered, its extremely high replication rate (Ho *et al.,* 1995). Traditionally, drug resistance in the infectious diseases area is reduced by ensuring that high blood levels of the drug [meaning multiples of the drug's 90% effective antiviral dose (ED_{90})] are rapidly achieved and safely maintained. Thus, again at an early point in the discovery program, it would be extremely important for the medicinal chemists to work closely with their drug development colleagues to ensure not only that the inherent potency of the series was being enhanced but also that pharmacokinetic parameters and safety were optimized. A second paradigm that has been useful in preventing drug resistance, particularly in the infectious diseases as well as the oncology fields, is the use of combination chemotherapy. Thus, we assumed that potential drugs would probably eventually need to be combined with other agents to get the most durable therapeutic effect *in vivo.* Thus, it would be important to have compounds that were at least additive, if not synergistic, with other anti-HIV agents in their antiviral activity. Furthermore, it would be advantageous not to have overlapping toxicities with other anti-HIV agents that patients might be taking. In fact, the study of drug–drug interactions in general, as HIV-infected patients are usually being treated with a variety of medications, would be a particularly important issue with any new drug and merited serious consideration early rather than later in the drug development process.

As detailed information about HIV was elucidated, the reverse transcriptase (RT) enzyme of HIV-1 became a prime target for antiviral therapy. This enzyme catalyzes the conversion of the viral genomic RNA into double-stranded DNA by a process involving RNA-directed DNA polymerization, cleavage of the RNA strand of the resulting RNA–DNA strand by the ribonuclease H activity, and finally DNA-directed DNA polymerization. There were a number of strategic reasons for selecting HIV-1 RT as a drug target: (1) there was no known closely related human homologue to RT, so selectivity might be possible; (2) RT was clearly essential for viral replication; (3) RT was required early in the life cycle of the virus; and (4) there were multiple enzymatic activities to inhibit.

Importantly, the validity of targeting HIV-1 RT was supported with the report in 1987 (Fischl *et al.*, 1987) that the RT inhibitor azidothymidine (AZT) was clinically effective. At that time a number of other nucleoside RT inhibitors were also in development and beginning to look promising. However, we decided to search for nonnucleoside inhibitors of RT because it was apparent that the administration of the nucleoside inhibitors was limited by various toxicities (presumably arising from their recognition by normal cellular polymerases) and the development of viral resistance (DeClercq, 1994). Thus, it appeared that combinations of RT inhibitors would be required for the prolonged, effective therapy of HIV infection, and our goal became the discovery and development of an orally efficacious nonnucleoside HIV RT inhibitor that could be used in combination with nucleosides such as AZT.

2. DISCOVERY OF INITIAL LEAD (PNU-80493E)

Our basic strategy utilized a computer-directed dissimilarity analysis of the Pharmacia & Upjohn chemical library to select compounds for screening against HIV-1 RT. The initial dissimilarity set of compounds consisted of approximately 1500 structurally diverse compounds that were selected for primary screening against the recombinant RT enzyme. This led to the identification of about 100 inhibitors with some degree of RT inhibitory activity. These compounds were then evaluated for their selectivity by determining whether they inhibited normal DNA polymerases such as human polymerases α and δ, and for their antiviral activity and cytotoxicity in HIV-1-infected human lymphocytes. Structurally attractive compounds that exhibited anti-HIV-1 activity in this initial assay at noncytotoxic concentrations were tested further in additional antiviral cell culture assays using other cell types and viral strains. Lead compounds were selected for further optimization of potency, selectivity, and pharmaceutical properties (e.g., pharmacokinetic characteristics and toxicology).

Resulting from this strategy was the identification of the lead template PNU-80493E (Fig. 1), an arylpiperazine with modest activity against HIV-1 RT (IC_{50} =

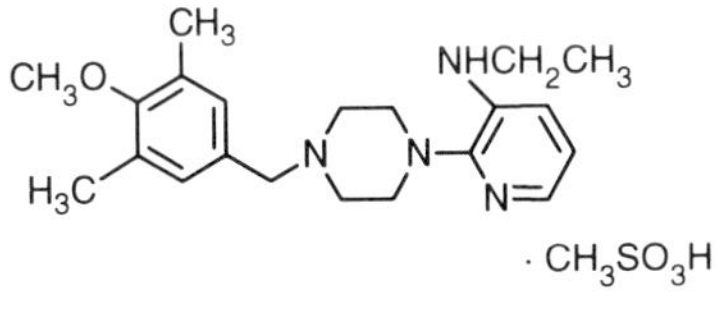

Figure 1. PNU-80493E.

20 μM versus IC_{50} = 0.15 μM for the triphosphate of AZT), but nevertheless at least a 30-fold selectivity at the enzyme level (IC_{50} = 600 μM for pol α and >1250 μM for pol δ) (Romero *et al.,* 1991). Similarly, PNU-80493E was weakly active against the virus (ED_{50} = 2 μM versus ED_{50} = 0.07 μM for AZT) in infected MT-2 cells with a narrow but significant therapeutic window (the cytotoxic concentration, CC_{50} = 15 μM). Thus, PNU-80493E was a bona fide RT inhibitor and anti-HIV agent. Furthermore, it was structurally attractive as all parts of the molecule could be readily varied, and analogues easily prepared in only about four to six chemical steps (Romero *et al.,* 1994).

3. SELECTION OF FIRST-GENERATION CANDIDATE (PNU-87201)

We undertook a synthetic program to systematically explore the relationship between structure and anti-HIV activity. This work resulted in the identification of a series of indol-2-yl substituted arylpiperazines with good activity (Table I; Romero *et al.,* 1994). Consideration of the propensity of indoles to undergo metabolism via hydroxylation at the 5-position led to the synthesis of the 5-methoxy and 5-fluoro congeners. Similarly, the known propensity for oxidative *N*-dealkylation of secondary amines led to the synthesis of compounds containing an *N*-isopropylamine as it was more sterically hindered than an *N*-ethylamine and it was thought the additional steric bulk might decrease the rate of *N*-dealkylation, should it occur. Synthesis of all possible combinations resulting from the variation in the 5-indole substitution and variation of the *N*-alkyl substituent led to six compounds that possessed good antiviral activity. Thus, all six were considered as possible first-generation clinical candidates (Table I).

At this point, a multidisciplinary team was assembled and charged with determining which of these six compounds would make the optimal first-generation clinical candidate. Selection criteria included assessment of aqueous solubility at pH 6.0, i.v. clearance, and, in some cases, absolute oral bioavailability in the rat, multiple-dose toxicity in the rat, manufacturability, and physical stability (Table II). In order to efficiently select the most appropriate drug candidate, many of the activities described below were conducted in parallel. One of the first activities involved the preparation of hydrochloride and mesylate salts of the lead compounds, which were evaluated for their aqueous solubility and physical stability. The physical stability was acceptable in every case, although the exceedingly poor solubility of PNU-88353 and its salts resulted in its elimination from further consideration.

The i.v. clearance and, in some cases, the oral bioavailability of the remaining compounds were evaluated in rats, and on the basis of these experiments it was determined that PNU-88141 and PNU-87201 had low to moderate i.v. clearances

Table I

Antiviral Evaluation of Lead RT Inhibitors in Human Lymphocytes

PNU-number	Substituents		MT-2/IIIb[a]		PBMC/D34[a]		H9/IIIb[a]	
	X	Y	ED_{50} (μM)	CC_{50}[b] (μM)	ED_{50} (μM)	CC_{50} (μM)	ED_{50} (μM)	CC_{50} (μM)
85961	H	Et	0.3	>30	0.01	>10	0.06	>10
88141[c]	F	Et	<0.3	>27	0.001	>10	ND[d]	ND
87201[c]	OCH_3	Et	<0.2	>20	0.001	>10	0.04	>10
88204[c]	H	*i*-Pr	0.3	>27	0.001	>10	0.04	>10
88352[c]	F	*i*-Pr	0.3	>26	0.003	>10	0.05	>10
88353[c]	OCH_3	*i*-Pr	0.3	>25	0.001	>10	0.06	63
AZT	NA[e]	NA	0.07	12	0.001	10	0.03	63
89227[f]	OHC_3	*t*-Bu	0.22—2.2	>22	0.01	10	ND	ND

[a]HIV-1 infectivity and drug cytotoxicity studies were conducted in MT-2 and H9 cells with the HIV-1_{IIIb} isolate and PBMC with the HIV-1_{D34} isolate as previously described (Romero *et al.*, 1991).

[b]CC_{50} is the drug concentration required to decrease cell viability compared with uninfected controls; in most cases CC_{50} is estimated because cell viability was > 50% at the highest concentrations tested.

[c]Tested as the mesylate salt.

[d]ND, not done.

[e]NA, not applicable.

[f]Tested as the hydrochloride salt.

Table II
Selection Criteria and Comparative Data for the First-Generation HIV-1 RT Inhibitor Clinical Candidate

Criteria	PNU-85961	PNU-88141	PNU-87201	PNU-88204	PNU-88352	PNU-88353
i.v. clearance	+	+++	++	++	–	
Oral bioavailability	++	++++	+++	+++	+	ND
Solubility (salt form)	++	ND[a]	+++	++	+	poor
	(mesylate)		(HCl)	(mesylate)	(HCl)	(mesylate)
Toxicity						
7-day rat	negative	ND	ND	negative	negative	ND
14-day rat	ND	ND	negative	negative	ND	ND
Ames essay	ND	ND	negative	negative	ND	ND
P450 induction	ND	ND	no	yes	ND	ND

[a]ND, not done.

(13.0 ± 1.3 and 19.9 ± 1.5 ml/min per kg, respectively) and PNU-87201, PNU-88204, and PNU-88141 possessed acceptable oral bioavailability (62, 45, and 79%, respectively) (Schwende *et al.,* unpublished data). A 7-day non-GLP safety study in rats with PNU-85961, PNU-88204, and PNU-88352 was conducted in parallel with the above pK studies; results indicated that these compounds possessed similar toxicity profiles (i.e., all were well tolerated). The major structural difference between the three remaining compounds was the 5-indole substituent (e.g., unsubstituted, methoxy, or fluoro). Consideration of the manufacturability and cost of goods raised a concern regarding supply of the 5-fluoroindole-2-carboxylic acid required to manufacture PNU-88141 and led to its elimination. To summarize, PNU-87201 and PNU-88204 had similar oral bioavailability, aqueous solubility, and bulk drug stability properties. It appeared as though the 5-methoxyindole-2-carboxylic acid required for the synthesis of PNU-87201 would cost about two times that of the unsubstituted indole required for PNU-88204. On the other hand, the i.v. clearance of PNU-88204 (42 ± 7 ml/min per kg) in rats was approximately two times higher than that of PNU-87201 (19.9 ± 1.5 ml/min per kg) and the oral bioavailability of PNU-88204 (45%) appeared to be lower than that of PNU-87201 (62%). The team decided to further distinguish between the two compounds by conducting 14-day non-GLP rat toxicity and Ames studies, and *in vitro* metabolism and metabolite stability studies in hepatic microsomal metabolism studies.

Throughout the period of time the bis(heteroaryl)piperazines (BHAPs) described above were being evaluated for their suitability as clinical candidates, the medicinal chemists continued synthesizing more analogues. The information obtained from these evaluations was used in the design of new analogues with the aim of identifying compounds with better properties than the compounds being considered above. For example, early work delineated a major metabolic pathway of the BHAPs in rats, which was *N*-dealkylation of the alkylamine substituent on the pyridine ring (Voorman *et al.,* unpublished data). Attempts to suppress oxidation of the α-carbon by replacing the hydrogen substituents with methyl groups led to the synthesis of PNU-89227 (Romero *et al.,* 1994), the *t*-butyl analogue of PNU-87201E (Table I). This alteration resulted in a longer half-life for this analogue in *in vitro* hepatic microsomal preparations, and thus PNU-89227 was briefly considered in the quest for the first-generation clinical candidate. Unfortunately, its poor aqueous solubility profile led to its elimination from serious consideration.

Meanwhile, both PNU-87201 and PNU-88204 were shown to be negative in the Ames assay. Comparable toxicity was observed in rats treated with PNU-88204 and PNU-87201 for 14 days, although higher serum concentrations were attained in the PNU-87201-treated rats. In addition, PNU-88204 induced cytochrome P450 while PNU-87201 did not. Therefore, PNU-87201 was selected as the first-generation drug candidate.

4. DEVELOPMENT OF PNU-87201E (ATEVIRDINE MESYLATE)

Selection of the appropriate salt form was based on the work conducted during the early lead finding evaluation described above, wherein it was determined that the mesylate salt (PNU-87201E) possessed a higher dissolution rate than either the hydrochloride or hydrobromide salts. The suitability of atevirdine mesylate for development as an orally administered drug was subsequently confirmed by an oral bioavailability study conducted in the dog. Potential issues identified during these early studies included the low aqueous solubility of PNU-87201, which might make development of formulations difficult and may have contributed to the variable serum concentrations observed in animal studies. In addition, data from the early toxicokinetic studies indicated that systemic concentrations of PNU-87201 increased less than proportionally with drug doses administered in the toxicity studies. Because of an apparent plateauing of serum concentrations and the highly variable serum concentrations observed in animals, extensive toxicokinetic monitoring was completed in toxicity studies. Metabolism of PNU-87201 appeared straightforward in that the major metabolite in hepatic microsomal preparations and *in vivo* in rats was identified as the *N*-desethyl compound. These studies further suggested that PNU-87201 inhibited, at least in part, its own metabolism. Four-week toxicity studies in rats and dogs were completed to support initial human clinical trials. Phase I studies of PNU-87201E (atevirdine mesylate) began by studying its tolerance in normal male volunteers and were subsequently followed by studies to determine its effects in asymptomatic HIV-positive male patients when it was administered alone or in combination with AZT. In these early clinical studies, atevirdine mesylate was very well tolerated (Mieke *et al.,* 1995).

5. GOALS FOR SECOND-GENERATION CANDIDATE

With the identification of atevirdine mesylate as the first-generation BHAP clinical candidate, we turned our attention to the discovery of a second-generation compound that would retain the desirable attributes of atevirdine mesylate while trying to improve on several criteria employed in the initial selection process described above. In particular, we sought analogues with increased potency and improved pharmacokinetic properties. Improved potency should translate into an optimized antiviral effect, lower doses for minimization of side effects, and minimization of the potential for resistance development. Central to this aspect of the program was the requirement to maintain high selectivity for HIV RT versus human polymerases, synergy with other HIV inhibitors (e.g., protease inhibitors and nucleoside reverse transcriptase inhibitors), and activity versus nucleoside-resistant HIV strains. Two approaches were explored to provide analogues with pharmacokinetic characteristics superior to atevirdine mesylate. We first sought compounds with improved intrinsic metabolic stability, low i.v. clearance, and high oral bioavailability, whereas the second focused on enhanced absorption through increased aqueous solubility and optimized formulations. Maintenance of the excellent safety profile, facile chemical synthesis, chemical stability, and appropriate physical properties of atevirdine mesylate were also deemed essential for a second-generation candidate.

6. SELECTION PROCESS

This was a daunting list of properties to be considered, especially in light of the large number of compounds to be evaluated and the interrelationship of many of the parameters. A modified testing funnel was established, which took advantage of the higher volume and more critical assays to prioritize analogues for further testing, to ensure the orderly and efficient evaluation of the analogues prepared by the chemistry team. All compounds prepared by the chemists were immediately evaluated for activity versus recombinant HIV-1 RT followed by antiviral testing in cell culture when appropriate. Only those compounds approaching the target activity level were carried forward, with the exception of an occasional analogue designed to test a hypothesis in one of the other assays. Operating in parallel to the HIV inhibition assays was a high-volume aqueous solubility screen that served as a marker for compounds with anticipated improvements in oral absorption. Information garnered from these assays was employed in the selection of compounds for further evaluation and in the design of additional analogues for iterative rounds of optimization. Compounds emerging from the initial stage of the testing funnel were evaluated *in vitro* using rat hepatic microsomal preparations to determine their metabolic stability. Experience with the original BHAP analogues and results gathered during the course of the second-generation search indicated that there was a relatively good correlation between *in vitro* metabolic stability in rat microsomes (Voorman *et al.*, unpublished data) and observed *in vivo* i.v. clearance in rats (Adams *et al.*, unpublished data). A solution formulation was selected that permitted i.v. and p.o. dosing of compounds for the determination of the i.v. clearance and oral bioavailability of the compounds in crossover design studies in the rat. Using a common dosage vehicle for all compounds in these studies minimized the possibility of confounding results related to formulation differences. Compounds surviving these stages of the testing funnel were next evaluated in 14-day rat toxicity studies. Ames and unscheduled DNA synthesis assays were employed to assess potential genotoxicity. Crossover design i.v. clearance and oral bioavailability studies of solution and solid dosage formulations were then conducted in the dog. Finally, the more compound-intensive 14-day toxicity studies in dogs completed the testing scheme.

Employing knowledge of BHAP structure–activity relationships from earlier

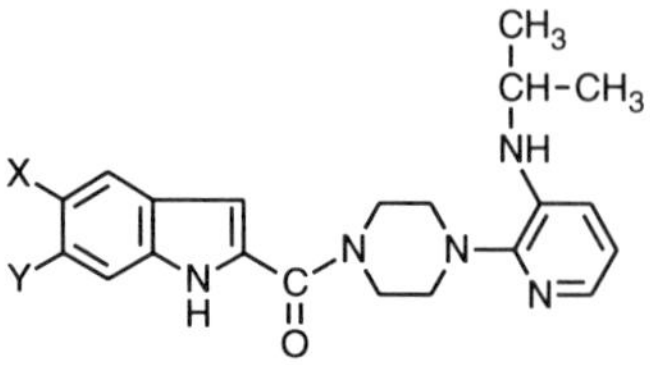

PNU-89388 X = H, Y = $(CH_3)_2NH$-
PNU-90152 X = CH_3SO_2NH-, Y = H

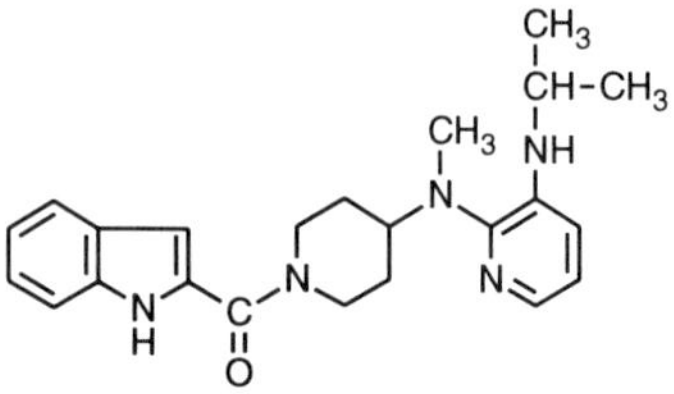

PNU-90328

Figure 2. PNU-89388, PNU-90328, and PNU-90152.

studies and information obtained at all stages of the testing funnel, the medicinal chemistry team synthesized several hundred analogues for evaluation. An exhaustive description of the rationale for analogue design and the resulting SAR is beyond the scope of this review and only a small selection of analogues will be presented to highlight the various aspects of the evaluation scheme. A more complete description of these studies has been presented elsewhere (Romero *et al.,* 1993).

The structure–activity studies that led to the selection of atevirdine mesylate showed that, in general, modification of the indole ring was well tolerated and could have a dramatic influence on the antiviral activity. Additional work conducted at that time focused on developing the SAR of the indole substituent as well as that of the central spacer and right-hand heterocycle. Such variation of the central spacer led to the discovery of a highly active series of (alkylamino)piperidine analogues, exemplified by PNU-90328 (Romero *et al.,* 1996). Numerous analogues from both series were evaluated in the testing scheme and three were selected as class representatives for 14-day toxicity studies conducted in rats with toxicokinetic monitoring (Fig. 2): the 5-substituted indole PNU-90152, the 6-substituted indole PNU-89388, and the (alkylamino)piperidine PNU-90328. (Solution formulations were employed in these studies, and PNU-89388 and PNU-90152 were tested as the hydrochloride salts.)

PNU-89388 was more toxic and had lower serum concentations than an equivalent dose of PNU-90152. PNU-90328 was well tolerated but very low serum concentrations were attained. Of the three compounds evaluated, PNU-90152 had the highest serum concentations and was well tolerated. Therefore, the team decided to focus on the PNU-90152-type template for further optimization.

7. WATER-SOLUBLE COMPOUNDS

Throughout the evaluation of the BHAPs, we encountered numerous examples of compounds with suboptimal pharmacokinetic performance (e.g., PNU-90328) because of high presystemic clearance or malabsorption. The high presystemic clearance was related, in large part, to extensive oxidative metabolism. A hypothesis was proposed linking the low oral bioavailability of compounds having low i.v. clearances to their poor aqueous solubility, which resulted in the precipitation of the compounds in the gut following p.o. administration of solution formulations in the rat. Several analogue series were designed to address the solubility issue. With the knowledge that activity was retained with substitution at the 5-position of the indole ring, various surrogates for the methanesulfonamide group of PNU-90152 bearing basic nitrogen atoms were designed and synthesized, including ureas (e.g., PNU-93923), substituted sulfonamides (e.g., PNU-93898, PNU-93750, and PNU-94423), and sulfamides (e.g., PNU-90781). In general, all analogues in these series retained the desired antiviral activity profile (Table III). Aqueous solubilities of these analogues were 100- to 400-fold higher at pH 6.0 than PNU-87201.

In vitro metabolic stability studies indicated that the half-life of these compounds in rat hepatic microsomal preparations increased in the rank order PNU-93750 < PNU-93898 < PNU-94423 < U-90152 < U-93923 (Voorman *et al.*, unpublished data), and suggested that only PNU-93923, and possibly PNU-94423, had adequate metabolic stability to warrant further study (Voorman, unpublished data). *In vivo* pharmacokinetic studies in rats indicated that the i.v. clearance of these compounds decreased in the rank order PNU-93750 > PNU-90781 > PNU-94423 > PNU-93898 > PNU-93923, with only PNU-93923 having an i.v. clearance comparable to that of PNU-90152. The absolute oral bioavailability of this series of compounds increased in the rank order PNU-93750 < PNU-94423 < PNU-93898 < PNU-90781 < PNU-93923 (Adams *et al.*, unpublished data). Thus, the *in vitro*/*in vivo* correlation for this series of compounds was quite good, with their *in vitro* metabolic stability being nearly as good as the i.v. clearance in predicting the rank order oral bioavailability.

In addition to exploring the effect of the indole substituent, we also explored alterations in the pyridine ring to look for enhanced performance (Table IV). Substituting a pyrazine ring (PNU-93486) for the pyridine ring did not offer any advantages in terms of aqueous solubility. However, a pyridazine ring (PNU-91580 or PNU-94160) enhanced the aqueous solubility 10- to 100-fold. Although the antiviral activity of PNU-91580 was not as good as targeted, it was submitted for *in vivo* pharmacokinetic evaluation in the hope that the information obtained would aid in the design of future analogues. Furthermore, the *in vitro* metabolic stability of PNU-91580 was nearly as good as that of PNU-93923, and it would provide additional data to determine whether there was a good correlation between *in vitro* metabolic stability and *in vivo* i.v. clearance/absolute oral bioavailability. The i.v. clearance of PNU-91580 was only slightly higher than that of PNU-93923 (18.9 $\pm$ 1.5 versus 16.1 $\pm$ 1.0 ml/min per kg) and its oral bioavailability was comparable to that of PNU-93923 (73 $\pm$ 6 versus 70 $\pm$ 8%). Thus, these results provided further data to confirm that there was a good correlation between *in vitro* metabolic stabiltiy in rat hepatic microsomal preparations and i.v. clearance/absolute oral bioavailability. These results also indicated that the i.v. clearances and ab-

Table III

Antiviral Evaluation and Aqueous Solubility of Selected Analogues Synthesized with the Aim of Improving the Properties of Atevirdine Mesylate (PNU-87201E)

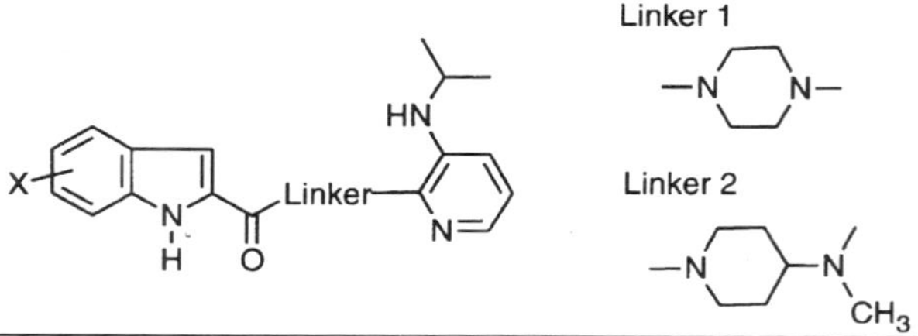

Linker 1

Linker 2

PNU-number	X	Linker	PBMC/D34 ED_{50} (nM)	Solubility[a] (μg/ml)	Dose[b] (mg/kg) i.v.	Dose[b] (mg/kg) p.o.	Absolute oral bioavailability[b] (%)	i.v. clearance[b] (ml/min/kg)
87201E	5-OCH_3	1	1	0.5	12	23	75 ± 14	13.0 ± 1.3
90152	5-$NHSO_2CH_3$	1	0.1	0.4	14	15	64 ± 8	13.6 ± 0.3
89388	6-$N(CH_3)_2$	1	0.01[b]	2.6[c]	—	—	ND	ND
90328	H	2	<0.1	10.3[c]	—	—	ND	ND

93923	CH_3N(piperazine)$N(CO)HN-$	1	10–100	198	15	29	70 ± 8	16.1 ± 10
93898	(piperidine)$N-(CH_2)_2SO_2HN-$	1	0.1–1	47	15	30	18 ± 4	33 ± 8
93750	$(CH_3)_2N(CH_2)_2SO_2HN-$	2	0.1–1	204	15	31	4.6 ± 0.5	54 ± 6
94423	(piperidine)$N-(CH_2)_3SO_2HN-$	1	10–100	190	16	28	12 ± 3	41.3 ± 21
90781	CH_3N(piperazine)NSO_2HN-	1	0.1–1	160	14	31	41 ± 28	

[a]Solubility determined after 24 hr in pH 6.1 MES buffer, unless noted otherwise.
[b]In male Sprague–Dawley rats. Tested as the free base.
[c]Solubility determined after 24 hr at pH 7.5.

Table IV
Antiviral Evaluation and Aqueous Solubility and Pharmacokinetic Characteristics of Selected Pyridazine and Pyrazine Analogues Synthesized with the Aim of Improving Pharmacokinetic Properties

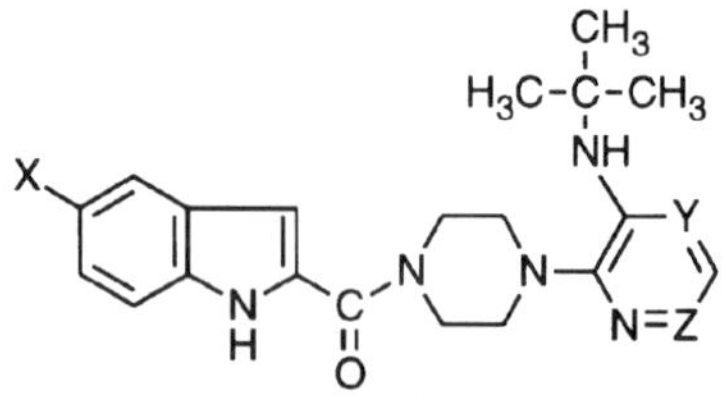

PNU-number	X	Y	Z	PBMC/D34 ED_{50} (nM)	Solubility[a] (μg/ml)	Oral bioavailability (%)	Clearance (ml/min/kg)
93486	$NHSO_2CH_3$	N	CH	ND	0.8	ND	ND
91580	H	CH	N	>0.1	71	73	19
94160	$NHSO_2CH_3$	CH	N	ND	9.4	ND	ND

[a]Solubility determined after 24 hr in pH 6.1 MES buffer.

solute oral bioavailabilities of PNU-93923 and PNU-91580 (Adams *et al.*, unpublished data) did not different from those of the much less soluble PNU-87201. Hence, the much higher aqueous solubilities of PNU-93923 and PNU-91580 did not have a major impact on the oral bioavailability of these compounds when solution formulations were administered at low doses. These results clearly indicated that compounds with excellent antiviral activity should be further evaluated in the *in vitro* metabolic stability and/or *in vivo* pharmacokinetic screen even if they had low aqueous solubility.

A comparative pharmacokinetics and oral bioavailability study of PNU-93923 and PNU-90152 was conducted in the beagle dog to determine whether the more soluble PNU-93923 had better pharmacokinetic properties than the much less soluble PNU-90152 when administered as solution and solid dosage formulations. This study indicated that the higher aqueous solubility of PNU-93923 provided no pharmacokinetic advantage over PNU-90152 in terms of oral bioavailability or interanimal variability in systemic drug concentrations (Adams *et al.*, unpublished data). In addition, PNU-93923 was cleared very rapidly by oxidative metabolism in the dog. Preliminary investigations revealed that a major pathway of metabolism of PNU-93923, both *in vitro* and *in vivo,* was on the piperazine of the 5-indole substituent to form an *N*-oxide (Voorman *et al.*, unpublished data). Therefore, we attempted to slow the metabolism of PNU-93923 by synthesizing analogues (PNU-94749, PNU-94169, and PNU-95109) with variations in the culprit piperazine (Table V). *In vitro* metabolic stability studies indicated that PNU-94749 was much more stable and PNU-94169 had equivalent stability compared to the metabolic stability of PNU-93923 (Voorman, unpublished data). *In vivo*

Table V

Activity, Aqueous Solubility at pH 6.0, and Pharmacokinetic Characteristics of Soluble Analogues of PNU-93923

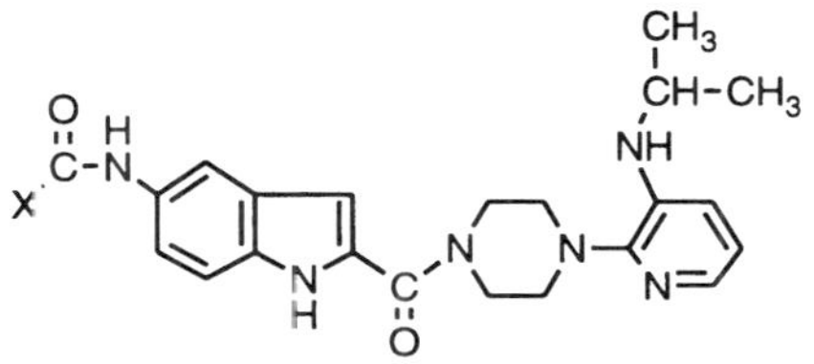

PNU-number	X	PBMC/D34 ED_{50} (nM)	$c \log P$	$c \log P_{app}$ [a]	Solubility (μg/ml)	Dose (mg/kg) i.v.	Dose (mg/kg) p.o.	Absolute oral bioavailability (%)	i.v. clearance (ml/min/kg)
93923	 CH_3N N—	10–100	3.59	0.3	198	15	29	70 ± 8	16.1 ± 1.0
94749	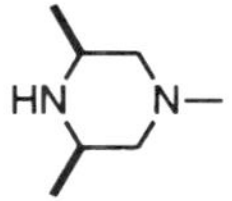 HN N—	ND	3.77	−0.2	191	15	29	14 ± 6	19.9 ± 1.3
94169	 HN N—	>100	2.73	−1.2	216	16	28	0.89 ± 0.27	14.4 ± 1.3
95109	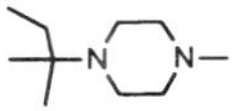 N N—	10	5.35	1.8	23	14	31	47 ± 14	27 ± 5

[a] $c \log PC_{app_{pH\,6}} = c \log P - \log(1 + 10^{pK_a - 6})$.

pharmacokinetic studies in the rat indicated that the i.v. clearances of PNU-94749 and PNU-94169 were comparable to the i.v. clearance of PNU-93923 whereas the more lipophilic PNU-95109 had a substantially higher i.v. clearance than PNU-93923. However, the absolute oral bioavailabilities of these compounds were markedly different and indicated that the absorption of these compounds differed greatly. The low oral bioavailabilities of PNU-94169 and PNU-94749 relative to PNU-93923 were attributed to the much lower apparent partition coefficients of PNU-94169 and PNU-94749 compared with PNU-93923 (Adams *et al.*, unpublished data).

Because the clearance of these compounds is low, the extent of absorption approximates the oral bioavailability and a correlation between lipophilicity at pH 6, the pH of the small intestine, and bioavailability is expected. The clog P_{app} values of PNU-94169 and PNU-94749 are low, about -1.2 and -0.2, respectively. Therefore, poor oral absorption of PNU-94169 and PNU-94749 is expected and this is in accordance with their observed bioavailabilities of only 0.89 and 14% respectively. The other compounds in Table V have clog P_{app} values greater than 0 and more efficient oral absorption is expected and was observed. These results indicate that attempts to increase the aqueous solubility of the BHAP class of RT inhibitors by introducing ionizable amino groups into the BHAP moiety can result in reduced oral bioavailability if the apparent partition coefficient at pH 6 is too low. As the more water-soluble analogues did not seem to offer any advantage, PNU-90152 was selected for development as a second-generation drug candidate.

8. DEVELOPMENT OF PNU-90152T (DELAVIRDINE MESYLATE)

8.1. Pharmacology

PNU-90152 is a potent and selective inhibitor of the HIV-1 RT ($IC_{50} = 0.26$ μM versus IC_{50}s > 440 μM for pol alpha and delta; Dueweke *et al.*, 1993a). In PBMC the mean ED_{50} versus a panel of 25 primary HIV-1 isolates, many of which were highly AZT-resistant, was determined to be 0.066 ± 0.137 μM; moreover, PNU-90152 has low cellular cytotoxicity, causing less than 8% reduction in human lymphocyte viability at 100 μM. In experiments assessing inhibition of the spread of HIV-1 in cell culture, PNU-90152 was much more effective than AZT: While AZT only slightly delayed the spreading HIV-1 infection, PNU-90152 totally prevented the infection. A considerable amount of *in vitro* work was directed at defining the molecular basis of HIV-1 resistance to PNU-90152. These studies indicated that the RT mutations that resulted in resistance to PNU-90152 were distinct from those observed with other antivirals; indeed, in some cases PNU-90152-resistant HIV-1 was found to be more sensitive to the antiviral activity of other compounds compared with parenteral or wild-type HIV-1 strains (Dueweke *et al.*, 1993b). Finally, in several combination experiments, PNU-90152 has exhibited synergistic antiviral activity with protease inhibitors, nucleoside analogue inhibitors, and immune modulating agents (Chong *et al.*, 1994). Collectively, the above antiviral properties of PNU-90152 make it an excellent candidate for HIV-1 combination therapy.

8.2. Formulation/Salt Selection/Crystal Form

8.2.1. pH–SOLUBILITY PROFILE AND BEHAVIOR OF DELAVIRDINE IN WATER

The pH–solubility profile (Fig. 3) of crystalline anhydrous delavirdine free base in unbuffered water at room temperature revealed a pK_a of 4.56 and an intrinsic solubility of 0.86 μg/ml. At pH 1, the solubility of crystalline anhydrous

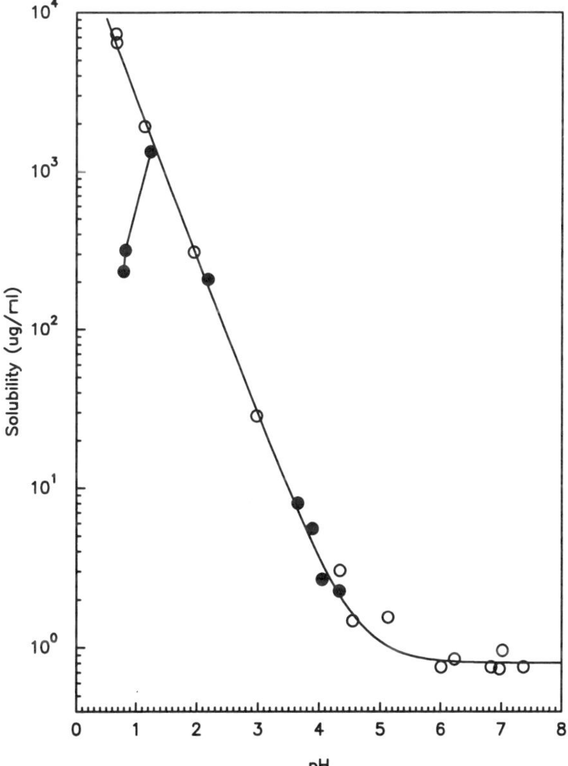

Figure 3. Equilibrium solubility of crystalline anhydrous delavirdine free base in water at room temperature using hydrochloric acid (●), methanesulfonic acid (○), or sodium hydroxide (○, > pH 5).

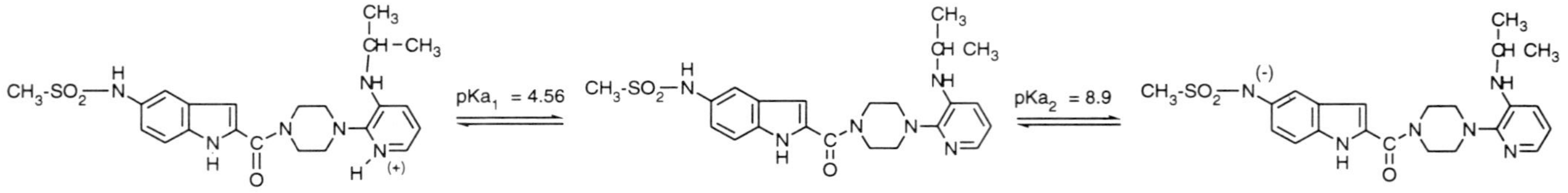

$pKa_1 = 4.56$

$pKa_2 = 8.9$

Figure 4. Ionization of delavirdine.

delavirdine free base is 2.94 mg/ml (or 3.56 mg/ml calculated as the mesylate salt of delavirdine), and at pH 6.0, the approximate pH of the small intestine, the solubility is about 1 μg/ml. The hydrochloride salt of delavirdine has a maximum solubility of 1.34 mg/ml at pH 1.2 (Morozowich *et al.*, unpublished data).

Delavirdine has a second pK_a at 8.9 as determined from partitioning data. The ionization scheme for delavirdine is shown in Fig. 4.

The *n*-octanol water log *PC* of delavirdine is 2.84, indicating adequate lipophilicity for absorption and membrane transport.

8.2.2. SALT SELECTION

In attempts to identify a rapidly dissolving acidic salt of delavirdine, intrinsic dissolution rates of a number of delavirdine salts were determined using constant surface pellets. Table VI shows that the intrinsic dissolution rate of delavirdine mesylate is about 255 times faster than that of delavirdine free base. The tosylate, HBr and HCl salts of delavirdine dissolved much slower than the mesylate salt. As a result, the mesylate salt of delavirdine was selected for development.

8.2.3. CRYSTAL FORMS OF DELAVIRDINE MESYLATE

Initially, two crystalline solid forms of delavirdine mesylate (PNU-90152E) were isolated (Forms I and II). These forms were hygroscopic and deliquescent at high humidities and a more stable crystal form was clearly desirable. An extensive study of the crystallization behavior of delavirdine mesylate was undertaken and an anhydrous, nonhygroscopic crystal form was discovered (Bergren *et al.*, 1996). This form, designated Form VIII of delavirdine mesylate (PNU-90152E), was giv-

Table VI
Intrinsic Dissolution Rate of Delavirdine (PNU-90152) and Atevirdine (PNU-87201) and Salts at pH 2 (0.01 N HCl) at 25°C and 300 rpm

PNU-number	Dissolution rate (μg/sec/cm^2)
87201E (mesylate)	255
87201 (free base)	0.8
90152E (mesylate)	258
90152F (tosylate)	11.8
90152B (hydrobromide)	4.8
90152A (hydrochloride)	4.2
90152 (free base)	1.0

en the distinct code designation PNU-90152S. PNU-90152S was employed in early delavirdine mesylate toxicology and clinical trials. In early process crystallizations, however, PNU-90152S generally contained several percent of Form II. An alternate crystallization process was developed to eliminate Form II. During this development, a new crystal form of delavirdine mesylate, designated Form XI, was discovered. Form XI proved to be the most thermodynamically stable anhydrate. Once Form XI had been crystallized in process equipment, subsequent large-scale crystallizations reliably yielded Form XI but could not reliably yield Form VIII. Form XI was given the unique code designation PNU-90152T. After the bio-equivalence of Forms VIII and XI was demonstrated in the toxicology and clinical formulations, Form XI (PNU-90152T) was selected for final formulation development.

8.2.4. BEHAVIOR OF DELAVIRDINE MESYLATE IN WATER

Delavirdine mesylate dissolves rapidly in water and highly concentrated supersaturated solutions (>100 mg/ml) can be generated temporarily. Within a few minutes, precipitation of delavirdine free base occurs and the resulting equilibrium solubility is dictated by the pH that is generated and this, in turn, is determined by the amount of the delavirdine salt used to saturate the solution. The resulting equilibria are shown in Fig. 5 where $N\text{-}H^+$ and N^o represent the protonated (salt form) and the free base form of delavirdine, respectively, and MSA^- and MSA-H represent the methanesulfonate ion and methanesulfonic acid, respectively.

Delaviridine free base trihydrate is precipitated from water and the solubility of the trihydrate (1.9 μl/ml) is about twice that of anhydrous crystalline delavirdine free base. The short-term solubility of delavirdine mesylate in water is greater than 100 mg/ml using a large excess of drug but delavirdine free base will eventually precipitate from these solutions. Delavirdine mesylate forms supersaturated solutions *in vitro* as well as *in vivo*. Oral administration of a 100-mg delavirdine mesylate tablet to duodenal fistulated dogs resulted in concentrations up to about 400 μg/ml in the duodenal fluid at pH 4–6. This ability to generate high duodenal concentrations of delavirdine may be implicated in the efficient absorption of the drug.

$$N\text{-}H^+_{Solid} \longrightarrow N\text{-}H^+_{Sol'n} + H_2O \underset{}{\overset{Hydrolysis}{\rightleftharpoons}} H_3O^+ + N^o_{Sol'n} \overset{Precipitation}{\rightleftharpoons} N^o_{Solid}$$

$$\searrow MSA^- + H_2O \overset{Hydrolysis}{\rightleftharpoons} OH^- + MSA\text{-}H$$

Figure 5. Equilibria on contact of delavirdine mesylate with water.

8.2.5. DELAVIRDINE MESYLATE FORMULATION FOR ANIMAL STUDIES

Delavirdine mesylate was formulated at concentrations up to about 150 mg/ml in 80% propylene glycol containing 3 μl/ml of methanesulfonic acid and this solution formulation, with a pH of about 1.2, was employed in animal safety and pharmacokinetic studies, with administration by oral and i.v. routes.

8.3. Absorption, Distribution, Metabolism, and Excretion

The absorption, distribution, metabolism, and excretion (ADME) of delavirdine were determined following single i.v. dose administration in rat, dog, and monkey, single and multiple p.o. dose administration in mouse, rat, dog, monkey, rabbit, and human, and multiple i.v. dose administration in dog and monkey. *In vitro* studies were conducted to characterize, among other things, the microsomal metabolism, P450 isoforms responsible for the metabolism of the drug, and plasma/serum protein binding. These studies were designed to support preclinical drug safety studies and clinical investigations of delavirdine mesylate. The species, drug doses, and formulations used in the animal ADME studies generally reflected those employed in preclinical pharmacology and toxicology evaluations. Single- and multiple-dose pharmacokinetic data were obtained as an integral part of the toxicological evaluation of the drug. The doses administered to animals in the drug safety studies were much higher (on a mg/kg dosage basis) than the recommended clinical dose of delavirdine mesylate. Specific high-performance reversed-phase liquid chromatographic methods that utilized ultraviolet or fluorescence detection were used for the quantitation of the intact compound (Hosley *et al.*, unpublished data) and for the quantitation of the intact compound and its major metabolite (Staton *et al.*, 1995) in systemic circulation. Delavirdine-related radioactivity in biological samples was determined by liquid scintillation counting techniques following administration of radiolabeled [^{14}C-carboxamide]delavirdine mesylate or [2-^{14}C-pyridine]delavirdine mesylate (Hsi *et al.*, unpublished data). The results of some of these studies and their impact on the delavirdine mesylate development program are briefly described below.

8.3.1. SINGLE-DOSE PHARMACOKINETICS

The pharmacokinetics of delavirdine was determined following single i.v. and p.o. dose administration of nonradiolabeled drug in male rats, male dogs, and male and female monkeys, and following single p.o. dose administration of radiolabeled drug in male and female mice, male and female rats, male dogs, male and female monkeys, female rabbits, and male humans. The nonradiolabel studies in rats and dogs (Adams *et al.,* unpublished data) were conducted during the drug discovery phase of the project, as previously noted. The pharmacokinetics of delavirdine was nonlinear in rats, and appeared to be nonlinear in dogs. Systemic clearances in the dog, at a dose of 10 mg/kg, were approximately 20-fold lower than in the rat, whereas the clearances in monkeys (Adams *et al.,* unpublished data) were comparable to those in rats. Absolute p.o. bioavailabilities increased in the order monkey (30%) < rat (65%) < dog (100%) when equivalent i.v. and p.o. doses were administered.

Oral bioavailability studies conducted in the dog indicated that the bioavailabilities of solution and milled- or micronized-quality delavirdine mesylate were equivalent, whereas the bioavailabilities of the free base suspension and milled-quality delavirdine hydrochloride were two-fold and five-fold lower, respectively, than the solution formulation. These results indicated that the mesylate salt should be developed for clinical use. Several different crystalline forms of the mesylate salt (E, S, and T forms) were also evaluated in preliminary relative oral bioavailability studies in dogs and were found to be bioequivalent (Adams, unpublished data), confirming that different crystalline forms of the mesylate salt could be interchangeably used in the toxicological evaluation of the drug. Concurrent administration of food (ad libitum) to male rats (Rodríquez *et al.,* unpublished data) and dogs (Adams *et al.,* unpublished data) reduced systemic exposure to delavirdine by two-fold and five-fold, respectively, relative to fasted animals. This information led to the control of animal feeding schedules in the dog safety studies.

Following single-dose p.o. administration of radiolabeled drug, dose-normalized systemic exposure to delavirdine (on a mg/kg dosing basis) increased in the order mouse < rabbit < monkey < human < dog, and was confirmatory of the single-dose nonradiolabel studies (Chang *et al.,* unpublished data) . The relative exposure to the main metabolite in systemic circulation was dose-dependent. The majority of delavirdine-related material was excreted in the feces. Urinary excretion of delavirdine-related material was highest in the human and lowest in the dog.

8.3.2. MU+SE PHARMACOKINETICS

Multiple-dose pharmacokinetics studies of delavirdine were conducted following p.o. dose administration in the mouse, rat, dog, monkey, rabbit, and human, and following i.v. dose administration in the dog and monkey. The nonradiolabel multiple-dose studies were conducted in conjunction with subchronic, chronic, and reproductive toxicity studies. The pharmacokinetics was nonlinear in all species, with most, if not all, of the nonproportional increase in systemic exposure to delavirdine being related to the fact that the biotransformation of delavirdine to *N*-desisopropyldelavirdine was, at least in part, inhibitable or capacity-limited (Adams *et al.,* unpublished data). Because the biotransformation of delavirdine was in part inhibitable or capacity-limited, systemic concentrations of *N*-desisopropyldelavirdine, compared with simultaneous delavirdine concentrations, were higher when concentrations of delavirdine were low and then achieved asymptotic values at higher delavirdine concentrations. Comparison of the pharmacokinetic data from the toxicity studies indicated that the systemic exposure to delavirdine (on a mg/kg per day dosage basis) increased in the order mouse $<$ rabbit $\leq$ monkey $\leq$ rat $<$ dog (Adams *et al.*, unpublished data). These data were consistent with the higher toxicity of the drug (on a mg/kg per day dosage basis) in the dog relative to the other species (Jensen *et al.*, unpublished data).

Administration of multiple p.o. doses of radiolabeled delavirdine indicated that systemic exposure to delavirdine (on a mg/kg per day dosage basis) increased in the order monkey $<$ rat $<$ human $<$ dog (Chang *et al.*, unpublished data). These results were consistent with the multiple-dose nonradiolabel studies and the single-dose radiolabel studies. The majority of delavirdine-related radioactivity was excreted in the feces, as was found in the single-dose studies.

8.3.3. DISTRIBUTION

Tissue distribution of delavirdine was investigated in mouse, rat, dog, and human. Distribution of delavirdine-related radioactivity into most rat tissues, with the exception of brain, was rapid and extensive, with the highest concentrations observed in liver, kidney, and adrenal glands (Chang *et al.*, unpublished data). The brain penetration of delavirdine in mice, rats, and dogs was limited to approximately 1 to 6% of simultaneous plasma concentrations. After single p.o. dose administration of radiolabeled delavirdine mesylate to lactating rats, concentrations of delavirdine-related material and delavirdine were significantly higher in milk than in plasma. The *in vitro* protein binding of delavirdine in rat, dog, monkey, and human serum indicated binding of greater than 96%, with albumin being the major serum protein contributing to delavirdine binding (Bombardt *et al.*, unpublished data). Binding of delavirdine in plasma from HIV^+ patients was consistent with the *in vitro* protein binding and was approximately 98%. The binding capacity of human serum for delavirdine was approximately two-fold greater than in rat, beagle dog, and monkey serum, and about equivalent with mongrel dog serum. The *in vitro* distribution of delavirdine in human PBMC indicated that distribution was proportional to free drug concentration (Zhao *et al.*, unpublished data). Although delavirdine was extensively bound to plasma or serum proteins, the *in vitro* antiviral activity of delavirdine, as determined by a cell culture model using human plasma albumin or α_1-acid glycoprotein, was minimally affected by protein binding.

8.3.4. BIOTRANSFORMATION

The *in vivo* metabolism of delavirdine was investigated in the mouse, rat, dog, monkey, rabbit, and human (Chang *et al.*, unpublished data). Metabolite profiling of plasma and urine was conducted in the mouse, dog, monkey, rabbit, and human following single p.o. dose administration of radiolabeled delavirdine mesylate; in the mouse, rat, dog, monkey, rabbit, and human following multiple p.o. dose administration of radiolabeled delavirdine mesylate; and in the rat following single i.v. dose administration of radiolabeled delavirdine mesylate. In addition, the metabolite profiles of urine and bile were investigated in male rats after single and multiple p.o. dose and single i.v. dose administration of radiolabeled delavirdine mesylate. Metabolite profiling of feces was conducted in the mouse, rat, dog, rabbit, and human following single p.o. dose administration of radiolabeled delavirdine mesylate; and in the mouse, rat, rabbit, and human following multiple p.o. dose administration of radiolabeled delavirdine mesylate.

The metabolism of delavirdine involves four pathways (Chang *et al.*, unpublished data): first, *N*-desalkylation to desalkyl delavirdine, followed by conjugation with sulfate or with *N*-acetylglucosamine; second, hydroxylation of *N*-desalkyl delavirdine at the pyridine ring C-4′ or C-6′ and subsequent conjugation with sulfate; third, cleavage of the amide bond to give indole carboxylic acid and *N*-isopropylpyridinepiperazine, with subsequent *N*-desalkylation to form aminopyridinepiperazine, and conjugation of the indole carboxylic acid. Amide bond cleavage with release of *N*-isopropylpyridinepiperazine is observed as a significant pathway in mice only. In a fourth pathway observed in all species, delavirdine is hydroxylated at the pyridine ring C-6′ to 6′-pyridinol delavirdine. Subsequent conjugation with glucuronic acid or sulfate gives 6′-*O*-glucuronide delavirdine and 6′-*O*-sulfate delavirdine, respectively, or *N*-desalkylation and conjugation yield 6′-*O*-sulfate desalkyl delavirdine. Alternatively, the pyridine ring in 6′-pyridinol delavirdine is cleaved to give pyridine-cleaved delavirdine and the pyridine ring-opened metabolite. Further conjugation of pyridine-cleaved delavirdine also occurs. In all species, *N*-desisopropyldelavirdine was the the principal metabolite in circulation. *N*-isopropylpyridinepiperazine was observed as a minor metabolite in circulation in the mouse only, but higher concentrations of this metabolite were found in mouse brain. With the exception of the mouse, metabolite profiles in animals used in drug safety studies were qualitatively similar to the profiles in humans.

The *in vitro* metabolism of delavirdine was investigated using liver microsomes from mouse, rat, dog, monkey, rabbit, and human (Voorman *et al.*, unpublished data). In all species, the primary microsomal metabolite was *N*-desisopropyldelavirdine. 6′-Pyridinol delavirdine and a conjugate or isomer of 6′-pyridinol delavirdine were tentatively identified as minor microsomal metabolites. With human liver microsomes, delavirdine was metabolized primarily by cytochrome P450 3A (CYP3A), which catalyzed both delavirdine *N*-desalkylation and 6′-hydroxylation. Delavirdine was also metabolized by CYP2D6, which catalyzed only *N*-desalkylation and was probably a lower-capacity pathway than CYP3A.

8.4. Safety/Toxicokinetics

A 2-week p.o. dose toxicity study in dogs was completed to characterize the toxicity and toxicokinetics of delavirdine mesylate in a second species so as to determine the acceptability of delavirdine mesylate as a second-generation drug candidate. Results of toxicokinetic analyses indicated that very high and persistent serum concentrations of delavirdine were achieved in dogs and that serum concentrations of delavirdine increased more than expected following multiple-dose administration, suggesting that the clearance of PNU-90152 was capacity-limited or inhibited. Toxicity in dogs was associated with very high concentrations of delavirdine. The toxicokinetic and toxicity profiles of delavirdine mesylate were determined to be acceptable in rats and dogs in these preliminary studies (Jensen *et al.*, unpublished data).

To support clinical trials and registrations of delavirdine mesylate, definitive p.o. dose toxicity studies were completed in rats and dogs through 6 and 12 months' duration, respectively, and in cynomolgus monkeys through 3 months' duration. Also conducted were a standardized panel of reproductive toxicity studies in rats and rabbits, and carcinogenicity studies in rats and mice. In these studies, extensive toxicokinetic monitoring was conducted to correlate toxicity with systemic exposure to delavirdine.

8.5. Clinical Summary

Delavirdine mesylate entered phase I clinical development in April 1993 using a 100-mg tablet formulation (Rescriptor®, Pharmacia & Upjohn). The strategy for the clinical evaluation was to focus on trials to support a combination therapy indication. A total of five phase I and II trials that investigated delavirdine mesylate safety and pharmacokinetics in humans have been conducted. These trials collectively demonstrated that the compound is extremely well tolerated and exhibits an antiviral effect as evidenced by increased levels of $CD4^+$ lymphocytes and reductions in plasma viral RNA (viral load) and viral p24 antigen in diverse populations of HIV-1-infected patients (Davey, 1996).

The major clinical toxicity noted in some AIDS patients was a mild to moderate rash, which gradually resolved. The majority of patients experiencing the rash could continue delavirdine mesylate treatment without dose interruption as this did not appear to cause any additional complications or delay resolution of the rash (Davey *et al.*, 1996). Based on the promising early clinical results with delavirdine mesylate combinations, two pivotal phase III trials were begun in May 1994. These trials evaluated the drug in combination with AZT or ddI versus AZT or ddI monotherapy. The trials were designed as clinical endpoint studies of greater than 1 year's duration to correlate the surrogate marker response with clinical efficacy, defined as the time to and incidence of AIDS-defining illnesses or death. In addition, the trials were powered to allow an analysis of surrogate marker response after 6 months of patient treatment.

An analysis of delavirdine mesylate safety and efficacy based on the data collected in these trials was completed in late 1995. The analysis was performed maintaining the blind of individual patients and their assigned therapy; thus, the effect of antiviral treatment on clinical outcome has not been fully examined and remains blinded. Nevertheless, the analysis indicated that baseline $CD4^+$ lymphocyte levels and plasma viral load as well as reduction in viral load in patients receiving antiviral therapy were significantly correlated to clinical progression. In contrast, a change in $CD4^+$ lymphocyte levels or p24 levels was much less strongly associated with the risk of clinical progression (Freimuth *et al.*, unpublished data). In both trials, patients administered high levels of delavirdine mesylate (1200 mg daily) exhibited positive surrogate marker responses. For example, patients receiving AZT and delavirdine mesylate showed a significantly greater increase in $CD4^+$ cells and $CD4^+$ cell percentage levels, greater reductions in viral load and antigen for up to 60 weeks when compared with AZT monotherapy.

In summary, data from over 2000 patients in two double-blind, randomized, comparative phase III studies provide the primary data demonstrating the safety and efficacy of delavirdine mesylate in HIV-1 disease. Data from five phase I and II studies provide supportive evidence for the compound's efficacy. Taken together, these data provide strong evidence that delavirdine mesylate when used in combination with other antiviral therapy is safe and efficacious in HIV-1-infected patients. As a result of these positive findings, Pharmacia & Upjohn filed a New Drug Application with the U.S. FDA in July 1996.

9. CONCLUSIONS

A program directed toward the discovery and development of novel and efficacious nonnucleoside HIV RT inhibitors involved a close collaboration between medicinal chemists and biologists and their colleagues in drug delivery, drug metabolism, and drug safety. An initial clinical candidate, PNU-87201E (atevirdine mesylate), helped pave the way for a far more potent and effective second-generation BHAP candidate, delavirdine mesylate (PNU-90152T). Emerging results from clinical trials suggest that delavirdine mesylate is a promising new agent for the treatment of AIDS.

ACKNOWLEDGMENTS

We thank Mike Bergren for contributing the section on crystal forms of delavirdine mesylate. We also thank the many dedicated and talented individuals who contributed to the successful discovery and development of delavirdine mesylate.

REFERENCES

Bergren, M. S., Chao, R. S., Meulman, P. A., Sarver, R. W., Lyster, M. A., Havens, J. L., and Hawley, M., 1996, Solid phases of delavirdine mesylate, *J. Pharm. Sci.* **85:**834–841.

Chang, M., Sood, V. K., Kloosterman, D. A., Hauer, M. J., Gaerness, P. E., Sanders, P. E., and Vrbanac, J. J., 1997, Identification of the metabolites of the HIV-1 reverse transcriptase inhibitor delavirdine in monkeys, *Drug Metab. Dispos.* **25:**828–839.

Chang, M., Sood, V. K., Wilson, G. J., Kloosterman, D. A., Sanders, P. E., Hauer, M. J., and Fagerness, P. E., 1997, Metabolism of the HIV-1 reverse transcriptase inhibitor delavirdin in rats, *Drug Metab. Dispos.* **25:**228–242.

Chang, M., Sood, V. K., Wilson, G. J., Kloosterman, D. A., Sanders, P. E., Hauer, M. J., Zhang, W., and Granstetter, D. G., 1997, Metabolism of the HIV-1 reverse transcriptase inhibitor delavirdine in mice, *Drug Metab. Dispos.* **25:**814–827.

Chong, K. T., Pagano, P. J., and Hinshaw, R. R., 1994, Bisheteroaryl piperazine reverse transcriptase inhibitor in combination with 3′-azido-3′-deoxythymidine or 2′,3′-dideoxycytidine synergistically inhibits human immunodeficiency virus type I replication in vitro, *Antimicrob. Agents Chemother.* **38:**288–293.

Davey, R. T., Chaitt, D. G., Reed, G. F., Freimuth, W. W., Herpin, B. R., Metcalf, J. A., Eastman, P. S., Falloon, J., Kovacs, J. A., Polis, M. A., Walker, R. E., Masur, H., Boyle, J., Coleman, S., Cox, S. R., Wathen, L., Daenzer, C. L., and Lane, H. C., 1996, Randomized, controlled phase I/II trial of combination therapy with delavirdine (U-90152S) and conventional nucleosides in human immunodeficiency virus type 1-infected patients, *Antimicrob. Agents Chemother.* **40:**1657–1664.

DeClercq, E., 1994, HIV resistance to reverse transcriptase inhibitors, *Biochem. Pharmacol.* **47:**155 169.

Dueweke, T. J., Poppe, S. M., Romero, D. L., Swaney, S. M., So, A. G., Downey, K. M., Althaus, I. W., Reusser, F., Busso, M., Resnick, L., Mayers, D. L., Lane, J., Aristoff, P. A., Thomas, R. C., and Tarpley, W. G., 1993a, U-90152, a potent inhibitor of human immunodeficiency virus type I replication, *Antimicrob. Agents Chemother.* **37:**1127–1131.

Dueweke, T. J., Pushkarskaya, T., Poppe, S. M., Swaney, S. M., Zhao, J. Q., Chen, I. S. Y., Stevenson, M., and Tarpley, W. G., 1993b, A mutation in reverse transcriptase in bis(heteroaryl)piperazine-resistant human immunodeficiency virus type I that confers sensitivity to other nonnucleoside inhibitors, *Proc. Natl. Acad. Sci. USA* **90:**4713–4717.

Fischl, M. A., Richman, D. D., Grieco, M. H., Gottlieb, M. S., Volberding, P. A., Lasking, O. L., Lecdom, J. M., Groopman, J. E., Mildvan, D., Schooley, R. T., Jakson, G. G., Durack, D. T., King, D., and the AZT Collaborative Working Group, 1987, The efficacy of azidothymidine (AZT) in the treatment of patients with AIDS and AIDS-related complex. A double-blind, placebo-controlled trial, *N. Engl. J. Med.* **317:**185–191.

Gottlieb, M. S., Schroff, R., Shanker, H. M., Weisman, J. D., Fan, P. T., Wolf, R. A., and Saxon, A., 1981, *Pneumocystis carinii* pneumonia and mucosal candidiasis in previously healthy homosexual men: Evidence of a new acquired cellular immunodeficiency, *N. Eng. J. Med.* **305:**1425–1431.

Ho, D. D., Neumann, A. G., Perelson, A. S., Chen, W., Leonard, J. M., and Markowitz, M., 1995, Rapid turnover of plasma virions and CD4 lymphocytes in HIV-1 infection, *Nature* **373:**123–126.

Mieke, A., Been-Tiktak, M., Vrehen, H. M., Schneider, M. M. E., van der Feltz, M., Branger, T., Ward, P., Cox, S. R., Harry, J. D., and Borleffs, J. C., 1995, Safety, tolerance, and pharmacokinetics of atevirdine mesylate (U-87201E) in asymptomatic human immunodeficiency virus-infected patients, *Antimicrob. Agents Chemother.* **39:**602–607.

Quinn, T. C., 1996, Global burden of the HIV pandemic, *Lancet* **348:**99–105.

Romero, D. L., Busso, M., Tan, C. K., Reusser, F., Palmer, J. R., Poppe, S. M., Aristoff, P. A., Downey, K. M., So, A. G., Resnick, L., and Tarpley, W. G., 1991, Nonnucleoside reverse transcriptase inhibitors that potently and specifically block human immunodeficiency virus type I replication, *Proc. Natl. Acad. Sci. USA* **88:**8806–8810.

Romero, D. L., Morge, R. A., Genin, M. J., Biles, C., Busso, M., Resnick, L., Althaus, I. W., Reusser, F., Thomas, R. C., and Tarpley, W. G., 1993, Bis(heteroaryl)piperazine (BHAP) reverse transcriptase inhibitors: Structure–activity relationships of novel substituted indole analogues and the identification of 1-[(5-methanesulfonamido-1*H*-indol-2-yl)carbonyl]-4-[3-[(1-methylethyl)-amino]pyridinyl]piperazine monomethanesulfonate (U-90152S), a second-generation clinical candidate, *J. Med. Chem.* **36:**1505–1508.

Romero, D. L., Morge, R. M., Biles, C., Berrios-Pena, N., Max, P. D., Smith, H. W., Busso, M., Tan, C. K., Voorman, R. L., Reusser, F., Althaus, I. W., Downey, K., So, A. G., Resnick, L., Tarpley, W. G., and Aristoff, P. A., 1994, Discovery, synthesis, and bioactivity of bis(heteroaryl)piperazines. I. A novel class of non-nucleoside HIV-1 reverse transcriptase inhibitors, *J. Med. Chem.* **37**:999–1014.

Romero, D. L., Olmsted, R. A., Poel, T. J., Morge, R. A., Biles, C., Keiser, B. J., Kopta, L. A., Friis, J. M., Hosley, J. D., Stefanski, K. J., Wishka, D. G., Evans, D. B., Morris, J., Stehle, R. G., Sharma, S. K., Yagi, Y., Voorman, R. L., Adams, W. J., Tarpley, W. G., and Thomas, R. C., 1996, Targeting delavirdine/atevirdine resistant HIV-1: Identification of (alkylamino)piperidine-containing bis(heteroaryl)piperazines as broad spectrum HIV-1 reverse transcriptase inhibitors, *J. Med. Chem.* **39**:3769–3789.

Staton, B. A., Johnson, M. G., Friis, J. M., and Adams, W. J., 1995, Simple, rapid and sensitive high-performance liquid chromatographic determination of delavirdine and its N-desisopropyl metabolite in human plasma, *J. Chromatogr. B* **668**:99–106.

Chapter 14

Famciclovir

Discovery and Development of a Novel Antiherpesvirus Agent

Richard L. Jarvest, David Sutton, and R. Anthony Vere Hodge

1. INTRODUCTION

Famciclovir (**1**) is a new oral antiviral agent. The active circulating metabolite following oral administration of famciclovir is penciclovir (**2**), which has antiviral activity against members of the human herpesvirus family such as herpes simplex virus types 1 and 2 (HSV-1 and HSV-2), varicella zoster virus (VZV), and Epstein–Barr virus (EBV) and also against hepatitis B virus (HBV). Famciclovir has been licensed as Famvir™ in over 50 countries for the treatment of herpes zoster (shingles) and in over 25 countries for the acute treatment of genital herpes. A topical formulation of penciclovir itself has been licensed in the United Kingdom as Vectavir™ and in the United States as Denavir™ for the treatment of cold sores.

In order to successfully identify and characterize the most appropriate derivative of penciclovir for oral administration and to evaluate the best compound for topical and i.v. administration, the role of bioavailability and metabolism studies in the discovery phase was essential. In addition, biochemical mode of action studies were essential to understand the antiviral selectivity against VZV, HSV-1 and

---

Richard L. Jarvest, David Sutton, and R. Anthony Vere Hodge • SmithKline Beecham Pharmaceuticals, Harlow, Essex CM19 5AW, England.

Integration of Pharmaceutical Discovery and Development: Case Studies, edited by Borchardt *et al.*, Plenum Press, New York, 1998.

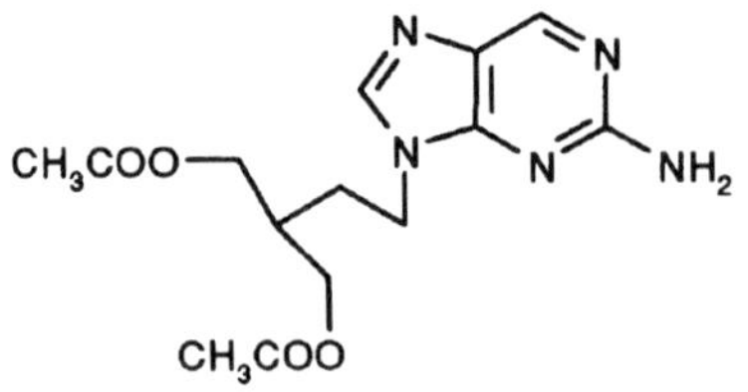

1 Famciclovir (FCV, BRL 42810)

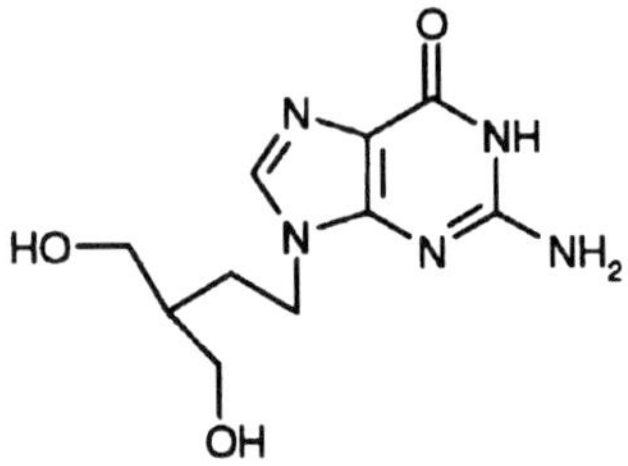

2 Penciclovir (PCV, BRL 39123)

HSV-2, and HBV and had a significant impact on the design of clinical studies and target therapeutic profile.

1.1. Identification of Penciclovir as an Antiherpesvirus Agent

Penciclovir (**1**) was synthesized in these laboratories in a pure form by a short five-step route in 1983 (Harnden and Jarvest, 1985; Harnden *et al.,* 1987). When tested in a primary assay for activity against HSV-1, penciclovir had potent selective antiviral activity and thus became the focus of detailed evaluation. Synthetic programs investigating systematic modification of the molecular structure were undertaken. The antiviral activity of penciclovir proved to be very structure specific and it was found that a variety of molecular changes resulted in a significant decrease in antiviral potency (Bailey and Harnden, 1988; Harnden and Jarvest, 1988a, 1989; Harnden *et al.,* 1988a; Geen *et al.,* 1991). However, a novel class of compounds in which the first atom of the alkyl side chain was replaced by a heteroatom was found to possess antiherpesvirus activity (Harnden and Jarvest, 1988b; Harnden *et al.,* 1988b, 1990b). Of these, compound **3** had five-fold increased activity against the HSVs and in cell culture it was the most potent acyclonucleoside that we tested (Harnden *et al.,* 1988b, 1990b). While ours was the

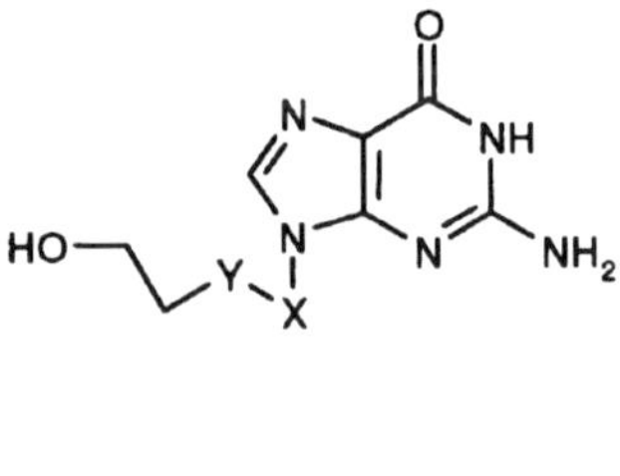

3 X = O, Y = CH_2
4 X = CH_2, Y = O

first patent publication relating to pure penciclovir and its antiviral properties, others also reported its antiherpesvirus activity (Tippie *et al.*, 1984; Larsson *et al.*, 1986; MacCoss *et al.*, 1986).

1.2. Antiviral Activity and Spectrum of Activity

As part of the initial evaluation of penciclovir, laboratory strains of various human and animal herpesviruses were tested for susceptibility to penciclovir and acyclovir. Both compounds were very active against HSV-1 and HSV-2, and had good activity against VZV. However, activity against cytomegalovirus (CMV) was limited (Boyd *et al.*, 1987). Whereas penciclovir and acyclovir had similar activity against laboratory strains of human herpesviruses, there were marked differences in potency between the two compounds among animal herpesviruses. For example, penciclovir was 10-fold more active than acyclovir against feline herpesvirus type 1, but was inactive against simian varicella virus (Boyd *et al.*, 1987, 1993).

The activity of penciclovir against clinical isolates of human herpesviruses in a variety of cell lines has been examined in more detail (Boyd *et al.*, 1993). The cell line was found to influence both absolute and relative potencies of both penciclovir and acyclovir (Bacon and Howard, 1996). Using a plaque reduction assay in MRC-5 cells, penciclovir was significantly less active than acyclovir against both HSV-1 and HSV-2 (Table I; Boyd *et al.*, 1987, 1993), whereas in WISH cells penciclovir was significantly more active than acyclovir. The assay type was also

Table I
Spectrum of Activity of Penciclovir in Human Cells

Virus	Assay method	Cell line	No. of strains tested	EC_{50} (μg/ml)[a]	
				Penciclovir	Acyclovir
HSV-1	Plaque reduction[b]	MRC-5	19	0.4	0.2
HSV-2	Plaque reduction[b]	MRC-5	22	1.8	0.7
VZV	Plaque reduction[c]	MRC-5	29	3.8	4.2
EBV	DNA inhibition[d]	P3HR-1	1	2.3	2.2
CMV	Plaque reduction[e]	MRC-5	3	18.0	7.4
HBV	DNA inhibition[f]	2.2.15	1	0.05[g]	28[h]

[a]Mean values except for VZV where median is shown.
[b]Boyd *et al.* (1993).
[c]Cell-free virus preparations (Bacon *et al.*, 1996).
[d]Bacon and Boyd (1995).
[e]Unpublished data for three clinical isolates (T. H. Bacon, personal communication).
[f]Inhibition of the production of extracellular virus.
[g]Korba and Boyd (1996).
[h]B. Korba (Georgetown University, personal communication).

found to affect the outcome of antiviral assays. When a virus yield reduction assay was employed using MRC-5 cells infected with a high virus inoculum, penciclovir and acyclovir had similar activity against HSV-1 and HSV-2 (Boyd *et al.*, 1987). However, when a lower virus-to-cell ratio and longer incubation time were used, penciclovir was significantly more active than acyclovir against both types of virus (Bacon *et al.*, 1996b). The effect of assay conditions on the relative potency of penciclovir and acyclovir against HSV has been extensively reviewed (Bacon, 1996).

In plaque reduction assays in MRC-5 cells, penciclovir and acyclovir had similar activity against VZV, but were less active against HSV (Table I; Bacon *et al.*, 1996a). However, in Hs68 cells, both compounds were approximately five-fold more active than in MRC-5 cells (Bacon *et al.*, 1996a). EBV was equally sensitive to penciclovir and acyclovir in P3HR-1 cells (Table I; Bacon and Boyd, 1995). CMV was much less sensitive to both penciclovir and acyclovir than the other herpesviruses tested (Table I; Boyd *et al.*, 1987). Recent clinical isolates appeared to be more sensitive than laboratory strains to both drugs; the EC_{50}s for penciclovir were 18 and 52 μg/ml respectively and 7 and 25 μg/ml for acyclovir (Table I).

Penciclovir was found to cause more prolonged inhibition of HSV replication than acyclovir following removal of extracellular drug (Boyd *et al.*, 1987; Bacon and Schinazi, 1993). This effect is of great significance for the target pharmacokinetics of penciclovir and its mechanistic basis was thoroughly elucidated (Section 1.3).

Unexpectedly, penciclovir was found to have antiviral activity against HBV. Cell culture experiments demonstrated that penciclovir was extremely effective at inhibiting both the extracellular and intracellular levels of human HBV DNA in transfected HepG2 cells (Table I; Korba and Boyd, 1996). Acyclovir was much less effective than penciclovir in this system (Table I). Penciclovir has also been demonstrated to inhibit duck HBV DNA synthesis in primary duck hepatocytes (Shaw *et al.*, 1994) and in chronically infected ducks (Tsiquaye *et al.*, 1996), an accepted animal model of HBV infection.

The antiviral activity of penciclovir in cell culture is highly selective. The exceptional lack of cytotoxicity of penciclovir was demonstrated in a range of 12 human cell lines. The IC_{50} was at least 100 μg/ml in all but one cell line (where it was 60 μg/ml) (Boyd *et al.*, 1987, 1993).

1.3. Mechanism of Action

The early evaluation of penciclovir in cell culture antiviral assays had suggested that the drug is selectively activated in herpesvirus-infected cells, as it was inactive against thymidine kinase-negative strains of herpesviruses. In this respect, penciclovir seemed to be similar to acyclovir. However, in contrast to acyclovir, penciclovir had prolonged antiviral activity after its removal from the cell culture medium. Consequently, biochemical mode of action studies were undertaken initially to provide a rationale for the selective activity of penciclovir in cell culture against HSV-1, HSV-2, and VZV. Unexpectedly, these studies revealed major differences between penciclovir and acyclovir, which influenced both the choice of prodrug and the design of clinical trials.

Penciclovir rapidly enters both uninfected and herpesvirus-infected cells. However, in uninfected cells, because there is no viral thymidine kinase, penciclovir remains essentially unchanged, although very low levels of penciclovir triphosphate (about 0.04 μM) have been detected (Lowe *et al.,* 1995). However, penciclovir triphosphate only inhibits cellular DNA polymerase alpha at much higher concentrations, more than 4000-fold those detected in uninfected cells (K_i 200 μM) (Earnshaw *et al.,* 1992; Ilsley *et al.,* 1995). The virtual lack of phosphorylation of penciclovir in uninfected cells and its low affinity for the cellular polymerases account for the fact that it is exceptionally nontoxic to replicating cells in culture and for its safety profile in clinical therapy (see Sections 1.2 and 4.2).

Within cells infected with HSV-1, HSV-2, or VZV, the viral thymidine kinase has a high affinity for penciclovir (100 times greater than acyclovir in the case of HSV-1; Larsson *et al.,* 1986; Datema *et al.,* 1987) and this viral enzyme phosphorylates penciclovir to its monophosphate. In this form, penciclovir is trapped within the infected cell and is further phosphorylated by cellular enzymes to the active triphosphate (Vere Hodge and Perkins, 1989; Earnshaw *et al.,* 1992, Vere Hodge, 1993). During incubation of HSV-1-infected MRC-5 cells with 10 μM penciclovir for 4 hr, penciclovir triphosphate continued to increase up to about 300 μM (Vere Hodge and Perkins, 1989). By comparison, the maximum concentration of acyclovir triphosphate was about 2.5 μM. Recently, Lowe *et al.* (1995) have reported similar results. Both acyclovir and penciclovir are phosphorylated very selectively in herpesvirus-infected cells, with the greater selectivity of penciclovir being related to its higher affinity for viral thymidine kinases.

Penciclovir has a prochiral structure and phosphorylation of the molecule results in a new chiral center. To determine the stereospecificity and absolute configuration of the penciclovir phosphates formed intracellularly, we synthesized penciclovir in isotopically chiral form, with known absolute configuration, by incorporating ^{13}C into one of the hydroxymethyl groups (Jarvest *et al.,* 1990; Sime *et al.,* 1992). The isotopically chiral penciclovir was incubated in the appropriate biological system, the resulting phosphates were isolated, and ^{13}C NMR was used to determine whether the phosphoryl group was adjacent to the ^{13}C label.

HSV-1 thymidine kinase phosphorylated [4′-^{13}C]penciclovir to give 75% of the (*S*)- and 25% of the (*R*)-penciclovir monophosphate (Fig. 1) (Vere Hodge *et al.,* 1993a). HSV-2 thymidine kinase gave at least 70% of the (*S*)-enantiomer (Ertl *et al.,* 1995). When penciclovir triphosphate was extracted from HSV-1-infected cells, the absolute configuration of the triphosphate was (*S*) with an enantiomeric purity greater than 95% (Fig. 1) (Vere Hodge *et al.,* 1993a). Whereas none of the

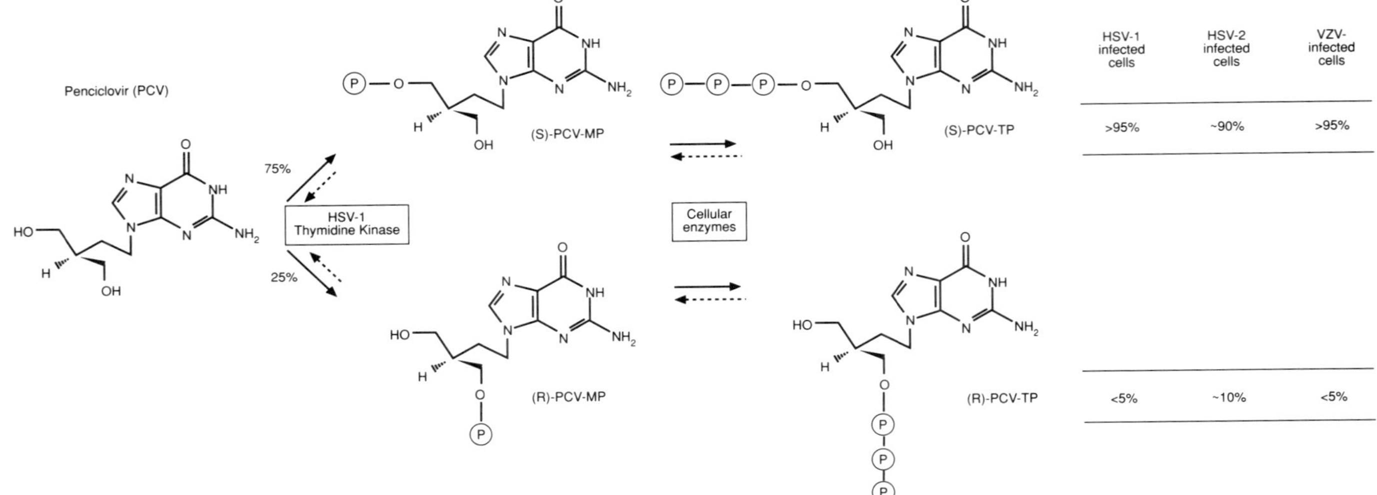

Figure 1. Summary of the chirality of penciclovir phosphorylation. Reproduced with permission from Vere Hodge *et al.* (1993a).

(*R*)-penciclovir triphosphate was detected in HSV-1-infected cells, there was about 10% of the (*R*)-enantiomer in HSV-2-infected cells (Vere Hodge *et al.,* 1993). This suggests that the HSV-2-encoded thymidine kinase is less specific than the HSV-1 enzyme. As for HSV-1-infected cells, only (*S*)-penciclovir triphosphate (>95%) was detected in VZV-infected cells (Bacon *et al.,* 1996a).

(*S*)-penciclovir triphosphate, isolated from HSV-1-infected cells, was compared with chemically synthesized racemic penciclovir triphosphate for their inhibitory effects on the viral DNA polymerases. For both HSV-1 and HSV-2 DNA polymerases, (*S*)-penciclovir triphosphate has a higher affinity (K_i 8.5 and 5.8 μM, respectively) than the racemic triphosphate (Earnshaw *et al.,* 1992). Although not compared with racemic penciclovir triphosphate, the (*S*)-enantiomer inhibits VZV DNA polymerase (K_i 7.5 and 1.6 μM; Bacon *et al.,* 1996a, and Ertl *et al.,* 1995, respectively). The K_i values for (*S*)-penciclovir triphosphate for HSV-1, HSV-2, and VZV DNA polymerases were about 100-fold higher than for acyclovir triphosphate (Earnshaw *et al.,* 1992; Ertl *et al.,* 1995; Bacon *et al.,* 1996a). Although penciclovir triphosphate is a less powerful inhibitor of the viral DNA polymerases, it is a highly effective inhibitor of viral DNA extension when present at the high concentrations formed in virally infected cells. When enzyme assay conditions were designed to represent virus-infected cells, with dGTP and the nucleotide analogues at their respective concentrations, penciclovir triphosphate was more effective than acyclovir triphosphate in inhibiting viral (HSV-2) DNA polymerase-mediated DNA chain elongation (Earnshaw and Vere Hodge, 1992; Vere Hodge and Cheng, 1993). The concentrations of the nucleoside triphosphates were 12 μM dGTP (Karlsson *et al.,* 1986), 300 μM PCV-TP, and 18 μM ACV-TP (Earnshaw *et al.,* 1992). As predicted by the work of Reardon and Spector (1989), acyclovir triphosphate did not prevent the incorporation of dGTP into many of the individual DNA strands, thereby permitting further DNA chain elongation (about 40 to 70 nucleotides added to the primer). In contrast, penciclovir triphosphate gave striking inhibition of DNA synthesis, preventing any detectable elongation. Hence, penciclovir has been called a "short DNA chain terminator." Clearly, termination of viral DNA extension within either 20 nucleotides by penciclovir triphosphate or 70 nucleotides by acyclovir triphosphate will give highly effective inhibition of viral replication.

Penciclovir triphosphate is highly stable within the infected cell and so can exert its antiviral actions for long periods (Vere Hodge and Perkins, 1989; Earnshaw *et al.,* 1992; Vere Hodge and Cheng, 1993; Bacon *et al.,* 1996a). The half-lives of the triphosphates within cells infected by HSV-1, HSV-2, or VZV are shown in Table II. The high intracellular concentrations of penciclovir triphosphate together with its high stability led to the anticipation that it could continue to exert its antiviral effects after the blood concentrations of the compound have fallen.

VZV remains latent in the satellite cells that surround the nerve cells. The pain experienced by shingles patients is thought to be related, at least in part, to

Table II
Half-Lives of the Triphosphates of Penciclovir and Acyclovir in Herpesvirus-Infected Cells

	Half-life (hr)			
	HSV-1	HSV-2	VZV	References
Penciclovir	10	20	9	Vere Hodge and Perkins (1989), Earnshaw *et al.* (1992), Bacon *et al.* (1996)
			17	Bebault *et al.* (1995)
Acyclovir	0.7	1	0.8	Vere Hodge and Perkins (1989), Earnshaw *et al.* (1992), Bacon *et al.* (1996)
			1	Bebault *et al.* (1995)

virus-induced damage to the Schwann cells surrounding the axon of nerve cells. Therefore, the finding that penciclovir is converted efficiently to its triphosphate in a VZV-infected human schwannoma cell line but not in uninfected cells (Sacks *et al.*, 1994; Bebault *et al.*, 1995) may be of clinical significance. Triphosphate concentrations reach similar levels to those found in VZV-infected fibroblasts and penciclovir triphosphate has similar stability (half-life 14 hr). In contrast, the half-life of acyclovir triphosphate is only 1 hr. The rapid formation and persistence of penciclovir triphosphate in human schwannoma cells may help to explain the beneficial effects of famciclovir on postherpetic neuralgia (Section 4.2).

The differing affinities of penciclovir and acyclovir for the viral enzymes, thymidine kinase and DNA polymerase, may influence their cross-resistance. Although both compounds are inactive against HSV strains that lack the thymidine kinase, such strains have a reduced ability for reactivation and are generally less pathogenic than strains that have become resistant to acyclovir via a change in the thymidine kinase or DNA polymerase. Certain thymidine kinase-altered strains of HSV and VZV resistant to acyclovir retain sensitivity to penciclovir whereas others are cross-resistant (Talarico *et al.*, 1993). Chiou *et al.* (1995) reported on five DNA polymerase mutants resistant to acyclovir. One strain was two-fold hypersensitive to penciclovir, and the other four strains retained sensitivity (within three-fold). Whether an acyclovir pol mutant can confer high resistance to penciclovir remains to be seen.

The activity of penciclovir against HBV derives from the potent activity of penciclovir triphosphate against HBV DNA polymerase, both when this enzyme is making the primer for DNA synthesis and during the reverse transcriptase stage (Mok *et al.*, 1995; Korba and Boyd, 1996; Dannaoui *et al.*, 1997). Penciclovir does not appear to be selectively phosphorylated in HBV-infected cells and although only very small amounts of penciclovir triphosphate are formed, its affinity for HBV DNA polymerase is so great that there is sufficient triphosphate to cause efficient inhibition. Antiviral selectivity is achieved because the affinity of penciclovir triphosphate for HBV DNA polymerase is over 4000-fold higher than that for cellular DNA polymerase alpha.

1.4. Oral Bioavailability

In common with acyclovir (de Miranda *et al.*, 1981; de Miranda and Blum, 1983) and other guanine nucleoside analogues such as ganciclovir (Jacobson *et al.*, 1987), the oral absorption of penciclovir was found to be very low. In studies carried out in mice (Boyd *et al.*, 1988a; Harnden *et al.*, 1989; Sutton and Kern, 1993) and rats (Vere Hodge *et al.*, 1989), the oral bioavailability was approximately 1–3%. For comparison, the oral bioavailability of acyclovir in the mouse and rat was 43 and 19%, respectively (de Miranda *et al.*, 1981). The oral bioavailability of acyclovir in humans varies between 10 and 20% in a manner inversely dependent on the dosage (de Miranda and Blum, 1983; Weller *et al.*, 1993). As the oral absorption of penciclovir in experimental animals was less than that of acyclovir, it was predicted that the oral absorption of penciclovir in humans would be inadequate.

2. PRODRUG FORMS OF PENCICLOVIR

2.1. Strategy and Evaluation of Oral Bioavailability

The poor oral absorption of penciclovir was a key feature that we believed needed to be improved for the successful development of the compound as an antiherpesvirus agent. As poor oral bioavailability is a feature common to other guanine acyclonucleosides and the antiviral activity was very specific to the penciclovir structure, it was decided to attempt to improve the bioavailability of penciclovir by synthesis and evaluation of potential prodrugs.

In order to rapidly screen a high number of potential oral prodrugs of penciclovir, a mouse oral bioavailability test was used (Harnden *et al.*, 1989). Compounds were administered at 0.2 mole/kg (equivalent to 50 mg/kg penciclovir) by discrete oral gavage and oral penciclovir acted as the control in each test. Blood samples were collected from each of three mice 15, 60, and 180 min after dosing. Pooled samples were treated with 16% trichloroacetic acid to precipitate proteins prior to high-performance liquid chromatography (HPLC) analysis of the supernatant for penciclovir, the parent compound, and potential metabolic precursors. The acid stability of each compound was tested in parallel to the biological test to assist in the interpretation of the blood level data. In the event of an acid-labile compound giving a significant improvement in penciclovir blood levels, the oral bioavailability test was repeated using ethanol as an alternative protein precipitant.

The first chemical approach to penciclovir prodrugs was to prepare simple mono- and dicarboxylic esters of penciclovir, **5** and **6,** respectively (Harnden *et al.*, 1987). None of these compounds resulted in improved blood levels of penciclovir after oral administration (Harnden *et al.*, 1989). The physicochemical properties of penciclovir are dominated by the polar guanine ring and it was reasoned that this would have to be modified to improve absorption. Three main strategies for modification of the 6-position of the purine ring were pursued: potential substrate moieties for adenosine deaminase; higher alkoxy groups; and 6-unsubstituted as a potential oxidase substrate.

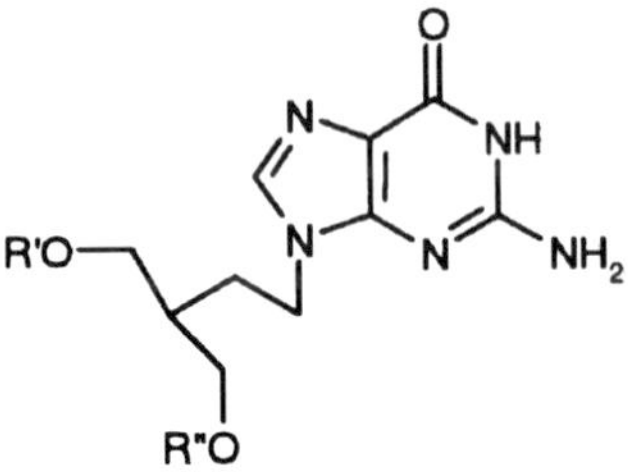

5 R' = H, R" = RCO

6 R' = R" = RCO

The enzyme adenosine deaminase hydrolyzes small polar groups at the 6-position of a number of purine nucleoside analogues. The 6-amino (**7**), 6-chloro (**8**), and 6-methoxy (**9**) analogues of penciclovir were thus prepared as potentially adenosine deaminase-activated prodrugs. All three compounds gave poor blood levels of penciclovir, partly related to the fact that they proved to be rather poor substrates for the deaminase (Harnden *et al.*, 1989).

A wide range of higher 6-alkoxy compounds such as **10** and **11** were synthesized and a number of these were found to be more efficiently absorbed than penciclovir (Harnden *et al.*, 1989). The total concentration of penciclovir and parent compound at the 15-min time point in the mouse showed a bell-shaped correlation with calculated log *P* values, maximum absorption occurring with ethoxy (**10**) and isopropoxy (**11**). However, although the maximum total concentration of parent and penciclovir were increased up to 11-fold over the equivalent dose of penciclovir, the plasma levels of penciclovir were increased only a maximum of 2-fold. It was evident that metabolic conversion rather than absorption from the gastrointestinal tract was the limiting factor in delivering circulating penciclovir from this class of prodrug. The higher 6-alkoxy group was a novel prodrug moiety for guanine derivatives and although the mechanism of conversion was not characterized, it was presumed to be an oxidative dealkylation. It seemed unlikely that the metabolic conversion could be significantly enhanced so the series was not pursued further.

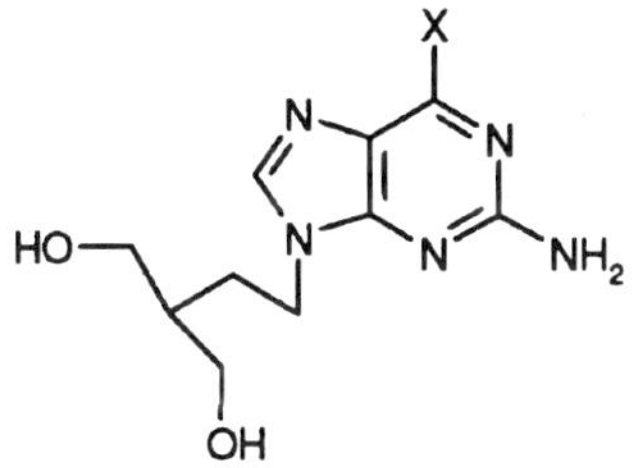

7 X = NH_2
8 X = Cl
9 X = CH_3O
10 X = CH_3CH_2O
11 X = $(CH_3)_2CHO$

In contrast, the 6-deoxy analogue of penciclovir (**12;** BRL 42359) was only moderately well absorbed in mice but was efficiently converted to penciclovir, affording a three-fold higher plasma concentration than that obtained from dosing penciclovir (Harnden *et al.,* 1989). The 6-deoxy congener of acyclovir had been described as a xanthine oxidase-activated prodrug form of acyclovir (Krenitsky *et al.,* 1984), and it was shown that bovine xanthine oxidase was able to oxidize the 6-deoxy compound to penciclovir (Harnden *et al.,* 1989) (however, see Section 3.3 for the situation in human tissues).

As absorption rather than metabolism appeared to be limiting for **12,** a series of mono- and diesters were prepared with the aim of further enhancing absorption. This strategy was successful, resulting in an increase of 7- to 16-fold in circulating levels of penciclovir (Harnden *et al.,* 1989). For six mono- and diesters of **12,** the total concentration of acyclonucleoside present in the blood at the 15-min time point showed some degree of correlation with calculated log P ($r = 0.79$) indicating that lipophilicity is an important determinant of the degree of absorption in this series. However, a much more significant correlation was obtained for an inverse

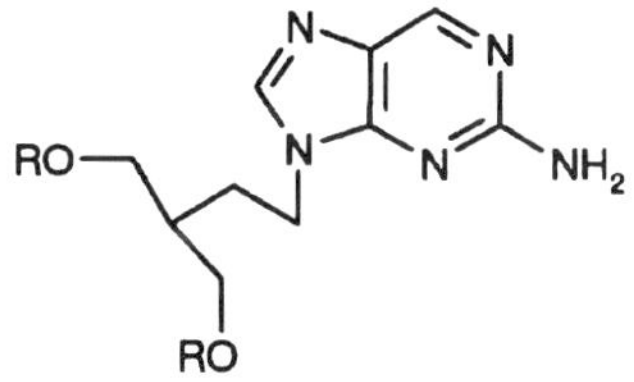

12 R = H (BRL 42359)
13 R = CH_3CH_2CO

relationship with melting point ($r = 0.99$). Melting point may effectively act as a marker of lattice energy in certain high-melting crystalline compounds and it has been postulated that disruption of intermolecular hydrogen bonding via creation of bioreversible adducts may lower lattice energy, increase lipid solubility, and result in improved absorption (the melting point of some phenytoin derivatives was correlated with lipid solubility; Yamaoka *et al.*, 1983). The lower lattice energy of the diacetyl ester, famciclovir (**1**), is also reflected in increased solubility relative to **2** and **12** both in aqueous buffer and in organic solvents. When the crystal structures of penciclovir and famciclovir were determined, they afforded a visual insight into the lattice structure that results in their different physicochemical properties and ultimately oral absorption (Fig. 2) (Harnden *et al.*, 1990a). Examination

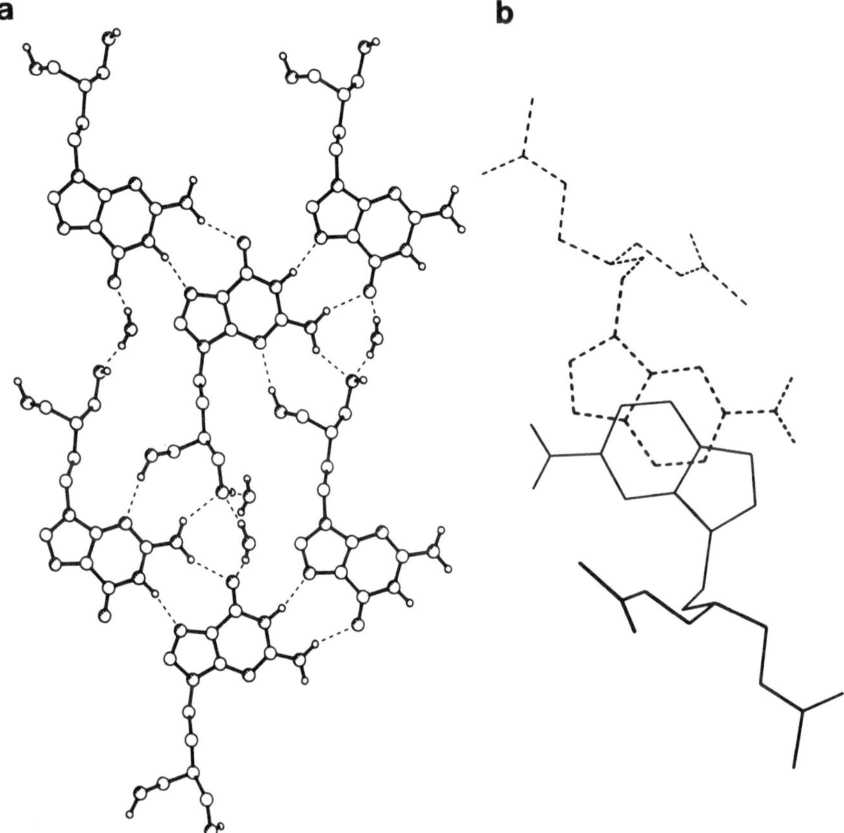

Figure 2. Crystal structures of penciclovir and famciclovir. (a) Part of one of the continuous sheets of hydrogen-bonded molecules in the structure of penciclovir monohydrate. (b) Parallel overlap of symmetry-related pairs of molecules in the structure of famciclovir monohydrate. Reproduced with permission from Harnden *et al.* (1990a).

of the structure of penciclovir shows that all of the available heteroatoms and exchangeable hydrogen atoms are involved in an extensive sheetlike hydrogen-bonding network. In contrast, famciclovir does not have this strong hydrogen-bonding network, the main interaction being a relatively weak π–π stacking interaction of the purine rings of a pair of symmetry-related molecules.

The diacetate (famciclovir) and dipropionate (**13;** BRL 43599) esters of 6-deoxypenciclovir were selected for further evaluation in other animal models of oral absorption and for metabolic studies in rodent and human tissue homogenates.

Secondary evaluation of oral bioavailability was carried out in rats (Vere Hodge *et al.,* 1989). By using the rat, it was possible to collect sequential blood samples from individual animals. The rat studies both confirmed and extended the results of the primary mouse screen (Harnden *et al.,* 1989). As in the mouse, the 6-deoxy derivative of penciclovir gave only a modest increase in blood levels of penciclovir relative to penciclovir itself (Fig. 3). The oral bioavailability of oral penciclovir itself was between 1 and 2%. The 6-deoxy prodrug increased this value by approximately six-fold to 9%. However, both the diacetate, famciclovir, and the dipropionate, **13,** were quickly absorbed and efficiently metabolized to penciclovir. The oral bioavailability of penciclovir following administration of famciclovir and BRL 43599 was 41 and 27%, respectively (Fig. 3; Vere Hodge *et al.,* 1989). The major metabolic intermediate from each prodrug was 6-deoxy penciclovir indicating that the rate-determining step in the metabolism of these prodrugs

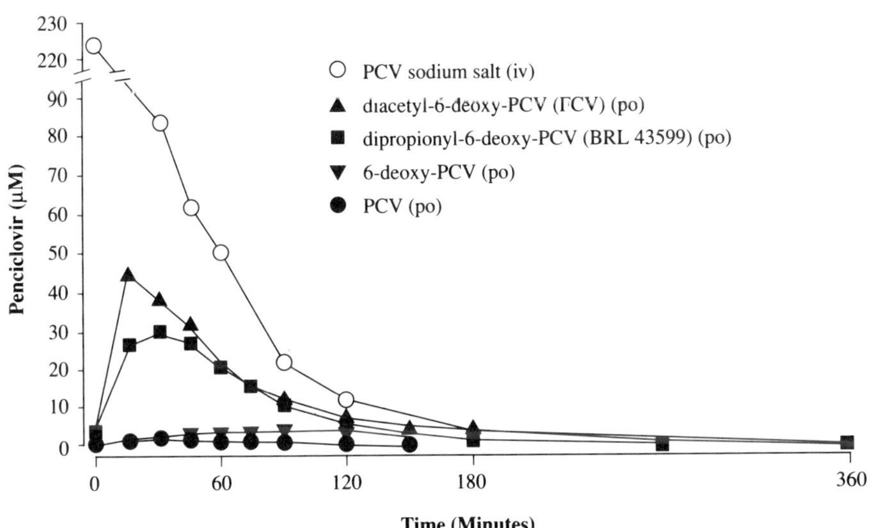

Figure 3. Mean concentrations of penciclovir in blood after administration of penciclovir and prodrugs (0.2 mmole/kg) to rats. Reproduced with permission from Vere Hodge *et al.* (1989).

to penciclovir was oxidation at the 6-position of the guanine ring. The metabolism of these compounds was then studied in more detail in rat and human tissues.

2.2. Evaluation of Metabolic Conversion in Human Body Fluids and Tissues

The primary aim of the work with body fluids and tissues was to show that the potential prodrugs, famciclovir and BRL 43599, would be converted to the antiviral drug by human enzymes, not just those from rats and mice. Also, these studies provided a key piece of information for making the choice between the candidate compounds. The strategy was to model the metabolic process during absorption by using duodenal contents, extracts from intestinal wall and liver, and blood. Our aim was to select a prodrug that was relatively stable in human duodenal contents, to allow time for absorption, yet be efficiently converted to penciclovir by the combined effects of the other human tissues (Vere Hodge *et al.,* 1989).

The rate of metabolism of compounds in tissue homogenates and extracts is commonly much slower than occurs *in vivo,* even when the tissues are freshly collected. Therefore, it was important to include appropriate controls. The metabolism of the compounds in human fluids and tissue extracts was compared with that in rat samples, thus providing a link with *in vivo* data. Also, we included in the tests 6-deoxyacyclovir, which was known to be efficiently converted to acyclovir in humans. The prodrugs, intermediate metabolites, and penciclovir itself were assayed by HPLC using a single-gradient elution cycle to assay all compounds (Vere Hodge *et al.,* 1989).

In rat duodenal contents, diluted 10-fold, BRL 43599 was very quickly hydrolyzed, the half-life being less than 2 min. This result confirmed the impression gained from the *in vivo* assay that metabolism of BRL 43599 was competing with, and limiting, the prodrug absorption, which took about 15–30 min. Similarly in human duodenal contents (undiluted), BRL 43599 was rapidly metabolized with a half-life of 7 min. In contrast, famciclovir had good stability in the duodenal contents of both rats (half-life 35 min) and humans (half-life about 6.5 hr) (Vere Hodge *et al.,* 1989).

With intestinal wall extract, both rat and human, one ester group was hydrolyzed much faster than the other. The potential stereochemical consequences of this monohydrolysis were subsequently investigated (Section 3.2).

The liver extracts, relative to the other tissues, had the greater metabolic activity. Both ester groups were hydrolyzed to give 6-deoxypenciclovir, which was converted further to penciclovir. The oxidation step was clearly the rate-limiting step but the rate in human liver extract was slightly greater (about 1.5-fold) than for 6-deoxyacyclovir. The enzyme responsible for the oxidation of the purine in human liver was subsequently identified (Section 3.3). Penciclovir appeared to be stable in the liver extracts and also in the other body fluids/extracts. A possible metabolite, 8-hydroxypenciclovir, was sought but not detected.

2.3. Selection of Preferred Oral Candidate: Famciclovir

From the studies of oral bioavailability in mice, the esters of 6-deoxypenciclovir were identified as the best class of prodrugs for penciclovir. Two compounds, famciclovir and BRL 43599, were selected for detailed study in rats. The ability to take sequential samples from individual rats and to analyze these for all of the intermediate metabolites in addition to penciclovir itself, proved to be invaluable as it afforded data on the rates of absorption and conversion and on the variation between individual animals. From these studies, famciclovir was found to be the preferred prodrug of penciclovir in rats.

From the package of tests in human body fluids and tissues, it appeared that the dipropionyl ester was too readily hydrolyzed in human duodenal contents to give reliable, consistent absorption. In contrast, famciclovir had sufficient stability (half-life about 6.5 hr) to ensure time for absorption prior to metabolism under varying conditions, such as change in dose, patient-to-patient individuality, young or old, before or after food. We proposed that the metabolic conversion starts during passage through the intestinal wall and that conversion to penciclovir would be completed mainly in the liver. The rate-limiting step would be the oxidation of the purine but, as this step was slightly faster than for 6-deoxyacyclovir, this would be sufficiently fast to allow efficient conversion of famciclovir to penciclovir. These conclusions, summarized schematically in Fig. 4a (Vere Hodge *et al.*, 1989), were later fully confirmed by clinical evaluation of famciclovir (Section 4.1). We conclude that the use of human tissues was of critical importance in the early identification of the preferred prodrug.

2.4. Other Routes of Administration

Penciclovir cream (5% w/w) was found to have topical efficacy against a cutaneous HSV-1 infection in guinea pigs (Boyd *et al.*, 1988a). Subsequent work demonstrated that penciclovir was highly effective even if the start of therapy was delayed until lesions were present (Sutton and Kern, 1993). Topical famciclovir was also highly effective in this model and *in vitro* studies using human skin showed that famciclovir was able to readily penetrate human skin. However, there was insufficient oxidase activity in human skin *in vitro* to metabolize famciclovir to penciclovir. Therefore, topical formulations of penciclovir were progressed to clinical studies.

An intravenous form of penciclovir needed to be identified for potential use

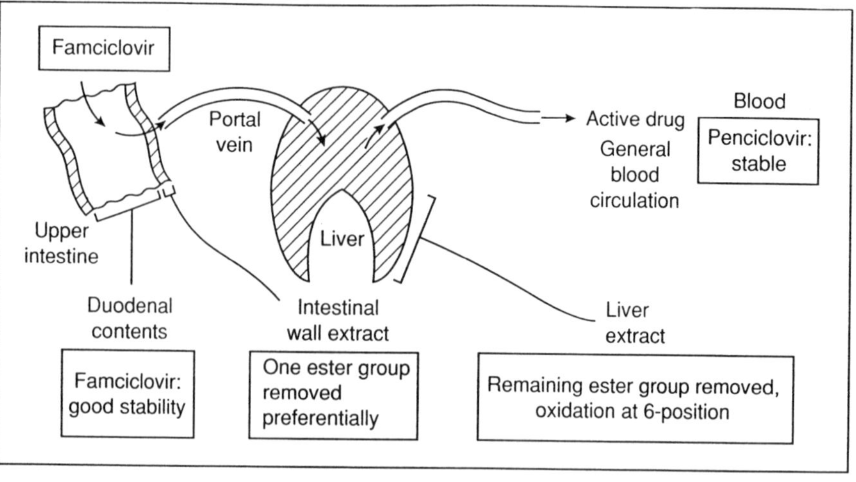

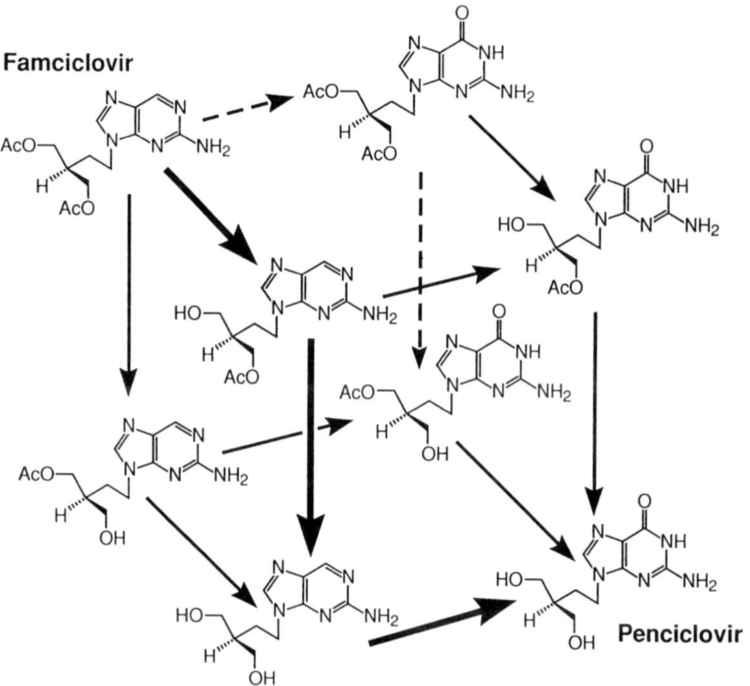

Figure 4. Conversion of famciclovir to penciclovir. Both in extracts of human tissues (intestinal wall and liver) and after administration to healthy subjects, the major metabolic route is by deacetylation followed by oxidation of the purine. (a) Schematic diagram showing the expected conversion of the prodrug, famciclovir, and the dipropionyl compound (BRL 43599) into penciclovir.(b) The chirality of the monoacetates is shown as determined by incubation of [4′-^{13}C] famciclovir in an extract of human intestinal wall. Reproduced with permission from Vere Hodge *et al.* (1989).

in seriously ill patients. The modest solubility of penciclovir itself limits its use for intravenous administration. In contrast, the sodium salt of penciclovir is highly water-soluble. However, the resulting solution is alkaline. Pharmacokinetic studies were carried out in both mice and rats using the sodium salt (Vere Hodge *et al.*, 1989; Sutton and Kern, 1993). Intravenous administration of famciclovir to rats demonstrated that although there was extensive metabolism to penciclovir, the actual systemic exposure to penciclovir was much reduced. This was the result of excretion of significant amounts of metabolic intermediates prior to their metabolism to penciclovir. The sodium salt of penciclovir was thus chosen as the preferred form for intravenous administration.

3. PRECLINICAL EVALUATION OF FAMCICLOVIR

3.1. Animal Models of Infection

Initial studies using penciclovir demonstrated that the potent antiherpesvirus activity seen in cell culture was reflected by efficacy against HSV-1 and HSV-2 in murine models of infection by both oral and systemic routes. These data have been extensively reviewed by Sutton and Kern (1993).

In a direct comparison of penciclovir and famciclovir given orally at 10 mg/kg per dose twice daily for 4 days to mice infected on the flank with HSV-1, famciclovir was superior to penciclovir in reducing the severity of the zosteriform lesions (Sutton and Kern, 1993). In a further independent study carried out using a murine model of herpes encephalitis, compounds were administered via the drinking water from day 1 postinfection. Famciclovir administered at 0.2 mg/ml was highly effective at reducing mortality of infected mice. In the treated group, 70% of animals survived compared with 0% in the control group. This is in contrast to the 50% survival in mice receiving penciclovir at a fivefold higher concentration (1 mg/ml) (Goldthorpe *et al.*, 1992). These studies demonstrate that the increased bioavailability of penciclovir from famciclovir is associated with improved efficacy.

Further *in vivo* evaluation of the oral efficacy of famciclovir was carried out in comparative studies with the competitors, acyclovir or valaciclovir. Use was made of an intraperitoneal HSV-1 infection model in mice that gave a direct quantitative measure of antiviral effect. Famciclovir was found to be highly effective following oral administration. In a study that compared multiple doses given over a 16-hr period starting 24 hr postinfection, famciclovir was significantly more effective than acyclovir at reducing peritoneal virus replication. A single dose of famciclovir (10 mg/kg) was significantly more active than four doses of acyclovir (10 mg/kg). Famciclovir was still active if the dose was reduced to 5 mg/kg, whereas acyclovir was inactive at this dose even after five doses (Ashton *et al.*, 1994).

In a model of cutaneous HSV infection in normal or immunosuppressed mice, famciclovir has been found to be superior to valaciclovir in moderating both clinical signs and viral replication in both skin and neural tissue (Field *et al.,* 1995; Field and Thackray, 1995; Thackray and Field, 1996a,b). Following cessation of valaciclovir therapy, there is a transient but reproducible recurrence of infectious virus. No such rebound has been observed when famciclovir treatment is stopped. Prolonged suppression of viral replication by penciclovir, but not acyclovir, has also been reported previously in a model of systemic HSV infection in mice (Sutton and Boyd, 1993). Intriguingly, famciclovir therapy during the acute infection has been found to reduce the subsequent recovery of latent virus from explanted ganglia; valaciclovir has no such effect (Thackray and Field, 1996a,b). The authors have suggested that this activity of famciclovir may reflect the high affinity of penciclovir for the viral thymidine kinase.

Studies in ducks chronically infected with DHBV demonstrated that both oral penciclovir and famciclovir reduced plasma DHBV DNA and DNA polymerase levels to below the limit of detection while treatment continued (Tsiquaye *et al.,* 1996). After cessation of therapy, there was a delay of 2 to 8 days before plasma levels of DHBV DNA and DNA polymerase began to increase. These data clearly demonstrated that famciclovir has *in vivo* efficacy against DHBV. This has been confirmed by Lin *et al.* (1996).

3.2. Chirality of Metabolic Products from Famciclovir

Like penciclovir itself, famciclovir is prochiral. The monoacetylated metabolites, monoacetyl-6-deoxypenciclovir and monoacetyl-penciclovir, formed during the metabolic conversion of famciclovir to penciclovir, are both chiral. The preferential removal of just one acetyl group by human intestinal wall extract (Section 2.2) had suggested that there may be enantioselectivity associated with this step, and to fully characterize the route of metabolic conversion it was necessary to determine the stereospecificity and absolute configuration of these metabolites. Isotopically chiral famciclovir was synthesized in a similar way to penciclovir by incorporating ^{13}C into one of the acetoxymethyl groups (Jarvest *et al.,* 1990; Sime *et al.,* 1992).

[4′-^{13}C]-Famciclovir was incubated in an extract from human intestinal wall. It was found that the human esterase(s) present in intestinal wall hydrolyze the acetyl group preferentially from the pro-(*S*)-acetoxymethyl group of famciclovir (Vere Hodge *et al.,* 1993b). The specificity of the esterase action in forming monoacetyl-6-deoxypenciclovir and monoacetyl-penciclovir was about 77 and 72%, respectively. Thus, all of the metabolites shown in Fig. 4b were detected but the major route is indicated by the bold arrows. Importantly, all routes lead to penciclovir.

3.3. Identification of Enzymatic Oxidation in Humans

The 6-deoxypenciclovir generated *in vivo* by esterase activity was initially expected to be oxidized to the guanine by the molybdenum-dependent enzyme xanthine oxidase and it was shown that 6-deoxypenciclovir was efficiently oxidized to penciclovir by bovine xanthine oxidase (Harnden *et al.*, 1989). In order to determine the enzyme responsible for the oxidation in humans, the metabolism of 6-deoxypenciclovir in human liver cytosol was examined in the presence or absence of inhibitors of xanthine oxidase or the related enzyme aldehyde oxidase (Clarke *et al.*, 1995). It was found that the inhibitors of aldehyde oxidase, menadione and isovanillin, impaired the oxidation to penciclovir but that allopurinol, an inhibitor of xanthine oxidase used clinically, had no effect. This suggests that aldehyde oxidase is the main enzyme responsible for oxidative step in the metabolism of famciclovir in humans. This conclusion was supported by drug interaction studies in human volunteers (Section 4.1).

4. CLINICAL EVALUATION

4.1. Metabolism and Pharmacokinetics

For initial evaluation of tolerance and pharmacokinetics, penciclovir was administered to volunteers intravenously as the sodium salt. In the dosage range of 10–20 mg/kg, both maximal blood concentration of penciclovir (C_{max}) and the area under the curve (AUC) increased proportionately with dose (Fowles *et al.*, 1992). Penciclovir was extensively distributed and rapidly excreted. No metabolites of penciclovir were detected in plasma and approximately 70% of the dose was excreted unchanged in the urine. When penciclovir was administered intravenously to healthy elderly subjects, both C_{max} and $t_{1/2}$ increased as would be expected for an age-related decrease in renal clearance (Pratt *et al.*, 1993). For a t.i.d. dosing schedule, the change in the pharmacokinetic parameters in the elderly was considered unlikely to result in significant accumulation and it was considered not necessary to adjust the dosage.

As anticipated from the animal studies, bioavailability of penciclovir from oral administration of penciclovir in human volunteers proved to be poor. From an oral dose of 5 mg/kg, both the AUC and the urinary excretion of penciclovir were between 20- and 25-fold lower than the values obtained from intravenous administration of the sodium salt (Fig. 5a; Boyd *et al.*, 1988b).

With oral famciclovir there was a dramatic improvement in the bioavailability of penciclovir, from about 4% to 75%. The AUC was increased 18-fold and the urinary excretion was increased 13-fold (Fig. 5a; Boyd *et al.*, 1988b). The termi-

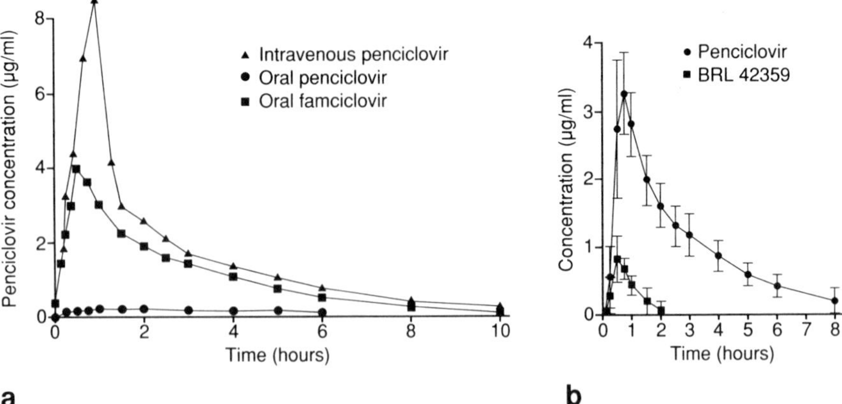

Figure 5. Mean plasma concentration–time data in healthy volunteers. (a) Penciclovir following oral administration of famciclovir or oral or intravenous administration of penciclovir at doses equivalent to 5 mg/kg penciclovir. (b) Penciclovir and 6-deoxypenciclovir (BRL 42359) following oral administration of 500 mg famciclovir. Reproduced with permission from Pue and Benet (1993).

nal half-life of penciclovir derived from oral administration was essentially identical to that obtained for intravenous administration of the sodium salt. This was the first indication that the strategy for evaluation of the absorption and metabolism of penciclovir prodrugs had produced a highly effective compound for oral administration to humans.

The pharmacokinetics of penciclovir following oral administration of fixed doses of famciclovir has been studied in detail (Pue and Benet, 1993; Pue *et al.*, 1994). Famciclovir (500 mg) was rapidly absorbed and maximum plasma concentrations of penciclovir were obtained at a median time of 0.75 hr (Fig. 5b). The only other metabolite consistently observed was 6-deoxypenciclovir, which was present in plasma for a relatively short period of time at concentrations up to one-third those of penciclovir (Fig. 5b). Approximately 65% of the administered dose was excreted in the urine, of which 60% was penciclovir and the remaining 5% 6-deoxypenciclovir (Pue *et al.*, 1994). The absolute bioavailability of penciclovir was 77% with a 95% confidence interval of 72 to 83%, indicating a relatively small degree of variability of absorption and metabolism. The 500-mg dose of famciclovir afforded a mean C_{max} for penciclovir of 3.3 μg/ml and a mean elimination $t_{1/2}$ of 2.3 hr. The dose dependency of penciclovir pharmacokinetics was measured for a range of oral famciclovir doses between 125 and 750 mg (Pue *et al.*, 1994). Both the maximal blood concentration and the area under the curve of penciclovir were shown to be dose proportional over this range. Following oral administration of famciclovir, the pharmacokinetics of penciclovir is similar in male and female volunteers (Pratt *et al.*, 1994a).

There are no clinically significant drug interactions with a number of drugs that have been studied for their potential pharmacokinetic interaction with famciclovir (Daniels and Schentag, 1993). Consistent with the *in vitro* demonstration that in human liver the oxidative step in the metabolism of famciclovir to penciclovir is catalyzed by aldehyde oxidase rather than xanthine oxidase, it was found that allopurinol (a xanthine oxidase inhibitor used clinically in the treatment of hyperuricemia and gout) at therapeutic doses had no effect on the metabolism of famciclovir to penciclovir in human volunteers (Fowles *et al.,* 1994). There was also no significant interaction of famciclovir with cimetidine, a nonspecific inhibitor of cytochrome P450-mediated drug metabolism, confirming that the cytochrome P450 isozymes inhibited by cimetidine are not of importance in the metabolism of famciclovir to penciclovir (Pratt *et al.,* 1991).

The effect of food on the bioavailability and pharmacokinetics of penciclovir following administration of famciclovir has also been investigated. In fasting volunteers, maximum penciclovir plasma concentrations were attained within 1 hr of administration of famciclovir, whereas in fed volunteers absorption was delayed but the bioavailability of penciclovir was not reduced (Fowles *et al.,* 1990, 1991). Both cell culture studies (Pratt *et al.,* 1994b) and biochemical considerations (Vere Hodge and Cheng, 1993) suggest that for penciclovir the AUC is more important than initial peak concentrations in determining antiviral efficacy in patients. Thus, in HSV-2-infected MRC-5 cells, concentration–time profiles of penciclovir representative of fed and fasted treatments were shown to be similar in efficacy (Pratt *et al.,* 1994b) and in phosphorylation studies it was shown that the rate of formation of penciclovir triphosphate is proportional both to the concentration of penciclovir outside the cell and to the incubation time (Vere Hodge and Perkins, 1989; Earnshaw *et al.,* 1992). Also, as penciclovir triphosphate has a long half-life in virus-infected cells, the antiviral effect is maintained between doses while the plasma penciclovir concentrations are low.

4.2. Efficacy

4.2.1. HERPES ZOSTER (SHINGLES)

There have been two large double-blind clinical trials of the efficacy of famciclovir in herpes zoster, one comparing famciclovir with placebo and the other with acyclovir. For both trials, famciclovir was dosed at a lower frequency (only three times daily) than acyclovir (standard therapy five times daily) as the preclinical studies had indicated that famciclovir should have a prolonged antiviral effect. Patients were enrolled within 72 hr of rash onset and the treatment was for 7 days. In these trials, the safety profile of famciclovir was similar to those of placebo and acyclovir (Saltzman *et al.,* 1994).

In the placebo-controlled study, the efficacy of famciclovir, 500 mg or 750 mg t.i.d., was demonstrated in reducing not only the duration of the acute symptoms of shingles but also of the long-term pain known as postherpetic neuralgia (PHN) (Tyring *et al.*, 1995). PHN is the most common complication of shingles and its incidence increases with age such that about one-half of patients more than 60 years old have this complication. PHN is clearly the most distressing aspect of shingles for both the patient and the physician.

Famciclovir recipients stopped shedding virus about two times faster than the placebo group, relative ratio (RR) 2.0 ($p = 0.0005$) for the 500-mg group and 2.3 ($p = 0.0001$) for the 750-mg group (Tyring *et al.*, 1995). In patients with severe rash at enrollment, the acute pain resolved faster than placebo in both famciclovir groups (RR = 2.9, $p = 0.003$; RR = 2.0, $p = 0.01$). However, the clinically most important result is that PHN resolved about two times faster in the famciclovir groups than placebo (RR = 1.9, $p = 0.01$; RR = 2.3, $p = 0.0007$ for the 500- and 750-mg groups, respectively), resulting in a 2-month reduction in the median duration of PHN. The effect is even more marked in older patients, age 50 years or older. In these patients, famciclovir reduced the duration of PHN by almost three times (RR = 2.6, $p = 0.004$; RR = 2.8, $p = 0.003$ for 500- and 750-mg groups, respectively), resulting in a 3.5-month reduction in the median duration of PHN (Fig. 6).

In the acyclovir controlled trial, three doses of famciclovir (250, 500, and 750 mg t.i.d.) were compared with acyclovir, 800 mg five times daily (Degreef *et al.*, 1994). When all patients are considered, resolution of zoster-associated pain, defined as pain from enrollment to last cessation of pain, occurred at a faster rate in

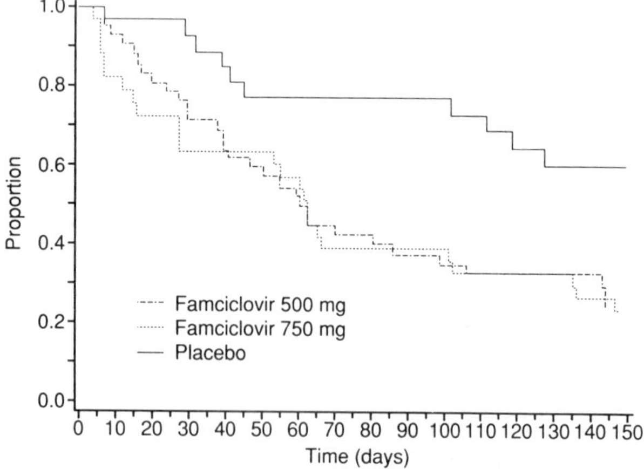

Figure 6. Time to resolution of postherpetic neuralgia for patients 50 years of age or older (Kaplan–Meier survival curves). Reproduced with permission from Tyring *et al.* (1995).

famciclovir-treated groups than in the acyclovir group, although this was significant only for the 500-mg famciclovir group. This difference was more apparent in patients treated within 48 hr of rash onset (Degreef *et al.,* 1994). When covariate-adjusted analysis was performed on the efficacy-evaluable population to take account of age, pain, and rash severity at presentation, factors that have been shown previously to affect outcome, loss of zoster-associated pain was significantly faster in all famciclovir groups compared with acyclovir (250 mg: 1.7 times faster, $p = 0.01$; 500 mg: 1.8 times faster, $p = 0.01$; 750 mg: 1.5 times faster, $p = 0.04$) (Carrington, 1996).

4.2.2. GENITAL HERPES

Primary genital herpes often causes severe symptoms with virus shedding lasting about 1 to 2 weeks and complete healing taking up to 3 weeks. Recurrent genital herpes is usually a milder disease with cessation of virus shedding and healing within a week. The standard approved acyclovir therapy has been with 200-mg doses five times daily. As for the clinical trials with famciclovir for zoster, the dosing frequency has been reduced, relative to acyclovir, to three times daily for the primary infection and to only twice a day for the recurrent episodes.

For first-episode genital herpes, famciclovir 250, 500, and 750 mg t.i.d. were compared with acyclovir 200 mg five times daily, each given for 5 days. All of the famciclovir groups were comparable to each other and to the acyclovir group in their effects on virus shedding, healing, and symptoms (Loveless *et al.,* 1995). The median time to stopping virus shedding was 2 to 3 days for famciclovir and 3 days for acyclovir.

There have been two similar double-blind placebo-controlled trials with famciclovir (125, 250, and 500 mg) for episodic therapy of recurrent genital herpes, one in which therapy was initiated in the clinic (Sacks *et al.,* 1994) and the other by the patient (Sacks *et al.,* 1996). In the clinic-initiated trial, famciclovir significantly reduced the duration of virus shedding, time to lesion healing, and the number of patients with new lesion formation compared with placebo. Additional significant benefits of famciclovir therapy included reduced time to loss of vesicles, ulcers, and crusts and in the relief of symptoms of tenderness, pain, and itching (Perry and Wagstaff, 1995). Similar results were obtained in the patient-initiated trial. Treatment was started after the patient had taken a swab for viral culture. Duration of virus shedding was significantly shorter in the famciclovir groups (Fig. 7). Among those not shedding virus at enrollment, significantly fewer went on to shed virus later and in these cases, famciclovir aborted virus shedding. Famciclovir therapy also resulted in a significantly shorter duration of lesion-associated symptoms and time to complete lesion healing than placebo. Thus, famciclovir, 125 mg given twice daily, is highly effective for the episodic therapy of genital herpes.

For patients with frequent episodes of recurrent genital herpes, suppressive

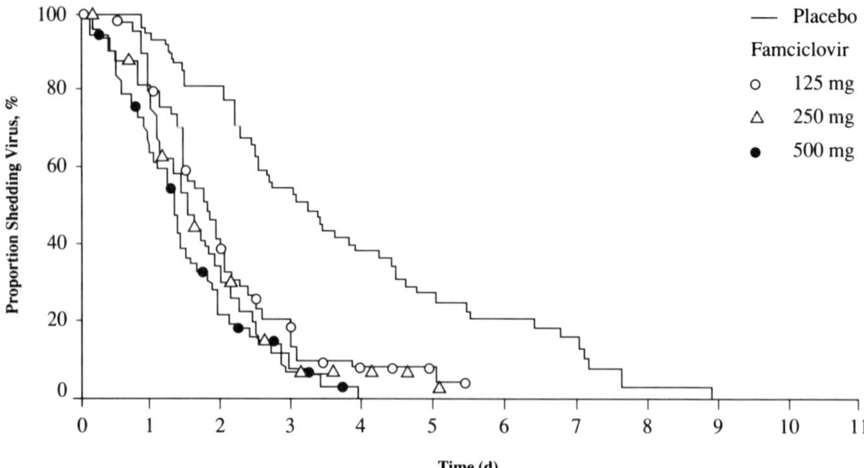

Figure 7. Time to cessation of virus shedding in genital herpes. Kaplan–Meier survival curves comparing famciclovir with placebo twice daily for 5 days. The symbols are for identification of the curves and do not represent data points. Reproduced with permission from Sacks *et al.* (1996).

antiviral therapy is preferred by patients. In a 4-month double-blind placebo-controlled trial, a range of doses of famciclovir were evaluated. Over the 4 months, virologically confirmed recurrences were seen in half of the patients on placebo but only 10% of those with famciclovir, 250 mg b.i.d. In a multicenter, placebo-controlled trial, patients with a history of frequently recurring genital herpes were treated with famciclovir (125 mg t.i.d., 250 mg b.i.d., 250 mg t.i.d.) or placebo for 1 year (Diaz-Mitoma *et al.,* 1996). The median time to the first virologically confirmed recurrence was 2.7 months for placebo and more than 1 year for the famciclovir groups. The time to the first clinically confirmed lesion episode was also significantly prolonged with famciclovir therapy, the median times being 8, 10, and 11 months, respectively, for the famciclovir groups compared with 1.5 months for placebo. These data demonstrate that long-term administration of famciclovir is an effective and well-tolerated treatment for the suppression of recurrent genital herpes.

A recent clinical observation suggests that famciclovir therapy of primary genital herpes may reduce the chance of recurrence (Ahmed and Woolley, 1996). This observation is consistent with the findings of Thackray and Field (1996a,b) discussed in Esection 3.1.

4.2.3. HERPES LABIALIS (COLD SORES)

Penciclovir as a 1% cream significantly shortened the time to loss of cold sore lesions, associated pain, and virus shedding in a double-blind placebo-controlled trial (Spruance, 1996). Patients who received penciclovir lost lesions 33% faster than placebo-treated patients. In addition, penciclovir conferred a benefit regardless of whether treatment had started "early" (during the prodrome or erythema stage) or "late" (during the papule, vesicle, or ulcer stages). A second trial gave similar results (Raborn *et al.*, 1996). Penciclovir cream is the first topical antiviral treatment to convincingly impact the clinical course of recurrent herpes labialis.

4.2.4. HEPATITIS B VIRUS

In liver transplant patients, HBV reinfection is a frequent and sometimes fatal complication. First experience in an HBV patient indicated that famciclovir was effective in suppressing viral replication of HBV (Boker *et al.*, 1994). In a compassionate use study, famciclovir (usually 500 mg t.i.d.) was well tolerated in patients treated for recurrent HBV after orthotopic liver transplantation (Kruger *et al.*, 1994, 1996). A median reduction in serum HBV DNA of 88% was observed in 10 of 11 patients within 8 weeks after starting famciclovir treatment. A median reduction of alanine aminotransferase (ALT) levels of 73% was observed in 8 of 11 patients. In the patient who has been under treatment for the longest time (112 weeks), the dose was tapered to 125 mg t.i.d., as HBV DNA is negative by the sensitive PCR assay, ALT levels are normal, and seroconversion to anti-HBe was noted. Long-term famciclovir treatment (over 1 year) is well tolerated in these immunocompromised patients.

Large-scale trials of famciclovir for the treatment of chronic HBV infection are ongoing.

5. CONCLUSION

Penciclovir is a highly selective antiviral agent with a prolonged mechanism of action against members of the herpesvirus family and against HBV. In the discovery phase, biochemical mode of action studies identified key differences to the existing therapy, acyclovir, which impacted the strategy of the subsequent clinical evaluation. A key issue for the research program was the identification of a suitable orally bioavailable prodrug. In addition to oral bioavailability studies in animals, studies of metabolism in human tissue and body fluids played a crucial role in the choice of the preferred prodrug, famciclovir. Clinical evaluation of the human pharmacokinetics of famciclovir fully confirmed its utility for the oral delivery of penciclovir. The good blood levels of penciclovir from oral administration of famciclovir combined with the prolonged mechanism of action result in a clinically effective agent with significant reductions in both size and frequency of dose relative to previously available therapy. Overall, an early and prolonged interaction of the research program with the needs of development, clinical investigators, and regulatory authorities greatly influenced the progression of penciclovir and famciclovir into clinical usage.

ACKNOWLEDGMENTS

The authors wish to thank their many colleagues who have made significant contributions to the work described in this chapter. We also wish to record our appreciation of all those who have progressed famciclovir and topical penciclovir from research compounds into widely used therapies.

REFERENCES

Ahmed, A., and Wooley, P. D., 1996, Comparison of famciclovir and aciclovir in first episode genital herpes: Possible clinical effect on latency (abstract), European Congress on STDs, Paris.

Ashton, R. J., Abbott, K. H., Smith, G. M., and Sutton, D., 1994, Antiviral activity of famciclovir and acyclovir in mice infected intraperitoneally with herpes simplex virus type 1 SC16, *J. Antimicrob. Chemother.* **34:**287–290.

Bacon, T. H., 1996, Famciclovir, from the bench to the patient—A comprehensive review of preclinical data, *Int. J. Antimicrob. Agents* **7:**119–134.

Bacon, T. H., and Boyd, M. R., 1995, Activity of penciclovir against Epstein-Barr virus, *Antimicrob. Agents Chemother.* **39:**1599–1602.

Bacon, T. H., and Howard, B. A., 1996, Further characterisation of the inhibition of herpes simplex virus replication in human cell lines by penciclovir and aciclovir, *Antiviral Chem. Chemother.* **7:**128–137.

Bacon, T. H., and Schinazi, R. F., 1993, An overview of the further evaluation of penciclovir against herpes simplex virus and varicella-zoster virus in cell culture highlighting contrasts with acyclovir, *Antiviral Chem. Chemother.* **4**(Suppl. 1)**:**25–36.

Bacon, T. H., Gilbart, J., Howard, B. A., and Standring-Cox, R., 1996a, Inhibition of varicella-zoster virus by penciclovir in cell culture and mechanism of action, *Antiviral Chem. Chemother.* **7:**71–78.

Bacon, T. H., Howard, B. A., Spender, L. C., and Boyd, M. R., 1996b, Activity of penciclovir in antiviral assays against herpes simplex virus, *J. Antimicrob. Chemother.* **37:**303–313.

Bailey, S., and Harnden, M. R., 1988, Analogues of the antiviral acyclonucleoside 9-(4-hydroxy-3-hydroxymethylbutyl)guanine. Part 2. Substitutions on C-1′ and C-3′ of the acyclic N-9 substituent, *J. Chem. Soc. Perkin Trans. 1* **1988:**2767–2775.

Bebault, G. M., Wall, R. A., Ronnie, B. A., and Sacks, S. L., 1995, Penciclovir (PCV) triphosphate (PCVTP) and acyclovir (ACV) triphosphate (ACVTP) in a human schwannoma continuous cell line (SW) infected with varicella zoster virus (VZV), *Can. J. Infect. Dis.* **6**(Suppl. C)**:**287C.

Boker, K. H., Ringe, B., Krüger, M., Pichlmayr, R., and Manns, P., 1994, Prostaglandin E plus famciclovir—A new concept for the treatment of severe hepatitis B after liver transplantation, *Transplantation* **57:**1706–1708.

Boyd, M. R., Bacon, T. H., Sutton, D., and Cole, M., 1987, Antiherpesvirus activity of 9-(4-hydroxy-3-hydroxymethylbut-1-yl)guanine (BRL 39123) in cell culture, *Antimicrob. Agents Chemother.* **31:**1238–1242.

Boyd, M. R., Bacon, T. H., and Sutton, D., 1988a, Antiherpesvirus activity of 9-(4-hydroxy-3-hydroxymethylbut-1-yl)guanine (BRL 39123) in animals, *Antimicrob. Agents Chemother.* **32:**358–363.

Boyd, M. R., Boon, R., Fowles, S. E., Pagano K., Sutton, D., Vere Hodge, R. A., and Zussman, B. D.,

1988b, Some biological properties of BRL 42810, a well absorbed oral prodrug of the antiherpesvirus agent BRL 39123, *Antiviral Res.* **9:**146.

Boyd, M. R., Safrin, S., and Kern, E. R., 1993, Penciclovir: A review of its spectrum of activity, selectivity and cross-resistance pattern, *Antiviral Chem. Chemother.* **4**(Suppl. 1)**:** 3–11.

Carrington, D., 1996, Reducing the duration of zoster-associated pain with famciclovir (abstract), in: *First European Congress of Chemotherapy,* Glasgow.

Chiou, H. C., Kumura, K., Hu, A., Kerns, K. M., and Coen, D. M., 1995, Penciclovir-resistance mutations in the herpes simplex virus DNA polymerase gene, *Antiviral Chem. Chemother.* **6:**281–288.

Clarke, S. E., Harrell, A. W., and Chenery, R. J., 1995, The role of aldehyde oxidase in the *in vitro* conversion of famciclovir to penciclovir in human liver, *Drug Metab. Dispos.* **23:**251–254.

Daniels, S., and Schentag, J. J., 1993, Drug interaction studies and safety of famciclovir in healthy volunteers: A review, *Antiviral Chem. Chemother.* **4**(Suppl. 1)**:**57–64.

Dannaoui, E., Trépo, C., and Zoulim, F., 1997, Inhibitory effect of penciclovir-triphosphate on duck hepatitis B virus reverse transcription, *Antimicrob. Agents Chemother.* **8:**38–46.

Datema, R., Ericson, A.-C., Field, H. J., Larsson, A., and Stenberg, K., 1987, Critical determinants of antiherpes efficacy of buciclovir and related acyclic guanosine analogs, *Antiviral Res.* **7:**303–316.

Degreef, H., and Famciclovir Herpes Zoster Clinical Study Group, 1994, Famciclovir, a new oral antiherpes drug: Results of the first controlled clinical study demonstrating its efficacy and safety in the treatment of uncomplicated herpes zoster in immunocompetent patients, *Int. J. Antimicrob. Agents* **4:**241–246.

de Miranda, P., and Blum, M. R., 1983, Pharmacokinetics of acyclovir after intravenous and oral administration, *J. Antimicrob. Chemother.* **12**(Suppl. B)**:**29–37.

de Miranda, P., and Good, S. S., 1992, Species differences in the metabolism and disposition of antiviral nucleoside analogues: 1. Acyclovir, *Antiviral Chem. Chemother.* **3:**1–8.

de Miranda, P., Krasny, H. C., Page, D. A., and Elion, G. B., 1981, The disposition of acyclovir in different species, *J. Pharmacol. Exp. Ther.* **219:**309–315.

Diaz-Mitoma, F., Sibbald, R. G., Shafran, S. D., and the Collaborative Famciclovir Genital Herpes Suppression Group, 1996, Famciclovir suppression of recurrent genital herpes, in: *Abstracts of the 36th Interscience Conference on Antimicrobial Agents and Chemotherapy,* New Orleans, American Society for Microbiology.

Earnshaw, D. L., and Vere Hodge, R. A., 1992, Effective inhibition of herpesvirus DNA synthesis by (*S*)-penciclovir-triphosphate, in: *Program and Abstracts of the 32nd Interscience Conference on Antimicrobial Agents and Chemotherapy,* Anaheim, American Society for Microbiology.

Earnshaw, D. L., Bacon, T. H., Darlison, S. J., Edmonds, K., Perkins, R. M., and Vere Hodge, R. A., 1992, Mode of antiviral action of penciclovir in MRC-5 cells infected with herpes simplex virus type 1 (HSV-1), HSV-2 and varicella-zoster virus, *Antimicrob. Agents Chemother.* **36:**2747–2757.

Ertl, P., Snowden, W., Lowe, D., Miller, W., Collins, P., and Littler, E., 1995, A comparative study of the *in vitro* and *in vivo* antiviral activities of acyclovir and penciclovir, *Antiviral Chem. Chemother.* **6:**89–97.

Field, H. J., and Thackray, A. M., 1995, The effects of delayed-onset chemotherapy using famciclovir or valaciclovir in a murine immunosuppression model for HSV-1, *Antiviral Chem. Chemother.* **6:**210–216.

Field, H. J., Tewari, D., Sutton, D., and Thackray, A. M., 1995, Comparison of efficacies of famciclovir and valaciclovir against herpes simplex virus type 1 in a murine immunosuppression model, *Antimicrob. Agents Chemother.* **39:**1114–1119.

Fowles, S. E., Pierce, D. M., Prince, W. T., and Thow, J. C., 1990, Effect of food on the bioavailability and pharmacokinetics of penciclovir, a novel antiherpes agent, following oral administration of the pro-drug, famciclovir, *Br. J. Clin. Pharmacol.* **29:**620P–621P.

Fowles, S. E., Fairless, A. J., Pierce, D. M., and Prince, W. T., 1991, A further study on the effect of food on the bioavailability and pharmacokinetics of penciclovir after oral administration of famciclovir, *Br. J. Clin. Pharmacol.* **32:**657P.

Fowles, S. E., Pierce, D. M., Prince, W. T., and Staniforth, D., 1992, The tolerance to and pharmacokinetics of penciclovir (BRL 39123A), a novel antiherpes agent, administered by intravenous infusion to healthy subjects, *Eur. J. Clin. Pharmacol.* **43:**513–516.

Fowles, S. E., Pratt, S. K., Laroche, J., and Prince, W. T., 1994, Lack of a pharmacokinetic interaction between oral famciclovir and allopurinol in healthy volunteers, *Eur. J. Clin. Pharmacol.* **46:**355–359.

Geen, G. R., Harnden, M. R., and Parratt, M. J., 1991, Synthesis of 9-[2,2-bis(hydroxymethyl)cycloprop-1-yl]guanine as a potential antiviral agent, *BioMed. Chem. Lett.* **1:**347–348.

Goldthorpe, S. E., Boyd, M. R., and Field, H. J., 1992, Effects of penciclovir and famciclovir in a murine model of encephalitis induced by intranasal inoculation of herpes simplex virus type 1, *Antiviral Chem. Chemother.* **3:**37–47

Harnden, M. R., and Jarvest, R. L., 1985, An improved synthesis of the antiviral acyclonucleoside 9-(4-hydroxy-3-hydroxymethylbut-1-yl)guanine, *Tetrahedron Lett.* **26:**4265–4268.

Harnden, M. R., and Jarvest, R. L., 1988a, Analogues of the antiviral acyclonucleoside 9-(4-hydroxy-3-hydroxymethylbutyl)guanine. Part 3. Modification of a 3′-hydroxymethyl group, *J. Chem. Soc. Perkin Trans. 1* **1988:**2777–2784.

Harnden, M. R., and Jarvest, R. L., 1988b, Synthesis of 9-(3-hydroxyalkylamino)guanines, novel antiviral acyclonucleosides, *Tetrahedron Lett.* **29:**5995–5998.

Harnden, M. R., and Jarvest, R. L., 1989, Analogues of the antiviral acyclonucleoside 9-(4-hydroxy-3-hydroxymethylbutyl)guanine. Part 4. Substitution on the 2-amino group, *J. Chem. Soc. Perkin Trans. 1* **1989:**2207–2213.

Harnden, M. R., Jarvest, R. L., Bacon, T. H., and Boyd, M. R., 1987, Synthesis and antiviral activity of 9-[4-hydroxy-3-(hydroxymethyl)but-1-yl]purines, *J. Med. Chem.* **30:**1636–1642.

Harnden, M. R., Parkin, A., and Wyatt, P. G., 1988a, Analogues of the antiviral acyclonucleoside 9-(4-hydroxy-3-hydroxymethylbutyl)guanine. Part 1. Substitution on C-2′ of the acyclic N-9 substituent, *J. Chem. Soc. Perkin Trans. 1* **1988:**2757–2765.

Harnden, M. R., Parkin, A., and Wyatt, P. G., 1988b, Synthesis of 9-(3-hydroxypropoxy)guanine, a novel antiviral acyclonucleoside, *Tetrahedron Lett.* **29:**701–704.

Harnden, M. R., Jarvest, R. L., Boyd, M. R., Sutton, D., and Vere Hodge, R. A., 1989, Prodrugs of the selective antiherpesvirus agent 9-(4-hydroxy-3-(hydroxymethyl)-but-1-yl)guanine (BRL 39123) with improved gastrointestinal absorption properties, *J. Med. Chem.* **32:**1738–1743.

Harnden, M. R., Jarvest, R. L., Slawin, A. M. Z., and Williams, D. J., 1990a, Crystal and molecular structure of the antiviral acyclonucleoside 9-[4-hydroxy-3-(hydroxymethyl)butyl]guanine (BRL 39123, penciclovir) and its prodrug 9-[4-acetoxy-3-(acetoxymethyl)butyl]-2-aminopurine (BRL 42810, famciclovir), *Nucleosides Nucleotides* **9:**499–513.

Harnden, M. R., Wyatt, P. G., Boyd, M. R., and Sutton, D., 1990b, Synthesis and antiviral activity of 9-alkoxypurines. 1. 9-(3-hydroxypropoxy)- and 9-[3-hydroxy-2-hydroxymethyl)propoxy]-purines, *J. Med. Chem.* **33:**187–196.

Ilsley, D. D., Lee, S.-K., Miller, W. H., and Kuchta, R. D., 1995, Acyclic guanosine analogues inhibit DNA polymerases α, δ, and ϵ with very different properties and have unique mechanisms of action, *Biochemistry* **34:**2504–2510.

Jacobson, M. A., de Miranda, P., Cederberg, D. M., Burnett, T., Cobb, E., Brodie, H. R., and Mills, J., 1987, Human pharmacokinetics and tolerance of oral ganciclovir, *Antimicrob. Agents Chemother.* **31:**1251–1254.

Jarvest, R. L., Barnes, R. D., Earnshaw, D. L., O'Toole, K. J., Sime, J. T., and Vere Hodge, R. A., 1990, Synthesis of isotopically chiral [^{13}C] penciclovir (BRL 39123) and its use to determine the absolute configuration of penciclovir triphosphate formed in herpes virus infected cells, *J. Chem. Soc. Chem. Commun.* **1990:**555–556.

Karlson, A. H. J., Harmenberg, J. G., and Wahren, B. E., 1986, Influence of acyclovir and bucyclovir on nucleotide pools in cells infected with herpes simplex virus type 1, *Antimicrob. Agents Chemother.* **29:**821–824.

Korba, B. E., and Boyd, M. R., 1996, Penciclovir is a selective inhibitor of hepatitis B virus replication in cultured human hepatoblastoma cells, *Antimicrob. Agents Chemother* **40:**1282–1284.

Krenitsky, T. A., Hall, W. W., de Miranda, P., Beauchamp, L. M., Schaeffer, H. J., and Whiteman, P. D., 1984, 6-Deoxyacyclovir: A xanthine oxidase-activated prodrug of acyclovir, *Proc. Natl. Acad. Sci. USA* **81:**3209–3213.

Krüger, M., Tillmann, H. L., Trautwein, C., Bode, V., Oldhafer, K., Boker, K. H. W., Pichlmayr, R., and Manns, M. P., 1994, Treatment of hepatitis B virus reinfection after liver transplantation with famciclovir, in: *45th Meeting of the American Association for the Study of Liver Diseases,* Chicago (Poster).

Krüger, M., Tillmann, H. L., Trautwein, C., Bode, V., Oldhafer, K., Maschek, H., Boker, H., Broelsch, C. E., Pichlmayr, R., and Manns, M. P., 1996, Famciclovir treatment of hepatitis B virus recurrence after liver transplantation: A pilot study, *Liver Transplant. Surg.* **2:**253–262.

Larsson, A., Stenberg, A.-C., Ericson, U., Haglund, W.-A., Yisak, N.-G., Johansson, B., Öberg, B., and Datema, R., 1986, Mode of action, toxicity, pharmacokinetics and efficacy of some new antiherpesvirus guanosine analogs related to buciclovir, *Antimicrob. Agents Chemother.* **30:**598–605.

Lin, E., Luscombe, C., Wang, Y. Y., Shaw, T., and Locarnini, S., 1996, The guanine nucleoside analogue penciclovir is active against chronic duck hepatitis B virus infection *in vivo, Antimicrob. Agents Chemother.* **2:**413–418.

Loveless, M., Harris, W., and Sacks, S., 1995, Treatment of first episode genital herpes with famciclovir, in: *Abstracts of the 35th Interscience Conference on Antimicrobial Agents and Chemotherapy,* San Francisco, American Society for Microbiology.

Lowe, D. M., Alderton, W. K., Ellis, M. R., Parmar, V., Miller, W. H., Roberts, A. B., Fyfe, J. A., Gaillard, R., Ertl, P., Snowden, W., and Littler, E., 1995, Mode of action of (R)-9-[4-hydroxy-2-(hydroxymethyl)butyl]guanine against herpesvirus, *Antimicrob. Agents Chemother.* **39:**1802–1808.

MacCoss, M., Tolman, R. L., Ashton, W. T., Wagner, A. F., Hannah, J., Field, A. K., Karkas, J. D., and Germershausen, J. I., 1986, Synthetic, biochemical and antiviral aspects of selected acyclonucleosides and their derivatives, *Chem. Scri.* **26:**113–121.

Mok, S. S., Shaw, T., and Locarnini, S., 1995, Preferential inhibition of hepatitis B virus (HBV) DNA polymerase by the (R)-enantiomer of penciclovir triphosphate, in: *Abstracts of the 35th Interscience Conference on Antimicrobial Agents and Chemotherapy,* San Francisco, American Society for Microbiology.

Perry, C. M., and Wagstaff, A. J., 1995, Famciclovir, a review of its pharmacological properties and therapeutic efficacy in herpes virus infection, *Drugs* **50:**396–415.

Pratt, S. K., Fowles, S. E., Pierce, D. M., and Prince, W. T., 1991, An investigation of the interaction between cimetidine and famciclovir in non-patient volunteers, *Br. J. Clin. Pharmacol.* **32:**656P–657P.

Pratt, S. K., Pue, M. A., Fairless, A J., Fowles, S. E., Laroche, J., Bygate, E., Glenny, H., Daniels, S., and Freedman, P. S., 1993, The pharmacokinetics of penciclovir following intravenous administration to healthy elderly subjects, in: *Program and Abstracts of the 33rd Interscience Conference on Antimicrobial Agents and Chemotherapy,* New Orleans, American Society of Microbiology.

Pratt, S. K., Pue, M. A., Fairless, A. J., Fowles, S. E., Laroche, J., Kumar, R., and Prince, W. T., 1994a, Lack of an effect of gender on the pharmacokinetics of penciclovir, following single oral doses of famciclovir, *Br. J. Clin. Pharmacol.* **37:**493P.

Pratt, S. K., Standring-Cox, R., Writer, D., Brooks, S., Fowles, S. E., Fiala, S., Schubert, C., and Hust, R., 1994b, Penciclovir pharmacokinetics in fed and fasted subjects following oral famciclovir in relation to in-vitro antiviral activity, in: *Book of Abstracts of the 6th International Congress for Infectious Diseases,* Prague.

Pue, M. A., and Benet, L. Z., 1993, Pharmacokinetics of famciclovir in man, *Antiviral Chem. Chemother.* **4**(Suppl. 1)**:**47–55.

Pue, M. A., Pratt, S. K., Fairless, A. J., Fowles, S., Laroche, J., Georgiou, P., and Prince, W., 1994, Linear pharmacokinetics of penciclovir following administration of single oral doses of famciclovir 125, 250, 500 and 750 mg to healthy volunteers, *J. Antimicrob. Chemother.* **33:**119–127.

Raborn, G. W., and the Penciclovir Topical Collaborative Study Group, 1996, Penciclovir cream for recurrent herpes simplex labialis: An effective new treatment, in: *Abstracts of the 36th Interscience Conference on Antimicrobial Agents and Chemotherapy,* New Orleans, American Society for Microbiology.

Reardon, J. E., and Spector, T., 1989, Herpes simplex virus type 1 DNA polymerase. Mechanism of inhibition by acyclovir triphosphate, *J. Biol. Chem.* **264:**7405–7111.

Sacks, S. L., Bebault, G. M., Rennie, B. A., Wall, R. A., Vere Hodge, R. A., and Strauss, S. E., 1994, Virus-specified phosphorylation of penciclovir in a human schwannoma continuous cell line infected with varicella zoster virus, in: *Abstracts of the 34th Interscience Conference on Antimicrobial Agents and Chemotherapy,* Orlando, American Society for Microbiology.

Sacks, S. L., Aoki, F. Y., Diaz-Mitoma, F., Sellors, J., and Shafran, S. D., 1996, Patient initiated, twice daily oral famciclovir for early recurrent genital herpes: A Canadian multicenter trail, *J. Am. Med. Assoc.* **276:**44–49.

Saltzman, R., Jurewicz, R., and Boon, R., 1994, The safety of famciclovir in patients with herpes zoster virus and genital herpes virus, *Antimicrob. Agents Chemother.* **38:**2454–2457.

Shaw, T., Amor, P., Civitico, G., Boyd, M., and Locarnini, S., 1994, In vitro antiviral activity of penciclovir, a novel purine nucleoside, against duck hepatitis B virus, *Antimicrob. Agents Chemother.* **38:**719–723.

Sime, J. T., Barnes, R. D., Elson, S. W., Jarvest, R. L., and O'Toole, K. J., 1992, Chemoenzymatic approach to the synthesis of the antiviral agents penciclovir and famciclovir in isotopically chiral [^{13}C] labelled form, *J. Chem. Soc. Perkin Trans. I* **1992:**1653–1658.

Spruance, S. L., 1996, Penciclovir cream: A new and effective treatment for recurrent herpes simplex labialis (abstract), in: *Congress of Clinical Dermatology 2000,* Vancouver, Canada.

Sutton, D., and Boyd, M. R., 1993, Comparative activity of penciclovir and acyclovir in mice infected intraperitoneally with herpes simplex virus type 1 SC16, *Antimicrob. Agents Chemother.* **37:**642–645.

Sutton, D., and Kern, E. R., 1993, Activity of famciclovir and penciclovir in HSV-infected animals: A review, *Antiviral Chem. Chemother.* **4**(Suppl. 1)**:**37–46.

Talarico, C. L., Phelps, W. C., and Biron, K. K., 1993, Analysis of the thymidine kinase genes from acyclovir-resistant mutants of varicella zoster virus isolated from patients with AIDS, *J. Virol.* **67:**1024–1033.

Thackray, A. M., and Field, H. J., 1996a, Differential effects of famciclovir and valaciclovir on the pathogenesis of herpes simplex virus in a murine infection model including reactivation from latency, *J. Infect. Dis.* **173:**291–299.

Thackray, A. M., and Field, H. J., 1996b, Comparison of the effects of famciclovir and valaciclovir on the pathogenesis of herpes simplex virus type 2 in a murine infection model, *Antimicrob. Agents Chemother.* **40:**846–851.

Tippie, M. A., Martin, J. C., Smee, D. F., Matthews, T. R., and Verheyden, J.P.H., 1984, Antiherpes simplex virus activity of 9-[4-hydroxy-3-(hydroxymethyl)-1-butyl]guanine, *Nucleosides Nucleotides* **3:**525–535.

Tsiquaye, K. N., Slomka, M. J., and Maung, M., 1994, Oral famciclovir against duck hepatitis B virus replication in hepatic and nonhepatic tissues of ducklings infected in ovo, *J. Med. Virol.* **42:**306–310.

Tsiquaye, K. N., Sutton, D., Maung, M., and Boyd, M. R., 1996, Antiviral activities and pharmacokinetics of penciclovir and famciclovir in Peking ducks chronically infected with duck hepatitis virus, *Antiviral Chem. Chemother.* **7:**153–159.

Tyring, S., Barbarash, R. A., Nahlik, J., *et al.,* 1995, Famciclovir for the treatment of acute herpes zoster: Effects on acute disease and postherpetic neuralgia. A randomized, double-blind, placebo-controlled trial, *Ann. Intern. Med.* **123:**89–96.

Vere Hodge, R. A., 1993, Famciclovir and penciclovir. The mode of action of famciclovir including its conversion to penciclovir, *Antiviral Chem. Chemother.* **4:**67–84.

Vere Hodge, R. A., and Cheng, Y.-C., 1993, Mode of action of penciclovir, *Antiviral Chem. Chemother.* **4**(Suppl. 1):13–24.

Vere Hodge, R. A., and Perkins, R. M., 1989, Mode of action of 9-(4-hydroxy-3-hydroxymethylbut-1-yl)guanine (BRL 39123) against herpes simplex virus in MRC-5 cells, *Antimicrob. Agents Chemother.* **33:**223–229.

Vere Hodge, R. A., Sutton, D., Boyd, M. R., Harnden, M. R., and Jarvest, R. L., 1989, Selection of an oral prodrug (BRL 42810; famciclovir) for the antiherpesvirus agent BRL 39123 [9-(4-hydroxy-3-(hydroxymethyl)but-1-yl)guanine; penciclovir], *Antimicrob. Agents Chemother.* **33:**1765–1773.

Vere Hodge, R. A., Darlison, S. J., Earnshaw, D. L., and Readshaw, S. A., 1993a, Use of isotopically chiral [4′-^{13}C]penciclovir and ^{13}C NMR to determine the specificity and absolute configuration of penciclovir phosphate esters formed in HSV-1 and HSV-2 infected cells and by HSV-1 encoded thymidine kinase, *Chirality* **5:**583–588.

Vere Hodge, R. A., Darlison, S. J., Earnshaw, D. L., and Readshaw, S. A., 1993b, Use of isotopically chiral [4′-^{13}C]famciclovir and ^{13}C NMR to identify the chiral monoacetylated intermediates in the conversion of famciclovir to penciclovir by human intestinal wall extract, *Chirality* **5:**577–582.

Weller, S., Blum, M. R., Doucette, M., Burnette, T., Cederberg, D. M., de Miranda, P., and Smiley, M. L., 1993, Pharmacokinetics of the acyclovir pro-drug valaciclovir after escalating single- and multiple-dose administration to normal volunteers, *Clin. Pharmacol. Ther.* **54:**595–605.

Yamaoka, Y., Roberts, R. D., and Stella, V. J., 1983, Low-melting phenytoin prodrugs as alternative oral delivery modes for phenytoin: A model for other high-melting sparingly water-soluble drugs, *J. Pharm. Sci.* **72:**400–405.

Chapter 15

The Use of Esters as Prodrugs for Oral Delivery of β-Lactam Antibiotics

Linda Mizen and George Burton

1. INTRODUCTION

Orally administered antibiotics are particularly suitable for community practice, and there are a number of β-lactam antibiotics that demonstrate satisfactory oral bioavailability in their own right including for example the penicillins, amoxicillin, penicillin V, cloxacillin, and nafcillin (Bergan, 1978), and the cephalosporins cefaclor and cephalexin (Brogard *et al.,* 1978). However, over the years there has been a continuing need for the development of new β-lactam antibiotics to increase potency, to broaden the spectrum of antibacterial activity, and to combat the development of bacterial resistance. In this search, many excellent injectable β-lactams have been discovered and developed for the clinic. These include carbenicillin and ticarcillin, ureido penicillins, temocillin (Hampel *et al.,* 1985), the carbapenems, monobactams, penems (Bergan, 1978), the trinem (DiModugno *et al.,* 1994), and many cephalosporins including cefotaxime, cefuroxime, ceftazidime, and ceftriaxone (Brogard *et al.,* 1978). A number of injectable antibiotics have displayed limited oral bioavailability and one approach to increase this absorption has been to develop esters that can enhance uptake from the gastrointestinal tract.

The development of prodrug esters of these antibiotics has not been straightforward and not all β-lactam antibiotics are amenable to the prodrug approach as a result either of physicochemical properties of the intact ester (Ferres, 1983) or

Linda Mizen and George Burton • SmithKline Beecham Pharmaceuticals, Collegeville, Pennsylvania 19426-0989.

Integration of Pharmaceutical Discovery and Development: Case Studies, edited by Borchardt *et al.,* Plenum Press, New York, 1998.

of the pharmacokinetic properties of the parent β-lactam. This latter point is exemplified by the observations of Mizen *et al.* (1995) for the (*Z*)-alkyloxyimino penicillins. The bioavailability of these penicillins was poor after oral administration, but it was found that this was related to extensive biliary clearance rather than poor uptake from the gastrointestinal tract.

Instead of focusing on one particular β-lactam antibiotic and the development of a suitable prodrug, this chapter provides an overview of what could be regarded as a series of case histories, as this approach to the enhancement of the oral bioavailability of β-lactam antibiotics has been in use for many years and is similar for most compounds. The development of ester prodrugs is a multidisciplinary process requiring expertise in microbiology, physiology and pharmacology, biochemistry, organic chemistry and last but not least pharmaceutical science. Go/no-go decisions are required throughout the process and the ability to make the right decisions has always depended on the extent of the predictability of testing methods.

Studies in laboratory animals are of value in all selection processes at the research stage. However, species differences in pharmacokinetics can exist not only for the intact prodrug ester and its conversion to the active antibiotic but also for the parent β-lactam. The prodrug esters listed in Tables I and II are not all marketed despite promising results for oral bioavailability in preclinical studies. Clinically acceptable bioavailability is based on concentrations achieved in human serum and their duration relative to minimum inhibitory concentrations (MIC) against the bacterial pathogens important for the identified clinical indication. In general, penicillins are at their most effective when concentrations in serum exceed MIC values for around 40% of the dosing interval (Craig, 1995). Increasing the dose or altering formulations are obvious ways to improve results. However,

Table I
Examples of Penicillin, Penem, and Trinem Ester Prodrugs

Prodrug ester	Antibiotic	Ester group	Reference
Penicillins			
pivampicillin	ampicillin	pivaloyloxymethyl	Sjovall *et al.* (1978)
talampicillin		phthalidyl	Clayton *et al.* (1976)
bacampicillin		ethoxycarbonyloxymethyl	Sjovall *et al.* (1978)
lenampicillin		daloxylate	Sakamoto *et al.* (1984)
pivmecillinam	mecillinam	pivaloyloxymethyl	Parsons *et al.* (1977)
carfecillin	carbenicillin	phenyl	Clayton *et al.* (1975)
carindacillin		indanyl	Butler *et al.* (1973)
Penems			
FCE 22891	FCE 22101	acetoxymethyl	Webberley *et al.* (1988)
—	CP-65,207	pivaloyloxymethyl	Gootz *et al.* (1990)
Trinem			
GV 118819	trinem (GV 104326)	1-cyclohexyloxycarbonyl oxyethyl (hexetil)	DiModugno *et al.* (1994)

Table II
Examples of Cephalosporin Prodrug Esters

Prodrug ester	Cephalosporin	Ester group	Reference
—	cephaloglycine	acetoxymethyl pivaloyloxymethyl	Binderup *et al.* (1971)
cefpodoxime proxetil	cefpodoxime	1-(isopropoxycarbonyloxy)ethyl (proxetil)	Frampton *et al.* (1992)
cefetamet pivoxil (® Globocef)	cefetamet	pivaloyloxymethyl	Stoeckel *et al.* (1989)
cefuroxime axetil	cefuroxime	1-(acetoxy)ethyl	Harding *et al.* (1984)
cefteram pivoxil (T2588) (® Tomiron)	cefteram (T2525)	pivaloyloxymethyl	Goto *et al.* (1986)
cefdaloxime pentexil (HR 916K)	cefdaloxime (RU 29246)	1-(pivaloyloxy)ethyl	Mendes *et al.* (1992)
ceftrazonal (Ro 41-3399)	Ro 40-6890	isobutoxycarbonylpentenyl	Weber *et al.* (1992), Angehrn *et al.* (1992)
cefcamate pivoxil (S1108)	cefcamate (S1006)	pivaloyloxymethyl	Nakashima *et al.* (1989), Matsuura *et al.* (1989)
E1100	E1101	1-(isopropoxycarbonyloxy)ethyl	Nakashima *et al.* (1994)

cost of goods and patient compliance are important considerations here, together with potential safety issues.

It is therefore important to consider a number of factors at the preclinical stage when developing prodrug esters. The properties of the intact prodrug ester, including enzymatic hydrolysis, stability, lipid and aqueous solubility, together with species differences in metabolism, are important considerations in the research and development of prodrug esters.

This overview is based on our experiences and those of others in the literature. An understanding of the information that has accumulated in this area of research is important for the future of prodrug esters and the rewards from successful development are such that these challenges need to be overcome.

2. CHEMICAL OVERVIEW

In the early history of penicillin, efforts to improve oral absorption were directed at increasing the lipophilicity of penicillin G by esterification of the C-3 carboxylic acid moeity (Richardson *et al.*, 1945). However, simple alkyl and arylalkyl esters of penicillin (e.g., **1**) were found to be devoid of antibacterial activity *in vitro* and in human where, unlike in rats and mice, esterases of the necessary specificity to hydrolyze the esters to the active parent are absent (Barnden *et al.*, 1953). In an alternative explanation for the failure of simple esters in human, Neilsen and Bundgaard (1988) suggested the presence of an enzyme in human serum that attacked the β-lactam ring rather than the ester function of these penicillin esters.

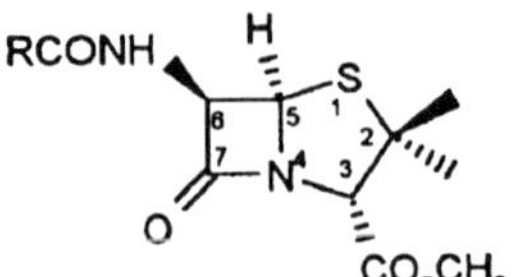

One early discovery, which demonstrated that some penicillin esters could be hydrolyzed, was the dialkylaminoethyl esters (e.g., **2**). These underwent nonenzymatic hydrolysis in the gastrointestinal tract and were not suitable for oral drug delivery, although they found use as parenteral agents (Ungar and Muggleton, 1952) as they demonstrated improved distribution in tissues. The search continued for suitable esters that were stable to nonenzymatic hydrolysis, but labile to hydrolytic enzymes in humans. This resulted in the methylenedioxy diesters or "*gem*-diol double esters" (**3**) (Jansen and Russell, 1965), which overcame the steric hindrance around the carboxylic acid of the penicillins and the steric and electronic effects in cephalosporins, penems, and carbapenems. The double esters (A) achieve this with a second ester function remote from the β-lactam that is susceptible to the required enzyme attack. Hydrolysis of this second ester gives the carboxylic acid (B) and the intrinsically unstable hydroxymethyl ester of the β-lactam (C) which collapses to the free antibiotic (D) and formaldehyde (E) (Agersborg *et al.*, 1966; Clayton *et al.*, 1976).

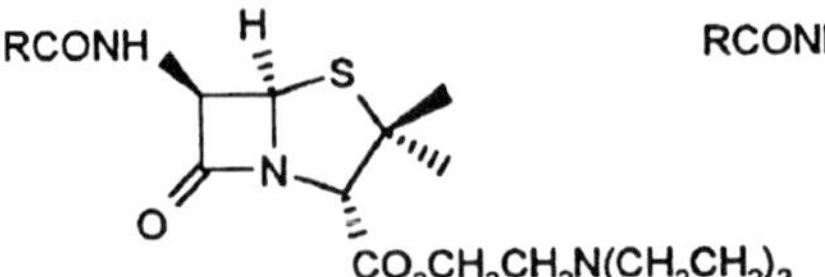

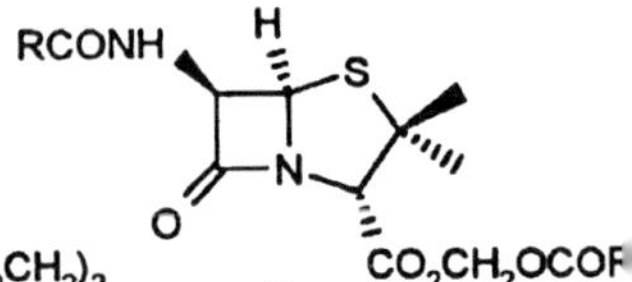

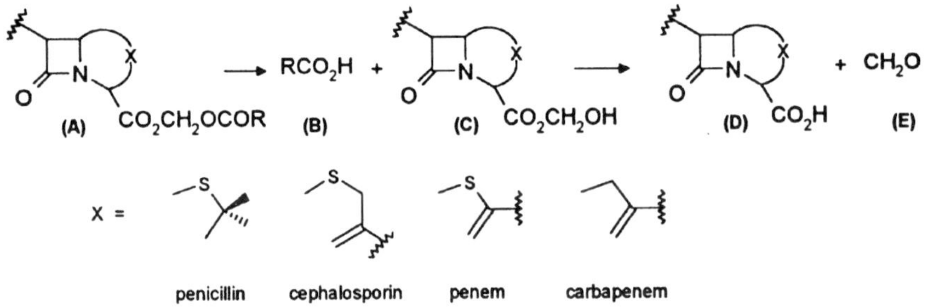

From the examination (Daehne *et al.*, 1970) of a range of these double esters, pivampicillin (**4**), the pivaloyloxymethyl ester of ampicillin, was identified, which was absorbed far more efficiently than ampicillin itself and was marketed by Leo Pharmaceuticals in 1972 (Sjovall *et al.*, 1978).

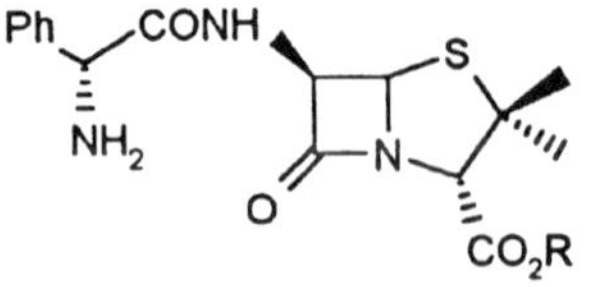

4 R = $CH_2OCOC(CH_3)_3$

5 R =

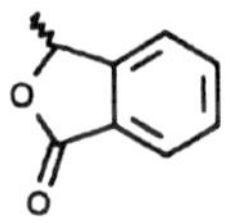

6 R = $CH(CH_3)OCO_2CH_2CH_3$

The success of pivampicillin stimulated much interest in the use of these prodrugs and further structural variations rapidly appeared. Talampicillin (**5**) (Clayton *et al.,* 1974), the phthalide ester, in which the carboxylic acid of the R group is cyclized onto the linking methylene, was marketed by Beecham Pharmaceuticals in 1975, and bacampicillin (**6**) (Bodin *et al.,* 1975), the ethoxycarbonyloxyethyl ester, by Astra Pharmaceuticals in 1977. The two latter esters are both examples in which the linking methylene is substituted and so have the disadvantage of existing as mixtures of diastereomers, each with the potential for different oral bioavailabilities.

These three esters of ampicillin (**4, 5,** and **6**) serve as examples of the main types of esters used for the β-lactam antibiotics (Tables I and II). Two other structures are worthy of mention, the cyclic carbonate (**7**) or daloxate ester (Sakamoto *et al.,* 1984) and the 2-(alkyloxycarbonyl)-2-alkylideneethyl esters (**8**) described by workers from F. Hoffmann-La Roche Ltd. (Hubschwerlen *et al.,* 1992).

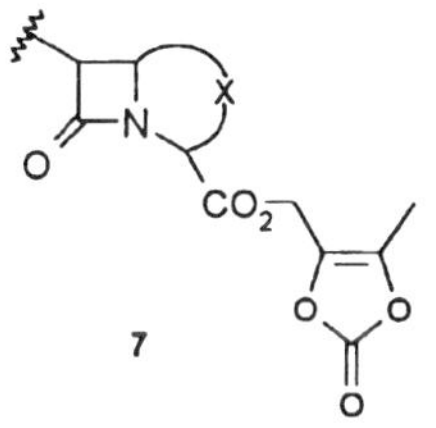

X N O CO_2 O O O 7

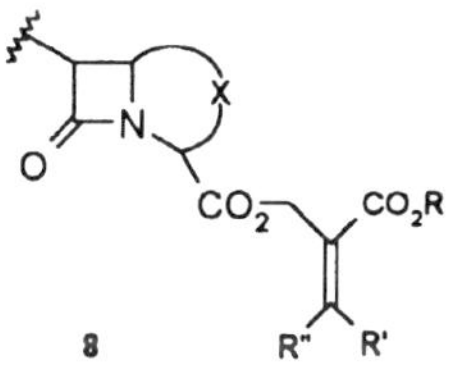

X N O CO_2 CO_2R R" R' 8

Esters of cephalosporins (**9**), unlike the acids, have the potential to isomerize to the antibacterially inactive Δ-2 isomers *in vivo* (**10**) (Cocker *et al.,* 1966). This isomerization increases with increasing pH (Saab *et al.,* 1988; Richter *et al.,* 1990) and is also influenced by substitution at the C-3 position (Miyauchi *et al.,* 1989). This represents a further factor that may be responsible for species differences in the levels of active cephalosporins found *in vivo* after dosing prodrug esters.

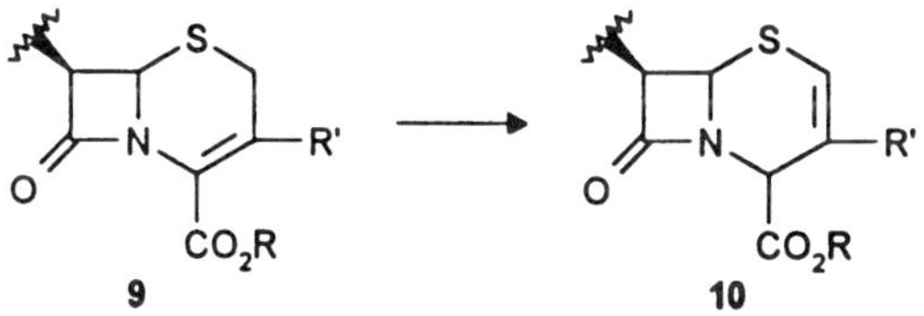

S N R' O CO_2R 9 → S N R' O CO_2R 10

Carbenicillin (**11**), an important penicillin analogue introduced in the 1960s with bioactivity against *Pseudomonas aeruginosa* and the indole-positive *Proteus* (Bergan, 1978), is administered parenterally because of its poor gastrointestinal absorption. Additionally, it is rendered inactive in the gastric contents as a result of acid lability associated with the C-6 acylamido group; therefore, prodrug approaches concentrated on the side chain rather than the C-3 carboxylic acid.

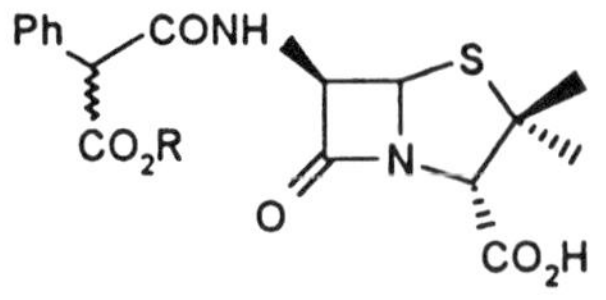

11 R = H

12 R =

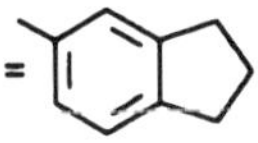

13 R = Ph

In contrast to the C-3 carboxylic acid, it was demonstrated here that simple alkyl and aryl esters of the carboxylic acid in the C-6 acylamido side chain give acid-stable esters that are orally absorbed and enzymatically hydrolyzed in humans. As a result of this approach, two derivatives, carindacillin (Butler *et al.*, 1973) (**12**) and carfecillin (Clayton *et al.*, 1975) (**13**), were marketed as oral forms of carbenicillin for the treatment of urinary tract infections.

3. ANIMAL BIOAVAILABILITY STUDIES AND SELECTION

Laboratory animal species, usually the mouse or rat, figure significantly in the earlier stages of selection of a prodrug ester auxiliary because of their small size relative to the availability of compound, and selection is based on comparative bioavailability by the oral route. The determination of absolute oral bioavailability, i.e., expressing the area under plasma level curve (AUC) after oral administration as a percentage of that after intravenous dosing of the parent antibiotic, is not common as these intravenous data are not often available. The percentage of dose recovered in urine after oral administration has been used for comparison by some (e.g., Daehne *et al.*, 1970; Fujimoto *et al.*, 1987), but this figure is not a true measure of absolute oral bioavailability unless compared with urinary recovery data of the parent after intravenous dosing. The comparison of the rank order of concentrations in serum or blood either as peak concentration or as AUC has been the most frequently used method to assess the clinical potential of prodrug esters. For studies in laboratory animals to be truly predictive, therefore, the rank order of antibiotic concentrations in serum would need to be the same in animals and human. Examples below will illustrate that this is seldom the case. Species differences in absorption, metabolism, and excretion can be significant factors in experimental studies (Mizen and Woodnutt, 1986), and the different dosing vehicles and dose levels that are used in experimental studies may influence results. In order to assess the influence of species differences in prodrug ester evaluation, data from the literature and from our own studies measuring the oral bioavailabilities for ester prodrugs of penicillins, cephalosporins, penems, and the trinem have been utilized.

3.1. Penicillins, Penems, Trinem

The bioavailability results obtained for a number of simple alkyl and aryl esters of the carboxylic acid of the C-6 acylamido group of carbenicillin in mouse, rat, rabbit, dog (beagle), rhesus monkey (RM), squirrel monkey (SM), pig (unpublished data), and human (Clayton *et al.,* 1976) are shown in Fig. 1. No significant rank correlations of maximum observed concentrations in serum (C_{max}) were demonstrated between human and each of the animal species tested. However, despite the lack of direct correlation, the indanyl and phenyl esters were selected for further study on the basis of animal data, and both gave rise to therapeutic concentrations of carbenicillin after oral administration to humans.

In Table III, results for three acyloxymethyl esters of the penicillin ampicillin—the pivaloyloxymethyl (pivampicillin), ethoxycarbonyloxyethyl (bacampicillin), and phthalidyl (talampicillin) esters—have been compared in laboratory animal species and humans. Ampicillin is stable to stomach acid and is absorbed by the oral route in all species but incompletely and in human only 30–40% is excreted into urine (Ferres, 1983). The aim was to increase this bioavailability by the use of prodrug esters. All three esters gave rise to higher concentrations of ampicillin and in this instance there was good correlation between animals and human (Table III). Clayton *et al.* (1976) also reported on the bioavailability in the squirrel monkey of phthalidyl esters of other β-lactam antibiotics that are available by the oral route, i.e., penicillin V and cloxacillin. The oral bioavailability of these compounds was decreased, not increased, and it was reported that this could be the result of inadequate aqueous solubility of the esters in the lumen of the gastrointestinal tract of the monkey. The dose used in these studies was 100 mg/kg and in fact no difference was observed between ampicillin and talampicillin (data not shown) at this dose level, whereas at 25 mg/kg, the ampicillin concentrations were higher after administering the ester than ampicillin itself (Table III). These dose-related effects may have been indicative of poor solubility limiting absorption at the higher dose, but interestingly, the results reported by Clayton *et al.* (1976) for another ester of ampicillin, the 5,6-dimethoxyphthalidyl ester, which was administered at 100 mg/kg, indicated that concentrations of ampicillin were higher after administration of this ester in both monkey and humans. The relative aqueous and lipid solubilities of these esters were not described.

The prodrug esters of penems and the trinem GV 104326 have not been compared in tabular form as data are incomplete. The hexetil ester of the trinem was shown to be absorbed in humans, yielding a C_{max} of 3.5 μg/ml after a single oral dose of 500 mg (Efthymiopoulus *et al.,* 1994); no animal data were available at the time of writing. However, experimental efficacy studies in mice indicate adequate bioavailability by the oral route in this species (DiModugno *et al.,* 1994). The pivaloyloxymethyl ester of the penem CP-65,207 was well absorbed in rats

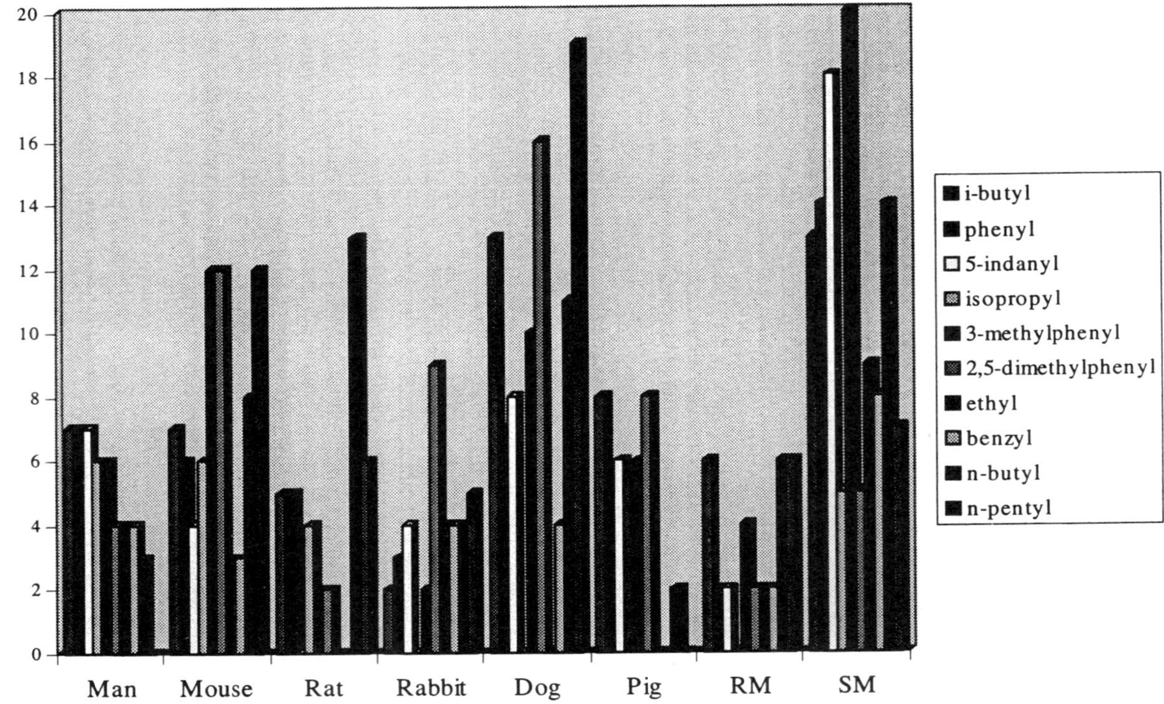

Figure 1. A comparison of maximum observed concentrations (C_{max}) in animal and human plasma after single oral doses of each prodrug ester of carbenicillin. All animals received the ester dosed in terms of pure free acid carbenicillin and concentrations of carbenicillin in plasma were assayed using hole-in-plate bioassay with *Pseudomonas aeruginosa* as the assay organism, minimum assayable concentration 1.5 μg/ml being depicted as 0 μg/ml.

Table III
Bioavailability of Prodrug Esters of Ampicillin in Humans and Animal Species

	Maximum observed concn. (μg ampicillin/ml serum)			
	Mouse (50 mg/kg)	Rat (100 mg/kg)	Squirrel monkey (25 mg/kg)[a]	Human[b]
Talampicillin (Clayton *et al.*, 1976)				
Ester	7.6	2.7	7.1	6.5 (250 mg)
Ampicillin	2.2	1.4	4.8	2.4 (250 mg)
			Dog (14 mg/kg)	(Sjovall *et al.*, 1978)
Bacampicillin (Bodin *et al.*, 1975)				
Ester	–	33.8	4–6.8	8.3 (278 mg)
Ampicillin	–	10.4	1.1–3.3	3.7 (278 mg)
			Dog (30 mg/kg)	
Pivampicillin (Daehne *et al.*, 1970)[c]				
Ester	–	2 hr[d] 4.9	2 hr 5.7	5.8 (250 mg)
Ampicillin		2 hr 0.84	2 hr 2.0	1.72 (250 mg)

[a]Unpublished data for squirrel monkey at this dose level.
[b]1 : 1 molar comparison of prodrug and ampicillin.
[c]Only 2-hr data were available.
[d]Two-hour sample only.

with a bioavailability of 53–68% (Gootz *et al.*, 1990) but no blood level data were given. It is noteworthy that the pivaloyloxymethyl prodrug auxiliary has been effective in enhancing bioavailability in at least three groups of β-lactam antibiotics (Tables I and II). Another penem prodrug ester, FCE 22101 (the acetoxymethyl ester), was administered to humans and the oral bioavailability was 29%, versus 47% in rats (Webberley *et al.*, 1988).

3.2. Cephalosporins

The bioavailabilities, as indicated by observed C_{max} values, for a number of cephalosporin prodrug esters in animals and humans have been collected from various sources in Table IV, together with results for two cephalosporins (cefaclor and cefixime) that are absorbed without the use of prodrug esters. The rank order of results has been compared between compounds in the different species and shows the variability that has been observed for cephalosporins.

The C_{max} values (Table IV) for cefpodoxime after oral administration of its proxetil ester at a dose of 400 mg to healthy human volunteers ranged from 3.72 to 4.5 μg/ml plasma with approximately 30% of the dose being excreted in urine (Frampton *et al.*, 1992). These values were similar to those for cefuroxime after administration of its axetil ester (500-mg dose). In contrast, in rat and mouse, the

Table IV
Comparative Bioavailabilities of Cephalosporins in Animals and Humans

Compound	Maximum observed concn. (μg parent compound/ml plasma) Mouse	Rat	Squirrel monkey	Human
		(dose mg/kg)		(dose mg)
cefpodoxime proxetil	(50) 28.3[a]	(100) 23.6[a]	(25) 2.2[a]	(400) 3.72–4.5[b]
	(40) 26.3[c]	(20) 9.4[c]		(200) 2.1[d]
cefuroxime axetil	(50) 8.6[a]	(100) 3.1[a]	(25) 1.6[a]	(500) 3.6–8.3[d]
	(40) 11.9[c]	(20) 6.7[c]		
cefdaloxime pentexil	(40) 31.1[c]	(20) 14.5[c]		(200) 3.95[m]
	(20) 26.7[e]			
cefcamate pivoxil (S1108)	(20) 4.7[f]	(20) 2.1[f]		(600) 5.7[g]
E1101	(20) 25.9[h]	(20) 18.2[h]		(200) 2.5[i]
cephaloglycine				
acetoxymethyl ester		(100) 6.8[j]		(200) 7.7[j]
pivaloyloxymethyl ester		(100) 3.1[j]		(200) 5.8[j]
Ro 41-3399	(20) 25–28[k]			(500) 1.64[l]
cefaclor	(50) 15.1[a]	(100) 9.1[a]		(500) 12.0[d]
	(20) 11.4[k]			
cefixime	(50) 5.8[a]	(100) 17.8[a]		(500) 4.0[d]
	(40) 10.4[c]	(20) 7.1[c]		
	(20) 9.0[c]			

[a]In-house data, unpublished, whole blood assay.
[b]Frampton *et al.* (1992).
[c]Klesel *et al.* (1992).
[d]Jones (1989).
[e]Isert *et al.* (1992).
[f]Matsuura *et al.* (1989).
[g]Nakashima *et al.* (1989).
[h]Uemura *et al.* (1994).
[i]Nakashima *et al.* (1994).
[j]Binderup *et al.* (1971).
[k]Angehrn *et al.* (1992).
[l]Hesse *et al.* (1992).
[m]Mendes *et al.* (1992).

cefpodoxime levels were in excess of those of cefuroxime. Concentrations of the two cephalosporins were closer, however, in the squirrel monkey and more representative of results in humans.

The results in mice and rats for cefdaloxime pentexil and the development cephalosporin E1101 (Table IV) looked particularly promising in comparison with cefcamate pivoxil (S1108) as concentrations were at least fivefold higher. In humans, however, although the doses varied, no such differences between concentrations were observed. In further comparisons of the bioavailabilities of the parent cephalosporins in mice and rats, cefdaloxime pentexil was shown to be superior to the prodrug esters cefuroxime axetil and cefpodoxime proxetil and to the orally absorbed cefaclor and cefixime; in humans, the differences were not as marked, although cefdaloxime pentexil was still superior overall. Although data are limited, it is of interest to observe that the rank order of results for the two prodrugs of cephaloglycine, the acetoxymethyl and pivaloyloxymethyl esters, was the same in rats and humans.

The concentrations of parent cephalosporin Ro 40-6890 in mice after administration of its prodrug Ro 41-3399 were high (Table IV) and of the same order as those of the parent cephalosporins delivered by cefpodoxime proxetil, cefdaloxime pentexil, and E1101. Despite these encouraging results, administration of this new prodrug ester to human volunteers resulted in only a low C_{max} (1.64 μg/ml) of Ro 40-6890 after a single oral dose of 500 mg: Only 9% of the dose was excreted in urine over 24 hr (data not shown). It is also seen that the concentrations of Ro 40-6890 in human plasma were less than observed not only for other cephalosporins from their prodrug esters but also for those cephalosporins not requiring a prodrug for oral absorption such as cefaclor and cefixime.

It is apparent that differences even within the rodent species exist. In their search to find novel ester groups for the new third-generation cephalosporin Ro 40-6890, Hubschwerlen *et al.* (1992) used both rats and mice for evaluation and compared AUC values for the parent cephalosporin. The mouse was used in the primary selection screen for the range of prodrug esters prepared (7a–k, Fig. 2). Cefetamet pivoxil (Globocef®) was included as an example of an oral cephalosporin prodrug in late development. In the rat the absolute oral bioavailabilities (on the basis of AUC values after oral and subcutaneous dosing) of the selected compounds were determined and compared. The pivaloyloxymethyl ester of Ro 40-6890 was included for comparison. The rank order of AUC values for the two species is shown in Fig. 2. In contrast to results in mice, the best ester in the rat was 7k, the isobutoxycarbonyl-2-propylidene ethyl ester (Ro 41-3399), and this entered development. Differences between rats and mice were also observed for the pivaloyloxymethyl ester of Ro 40–6890. In mice the AUC (20.7 μg·hr/ml) was similar to that for 7k (19.9 μg·hr/ml) whereas in rats the AUC for 7k (22.8 μg·hr/ml) was higher than that observed for the pivaloyloxymethyl ester (17.3 μg·hr/ml) as was the absolute oral bioavailability, 61 versus 41%.

In addition, a comparison of the rank order of results for cefetamet pivoxil and Ro 41-3399 (7k) in the mouse (Hubschwerlen *et al.*, 1992) with those in human volunteers was a further example of the lack of correlation between animal and human data. In the mouse the AUC values for 7k (19.9 μg·hr/ml Ro 40-6890) were higher than those for cefetamet (16.2 μg·hr/ml) after an oral dose of 20 mg/kg cefetamet pivoxil whereas the C_{max} of cefetamet in humans (4.11 μg/ml) (Tam *et al.*, 1989) was higher than observed for Ro 41-3399 (7k) (1.6 μg/ml) (Hesse *et al.*, 1992) after doses of 500 mg of the prodrug esters.

In summary, the rank order of bioavailability data in laboratory animals for any one series of analogues can differ and for some esters the differences were par-

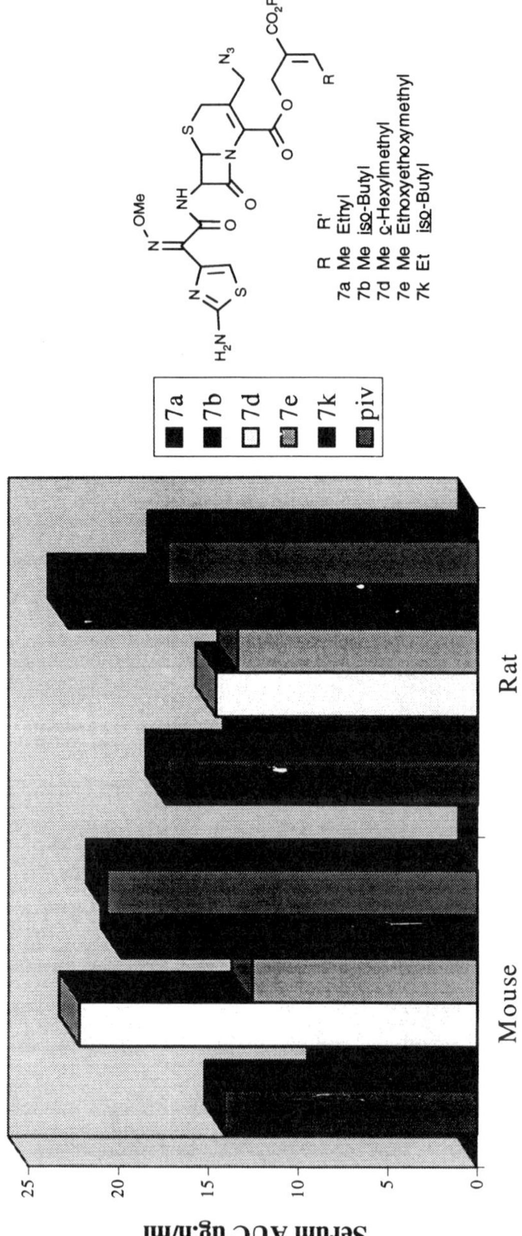

Figure 2. A comparison of area under the serum curve (AUC) in rats and mice after oral administration of a series of alkyloxycarbonyl-2-alkylidene ethyl esters (7a–7k) and the pivaloyloxymethyl ester of the cephalosporin Ro 40-6890 (dose: 20 mg Ro 40-6890 equiv./kg).

ticularly marked. The studies identified above were collected from different sources and do not take into account the potential influences of different dosing vehicles in either animals or humans. These data serve to illustrate the problems that arise in selection of a candidate for development. With the present-day requirements for safety evaluation and the time scales involved in development, it would not be possible to test a large number of esters in humans in order to select the best candidate and this has necessitated reliance on animal studies. However, despite the lack of correlation illustrated above, prodrugs have been marketed (Tables I and II), although the best compound may not have been selected for development and an alternative could have been missed at the primary selection stage in animals. Solubility, stability, and comparative hydrolysis rates are additional factors that can be usefully considered in conjunction with bioavailability studies with a view to understanding the mechanisms involved in achieving adequate oral bioavailability.

4. HYDROLYSIS RATES AND PHYSICOCHEMICAL PROPERTIES

The extent of hydrolysis in gastrointestinal contents and wall, blood, and liver may, either together or separately, influence the concentrations of the parent compound observed in serum after oral administration of an ester prodrug. Species differences in hydrolysis may account for the differences that have been observed for oral bioavailability in terms of concentrations in serum. An assessment of the relationship of hydrolysis rates to serum concentrations of antibiotic measured has been viewed as a means to obtain more comprehensive and therefore more predictive data.

4.1. Hydrolysis by Liver

In the selection process for carfecillin, Clayton *et al.* (1975) compared rates of hydrolysis of a series of esters of carbenicillin in squirrel monkey and human liver homogenates (2 and 4% w/v, respectively, a dilution that allowed rates of hydrolysis to be measured and compared). The results in Fig. 3 for hydrolysis in liver showed little correlation between humans and squirrel monkey; in addition, the rates of hydrolysis did not directly correlate with the rank order of peak concentrations observed in the plasma of those species. It was found, however, that the aryl esters hydrolyzed more rapidly than the alkyl esters in the tissues of both species and this correlated with the observation that some unhydrolyzed alkyl esters had been detected in human urine after oral administration.

4.2. Hydrolysis by Small Intestine

The rates of hydrolysis of esters of carbenicillin (Clayton *et al.,* 1975) in homogenized small intestine of the squirrel monkey (1 mg/ml at 2% w/v) were in the same rank order as found in liver homogenate (Fig. 3). In both squirrel monkey and human tissues the phenyl ester (carfecillin) was one of the most rapidly hydrolyzed esters. Hydrolysis in rat tissues was also rapid, but not in dog tissues. Humphrey *et al.* (1980) found significant levels of intact carfecillin in dog portal vein after duodenal administration, unlike the case in rats. The work of Jeffery *et al.* (1978) on the metabolism of talampicillin also supported the suggestion of low esterase activity of dog intestine. This may have resulted in intact ester reaching the liver via the hepatic portal vein, offering an explanation for hepatocellular changes seen in toxicity studies in the dog but not in the rat.

The observation of little or no ester in portal blood indicates that hydrolysis in the gut wall is a significant factor in the uptake of prodrug esters from the gastrointestinal tract. Esterases are present within the mucosal epithelium of the small intestine (Inoue *et al.,* 1979; Friedman *et al.,* 1966), and the elegant work of Shindo *et al.* (1973) with [^{14}C] pivampicillin showed clearly that the site of hydrolysis was primarily in the apical cytoplasm. The esterase activity was associated with the rough and smooth endoplasmic reticulum, while the nucleus and basal cytoplasm were unreactive. Using *in situ* gut perfusion techniques, Shindo *et al.* (1973) found that the newly formed ampicillin accumulated only transiently within the cells and then diffused into portal blood. Of interest here was the observation that there were species differences in the distribution and activity of esterases. Human and rhesus monkey intestine showed the same strong activity of esterases but these were not distributed throughout the cytoplasm as found in the rat, but rather localized only at the membranes of the smooth and rough endoplasmic reticulum and also the lysosomes. In contrast, they found that the dog displayed only weak esterase activity and this observation confirms the work of others (Inoue *et al.,* 1979; Williams, 1985, 1987) and supports the studies of Humphrey *et al.* (1980) with esters of carbenicillin and those of Jeffery *et al.* (1978) with talampicillin.

4.3. Hydrolysis by Blood

Talampicillin was rapidly hydrolyzed in mouse, rat, and human blood (Clayton *et al.,* 1976) and the half-life for this hydrolysis in 90% blood was less than 2.0 min. Shiobara *et al.* (1974) also found a rapid hydrolysis rate of talampicillin in mouse and rat blood and this was faster than in dog blood. The comparative rates of hydrolysis of the three marketed ampicillin esters, talampicillin, pivampicillin, and bacampicillin, were different in both human blood and tissue homogenates

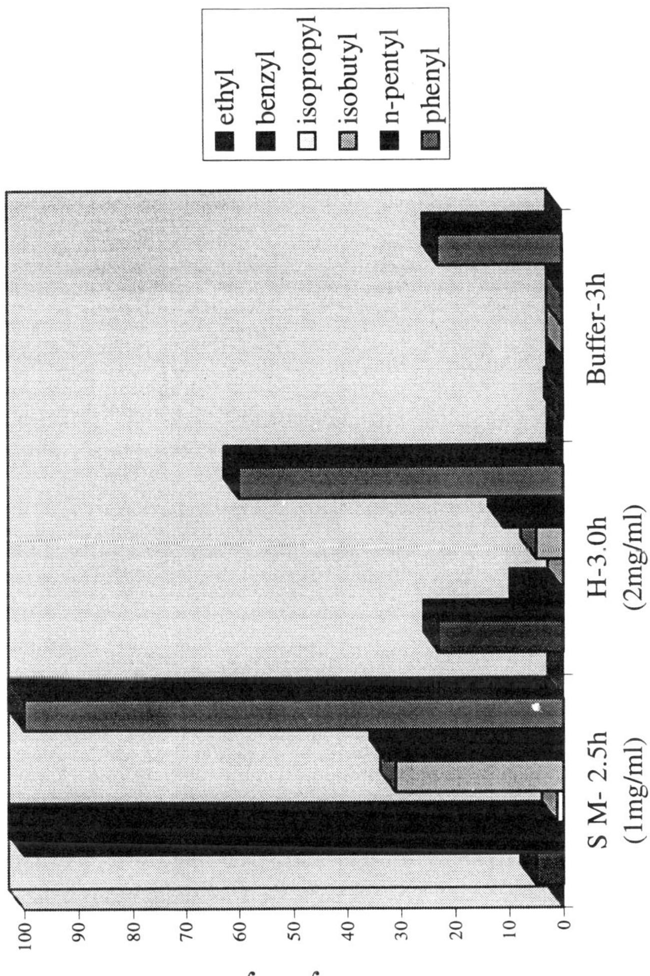

Figure 3. The extent of hydrolysis of a series of simple alkyl and aryl ester prodrugs of carbenicillin in human (H) and squirrel monkey (SM) liver.

(Ekstrom *et al.*, 1977). The hydrolysis of pivampicillin was the slowest and Daehne *et al.* (1970) found about 1–2% of intact ester in human blood. Results for hydrolysis in blood can be influenced by factors other than esterase activity. The *N,N*-diethylaminoethyl ester of methicillin was hydrolyzed at different rates in human, mouse, and squirrel monkey blood (90% weight volume) and after incubation at 37°C for 60 min the ester was completely hydrolyzed in mouse blood, 75% hydrolyzed in human blood, and only 47% hydrolyzed in squirrel monkey blood (Ferres, 1983). However, this ester was also completely hydrolyzed in the buffered saline control and so species differences may have been related more to differences in binding of the complete ester to serum proteins than to esterase activity.

4.4. Physicochemical Properties

The relationship of physicochemical properties to oral bioavailability of not only intact esters but also the parent β-lactams is unclear. Yoshimura *et al.* (1987) attempted to obtain a more rational design of prodrug esters of the cephalosporin cefotiam by measuring oral bioavailability in mice, water solubility, lipophilicity, hydrolysis in liver and small intestine, and isomerization to Δ-2 cefotiam for a series of 16 esters. They found no significant correlation between water solubility and rank order of bioavailability, probably because the esters were of sufficient water solubility for dissolution rate not to be critical. The bioavailability was, however, shown to be correlated with lipophilicity. A relationship between hydrolysis and bioavailability was also demonstrated provided that a contribution from the analysis of the steric hindrance of the ester auxiliary was included. The results showed that hydrolysis and isomerization may be parallel reactions *in vitro* and the ratio of the products would depend on both esterase activity and the susceptibility of the ester to isomerization. Yoshimura *et al.* suggested that sufficient information had been gained from these comparative studies to understand the basic requirements for oral absorption of cefotiam prodrug esters. It should be pointed out, however, that bioavailability studies were only performed in the mouse. Earlier, Yoshimura *et al.* (1985) reported similar comparative data for the pivaloyloxymethyl esters of a series of parenteral cephalosporins, which demonstrated a correlation with water solubility, but only if the lipophilicity and hydrolysis rates were sufficiently high. These studies of Yoshimura *et al.* did supply very useful baseline data for the rational design of prodrugs, but again, as the bioavailability was only examined in the mouse, the predictive quality of these data for human studies is unclear.

In conclusion, the foregoing analysis demostrates that for the selection of the optimal prodrug ester for each parent antibiotic, all factors must be considered in relationship to each other in order to select the most appropriate candidate for development and eventually for administration to healthy human volunteers.

5. DOSING VEHICLES AND FORMULATIONS

The interaction between the physicochemical properties of the prodrug ester and the choice of dosing vehicles at the research stage can influence the bioavailability. There are a number of suspending agents that can be used; the most commonly used ones (in varying dilutions) include dimethyl sulfoxide, dimethylformamide, gum arabic, the Tweens, cyclodextrins, various oils, ethanol, methyl cellulose, carboxymethyl cellulose, and polyethyleneglycol (PEG) 400, although not all are suitable for use in human volunteer studies. Usually the same diluent is used throughout the research selection process for any one series of analogues.

In the initial human volunteer studies a simple suspension is often used in an appropriate vehicle. If the concentrations in serum are considered inadequate after administration of the prodrug, the usual options are either to select another ester or to examine different formulations of the first ester. Both of these approaches depend on the failure being related to poor absorption of the prodrug and not to rapid metabolism and elimination of the parent antibiotic. This can only be clarified by intravenous administration of the parent antibiotic, which is not always feasible if it is not also being considered as an agent for parenteral use. Potential opportunities to improve the extent of uptake from the gastrointestinal tract include reducing particle size and the use of various excipients. Bioavailability studies in animals can help in this respect but no one animal is a true representative of the human gastrointestinal tract particularly for evaluating dosage forms. The pig has been suggested as the species nearest to human (Wilding *et al.*, 1994), but it would not be suitable for early stage studies in development because of the large compound requirement reflecting its size.

6. SUMMARY AND CONCLUSIONS

It is apparent that the sequence of events that has been followed in the approach to the discovery and development of a new β-lactam prodrug has been similar in many of the case histories we have studied and indeed similar to the approach we have followed. Initially, we select a suitable series of prodrug moieties, which either comprises totally novel structures or is deduced from the data bases available (bearing in mind reports of potential toxicity) or both. The successful preparation of these prodrugs and the studies undertaken to ensure they are of known purity and stability is not easy and, as would be expected, is the initial go/no-go decision. Usually, the next stage has involved the assessment of whether or not bioavailablity of the parent molecule is increased after administration of the prodrug ester by gavage to laboratory animal species. The selection of which species to use has very often been made according to which has the most information available in those particular laboratories and in the literature. It is this process that can be dishearteningly misleading as was demonstrated in Table IV and Fig. 1. Increasing the range of animal species does not lead to a better ability to predict bioavailability in humans. Hydrolysis studies are important to ensure that any novel prodrug will hydrolyze in human tissues, and also in the clarification of why a particular prodrug is not performing as expected in animals. After selection, it is essential to determine where and how rapidly hydrolysis takes place in the animal species to be used for safety evaluation prior to the first bioavailability studies in humans.

The assessment of absolute oral bioavailability has not always been undertaken. This would seem critical for studies in not only the selected animal species but also in humans. In the absence of these data it is difficult to judge whether oral uptake can be increased further by modifying the ester moieties and at the development stage to determine whether or not modifications in formulation could increase bioavailability. When the prodrug is being developed for an injectable β-lactam already available for humans, there would be no problem, but it would be an important consideration during the development of an entirely novel β-lactam antibiotic for which no parenteral data are available in humans. Animal data are not totally predictive.

The development of prodrugs is not easy, as a consequence of species differences in the properties of the prodrug superimposed on those of the parent compound during the evaluation. However, technical advances have enabled us to assay concentrations more precisely, determine basic physicochemical properties more efficiently, understand absorption processes by the use of *in vitro* systems, and analyze data far more comprehensively by the use of ever-evolving computer software.

The prodrug approach to increasing the oral bioavailability of β-lactam antibiotics has provided clinically valuable agents and continues. Despite the inherent difficulties, knowledge gained over the years, of the relationships between physicochemical and biological properties of the parent compound and the intact prodrug ester, has contributed to the design of novel prodrugs and a number of novel auxiliaries have been developed.

ACKNOWLEDGMENTS

The authors wish to thank John Bateson, Gary Woodnutt, and Michael Pearson for helpful discussions during preparation of the manuscript.

REFERENCES

Agersborg, H. P. K., Cambridge, G. W., and Rule, A. W., 1966, The pharmacology of penamecillin, *Br. J. Pharmacol.* **26:**649–655.

Angehrn, P., Hohl, P., Hubschwerlen, C., Page, M., and Then, R., 1992, Antibacterial properties of Ro 40-6890, a broad spectrum cephalosporin, and its novel orally absorbable ester, Ro 41-3399, *Antimicrob. Agents Chemother.* **36**(12)**:**2825–2834.

Barnden, R. L., Evans, R. M., Hamlet, J. C., Hems, B. A., Jansen, A. B. A., Trevett, M. E., and Webb, G. B., 1953, Some preparative uses of benzylpenicillinic ethoxyformic anhydride, *J. Chem. Soc.* **1953:**3733–3739.

Bergan, T., 1978, Penicillins, in: *Antibiotics and Chemotherapy,* Volume 25 (H. Schonefeld, ed.), pp. 1–122, Karger, Basel.

Binderup, E., Godtfredsen, W. O., and Roholt, K., 1971, Orally active cephaloglycine esters, *J. Antibiot.* **24**(11)**:**767–773

Bodin, N. O., Ekstrom, B., Forsgren, U., Jalar, L. P., Magni, L., Ramsay, C. H., and Sjoberg, B., 1975, Bacampicillin: A new orally well-absorbed derivative of ampicillin, *Antimicrob. Agents Chemother.* **8:**518–525.

Brogard, J. M., Compte, F., and Pinget, M., 1978, Pharmacokinetics of cephalosporin antibiotics, in: *Antibiotics and Chemotherapy,* Volume 25 (H. Schonefeld, ed.), pp. 123–162, Karger, Basel.

Butler, K., English, A. R., Briggs, B., Gralla, E., Stebbins, R. B., and Hobbs, D. C., 1973, Indanyl carbenicillin: Chemistry and laboratory studies with a new semisynthetic penicillin, *J. Infect. Dis.* **127**(Suppl. May)**:**97–104.

Clayton, J. P., Cole, M., Elson, S. W., Hardy, K. D., Mizen, L. W., and Sutherland, R., 1975, Preparation, hydrolysis, and oral absorption of alpha-carboxy esters of carbenicillin, *J. Med. Chem.* **18**(2)**:**172–177.

Clayton, J. P., Cole, M., Elson, S. W., Ferres, H., Hanson, J. C., Mizen, L. W., and Sutherland, R., 1976, Preparation, hydrolysis, and oral absorption of lactonyl esters of penicillins, *J. Med. Chem.* **19**(12)**:**1385–1391.

Cocker, J. D., Eardley, S., Gregory, G. I., Hall, M. E., and Long, A. G., 1966, Cephalosporanic acids. Part IV. 7-Acylamidoceph-2-em-4-carboxylic acids, *J. Chem. Soc. C* **1966:**1142.

Craig, W. A., 1995, Interrelationship between pharmacokinetics and pharmacodynamics in determining dosage regimens for broad spectrum cephalosporins, *Diagn. Microbiol. Infect. Dis.* **22** (2-2)**:**89–96.

Daehne, W. V., Frederiksen, E., Gundersen, E., Lund, F., Morch, P., Peterson, H. J., Roholt, K., Tybring, L., and Godtfredson, O., 1970, Acyloxymethyl esters of ampicillin, *J. Med. Chem.* **13:**607–612.

Di Modugno, E., Erbetti, I., Ferrari, L., Galassi, G., Hammond, S. M., and Xerri, L., 1994, In vitro activity of the tribactam GV104326 against gram positive, gram negative and anaerobic bacteria, *Antimicrob. Agents Chemother.* **38**(10)**:**2362–2368.

Efthymiopoulos, C., Capriati, A., Barrington, P., Patel, J., Shenoy, E. V. B., and Bye, A., 1994, Pharmacokinetics of GV 104326, a novel tribactam antibiotic, following single intravenous and oral (as its prodrug GV 118819X) administration in man, in: *Proc. 34th Interscience Conference on Antimicrobial Agents and Chemotherapy,* Orlando, F82 (abstract).

Ekstrom, B., Forsgren, U., Jalar, L. P., Magni, L., Sjoberg, B., and Sjovall, J., 1977, Preclinical studies with bacampicillin—A new orally well absorbed prodrug of ampicillin, *Drugs Exp. Clin. Res.* **3:**3–10.

Ferres, H., 1983, Prodrugs of beta-lactam antibiotics, *Drugs Today* **19:**499–538.

Frampton, J. E., Brogden, R. N., Langtry, H. D., and Buckley, M. M., 1992, Cefpodoxime proxetil: A review of its antibacterial activity, pharmacokinetic properties and therapeutic potential, *Drugs* **44**(5)**:**889–917.

Friedman, B., Strachan, D. S., and Dewey, M. M., 1966, Histochemical and biochemical analysis of the nonspecific esterases of the small intestine of the rat, *J. Histochem. Cytochem.* **14**(7)**:**560–566.

Fujimoto, K., Ishihara, S., Yanagisawa, H., Ide, J., Nakayama, E., Nakao, H., Sugawara, S., and Iwata, M., 1987, Studies on orally active cephalosporin esters, *J. Antibiot.* **40**(3)**:**370–384.

Gootz, T., Girard, D., Schelkley, W., Tensfedlt, T., Foulds, G., Kellogg, M., Stam, J., Campbell, B.,

Jasys, J., Kelbaugh, P., Volkmann, R., and Hamanaka, E., 1990, Pharmacokinetic studies in animals of a new parenteral penem CP-65-207 and its oral prodrug ester, *J. Antibiot.* **43**(4)**:**422.

Goto, S., Ogawa, M., and Tsuji, A., 1986, In vitro and in vivo antibacterial activities of T-2588 a new oral cephem antibiotic, *Chemotherapy (Tokyo)* **34**(Suppl. 2)**:**13–23.

Hampel, B., Feike, M., Koeppe, P., and Lode, H., 1985, Pharmacokinetics of temocillin in volunteers, *Drugs* **29**(5)**:**99–102.

Harding, S. M., Williams, P .E. O., and Ayrton, J., 1984, Studies on the pharmacology of cefuroxime axetil (cefuroxime as the 1-acetoxyethyl ester) in volunteers, *Antimicrob. Agents Chemother.* **25:**78–82.

Hesse, W. H., Hoogkamer, J., and Gower, M., 1992, New oral cephalosporin Ro41–3399: Single and multiple dose pharmacokinetics, in: *Proc. Interscience Conference on Antimicrobial Agents and Chemotherapy,* Anaheim, 204 (abstract).

Hubschwerlen, C., Charnas, R., Angehrn, P., Furlenmeier, A., Graser, T., and Montavon, M., 1992, Orally active 2-(alkyloxycarbonyl)-2-alkylideneethyl esters of cephalosporins, *J. Antibiot.* **45**(8)**:**1358–1364.

Humphrey, M. J., Filer, C. W., Jeffery, D. J., Langley, P. F., and Wadds, G. A., 1980, The availability of carfecillin and its phenol moiety in rat and dog, *Xenobiotica* **10**(10)**:**771–778.

Inoue, M., Morikawa, M., Tsuboi, M., and Sugiura, M., 1979, Species difference and characterisation of intestinal esterase on the hydrolyzing activity of ester-type drugs, *Jpn. J. Pharmacol.* **29:**9–16.

Isert, D., Fischer, G., Klesel, N., Limbert, M., Markus, A., Riess, G., and Seibert, G., 1992, Cefdaloxime pentexil tosilate (HR 916K): A diastereomerically pure novel oral cephalosporin ester with outstanding absorption characteristics, in: *Proc. 32nd Interscience Conference on Antimicrobial Agents and Chemotherapy,* Anaheim, 188 (abstract).

Jansen, A. B .A., and Russell, T. J., 1965, Some novel penicillin derivatives, *J. Chem. Soc.* **1965:** 2127–2132.

Jeffery, D. J., Jones, K. H., and Langley, P. F., 1978, The metabolism of talampicillin in rat, dog and man, *Xenobiotica* **8**(7)**:**419–427.

Jones, R. N., 1989, New oral antimicrobial agents, *Curr. Opin. Infect. Dis.* **2:**367–375.

Klesel, N., Adam, F., Isert, D., Limbert, M., Markus, A., Schrinner, E., and Seibert, G., 1992, RU29-246, the active compound of the cephalosporin prodrug ester HR 916: III. Pharmacokinetic properties and antibacterial activity *in vivo, J. Antibiot.* **45**(6)**:**922–931.

Matsuura, S., Yamano, Y., Nakashimizu, H., Matsubara, T., Kobayashi, F., and Kuwahara, S., 1989, S-1108, a new oral cephem, in: *Proc. 29th Interscience Conference on Antimicrobial Agents and Chemotherapy,* Houston, 369 (abstract).

Mendes, P., Meyer, W. H., Muller, P. O., Scholl, T., and Luus, H., 1992, Pharmacokinetics of cefdaloxime pentexil tosilate HR916K after a single oral dose in human volunteers. in: *Proc. Interscience Conference on Antimicrobial Agents and Chemotherapy,* Anaheim, 189 (abstract).

Miyauchi, M., Kurihara, H., Fujimoto, K., Kawamoto, I., Ide, J., and Nakao, H., 1989, Studies on orally active cephalosporin esters. III. Effect of the 3-substituent on the chemical stability of pivaloyloxymethyl esters in phosphate buffer solution, *Chem. Pharm. Bull.* **37**(9)**:**2375–2378.

Mizen, L., and Woodnutt, G., 1988, A critique of animal pharmacokinetics, *J. Antimicrob. Chemother.* **21:**273–278.

Mizen, L., Berry, V., and Woodnutt, G., 1995, The influence of uptake from the gastrointestinal tract and first pass effect on oral bioavailability of (Z)-alkyloxyimino penicillins, *J. Pharm. Pharmacol.* **47:**725–730.

Nakashima, M., Matsuno, S., Yoshida, T., Kimura, Y., Toguma, F., and Ishii, H., 1989, Pharmacokinetics and safety of S1108 in healthy volunteers, in: *Proc. Interscience Conference on Antimicrobial Agents and Chemotherapy,* Houston, 370 (abstract).

Nakashima, M., Uematsu, T., Uemura, K., Kosuge, K., Yasuda, S., Kawahara, M., Tomono, Y., Ohno, T., Okano, K., and Yamoto, C., 1994, Pharmacokinetics and safety of E1101 a new oral cephalosporin in healthy male volunteers, in: *Proc. Interscience Conference on Antimicrobial Agents and Chemotherapy,* Orlando, F89 (abstract).

Neilsen, N. M., and Bundgaard, H., 1988, Facile plasma-catalysed degradation of penicillin alkyl esters but with no liberation of the parent penicillin, *J. Pharm. Pharmacol.* **40:**506–509.

Parsons, R. L., Hossack, G. A., and Paddock, G. M., 1977, Pharmacokinetics of pivmecillinam, *Brt J. Clin. Pharmacol.* **4**(3)**:**267–273.

Richardson, A. P., Walker, H. A., Loeb, P., and Miller, I., 1945, Metabolism of methyl and benzyl esters of penicillin by different species, *Proc. Soc. Exp. Biol. Med.* **60:**272–276.

Richter, W. F., Chong, Y. H. and Stella, V. J., 1990, On the mechanism of isomerisation of cephalosporin esters, *J. Pharm. Sci.* **79:**185–186.

Saab, A. K., Dittert, L. W., and Hussain, A. A., 1988, Isomerisation of cephalosporin esters: Implications for the prodrug ester approach to enhancing the oral bioavailabilities of cephalosporins, *J. Pharm. Sci.* **77:**906–907.

Sakamoto, F., Ikeda, S., and Tsukamoto, G., 1984, Studies on prodrugs. II. Preparation and characterisation of (5-substituted 2-oxo-1,3-dioxolen-4-yl)methyl esters of ampicillin, *Chem. Pharm. Bull.* **32**(6)**:**2241–2248.

Shindo, H., Kawai, K., Fukuda, K., Matsumura, M., Tanaka, K., Tanaka, M., and Yokota, T., 1973, Absorption and metabolism of pivampicillin—site of ester hydrolysis and species difference, in: *Proc. 5th Symposium on Drug Metabolism and Action,* 49–66.

Shiobara, Y., Tachibana, A., Sasaki H., Watanabe, T., and Sado, T., 1974, Phthalidyl D-α-aminobenzylpenicillinate hydrochloride (PC-183), a new orally active ampicillin ester. I. Absorption, excretion and metabolism of PC-183 and ampicillin, *J. Antibiot.* **28**(9)**:**665–673.

Sjovall, J., Magni, L., and Bergan, T., 1978, Pharmacokinetics of bacampicillin compared with those of ampicillin, pivampicillin, and amoxycillin, *Antimicrob. Agents Chemother.* **13**(1)**:**90–96.

Stoeckel, K., Tam, Y. K., and Kneer, J., 1989, Pharmacokinetics of oral cefetamet pivoxil (Ro 15-8075) and intravenous cefetamet (Ro15–8074) in humans: A review, *Curr. Med. Res. Opin.* **11**(7)**:** 432–444.

Tam, Y. K., Kneer, J., Dubach, U. C., and Stoeckel, K., 1989, Pharmacokinetics of cefetamet pivoxil (Ro 15-8075) with ascending oral doses in normal healthy volunteers, *Antimicrob. Agents Chemother.* **33**(6)**:**957–959.

Uemura, Y., Tokumura, T., Hiruma, R., Ueno, J., and Yuzuriha, T., 1994, E1101, a new oral cephalosporin: Pharmacokinetics in laboratory animals, in: *Proc. 34th Interscience Conference on Antimicrobial Agents and Chemotherapy,* Orlando, F87 (abstract).

Ungar, J., and Muggleton, P. W., 1952, Accumulation of diethylaminoethyl ester of penicillin in inflamed lung tissue, *Brt Med. J.* **1:**1211–1213.

Webberley, J. M., Wise, R., Andrews, J. M., Ashby, J. P., and Wallbridge, D., 1988, The pharmacokinetics and tissue penetration of FCE 22101 following intravenous and oral administration, *J. Antimicrob. Chemother.* **21**(4)**:**445–450.

Weber, C., Hesse, W. H., and Hoogkamer, F., 1992, Pharmacokinetics of Ro 40-6890 following oral doses of its prodrug ester Ro 41-3399 in healthy male volunteers, in: *Proc. Interscience Conference in Antimicrobial Agents and Chemotherapy,* Anaheim, 202 (abstract).

Wilding, I. R., Davis, S. S., and O'Hagan, D. T., 1994, Optimising gastrointestinal delivery of drugs, Bailliere's *Clinical Gastroenterology* **8**(2)**:**255–270.

Williams, F. M., 1985, Clinical significance of esterases in man, *Clin. Pharmacokinet.* **10:**392–403.

Williams, F. M., 1987, Serum enzymes of drug metabolism, *Pharmacol. Ther.* **34:**99–109.

Yoshimura, Y., Hamaguchi, N., and Yashiki, T., 1985, Synthesis and relationship between physicochemical properties and oral absorption of pivaloyloxymethyl esters of parenteral cephalosporins, *Int. J. Pharm.* **23:**117–129.

Yoshimura, Y., Hamaguchi, N., and Yashiki, T., 1987, Synthesis and oral absorption of acyloxymethyl ester of 7β-(2-(2-aminothiazol-4-yl)acetamido)-3-(((1-(2-dimethylaminoethyl)-1H-tetrazol-5-yl)thio)-methyl)ceph-3-em-4-carboxylic acid (cefotiam), *Int. J. Pharm.* **38:**179–180.

Chapter 16

Hematoregulators

A Case History of a Novel Hematoregulatory Peptide, SK&F 107647

Pradip K. Bhatnagar, William F. Huffman, Andrew G. King, Dagfinn Løvhaug, Louis M. Pelus, William M. Potts, and Philip L. Smith

1. INTRODUCTION

Hematopoiesis is the process of lifelong blood-cell renewal and has been the subject of various literature reviews (Broxmeyer, 1995; Guillosson, 1996; Hunt and Foote, 1995; Kelley *et al.*, 1996; Klein, 1995; Lasky, 1996; Lowry, 1995; Orkin, 1995; Sachs, 1996; Shivdasani and Orkin, 1996; Weiss and Orkin, 1996; Yu, 1996). Mature blood cells are derived from stem cells that possess the ability to both self-renew and produce more mature progenitor cells that are committed to differentiation and proliferate within single or multiple lineages. In normal individuals, this system has the capability to increase the production of mature cells in response to

---

Pradip K. Bhatnagar and William F. Huffman • Department of Medicinal Chemistry, SmithKline Beecham Pharmaceuticals, King of Prussia, Pennsylvania 19406-0939. *Andrew G. King and Louis M. Pelus* • Department of Molecular Virology and Host Defense, SmithKline Beecham Pharmaceuticals, Collegeville, Pennsylvania 19426. *William M. Potts* • Department of Drug Metabolism and Pharmacokinetics, SmithKline Beecham Pharmaceuticals, King of Prussia, Pennsylvania 19406-0939. *Philip L. Smith* • Department of Drug Delivery, SmithKline Beecham Pharmaceuticals, Collegeville, Pennsylvania 19426. *Dagfinn Løvhaug* • Nycomed Imaging AS, Bioreg Research, Oslo N0371, Norway.

Integration of Pharmaceutical Discovery and Development: Case Studies, edited by Borchardt *et al.*, Plenum Press, New York, 1998.

different stimuli, such as infection or bleeding, as well as to maintain its reserve of pluripotent cells. Therefore, effective hematopoiesis requires orchestration by a complex network of autocrine and paracrine cytokines that impose positive and negative feedback on the system. In addition, extracellular matrix containing glycosaminoglycan, fibronectin, laminin, and collagen is involved in hematopoiesis. In the last decade, the availability of recombinant proteins and biological assays that can identify specific hematopoietic cell populations has resulted in a dramatic expansion of research in hematopoiesis and understanding of the regulatory mechanisms that control this complex organ system.

2. HEMATOPOIESIS, ENDOGENOUS REGULATORS, AND HOST DEFENSE MECHANISM

Pathogenic infections are the major cause of morbidity and mortality in patients who receive cytotoxic chemotherapy or who suffer from congenital neutropenia. AIDS patients and patients with transient immunosuppression resulting from antibiotics or steroid therapy, surgery, burn, and other septic shock-associated situations also suffer from increased episodes of infections. Strategies to prevent or reduce neutropenia and myelotoxicities as a means to potentially increase antimicrobial efficacy and reduce morbidity have been clinically realized through the use of myelopoietic growth factors (G-CSF, GM-CSF).

The hematopoietic growth factors play a dual role in (1) regulation of blood cell production, stimulating both proliferation and differentiation of pluripotent cells, as well as (2) priming and enhancing the functions of mature blood effector cells that are responsible for host defense mechanisms. The net result of their activity is to provide competent host defense against bacterial and fungal infections (Aviles *et al.,* 1996a,b; Metcalf, 1990). The colony stimulating factors (CSFs) have also demonstrated efficacy for decreasing the myelotoxic side effects of chemotherapy and radiation therapy (Chatta and Dale, 1996) and are being used to increase the therapeutic index of cytotoxic agents and irradiation employed in the treatment of various cancers. Table I shows the potential clinical applications of various hematopoietic factors.

Both GM-CSF and G-CSF are used clinically and are well tolerated during short-term administration. The most commonly reported adverse effects are bone pain, musculoskeletal pain, headache, mild rash, and low-grade fever (Frampton *et al.,* 1994). These adverse effects are rarely treatment-limiting. The data for long-term treatment with these factors are still being collected. The factors seem to be well tolerated, although adverse effects such as exacerbation of osteoporosis have been reported (Bonilla *et al.,* 1994). Investigations with other cytokines are still ongoing but adverse events have been reported for IL-1, IL-3, IL-6, and SCF (Vial and Descotes, 1996). Some of the reported adverse effects of hematopoietic proteins are shown in Table II.

Table I
Potential Clinical Applications of Hematopoietic Proteins

Factors	Potential clinical applications	References
G-CSF (Filgrastim)	Oncology-associated febrile neutropenia; neutropenia associated with BMT; febrile neutropenia in patients with congenital neutropenia; stem cell mobilization and dose-intensive chemotherapy	Hartung *et al.* (1995), Foster *et al.* (1995), Rackoff *et al.* (1996)
GM-CSF (Sargramostim)	Myeloid reconstitution postautologous BMT; BMT failure or engraftment delay; stem cell mobilization and neutropenia and reduction of infections	Jones (1994), Schiffer (1996), Meropol *et al.* (1996)
M-CSF	Augmentation of antitumor functions of macrophage	Nemunaitis *et al.* (1993)
Erythropoietin (Epogen)	Treatment of anemia	Eschbach *et al.* (1991), Danna *et al.* (1990)
Thrombopoietin	Development of megakaryocytes and platelets and expansion of the marrow CFU-MK pool	Zucker-Franklin and Kaushansky (1996), Miyazaki (1996)
IL-3	Reduction of duration and severity of thrombopenia and neutropenia caused by chemotherapy	Slenar *et al.* (1994), Gianella-Borradori (1994)
PIXY321	Attenuation of myeloid toxicity associated with dose-intensive chemotherapy	Vadhan-Raj *et al.* (1993), Jakubowski *et al.* (1992)
IL-1	Radioprotection, bone marrow expansion following stem cell harvest	Neta (1990), Dinarello (1992)
IL-6	Autologous bone marrow transplant	Barton (1996), Devine *et al.* (1994)
IL-11 (Neumega)	Bone marrow recovery and small intestinal epithelial cell regeneration postmyelotoxic therapy	Du *et al.* (1994)
IL 12	Restoration of defective NK activity of monocytes and the enhancement of cytolytic T cells from cancer patients	Soiffer *et al.* (1993), Gately and Mulqueen (1996)
SCF (C-kit ligand)	Mobilization of peripheral blood progenitor cells	Glaspy *et al.* (1995)
TGF-β	Immunosuppression and enhancement of the formation of extracellular matrix	Lawrence (1996), Maiti and Singh (1996)

3. UNMET NEEDS

Although various hematopoietic proteins have proven efficacious, they may not be suitable for extended chronic use because of issues associated with development of oral formulations of proteins, pharmacoeconomics, and patient compliance. Niven *et al.*, (1994) described aerosol delivery as an alternate approach for G-CSF. Development of this route of administration is being evaluated for a

Table II
Reported Adverse Effects of Hematopoietic Proteins

Factors	Reported adverse effects	References
G-CSF (Filgrastim)	Musculoskeletal pain, transient bone pain, headache, and mild rash	Frampton *et al.* (1994)
GM-CSF (Sargramostim)	Low-grade fever, flulike symptoms, myalgia, gastrointestinal disorders, and vascular leak syndrome	Grant and Heel (1992)
M-CSF	Thrombocytopenia and ocular toxicity, malaise, headache	Bukowski *et al.* (1994), Sanda *et al.* (1992)
Erythropoietin (Epogen)	Hypertension, headache, tachycardia, nausea, clotted vascular access, and hyperkalemia	Buckner *et al.* (1990)
Thrombopoietin	No reported adverse effects	
IL-1	Fever, flulike symptoms, dose-limiting hypotension, abdominal pain, neurotoxicity, myocardial infarction	Smith *et al.* (1992), Nemunaitis *et al.* (1994a,b)
IL-3	Low-grade fever, flulike symptoms, myalgia, gastrointestinal disorders, skin rash	Huhn *et al.* (1995)
IL-6	Fever, flulike symptoms, anemia, atrial fibrillation, hyperbilirubinemia, and hepatotoxicity	Veldhuis *et al.* (1995), van Gameren *et al.* (1994), Gordon *et al.* (1992), Kammüeller (1995)
IL-11	Stimulates leukemia and myeloma cells, myalgias/arthralgias and fatigue	Teramura *et al.* (1996), Gordon *et al.* (1996)
IL-12	Hematological toxicity, hepatotoxicity, and skeletal muscle degeneration (in mice)	Hendrzak and Brunda (1995)
SCF	Melanocyte hyperplasia, cutaneous hyperpigmentation, mast cell degranulation	Costa *et al.* (1996)
TGF-β	No reported adverse effects	

variety of proteins and appears to be a promising alternative delivery site (Patton and Platz, 1992; Niven, 1993; Wall, 1995). Takada *et al.* (1989) studied the oral bioavailability of recombinant human G-CSF formulated in polyoxyethylated castor oil derivative (HCO-60) in rats. Although a measurable effect on the number of blood leukocytes was observed, the level of G-CSF in the plasma was not reported suggesting that the bioavailability of recombinant human G-CSF is low. A recent review by Rollwagen and Baqar (1996) describes the local and systemic biological activities of orally administered cytokines. Reproducibility, safety, and cost-effectiveness of these approaches remain the major issues. Thus far, small molecules that affect hematopoiesis and host defense mechanisms have not advanced far enough to be rigorously scrutinized. However, safe and efficacious small compounds may better address the needs of oral bioavailability and cost-effectiveness.

4. NONPROTEINACEOUS HEMATOREGULATORS

4.1. Polymeric Carbohydrate: Betafectin

Betafectin (PGG-glucan) (poly-[1–6]-β-D-glucopyranosyl-[1–3]-β-D-glucopyranose) is a genetically engineered, yeast-derived carbohydrate activator of nonspecific host defenses. It is a β(1–3)/β(1–6)-linked glucose polymer. The target for this immunostimulator is the beta-glucan receptor on white blood cells. In animal models of infection, betafectin-treated animals showed increases in total leukocyte counts and enhanced bacterial clearance from blood (Cisneros *et al.*, 1996). A randomized phase I/II trial of betafectin in high-risk surgical patients showed significant reduction in the number and severity of postoperative infections (Babineau *et al.*, 1994). Betafectin is now in phase III clinical trials for reduction of infections in high-risk surgical patients (Washburn *et al.*, 1996; Stashenko *et al.*, 1995).

4.2. Low-Molecular-Weight Hematoregulators

Relative to proteins, small molecules generally have better oral bioavailability, and there has thus been considerable interest in identification and development of small molecules that mimic the pharmacological activities of hematopoietic growth factors (Broxmeyer, 1993). The idea that small molecules can modulate hematopoiesis/myelopoiesis was recognized almost three decades ago when Rytömaa and Kiviniemi (1968a,b) proposed a "chalone hypothesis" suggesting that one of the regulatory mechanisms in granulocytopoiesis is based on a tissue-specific mitotic inhibitor, the "granulocytic chalone," which originates from the mature cells of the myeloid population. This chalone mechanism prevents the excessive multiplication of the progenitor cells. These investigators isolated a factor (~ 4 kDa) from the neutrophils of normal and chloroleukemic rats that suppressed granulocyte production *in vivo* (Rytömaa and Kiviniemi, 1968a,b; Vilpo *et al.*, 1973). Since then, several other endogenous and exogenous agents that modulate hematopoiesis have been identified and are the subject of intensive investigation for their clinical potential. Some of these nonproteinaceous agents (Table III) are briefly reviewed below.

4.2.1. AS 101 (OSSIRENE)

The immunomodulator AS 101 (Table III), has been shown to induce lymphocyte proliferation and to increase the secretion of various hematopoietic factors (e.g., IL-1, IL-2, CSFs, interferon, TNF) in murine and humans (Xu *et al.*,

Table III
Structures of Low-Molecular-Weight Nonproteinaceous Hematoregulators

AS 101
Ossirene, ammonium-trichloro(dioxoethylene-*O,O′*) tellurate

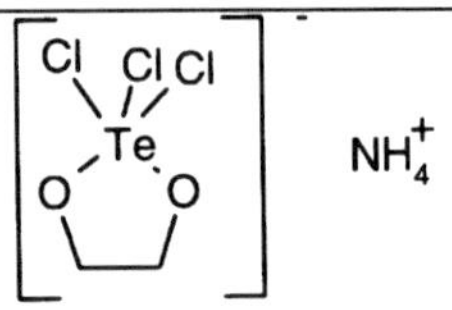

Y-25510
(±)-3-[4-(2-dimethylamino-1-methylethoxy)phenyl)]-1*H*-pyrazolo[3,4-*b*]pyridine 1-acetic acid

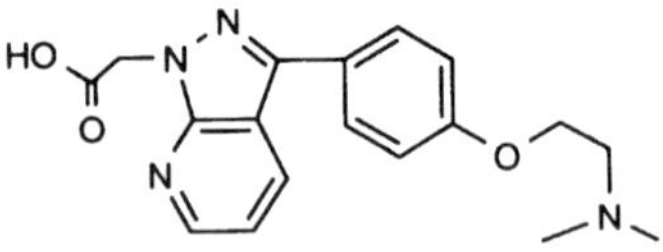

Goralatide
Seraspenide, N-acetyl-Ser-Asp-Lys-Pro

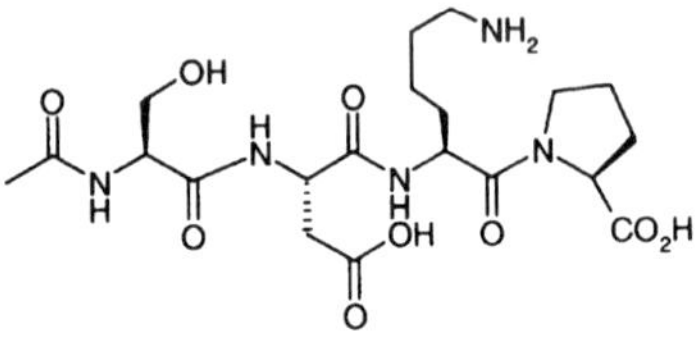

Lisofylline
2,6-dione-3,7-dihydro-1-(5-hydroxyhexyl)-3,7-dimetyl-1*H* purine

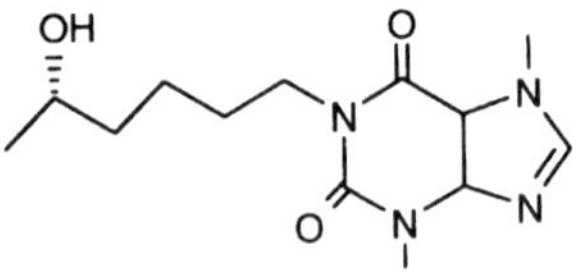

Bestatin
Ubenimex, 3-(*R*)-amino-2-(*S*)-hydroxy-4-phenylbutanoyl-L-leucine

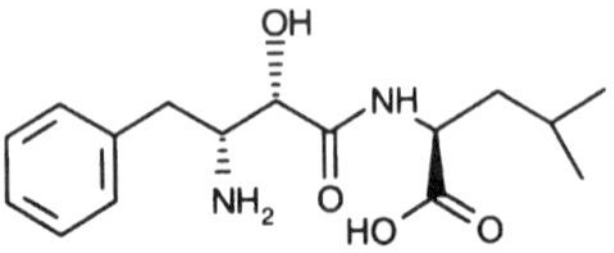

1996; Sredni *et al.,* 1996). Recent studies have shown that in mice oral administration of AS 101 at doses between 1 and 2 mg/kg alleviates hematopoietic suppression observed after treatment with sublethal doses of cyclophosphamide (CYP) and protects mice from the lethal effects of CYP (Sredni *et al.,* 1992). In addition, AS 101 administered orally confers a strong radioprotective effect in mice when given prior to irradiation (Sredni *et al.,* 1992). Treatment of tumor-bearing mice with AS 101 results in predominance of the Th-1 response with a concomitant Th-2 response. AS 101 also upregulated B7–1 on B-cell expression in a dose-dependent manner (Kalechman and Sredni, 1996). Clinical studies are ongoing with this agent in an effort to prevent chemotherapy-induced neutropenia and thrombocytopenia.

4.2.2. Y-25510

Y-25510 [(±)-3-[4-(2-dimethylamino-1-methylethoxy)phenyl)]-1H-pyrazolo [3,4-b] pyridine 1-acetic acid] (Table III) has been shown to prevent 5-fluorouracil-induced leukopenia in mice. Although the mechanism of the restorative action of Y-25510 is not known, it is believed to induce production of and/or potentiate the actions of stem cell growth factors that affect differentiation and proliferation of primitive multipotential hematopoietic stem cells. The molecular target of action for this molecule has not been identified (Hisadome *et al.,* 1992, 1996).

4.2.3. GORALATIDE (SERASPENIDE)

Seraspenide (N-acetyl-Ser-Asp-Lys-Pro-OH) (Table. III), a tetrapeptide, is a negative regulator of hematopoiesis and was first identified in calf bone marrow extracts (Lenfant *et al.,* 1989; Guigon and Bonnet, 1995; Watanabe *et al.,* 1996, Genevay *et al.,* 1996). It has been shown to inhibit the entry of murine colony-forming units–spleen (CFU-S) into S phase after chemotherapy or irradiation (Coutton *et al.,* 1994). Seraspenide also inhibits the proliferation of hematopoietic precursor cells from human marrow (Volkov *et al.,*1996 a,b); however, malignant cell lines, HL-60, K562, or precursor cells from chronic myeloid leukemic patients are not inhibited (Jackson *et al.,* 1996). The mechanism of action of this peptide is not known. It has been reported to cause inhibition of hypothalamic calmodulin-dependent phosphodiesterase stimulated by calmodulin and may achieve its biological affects by modulating the metabolism of cyclic nucleotides (Voelter *et al.,* 1995). It has been the subject of several clinical studies (Carde, 1994; Ezan *et al.,* 1994; Liozon *et al.,* 1995), and in Ara-C (arabinofuranosylcytosine)-treated patients Seraspenide seems to reduce the depth of granulocytes and platelet nadirs.

4.2.4. LISOFYLLINE

Lisofylline (Table III) is a metabolite of pentoxifylline and is under development as an immunomodulator to decrease toxicity of chemotherapy (Clarke *et al.,* 1996). It confers protection against chemo/radiotherapies by inhibiting formation of a phosphatidic acid that is linked to oxidative damage in tissues and is involved in intracellular signal transduction (Bursten *et al.,* 1996). It is currently in phase III clinical trials in patients with advanced hematological malignancies undergoing allogeneic bone marrow transplantation. The study was designed to evaluate its effect on infections resulting from neutropenia caused by high doses of chemotherapy (Bauer *et al.,* 1996), and preliminary evidence indicates protection from infection without acceleration of white blood cell (WBC) recovery.

4.2.5. BESTATIN (UBENIMEX)

Bestatin [3-(*R*)-amino-2-(*S*)-hydroxy-4-phenylbutanoyl-L-leucine] (Table III) has been shown to enhance the recovery of peripheral leukocytes in sublethally irradiated mice by inducing release of colony-stimulating factors by WBCs (Hourchi and Miyamoto, 1992). Its hematopoietic and therapeutic properties have been reviewed by Talmadge *et al.* (1991). Tsunogake *et al.* (1994) reported the effect of bestatin and its stereoisomers on the production of various cytokines from normal peripheral blood mononuclear cells (PBMNC). Bestatin stimulates production of G-CSF, GM-CSF, IL-6, IFN-γ, and TNF-α by PBMNC. Bestatin also exerts an inhibitory effect on hematopoiesis and tumor growth probably through stimulation of the production of TNF-α and IFN-γ. Its stereoisomers had no significant effect on hematopoiesis. Bestatin also stimulates the expression of stem cell factor mRNA in the stromal cells (KM 102). Bestatin is an orally bioavailable immunomodulator that utilizes the intestinal dipeptide transporter for absorption via the oral route (Saito *et al.,* 1996).

4.2.6. HEMATOREGULATORY PEPTIDE 5b (HP-5b)

In an attempt to isolate and characterize the active component of granulocyte chalone, Paukovits (1982) and Laerum and Paukovits (1984) published articles on the purification and chemical properties of an inhibitor of myelopoiesis. They isolated the inhibitor from the buffy coats of human blood lymphocytes. During purification of this factor it was noted that the active material was very sensitive to atmospheric oxidation. Exposure to oxygen resulted in the loss of the inhibitory effect, which was replaced, in time, with a stimulatory effect on myeloid colony formation. These observations were attributed to the presence of a thiol group in the molecule. This factor was found to be a peptide that contained only four amino acids, namely, aspartic acid, glutamic acid, cystine, and lysine. The N-terminus of this peptide was blocked. Treatment of the isolated factor with pyroglutamyl aminopeptidase resulted in loss of activity indicating that the N-terminus of the peptide was a pyroglutamic acid. Paukovits *et al.* (1987) assigned the sequence of the peptide as pyroGlu-Glu-Asp-Cys-Lys based on chemical synthesis to afford a synthetic peptide that was identical in biological profile to that of the natural factor. They named this factor HP-5b. To date it is not known if this hematoregulatory pentapeptide is part of a larger precursor molecule. A similar sequence motif has been shown to be a part of the effector domain of Gi alpha proteins but there is no surrounding sequence that would suggest that these proteins are the precursor of HP-5b (Laerum *et al.,* 1990). HP-5b demonstrated a dose-dependent inhibitory effect on myelopoietic colony formation (CFU-GM) on human and mouse cells at concentrations from 10^{-13} to 10^{-6} M. Air oxidation of this peptide resulted in the formation of a homodimer that stimulated myelopoietic cells (Fig. 1).

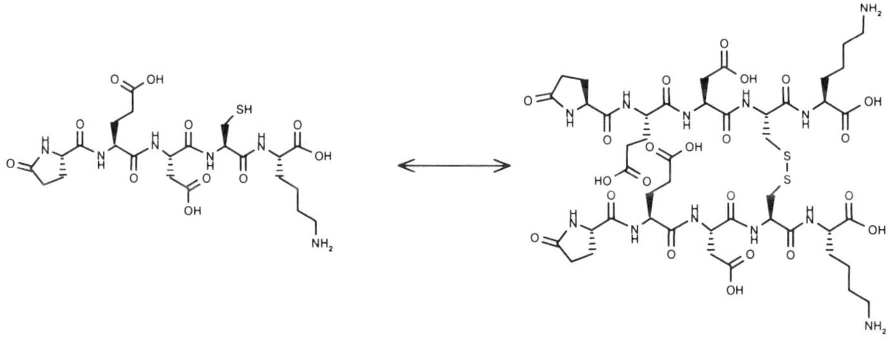

Figure 1. Interchangeable monomeric and dimeric isoforms of HP-5b.

Laerum *et al.* (1988) showed that female C3H mice when injected with a single dose of dimer of HP-5b produced a gradual increase in CFU-GM and CFU-S number in femur and spleen. Infusion of dimer for 6 days was followed by increased CFU-GM. On the other hand, infusion of monomer (HP-5b) resulted in 50% reduction of CFU-GM after 6 days and normal numbers of CFU-GM were reached after 13 days. The authors suggested that this monomer–dimer combination could constitute an efficient regulatory system for hematopoiesis *in vivo.* They also showed that the monomer–dimer combination could be used for protection from Ara-C- and N-mustard-induced myelotoxicity.

5. SK&F 107647 AND ANALOGUES

In an attempt to reproduce these results, we (King *et al.,* 1992) encountered some erratic results with this combination of monomer and dimer and attributed the problem to the redox chemistry in the cellular milieu. This property of these peptides made it particularly difficult to develop these agents as drugs.

Replacement of the thiol group in the monomer with a methyl group and replacement of the disulfide bond in the dimer with an ethylene group resulted in two stable and noninterchangeable peptides that acted as inhibitor and stimulator, respectively, of the hematopoietic system (Fig. 2).

These two peptides have been investigated for their potential clinical applications. The monomeric peptide (Fig. 2, SK&F 108636) has been shown to be a specific inhibitor of primitive stem cell proliferation and differentiation in mice and thus may be useful as a myeloprotectant protecting these crucial cells from being killed by chemotherapeutic agents (Veiby *et al.,* 1996).

The dimeric peptide, SK&F 107647, has been studied in detail. It is 10 times

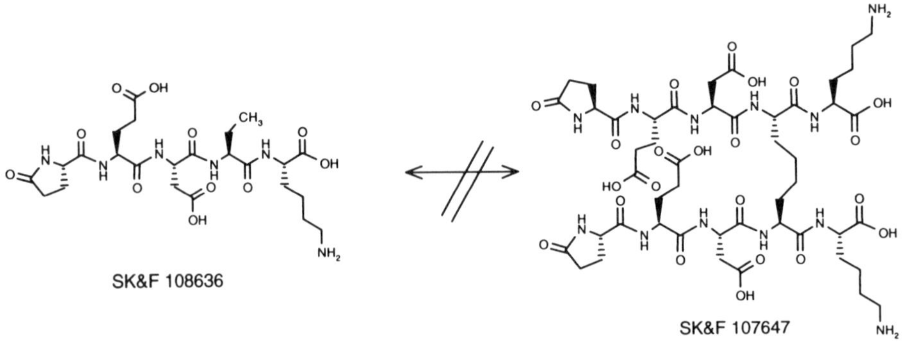

SK&F 108636

SK&F 107647

Figure 2. Structures of noninterchangeable monomeric (SK&F 108636) and dimeric (SK&F 107647) hematoregulatory peptides.

more potent than HP-5b dimer in eliciting colony-stimulating activity *in vitro* as well as *in vivo* (Pelus *et al.*, 1992, 1994). SK&F 107647 increases the CFU-GM cycle rate (King *et al.*, 1991). In addition to effects on early hematopoietic cells, this peptide also modulates the function of mature effector cells (Frey *et al.*, 1991). These properties potentially make SK&F 107647 a unique anti-infective agent that acts through host modulation. It has been proven to be efficacious in fungal and bacterial infection models (DeMarsh *et al.*, 1991, 1992, 1993). Recently it was discovered that the mechanism of action of SK&F 107647 and its analogues is to induce stromal cells to produce a factor that synergizes with the endogenous CSFs and directly activates mature PMN and monocytes (Frey *et al.*, 1991; King *et al.*, 1991, 1995a,b).

5.1. Structure–Activity Relationships of SK&F 107647

Because there is no known receptor for SK&F 107647, it was essential to ensure that the observed biological effects were related to some structural parameters of this peptide and did not reflect nonspecific effects. Studies delineating the structure–activity relationships of SK&F 107647 revealed that this molecule has very stringent requirements for biological activity (Bhatnagar *et al.*, 1996b). There are only a few allowable substitutions that result in either comparable or enhanced biological activity (Table IV). Most of the substitutions render the compound inactive. D-Amino acid substitutions or amino acid truncations were detrimental for colony-stimulating activity. The pGlu at position 1 could be substituted with heterocyclic carboxylic acids that contain a nitrogen atom α to the carboxylic acid. The Glu at position 2 could be replaced with Asp or Ser suggesting that a charged side chain is not required at this position. The stringent requirement of a negatively charged residue at position 3 suggested that this residue forms a critical salt bridge with some basic residue. The amino group of lysine at position 5 is very important for biological activity. The C-terminal carboxyl group of lysine could be replaced with carboxamide without any loss of activity. The number of methylene units spanning the diaminodicarboxylic acids at position 4 appeared to be critical. The di- and tetramethylene spacers were well tolerated whereas the mono-, tri-, penta-, and hexamethylene spacers were not. This suggested that not only the length of the span but the relative conformation of the methylene units was also important for biological activity. The vast difference in the EC_{50} of analogues containing even- and odd-membered spacers suggested that the distance and the relative orientation of the two peptide chains were crucial for the biological activity. The data indicated that both the net charge and the exact location of the charged groups were critical. The stringent structural requirements for this peptide and the availability of a panel of analogues with EC_{50}s ranging from micro- to picomolar strongly suggested that these compounds indeed interacted with a specific yet unidentified molecular target. The relative potency of some of these analogues are presented in Table IV.

Based on the SAR, a hypothesis for the pharmacophore of the peptide was developed (Fig. 3). The model suggests that the three residues of the C-termini act

Table IV
Relative Potency of SK&F 107647 Analogues

Structure	Relative potency in the CSA assay[a]
pGlu-Glu-Asp-Sub[c]-Lys (SK&F 107647)	1
pGlu-**Asp**-Asp-Sub-Lys	1
pGlu-**Ser**-Asp-Sub-Lys	10
pGlu-Glu-**Glu**-Sub-Lys	1
pGlu-Glu-Asp-**Cys**[d]-Lys	10^{-2}
pGlu-Glu-Asp-**Adp**[e]-Lys	10^{2}
pGlu-Glu-Asp-**Pim**[f]-Lys	na[b]
pGlu-Glu-Asp-**Aza**[g]-Lys	na
pGlu-Glu-Asp-**Asa**[h]-Lys	na
pGlu-Glu-Asp-Sub-**Arg**	na
pGlu-Glu-Asp-Sub-**Orn**	na
pGlu-Glu-Asp-Sub-**Lys-NH$_2$**	1
pGlu-Glu-Asp-Sub-**Lysinol**	na

[a]Relative potency = [EC_{50}(SK&F 107647)/EC_{50}(analogue)].
[b]na = relative potency $< 1 \times 10^{-6}$.
[c]Sub = (2*S*,7*S*)-2,7-diaminosuberic acid.
[d]Cys = cystine.
[e]Adp = (2*S*,5*S*)-2,5-diaminoadipic acid.
[f]Pim = (2*S*,6*S*)-2,6-diaminopimelic acid.
[g]Aza = (2*S*,8*S*)-2,8-diaminoazelaic acid.
[h]Asa = (2*S*,9*S*)-2,9-diaminosebacic acid.

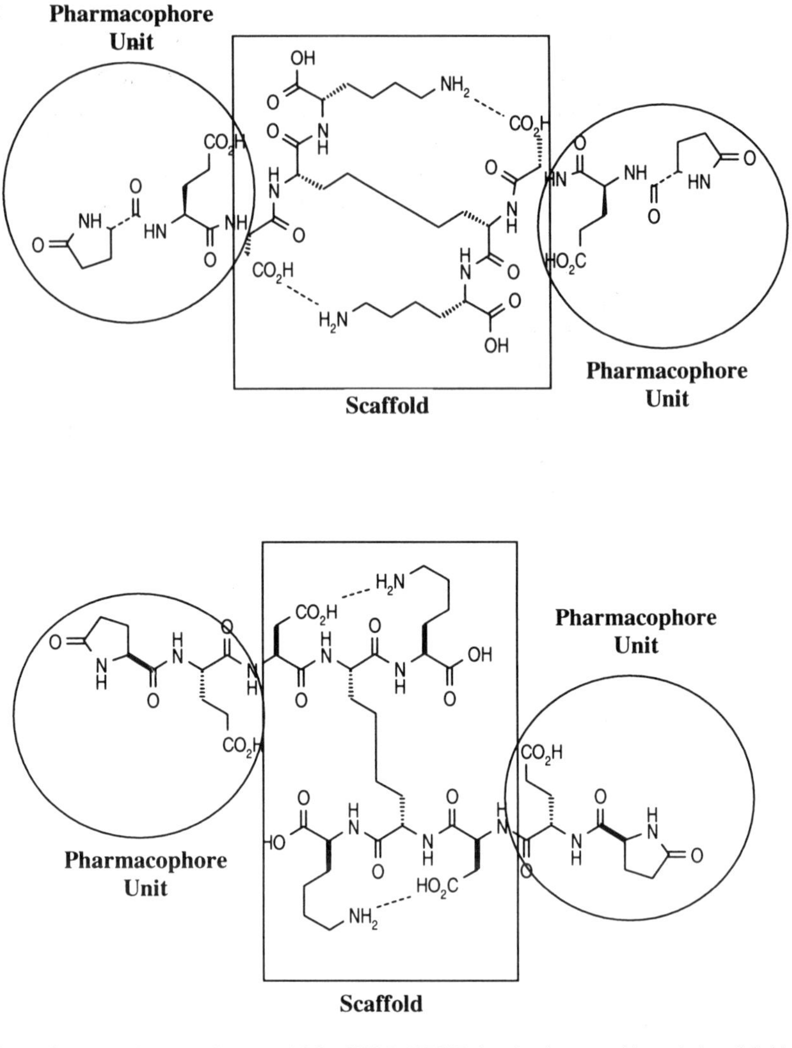

Figure 3. The pharmacophore model for SK&F 107647 showing intra- and interchain salt bridges.

as a scaffold and the N-termini interact with the putative target and act as the pharmacophore unit. This model was used to design biologically active peptidomimetic analogues of SK&F 107647 (Bhatnagar *et al.*, 1996a). These analogs further supported the idea that SK&F 107647 and related molecules manifested their biological activities through the interaction with an as yet unidentified target. The characterization of this target molecule might lead to the design of the next generation of hematoregulators.

5.2. Mechanism of Action

A great deal of effort has been devoted to elucidation of the mechanism of action of these peptides. The effects of this class of compounds on hematopoiesis and host defense mechanism appear to be indirect. So far, attempts to identify a receptor and/or secondary message for these peptides have not been successful. It has been shown that the peptides induce a hematopoietic synergistic factor (HSF) from the stromal cell. This factor has been purified from an SK&F 107647-treated murine stromal cell line C-6.4 and human cell line TF-274 (King *et al.,* 1995a,b). Amino acid sequence analysis identified the synergistic activity as the N-terminal truncated forms of the chemokine KC (KC-T) in murine and Gro-β (Gro-β-T) in humans. *In vitro* studies have shown that these synergistic factors mimic the hematopoietic activity of SK&F 107647 (King *et al.,* 1995a,b). These activities include increased CFU-GM number, fraction of CFU-GM in the S phase of the cell cycle, and augmentation of nonspecific host defense mechanisms *in vivo* defined by increased superoxide production and expression of CD11b/CD18 on PMN and monocytes.

5.3. Colony Stimulating Activity Induction Assay

To study these compounds *in vitro,* an assay widely used for monitoring hematopoietic activity of various growth factors was used (King *et al.,* 1992). Briefly, this assay utilizes an immortalized murine stromal cell line (C-6.4), derived from cultures of marrow stromal cells. In response to these compounds, C-6.4 cells produce factors in the conditioned media that stimulate colony forming activity (CSA) of murine bone marrow CFU-GM. Test compounds were added at various concentrations to confluent serum-free cultures of C-6.4. Cell-free supernatants were collected and sterile filtered after overnight incubation at 37°C in a humidified atmosphere and were evaluated for CSA. In this assay, SK&F 107647 and its analogues display a bell-shaped dose–response curve and have EC50s in the micro- to picomolar range (Bhatnagar *et al.,* 1996a,b).

5.4. Hematopoietic Synergistic Factor Assay

The analoguss of SK&F 107647 were also analyzed for their ability to induce hematopoietic synergistic factor (HSF) from the C-6.4 cell line. In this case, the above-mentioned CFU-GM assay was slightly modified. The endogenous CSFs produced by the C-6.4 cells were removed by filtration through a Centricon 30,000 MW cutoff membrane. The filtrates (which alone were unable to stimulate CFU-

GM formation) were analyzed for their ability to synergize with a suboptimal amount of exogenous M-CSF and enhance CFU-GM proliferation.

5.5. Preclinical Studies

As mentioned above, the dimer has been studied more extensively than the monomer. The following is a synopsis of steps that were required to bring this agent forward for development.

5.5.1. SYNTHESIS

The first step was to ensure that the peptide could be synthesized on a commercial scale in a reproducible manner. Large-scale solution- and solid-phase syntheses of SK&F 107647 were developed and it was shown that different preparations of peptide had identical biological activity (Alberts *et al.,* 1993; Bhatnagar *et al.,* 1996a,b). The synthesis of this peptide was complicated by the fact that it contained an unnatural amino acid, namely, 2,7-R,R-diaminosuberic acid in the sequence. This amino acid is not commercially available and an efficient large-scale synthesis had to be developed (Heibl and Rovenszky, 1996).

5.5.2. ANALYSIS OF ORAL ACTIVITY OF SK&F 107647

Oral activity is one of the most desired attributes in a clinical candidate. SK&F 107647 was evaluated for oral activity and shown to be orally bioactive (Pelus *et al.,* 1992) in various models of infection. However, drug absorption studies did not demonstrate the presence of SK&F 107647 in plasma, suggesting that this peptide was not orally bioavailable. Oral activity of SK&F 107647 without oral bioavailability presented an interesting paradigm. From *in vitro* studies employing intestinal tissues in a Ussing chamber, the permeability of SK&F 107647 was evaluated (Smith, 1996). In this *in vitro* approach, intestinal tissues obtained from rabbit are mounted between Lucite chambers and bathed on the luminal (mucosal) and serosal (blood) surfaces with buffer. Tissues remain viable for several hours with this technique and have been demonstrated to possess appropriate transport and barrier functions (Smith, 1996). Analysis of the mucosal and serosal bathing solutions indicated that following SK&F 107647 addition, very low amounts ($< 0.1\%$) of peptide were transported from either the mucosal to serosal or serosal to mucosal bathing solutions. These results suggested that the small intestine has little or no permeability to SK&F 107647. Evaluation of the mucosal and serosal bathing solutions from *in vitro* studies for HSF bioactivity indicated that the intestinal tissue itself is capable of producing HSF. Its production in small intestinal tissues appeared to be stimulated only when SK&F 107647 was added to the mucosal bathing solution. When SK&F 107647 was added to the mucosal bathing solution of small intestinal tissues, HSF activity could only be detected in the serosal bathing solution. HSF activity increased with time of incubation. No HSF bioactivity was ever detected in the mucosal bathing solution whether SK&F 107647 was added to the mucosal or serosal bathing solution. When SK&F 107647 was placed in the serosal bathing solution with small intestinal tissues, no detectable HSF was found in either bathing solution. These observations suggest that SK&F 107647 stimulates cells within rabbit small intestinal tissue to produce HSF, which is then secreted in a directional manner into the serosal bathing solution. These results may explain the oral bioactivity seen with SK&F 107647 in the absence of measurable plasma levels of the compound.

Together, these results support the hypothesis that SK&F 107647-induced production of HSF bioactivity from gastrointestinal tissues results in beneficial hematopoietic stimulation as well as augmentation of nonspecific host defense mechanisms *in vivo*.

5.5.3. PARENTERAL DOSAGE FORM EVALUATION

Following abdominal surgery, radiation therapy, or chemotherapy in patients, gastrointestinal physiology is dramatically altered affecting its absorptive function (Parsons, 1977). In such cases, development of a parenteral dosage form of the drug is imperative. Although SK&F 107647 conferred significant protection when dosed orally to animals that were infected with gram-negative (*E. coli*) or gram-positive (*S. aureus*) bacteria (the bacteria were dosed via a fibrin–thrombin clot, Ahrenholz and Simmons, 1980), demonstration of these effects following parenteral administration was also desirable.

Parenteral administration of SK&F 107647 to rats is complicated by a number of factors including: (1) the infection models required pretreatment for 6 days, and posttreatment for several days; (2) it was technically difficult to administer the drug to the mice intravenously over this time period; and (3) excessive handling of small rodents often results in immunomodulation, which may affect the conclusions of the experiment. Therefore, alternate approaches for parenteral administration of this peptide to rats were developed. In initial studies, Alzet® pumps were used to deliver the peptide. In some cases, it was observed that these pumps were not completely inert and did exert some immunomodulatory effects.

Recent commercial success in developing implantable biodegradable formulations containing peptides was the catalyst to use a similar preparation for SK&F 107647 delivery (DeMarsh *et al.,* 1996). Biodegradable microspheres have been used for a variety of purposes including sustained release of drugs and in taste masking (Uchida *et al.,* 1995). In addition, polylactide-coglycolide (PLGA) microspheres were used to deliver a decapeptide (Leuprolide) (Ogawa *et al.,* 1988).

The release of drug in this formulation is controlled by diffusion through channels in the spheres and hydrolysis of ester linkages. This formulation avoids the surgical treatment required for Alzet® pump implantation and also alleviates the need for daily injections. SK&F 107647 was formulated in biodegradable PLGA microspheres and the formulation of the peptide was shown to release uniform amounts of peptide for a period greater than 2 weeks (DeMarsh *et al.,* 1996). Animals received a single subcutaneous injection of biodegradable microspheres containing SK&F 107647 six days before infection. SK&F 107647 in this formulation was efficacious as evidenced by increased effector cell function measured by expression of CD11b on neutrophils and monocytes and by a 1000-fold reduction in bacterial count in the blood. As in previous studies, SK&F 107647 also showed a bell-shaped dose response (DeMarsh *et al.,* 1991, 1992, 1993).

5.5.4. DRUG DISPOSITION STUDIES

Because SK&F 107647 lacks a UV chromophore, conventional HPLC/UV analysis was not suitable for drug disposition studies. The first studies were therefore undertaken with radiolabeled material. A radiochemical synthetic strategy was devised whereby a diacetylenic precursor of the molecule was fully reduced with tritium to generate a high-specific-activity material for investigative studies. The first study of the disposition of [^{3}H]-SK&F 107647 was performed in the rat. The dose administered was selected based on consideration of the sensitivity of the radiometric analysis methods to be employed. The dosages thus selected for intravenous and oral administration were two orders of magnitude higher than the peak pharmacological dose. Analysis of the plasma concentration time data revealed that SK&F 107647 had moderate clearance in the rat with a terminal elimination half-life of approximately 20 min.

5.5.5. DEVELOPMENT OF ANALYTICAL METHODS

As mentioned above, conventional HPLC analysis was not suitable for pharmacokinetic and toxicokinetic studies and an ultrasensitive method of detection was required. Precolumn derivatization of this peptide was precluded by the presence of multiple sites of reactions and limited stability of the peptide in base. A postcolumn method of detection was devised in which the peptide was reacted with *o*-phthaldialdehyde after chromatographic separation on a reversed-phase column. This method was very sensitive and allowed detection of less than 10 ng/ml in 250 μl of plasma. The mean accuracy of detection ranged from 91.61 to 106.95%. The assay was validated over a range of 20 to 4000 ng/ml and was successfully used for the analysis of the plasma samples from preclinical studies in dogs and rats (Boppana and Miller-Stein, 1994).

The HPLC assay described above was used to analyze dog and rat plasma samples. However, much smaller doses were administered to humans and a more sensitive radioimmunoassay was developed for the quantification of SK&F 107647 in animal and human plasma. The sheep antiserum that cross-reacted with SK&F 107647 was produced by immunizing sheep with a conjugate of SK&F 107647 and ovalbumin. The immunogen was prepared by conjugating SK&F 107647 to albumin with glutaraldehyde. The antiserum was incubated with the radioiodinated analogue of SK&F 107647. The mixture was incubated with the test sample for 16–24 hr at 4°C and bound radiolabeled tracer was separated from unbound tracer by polyethylene glycol precipitation. A standard curve was obtained by using known concentrations of SK&F 107647 as test compounds and unknown samples were analyzed against this standard curve. The sensitivity of this method was 40 pg/ml in human plasma.

5.5.6. PHARMACOKINETIC STUDIES

Normally, peptides have a short half-life, high plasma clearance, restricted distribution to tissues, and, if small in size (MW < 5000), are filtered unchanged through the glomerulus (Humphrey and Ringrose, 1986). SK&F 107647 is no exception. It has been extensively studied for its pharmacokinetic properties in rat, dog, and human (Brocks *et al.*, 1996). In each species the plasma clearance is low in relation to hepatic blood flow. Its short half-life in plasma is attributed to a combination of high plasma clearance and restricted tissue distribution.

The relevance of the pharmacokinetic parameters of SK&F 107647 with respect to its pharmacodynamics is not clear. As mentioned above, SK&F 107647 causes induction of a protein factor (HSF) from stromal cells that in turn synergizes with hematopoietic growth factors and modulates effector cell functions. The length of exposure to SK&F 107647 that stromal cells require to produce HSF *in vivo* is not known; however, *in vitro* exposures of 1 or 24 hr did not change the induction patterns, suggesting that a short half-life may not be a detriment.

6. CONCLUSIONS

Natural host defenses against pathogens require efficient production of mature effector cells. Impairment of this function results in invasive infections in cancer patients or in other patients with suppressed immune systems such as burn or surgical patients. The hematopoietic factors, chemokines, and interleukins, which stimulate host defense mechanisms, offer new therapeutic venues. Since the late 1980s, clinical trials of CSFs have demonstrated that these agents can shorten the duration of neutropenia following intensive chemotherapy and reduce the episodes of neutropenia-associated infections. However, they are not completely risk-free. Various adverse effects, such as fever, flulike symptoms, dose-limiting hypotension, abdominal pain, neurotoxicity, and myocardial infarction, have been associated with some of these agents. The analysis of the pharmacoeconomics of these protein agents has yet to be completed. Recently, several small molecules that induce hematopoietic factors, chemokines, and interleukins have been identified, although their mechanisms of action are poorly understood. One such novel agent is SK&F 107647, which has been shown to be effective in various animal models of infections and is currently being evaluated for its potential clinical use. Developing such agents is a major challenge to pharmaceutical organizations, which are primarily attuned to the development of antipathogenic agents rather that host defense modifiers. Their development will require a greater understanding of the clinical pharmacology, understanding of chronobiology, and deconvolution of complicated pharmacodynamics. Nonetheless, in the current era of increasing incidence of resistant pathogens, host defense modifiers will have an important role in the new armamentarium against infections. Modulators of the hematopoietic system such as protein agents as well as orally active small molecules will provide novel therapeutic modalities in the future.

REFERENCES

Ahrenholz, D. H., and Simmons, R. L., 1980, Fibrin in peritonitis. I. Beneficial and adverse effects of fibrin in experimental *E. coli* peritonitis, *Surgery* **88:**41–47.

Alberts, D. P., Agner, E., Silvestri, J. S., Kwon, C., Newlander, K., King, A. G., Pelus, L. M., DeMarsh, P. L., Frey, C., Petteway, S. R., Huffman, W. F., and Bhatnagar, P. K., 1993, Synthesis of a novel hematoregulatory peptide SK&F 107647, *Am. Peptide Symp.* **13:**357–359.

Aviles, A., Guzman, R., Delgado, S., Nmabo, M. J., Gracia, E. L., and Diaz-Maqueo, J. C., 1996a, Intensive brief chemotherapy with hematopoietic growth factors as hematological support and adjuvant radiotherapy improve the prognosis in aggressive malignant lymphoma, *Am. J. Hematol.* **52:**275–280.

Aviles, A., Guzman, R., Gracia, E. L., Talavera, A., and Diaz-Maqueo, J. C., 1996b, Results of a randomized trial of granulocyte colony-stimulating factor in patients with infection and severe granulocytopenia, *Anticancer Drugs* **7:**392–397.

Babineau, T. J., Hackford, A., Kenler, A., Bistrian, B., Forse, R. A., Fairchild, P. G., Heard, S., Keroack, M., Caushaj, P., and Benotti, P., 1994, A phase II multicenter, double-blind, randomized, placebo-controlled study of three dosages of an immunomodulator (PGG-glucan) in high-risk surgical patients, *Arch. Surg.* **129:**1204–1210.

Barton, B. E., 1996, The biological effects of interleukin 6, *Med. Res. Rev.* **16:**87–109.

Bauer, G. J., Garcia, I., and Maier, R. V., 1996, Lisofylline attenuates GM-CSF enhancement of TNF-alpha production by alveolar macrophages in response to LPS, *Surg. Forum* **47:**99–101.

Bhatnagar, P. K., Alberts, D., Callahan, J. F., Heerding, D., Huffman, W. F., King, A. G., LoCastro, S., Pelus, L. M., and Takata, J. S., 1996a, Development of a pharmacophore model for a novel hematoregulatory peptide, *J. Am. Chem. Soc.* **118:**12862–12863.

Bhatnagar, P. K., Agner, E. K., Alberts, D., Arbo, B. E., Callahan, J. F., Cuthbertson, A. S., Engelsen, S. J., Fjerdingstad, H., Hartmann, M., Heerding, D., Hiebl, J., Huffman, W. F., Husbyn, M., King, A. G., Kremminger, P., Kwon, C., LoCastro, S., Lovhaug, D., Pelus, L. M., Petteway, S., and Takata, J. S., 1996b, Structure–activity relationships of novel hematoregulatory peptides, *J. Med. Chem.* **39:**3814–3819.

Bonilla, M. A., Dale, D., Zeidler, C., Last, L., Reiter, A., Ruggiero, M., Davis, M., Koci, B., Hammond, W., Gillio, A., and Welte, K., 1994, Long-term safety of treatment with recombinant human granulocyte colony-stimulating factor (r-metHuG-CSF) in patients with severe congenital neutropenias, *Br. J. Haematol.* **88:**723–730.

Boppana, V. K., and Miller-Stein, C., 1994, Determination of a novel hematoregulatory peptide in dog plasma by reversed-phase high performance liquid chromatography and an amine-selective o-phthaldialdehyde-thiol post column reaction with fluorescence detection, *J. Chromatogr. A* **676:**161–167.

Brocks, D. R., Freed, M. I., Martin, D. E., Sellers, T. S., Mehdi, N., Citerone, D. R., Boppana, V., Levitt, B., Davies, B. E., Nemunaitis, J., and Jorkasky, D. K., 1996, Interspecies pharmacokinetics of a novel hematoregulatory peptide (SK&F 107647) in rats, dogs, and oncologic patients, *Pharm. Res.* **13:**794–797.

Broxmeyer, H. E., 1993, Combination cytokine therapy or compound that may indirectly mimic such effects by stimulating production of multiple cytokines, *J. Exp. Hematol.* **20:**149–151.

Broxmeyer, H. E., 1995, Role of cytokines in hematopoiesis, in: *Human Cytokines* (B. B. Aggarwal and R. K. Puri, eds.), pp. 27–36, Blackwell, Oxford.

Buckner, F. S., Eschbach, J. W., Haley, N. R., Davidson, R. C., and Adamson, J. W., 1990, Hypertension following erythropoietin therapy in anemic hemodialysis patients, *Am. J. Hypertens.* **3:**947–955.

Bukowski, R. M., Budd, G. T., Gibbons, J. A., Bauer, R. J., Childs, A., Antal, J., Finke, J., Tuason, L., Lorenzi, V., McLain, D., Tubbs, R., Edinger, M., and Thomassen, M. J., 1994, Phase I trial of subcutaneous recombinant macrophage colony stimulating factor: Clinical and immunomodulatory effects, *J. Clin. Oncol.* **12:**97–106.

Bursten, S. L., Harris, W. E., and Rice, G. C., 1996, Selective inhibition of phosphatidic acid synthesis: A novel approach to the treatment of sepsis and the systemic inflammatory response syndrome, *Infect. Dis. Ther.* **19:**199–226.

Carde, P., 1994, Inhibitors of hematopoiesis: From physiology to therapy, *Bull. Acad. Natl. Med.* **178:**793–806.

Chatta, G. S., and Dale, D. C.,1996, Aging and hemopoiesis. Implications for treatment with haemopoietic growth factors, *Drugs Aging.* **9:**37–47.

Cisneros, R. L., Gibson, F. C., and Tzianabos, A. O., 1996, Passive transfer of poly-(1–6)-beta-glucotriosyl-(1–3)-beta-glucopyranose glucan protection against lethal infection in an animal model of intra-abdominal sepsis, *Infect. Immun.* **64:**2201–2205.

Clarke, E., Rice, G. C., Weeks, R. S., Jenkins, N., Nelson, R., Bianco, J. A., and Singer, J. W., 1996, Lisofylline inhibits transforming growth factor-β release and enhances trilineage hematopoietic recovery after 5-fluorouracil treatment in mice, *Cancer Res.* **56:**105–112.

Costa, J. J., Demetri, G. D., Harrist, T. J., Dvorak, A. M., Hayes, D. F., Merica, E. A., Menchaca, D. M., Gringeri, A. J., Schwartz, L. B., and Galli, S. J., 1996, Recombinant human stem cell factor (kit ligand) promotes human mast cell and melanocyte hyperplasia and functional activation *in vivo, J. Exp. Med.* **183:**2681–2686.

Coutton, C., Guigon, M., Bohbot, A., Ferrani, K., and Oberling, F., 1994, Photoprotection of normal human hematopoietic progenitors by the tetrapeptide N-AcSDKP, *Exp. Hematol.* **22:**1076–1080.

Danna, R. P., Rudnick, S. A., and Abels, R. I., 1990, Erythropoietin therapy for anemia associated with AIDS and AIDS therapy and cancer, in: *Erythropoietin in Clinical Applications—An International Perspective* (M. B. Garnick, ed.), pp. 310–324, Dekker, New York.

DeMarsh, P. L., Sucoloski, S. K., Wells, G. I., Frey, C. L., Bhatnagar, P. K., and Petteway, S. R., 1991, Efficacy of the hematopoietic peptide SK&F 107647 in normal and immunosuppressed mice challenged with *Candida albicans,* in: *XI Congress of the International Society for Human and Animal Mycology.*

DeMarsh, P. L., Sucoloski, S. K., Wells, G. I., Frey, C. L., Bhatnagar, P. K., and Petteway, S. R., 1992, Efficacy of the hematoregulatory peptide SK&F 107647 in mice challenged with Candida albicans treated with fluconazole or amphotericin (abstract), in: *Program and Abstracts of the 32nd Interscience Conference on Antimicrobial Agents and Chemotherapy,* New Orleans, American Society for Microbiology.

DeMarsh, P. L., Frey, C. L., Sucoloski, S. K., Henne, S. L., Barney, S., Bhatnagar, P. K., and Petteway, S. R., 1993, Efficacy of the hematoregulatory peptide SK&F 107647 in experimental herpes simplex II infection (abstract), in: *Program and Abstracts of the 33rd Interscience Conference on Antimicrobial Agents and Chemotherapy,* New Orleans, American Society for Microbiology.

DeMarsh, P. L., Wells, G. I., Lewandowski, T. F., Frey, C. L., Bhatnagar, P. K., and Ostovic, J. R., 1996, Treatment of experimental gram-negative and gram-positive bacterial sepsis with the hematoregulatory peptide SK&F 107647. *J. Infect. Dis.,* **173:**203–211.

Devine, S. M., Winton, E. F., Holland, H. K., Geller, R. B., Heffner, L. T., Hillver, C. D., Morris, L. E., Rodey, G. E., Beveridge, R., Lynch, J., Klein, L., and Dix, S. P., 1994, Simultaneous administration of interleukin-6 and Neupogen (rhG-CSF) following autologous bone marrow transplantation for breast cancer (abstract), *Blood* **84**(Suppl.):88a.

Dinarello, C. A., 1992, Role of interleukin-1 in infectious diseases, *Immunol. Rev.* **127:**119–146.

Du, X. X., Doerschuk, C. M., Orazi, A., and Williams, D. A., 1994, A bone marrow stromal-derived growth factor, interleukin-11, stimulates recovery of small intestinal mucosal cells after cytoablative therapy, *Blood* **83:**33–37.

Eschbach, J. W., Egrie, J. C., Downing, M. R., Browne, J. K., and Adamson, J. W., 1991, The safety of epoetin-alpha: Results of clinical trials in the United States, *Contrib. Nephrol.* **88:**72–80.

Ezan, E., Carde, P., Le Kerneau, J., Ardouin, T., Thomas, F., Isnard, F., Deschamps-de-Paillette, E., and Grognert, M., 1994, Pharmacokinetics in healthy volunteers and patients of N-Ac-SDKP (Seraspenide), a negative regulator of hematopoiesis, *Drug Metab. Dispos.* **22:**843–848.

Foster, P. F., Mital, D., Sankary, H. N., McChesney, L. P., Marcon, J., Koukoulis, G., Kociss, K., Leurgans, S., Whiting, J. F., and Williams, J. W., 1995, Use of granulocyte colony stimulating factor after liver transplantation, *Transplantation* **59:**1557–1563.

Frampton, J. E., Lee, C. R., and Faulds D., 1994, Filgrastim: A review of its pharmacological properties and therapeutic efficacy in neutropenia, *Drug* **48:**731–760.

Frey, C. L., DeMarsh, P. L., Sucoloski, S. K., Bhatnagar, P. K., and Pelus, L. M., 1991, The effect of the hematoregulatory peptide SK&F 107647 on murine peritoneal macrophage anti-Candida activity (abstract), in: *Program and Abstracts of the 31st Interscience Conference on Antimicrobial Agents and Chemotherapy,* New Orleans, American Society for Microbiology, 85.

Gately, M. K., and Mulqueen, M. J., 1996, Interleukin-12: Potential clinical applications in the treatment and prevention of infectious diseases, *Drug* **52**(Suppl. 2)**:**18–26.

Genevay, M. C., Mormont, C., Thomas, F., and Berthier, R., 1996, The synthetic tetrapeptide AcSDKP protects cells that reconstitute long-term bone marrow stromal cultures from the effects of mafosfamide, *Exp. Hematol.* **24:**77–81.

Gianella-Borradori, A., 1994, Present and future clinical relevance of interleukin-3, *Stem Cells* **12**(Suppl. 1)**:**241–248.

Glaspy, J., LeMaistre, C. F., Lill, M., Jones, R., Moore, R., Briddell, D., Menchaca, S., Turner, S., and Shpall, E. J., 1995, Dose-response of 7 day administration of recombinant methionyl human stem cell factor (SCF) in combination with Filgrastim (G-CSF) for progenitor cell mobilization in patients with stage II–IV breast cancer (abstract), *Blood* **86**(Suppl)**:**463a.

Gordon, M. S., Nemunaitis, J., Hoffman, R., Paquette, R., Samuel, S., Copper, R., Young, D., and Nimer, S., 1992, Phase I trials for subcutaneous recombinant human Il-6 in patients with myelodysplasia and thrombocytopenia, (abstract) *Blood* **80**(Suppl.)**:**249a.

Gordon, M. S., McCaskill-Stevens, W. J., Battiato, L. A., Loewy, J., Loesch, D., Breeden, E., Hoffman, R., Beach, K. J., Kuca, B., Kaye, J., and Sledge, G. W., 1996, A phase I trial of recombinant human interleukin-11 (Neumega rhIL-11 growth factor) in women with breast cancer receiving chemotherapy, *Blood* **87:**3615–3624.

Grant, S. M., and Heel, R. C., 1992, Recombinant granulocyte-macrophage colony stimulating factor. A review of its pharmacological properties and prospective role in the management of myelosuppression, *Drug* **43:**516–560.

Guigon, M., and Bonnet, D., 1995, Inhibitory peptides in hematopoiesis, *Exp. Hematol.* **23:**477–481.

Guillosson, J. J., 1996, Hematopoietic growth factors: General presentation, *Ann. Pharm. Fr.* **54:**145–150.

Hartung, T., Volk, H. D., and Wendel, A., 1995, G-CSF—an anti-inflammatory cytokine, *J. Endotox. Res.* **2:**195–201.

Hendrzak, J. A., and Brunda, M. J., 1995, Interleukin-12 . Biological activity, therapeutic utility, and role in disease, *Lab. Invest.* **72:**619–637.

Hiebl, J., and Rovenszky, F., 1996, Verfahren zur Herstellung substitulerter Diaminodicarbonsurederivate, EP 0691422-A1.

Hisadome, M., Fukuda, T., Terasawa, M., Oe, T., Takahata, H., Goto, K., Tsuru, S., and Nomoto, K., 1992, Enhancement of host defense by Y-25510, (±)-3-[4-(2-dimethylamino-1-methylethoxy)-phenyl]-1H-pyrazolo[3,4-b]pyridine-1-acetic acid, a novel synthetic compound. A comparison with recombinant human granulocyte colony-stimulating factor in 5-fluorouracil-treated mice, *Int. J. Immunopharmacol.* **14:**1195–1201.

Hisadome, M., Fukuda, T., Matsuyuki, H., Ikeda, Y., and Nomoto, K., 1996, Enhancement of in vivo production of IL-1 alpha and IL-6 in mice by Y-25510, a 1H-pyrazolo [3,4-b]pyridine-1-acetic acid derivative, *Int. J. Immunopharmacol.* **18:**379–384.

Hourchi, K., and Miyamoto, T., 1992, Radioprotective effects of Bestatin in Balb/c mice, *Int. J. Radiat. Biol.* **62:**73–80.

Huhn, R. D., Yurkow, E. J., Kuhn, J. G., Clarke, L., Gunn, H., Resta, D., Shah, R., Meyer, L. A., Seibold, J. R., Sturgill, M. G., Hoffman, R., Sheay, W., Cody, R., Philipp, C., Resta, D., and George, M., 1995, Pharmacodynamics of daily subcutaneous recombinant human interleukin-3 in normal volunteers, *Clin. Pharmacol. Ther.* **57:**32–41.

Humphrey, M. J., and Ringrose, P. S., 1986, Peptides and related drugs: A review of their absorption, metabolism and excretion, *Drug Metab. Rev.* **17:**383–410.

Hunt, P., and Foote, M. A., 1995, The new generation of recombinant human hematopoietic cytokines, *Curr. Opin. Biotechnol.* **6:**692–697.

Jackson, J. D., Yan, Y., Ewel, C., and Talmadge, J. E., 1996, Activity of N-acetyl-Ser-Asp-Lys-Pro (AcSDKP) on hematopoietic progenitor cells in short-term and long-term murine bone marrow cultures, *Exp. Hematol.* **24:**475–481.

Jakubowski, A., Rapits, G., Gilewski, T., Gabrilove, J., Shuster, S., Crown, J. Hudis, C., Seidman, A., Hoffman, R., and Caron, D., 1992, A phase I/II trial of PIXY-321 in patients receiving doxorubicin and thiotepa (abstract), *Blood* **80**(Suppl.)**:**88a.

Jones, T. C., 1994, Future use of granulocyte-macrophage colony-stimulating factor (GM-CSF), *Stem Cells* **12**(Suppl. 1)**:**229–240.

Kalechman, Y., and Sredni, B., 1996, Differential effect of the immunomodulator AS101 and B7-1 and B7-2 costimulatory molecules. Role in the antitumoral effects of AS101, *J. Immunol.* **157:**589–597.

Kammüeller, M. E., 1995, Recombinant human interleukin-6: Safety issues of a pleotropic growth factor, *Toxicology* **105:**91–107.

Kelley, K. W., Arkins, S., Minshall, C., Liu, Q., and Dantzer, R., 1996, Growth hormone, growth factors and hematopoiesis, *Horm. Res* **45:**38–45.

King, A. G., Bhatnagar, P., Balcarek, J., and Pelus, L. M., 1991, Modulation of bone marrow stromal cell production of colony stimulating activity by the synthetic peptide, SK&F 107647, *Exp. Hematol.* **19:**481.

King, A. G., Talmadge, J. E., Badger, A. M., and Pelus, L. M., 1992, Regulation of colony stimulating activity production from bone marrow stromal cells by the hematoregulatory peptide HP-5, *Exp. Hematol.* **20:**223–228.

King, A. G., Frey, C. L., Arbo, B., Scott, M., Johansen, K., Bhatnagar, P. K., and Pelus, L. M., 1995a, Hematoregulatory peptide, SK&F 107647, induced stromal cell production of KC [5–72] enhances CFU-GM growth and effector cell function (abstract) *Blood* **86:**(Suppl)**:**309a.

King, A. G., Scott, R., Wu, D. W., Strickler, J., McNulty, D., Scott, M., Johansen, K., McDevitt, D., Bhatnagar, P. K., Balcarek, J., and Pelus, L. M., 1995b, Characterization and purification of a stromal cell-derived hematopoietic synergistic factor induced by a novel hematoregulatory compound, SK&F 107647 (abstract), *Blood* **86**(Suppl.)**:**310a.

Klein, G., 1995, The extracellular matrix of the hematopoietic microenvironment, *Experientia* **51:**914–926.

Laerum, O. D., and Paukovits, W. R., 1984, Modulation of murine hematopoiesis in vivo by a synthetic hematoregulatory pentapeptide (HP5b), *Differentiation* **27:**106–112.

Laerum, O. D., Sletvold, O., Bjerknes, R., Eriksen, J. A., Johansen, J. H., Schanche, J. S., Tveteras, T., and Paukovits, W. R., 1988, The dimer of hemoregulatory peptide (HP5B) stimulates mouse and human myelopoiesis in vitro, *Exp. Hematol.* **16:**274–280.

Laerum, O. D., Frostad, S., Ton, H. I., and Kamp, D., 1990, The sequence of the hemoregulatory peptide is present in Gi alpha proteins, *FEBS Lett.* **269:**11–14.

Lasky, L. A., 1996, Hematopoiesis: Wandering progenitor cells, *Curr. Biol.* **6:**1238–1240.

Lawrence, D. A., 1996, Transforming growth factor-beta: A general review, *Eur. Cytol. Network* **7:**363–374.

Lenfant, M., Wdzieczak-Bakala, J., Guittet, E., Prome, J. C., Sotty, D., and Frindel, E., 1989, Inhibitor of hematopoietic pluripotent stem cell proliferation: Purification and determination of its structure, *Proc. Natl. Acad. Sci. USA* **86:**779–783.

Liozon, E., Volkov, L., Comte, L., Trimoreau, F., Pradelles, P., Bordessoule, D., Frindel, E., and Praloran, V., 1995, AcSDKP serum concentrations vary during chemotherapy in patients with acute myeloid leukaemia, *Br. J. Haematol.* **89:**917–920.

Lowry, P. A., 1995, Hematopoietic stem cell cytokine response, *J. Cell. Biochem.* **58:**410–415.

Maiti, S. K., and Singh, G R., 1996, Transforming growth-factor in bone remodeling, *Curr. Sci.* **71:**613–617.

Meropol, N. J., Petrelli, N. J., Lipman, B. J., Rodriguezbigas, M., Hicks, W., Douglass, H. O., Smith J. L., Rasey, M., Blumnenson, L. E., Vaickus, L., Hayes, F. A., and Agosti, J. M., 1996, Granulocyte-macrophage colony stimulating factor as an infection prophylaxis in high risk oncology surgery, *Am. J. Surg.* **72:**299–302.

Metcalf, D., 1990, The colony stimulating factors. Discovery development and clinical applications, *Cancer* **65:**2185–2195.

Miyazaki, H., 1996, Cloning of thrombopoietin and its therapeutic potential, *Cancer Chemother. Pharmacol.* **38**(Suppl.)**:**S74–S77.

Nemunaitis, J., Shannon-Dorcy, K., Appelbaum, F. R., Meyers, J., Owens, A., Day, R., Ando, D., O'Neil, C., Buckner, C. D., and Singer, J., 1993, Long-term follow-up of patients with invasive fungal disease who received adjunctive therapy with recombinant human macrophage colony-stimulating factor, *Blood* **82:**1422–1427.

Nemunaitis, J., Applebaum, F. R., Lilleby, K., Buhles, W. C., Rosenfeld, C., Zeigler, Z. R., Shadduck, R. K., Singer, J. W., Meyer, W., and Buckner, C. D., 1994a, Phase I study of recombinant interleukin-1 beta in patients undergoing autologous bone marrow transplant for acute myelogenous leukemia, *Blood* **83:**3473–3479.

Nemunaitis, J., Ross, M., Meisenberg, B., O'Reilly, R., Lilleby, K., Buckner, C. D., Appelbaum, F. R., Buhles, W., Singer, J., and Peters, W. P., 1994b, Phase I study of recombinant human interleukin-1 beta (rhIL-1 beta) in patients with bone marrow failure, *Bone Marrow Transplant* **14:**583–588.

Neta, R., 1990, Radioprotection and therapy of radiation injury with cytokines, *Prog. Clin. Biol. Res.* **352:**471–481.

Niven, R. W., 1993, Delivery of biotherapeutics by inhalation aerosols, *Pharm. Tech.* **17:**72–82.

Niven, R. W., Lott, F. D., and Cribbs, J. M., 1994, Pulmonary absorption of recombinant methionyl human granulocyte colony stimulating factor (r-huG-CSF) after intrathecal instillation to the hamster, *Pharm. Res.* **10:**1604–1610.

Ogawa, Y., Yamamoto, M., Okada, H., Yashiki, T., and Shimamoto, T., 1988, A new technique to efficiently entrap leuprolide acetate into microcapsules of poly lactic acid or copoly (lactic/glycolic) acid, *Chem. Pharm. Bull.* **36:**1095–1103.

Orkin, S. H., 1995, Hematopoiesis: How does it happen? *Curr. Opin. Cell Biol.* **7:**870–877.

Parsons, R. L., 1977, Drug absorption in gastrointestinal disease with particular reference to malabsorption syndromes, *Clin. Pharmacokinet.* **2:**45–60.

Patton, J. S., and Platz, R. M., 1992, Pulmonary delivery of peptides and proteins for systemic action, *Adv. Drug Del. Rev.* **8:**179–196.

Paukovits, W. R., 1982, Isolation and synthesis of a hematoregulatory peptide, *Z. Naturforsch.* **37C:**1297–1300.

Paukovits, W. R., Larrum, O. D., Paukovits, J. B., Guigon, M., and Scanche, J., 1987, Regulatory peptides inhibiting granulopoiesis, in: *The Inhibitors of Hematopoiesis* (A. Najman and M. Guigon, eds.), pp. 31–42, Colloque INSERM, John Libbey Eurotext.

Pelus, L. M., DeMarsh, P., King, A., Balcarek, J., Frey, C., Bhatnagar, P., Levin, R., and Scott, R., 1992, In vitro and in vivo hematopoietic activity of a novel synthetic hematoregulatory peptide, *J. Cell Biol.* **16C:**87.

Pelus, L. M., King, A. G., Broxmeyer, H. E., DeMarsh, P. L., Petteway, S. R., and Bhatnagar, P. K., 1994, In vivo modulation of hematopoiesis by a novel hematoregulatory peptide, *Exp. Hematol.* **22:**239–247.

Rackoff, W. R., Orazi, A., Robinson, C. A., Cooper, R. J., Alter, B. P., Freedman, M. H., Harris, R. E., and William, D. A., 1996, Prolonged administration of granulocyte colony-stimulating factor (Filgrastim) to patients with Fanconi anemia: A pilot study, *Blood* **88:**1588–1593.

Rollwagen, F. M., and Baqar, S., 1996, Oral cytokine administration, *Immunol. Today* **17:**548–550.

Rytömaa, T., and Kiviniemi, K., 1968a, Control of cell production in rat chloroleukemia by means of the granulocytic chalone, *Nature* **220:**136–137.

Rytömaa, T., and Kiviniemi, K., 1968b, Control of granulocyte production: Chalone and antichalone, two specific humoral regulators, *Cell Tissue Kinet.* **1:**329–340.

Sachs, L., 1996, The control of hematopoiesis and leukemia: From basic biology to the clinic, *Proc. Natl. Acad. Sci. USA* **93:**4742–4749.

Saito, H., Terada, T., Okuda, M, Sasaki, S., and Inui, K., 1996, Molecular cloning and tissue distribution of rat peptide transporter PEPT2, *Biochim. Biophys. Acta* **1280:**173–177.

Sanda, M. G., Yang, J. C., Topalian, S. L., Groves, E. S., Childs, A., Belfort, R., deSmet, M. D., Schwartzentruber, D. J., White, D. E., Lotze, M. T., and Rosenburg, S. A., 1992, Intravenous administration of human macrophage colony-stimulating factor to patient with metastatic cancer: A phase I study, *J. Clin. Oncol.* **10:**1643–1649.

Schiffer, C.A., 1996, Hematopoietic growth factors as adjuncts to the treatment of acute myeloid leukemia, *Blood* **88:**3675–3685.

Shivdasani, R. A., and Orkin, S. H., 1996, Review article: The transcriptional control of hematopoiesis, *Blood* **87:**4025–4039.

Slenar, I., Gianella-Borradori, A., and Jones, T. C., 1994, Update on clinical trials on the use of interleukin 3, in: *Cytokine Hemopoiesis, Oncology, Immunology,* III (M. Freund, ed.), pp. 23–30, Springer, Berlin.

Smith, J. W., Urba, W. J., Curti, B. D., Elwood, L. J., Steis, R. G., Janik, J. E., Sharfman, W. H., Miller, L. L., Fenton, R. G., Conlon, K. C., Sznol, M., Creekmore, S. P., Wells, N. F., Ruscetti, F. W., Keller, J. R., Hestdal, K., Shimizu, M., Rossio, J., Alvord, W. G., Oppenheim, J. J., and Longo, D. L., 1992, The toxic and hematological effects of interleukin 1 alpha administered in phase I trials to patients with advanced malignancies, *J. Clin. Oncol.* **10:**1141–1152.

Smith, P. L., 1996, Methods for evaluating intestinal permeability and metabolism *in vitro,* in: *Models for Assessing Drug Absorption and Metabolism* (R. T. Borchardt, P. L. Smith, and G. Wilson, eds.), pp. 13–34, Plenum Press, New York.

Soiffer, R. J., Roberston, M. J., Murray, C., Cochran, K., and Ritz, J., 1993, Interleukin-12 augment cytolytic activity of peripheral blood lymphocytes from patients with hematologic and solid malignancies, *Blood* **82:**2790–2796.

Sredni, B., Albeck, M., Kazimirsky, G., and Shalit, F., 1992, The immunomodulator AS101 administered orally as a chemoprotective and radioprotective agent, *Int. J. Immunopharmacol.* **14**(4)**:**613–619.

Sredni, B., Xu, R. H., Albeck, M., Grafter, U., Gal, R., Shani, A., Tichler, T., Shopira, J., Bruderman, I., Catane, R., Kaufman, B., Whisnant, J. K., Mettinger, K. L., and Kalechaman, Y., 1996, The protective role of the immunomodulator AS101 against chemotherapy-induced alopecia studies on human and animal models, *Int. J. Cancer* **65:**97–103.

Stashenko, P., Wang, C. Y., Riley, E., Wu, Y., Ostroff, G., and Niederman, 1995, Reduction of infection-stimulated periapical bone resorption by the biological response modifier PGG glucan, *J. Dent. Res.* **74:**323–330.

Takada, K., Tohyama, Y., Oohashi, M., Yoshikawa, H., Muranishi, S., Shimosaka, A., and Kaneko, T., 1989, Is recombinant human granulocyte colony-stimulating factor (G-CSF) orally available in rats? *Chem. Pharm. Bull.* **37:**838–839.

Talmadge, J. E., Pelus, L. M., Black, P. L., and Abe, F., 1991, Hematopoietic and therapeutic properties of bestatin in normal and myelosuppressed mice, *Biomed. Pharmacother.* **45:**61–69.

Teramura, M., Kobayashi, S., Yoshinaga, K., Iwabe, K., and Mizoguchi, H., 1996, Effect of interleukin 11 on normal and pathological thrombopoiesis, *Cancer Chemother. Pharmacol.* **38**(Suppl.)**:**99–102.

Tsunogake, S., Furusawa, S., Nagashima, S., Nakamura, Y., Enokihara, H., Shishido, H., Fujii, H., and Abe, F., 1994, Effect of aminopeptidase inhibitors on the production of various cytokines by peripheral blood mononuclear cells and stromal cells and on stem cell factor gene expression in stromal cells: Comparison of Ubenimex with its stereoisomers, *Int. J. Immunother.* **10:**41–47.

Uchida, T., Yoshida, K., Ninomiya, A., and Goto, S., 1995, Optimization of preparative condition of polylactide microspheres containing ovalbumin, *Chem. Pharm. Bull.* **43:**1569–1573.

Vadhan-Raj, S., Papadopoulos, N., Burgess, A, Patel, S., Linke, K., Hayes, C., Garrison, L., and Benjamin, R., 1993, PIXY321 (GM-CSF/IL-3) reduces chemotherapy (CT)-induced multilineage myelosuppression in patient with sarcoma, *Blood* **80**(Suppl. 1)**:**987.

van Gameren, M. M., Willemse, P. H., Mulder, N. H., Limburg, P. C., Groen, H. J., Vellenga, E., and deVries, E. G., 1994, Effects of recombinant human interleukin-6 in cancer patients: A phase I–II study, *Blood* **83:**1434–1441.

Veiby, O. P., LoCastro, S., Bhatnagar, P. K., and Olsen, W. M., 1996, Inhibition of enriched stem cells in vivo and in vitro by the hematoregulatory peptide SK&F 108636, *Stem Cells* **14:**215–224.

Veldhuis, G. J., Willemse, P. H., Sleijfer, D. T., van der Graaf, W. T., Groen, H. J., Limburg, P. C., Mulder, N. H., and de Vries, E. G., 1995, Toxicity and efficacy of escalating dosages of recombinant human interleukin-6 after chemotherapy in patients with breast cancer or non-small-cell lung cancer, *J. Clin. Oncol.* **13:**2585–2593.

Vial, T., and Descotes, J., 1996, Clinical toxicity of cytokines used as hematopoietic growth factors, *Drug Safety* **13:**371–406.

Vilpo, J. A., Kiviniemi, K., and Rytömaa, T., 1973, Inhibition of granulopoiesis by endogenous chalone study with the diffusion chamber technique, *Eur. J. Cancer* **9:**515–524.

Voelter, W., Kapuzniotu, A., Mihelic, M., Gurvits, B., Abrahamian, G., and Galoyan, A., 1995, The interaction of (1-4) fragment of thymosin beta-4 with calmodulin sensitive cAMP phosphodiesterase from hypothalamus, *Neurochem. Res.* **20:**55–59.

Volkov, L., Quere, P., Coudert, F., Comte, L., and Praloran, V., 1996a, The tetrapeptide AcSDKP, a physiological inhibitor of normal cell proliferation, reduces the S phase entry of continuous cell lines, *Exp. Cell Res.* **223:**112–116.

Volkov, L., Quere, P., Coudert, F., Comte, L., Antipov, Y., and Praloran, V., 1996b, The tetrapeptide

AcSDKP, a negative regulator of cell cycle entry, inhibits the proliferation of human and chicken lymphocytes, *Cell. Immunol.* **168:**302–306.

Wall, D. A., 1995, Pulmonary absorption of peptides and proteins, *Drug Deliv.* **2:**1–20.

Washburn, W. K., Otsu, I., Gottschalk, R., and Monaco, A. P., 1996, PGG-glucan, a leukocyte-specific immunostimulant, does not potentiate GVHD or allograft rejection, *J. Surg. Res.* **62:**179–183.

Watanabe, T., Brown, G. S., Kelsey, L. S., Yan, Y., Jackson, J. D., Ewel, C., Kessinger, A., and Talmadge, J. E., 1996, In vivo protective effects of tetrapeptide AcSDKP, with or without granulocyte colony-stimulating factor, on murine progenitor cells after sublethal irradiation, *Exp. Hematol.* **24:**713–721.

Weiss, M. J., and Orkin, S. H., 1996, In vitro differentiation of murine embryonic stem cells. New approaches to old problems, *J. Clin. Invest.* **97:**591–595.

Xu, R.H., Kalechman, Y., Albeck, M., Kung, H., and Sredni, B., 1996, Inhibition of B16 melanoma metastasis by the immunomodulator AS 101, *Int. J. Oncol.* **9:**319–325.

Yu, J., 1996, Regulation and reconstitution of human hematopoiesis, *J. Formosan Assoc.* **95:**281–293.

Zucker-Franklin, D., and Kaushansky, K., 1996, Effect of thrombopoietin on the development of megakaryocytes and platelets: An ultrastructural analysis, *Blood* **88:**1632–1638.

Chapter 17

Discovery and Development of GG745, a Potent Inhibitor of Both Isozymes of 5α-Reductase

Stephen V. Frye, H. Neal Bramson, David J. Hermann, Frank W. Lee, Achintya K. Sinhababu, and Gaochao Tian

1. INTRODUCTION

1.1. 5α-Reductases

The enzyme steroid 5α-reductase (5AR) has become the subject of significant biomedical research and drug discovery efforts largely because of its presumed role in the pathophysiology of the adult male. This connection to pathology and the function of 5AR in male sexual differentiation has led to advances in the understanding of the molecular biology, genetics, enzymology, and pharmacology of the 5AR enzymes. The utility of selective inhibitors of type 2 5AR as drugs has also been demonstrated. This chapter will focus on the discovery and early clinical development of a potent inhibitor of both isozymes of 5AR, GG745.

5AR catalyzes the NADPH-dependent reduction of $\Delta^{4,5}$ steroids and two isozymes of human 5AR have been cloned and characterized (Russell and Wilson,

---

Stephen V. Frye, H. Neal Bramson, David J. Hermann, Frank W. Lee, Achintya K. Sinhababu, and Gaochao Tian • Glaxo Wellcome Research and Development, Research Triangle Park, North Carolina 27709.

Integration of Pharmaceutical Discovery and Development: Case Studies, edited by Borchardt *et al.*, Plenum Press, New York, 1998.

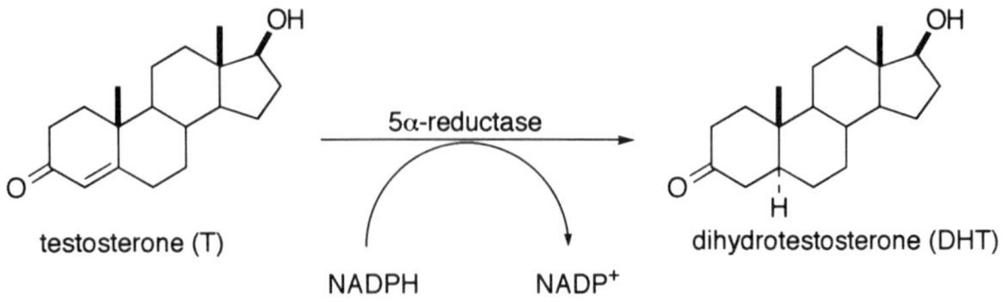

Figure 1. 5α-Reductase (5AR)-catalyzed conversion of testosterone to dihydrotestosterone. Adapted with permission from: Frye, S. V., 1996, Inhibitors of 5α-reductase, *Curr. Pharm. Des.* **2**:59–84. Copyright 1996 Bentham Science.

1994; Andersson and Russell, 1990; Jenkins *et al.,* 1992). The type 1 and 2 isozymes differ in their pH optima, sensitivity to inhibitors, and tissue distribution. The physiologically most well-characterized 5AR substrate, the androgen testosterone (T), and its more potent metabolite dihydrotestosterone (DHT; Fig. 1), are essential hormones responsible for male phenotypic sexual differentiation and maturation through their actions at the androgen receptor (Siiteri and Wilson, 1974; Wilson, 1989; Josso, 1994).

1.1.1. ROLE OF 5α-REDUCTASE IN NORMAL PHYSIOLOGY

The best-characterized function of 5AR is in the normal differentiation of the male reproductive tract. Initial male reproductive development requires production of testosterone and the antimüllerian hormone (AMH) by the fetal testes. The virilization of the external genitalia is dependent on the conversion of testosterone to DHT in the tissues of the urogenital sinus and a deficiency in type 2 5AR activity leads to an incomplete form of male pseudohermaphroditism (Imperato-McGinley *et al.,* 1974; Walsh *et al.,* 1974; Andersson *et al.,* 1991; Thigpen *et al.,* 1992). Affected males undergo varying degrees of virilization at puberty including penal enlargement, testicular descent, and development of male musculature. Facial and body hair is typically reduced in these men, temporal regression of the hairline is diminished, and the prostate remains small and is composed of exclusively stromal tissue (Imperato-McGinley *et al.,* 1992). This genetic phenotype demonstrates the dependency of the prostate on DHT for its development and suggests a possible role for DHT, and therefore 5AR, in diseases of aberrant prostate growth.

In addition to the unequivocal role of 5AR in male sexual development, there is evidence that 5AR serves a function in other physiological processes. There are high levels of 5AR activity in the liver, skin, and the tissues of central nervous system. Whereas the liver has been suggested to be a site where 5AR serves a catabolic function (Russell and Wilson, 1994), and the skin activity may mediate androgenic drive in that organ (Price, 1975; Darley, 1984; Imperato-McGinley *et al.*, 1993; Schweikert and Wilson, 1974; Diani *et al.*, 1992), the role of 5AR in the brain is less well understood. The distribution of 5AR activity throughout the central nervous system and the lack of sexual dimorphism in its expression are particularly intriguing (Pérez-Palacios *et al.*, 1975; Martini, 1982; Martini and Melcangi, 1991). It has recently been suggested that 5α-reduced metabolites of progesterone alter $GABA_A$ receptor function and play a part in sexual differentiation of the fetal brain (Lephart, 1993; Lephart and Husmann, 1993; Melcangi *et al.*, 1994).

1.1.2. BIOCHEMISTRY

The genetics, biochemistry, tissue distribution, and ontogeny of type 1 and 2 5AR have been reviewed recently by Russell and Wilson (1994) and only a brief summary of these topics will be presented here.

Because of the hydrophobicity and instability on chromatography of 5AR, neither isozyme has been purified to homogeneity to this day. Despite these difficulties, the existence of more than one 5AR was implied from the observed differences in pH optima of tissue-derived 5AR activity. A tremendous breakthrough came in 1989, when Russell and co-workers isolated the cDNA for a 5AR from rat liver using the technique of expression cloning (Andersson *et al.*, 1989). The homologous human 5AR was isolated from a prostate cDNA library by cross-hybridization and the enzyme is referred to as type 1 5AR. This 5AR possesses a neutral to basic pH optimum, is weakly inhibited by finasteride (a type 2 5AR-selective inhibitor; see Section 1.3), and has been shown not to be mutated in male pseudohermaphrodites suffering from 5AR deficiency. Human type 2 5AR was subsequently isolated via expression cloning from a prostate cDNA library and this isozyme had the sensitivity to finasteride and acidic pH optimum that had been anticipated based on studies carried out with prostatic tissue-derived 5AR activity. The gene coding for type 2 5AR was also shown to be mutated in subjects with 5AR deficiency (Andersson *et al.*, 1991).

The tissue distribution and biochemical characteristics of the isozymes of human 5AR are summarized in Table I (Russell and Wilson, 1994). Type 1 and 2 human 5AR are hydrophobic proteins of 259 and 254 amino acids, respectively, and their amino acid sequences are 50% identical. In man, type 2 5AR is primarily located in genital tissue and liver while the type 1 isozyme is found in the liver and skin (Levine *et al.*, 1996). The possibility that DHT produced by type 1 5AR can act as a circulating hormone, not just a paracrine hormone, and influence the growth and development of male sex organs is important to considerations of 5AR as a pharmacological target.

Table I
Biochemical Properties and Localization of Type 1 and 2 Human 5α-Reductase (Russell and Wilson, 1994)[a]

Properties	Type 1 5AR	Type 2 5AR
k_{cat}, sec^{-1} (testosterone)[b]	0.52	0.0001
K_m, μM (testosterone)[b]	6.3	0.0063
k_{cat}/K_m $M^{-1}sec^{-1}$[b]	8.2×10^4	1.6×10^4
pH optimum	6–8.5	5
Localization[c]		
Prostate, epididymis, seminal vesicle, genital skin[c]	absent	abundant
Testis, ovary, adrenal, brain, kidney	absent	absent
Liver	present	present
Nongenital skin	abundant	absent

[a]Adapted with permission from: Frye, S. V., 1996, Inhibitors of 5α-reductase, *Curr. Pharm. Des.* **2**:59–84. Copyright 1996 Bentham Science.
[b]See Tian *et al.* (1995a).
[c]Based on mRNA blot hybridization and immunoblotting to detect protein (Russell and Wilson, 1994).

In addition to testosterone, the 5ARs will reduce many other steroids with the 3-oxo-$\Delta^{4,5}$ structure and progesterone appears to be the optimal endogenous substrate. The mechanism of catalysis of rat 5AR has been well studied and has been used in the design of inhibitors and rationalization of their SAR. An ordered binding of substrates and release of products from the enzyme has been proposed as outlined in Fig. 2 (Levy *et al.*, 1990). The enolatelike transition state traversed during delivery of the hydride of NADPH is presumed to be stabilized by an electrophilic residue in the enzyme active site and most inhibitors of 5AR mimic this transition state in some fashion (Frye, 1996).

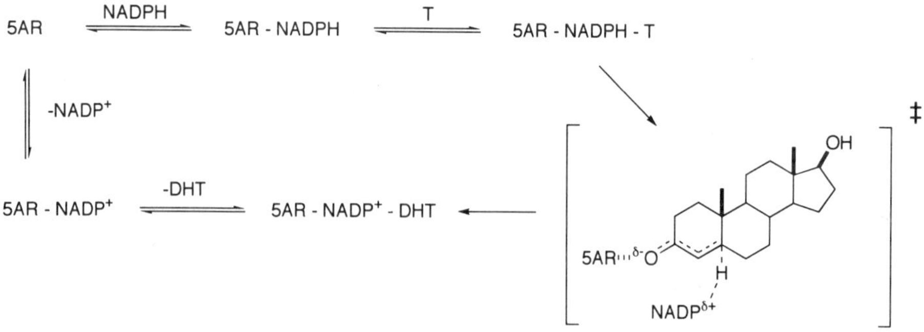

Figure 2. The kinetic mechanism of 5α-reductase and schematic transition state for reduction. Adapted with permission from: Frye, S. V., 1996, Inhibitors of 5α-reductase, *Curr. Pharm. Des.* **2**:59–84. Copyright 1996 Bentham Science.

1.2. Pathophysiology of DHT

Benign prostatic hyperplasia (BPH) and prostate cancer are major causes of morbidity and mortality in the aging male population (Isaacs, 1990; Geller, 1991; Arrighi *et al.*, 1991; Denis and Mahler, 1990; Geller, 1993). The clinical symptoms attributed to BPH occur in the majority of men over the age of 60 and microscopic hyperplasia is a nearly universal finding on autopsy. Prostate cancer is the most common cancer in men with more than 300,000 new cases diagnosed and more than 40,000 cancer-related deaths attributed each year in the United States. Both BPH and prostate cancer are rare in early adulthood but increase steadily from age 50 onward. BPH and prostate cancer are believed to be independent diseases that originate in different regions of the prostate although they share a dependence on androgens for growth.

BPH is a nonmalignant enlargement resulting from growth of both the stromal and glandular components of the prostate and is the most common neoplastic disease of man (Isaacs, 1990; Geller, 1991). As a consequence of the anatomical location of the prostate, surrounding the urethra just below the bladder, the growth of the prostate can produce difficulty in urination. Although the growth of the gland is associated with the symptoms of BPH, there is no direct correlation between size and symptomatology (Barry *et al.*, 1993).

Two prerequisites for the development of BPH are the presence of testes and aging (Geller, 1991). The dependence of BPH on testicular androgens has been known for some time and was investigated thoroughly by Moore (1944) who showed that absence of functioning testes prior to 40 years of age prevents both BPH and prostate cancer. Testosterone from the testes provides the substrate for 5AR-catalyzed production of DHT, the major androgen acting in the prostate. Given the phenotype of genetic 5AR type 2 deficiency, DHT is likely a necessary component in the development of BPH. Because benign or malignant disease of adjacent sex glands, such as the seminal vesicles, is practically unheard of, the special physiology of the prostate must contribute greatly to the development of disease (Griffiths *et al.*, 1991; Kreig *et al.*, 1993).

With the emergence of the role of DHT as the primary androgen in the prostate, Petrow and Padilla (1984) proposed that 5AR could be a target enzyme for prostate cancer. The relative role of testosterone and DHT in the hormone-dependent growth of prostatic carcinoma is critical to the potential success of 5AR inhibitor treatment (Gormley, 1991; Presti *et al.*, 1992). The possibility of preventing prostate cancer by treatment with finasteride, a type 2-selective 5AR inhibitor, is also under investigation (Brawley *et al.*, 1994).

In addition to BPH and prostate cancer, certain diseases of the skin may be dependent on DHT. Acne, idiopathic female hirsutism, and male pattern baldness have all been linked to increased 5AR activity in the areas of the skin affected (Price, 1975; Darley, 1984; Schweikert and Wilson, 1974; Brooks, 1986; Tenover,

1991). Studies to determine the efficacy of 5AR inhibitors in these conditions are now under way (Dallob *et al.*, 1994; Imperato-McGinley *et al.*, 1993; Diani *et al.*, 1992).

1.3. Finasteride: Clinical Effects of a Type 2-Selective 5α-Reductase Inhibitor

Workers from Merck described the first member of the 4-azasteroid class of 5AR inhibitors in 1981 and have pioneered many aspects of drug discovery in this area. Compound **1** (Fig. 3, 4MA) was described as a potent competitive, reversible inhibitor of rat prostatic 5AR (K_i = 5 nM) with some affinity for the rat androgen receptor (IC_{50} = 3 μM) (Liang and Heiss, 1981). Further characterization of 4MA ultimately led to the realization that it was a potent inhibitor of 3β-hydroxy-Δ^5-steroid dehydrogenase/3-keto-Δ^5-steroid isomerase (3BHSD) (Cooke and Robaire, 1986; Chan *et al.*, 1987; Brandt and Levy, 1989; Perron and Bélanger, 1994; Frye *et al.*, 1994), a critical enzyme for steroid biosynthesis (Potts *et al.*, 1978), in a number of species, including humans (Frye *et al.*, 1994). 4MA was ultimately not investigated clinically because of hepatotoxicity observed in the dog (McConnell, 1990).

Continued optimization of the 4-azasteroid class resulted in compound **2,** finasteride (Fig. 3), which was chosen for clinical development based on its *in vitro* and *in vivo* potency and selectivity (Rasmusson *et al.*, 1984, 1986). Finasteride is widely approved for the treatment of BPH and is by far the most studied inhibitor of 5AR (Peters and Sorkin, 1993; Sudduth and Koronkowski, 1993; Rittmaster, 1994). At the time finasteride was initially assessed clinically, the existence of two isozymes of human 5AR was unknown, as was finasteride's mechanism of inhibition. The pharmacodynamic results of phase I dose ranging studies examining the biochemical efficacy of finasteride foreshadowed both the discovery of type 1 5AR and the realization that finasteride is not a simple competitive inhibitor of 5AR (Vermeulen *et al.*, 1989; Gormley *et al.*, 1990; Ohtawa *et al.*,

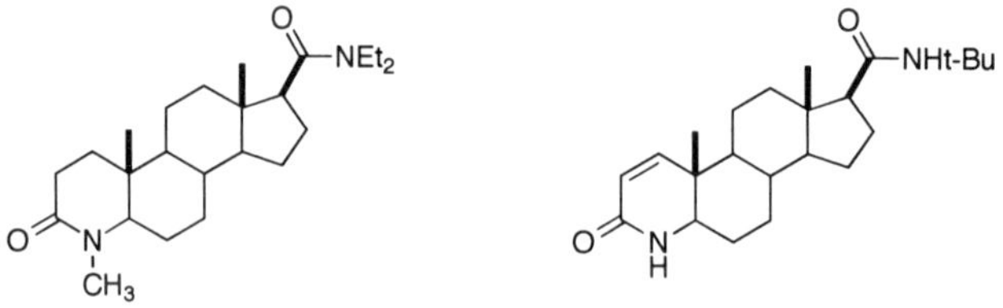

O NEt$_2$ O N CH$_3$

1, 4MA

O NHt-Bu O N H

2, finasteride

Figure 3. The 4-azasteroids, 4MA and finasteride.

1991; Vermeulen *et al.,* 1991; De Schepper *et al.,* 1991). Vermeulen (1991) reported a phase I study in which doses from 0.04 to 400 mg were examined. The resulting maximum reduction in plasma DHT achieved was 60–80% with up to 7 days required for return to baseline DHT levels, this despite finasteride's modest half-life of 6–8 hr in humans. The dose response observed in this study, depicted in Fig. 4 (see also Fig. 10, Section 3.2), is representative of phase I results with finasteride. The residual DHT observed in these and subsequent long-term studies (Mocellini *et al.,* 1993; Stoner, 1994) can now be attributed to the relatively slow rate of inhibition of type 1 5AR by finasteride and the long pharmacodynamic half-life is clearly related to the kinetics of its inhibition (Faller *et al.,* 1993; Tian *et al.,* 1994, 1995a; Tian, 1996; Bull *et al.,* 1996), as will be detailed in Section 2.1.

1.4. Potential Utility of a Dual 5α-Reductase Inhibitor

In clinical trials, finasteride has been shown to decrease plasma DHT, shrink the prostate, and result in a modest decrease in symptoms related to the disease (Peters and Sorkin, 1993; Sudduth and Koronkowski, 1993; Rittmaster, 1994). Overall, the results of finasteride therapy fall somewhat short of initial expectations and several reasons for this may be advanced (McConnell, 1990). Among these, the following obstacles will be faced by any inhibitor of 5AR: heterogeneity of the disease (Shapiro *et al.,* 1992); coincidental concurrence of BPH and symptoms unrelated to prostatic enlargement (Barry *et al.,* 1993); the role of residual, or rising testosterone levels in maintenance of prostate size (Grino *et al.,* 1990); and the possibility that the essential developmental role of DHT in the prostate is not mirrored in regression of the developed, hyperplastic gland on removal of DHT support (Geller, 1991). However, the residual circulating DHT in patients treated with finasteride (20–40% of baseline; see Fig. 4) (Vermeulen *et al.,* 1989, 1991; Gormley *et al.,* 1990; Ohtawa *et al.,* 1991; De Schepper *et al.,* 1991) is a clear target for possible improvement. A more effective dual inhibitor of type 1 and 2 human 5AR may lower circulating DHT to a greater extent than finasteride and show advantages in the treatment of BPH and other disease states that depend on DHT.

2. ENZYMOLOGY OF 5α-REDUCTASES

2.1. Time Dependence of Inhibition by Δ^1 4-Azasteroids

As discussed in Section 1.4. the results of clinical studies with finasteride were incompatible with finasteride being a competitive reversible inhibitor of the

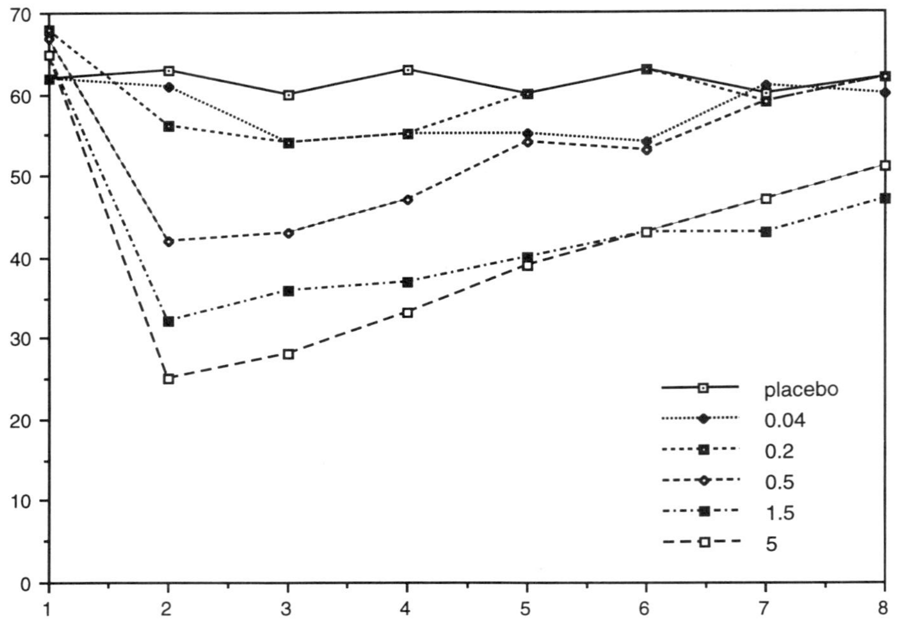

Figure 4. Plasma levels of dihydrotestosterone (DHT) (ng/dl) before and up to 7 days after a single dose of between 5 and 0.04 mg finasteride (dosed on day 2). Adapted with permission of S. Karger AG, Basel from: Vermeulen, A., Giagulli, V. A., De Schepper, P. J., and Buntinx, A., 1991, Hormonal effects of a 5α-reductase inhibitor (finasteride) on hormonal levels in normal men and in patients with benign prostatic hyperplasia, *Eur. Urol.* **20:**82–86. Copyright 1991 S. Karger AG.

two 5AR isozymes. Principle data not explained well by this mechanism were the unexpected potency of finasteride for lowering DHT concentrations by 60–80% and the discrepancy between the pharmacokinetics and pharmacodynamics of finasteride in man. The values of K_i for finasteride calculated according to the competitive reversible mechanism were in the 3–26 nM range for the type 2 5AR (Liang *et al.*, 1985; Andersson *et al.*, 1991; Jenkins *et al.*, 1992; Faller *et al.*, 1993), and more than 10-fold higher for the type 1 5AR (Andersson and Russell, 1990; Thigpen *et al.*, 1993). Contrary to predictions based on these potencies and pharmacokinetic measurements, single 1- to 10-mg doses of finasteride produce 60–80% suppression of DHT and maintain inhibition of 5AR for more than a week (Fig. 4) (Vermeulen *et al.*, 1989; Ohtawa *et al.*, 1991; Mocellini *et al.*, 1993). The discovery that finasteride is a time-dependent inhibitor of the type 2 (Faller *et al.*, 1993) and type 1 (Tian *et al.*, 1994) 5ARs provides an explanation for the potency and long-lasting effects of finasteride that were observed in phase I studies.

Unlike an inhibitor that binds to an enzyme active site at a diffusion-controlled rate and inhibits the enzyme by a classical competitive mechanism, the interactions between finasteride and the 5AR isozymes are described well by the two-step mechanism:

$$E + I \underset{}{\overset{K_i}{\leftrightarrows}} EI \overset{k_3}{\rightarrow} EI^* \tag{1}$$

where the equilibrium for the first step is established rapidly and the second step is slow (Faller *et al.*, 1993; Tian *et al.*, 1994, 1995a; Bull *et al.*, 1996). The presence of this second step gives rise to time-dependent inhibition as seen with finasteride and the 5AR isozymes (Fig. 5). Values for K_i, the inhibition constant for the initial binding step, and k_3, the rate constant for the second, time-dependent step, were obtained for each enzyme through progress curve analyses. Data for these experiments are summarized in Table II. Finasteride is an extremely fast time-dependent inactivator of the type 2 5AR, with a second-order rate constant k_3/K_i of $3 \times 10^5\ M^{-1}sec^{-1}$ (pH 7.0, 37°C), which is comparable to the k_{cat}/K_M for the reduction of testosterone (Tian *et al.*, 1995a; Bull *et al.*, 1996). Finasteride is a relatively slow time-dependent inactivator of the type 1 5AR, with a k_3/K_i of $4 \times 10^3\ M^{-1}sec^{-1}$ (pH 7.0, 37°C) (Tian *et al.*, 1995a; Bull *et al.*, 1996).

These time-dependent inhibition processes were shown to result from interactions of finasteride with 5AR active sites through preincubation studies in which the effects of known active-site reagents on the modification process were measured. This was accomplished by observing the effects of 4MA (**1,** Fig. 3), a simple competitive and reversible inhibitor of the human type 1 5AR, and the substrate progesterone (a more efficient 5AR substrate than testosterone) on finasteride-induced inactivations of 5ARs. The kinetic constants for the enzyme active-site reagents were indistinguishable from those determined in the absence of

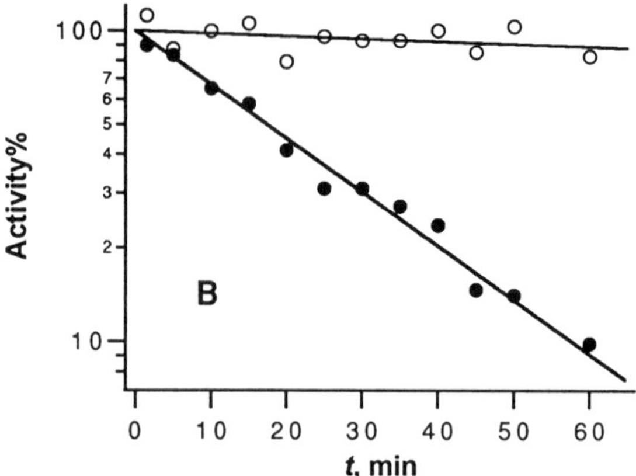

Figure 5. Time-dependent loss of 5AR 1 activity. The microsomal 5AR 1 was preincubated at 37°C with (●) or without (○) 1 μM finasteride. The activity versus time data were fit to a first-order decay function by nonlinear least-squares analysis. Reproduced with permission from: Tian, G., Stuart, J. D., Moss, M. L., Domanico, P. L., Bramson, H. N., Patel, I. R., Kadwell, S. H., Overton, L. K., Kost, T. A., Mook, R. A., Jr., Frye, S. V., Batchelor, K. W., and Wiseman, J. S., 1994, 17β-(N-tert-butylcarbamoyl)-4-aza-5α-androstan-1-en-3-one is an active site-directed slow time-dependent inhibitor of human steroid 5-alpha-reductase, *Biochemistry* **33:**2291–2296. Copyright 1994 American Chemical Society.

finasteride, demonstrating that the time-dependent inhibition by finasteride occurs at the enzyme active site.

The dialysis of [^{3}H]finasteride-inactivated type 1 or 2 5AR in denaturing solvents results in the release of more than 98% of labeled finasteride, suggesting that these enzymes are not covalently modified in the inactivated complexes (Bull *et al.*, 1996). The first-order rate constant K_{off} for finasteride dissociation from the inhibitory complex with the type 2 5AR was found to be (2.57 ± 0.03) × 10^{-7} sec^{-1} (pH 7.2, 37°C), by measuring the rate of exchange between radiolabeled fi-

Table II
Summary of Kinetic Parameters for the Inhibition of 5AR by Finasteride at pH 7.0 (Tian *et al.*, 1995a)

5AR	k_3 (sec^{-1})	K_i (nM)	k_3/K_i ($M^{-1}sec^{-1}$)[a]	T (°C)
Type 1	(6.7 ± 0.8) × 10^{-4}	360 ± 40	(1.9 ± 0.3) × 10^3	22
	(1.4 ± 0.1) × 10^{-3}	360 ± 40	(4.0 ± 0.6) × 10^3	37
Type 2	(5.1 ± 0.7) × 10^{-3}	69 ± 1[b]	(8.3 ± 0.5) × 10^4	22
	(4.9 ± 1.2) × 10^{-3}		(3.2 ± 0.4) × 10^5	37

[a] The standard errors were computed assuming that the errors from k_3 and K_i propagated independently.
[b] Determined from inhibition of initial rates.

nasteride in the enzyme inhibitory complexes and unlabeled finasteride in solution (Bull *et al.,* 1996). Thus, the half-life for finasteride dissociation from the enzyme–inhibitor complex is $\geq$ 31 days (pH 7.2, 37°C), and there is no evidence that the enzyme regains activity following the dissociation of inhibitor (Bull *et al.,* 1996). Similarly, the half-life for the dissociation of finasteride from the type 1 5AR is equivalent to 14 $\pm$ 2 days (pH 7.2, 37°C) (Bull *et al.,* 1996). The dissociation constants for dissociation of inhibitor from the inactivated enzyme complex at steady state is equal to k_{off}/k_{on}, and in this manner the K_is at steady state were calculated to be 2×10^{-10} and $\leq 3 \times 10^{-13}$ M (pH 7.0, 37°C) for the type 1 and 2 5ARs, respectively (Bull *et al.,* 1996). These data illustrate the thermodynamic driving force behind the potency of finasteride for inhibition of both 5ARs. However, the kinetics of inhibition renders finasteride essentially a type 2-selective 5AR inhibitor.

The nature of the inhibitor in the inactivated enzyme complexes was probed utilizing isotopic replacement of hydrogen to measure kinetic isotope effects. The possibility that k_3 [Eq. (1)] may describe a covalent reaction was probed through replacement of the C1 hydrogen atom with tritium, which would alter the rate of nucleophilic attack on this position. To perform this study, finasteride labeled with tritium at C-1 and C-2 positions and finasteride ^{14}C-labeled at the C-17 *tert*-butyl group were coincubated at pH 7.0 for 10 hr at 22°C with the type 1 or the type 2 5AR isozyme in the presence of 1 mM NADPH (Tian *et al.,* 1995b). The amount of label in the free and protein-bound finasteride were significantly different and a large inverse kinetic isotope effect was observed for both type 1 (0.71 $\pm$ 0.02) and type 2 (0.63 $\pm$ 0.02) isozymes (Tian *et al.,* 1995b). This suggests that the rate [k_3, Eq. (1)] of the finasteride-induced slow inactivation of the 5ARs results from a chemical transformation at the C-1 and/or C-2 positions of finasteride. Further, the inverse nature of the isotope effects is consistent with sp^2 to sp^3 rehybridization, and is therefore also consistent with nucleophilic addition at the Δ^1 double bond of finasteride. This, combined with the finding that finasteride is a time-dependent inhibitor of the type 1 and 2 5ARs suggests that the mechanism for slow inhibition by finasteride involves the attack of a nucleophile on the unsaturated C-1 position of the steroid *in vitro,* and possibly *in vivo* as well (Tian *et al.,* 1995b).

Resolution of the possible conflict between interpretations of the results of the dialysis experiments, which could be interpreted as ruling out a covalent interaction between 5AR and finasteride, and the kinetic isotope effect study was based on the finding that finasteride is released from the inactivated enzyme complexes as its 1,2-dihydrofinasteride metabolite (Bull *et al.,* 1996). However, dihydrofinasteride is a simple reversible inhibitor of the type 2 5AR, thus the reduction of finasteride does not account for the time-dependent inhibition of the 5ARs (Bull *et al.,* 1996), nor would it explain the inverse isotope effect (Tian *et al.,* 1995b). Based on the partitioning between organic and aqueous layers of the radiolabeled finasteride following its dissociation from the inactivated enzyme complex and physical data (mass spectrum of adduct), Bull *et al.* (1996) proposed that the ac-

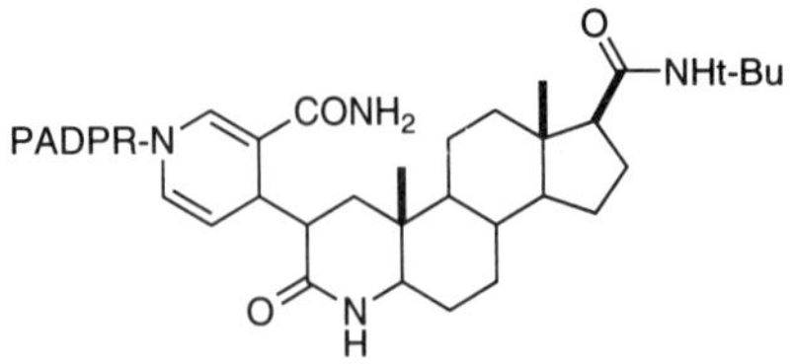

Figure 6. Proposed structure of the dihydrofinasteride–NADP adduct (Bull *et al.*, 1996).

tual potent inhibitor has the structure shown in Fig. 6. The inhibitory properties of the isolated potent inhibitor are also consistent with those of a bisubstrate analogue in the ordered bi–bi mechanism by which the 5ARs operate (Levy *et al.,* 1990; Bull *et al.,* 1996). Inhibition of the 5ARs by finasteride is extremely efficient, and the type 1 and 2 5ARs turn over 1.55 ± 0.05 and ≤ 1.07, respectively, finasteride molecules for each enzyme that is inactivated (Bull *et al.* 1996). In summary, all kinetic and mechanistic studies of finasteride's inhibition of 5ARs are consistent with nucleophilic addition of the hydride of NADPH to the Δ^1-double bond, followed by reaction between the enolate formed and $NADP^+$ to give the adduct of Fig. 6 (Bull *et al.,* 1996).

2.2. Modeling of the Clinical Effect of Finasteride

A model was developed (Tian, 1996) to enable quantitative analysis of time-dependent inhibition *in vivo* in order to understand why single doses of finasteride, a functionally irreversible inhibitor of both types of 5AR, only reduce plasma DHT levels 60–80% (Vermeulen *et al.,* 1989, 1991; Gormley *et al.,* 1990; Ohtawa *et al.,* 1991; De Schepper *et al.,* 1991). This two-compartment model provides a basis for understanding plasma DHT levels at 24 hr postdose when the effect of finasteride is near maximal. The model considered the partitioning of inhibitor between enzymatic inhibition and drug elimination, in addition to other *in vivo* factors such as distribution of inhibitor between the plasma and tissues. From the application of equations describing these features, it was estimated that a 5-mg dose of finasteride is sufficient to achieve a drug concentration in the prostate much greater than that needed to inhibit 95% of the type 2 5AR. In contrast, it was also predicted that the same dose would achieve only a 3% inhibition of type 1 5AR activities (Tian, 1996), a result deriving principally from the relatively slow rate of type 1 5AR inhibition. The accuracy of the two-compartment model was evaluated through the calculation of a theoretical fit through DHT levels measured at various finasteride clinical doses and reported by others (Vermeulen *et al.,* 1989, 1991; Gormley *et al.,* 1990; Ohtawa *et al.,* 1991; De Schepper et al., 1991). The theoretical fit is shown in Fig. 7 (Tian, 1996). This analysis suggests that single doses of finasteride are in-

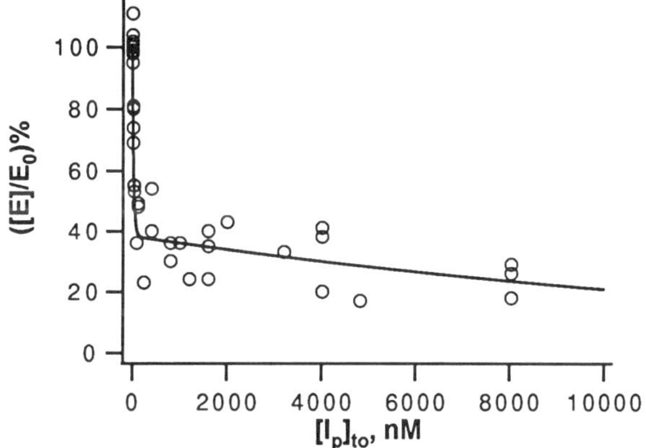

Figure 7. Effect of finasteride on DHT levels in plasma 24 hr after oral dosing (open circles) versus the theoretical fit based on two isoforms of 5AR that are differentially inhibited by finasteride. [E] is the remaining concentration of active 5AR enzyme, E_0 is the initial concentration of 5AR, and $[I_p]_{to}$ is related to the plasma concentration of inhibitor. [Reproduced with permission from: Tian, G. J., 1996, In vivo time-dependent inhibition of human steroid 5α-reductase by finasteride, *Pharm. Sci.* **85:**106–111. Copyright 1996 American Chemical Society.

sufficient to inhibit the type 1 5AR, and that this accounts for the residual DHT found after finasteride treatment (Tian, 1996; Mocellini *et al.*, 1993; Stoner, 1994).

3. DISCOVERY OF DUAL 5α-REDUCTASE INHIBITORS: 6-AZASTEROIDS

3.1. Medicinal Chemistry

Based on our analysis of the limited effect of finasteride on circulating DHT levels and coincident with the discovery of a second isozyme of human 5AR (Andersson *et al.*, 1991), we set out to discover a novel 5AR inhibitor framework. One of the initial targets we envisioned was the 6-azaandrost-4-en-3-one **3** (Table III). The hypothesis that **3** would be a 5AR inhibitor was based on the transition-state inhibitor paradigm (Wolfenden, 1972) whereby the ketoenamine functionality would mimic structural and charge-polarization features of the transition state for the enzyme-catalyzed transfer of hydride from NADPH to testosterone (see Fig. 2). In contrast to the 4-azasteroid nucleus, which could be considered to be rather productlike because of the 5α-configuration at C-5, the 6-azasteroid represents a more substratelike transition-state mimic because of the C-4–C-5 unsaturation. Unfortunately, as outlined below, the 6-azaandrost-4-en-3-one structure is for this

Table III
6-Azasteroid Inhibitors of 5AR[a]

No.	R, C-4 substituent (if other than H)	Type 1 5AR (K_i, nM)	Type 2 5AR (IC_{50}, nM)	3BHSD (K_i, nM)
3	-N(Et)$_2$	750	1.5	60
4	-N(Et)$_2$, C4-Cl	51	1.9	27
5	-OMe	150	3.2	12
6	-O-2-adamantyl	6.9	<0.1	180
7	-NHCH(Ph)$_2$	30	<0.1	150
8	-NHCH(4-chlorophenyl)$_2$	20	0.12	510
9	-NHCH(4-fluorophenyl)$_2$	20	0.16	320
10	-CH$_2$CH(Me)$_2$	9	<0.1	10
11	-NHPh	240	1.4	10
12	-NH[(2-*tert*-butyl)phenyl]	27	0.2	85
13	-NH[(5-bromo, 2-*tert*-butyl)phenyl]	4.2	<0.1	820
14	-NH[(2,5-bis-*tert*-butyl)phenyl]	5	<0.1	500
15	-NH[(2-*tert*-butyl, 5-trifluoromethyl) phenyl]	8.8	<0.1	1600
16	-NH[[2-*tert*-butyl, 5-(4-*tert*-butyl)phenyl] phenyl]	1.3	<0.1	8900
17	-NH[(2,5-bis-trifluoromethyl)phenyl]	4	<0.1	300
18	-NH[(3,5-bis-*tert*-butyl)phenyl]	8	<0.1	7.8
19	-NH[(3,5-bis-trifluoromethyl)phenyl]	26	0.2	58
20	-NH-N-[1-(4-chlorophenyl)cyclopentyl]	6.8	<0.1	1600
21	-NH-N-[1-(4-chlorophenyl)cyclopentyl], C4-Cl	0.6	<0.1	490
22	-NH[(2,5-bis-trifluoromethyl)phenyl], C4-Cl	0.2	<0.1	190

[a]Adapted with permission from: Frye, S. V., 1996, Inhibitors of 5α-reductase, *Curr. Pharm. Des.* **2**:59–84. Copyright 1996 Bentham Science.

reason a better mimic of the transition state of 3BHSD (Fig. 8) and selectivity versus this enzyme was a significant hurdle to overcome in this series (Brandt and Levy, 1989). The synthesis of the 6-azaandrost-4-en-3-one **3** followed the general strategy employed to introduce nitrogen at the 6-position of cholesterol (Lettré and Knof, 1960) and produced **3** in 12 steps from 3β-hydroxyetienic acid methyl ester (Frye *et al.*, 1993). Although **3** proved to be a potent inhibitor of type 2 5AR, with an IC_{50} of 1.5 nM, it was actually a more potent inhibitor of 3BHSD than type 1 5AR (60 nM versus 750 nM, IC_{50}). With a potent, novel 5AR inhibitor template in hand, extensive structure–activity relationship (SAR) studies were carried out to optimize the activity of the 6-azasteroids.

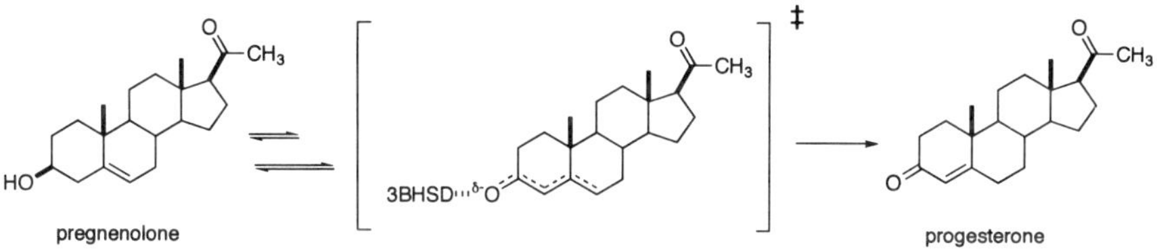

Figure 8. Schematic of the transition state in the isomerization reaction catalyzed by 3β-hydroxy-Δ^5-steroid dehydrogenase/3-keto-Δ^5-steroid isomerase (3BHSD). Adapted with permission from: Frye, S. V., 1996, Inhibitors of 5α-reductase, *Curr. Pharm. Des.* **2**:59–84. Copyright 1996 Bentham Science.

The initial investigation of the 6-azasteroids focused on exploring A- and B-ring substitutions, especially the relatively unexplored 6-position of the steroid (Frye *et al.,* 1994). Although a number of structural changes were investigated in this study, simple C-4 chloro substitution did the most to enhance type 1 5AR potency and selectivity versus 3BHSD. Chlorination of **3** produced **4,** which was roughly equipotent in its inhibition of type 1 5AR and 3BHSD, while maintaining potency versus type 2 5AR (Table III).

Based on early pharmacokinetic data, detailed in Section 3.2, and emerging SAR versus type 1 5AR and 3BHSD, exploration turned to C-17 of the steroid and ultimately led to the potency and selectivity required in the 6-azasteroid framework (Frye *et al.,* 1995). Compounds **5** and **6** (Table III) exemplify the differential sensitivity of type 1 5AR and 3BHSD to sterically demanding C-17 substituents as a change from a methyl ester to a 2-adamantyl ester transforms 10-fold 3BHSD selective **5** into a 25-fold selective type 1 inhibitor **6.** Bulky amides at C-17 also show some selectivity (**7–9**) whereas ketones do not (**10**). Additionally, compound **7** was the first 6-azasteroid to be equipotent to finasteride in an *in vivo* model of 5AR inhibitor-induced prostate shrinkage in the rat (Frye *et al.,* 1993). The C-17 anilides (**11–19**) display dramatic changes in selectivity with changes in substitution pattern. For example, introduction of a 2-*tert*-butyl group decreases 3BHSD potency 8-fold while increasing type 1 activity 10-fold (**11** versus **12**). Addition of a 5-substituent to the aniline further diminishes 3BHSD activity while increasing type 1 potency to give >100-fold selective dual 5AR inhibitors (**13–17**). Replacement of the 2-*tert*-butyl group of **15** with a trifluoromethyl group slightly increases type 1 potency, and results in a remarkably potent *in vivo* inhibitor as demonstrated in the castrated rat model of androgen-driven prostate growth (**17**) (Frye, 1996). Additionally (see Section 2.1) substitution of 2,5-bis(trifluoromethyl)aniline at C-17 results in a 6-azasteroid with a remarkable *in vivo* half-life of more than 30 hr and 100% bioavailability in the dog (Frye *et al.,* 1995).

Simply shifting the 2-substituent of the aniline in **14** and **17** to the 3-position converts these ~100-fold selective compounds to compounds with roughly equal potency versus type 1 5AR and 3BHSD (**18, 19**). A conformational model that is consistent with all of these results has been described (Frye *et al.,* 1995). Aryl-substituted cycloalkyl amines that satisfy this conformational model have also been utilized in the 6-azasteroids to produce potent, selective dual 5AR inhibitors (**20**). Combination of the favorable C-4 chloro substituent with the best C-17 groups produces picomolar dual 5AR inhibitors with up to 800-fold selectivity versus 3BHSD (**21, 22**).

3.2. Pharmacokinetic Studies: *In Vivo* and *in Vitro* Correlations

In order to identify a compound that would be suitable for once a day dosing, the half-lives of a selected group of 6-azasteroids containing C-17 anilides or alkyl amides were determined in dogs (Tippin *et al.,* 1995). The dog has been demonstrated to be a relevant animal model for studying 4-azasteroid pharmacokinetics in previous finasteride animal studies (Frye *et al.,* 1994; Carlin *et al.,* 1992a). Our initial lead compounds exhibited short $t_{1/2}$ after i.v. administration to dogs. For example, compound **23** (Fig. 9) was found to have $t_{1/2\beta}$ = 1.5 hr. Further studies with ^{14}C-labeled **23** revealed that only 5% of the administered radioactivity was excreted in urine as parent compound. No parent drug was detected in feces. These results suggested that the primary mode of clearance was by oxidative metabolic degradation. Multiple mono- or bis-hydroxylated metabolites were identified by HPLC–mass spectrometry and no single metabolite comprised a majority of the dose. Comparison of the rates of metabolism of **23** and its *N*6-demethylated analogue in dog liver microsomes suggested that the presence of the *N*6-methyl group was one of the major reasons for the metabolic instability of **23.** These early results suggested that removal of the *N*6-methyl group and variations at C-17 might be more fruitful in producing analogues with a longer half-life.

As the short $t_{1/2\beta}$ of **23** in the dog was primarily related to extensive oxidative metabolism, the utility of a dog liver microsomal assay to rapidly identify stable analogues of **23** was investigated. In addition, it was envisioned that if the metabolic profiles and rates of metabolism by dog liver microsomes are found to correlate with those by human liver microsomes, it would indicate that the dog is

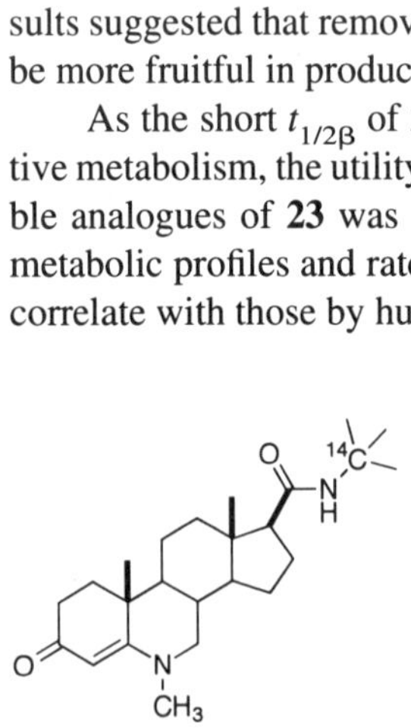

23

Figure 9. The first 6-azasteroid radiolabeled for metabolism studies.

a suitable model for humans in studying the metabolism of 6-azasteroids. Incubations with microsomes (1 mg protein/ml) in the presence or absence of NADPH were conducted at a compound concentration that reflected maximum concentrations that were observed for **23** and several of its analogues (2.5 μM) in dog blood after i.v. dosing. The rate constants for the disappearance of each test compound were obtained from the linear portion of the plots of the percentage of original test compound remaining versus time.

Initially, a total of 12 compounds were tested both *in vitro* and *in vivo* in dog and in human liver microsomes. Good correlation between dog *in vitro* half-life and dog *in vivo* $t_{1/2}\beta$ ($r^2 = 0.66$) was obtained. Thus, 6-azasteroids with half-lives less than 1.5 hr in dog liver microsomes were found to have short $t_{1/2}\beta$ (<3 hr) *in vivo* in dogs. An improved correlation ($r^2 = 0.78$) was obtained when only the C-17 anilide derivatives were considered (compounds **13–15, 17, 24–26;** see Fig. 10).

The correlation between human and dog *in vitro* half-lives was poor ($r^2 = 0.40$) when all 12 compounds were considered but was significant ($r^2 = 0.84$) among the anilides (compounds **13–15, 17, 24–26**). In addition, the metabolic profile produced from ^{14}C-labeled **23** in dog and human liver microsomes was qualitatively similar. Thus, the dog is a relevant animal model for predicting human metabolism of this class of compounds. These findings supported the use of the dog microsomal assay to rapidly identify 6-azasteroidal 5AR inhibitors that had the potential for improved *in vivo* half-life.

In terms of structure versus metabolic reactivity, in the C-17 anilide series, it was apparent that the presence of a halogen or a perfluoroalkyl group in the aniline ring (**13, 15, 17, 26**) improved metabolic stability significantly. Indeed analogues containing two trifluoromethyl groups (e.g., **17**) turned out to be among the most stable both *in vitro* and *in vivo* in dog.

In conclusion, metabolism studies suggested that (1) the dog *in vitro* microsomal assay is useful for screening 6-azasteroids prior to more labor-intensive *in*

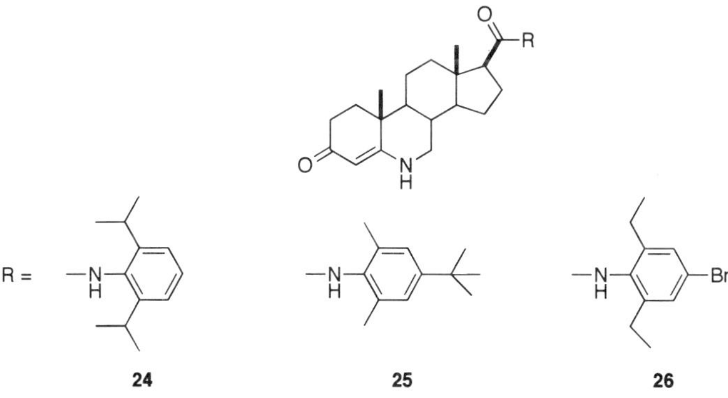

Figure 10. Compounds **24–26** of Table IV.

vivo dog studies; (2) the dog is a relevant animal model for predicting the human metabolism of this class of compounds; and (3) the C-17 anilide moiety containing two trifluoromethyl groups confers significant metabolic stability to the 6-aza-steroids.

4. DISCOVERY OF GG745

Effect of Optimal C-17 Substituents from the 6-Azasteroid Series on Other 5α-Reductase Inhibitors, *in Vitro, in Vivo,* and Pharmacokinetic Properties

Having discovered two novel series of C-17 amides (2,5-substituted anilines and cycloalkylarylamines) that significantly improve the potency, selectivity, and pharmacokinetic properties of 6-azasteroids, it seemed worthwhile to determine the effect of these substituents on other steroidal 5AR inhibitor frameworks (Frye *et al.,* 1995). This was of special interest in the Δ^1-4-azasteroid series where we (Tian *et al.,* 1995a) and others (Faller *et al.,*1993; Bull *et al.,* 1996) had shown that essentially irreversible inhibition of both 5ARs was a possibility. To that end, compounds **27** and **28** (Fig. 11), which bear representative C-17 substituents that lead to dual inhibition and appropriate pharmacokinetics in the 6-azasteroid framework (**17** and **20** in Tables III and IV), were prepared and assayed versus the 5AR isozymes and 3BHSD. Of these, **28** (GG745) proved to be the most potent dual 5AR inhibitor and the kinetic parameters regarding its inhibition are presented in Table V. Compared with finasteride, GG745 is 60-fold more potent in its initial K_i versus type 1 5AR and ~5-fold more rapid in inactivating the enzyme. Interestingly, Bakshi and co-workers have subsequently prepared and evaluated a series of C-17 anilides including GG745 and the kinetic data they determined are very consistent with the data of Table V (Bakshi *et al.,* 1995).

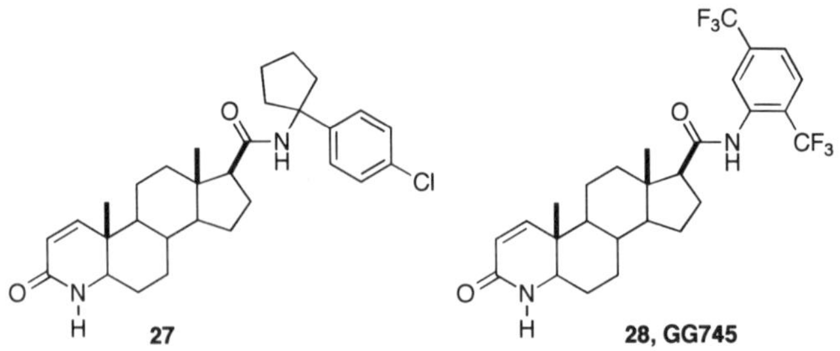

O N H 27 N H O Cl

O N H 28, GG745 N H O F3C CF3

Figure 11. Compounds **27** and **28.**

Table IV
Half-Lives of 6-Azasteroids Observed in Dog *in Vivo* and in Dog and Human Liver Microsomes[a]

	$t_{1/2}$ (hr)		
No.	Dog liver microsomes	Human liver microsomes	Dog *in vivo*
24	0.07	0.22	0.4
25	0.13	0.07	2.8
10	1.38	9.15	2.2
23	1.65	2.48	1.5
13	1.93	2.27	5.7
7	2.33	4.20	4.8
20	2.63	4.68	12.1
8	3.70	4.25	12.0
15	3.98	4.10	9.8
26	4.22	4.45	10.8
9	5.05	10.85	5.2
14	5.30	2.73	7.2
17	7.95	9.95	30.0

[a]See Fig. 10 for the structures of **24–26**.

When progressed to *in vivo* assessment in a model of prostate growth (Frye *et al.*, 1995), intact male rats treated daily with GG745 at 1, 10, or 100 mg/kg per day for 2 weeks had prostates about half as large as those of rats treated with vehicle alone (Table VI). There was no significant difference between the GG745 dose groups, indicating that the maximum effect in this model had been achieved. In contrast, finasteride produced a dose-related change in prostate volume. The top dose of finasteride (72 mg/kg per day, which is equimolar to the 100 mg/kg per day doses of GG745) produced similar effects on prostatic volume as those observed in the GG745 groups. Based on these data, it was assumed that GG745 was approximately 72 times (mg:mg) more potent than finasteride in this *in vivo* preclinical model. Blood samples collected from animals given similar doses indi-

Table V
Comparison of Kinetic Parameters for the Inhibition of 5AR by Finasteride (2) and GG745 (28) at 37°C (Tian *et al.*, 1995a)

5AR	k_3 (sec^{-1})	K_i (IC_{50}, nM)	k_3/K_i (M^{-1} sec^{-1})
Type 1			
2	1.4×10^{-3}	360	4×10^3
28	1.1×10^{-3}	6	1.8×10^5
Type 2			
2	2.2×10^{-2}	69	3.2×10^5
28	4.9×10^{-3}	7	6.8×10^5

Table VI
Effects of 14-Day Oral Administration of GG745 on Organ Weights and Body Weight of Intact Male Rats[a]

Treatment (dose, mg/kg/day)	Prostate (mg)	Seminal vesicles (mg)	Adrenals (mg)	Body weight (g)
Vehicle	154.1 ± 7.6	216.7 ± 13.0	38.0 ± 1.8	259.9 ± 10.1
Finasteride (0.7)	107.0 ± 4.7*	151.9 ± 18.0*	34.1 ± 0.09	232.4 ± 4.6
(7)	96.2 ± 6.1*	121.4 ± 8.5*	33.3 ± 1.2	234.7 ± 2.1
(70)	65.8 ± 3.7*	83.8 ± 12.8*	35.9 ± 1.3	210.6 ± 4.3
Vehicle	242.1 ± 14.3	367.6 ± 34.2	39.0 ± 1.4	274.4 ± 5.1
GG745 (1)	120.0 ± 6.9*	165.4 ± 22.1*	41.6 ± 1.5	275.3 ± 3.3
(10)	120.4 ± 9.9*	146.4 ± 12.7*	42.0 ± 1.0	265.5 ± 3.4
(100)	115.6 ± 20.7*	132.8 ± 15.9*	43.2 ± 0.9	264.8 ± 4.5

[a]Rats were treated with vehicle or test compound by the oral route for 14 days. Finasteride data are from a separate study from the GG745 data.
*Significantly different ($p < 0.05$) from the vehicle controls.

cated that on a concentration:concentration basis, GG745 was approximately 56 times more potent than finasteride.

Based on its outstanding *in vitro* and *in vivo* profile, detailed pharmacokinetic determinations in the rat and dog were carried out for GG745. Following i.v. infusion and oral administration of 5 mg/kg, the half-life ($t_{1/2}$), total body clearance (CL), volume of distribution at steady state (V_{ss}), and oral bioavailability (F) of GG745 in the dog were 65 hr, 0.5 ml/min per kg, 3 liters/kg, and 43%, respectively. The GG745 blood level reached its peak (745 ng/ml) at 2.5 hr after oral dosing. The GG745 blood concentration at the end of i.v. infusion (10 min) was 3430 ng/ml. The $t_{1/2}$, CL, and V_{ss} of finasteride in the dog were 3.9 hr, 4.9 ml/min per kg, and 1.6 liters/kg, respectively.

Following i.v. infusion and oral administration of 1 mg/kg, the $t_{1/2}$, CL, V_{ss}, and F of GG745 in the rat were 13.7 hr, 4.1 ml/min per kg, 4 liters/kg, and 100%, respectively. The GG745 blood level reached its peak (139 ng/ml) at 7 hr after oral dosing. The GG745 blood concentration was 200 ng/ml, 15 min after i.v. dosing. The $t_{1/2}$, CL, and V_{ss} of finasteride in the rat were 0.9 hr, 13.4 ml/min per kg, and 0.6 liter/kg, respectively.

Through an understanding of (1) the SAR for dual 5AR inhibition in the 6-azasteroid series, (2) the mechanism of inhibition of 5ARs by Δ^1-4-azasteroids, and (3) the effect of C-17 substituents on the pharmacokinetic parameters of steroidal 5AR inhibitors, GG745 (**28**) was discovered. Overall, GG745 is a remarkably potent dual inhibitor of 5AR, being as potent versus type 1 5AR as finasteride is versus type 2 (Table V) with 10,000-fold selectivity versus 3BHSD, outstanding *in vivo* potency, extended half-life in the dog and rat, and adequate bioavailability for clinical assessment. Based on these features, GG745 was selected for clinical development.

5. INITIAL CLINICAL STUDIES WITH GG745

5.1. Interspecies Scaling/Dose Selection

Pharmacokinetic and pharmacodynamic data generated in preclinical studies were used to estimate the dose of GG745 expected to produce a clinical effect as measured by changes in circulating DHT.

First, empirical allometric interspecies scaling was undertaken to obtain estimates of pharmacokinetic parameters in man. The principles of allometry (the study of size and its consequences) are well established as a method for estimating pharmacokinetic parameters in man from preclinical species (Boxenbaum, 1982; Mordenti, 1986; Ings, 1990; Ritschel *et al.*, 1992). This method relies on establishing a relationship between body size and pharmacokinetic parameters:

$$\text{Pharmacokinetic parameter in man} = a(\text{body weight})^{x} \quad (2)$$

Plotting pharmacokinetic parameters such as clearance and volume of distribution against weight on a log–log scale linearizes the above equation where a is the intercept and x is the slope parameter. From this relationship, pharmacokinetic parameters for a 70-kg human can be estimated. In this case, volume of distribution (log) and systemic clearance (log) were plotted against body weight (log) (Fig 12). Estimates of 12 ml/min and 180 liters were predicted for clearance and steady-

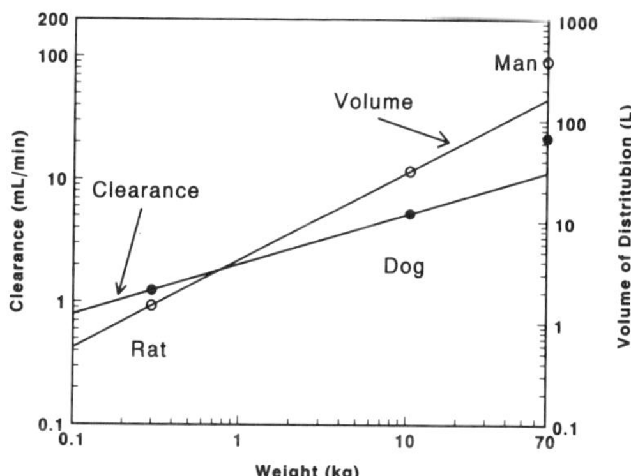

Figure 12. Log–log plot of pharmacokinetic parameters versus body weight. Solid line is the linear regression line through the rat and dog pharmacokinetic parameters determined after GG745 intravenous administration. Actual pharmacokinetic parameters determined from the oral single-dose GG745 evaluation are plotted for comparison. (Note: The reason observed values of clearance and volume of distribution for man are greater than predicted values is likely related to incomplete absorption; i.e., $CL_{oral} = C_{liv}/F$, where F is the absolute bioavailability.)

state volume of distribution for a 70-kg human. These estimates indicated that GG745 would have a long terminal half-life of approximately 180 hr.

Second, a target GG745 concentration was estimated for humans. This target was obtained by using the *in vivo* potency ratio determined for GG745 relative to finasteride in the rat experiments and indexing this ratio to published clinical literature on finasteride. The *in vivo* pharmacology experiments conducted in intact adult male rats indicated that GG745 was about 56 times more potent than finasteride on a concentration:concentration basis. In humans, the relationship between single doses of finasteride and maximal DHT suppression appeared to become asymptotic (approaching maximum effective exposure) at doses of 50–100 mg (De Schepper *et al.,* 1991). C_{max} observed following a single 100-mg dose of finasteride was approximately 836 ng/ml (Ohtawa *et al.,* 1991). Using the *in vivo* potency ratio of 56:1, a target GG745 concentration of about 15 ng/ml was estimated to be needed to reach the top part of the GG745 dose–response (DHT reduction) curve. Coincidentally, GG745 is also about 60-fold more potent versus type 1 5AR than finasteride so that consideration of either enzymology versus the human isozymes or rat pharmacology as a basis for dose selection gives similar predictions.

Third, a dose was estimated from the predicted pharmacokinetic parameters for a 70-kg man that would achieve the target GG745 concentrations of approximately 15 ng/ml. Absorption was considered to be rapid and complete. Based on these assumptions, a dose of 3 mg was estimated to provide peak GG745 concentrations of approximately 15 ng/ml and provide significant DHT suppression.

As GG745 was expected to have a long terminal half-life, a conservative starting dose of 0.01 mg was selected. This dose was approximately two orders of magnitude lower than the proposed clinically effective dose of 3 mg and was well below doses found to produce no toxicologically significant findings in long-term toxicology studies.

Subsequently, 48 healthy male subjects received single oral doses of GG745, placebo, or finasteride (5 mg) in a randomized, blinded, sequential cohort dose escalation study. GG745 doses of 0.01 to 40 mg were studied in cohorts consisting of 4 GG745, 1 placebo, and 1 finasteride subject. Doses were escalated in subsequent groups following an evaluation of safety. Serial serum samples were collected for determination of circulating DHT and GG745 concentrations. DHT samples were assayed via a GC-MS method with a limit of detection of 10 pg/ml and interday coefficient of variation of $\leq$9.5%. GG745 samples were assayed via an LC-MS method with a limit of detection of 0.1 ng/ml and interday coefficient of variation of $\leq$11.6% (Morris *et al.,* 1995).

5.2. Pharmacokinetic and Pharmacodynamic Results in Man

The pharmacokinetic parameters determined in man are in good agreement with predicted estimates from preclinical data (Fig. 12, Table VII). The reason ob-

Table VII
Predicted and Observed Values of Clearance and Volume of Distribution for GG745

Pharmacokinetic parameter	Predicted value	Observed value assuming 100% bioavailability	Observed value assuming 50% bioavailability
Clearance (ml/min)	12	22	11
Volume of distribution (liters)	180	385	193

served values of clearance and volume of distribution for man are greater than predicted values is likely related to incomplete absorption (i.e. $CL_{oral}=C_{liv}/F$, where F is the absolute bioavailability). Table VII compares observed pharmacokinetic parameters with estimates from interspecies scaling assuming 50 and 100% bioavailability. The absolute bioavailability of GG745 in man is unknown as an i.v. formulation is not available for administration to man. Absolute bioavailability of GG745 in preclinical studies ranged from 40 to 100%.

GG745 produced a dose-related decrease in DHT (Fig. 13). Little or no effect was observed at single oral doses below 0.1 mg. As predicted from the preclinical data, doses of approximately 3 mg produced significant reductions in serum DHT concentrations (Table VIII). Single oral GG745 doses of at least 5 mg decreased DHT significantly more than finasteride. The mean maximum decrease in DHT observed at the highest GG745 dose (40 mg) was 95%. In the present study, finasteride decreased DHT levels 80% from baseline. Single doses of 40 mg (Vermeulen *et al.*, 1989) and 100 mg (Ohtawa *et al.*, 1991) of finasteride produced negligible added reductions in serum DHT.

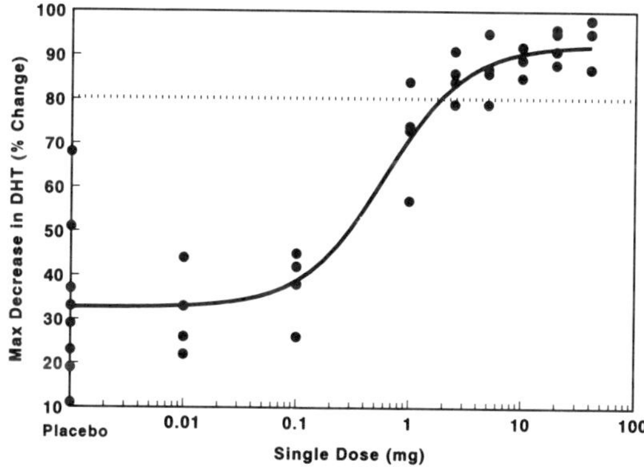

Figure 13. Relationship between (log) single oral dose and maximum DHT reduction. The horizontal line represents the mean maximum DHT reduction observed in the finasteride group. A sigmoidal E_{max} model was fitted to the GG745 dose–response data and is shown as a solid line.

Table VIII
Dose–Response Relationship for GG745

Treatment	n	DHT reduction (%)[a]	C_{max} (ng/ml)
Placebo	8	34	NA
Finasteride 5 mg	8	80	NA
GG745 0.01 mg	4	31	BQL[b]
0.1 mg	4	38	0.6
1.0 mg	4	72	5.4
2.5 mg	4	85	14.3
5.0 mg	4	87*	23.9
10 mg	4	89*	60.0
20 mg	4	92*	76.2
40 mg	4	95*	166

[a]Mean maximum percent decrease in DHT relative to baseline.
[b]BQL, below quantitation limits.
*$p < 0.05$ versus finasteride, t-test from a general linear model of the form: log(minimum DHT/baseline DHT) = treatment + log(baseline DHT).

In a multiple dose study, 53 BPH patients received daily oral doses of GG745, placebo, or finasteride (5 mg) for 28 days in a randomized, blinded, parallel group trial. GG745 doses of 0.1, 0.5, 2.5, 2.5 with a 40 mg loading dose, and 5 mg were studied. DHT measurements were taken before and after 28 days of study drug administration. GG745 groups were compared to placebo and finasteride using a general linear model with pairwise comparisons. The results after 28 days of treatment are presented in Table IX. Dual inhibition of 5AR with GG745 produced significantly greater reductions in serum DHT compared to finasteride. At doses of 2.5 mg/day and up, GG745 decreased DHT by 95% or greater.

Table IX
Twenty-Eight-Day Multiple-Dose–Response Relationship for GG745 versus finasteride (mean ± sd)

Group	n	DHT (% reduction)
Placebo	6	3 ± 21
Finasteride 5 mg	6	76 ± 7[a]
GG745 0.1 mg	8	72 ± 7[a]
GG745 0.5 mg	8	90 ± 2[a,b]
GG745 2.5 mg	8	95 ± 3[a,b]
GG745 40 + 2.5 mg	8	96 ± 1[a,b]
GG745 5.0 mg	9	97 ± 1[a,b]

[a]$p < 0.05$ vs. placebo; [b]$p < 0.05$ vs. finasteride.

Results from *in vitro* enzyme kinetic studies indicated that finasteride and GG745 inhibit both type 1 and 2 5AR. However, through modeling using the kinetics observed in these *in vitro* systems and accounting for pharmacokinetic characteristics, we showed that finasteride would be expected to only partially inhibit type 1 5AR whereas GG745 would be expected to effectively inhibit both isozymes (Section 2.2) (Tian, 1996). These data clearly demonstrate that GG745, a potent dual inhibitor of both human 5ARs, is more effective than finasteride, a type 2 5AR selective inhibitor, at reducing serum DHT levels in man. Further clinical trials will determine whether this further reduction in serum DHT offers added clinical benefit.

The path from discovery to development for GG745 depended on close coordination between medicinal chemistry, enzymology, pharmacology, drug metabolism, and clinical pharmacokinetics. The simultaneous, as opposed to sequential, determination of the influence of inhibitor structural change on enzyme potency, selectivity, and metabolism/pharmacokinetics was essential to the rapid discovery and development of GG745.

ACKNOWLEDGMENTS

The authors gratefully acknowledge the contributions of the Glaxo Wellcome 5AR project team: Curt D. Haffner, Patrick R. Maloney, Robert A. Mook, Jr., Roger N. Hiner, George F. Dorsey, Jr., Robert A. Noe, Rayomand J. Unwalla, Kenneth W. Batchelor, J. Darren Stuart, Stephanie L. Schweiker, John van Arnold, D. Mark Bickett, Marcia L. Moss, Timothy K. Tippin, Arthur Moseley, Michael K. James, Mary K. Grizzle, James E. Long, and Dallas K. Croom.

REFERENCES

Andersson, S., and Russell, D. W., 1990, Structural and biochemical properties of cloned and expressed human and rat steroid 5α-reductases, *Proc. Natl. Acad. Sci. USA* **87:**3640–3644.

Andersson, S., Bishop, R. W., and Russell, D. W., 1989, Expression cloning and regulation of steroid 5α-reductase, an enzyme essential for male sexual differentiation, *J. Biol. Chem.* **264:**16249–16255.

Andersson, S., Berman, D. M., Jenkins, E. P., and Russell, D. W., 1991, Deletion of steroid 5α-reductase 2 gene in male pseudohermaphroditism, *Nature* **354:**159–161.

Arrighi, H. M., Metter, E. J., Guess, H. A., and Fozzard, J. L., 1991, Natural history of benign prostatic hyperplasia and risk of prostatectomy, *Urology Suppl.* **38:**4–8.

Bakshi, R. K., Rasmusson, G. H., Patel, G. F., Mosley, R. T., Chang, B., Ellsworth, K., Harris, G. S., and Tolman, R. L., 1995, 4-Aza-3-oxo-5α-androst-1-ene-17β-N-aryl-carboxamides as dual inhibitors of human type 1 and 2 steroid 5α-reductases. Dramatic effect of N-aryl substituents on type 1 and type 2 5α-reductase inhibitory potency, *J. Med. Chem.* **38:**3189–3192.

Barry, M. J., Cockett, A. T. K., Holtgrewe, H. L., McConnell, J. D., Sihelnik, S. A., and Winfield, H. N., 1993, Relationship of symptoms of prostatism to commonly used physiological and anatomical measures of the severity of benign prostatic hyperplasia, *J. Urol.* **150:**351–358.

Boxenbaum, H., 1982, Interspecies scaling, allometry, physiological time, and the ground plan of pharmacokinetics, *J. Pharmacokinet. Biopharm.* **10**(2)**:**201–227.

Brandt, M., and Levy, M. A., 1989, 3β-Hydroxy-Δ^5-steroid dehydrogenase / 3-keto- Δ^5-steroid isomerase from bovine adrenals: Mechanism of inhibition by 3-oxa-4-aza steroids and kinetic mechanism of the dehydrogenase, *Biochemistry* **28:**140–148.

Brawley, O. W., Ford, L. G., Thompson, I., Perlman, J. A., and Kramer, B. S., 1994, 5-α-Reductase inhibition and prostate cancer prevention, *Cancer Epidemiol. Biomarker Prev.* **3:**177–182.

Brooks, J. R., 1986, Treatment of hirsutism with 5α-reductase inhibitors, *Clin. Endocrinol. Metab.* **15:**391–405.

Bull, H. G., Garcia-Calvo, M., Anderson, S., Baginsky, W. F., Chan, H. K., Ellsworth, D. E., Miller, R. R., Stearns, R. A., Bakshi, R. K., Rasmusson, G. H., Tolman, R. L., Myers, R. W., Kozarich, J. W., and Harris, G. S., 1996, Mechanism-based inhibition of human steroid 5α-reductase by finasteride: Enzyme-catalyzed formation of NADP-dihydrofinasteride, a potent bisubstrate analog inhibitor, *J. Am. Chem. Soc.* **118:**2359–2365.

Carlin, J. R., Christofalo, P., Arison, B. H., VandenHeuvel, W. J. A., Miller, R. R., Ellsworth, R. L., and Chiu, S.-H. L., 1992a, The disposition and pharmacokinetics of finasteride in the dog, *ISSX Proc.* **2:**197.

Carlin, J. R., Hoglund, P., Eriksson, L.-O., Christofalo, P., Gregoire, S. L., Taylor, A. M., and Anderson, K.-E., 1992b, Disposition and pharmacokinetics of [^{14}C]-finasteride after oral administration in humans, *Drug Metab. Dispos.* **20:**148–155.

Chan, W. K., Fong, C. Y., Tiong, H. H., and Tan, C. H., 1987, The inhibition of 3βHSD activity in porcine granulosa cells by 4-MA, a potent 5α-reductase inhibitor, *Biochem. Biophys. Res. Commun.* **144:**166–171.

Cooke, G. M., and Robaire, B. J., 1986, The effects of diethyl-4-methyl-3-oxo-4-aza-5α-androstane-17β-carboxamide (4MA) and (4R)-5,10-seco-19-norpregna-4,5-diene-3,10,20-trione (SECO) on androgen biosynthesis in the rat testis and epididymis, *Steroid Biochem.* **24:**877–886.

Dallob, A. L., Sadick, N. S., Unger, W., Lipert, S., Geissler, L. A., Gregoire, S. L., Nguyen, H. H., Moore, E. C., and Tanaka, W. K., 1994, The effect of finasteride, a 5α-reductase inhibitor, on scalp skin testosterone and dihydrotestosterone concentrations in patients with male pattern baldness, *J. Clin. Endocrinol. Metab.* **79:**703–706.

Darley, C. R., 1984, Recent advances in hormonal aspects of acne vulgaris, *Int. J. Dermatol.* **23:**539–541.

Denis, L., and Mahler, C., 1990, Prostatic cancer: An overview, *Rev. Oncol.* **3:**665–677.

De Schepper, P. J., Imperato-McGinley, J., Van Hecken, A., De Lepeleire, I., Buntinx, A., Carlin, J., Gressi, M. H., and Stoner, E., 1991, Hormonal effects, tolerability, and preliminary kinetics in men of MK-906, a 5α-reductase inhibitor, *Steroids* **56:**469–471.

Diani, A. R., Mulholland, M. J., Shull, K. L., Kubicek, M. F., Johnson, G. A., Schostarez, H. J., Brunden, M. N., and Buhl, A. E., 1992, Hair growth effects of oral administration of finasteride, a steroid 5α-reductase inhibitor, alone and in combination with topical minoxidil in the balding stumptail macaque, *J. Clin. Endocrinol. Metab.* **74:**345–350.

Faller, B., Farley, D., and Nick, H., 1993, Finasteride: A slow-binding 5α-reductase inhibitor, *Biochemistry* **32:**5705–5710.

Frye, S. V., 1996, Inhibitors of 5α-reductase, *Curr. Pharm. Des.* **2:**59–84.

Frye, S. V., Haffner, C. D., Maloney, P. R., Mook, R. A., Jr., Dorsey G. F., Jr., Hiner, R. N., Batchelor, K. W., Bramson, H. N., Stuart, J. D., Schweiker, S. L., van Arnold, J., Bickett, D. M., Moss, M. L., Tian, G., Unwalla, R. J., Lee, F. W., Tippin, T. K., James, M. K., Grizzle, M. K., Long, J. E., and Schuster, S. V., 1993, 6-Azasteroids: Potent dual inhibitors of human type 1 and 2 steroid 5α-reductase, *J. Med. Chem.* **36:**4313–4315.

Frye, S. V., Haffner, C. D., Maloney, P. R., Mook, R. A., Jr., Dorsey, G. F., Jr., Hiner, R. N., Cribbs, C. M., Wheeler, T. N., Ray, J. A., Andrews, R. C., Batchelor, K. W., Bramson, H. N., Stuart, J. D., Schweiker, S. L., van Arnold, J., Croom, S., Bickett, D. M., Moss, M. L., Tian, G., Unwalla, R. J., Lee, F. W., Tippin, T. K., James, M. K., Grizzle, M. K., Long, J. E., and Schuster, S. V., 1994,

6-Azasteroids: Structure activity relationships for inhibition of type 1 and 2 human 5α-reductase and human adrenal 3β-hydroxy-Δ^5-steroid dehydrogenase/3-keto-Δ^5-steroid isomerase, *J. Med. Chem.* **37:**2352–2360.

Frye, S. V., Haffner, C. D., Maloney, P. R., Hiner, R. N., Dorsey, G. F., Jr., Noe, R. A., Unwalla, R. J., Batchelor, K. W., Bramson, H. N., Stuart, J. D., Schweiker, S. L., van Arnold, J., Bickett, D. M., Moss, M. L., Tian, G., Lee, F. W., Tippin, T. K., James, M. K., Grizzle, M. K., Long, J. E., and Croom, D. K., 1995, Structure–activity relationships for inhibition of type 1 and 2 human 5α-reductase and human adrenal 3β-hydroxy-Δ^5-steroid dehydrogenase/3-keto-Δ^5-steroid isomerase by 6-azaandrost-4-en-3-ones: Optimization of the C17 substituent, *J. Med. Chem.* **38:**2621–2627.

Geller, J., 1991, Benign prostatic hyperplasia: Pathogenesis and medical therapy, *J. Am. Geriatr. Soc.* **39:**1208–1216.

Geller, J., 1993, Basis for hormonal management of advanced prostate cancer, *Cancer Suppl.* **71:**1039–1045.

Gormley, G. J., 1991, Role of 5α-reductase inhibitors in the treatment of advanced prostatic carcinoma, *Urol. Clin. N. Am.* **18:**93–98.

Gormley, G. L., Stoner, E., Rittmaster, R. S., Gregg, H., Thompson, D. L., Lasseter, K. C., Vlasses, P. H., and Stein, E. A., 1990, Effects of finasteride (MK-906), a 5α-reductase inhibitor, on circulating androgens in male volunteers, *J. Clin. Endocrinol. Metab.* **70:**1136–1141.

Griffiths, K., Eaton, C. L., Harper, M. E., Peeling, B., and Davies, P., 1991, Steroid hormones and the pathogenesis of benign prostatic hyperplasia, *Eur. Urol.* **20:**68–77.

Grino, P. B., Griffin, J. E., and Wilson, J. D., 1990, Testosterone at high concentrations interacts with the human androgen receptor similarly to dihydrotestosterone, *Endocrinology* **126:**1165–1172.

Imperato-McGinley, J., Guerrero, L., Gautier, T., and Peterson, R. E., 1974, Steroid 5α-reductase deficiency in man: An inherited form of male pseudohermaphroditism, *Science* **186:**1213–1215.

Imperato-McGinley, J., Gautier, T., Zirinsky, K., Hom, T., Palomo, O., Stein, E., Vaughan, E. D., Markisz, J. A., Arellano, E. R., and Kazam, E., 1992, Prostate visualization studies in males homozygous and heterozygous for 5α-reductase deficiency, *J. Clin. Endocrinol. Metab.* **75:**1022–1026.

Imperato-McGinley, J., Gautier, T., Cai, L.-Q., Yee, B., Epstein, J., and Pochi, P., 1993, The androgen control of sebum production. Studies of subjects with dihydrotestosterone deficiency and complete androgen insensitivity, *J. Clin. Endocrinol. Metab.* **76:**524–528.

Ings, R. M. J., 1990, Interspecies scaling and comparisons in drugs development and toxicokinetics, *Xenobiotica* **20:**1201–1231.

Isaacs, J. T., 1990, Importance of the natural history of benign prostatic hyperplasia in the evaluation of pharmacologic intervention, *Prostate Suppl.* **3:**1–7.

Jenkins, E. P., Anderson, S., Imperato-McGinley, J., Wilson, J. D., and Russell, D. W., 1992, Genetic and pharmacological evidence for more than one human steroid 5α-reductase, *J. Clin. Invest.* **89:**293–300.

Josso, N., 1994, Anatomy and endocrinology of fetal sex differentiation, in: *Endocrinology,* Volume 2 (L. J. DeGroot, ed.), pp. 1888–1900, Saunders, Philadelphia.

Kreig, M., Nass, R., and Tunn, S., 1993, Effect of aging on endogenous level of 5α-dihydrotestosterone, testosterone, estradiol, and estrone in epithelium and stroma of normal and hyperplastic human prostate, *J. Clin. Endocrinol. Metab.* **77:**375–381.

Lephart, E. D., 1993, Brain 5α-reductase: Cellular, enzymatic and molecular perspectives and implications for biological function, *Mol. Cell. Neurosci.* **4:**473–484.

Lephart, E. D., and Husmann, D. A., 1993, Altered brain and pituitary androgen metabolism by prenatal, perinatal or pre- and postnatal finasteride, flutamide or dihydrotestosterone treatment in juvenile male rats, *Prog. Neuro-Psychopharmacol. Biol. Psychiatry* **17:**991–1003.

Lettré, H., and Knof, L., 1960, 6-Aza-cholesterol and derivatives, *Chem. Ber.* **93:**2860–2864.

Levine, A. C., Wang, J.-P., Ren, M., Eliashvilli, E., Russell, D. W., and Kirschenbaum, A., 1996, Immunohistochemical localization of steroid 5-α-reductase 2 in human male fetal reproductive tract and adult prostate, *J. Clin. Endocrinol. Metab.* **81:**384–389.

Levy, M. A., Brandt, M., and Greway, A. T., 1990, Mechanistic studies with solubilized rat liver steroid 5α-reductase: Elucidation of the kinetic mechanism, *Biochemistry* **29:**2808–2815.

Liang, T., and Heiss, C. E., 1981, Inhibition of 5α-reductase, receptor binding, and nuclear uptake of androgens in the prostate by a 4-methyl-4-aza-steroid, *J. Biol. Chem.* **256:**7998–8005.

Liang, T., Cascieri, M. A., Cheung, A. H., Reynolds, G. F., and Rasmusson, G. H., 1985, Species differences in prostatic steroid 5α-reductases of rat, dog, and human, *Endocrinology* **117:**571–579.

Martini, L., 1982, The 5α-reduction of testosterone in the neuroendocrine structures. Biochemical and physiological implications, *Endocr. Rev.* **3:**1–25.

Martini, L., and Melcangi, R. C., 1991, Androgen metabolism in the brain, *J. Steroid Biochem. Mol. Biol.* **39:**819–828.

McConnell, J. D., 1990, Androgen ablation and blockade in the treatment of benign prostatic hyperplasia, *Urol. Cin. North Am.* **17:**661–670.

Melcangi, R. C., Celotti, F., and Martini, L., 1994, Progesterone 5-α-reduction in neuronal and in different types of glial cell cultures: Type 1 and 2 astrocytes and oligodendrocytes, *Brain Res.* **639:**202–206.

Mocellini, A. I., Gardiner, R., Marshall, V., Johnson, W., Bartsch, G., Schmidbauer, C. P., Mossing, H., Van Cangh, P. J., Denis, L. J., Arap, S., Freire, G. C., De Latorre, D., Botto, H., Richard, F., Devonec, M., Teillac, P., Vallancien, G., Bzaf, Z., DiSilverio, F., Miano, L., Pagano, F., Gabilondo, F., DeBuyne, F., Janknegt, R. A., Schroeder, F. H., Nacey, J., Furtado, L. A., Carretero, P., Jimenez, F. C., Hauri, D., Otto, U., Albrecht, J., Altwein, J. E., Egghart, G., Engelmann, U., Jacobi, G. H., Kreyes, G., Panijel, M., Riedasch, G., Rugendorff, E. W., Fabricius, P., O'Boyle, P. J., Gingell, C., Buck, A. C., Charig, C., Grino, P., Ferguson, D., Round, E., Shih, J., and Stoner, E., 1993, Finasteride (MK-906) in the treatment of benign prostatic hyperplasia, *Prostate* **22:**291–299.

Moore, R. A., 1944, Benign hypertrophy and carcinoma of the prostate, occurrence and experimental production in animals, *Surgery* **16:**152.

Mordenti, J., 1986, Man versus beast: Pharmacokinetic scaling in mammals, *J. Pharm. Sci.* **75:**1028–1040.

Morris, D. M., Grosse, C. M., and Selinger, K. A., 1995, Determination of GI198745 in human serum by LC/APCI/MS, Abstract, The 6th International Symposium for Pharmaceutical and Biomedical Analysis, St. Louis, M-P/A9.

Ohtawa, M., Morikawa, H., and Shimazaki, J., 1991, Pharmacokinetics and biochemical efficacy after single and multiple oral administration of N-(2-methyl-2-propyl)-3-oxo-4-aza-5α-androst-1-ene-17β-carboxamide, a new type of specific competitive inhibitor of testosterone 5α-reductase, in volunteers, *Eur. J. Drug Metab. Pharmacokin.* **16:**15–21.

Pérez-Palacios, G., Larsson, K., and Beyer, C., 1975, Biological significance of the metabolism of androgens in the central nervous system, *J. Steroid Biochem.* **6:**999–1006.

Perron, S., and Bélanger, A., 1994, Effects of 4MA, a potent inhibitor of 5α-reductase, on 3β-hydroxysteroid dehydrogenase / Δ^5 - Δ^4 -isomerase activity in guinea pig adrenals, *Steroids* **59:**371–376.

Peters, D. H., and Sorkin, E. M., 1993, Finasteride: A review of its potential in the treatment of benign prostatic hyperplasia, *Drugs* **46:**177–208.

Petrow, V., and Padilla, G. M., 1984, 5α-reductase: A target enzyme for prostatic cancer, in: *Novel Approaches to Cancer Chemotherapy* (P. S. Sunkara, ed.), pp. 269–305, Academic Press, Orlando.

Potts, G. O., Creange, J. E., Harding, H. R., and Schane, H. P., 1978, Trilostane, an orally active inhibitor of steroid biosynthesis, *Steroids* **32:**257–267.

Presti, J. C., Fair, W. R., Andriole, G., Sogani, P. C., Seidmon, E. J., Ferguson, D., Ng, J., and Gormley, G. J., 1992, Multicenter, randomized, double-blind, placebo controlled study to investigate the effect of finasteride (MK-906) on stage D prostate cancer, *J. Urol.* **148:**1201–1204.

Price, V. H., 1975, Testosterone metabolism in the skin, *Arch. Dermatol.* **111:**1496–1502.

Rasmusson, G. H., Reynolds, G. F., Utne, T., Jobson, R. B., Primka, R. L., Berman, C., and Brooks, J. R., 1984, Azasteroids as inhibitors of rat prostatic 5α-reductase, *J. Med. Chem.* **27:**1690–1701.

Rasmusson, G. H., Reynolds, G. F., Steinberg, N. G., Walton, E., Patel, G. F., Liang, T., Cascieri, M. A., Cheung, A. H., Brooks, J. R., and Berman, C., 1986, Azasteroids: Structure–activity relationships for inhibition of 5α-reductase and of androgen receptor binding, *J. Med. Chem.* **29:**2298–2314.

Ritschel, W. A., Vachharajani, N. N., Johnson, R. D., and Hussain, A. S., 1992, The allometric approach for interspecies scaling of pharmacokinetic parameters, *Comp. Biochem. Physiol. C Comp. Pharmacol. Toxicol.* **103C:**249–253.

Rittmaster, R. S., 1994, Finasteride, *N. Engl. J. Med.* **330:**120–125.

Russell, D. W., and Wilson, J. D., 1994, Steroid 5α-reductase: Two genes/two enzymes, *Annu. Rev. Biochem.* **63:**25–61.

Schweikert, H. U., and Wilson, J. D., 1974, Regulation of human hair growth by steroid hormones. I. Testosterone metabolism in isolated hairs, *J. Clin. Endocrinol. Metab.* **38:**811–819.

Shapiro, E., Becich, M. J., Hartanto, V., and Lepor, H., 1992, The relative proportion of stromal and epithelial hyperplasia is related to the development of symptomatic benign prostatic hyperplasia, *J. Urol.* **147:**1293–1297.

Siiteri, P. K., and Wilson, J. D., 1974, Testosterone formation and metabolism during male sexual differentiation in the human embryo, *J. Clin. Endocrinol. Metab.* **38:**113–125.

Stoner, E., 1994, Maintenance of clinical efficacy with finasteride therapy for 24 months in patients with benign prostatic hyperplasia, *Arch. Intern. Med.* **154:**83–88.

Sudduth, S. L., and Koronkowski, M. J., 1993, Finasteride: The first 5α-reductase inhibitor, *Pharmacotherapy* **13:**309–329.

Tenover, J. S., 1991, Prostates, pates and pimples: The potential medical uses of steroid 5α-reductase inhibitors, *Endocrinol. Metab. Clin. North Am.* **20:**893–909.

Thigpen, A. E., Davis, D. L., Milatovich, A., Mendonca, B. B., Imperato-McGinley, J., Griffin, J. E., Francke, U., Wilson, J. D., and Russell, D. W., 1992, Molecular genetics of steroid 5α-reductase 2 deficiency, *J. Clin. Invest.* **90:**799–809.

Thigpen, A. E., Silver, R. I., Guileyardo, J. M., Casey, M. L., McConnell, J. D., and Russell, D. W., 1993, Tissue distribution and ontogeny of steroid 5α-reductase isozyme expression, *J. Clin. Invest.* **92:**903–910.

Tian, G. J., 1996, In vivo time-dependent inhibition of human steroid 5α-reductase by finasteride, *Pharm. Sci.* **85:**106–111.

Tian, G., Stuart, J. D., Moss, M. L., Domanico, P. L., Bramson, H. N., Patel, I. R., Kadwell, S. H., Overton, L. K., Kost, T. A., Mook, R. A., Jr., Frye, S. V., Batchelor, K. W., and Wiseman, J. S., 1994, 17β-(N-tert-butylcarbamoyl)-4-aza-5α-androstan-1-en-3-one is an active site-directed slow time-dependent inhibitor of human steroid 5-alpha-reductase, *Biochemistry* **33:**2291–2296.

Tian, G., Mook, R. A., Jr., Moss, M. L., and Frye, S. V., 1995a, Mechanism of time-dependent inhibition of 5-alpha-reductases by Δ^1-4-azasteroids: Toward perfection of rates of time-dependent inhibition by using ligand-binding energies, *Biochemistry* **34:**13453–13459.

Tian, G., Chen, S.-Y., Facchine, K. L., and Prakash, S. R., 1995b, Chemical mechanism of the covalent modification of 5α-reductases by finasteride as probed by secondary tritium isotope effects, *J. Am. Chem. Soc.* **117:**2369–2370.

Tippin, T. K., Sinhababu, A. K., Lee. F. W., Haffner, C., Maloney, P., Hiner, R. and Frye, S. V., 1995, Application of microsomal metabolism rates to identify 6-azasteroidal 5α-reductase inhibitors with optium in vivo half-life, *ISSX Proc.* **8:**46.

Vermeulen, A., Giagulli, V. A., De Schepper, P., Buntinx, A., and Stoner, E., 1989, Hormonal effects of an orally active 4-azasteroid inhibitor of 5α-reductase in humans, *Prostate* **14:**45–53.

Vermeulen, A., Giagulli, V. A., De Schepper, P. J., and Buntinx, A., 1991, Hormonal effects of a 5α-reductase inhibitor (finasteride) on hormonal levels in normal men and in patients with benign prostatic hyperplasia, *Eur. Urol.* **20:**82–86.

Walsh, P. C., Madden, J. D., Harrod, M. J., Goldstein, J. L., MacDonald, P. C., and Wilson, J. D., 1974,

Familial incomplete male pseudohermaphroditism, type 2. Decreased dihydro-testosterone formation in pseudovaginal perineoscrotal hypospadias, *N. Engl. J. Med.* **291**:944–949.

Wilson, J. D., 1989, Sexual differentiation of the gonads and of the reproductive tract, *Biol. Neonate* **55**:322–330.

Wolfenden, R., 1972, Analog approaches to the structures of the transition state in enzyme reactions, *Acc. Chem. Res.* **5**:10–18.

Chapter 18

Discovery of a Potent and Selective α_{1A} Antagonist

Utilization of a Rapid Screening Method to Obtain Pharmacokinetic Parameters

Kimberly K. Adkison, Kathy A. Halm, Joel E. Shaffer, David Drewry, Achintya K. Sinhababu, and Judd Berman

1. INTRODUCTION

1.1. Benign Prostatic Hyperplasia

Benign prostatic hyperplasia (BPH) is a disease with an ever-increasing prevalence in men as they age. The human prostate is contained within a fairly rigid capsule and the increased proliferation of epithelial cells and/or the decreased apoptosis of existing cells leads to increased pressure on the prostatic urethra to produce the symptoms of BPH. The symptomatology is described in terms of "irritancy" and "obstructiveness." Irritancy refers to the increased frequency of urination in general as well as the frequency of nocturnal urination. The obstructive symptoms relate to the time to begin urination once the patient is ready, the rate of urine flow,

---

Kimberly K. Adkison, Kathy A. Halm, Joel E. Shaffer, David Drewry, Achintya K. Sinhababu, and Judd Berman • Glaxo Wellcome Research and Development, Research Triangle Park, North Carolina 27709.

Integration of Pharmaceutical Discovery and Development: Case Studies, edited by Borchardt *et al.*, Plenum Press, New York, 1998.

the time to void, the extent of "dribbling" at the end of urination, and the degree of complete bladder emptying. Historically, these symptoms were most commonly seen in men with enlarged prostates. However, more recent data suggest that the degree of enlargement does not necessarily correspond with the severity of symptoms.

Surgical removal or transurethral resection of the prostate has been the standard therapy for BPH in the United States with an annual health care cost of about $5 billion (Kirby and Christmas, 1993). Pharmacological alternatives to treat BPH have been developed. For example, 5α-reductase enzyme inhibitors such as finasteride block the conversion of testosterone to the more potent androgen, dihydrotestosterone. Dihydrotestosterone regulates growth of the prostate. Inhibition of dihydrotestosterone formation leads to prostate shrinkage, increases in flow rate, and an improvement in BPH symptoms in some patients (Stoner, 1994). Pioneering studies by Caine, Raz, and co-workers led to the observation that α-adrenergic blocking agents might also have a beneficial effect in patients with BPH symptoms (Raz *et al.*, 1973; Caine *et al.*, 1973, 1978).

1.2. Therapeutic Use of α_{1A}-Selective Antagonists

The prostate and prostatic urethra are innervated by the parasympathetic and sympathetic nerves. Nerve stimulation can result in contraction of the prostate and an increase in pressure on the prostatic urethra resulting in the obstructive symptoms of BPH (Caine, 1988). Studies have shown that the contractile activity of the prostatic tissue is primarily mediated by α_1 adrenoceptors (Lepor and Shapiro, 1984; Heible *et al.*, 1985). α_1 antagonists, like terazosin (Lepor *et al.*, 1992) and doxazosin (Gerber *et al.*, 1996), increase urinary flow rates and decrease the symptoms of BPH in most patients and have recently received approval for the treatment of BPH. However, these non-subtype-selective α-1-adrenoceptor antagonists have dose-limiting side effects such as dizziness, hypotension, orthostatic hypotension, lethargy, nasal stuffiness, and impotence.

Three subtypes of α-$_1$ receptors have been identified: α_{1A}, α_{1B}, and α_{1D}. The α_{1A} adrenoceptor is the predominant subtype found in prostatic smooth muscle (Price *et al.*, 1993). Consequently, blockade of the other two known subtypes may offer no additional therapeutic advantages, but may contribute to the incidence of undesirable side effects. It is hypothesized that an α_{1A}-selective antagonist would be effective in the treatment of the symptoms of BPH without the undesirable side effects associated with antagonism of the α_{1D} and α_{1B} receptors. Therefore, such a compound may be dosed higher to obtain greater efficacy or at least have a lower incidence of side effects at lower efficacious doses making it a more tolerable agent (Forray *et al.*, 1994; Goetz *et al.*, 1994). The only way to determine the reality of these contentions would be to make a highly selective compound and test the hypothesis.

1.3. Project Goal

The goal of the α_{1A} research project team was to rapidly discover a potent and selective α_{1A} adrenoceptor antagonist that demonstrated activity in the prostate over other tissues. The compound should provide symptomatic relief of BPH without the undesirable side effects of blood pressure lowering and lethargy to have a competitive advantage over commercially available nonselective agents. In addition, the compound should be orally bioavailable and have a half-life suitable for once-daily dosing.

The critical path that we originally put forth to evaluate and select molecules for advancement involved *in vitro* α-adrenoceptor binding assays, *in vivo* dog efficacy and functional selectivity studies, *in vivo* dog orthostasis risk studies, and pharmacokinetic evaluation of lead compounds. When we learned the SAR around α_{1A} receptor binding and selectivity and were consistently making potent and subtype-selective compounds, the project team ran into a bottleneck in evaluating compounds for their pharmacokinetic or pharmacodynamic characteristics. At that time we modified our compound progression scheme and used a risky pharmacokinetic approach to speed up the process of selecting a potent and α_{1A}-selective compound with appropriate pharmacokinetic characteristics for drug development.

In the remainder of this chapter we will present our strategy and highlight the ways in which calculated risk-taking and close integration and teamwork of drug discovery and development scientists during the early drug discovery stage enabled us to select a clinical candidate in a shorter time period than otherwise would have been the case.

2. RESEARCH STRATEGY

2.1. Compound Progression and Critical Path

Following compound synthesis or retrieval from our compound data base, a compound was entered into α-adrenoceptor binding assays using human α_{1A}, α_{1B}, and α_{1D} receptors expressed in rat-1 fibroblasts. The eventual goal was to find a potent and α_{1A}-selective antagonist with a pK_i (determined by inhibition of [^{125}I]-HEAT binding) for the α_{1A} receptor greater than 8 and a ratio of K_is for the α_{1B} and α_{1D} receptor versus the α_{1A} receptor of at least 100.

Potent and selective compounds were then advanced to an *in vivo* dog model to assess efficacy and functional selectivity. Anesthetized dogs were instrumented to measure prostatic urethral pressure in response to electrical stimulation of the hypogastric nerve and systemic blood pressure in response to intravenous phenylephrine. Doses of the α_1 antagonist were escalated and the ability of the antagonist to attenuate the response of the prostate to nerve stimulation was compared with its ability to inhibit systemic blood pressure response to phenylephrine. A nonselective antagonist (e.g., terazosin) would have similar ED_{50}s for both urethral pressure and blood pressure lowering. An α_{1A}-selective agent should have a lower ED_{50} for urethral pressure inhibition than for blood pressure lowering. The goal was to achieve a ratio of blood pressure ED_{50} to the urethral pressure ED_{50} greater than 100 (i.e., it should take a 100-fold greater dose to cause blood pressure lowering than urethral relaxation). We chose to use activation of the sympathetic nerves to the prostate to stimulate prostatic contractions because at the time it was not necessarily clear which type of adrenoceptor played the dominant role in mediating contraction of the prostate. We had two measures of prostate contraction, the nerve-stimulated contraction and the more modest contraction seen with systemic administration of phenylephrine. This ended up being critical to identifying compounds because the phenylephrine response was not as robust as the nerve-stimulated response, and compounds not selective for the α_{1A} subtype, but more selective for the other subtypes, frequently showed some degree of block on phenylephrine, but not against nerve-stimulated prostatic contractions.

Once compounds showed *in vitro* subtype specificity and *in vivo* potency and selectivity, the compounds entered several studies designed to assess side effect potential. Compounds were profiled for activity at other pharmacologically important targets [e.g., other seven-transmembrane (7TM) receptors and ion channels] by contract with Novascreen (Hanover, MD). Clean compounds (100-fold selectivity) were tested for their ability to induce orthostatic hypotension in the upright tilt conscious dog model. Terazosin and other nonselective compounds were shown to inhibit reflex-induced increases in blood pressure related to upright tilt. If the orthostatic hypotension is related to antagonist activity at the α_{1B} or α_{1D} receptors, then an α_{1A}-selective compound should exhibit less orthostatic hypotensive effects.

Compounds that met *in vitro* and *in vivo* efficacy and tolerability criteria were then evaluated for their pharmacokinetic parameters. At the time we felt we needed a compound with a low metabolic clearance, a low volume of distribution (to minimize CNS penetration and CNS effects), and a long terminal elimination half-life suitable for once-daily dosing. We chose dogs as our primary pharmacokinetic species because dogs were used in the pharmacology efficacy and safety studies and because we knew the pharmacokinetics of terazosin in the dog. Our goal was to find a compound with a half-life suitable for once-daily dosing and oral bioavailability greater than 30% in the dog. Compounds that met these criteria would then be tested in a second species for interspecies scaling comparisons.

2.2. Discovery of α_{1A}-Selective Oxazole-Containing Antagonists

Our early medicinal chemistry efforts resulted in the discovery of several novel and structurally distinct series of α_1 antagonists. We decided to focus our efforts on one series based on the following parameters: *in vitro* potency and selectivity, *in vivo* potency and selectivity, (relatively) clean profile against other 7TM receptors, chemical novelty, and the ability to build a strong chemical program around the molecule. The oxazole series, exemplified by compound **1** (Table I), was chosen based on this analysis. Compound **1** had low nanomolar affinity for the α_{1A} receptor, and although the *in vitro* selectivity was just moderate, the *in vivo* selectivity was good. This compound also had a reasonable profile against other 7TM receptors, with the next highest affinity for 5HT2 receptors (about 30-fold selective). The novelty of the structure and the ability to take advantage of several substitution positions on the oxazole also played into our decision to make this our lead chemical series and explore the SAR around it in depth.

2.2.1. ENHANCED SELECTIVITY WITH OXAZOLE 4-POSITION MODIFICATIONS

We eventually were led to substitution at the 4-position of the oxazole nucleus. Replacement of the proton with straight chain alkyls at the 4-position of the oxazole improved potency and selectivity. Chemistry was developed that allowed for easy incorporation of ethers into the 4-position (Table I). Increasing the length of the substituent led to an increase in potency and selectivity. The propyl ether

Table I
Effect of 4-Position Oxazole Modifications on Human α_{1A} Receptor Potency and Selectivity

O—CH3
O
S—NH2
O
N
O
N
R1

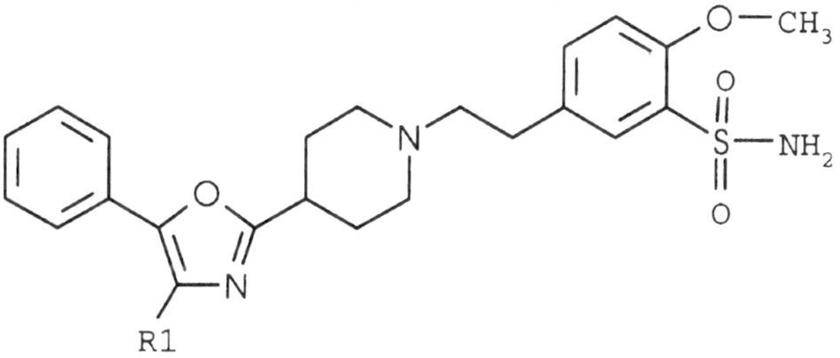

Compound	R1	α_{1A} pK_i	B/A	D/A
1	H	8.3	4	10
2	CH_2OCH_3	8.2	5	8
3	$CH_2OCH_2CH_3$	8.4	16	32
4	$CH_2OCH_2CH_2CH_3$	8.6	20	40
5	$CH_2OCH_2CH_2CH_2CH_3$	7.9	4	8
6	$CH_2OCH_2CH_2F$	8.9	50	80
7	$CH_2OCH_2CF_3$	9.3	112	145
8	$CH_2OCH_2CF_2CF_3$	8.4	16	25

(compound **4**) had a pK_i value of 8.6, an α_{1B} K_i-to-α_{1A} K_i selectivity ratio (B/A) of 20 and α_{1D} K_i-to-α_{1A} K_i selectivity ratio (D/A) of 40, representing a fourfold increase in B/A selectivity and a fivefold increase in D/A selectivity over the original 4-unsubstituted oxazole. It appeared that having a lipophilic chain four or five atoms in length at the 4-position conferred the desired α_{1A} selectivity. In order to probe this region of space, a set of fluorinated alkyls were synthesized. This fluorinated series gave rise to the highly potent and selective compound **7,** which retained the potency of the original lead molecule, and showed dramatically more selectivity (B/A selectivity improves from 5 to 112, and D/A selectivity improves from 10 to 145).

2.2.2. ENHANCED SELECTIVITY WITH SULFONAMIDE SUBSTITUTIONS

The SAR around the right-hand side of the molecule (aryl sulfonamide portion) was also explored, with work initially focusing on sulfonamide removal and substitution (Table II). The sulfonamide seemed to make an important receptor interaction, as its removal resulted in a compound with lower α_{1A} potency and lower B/A selectivity (compound **9**). The dimethyl sulfonamide (compound **10**) also lost potency and selectivity against the α_{1D} receptor. Monosubstitution of the sulfonamide with groups capable of hydrogen bonding met with more success. Compound **13,** made from the *N*-methyl amide of sarcosine, had a pK_i value of 9.3 for the α_{1A} receptor, B/A selectivity of 288, and D/A selectivity of 347. Other compounds of this series, such as compound **11,** made from glycine amide, and compound **17,** made from *N*-acetyl ethylenediamine, also met potency and selectivity criteria. A pharmacokinetic evaluation of compounds **7** and **11** in dogs revealed that these compounds had short half-lives and were metabolically eliminated (Table II).

2.2.3. STRUCTURE VERSUS *IN VITRO* METABOLIC REACTIVITY RELATIONSHIPS

LC/MS/MS analysis of bile collected from a dog dosed with compound **11** revealed the following major sites of metabolism: left-hand-side phenyl ring hydroxylation, piperidine *N*-dealkylation, and *N*-dealkylation of the alkylsulfonamide group. As the major metabolites formed were likely cytochrome P450 products, *in vitro* metabolism studies in dog liver microsomes were initiated in order to determine the relationship between structure and metabolic reactivity and metabolic pathways of a series of α_{1A} antagonists in the oxazole series. Another reason for conducting the *in vitro* studies was to determine if the major oxidative pathways of metabolism of this class of compounds were similar in dog and human.

The general procedure for conducting *in vitro* metabolism studies involved

Table II

Effect of Sulfonamide Substitutions on Human α_{1A} Receptor Potency and Selectivity and Pharmacokinetic Parameters in the Dog

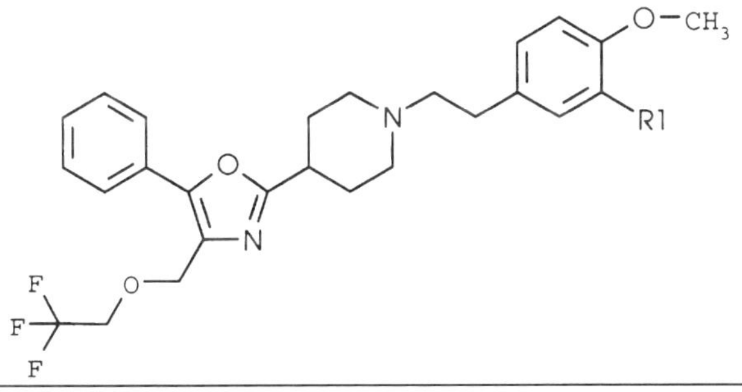

Compound	R1	α_{1A} pK_i	B/A	D/A	Clearance (ml/min/kg)	$t_{1/2}$ (hr)	Volume (ml/kg)
9	H	8.7	55	135	14	7.5	785
7	SO_2NH_2	9.3	112	145	5.0	1.4	382
10	$SO_2N(Me)_2$	8.8	76	63	11	2.0	1102
11	$SO_2NHCH_2CONH_2$	9.0	79	102	18	2.3	3175
12	$SO_2NHCH_2CONHMe$	9.0	141	123	22	2.4	3450
13	$SO_2NCH_2CONHMe$ Me	9.3	288	347	29	2.1	3464
14	$SO_2NHCHCONH_2$ Me	9.3	73	416	22	1.8	1241
15	$SO_2NH(CH_2)_2NHSO_2Me$	9.2	115	89	2.0	2.4	151
16	$SO_2NH(CH_2)_2NHCO_2Me$	8.8	115	66	9.1	2.4	647
17	$SO_2NH(CH_2)_2NHCOMe$	9.2	263	145	10	2.6	600

incubation of the α_{1A} antagonists at 10 μM concentrations in dog or human liver microsomes (1 mg protein/ml) at 37°C in the presence or absence of NADPH for various periods of time. The α_{1A} antagonist concentration was chosen based on maximum plasma concentrations observed in dog pharmacokinetic studies of several compounds. HPLC profiles of the reaction mixtures were generated by elution from a BDS Hypersil C8 column with a mixture of acetonitrile and ammonium acetate buffer under gradient conditions and fluorescence detection. The rate of metabolism was determined for each compound from the linear portion of the plot of percent parent compound remaining versus time. In a preliminary experiment the rates of metabolism (Fig. 1) and the metabolic profiles (data not shown) of eight analogues were determined. Analysis of the data suggested the following:

1. All of the test compounds underwent relatively rapid and extensive metabolism. The *in vitro* half-lives varied from 3 min for compound **13** to

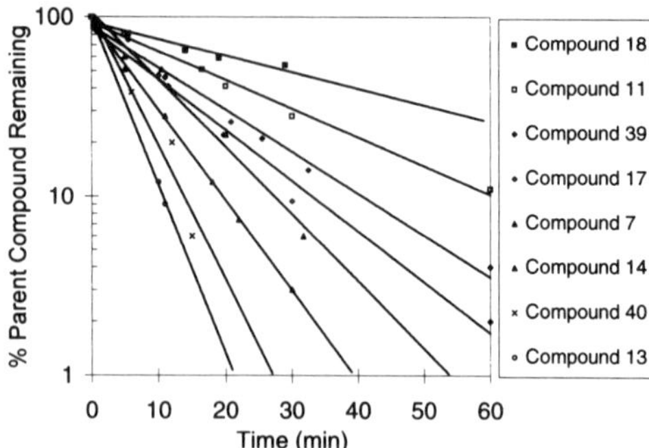

Figure 1. A comparison of the rates of oxidative metabolism of eight α_{1A} antagonists by dog liver microsomes.

33 min for compound **18,** suggesting that rates of metabolism of this class of compounds were dependent on their structures.

2. Left-hand-side phenyl ring hydroxylation was a major metabolic pathway. Compound **18** was designed to prevent phenyl ring hydroxylation with addition of a fluorine in the *para*-position. Presence of fluorine in the *para*-position of the phenyl ring blocked this pathway completely with a concomitant increase in *in vitro* half-life (compound **7,** half-life = 8 min versus its *para*-fluoro derivative compound **18,** half-life = 33 min).
3. Piperidine *N*-dealkylation, i.e., the loss of the right-hand-side arylethyl moiety, was a major metabolic pathway for all of the test compounds. It was virtually the only reaction that took place with compound **18.** Substitution on the sulfonamide moiety appears to influence the rate of piperidine *N*-dealkylation. In general, bulkier substituents on the sulfonamide nitrogen led to faster *N*-dealkylation.
4. Alkyl substituents on the sulfonamide nitrogen were also susceptible to metabolic dealkylation (compounds **11, 13,** and **17**). However, this dealkylation appeared to be blocked when the site of *N*-dealkylation was substituted with a methyl group as in compound **14.**

Compounds **18** (a compound with one of the longest *in vitro* half-lives and new lead), **13** (a compound with one of the shortest *in vitro* half-lives that gave rise to at least 12 metabolites), and **38** (the left-hand-side *para*-fluorophenyl analogue of compound **13**) were selected for metabolism by pooled human liver microsomes. The rates of metabolism of each of these compounds in dog and human liver microsomes were nearly identical. In addition, all of the metabolites produced by human liver microsomes were also produced by dog liver microsomes (data not shown). These results suggested that the dog might be a suitable model for human metabolism.

Twenty-nine additional compounds were subsequently tested in the *in vitro* metabolism assays. Results with the additional analogues confirmed the general conclusions reached with the initial set of eight. *In vitro* dog liver microsomal metabolism studies were influential in the selection of the lead compound (compound **18**) for further study and provided metabolism-based SAR that was useful in the design of novel analogues.

2.2.4. EFFECT OF PHENYL RING SUBSTITUTIONS ON POTENCY AND SELECTIVITY

Because the *para*-fluoro substituent increased metabolic stability, a series of compounds with halogen substitution on the aryl ring were prepared (Table III). Whereas compound **18** exhibited a threefold increase in potency over compound **7**, accompanied by a slight erosion of selectivity, the *para*-chloro compound showed a decrease in potency with a significant increase in selectivity. Di- and trihalogen-substituted compounds were usually more selective than compound **7**.

Table III
Effect of Phenyl Ring Substitutions on Human α_{1A} Receptor Potency and Selectivity and *in Vivo* Pharmacokinetic Parameters in the Dog

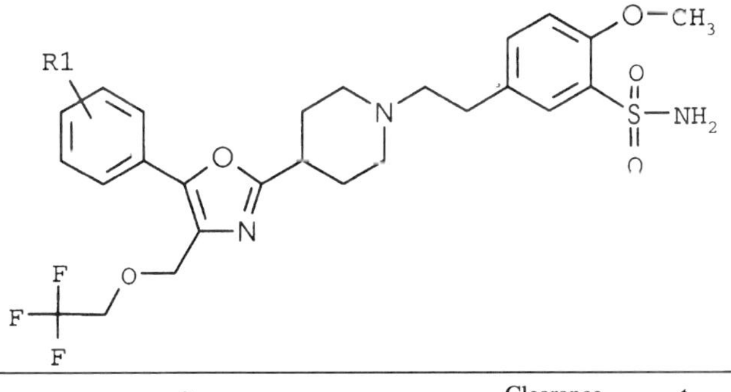

R1
O—CH$_3$
O
S—NH$_2$
O
N
O
N
O
F
F
F

Compound	R1	α_{1A} pK_i	B/A	D/A	Clearance (ml/min/kg)	$t_{1/2}$ (hr)	Volume (ml/kg)
7	4-H	9.3	125	160	5.0	1.4	382
18	4-F	9.7	81	112	0.8	4.1	202
19	4-Cl	8.7	263	204	0.7	4.9	263
20	3-F	8.6	182	59	5.0	1.5	523
21	2,3-F	8.1	708	339	2.6	0.8	242
22	2,4-F	9.3	170	145	1.2	4.4	340
23	2,5-F	8.1	339	339	4.6	2.5	631
24	3,4-F	8.9	200	209	1.9	4.7	527
25	2,3,4-F	8.5	575	324	0.6	7.2	240
26	2,4,5-F	8.3	331	295	2.3	2.3	372

The potency typically started to fall off for these multisubstituted analogues, and none approached the potency of the simple *para*-fluoro compound.

2.2.5. EFFECT OF REDUCED RIGHT-HAND-SIDE BULK ON POTENCY AND SELECTIVITY

In order to simplify the molecules, and explore a slightly different region of chemical space, a series of molecules were made that contained only *para*-substitution on the right-hand-side phenyl ring (Table IV). These molecules showed quite good potency and subtype selectivity. An unsubstituted phenyl ring (compound **27**) on the right-hand-side had a pK_i value of 9.1 for the α_{1A} receptor, B/A selectivity of 85, and D/A selectivity of 245. Both simple electron-donating groups (OMe, NH_2, NHMe, NMe_2) and electron-withdrawing groups (F, SO_2NH_2, $CONH_2$, CONHMe, SO_2Me) in the *para*-position gave potent and selective compounds. In summary, once we fixed the 4-position of the oxazole as the trifluoroethyl ether, we were able to modify both the left-hand side phenyl and the right-hand side of the molecule and create an extensive series of molecules with suitable potency and subtype selectivity *in vitro* and *in vivo*.

Table IV

Effect of Reduced Right-Hand-Side Bulk on Human α_{1A} Potency and Selectivity and Pharmacokinetic Parameters in the Dog

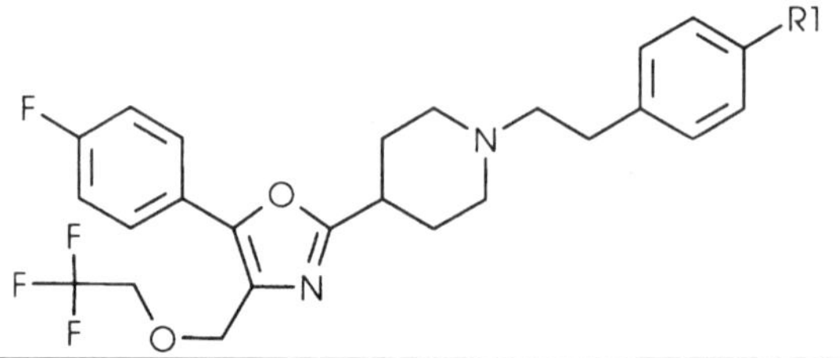

Compound	R1	α_{1A} pK_i	B/A	D/A	Clearance (ml/min/kg)	$t_{1/2}$ (hr)	Volume (ml/kg)
27	H	9.1	85	245	4.1	12.0	2743
28	F	9.1	55	138	7.4	1.8	884
29	OMe	9.2	60	186	7.9	5.3	2677
30	SO_2NH_2	9.1	76	76	35	5.9	9670
31	$CONH_2$	9.6	91	155	28	1.9	3655
32	CONHMe	9.3	115	302	33	2.1	5085
33	$CON(Me)_2$	8.1	62	100	8.4	4.2	2102
34	NH_2	9.6	85	295	4.6	14.5	9698
35	NHMe	9.3	93	257	6.6	14.9	5584
36	CH_2NHSO_2Me	9.1	234	275	5.0	12.2	4566
37	$CH_2NHCOMe$	8.9	151	166	5.7	8.1	3350

3. PHARMACOKINETIC/PHARMACODYNAMIC STRATEGY

Pharmacokinetic and oral bioavailability studies of some early molecules were conducted in dogs to gain some general knowledge about the absorption, distribution, and elimination of these compounds. In general, the compounds were well absorbed (also supported by permeability values across a Caco-2 cell monolayer $>75 \times 10^{-7}$ cm/sec), highly plasma protein bound (94–99%), and primarily eliminated by metabolism (generally <2% of the dose excreted as unchanged parent compound in urine or bile). However, the compounds had short half-lives (<2 hr) primarily because of a high systemic clearance.

Eventually, the left-hand-side *para*-fluorinated compound (**18**) emerged as the lead compound. Compound **18** had a half-life in dogs of 4 hr (twice that of any compound previously studied), a very low clearance (0.9 ml/min per kg), and an excellent oral bioavailability (90%). Also, it was very potent and selective in *in vitro* receptor binding assays and had a good *in vivo* potency (ED_{50}=13 μg/kg) and selectivity (BP ED_{50}/UP ED_{50} > 500) profile. The pharmacokinetics of compound **18** was studied in a second species (rat) to allow for interspecies scaling and prediction of half-life in human. Both the clearance (8.2 ml/min per kg) and steady-state volume of distribution (2274 ml/kg) of compound **18** were greater in rats than in dogs. The half-life (4 hr) was identical to that observed in dogs. Allometric scaling of clearance and volume from rats and dogs to humans predicted a half-life of roughly 5 hr in humans. A 5-hr half-life was believed to be too short for once-daily dosing.

3.1. *In Vitro* Metabolism Screening Prior to Pharmacokinetic Studies

The chemists were rapidly synthesizing potent and selective molecules and a substantial bottleneck had now been created at the *in vivo* pharmacokinetic testing stage. The preliminary *in vitro* metabolism work with eight compounds showed that there was some relationship between *in vitro* metabolism rates and *in vivo* pharmacokinetic clearances and elimination half-lives. For example, compound **7** had an *in vitro* half-life of 8 min, an *in vivo* clearance of 5 ml/min per kg, and an *in vivo* half-life of 1.4 hr. Compound **18** had a longer *in vitro* half-life than compound **7** and a lower clearance and longer half-life *in vivo*. Based on these results we thought the *in vitro* metabolism studies could be used to quickly screen out compounds that were rapidly metabolized prior to advancing compounds to more resource-intensive pharmacokinetic studies. Unfortunately, after screening 22 compounds and studying their pharmacokinetics in dogs we found a poor correlation between the *in vitro* rates of metabolism in dog liver microsomes and the *in vivo* half-lives of the compound in dogs. Structural changes that led to reduced metabolic rates also led to compensating decreases in volume of distribution and, consequently, no change in half-life. There was also a poor correlation between *in vitro* metabolic rates and total body clearance in the dog because of a 10-fold variability in plasma free fraction among compounds and variable contributions of nonoxidative metabolic pathways such as glucuronidation. Indeed, analysis of the urine of dogs following administration of selected compounds that were outliers in the *in vitro–in vivo* correlation analysis showed the presence of a significant fraction of the administered dose as glucuronide of parent. We realized that *in vitro* metabolism studies were not useful for rapidly identifying compounds with optimal *in vivo* half-life.

At that time, nearly 50 potent and α_{1A}-selective compounds had been synthesized, but remained untested for pharmacokinetics. The project team had a decision to make—whether to send compound **18** to development immediately or wait until the remaining compounds were tested to see if a compound with a longer pharmacokinetic half-life existed. It would take roughly 6 months to determine the pharmacokinetic properties of 50 compounds using conventional dosing and analytical techniques. The project team challenged the Bioanalysis and Drug Metabolism Department to screen this group of potent and selective compounds for their pharmacokinetic properties within a month. Compounds with pharmacokinetic properties better than compound **18** in dog would then be tested in the hypogastric nerve dog model and the orthostatic hypotension model to confirm *in vitro* potency and selectivity results. If confirmed, the compound would be moved into development. Otherwise, compound **18** would be moved into development.

3.2. Improved Pharmacokinetic Throughput: Mixture Dosing Coupled with LC/MS Analysis

The project team's goal of screening 50 compounds for their pharmacokinetic parameters was achieved by concomitantly dosing mixtures of compounds to one dog and analyzing the compounds simultaneously by LC/MS. The mixture dosing approach, which we refer to as N-in-One dosing (where N is the number of compounds dosed simultaneously to one animal), was initially validated by comparing the pharmacokinetics of five α_{1A} antagonists (compounds **7, 13, 17, 18, 38**) obtained from mixture dosing and sample analysis by LC/MS/MS in the selected reaction monitoring (SRM) mode to the pharmacokinetics obtained after individual dosing and sample analysis with HPLC fluorescence (Halm *et al.,* 1996). The five compounds were selected for validation of the approach because they demonstrated a wide range of clearances (0.9 to 28.9 ml/min per kg) and volumes of distribution (243 to 3464 ml/kg) in the individual studies. A good correlation was noted in the half-lives, clearances, and steady-state volumes of distribution suggesting that no significant compound–compound interactions altered the pharmacokinetics of the five α_{1A} antagonists (Table V). We subsequently went on to

Table V
Comparison of the Pharmacokinetic Parameters of Five α_1 Antagonists Obtained Following 5-in-One or Individual Dosing

	Half-life (hr)		Clearance (ml/min/kg)		Volume (ml/kg)	
Compound	5-in-One	Individual	5-in-One	Individual	5-in-One	Individual
7	1.5	1.4	5.9	5.0	473	382
13	2.4	2.1	17.3	28.9	2757	3464
17	2.9	2.6	18.7	10.3	2833	600
38	3.0	3.4	11.2	11.6	2857	2907
18	4.9	4.2	0.86	0.90	330	243

screen the potent and selective compounds by dosing mixtures of 12–22 compounds simultaneously with sample analysis by LC/MS.

3.2.1. LC/MS METHOD DEVELOPMENT

The plasma samples were prepared by protein precipitation with acetonitrile containing an internal standard. Composite plasma calibration standard curves were prepared over the range of 5–2500 ng/ml. Atmospheric pressure chemical ionization (APCI) LC/MS was utilized for plasma sample analysis rather than HPLC with fluorescence detection because of the inherent detection selectivity, specificity, and sensitivity of LC/MS for trace level quantitation of compounds from complex matrices. LC/MS methods were developed prior to dosing because grouping the compounds by their chromatographic characteristics and molecular weights facilitated sample processing and analysis. All analysis was done on a Finnigan TSQ-700 mass spectrometer in the positive ion mode at unit resolution. The compounds studied exhibited good mass spectrometric response in the positive ion mode using either APCI or electrospray ionization (ESI). APCI MS was chosen for analysis because of its compatibility with conventional HPLC column flow rates and relative insensitivity to ionic strength of the samples. A more generic approach using MS rather than MS/MS was used for quantitation of larger mixtures (12–22 compounds) by scanning in the SIM mode rather than in the SRM mode. The LC/MS dwell time was based on the number of compounds in the mixture with a total scan time of 2.5 sec. The molecular weights of the compounds to be studied ranged from 360 to 767. Isobars were included in a mixture if they could be chromatographically resolved with minimal effort. Enantiomers were excluded from being in the same mixture. Potential interferences in molecular weight redundancy from naturally occurring 13C isotope abundances, halogen isotope abundances, or possible common metabolite molecular weights such as the addition of oxygen (M+16) were taken into account during the pooling process.

The HPLC mobile phase consisted of acetonitrile and ammonium acetate buffer. A switching valve was used to divert the first 2 min of HPLC effluent to waste to minimize contamination of the mass spectrometer source with biological salts. For the larger mixture studies, the effects of compound coelution on suppression of ionization, in addition to the increase in the number of potential coeluting metabolites, were unknown. The HPLC method was modified from the 5-in-One study by using longer BDS Hypersil C18 columns (250 × 4.6 mm, 5-μm particle size or 150 x 4.6 mm, 3-μm particle size) to increase the selectivity of the isocratic HPLC separation and afford more specificity for LC/MS (SIM) detection. The compounds showed a wide variation in retention time. Compounds were included in a mixture if they could be retained from the column void volume and eluted within a capacity factor range of about 2 to 10. The percentage of acetonitrile in the mobile phase was adjusted accordingly. Mixtures of more diverse compounds would have required gradient elution.

3.2.2. *IN VIVO* N-IN-ONE DOSING AND SAMPLING

Compound groups for dosing were based on analytical requirements as described above and solubility issues. The α_{1A} antagonists for individual dosing and for the 5-in-One study were dissolved in 10–20 ml of 50 mM sodium acetate buffer (pH 4.5). However, when larger numbers of compounds were mixed, the compounds were insoluble in sodium acetate buffer. A formulation scientist from the Pharmaceutics Department helped solubilize the compounds for intravenous dosing in 20 ml of 100 mM sodium acetate buffer containing 30% propylene glycol and 1% Tween 80. Several compounds were eliminated from the groups because of solubility problems. Because the pharmacokinetics of compound **18** was known, compound **18** was dosed as a control compound in all studies to assess the validity of the mixture dosing approach in each animal.

The dose solutions were administered to three dogs as a 10-min infusion via a cephalic vein cannula. Each compound was dosed at 0.5 mg(base)/kg body wt (22-in-One) or 0.3 mg(base)/kg body wt (13 and 12-in-One). The dose of each compound was chosen based on analytical method sensitivity and consideration of potential adverse effects of a high combination dose. The dog from the 22-in-One study was ventilated and anesthetized with isofluorane during dosing so that potential adverse effects of the high dose of α_{1A} antagonists could be monitored. No adverse effects were noted from concurrent administration of 22 compounds so the anesthesia was stopped 15 min after dosing and the dog regained consciousness. Dogs in subsequent studies were dosed while conscious. Blood samples were withdrawn via a second cephalic vein cannula into a heparinized syringe at predetermined time points from 0 to 24 hr and the resulting plasma samples were quantitated by APCI LC/MS. The pharmacokinetic parameters of clearance, steady-state volume of distribution, and half-life were determined for each compound by model independent methods.

3.2.3. RESULTS OF N-IN-ONE STUDIES

Plasma concentration data were calculated using the ratio of the peak area of each compound to the peak area of the internal standard. Linear regression analysis with $1/x$ weighting was used to fit all of the plasma calibration standard curves. The correlation coefficients were greater than 0.99 and the intercepts of the lines were not statistically different from zero. Overall, the intrarun linearity and accuracy of the plasma calibration standards were acceptable for all but two compounds. The nonlinearity observed for two compounds that eluted close to the HPLC column void volume may have been the result of suppression of ionization. The limit of quantitation for each compound was 5–10 ng/ml. The percent differences from the line for the lowest concentration used in the curves was less than 26% for all compounds.

The pharmacokinetic parameters of some compounds studied in the three N-in-One studies are presented in Tables II–IV. The pharmacokinetic parameters varied widely across this structurally related series of compounds. The half-lives ranged from a low of 0.8 hr to a high of 14.9 hr. The clearances ranged from 0.4 to 43.6 ml/min per kg and the steady-state volume of distribution ranged from 69 to 18,685 ml/kg. The half-life of compound **18** (the control compound) varied between studies; however, the time-averaged parameters of clearance and volume for compound **18** were consistent within N-in-One studies and with values previously obtained in individually dosed dogs. Eight new compounds were identified with a half-life longer than 5 hr and a systemic clearance less than 7 ml/min per kg.

Six of the more potent and selective compounds identified as having a longer half-life than compound **18** and three compounds with short half-lives were retested by individual compound administration to dogs and sample analysis by HPLC with fluorescence detection. Figure 2 compares the half-lives of the nine compounds after N-in-One and individual dosing. There was an excellent correlation ($r^2 = 0.9$) in the half-lives of the nine compounds obtained from the N-in-One dosing to that obtained from individual dosing and the slope of the regression line was near unity. There was also a good correlation in the clearance and volume of distribution ($r^2 = 0.9$ and 0.9, respectively) between the two approaches. However, the slopes of the regression lines were less than unity, which indicates that the clearance and volume of distribution parameters obtained in the N-in-One experiment tended to be greater than those obtained in the individual experiments. This was particularly true for high-volume-of-distribution and high-clearance compounds. Nonetheless, the rank order in parameters was always the same and allowed us to use cutoff half-life and clearance criteria to determine which com-

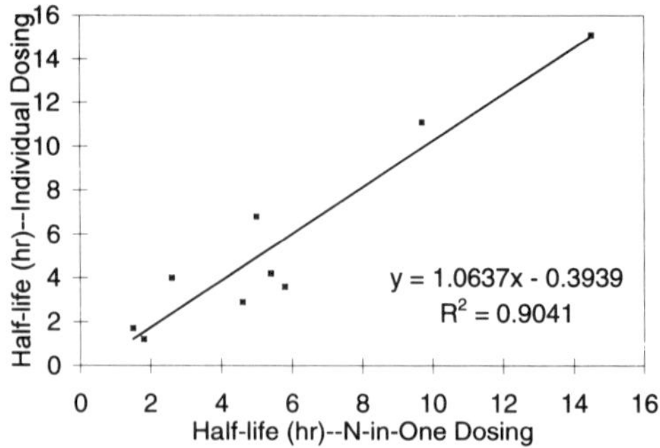

Figure 2. Comparison of the half-lives of nine compounds between N-in-One dosing and individual dosing.

pounds would be administered individually to confirm N-in-One pharmacokinetic data.

3.2.4. ADVANTAGES AND RISKS OF N-IN-ONE DOSING WITH LC/MS ANALYSIS

The N-in-One dosing approach offers several advantages to drug discovery programs. The method is very rapid relative to traditional, one-compound–one-dog studies by virtue of greatly reduced in-life experimental time and sample preparation time. In our case, pharmacokinetic parameters were obtained for roughly 15 compounds in about 7 days (this includes method development, dose solution preparation, dosing and sampling, sample processing, quantitation, and pharmacokinetic analysis). We also found that N-in-One dosing was a useful approach for rapidly identifying a backup candidate with suitable pharmacokinetic parameters. For example, 80 additional potent and selective compounds were dosed in seven more dog studies during the backup phase of the α_{1A} project.

N-in-One dosing has also enhanced our ability to generate *in vivo* SAR around pharmacokinetic parameters as a result of the increased number of compounds studied. For example, as we determined the pharmacokinetic parameters of about 120 compounds in this project, we learned that compounds with a *para*-fluoro on the left-hand-side phenyl had the lowest *in vivo* clearances. Compounds with no *meta*-substitution on the right-hand-side phenyl tended to have high volumes of distribution and compounds that had a nonsubstituted sulfonamide in the *meta*-position tended to have low volumes of distribution.

LC/MS techniques were critical to the success of N-in-One dosing for the α_1 antagonists. Mass spectrometry provided the necessary detection selectivity, specificity, and sensitivity. APCI LC/MS also offered the advantages of high sample throughput with minimal sample preparation (e.g., no need for sample derivatization as with traditional GC/MS) and decreased analytical method development time. Running calibration curves allowed us to quantitate the plasma concentrations of each compound and calculate clearance and volume of distribution for each compound. Knowledge of clearance and volume parameters better enabled us to select which compounds to redose individually (e.g., we chose to look at low-clearance compounds first to avoid first-pass liver extraction and optimize oral bioavailability). However, if the goal were to simply look for compounds with a long half-life, this technique could yield even faster results as there would be no need for calibration and quantitation.

We recognize potential limitations of the N-in-One approach. For example, there may be compound–compound interactions that alter metabolism, distribution (either plasma or tissue protein binding), or renal/biliary excretion. We were not concerned with interactions in renal/biliary excretion because these α_{1A} antagonists were not excreted to a significant extent in the urine or bile. We did not see any evidence for inhibition of enzymatic metabolism for the compounds that were also individually dosed. Inhibition of metabolic pathways or alterations in distribution by one molecule on another were probably avoided in our studies because of the relatively low plasma concentrations achieved. Generally, plasma concentrations in the nanomolar range are not high enough to inhibit enzymes involved in xenobiotic metabolism or to cause plasma protein binding displacement (Rowland and Tozer, 1995). Pharmacological or toxicological events may limit the total dose or number of molecules that can be coadministered. Fortunately, we observed no adverse effects from coadministration of these adrenergic blocking agents, probably reflecting the high degree of receptor selectivity of the compounds. Although mass spectrometry can provide selective detection for the analysis of multiple compounds in plasma, it is important to recognize that analytical complications could arise from redundancy in molecular weight. For example, metabolites could coelute with analytes of identical mass leading to inaccurate plasma concentrations. Finally, there are solubility considerations in formulating an intravenous dose containing more than one compound. Our approach to detecting problems with the data from a given study was to dose a "control" molecule in all N-in-One studies. Any gross differences in the pharmacokinetics of the control compound might alert us to any of the possible problems discussed above.

3.3. Pharmacokinetic Evaluation of Other Leads

Several compounds identified in the N-in-One studies were redosed individually to dogs to confirm the pharmacokinetic parameters and to study oral bioavailability. Of these compounds, compound **34** emerged as a lead candidate based on its good *in vitro* potency and selectivity, its concentration–time profile, long pharmacokinetic half-life (15 hr), low clearance (4.6 ml/min per kg), and good oral bioavailability (75%). This compound was tested in the hypogastric nerve dog model and was comparable in potency and selectivity to compound **18.** However, unlike compound **18** and other compounds in this series, compound **34** had profound adverse CNS effects in conscious dogs undergoing the upright tilt studies for assessment of orthostatic hypotension. For this reason, compound **34** along with several other molecules with long half-lives that were suspected to be metabolic precursors of compound **34** were dropped from further consideration.

Compound **18** remained the lead compound. However, concerns about its short half-life in animals remained. If the compound had a short half-life in humans as predicted, it would have to be administered several times per day or be administered at a high dose to keep the plasma concentration from falling below therapeutic levels. Multiple daily doses may compromise patient compliance, whereas the administration of a high daily dose may increase the risk of adverse effects resulting from greater exposure to the compound. It therefore became of interest to examine the pharmacokinetic/pharmacodynamic relationship of compound **18** in the dog. If the duration of pharmacological effect of compound **18** was sufficiently long, the need for a compound with a long pharmacokinetic half-life would be reduced.

3.4. Pharmacodynamics of the Lead Compound

A dose escalation study was done first to help pick a suitable dose for the duration of action study. Six dogs were anesthetized and instrumented to measure urethral pressure in response to phenylephrine stimulation. Compound **18** was administered as a series of 5-min intravenous infusions at 1, 3, 10, 30, 100 μg/kg (cumulative) doses spaced 15 min apart. Urethral pressure response to phenylephrine was measured before and after compound **18** administration. The dose–response curve generated using phenylephrine stimulation was identical to that observed for hypogastric nerve stimulation over the same dose range with an ED_{50} of ~13 μg/kg. Blood samples were collected immediately following the urethral pressure measurements and the plasma was analyzed for compound **18** content by positive ion APCI LC/MS in SIM mode.

In the duration of action study, a single 50 μg/kg dose of compound **18** was administered intravenously to 12 dogs. The dose selected was near the ED_{80} from the dose escalation study and was selected because it was a submaximal dose on the linear portion of the dose–response curve. At 5 min and 8, 17, 21, and 26 hr after compound administration, groups of dogs were anesthetized and instrumented as described above for phenylephrine stimulation and urethral pressure monitoring. Blood samples were collected immediately following all urethral pressure measurements and the resulting plasma samples were assayed for compound **18.**

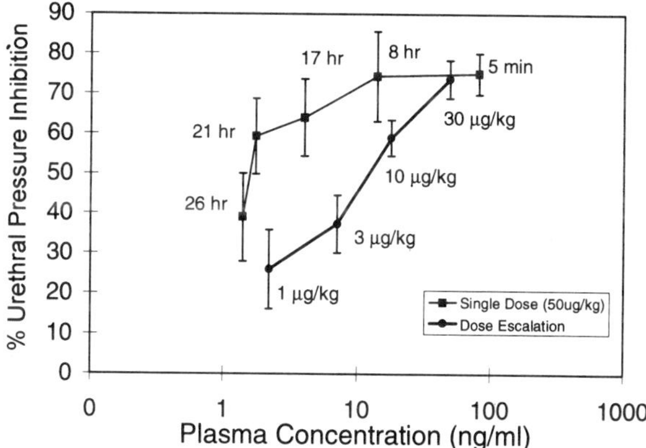

Figure 3. Plasma concentration–effect profiles of compound **18** during dose escalation and during single-dose duration experiments. Error bars represent the SEM.

The plasma concentration–effect relationships after dose escalation and the single-dose duration studies are shown in Fig. 3. A counterclockwise hysteresis was observed. The EC_{50}s differed by an order of magnitude (1 versus 13 ng/ml), which indicated that it took less compound in plasma to elicit a 50% inhibition during the single-dose duration study than it did during the dose escalation study. The major observation in this study was the prolonged efficacy after a single low dose of compound **18.** The plasma concentrations of compound **18** fell nearly 50-fold over the 26-hr period while the urethral pressure responses to phenylephrine stimulation only dropped 2-fold. If compound **18** also demonstrates a prolonged pharmacological effect in humans, then it could be dosed once daily in humans as well.

The pharmacokinetic/pharmacodynamic profile of compound **18** may be a result of delayed distribution of the compound from the plasma to the site of action, the formation of one or more active metabolites, or tight irreversible binding of the compound to the α_{1A} receptor. Slow distribution to the active site seemed unlikely because the prostate is a well-perfused organ. Receptor binding assays with tritiated compound **18** demonstrated reversible binding of compound **18** to the α_{1A} receptor with a rapid on/off rate. A search for metabolites in dog urine and plasma turned up several potent metabolites; however, the pharmacokinetic time profiles of these metabolites were never studied in the dog and the significance of their contribution to the pharmacodynamic profile remains unknown. Nonetheless, knowledge of the preclinical pharmacodynamic profile of compound **18** alerted the clinical pharmacologists to monitor for the presence of potential active metabolites as early as the healthy volunteer studies. Also, knowledge of effective plasma concentrations helped the clinical pharmacokineticists plan dosage regimens in phase I studies.

4. ADVANCEMENT OF COMPOUND 18 TO EXPLORATORY DEVELOPMENT

The project team decided to recommend compound **18** for Exploratory Development based on its chemical novelty, relative ease of synthesis, good *in vitro* α_{1A} receptor potency and selectivity, good *in vivo* potency and selectivity in the dog, lack of orthostatic hypotension effects up to 10 mg/kg, relative selectivity for α_{1A} over other 7TM receptors, good oral bioavailability, and long pharmacological half-life. Some additional studies were conducted with compound **18** to convince the team of its developability. These included a screening Ames test, a 7-day rat toxicology study at doses from 1 to 100 mg/kg, safety pharmacology studies, and salt selection and solid form bioavailability studies. No significant findings were observed in the toxicology studies and the hydrochloride salt was chosen as the salt form.

Compound **18** was formally accepted as an exploratory development candidate less than 6 months after it was first synthesized and entered the first healthy volunteer trials 7 months after that. A dedicated team of chemists, biologists, and "development" scientists (drug metabolism scientists, pharmacokineticists, formulation scientists, and toxicologists) helped in the selection of a candidate that not only met efficacy criteria but was rapidly progressed through the preclinical development hurdles necessary for first-time dosing in humans.

Acknowledgments

The authors would like to acknowledge the chemists in the Glaxo Wellcome Medicinal Chemistry departments for compound synthesis; David Saussy and Aaron Goetz for their receptor binding work; Ba-Jin Han, Don Anderson, and Jim Liacos for their *in vivo* pharmacology work; Frank Lee, Steve Unger, Arthur Moseley, Dhiren Thakker, Michelle Brosnan, Deanna Garrison, and Caroline Stafford for their drug metabolism and pharmacokinetics work; and Michael Jozwiakowski and W. Q. Tony Tong for their formulation work.

REFERENCES

Caine, M., 1988, Alpha-adrenergic mechanisms in the dynamics of benign prostatic hypertrophy, *Urology* **32:**16–20.

Caine, M., Peal, A., and Perlberg, S., 1973, The use of alpha-adrenergic blockers in benign prostatic obstruction, *Br. J. Urol.* **45:**663–667.

Caine, M., Perlberg, S., and Meretyk, S., 1978, A placebo-controlled double-blind study of the effect of phenoxybenzamine in benign prostatic obstruction, *Br. J. Urol.* **50:**551–554.

Forray, C., Bard, J. A., Wetzel, J. M., Chiu, G., Shapiro, E., Tang, R., Lepor, H., Hartig, P. R., Weinshank, R. L., Branchek, T. A., and Gluchowski, C., 1994, The α_1-adrenergic receptor that mediates smooth muscle contraction in the human prostate has the pharmacological properties of the cloned human α_{1c} subtype, *Mol. Pharmacol.* **45:**703–708.

Gerber, G. S., Kim, J. H., Contreras, B. A., Steinberg, G. D., and Rukstalis, D. B., 1996, An observational urodynamic evaluation of men with lower urinary tract symptoms treated with doxazosin, *Urology* **47:**840.

Goetz, A. S., Lutz, M. W., Rimele, T. R., and Saussy, D. L., 1994, Characterization of alpha-1 adrenoceptor subtypes in human and canine prostate membranes, *J. Pharmacol. Exp. Ther.* 271:1228–1233.

Halm, K. A., Adkison, K. K., Berman, J., and Shaffer, J. E., 1996, N-in-One dosing in the dog: LC/MS as a tool for higher throughput *in vivo* pharmacokinetic screening of drug discovery lead candidate mixtures, in: *IBC Molecular Diversity and Combinatorial Chemistry Conference,* San Diego, January 24–26, 1996.

Heible, J. P., Caine, M., and Zalaznik, E., 1985, In vitro characterization of the α-adrenoceptors in human prostate, *Eur. J. Pharmacol.* **107:**111–117.

Kirby, R. S., and Christmas, T. J., 1993, *Benign Prostatic Hyperplasia,* Raven Press, New York.

Lepor, H., and Shapiro, E., 1984, Characterization of alpha1 adrenergic receptors in human benign prostatic hyperplasia, *J. Urol.* **132:**1226–1229.

Lepor, H., Auerbach, S., Puras-Baez, A., Narayan, P., Soloway, M., Lowe, F., Moon, T., Leifer, G., and Madsen, P., 1992, A randomized, placebo-controlled multicentered study of the efficacy and safety of terazosin in the treatment of benign prostatic hyperplasia, *J. Urol.* **148:**1467–1474.

Price, D. T., Schwinn, D. A., Lomasney, J. W., Allen, L. F., Caron, M. G., and Lefkowitz, R. J., 1993, Identification, quantification and localization of the mRNA for the three distinct alpha1 adrenergic receptor subtypes in human prostate, *J. Urol.* **150:**546–551.

Raz, S., Zeigler, M., and Conti, M., 1973, Pharmacological receptors in the prostate, *Br. J. Urol.* **45:**663–667.

Rowland, M., and Tozer, T., 1995, *Clinical Pharmacokinetics: Concepts and Applications,* 3rd ed., Williams & Wilkins, Philadelphia.

Stoner, E., 1994, Three-year safety and efficacy data on the use of finasteride in the treatment of benign prostatic hyperplasia, *Urology* **43:**284–292.

Chapter 19

Discovery of Bioavailable Inhibitors of Secretory Phospholipase A_2

Steven G. Blanchard, Robert C. Andrews, Peter J. Brown, Liang-Shang L. Gan, Frank W. Lee, Achintya K. Sinhababu, and Thomas N. Wheeler

1. INTRODUCTION

1.1. Therapeutic Target

Phospholipase A_2s (PLA_2s) are enzymes that hydrolyze the C-2 fatty acid ester of phospholipids to liberate arachidonic acid and lysophospholipid. Arachidonic acid is converted by cyclooxygenases and lipoxygenases to proinflammatory prostaglandins and leukotrienes whereas the lysophospholipids can be acetylated to give the proinflammatory mediator platelet-activating factor.

Human secretory PLA_2 ($sPLA_2$), one of three major mammalian PLA_2s, is a low-molecular-mass (~15,000 kDa) enzyme secreted in response to inflammatory stimuli, e.g., by synoviocytes when stimulated by interleukin-1. The enzyme is found in the synovial fluid of arthritic joints, and enzyme levels correlate with the severity of the disease. $sPLA_2$ is found in high levels in the serum of endotoxic shock patients and the enzyme produces a local inflammatory response when injected *in vivo*.

Steven G. Blanchard, Robert C. Andrews, Peter J. Brown, Liang-Shang L. Gan, Frank W. Lee, Achintya K. Sinhababu, and Thomas N. Wheeler • Glaxo Wellcome Research and Development, Research Triangle Park, North Carolina 27709.

Integration of Pharmaceutical Discovery and Development: Case Studies, edited by Borchardt *et al.*, Plenum Press, New York, 1998.

Nonsteroidal anti-inflammatory drugs (NSAIDs) and steroid therapy for acute inflammation and rheumatoid arthritis are not effective at retarding progression of the disease and have potent gastrointestinal or immunosuppressive side effects. Based on the occurrence of the PLA_2-catalyzed reaction at the beginning of the eicosanoid manifold of proinflammatory mediators, it was thought that PLA_2 inhibitors may represent an alternative approach for development of anti-inflammatory agents.

1.2. Program Objective

Based on the above rationale, the objective of the program was the discovery of potent, selective, orally active inhibitors of human $sPLA_2$. It was thought that such inhibitors might present significant improvement over the currently available NSAID therapies for rheumatoid arthritis with a further potential for utility in other disease states involving acute inflammation.

2. *IN VITRO* IDENTIFICATION OF ACTIVE-SITE INHIBITORS OF $sPLA_2$

2.1. "Dual Substrate" Strategy for Inhibitor Discovery

In nature, essentially all phospholipid substrates are present in the form of lipid aggregates and PLA_2s have evolved to preferentially hydrolyze substrate in this form. As a result, PLA_2s exhibit complicated kinetics that are a reflection both of substrate hydrolysis and of enzyme partitioning between the aqueous environment and the lipid–aqueous interface. An apparent inhibition of enzyme activity may therefore occur by blocking either (1) binding of enzyme to the lipid–aqueous interface or (2) substrate binding to the active site (a specific event). The membrane binding step may be modulated either by a specific interaction with the enzyme or, more commonly (Jain *et al.,* 1991), by a nonspecific mechanism of perturbation of the organized structure of the substrate membrane. As a result of these complexities, agents that act to perturb the ordered structure of phospholipid substrate (e.g., detergents) can cause inhibition in an *in vitro* assay even though they have no direct effect on the enzyme.

It was therefore important to establish a testing strategy that could distinguish between such nonspecific "inhibition of the assay" and direct, specific enzyme inhibition. A number of different strategies have been employed to identify inhibitors that act via a direct, specific interaction with PLA_2. The first utilizes substrate dispersed in the form of mixed micelles with a nonhydrolyzable "carrier," i.e., a detergent such as Triton X-100 (Dennis, 1973; Deems *et al.,* 1975) or a lipid analogue (Jain *et al.,* 1991). Under conditions where the "carrier" is present in excess over substrate and inhibitor, addition of inhibitor has little effect on the overall structure of the micelle (Reynolds *et al.*, 1991) and observed inhibition may be attributed to an effect on enzyme rather than to a nonspecific effect on substrate organization. A second strategy is analysis of inhibition under conditions where enzyme is operating in the "scooting mode" of catalysis (Jain and Berg, 1989; Jain and Gelb, 1991; Jain *et al.*, 1991) where enzyme remains tightly bound to the substrate surface so that there is no partitioning of enzyme between solution and surface-bound forms.

In the studies reported here, we have utilized a different strategy in which concentration–response curves of potential inhibitors were determined in two different assays using substrates having different aggregation states. Both assays were performed under conditions where substrate concentration was well below the apparent K_m. Under these conditions, we reasoned that molecules that inhibited via a specific interaction with PLA_2 should show equivalent potency when tested against the two different substrates. A similar strategy was reported by Bennion *et al.* (1992), but their method does not allow direct comparison of inhibitor potency in different assay systems. Rather, parallel rank orders of potency for series of compounds were observed in the two assays employed, but the absolute inhibition constants differed for the different assay methods.

In the present study, the first method utilized for evaluation of $sPLA_2$ inhibition was a fluorescence assay that monitored hydrolysis of the aggregated substrate 1-acyl-2-(*N*-4-nitrobenzo-2-oxo-1,3-diazole)aminododecanoyl phosphatidylethanolamine (Blanchard *et al.*, 1994). The second method utilized di-1,2-hexanoylthio-glycerophosphatidylmethanol at concentrations below its critical micelle concentration. Although microaggregation of the substrate induced by $sPLA_2$ could not be ruled out, no deviations of the enzyme kinetics expected for a soluble substrate were observed. Hydrolysis of this substrate was based on spectrophotometric detection of thiol release using the chromogenic reagent 5,5-dithiobis-(2-nitrobenzoic acid) (Yuan *et al.*, 1990).

2.2. *In Vitro* Profile of Substrate Analogue PLA_2 Inhibitors

In the course of our studies of the substrate specificity of $sPLA_2$ (Wheeler *et al.*, 1994), we prepared GW 1763 (Fig. 1) and discovered that it was an inhibitor,

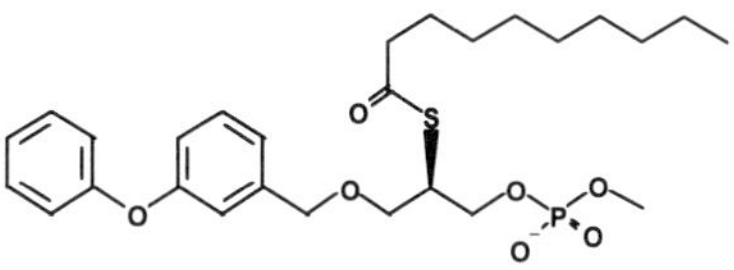

GW 1763

Figure 1. Structure of GW 1763.

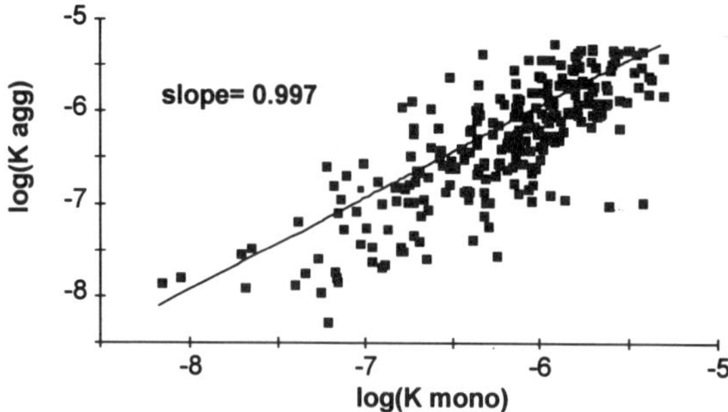

Figure 2. Correlation of inhibition constants for $sPLA_2$ measured against monomeric and aggregated substrate.

rather than a substrate, of $sPLA_2$. *In vitro* testing of this compound yielded IC^{50}(aggregated) = IC^{50}(monomeric) = 0.3 μM. Further evidence that GW 1763 bound directly to the enzyme was obtained from studies of binding of this compound to the PLA_2 from *Crotalus durrisus terrifficus* by monitoring of ligand-induced changes of the protein's intrinsic tryptophan fluorescence. A binding constant of 0.12 μM was obtained in these studies in good agreement with the values obtained by enzyme inhibition. As expected for an inhibitor binding to the active site of the enzyme, binding required calcium and occurred in the absence of a phospholipid membrane. Based on these initial findings, a study was undertaken to elucidate the structure–activity relationship for $sPLA_2$ inhibition. Figure 2 shows a plot of correlation between the inhibition constants observed for this series of inhibitors against the aggregated and monomeric substrates. For any given compound, the difference between the K_i(aggregated) and K_i(monomeric) could be expressed as the absolute value of the difference between the logarithms. For the data shown, the mean value of this parameter was 0.294 (for 110 compounds). This corresponds to an average twofold difference between the K_is determined versus the two substrates and represents an experimental verification of the definition of K_i(aggregated) = K_i(monomeric).

3. *IN VIVO* ANTI-INFLAMMATORY ACTIVITY OF INITIAL CANDIDATES

3.1. Choice of Animal Model

Initial *in vivo* evaluation of candidate $sPLA_2$ inhibitors for anti-inflammatory activity was in the rat carrageenan paw edema model. This is a well-characterized model of acute inflammation that is rapid and requires modest amounts of test compound. The model was chosen for evaluation of $sPLA_2$ inhibitors because increases in PLA_2 activity have been reported to be associated with administration of carrageenan and this activity is neutralized by antibodies directed against $sPLA_2$ (Murakami *et al.*, 1990). Further, nonsteroidal anti-inflammatory agents are active in this model. Disadvantages of the model include steep dose–response curves for anti-inflammatory agents and a limited therapeutic "window." Inhibition of the $sPLA_2$-sensitive component of carrageenan edema by, e.g., nonsteroidal agents typically gives only 50% inhibition of the observed signal. As a result, the model was used to give a qualitative, rather than quantitative, assessment of the activity of $sPLA_2$ inhibitors.

The protocol for compound testing consisted of injection of 0.1 ml of a 1% carrageenan solution into the hind paw of rats. Test compound was administered either at the time of carrageenan injection in the case of i.v. administration, or from 1 to 3 hr before carrageenan injection for oral tests. Edema was assessed by measurement of paw diameter 3 hr after carrageenan injection. The contralateral paw acted as control.

3.2. *In Vivo* Activity Is Dependent on Formulation of the Test Compound

Although some compounds showed modest anti-inflammatory activity in initial experiments, results for some compounds were highly variable. We noted, however, that the observed anti-inflammatory activity seemed to correlate with the appearance of the dose solutions. For instance, the dose preparations of GW 6209, an active compound, were clear whereas inactive compounds were cloudy. Laser light scattering of GW 6209 showed that the compound was present as small aggregates (average size ~ 15 nm). This finding was not unexpected as the inhibitors are glycerophospholipid analogues. We therefore reasoned that preparation of test compounds using standard laboratory methods for dispersion of phospholipids in aqueous buffers would maximize the surface area of compound exposed to the aqueous environment and the stability of the dose preparations. Accordingly, a routine protocol was developed for compound preparation in which the compounds were dissolved in ethanol, followed by removal of the solvent with a stream of dry nitrogen gas to leave a thin film of compound on the walls of the sample vessel. Buffer was added to the vessel, the film was allowed to hydrate for 1 hr with occasional vortexing of the sample, followed by extensive sonication. The final dose solutions were characterized by laser light scattering. Although the average particle size varied from compound to compound, little variation was observed for multiple preparations of a single compound. Ultraviolet spectroscopy and/or radiochemical detection were utilized to verify quantitative sample recovery.

3.3. Activity in the Rat Carrageenan Paw Edema Model

3.3.1. *IN VITRO* K_i IS NOT A USEFUL PREDICTOR OF *IN VIVO* ACTIVITY

Compounds were chosen for testing based solely on their *in vitro* potency as measured by enzyme inhibition. Although only weak activity was observed on oral administration, i.v. potency on the order of that observed for indomethacin could be achieved (e.g., GW 8219, Fig. 3). No apparent correlation was observed for the compounds tested between their *in vitro* K_i for PLA_2 inhibition and their *in vivo* activity. The calculated log of the octanol:water partition coefficient, clog *P*, was also not predictive of a compound's *in vivo* activity. The lack of concordance between *in vivo* and *in vitro* activity is best illustrated by the finding that even small changes in compound structure resulted in large changes in *in vivo* activity. GW 9624 and GW 8219 showed similar *in vitro* K_is and differed only by a single oxygen-to-sulfur change (Fig. 3A). Despite their similarity, however, these two compounds showed vastly differing potency in the carrageenan paw edema model (Fig. 3B). GW 9624 was essentially inactive whereas GW 8219 showed potency approximating that observed for a maximal dose of indomethacin.

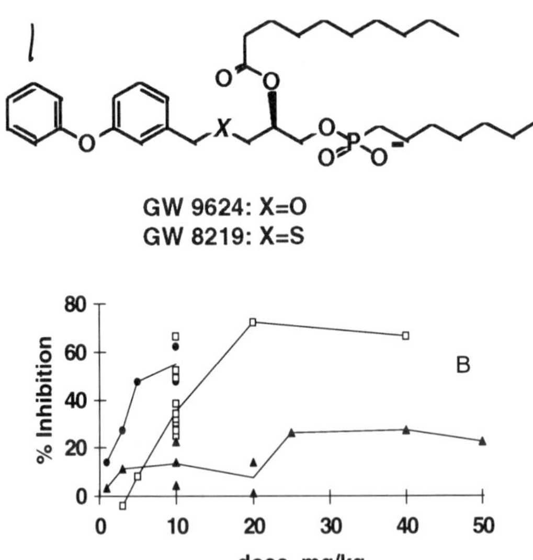

Figure 3. Inhibition of carrageenan paw edema by inhibitors of $sPLA_2$. (A) Structures of GW 9624 and GW 8129. (B) The inhibition of paw edema observed as a function of i.v. dose administered is shown for indomethacin (●), GW 8219 (□), and GW 9624 (▲). Each data point is the mean for eight animals as described in the text. Multiple data points at a single dose are for different experiments and serve to illustrate the range of day-to-day variation observed in these experiments.

3.3.2. BOTH *IN VITRO* AND *IN VIVO* ACTIVITIES ARE STEREOSELECTIVE

One possible explanation for the observed divergence between the ability of a compound to inhibit $sPLA_2$ *in vitro* and its *in vivo* anti-inflammatory effect was that the anti-inflammatory effect was related to an action of the test compound on a molecular target distinct from $sPLA_2$. PLA_2s preferentially hydrolyze phospholipid substrates with *sn*-2 stereochemistry at the asymmetric carbon of the glycerol backbone and stereoselective inhibition by substrate-mimetic inhibitors has also been reported (Yuan *et al.,* 1990). We reasoned, therefore, that the *in vivo* activity should reflect the known *in vitro* stereoselectivity of candidate inhibitors if the mechanism of the anti-inflammatory effect were via PLA_2 inhibition. In order to test this hypothesis, GW 6209 (active *in vitro*) and its enantiomer GW 4366 (inactive *in vitro*) were prepared and the stereoselective inhibition of PLA_2 was verified by *in vitro* testing (Table I). When tested in the carrageenan edema model via i.v. administration (50 mg/kg), GW 6209 showed anti-inflammatory activity whereas its inactive enantiomer GW 4366 did not. Although we cannot rule out stereoselective inhibition of another molecular target, these findings are consistent with the idea that the anti-inflammatory effect is mediated through inhibition of $sPLA_2$.

3.3.3. POSSIBLE EXPLANATIONS FOR THE DIFFERENCES IN *IN VITRO* AND *IN VIVO* ACTIVITY

The *in vitro* structure–activity relationship of the inhibitors described here was derived by testing against cloned human enzyme. In contrast, however, the anti-inflammatory activity was assessed in a rodent model and it was necessary to rule out differences in the structure–activity relationship for inhibition of human and rat enzymes as a possible explanation for the difference observed between *in vitro* and *in vivo* activity. As shown in Table II, a number of compounds were evaluated as inhibitors of both the rat and the human enzymes using the fluorescent enzyme assay. No significant differences were observed in inhibition of the enzymes

Table I
Stereoselective Inhibition of $sPLA_2$

Compound	K_i (μM) Monomeric	K_i (μM) Aggregated	*In vivo* activity[a] (50 mg/kg)
GW 6209	0.64	0.83	36%*
GW 4366	>5	2.5	18%

[a]Activity was measured in the carrageenan paw edema assay and is expressed as % inhibition of edema with respect to the vehicle control. Indomethacin (10 mg/kg) was used as a standard and gave 30% inhibition.
*$p < 0.05$.

Table II
Inhibition of Human and Rat $sPLA_2$ by Selected Compounds

Compound	K_i (μM)	
	Human	Rat
GW 7027	0.54	0.39
GW 0900	1.0	0.33
GW 4326	1.25	0.64
GW 5051	0.33	0.69
GW 8338	1.6	2.0

from the two species. Based on the available evidence, therefore, we hypothesized that differences in drug metabolism and/or other pharmacokinetic parameters may be dominant factors controlling the observed *in vivo* activity of this series of PLA_2 inhibitors. Our strategy to test this hypothesis was to radiolabel compounds of interest, determine their *in vitro* and *in vivo* stability, and compare these data with *in vivo* efficacy results.

4. PHARMACOKINETIC AND METABOLIC FATE OF CANDIDATE INHIBITORS

4.1. Plasma Levels and Metabolic Profiles after i.v. and p.o. Dosing

Three tritium-labeled PLA_2 inhibitors were prepared and their pharmacokinetic parameters and oral bioavailability were determined in rats. Compounds containing both amide (GW 7027) and ester (GW 9624, 8219) linkages at the C-2 position of the glycerol backbone were examined. The results are summarized in Table III. GW 9624 and GW 7027, the first two molecules examined, were rapidly metabolized following either an i.v. bolus or an oral administration. High concentrations of volatile radioactive metabolites were present in the circulation. As shown in Fig. 4 for GW 7027, the concentration of volatile metabolites was more than 20-fold higher than parent drug. Because both were radiolabeled in the C-2 side chains, the formation of volatile metabolites indicated oxidation of the C-2 decanoate moiety. Detailed information concerning metabolic pathways was lost, however, via loss of radioactive label. In order to minimize volatile metabolite formation and the resulting loss of information, radiolabel was incorporated into GW 8219 at an alternative site of the molecule. As indicated in Fig. 5, there was an approximately 100-fold decrease in the amount of volatile radioactive metabolite formed. A substantial amount of radioactivity was recovered (84% of

Table III
Pharmacokinetic Parameters for $sPLA_2$ Inhibitors in the Rat

Parameter	GW 9624	GW 7027	GW 8219
Dose (mg/kg)			
i.v.	11	9.7	10
p.o.	11	10	9.5
Cl_s (ml/min/kg)	11.4	13	24
V_{ss} (ml/kg)	170	472	1024
$t_{1/2}$ (min)	33	39	65
C_{max} (ng/ml), p.o.	465	3261	322
Oral bioavailability (%)	4	90	8
i.v. % dose in urine	8	5	19
p.o. % dose in urine	9	4	36
i.v. % dose in feces	2	5	65
p.o. % dose in feces	6	18	51
% parent drug in bile	ND[a]	ND	11

[a]Not determined.

dose) in urine and feces following i.v. administration of [^{3}H]-GW 8219, as compared with the low recovery (10% of dose) following i.v. administration of [^{3}H]-GW 9624. GW 8219 showed a small, 2-fold increase in the half-life observed for GW 9624. This could be related to the higher V_{ss} of GW 8219 in rats. The oral bioavailability of these C-2 esters (4–8%) was, however, much lower than that observed for the amide GW 7027 (90%).

4.2. *In Vitro* Studies

4.2.1. ACID STABILITY

An *in vitro* stability study showed that GW 9624 and GW 8219 were not stable in simulated gastric fluid (0.1 N HCl) whereas GW 7027 was stable under the same conditions. These results are consistent with the greater stability predicted for an amide versus an ester to acid hydrolysis. These *in vitro* studies indicate that the poor bioavailability of the two esters is, in part, a result of hydrolytic loss of the parent esters in the stomach after oral dosing.

4.2.2. *IN VITRO* ESTIMATION OF INTESTINAL TRANSPORT AND METABOLIC STABILITY IN Caco-2 CELLS

The transport of GW 9624, GW 8219, and GW 7027 across Caco-2 cell monolayers was examined in order to obtain an *in vitro* estimate of the possible intestinal absorption of these compounds. The apparent permeability coefficients

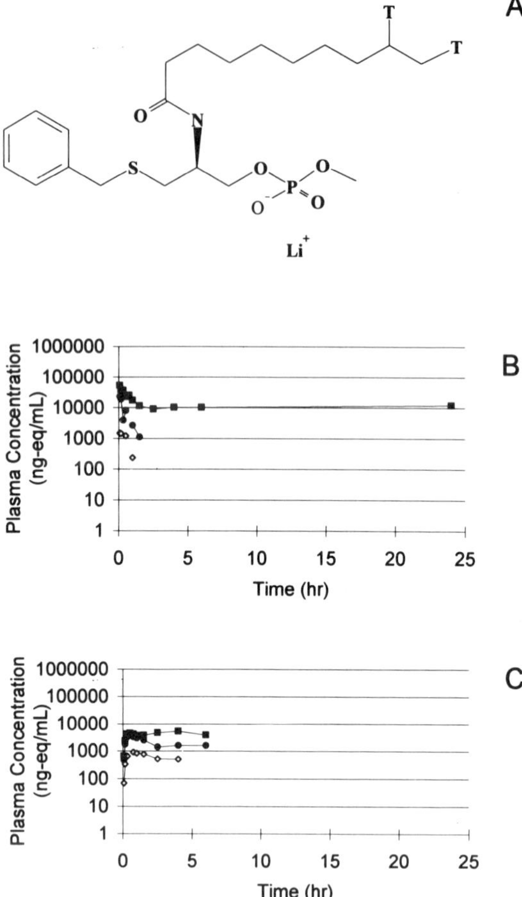

Figure 4. GW 7027 plasma concentration profiles in the rat. (A) Structure of [^{3}H]-GW 7027. The position of the tritium labels is indicated. The plasma concentration profiles following intravenous (B) and oral (C) doses of 10 mg/kg are shown. Concentration versus time profiles for parent compound (●), the major nonvolatile metabolite (◇), and the volatile metabolite (■) are shown for both methods of administration.

were determined to be 1×10^{-5}, 2×10^{-6}, and 5.6×10^{-7} cm/sec for GW 9624, GW 8219, and GW 7027, respectively. The esters GW 9624 and GW 8219 were rapidly metabolized in Caco-2 cells, presumably via hydrolysis of the C-2 ester linkage by esterases. In contrast, GW 7027 was stable in the presence of Caco-2 cell monolayers and transport of compound across the cell monolayer was observed. These results are consistent with the oral bioavailability for this molecule observed *in vivo*.

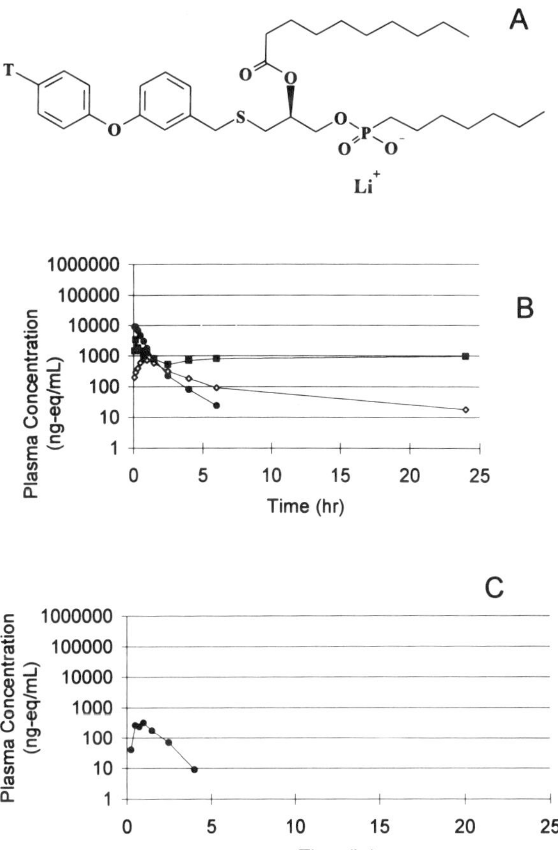

Figure 5. GW 8219 plasma concentration profiles in the rat. (A) Structure of [^{3}H]-GW 8219. The position of the tritium label is indicated. The plasma concentration profiles following intravenous (B) and oral (C) doses of 10 mg/kg are shown. Concentration versus time profiles for parent compound (●), the major nonvolatile metabolite (◇), and the volatile metabolite (■) are shown for both methods of administration.

4.2.3. METABOLISM BY TISSUE HOMOGENATES

Because the major route of elimination of the $sPLA_2$ inhibitors *in vivo* appeared to be via metabolism, *in vitro* metabolism studies were undertaken in order to elucidate the metabolic fate of selected $sPLA_2$ inhibitors. Stability of the inhibitors in rat blood and/or plasma as well as in the presence of liver microsomes and/or liver homogenates was studied. The aim was to understand the source(s) of metabolic instability and to use this information to design inhibitors with improved metabolic profile.

Although GW 7027 underwent rapid metabolism *in vivo* ($t_{1/2}$ = 39 min, Table III) in the rat with rapid appearance of volatile radioactivity, the metabolism of this compound was surprisingly slow in rat blood, plasma, liver homogenates, or liver microsomes. The extent of loss of parent drug after 2-hr incubation in blood and liver microsomes was less than 6 and 16%, respectively. The major metabolite observed in blood, liver homogenates, and liver microsomes displayed a very similar retention time to the authentic sulfoxide of GW 7027.

In contrast to the results obtained for GW 7027, both GW 9624 and GW 8219 underwent considerable *in vitro* metabolism when exposed to blood, plasma, liver homogenates, and liver microsomes. In the case of GW 9624, the metabolic profiles observed in blood were virtually identical to those observed in plasma, but completely different from those observed in liver homogenates or microsomes. The major metabolite in blood and plasma was suspected to be [^{3}H]decanoate, which would arise by the action of esterases present in blood or plasma on the parent compound, GW 9624. Unfortunately, [^{3}H]decanoic acid was not available as a standard. Nevertheless, indirect evidence suggested that the major metabolite seen in blood and plasma from [^{3}H]-GW 9624 was indeed [^{3}H]decanoate. Thus, when the 5-min plasma sample derived from a rat after i.v. administration of GW 9624 or a sample from the *in vitro* incubation of GW 9624 in plasma was analyzed by HPLC/MS, the compound corresponding to loss of the decanoate moiety from parent was identified as a metabolite.

As discussed in Section 3.3.1, GW 8219 differed from GW 9624 in that it contained sulfur in place of oxygen at C-3. In addition, the radiolabeled version of GW 8219 differed from that of GW 9624 in that the former was tritium labeled in the diphenyl ether moiety whereas the latter was labeled at the ω and ω-1 positions of the decanoate moiety (Fig. 5A). These differences led to remarkable differences in tissue-specific metabolism as well as in the appearance of the metabolite profiles *in vitro*. In contrast to the results observed for [^{3}H]-GW 9624, there was virtually no metabolism of [^{3}H]-GW 8219 in blood at 2 hr. However, there was considerable metabolism in liver homogenate (22% loss of parent in 2 hr) in the presence of NADPH giving rise to at least six metabolites, all of which were more polar than the parent. The major metabolite formed from [^{3}H]-GW 8219 in both the presence and absence of NADPH had the same retention time as an authentic sample of the compound corresponding to loss of the decanoate moiety from the parent.

A striking feature of the radiochromatographic profile of metabolites formed in the presence of NADPH was that the retention time separating any pair of adjacent peaks was virtually the same. Occurrence of such a profile of more polar metabolites suggested that the metabolites arose via the successive loss of a constant structural unit. Such fragmentation could arise by the involvement of the fatty acid oxidation pathway with the fragment lost being acetate. Although fatty acid oxidation could act at the ω positions of either the decanoate or the phosphonate groups of GW 8219, the constancy of time separating adjacent peaks and the number of metabolites formed suggested that fatty acid oxidation of only one of these groups occurred. Because no more than three rounds of fatty acid oxidation are possible with the phosphonate moiety, it was concluded that oxidation must have occurred at the decanoate moiety.

4.3. Conclusions Based on Metabolism Studies

Based on the observations and arguments presented in Section 4.2.3. the pathways shown in Fig. 6 were proposed for the metabolism of [^{3}H]-GW 8219 in rat liver homogenates. It should be emphasized that the involvement of the fatty acid oxidation pathway was proposed entirely on circumstantial evidence and that none of the postulated metabolites unique to this pathway were characterized. The first step in the proposed pathway is the cytochrome P450-mediated hydroxylation of the ω position of the decanoate moiety. Oxidation to the ω-COOH metabolite via the corresponding aldehyde intermediate and subsequent cleavage of the elements of acetic acid are generally catalyzed by enzymes other than cytochrome P450s (Schultz, 1991; Stryer, 1988). It should be noted that the major metabolite formed in the presence of NADPH probably arises from the loss of the decanoate moiety

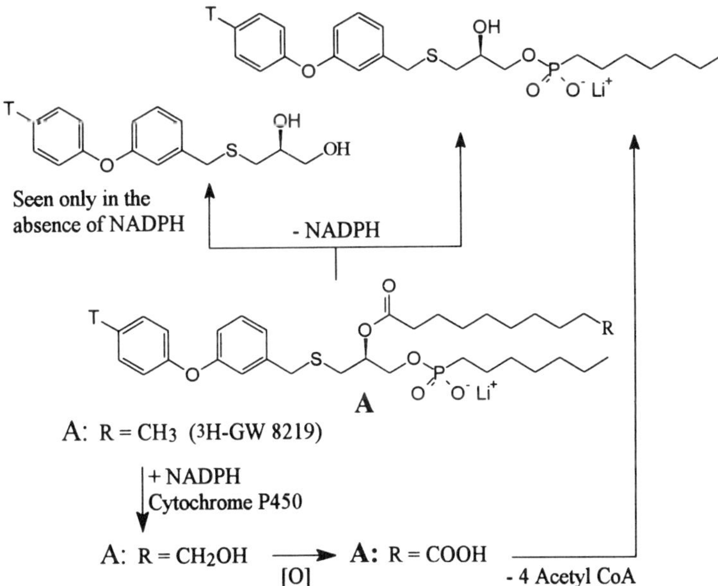

Figure 6. Hepatic metabolism of [^{3}H]-GW 8219 in rat: proposed pathways.

both from esterase mediated cleavage and as the terminal product of the fatty acid oxidation pathway.

Based on the results obtained with [^{3}H]-GW 8219 in terms of metabolism *in vitro,* we could make the following retrospective predictions on the metabolism of [^{3}H]-GW 9624 and [^{3}H]-GW 7027 *in vivo,* as well rationalizations of the metabolism *in vitro*. Had GW 9624 been labeled with tritium in the diphenyl ether moiety, metabolite profiles similar to those of [^{3}H]-GW 8219 would likely have been observed both *in vivo* and *in vitro*. Because the tritiums in [^{3}H]-GW 9624 were present in the ω and ω-1 positions, and these carbons were the first to be lost, metabolites arising from this pathway could not be detected after the first round of fatty acid oxidation. As mentioned in Section 4.1, both [^{3}H]-GW 9624 and [^{3}H]-GW 7027 lost a substantial fraction of radioactivity as a rapidly appearing volatile metabolite *in vivo*. Thus, it appeared that the fatty acid oxidation pathway was operative for both of these compounds *in vivo*.

In summary, the *in vivo* and *in vitro* metabolism data obtained were revealing in terms of the metabolic fate of this class of compounds. Molecules containing a C-2 ester linkage were susceptible to cleavage by both acidic pH and esterases *in vivo* and *in vitro*. Based on the total metabolic data obtained, two major recommendations for improving metabolic stability of this class of compounds were made. The recommendations were: (1) modify the C-2 ester moiety in order to minimize direct cleavage at this site and (2) devise ways to block ω (and, perhaps, ω-1) hydroxylation of the decanamide moiety. Section 5 describes the implementation of these recommendations and the data obtained for the resulting molecules.

5. PREPARATION OF INHIBITORS DESIGNED TO ADDRESS THE OBSERVED METABOLIC INSTABILITY

5.1. Synthesis and *in Vitro* Evaluation of Inhibitory Activity

Based on the findings described in Section 4, the decision was made to concentrate further analogue synthesis on modification of C-2 amides in order to minimize metabolism at this site. In an attempt to minimize or eliminate metabolism via the fatty acid oxidation pathway, synthesis of a number of analogues containing various degrees of fluorination of the decanamide moiety was undertaken. In general, the fluorinated compounds retained activity when tested as inhibitors of $sPLA_2$. Table IV shows the inhibition constants of two of these molecules, GW 8338 and GW 4776, in the aggregated and monomeric assays of $sPLA_2$ inhibition. GW 4776 was tritium labeled in positions that were not expected to undergo significant metabolism, and its *in vitro* and *in vivo* metabolism and pharmacokinetic parameters (Table V) were evaluated.

Table IV
Inhibition of $sPLA_2$ by Fluorinated Inhibitors

Compound[a]	K_i (μM)	
	Monomeric	Aggregated
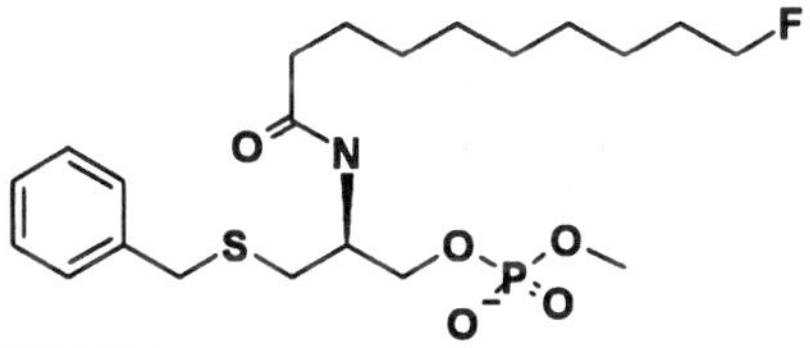 GW 8338	1.6	2.7
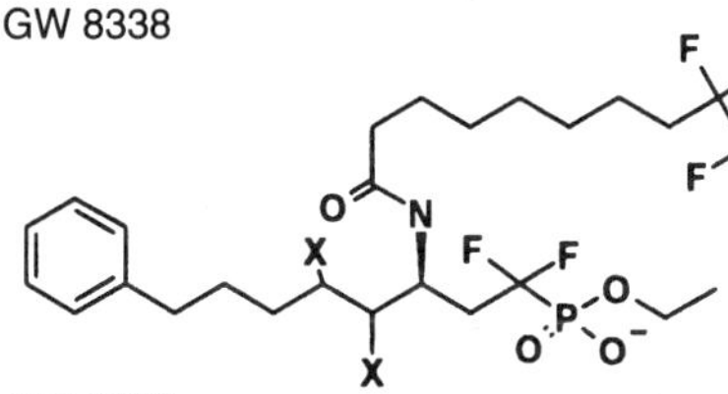 GW 4776	0.43	0.76

[a]For GW 4776, X is hydrogen for the unlabeled compound and tritium in the case of [^{3}H]-GW 4776.

Table V
Pharmacokinetic Parameters for Fluorinated $sPLA_2$ Inhibitors in the Rat

Parameter	GW 4776
Dose (mg/kg)	
i.v.	10
p.o.	10/50[a]
Cl_s (ml/min/kg)	20
V_{ss} (ml/kg)	6850
$t_{1/2}$ (min)	378
C_{max} (ng/ml), p.o.	1100/3850[a]
Oral bioavailability (%)	100/100[a]
i.v. % dose in urine	0.4
p.o. % dose in urine	0.3/0.4[a]
i.v. % dose in feces	83
p.o. % dose in feces	86/86[a]
% parent drug in bile	84

[a]Data obtained after oral administration of low (10 mg/kg) and high (50 mg/kg) doses of GW 4776.

5.2. Evaluation of *In Vitro* Stability

[^{3}H]-GW 4776 was stable in the presence of 0.1 N hydrochloric acid, Caco-2 cells, and rat blood. The extent of metabolism in 2 hr observed in rat liver homogenates in the presence of NADPH was small and amounted to approximately 15% for [^{3}H]-GW 4776. The major metabolite corresponded to a compound with a molecular weight 16 amu higher than the parent as assessed by LC/MS analysis. Metabolism did not occur in the absence of NADPH.

5.3. Pharmacokinetic Studies

Consistent with the observed *in vitro* stability, [^{3}H]-GW 4776 showed improved pharmacokinetic parameters relative to the earlier molecules. Table V summarizes the data obtained for [^{3}H]-GW 4776 in the rat. Whereas [^{3}H]-GW 7027, [^{3}H]-GW 9624, and [^{3}H]-GW 8219 all had half-lives in the range of 30 to 60 min (Table III), the corresponding value for [^{3}H]-GW 4776 was 378 min. [^{3}H]-GW 4776 was found to have an oral bioavailability of 100% at both low (10 mg/kg) and high (50 mg/kg) dose levels. It should be noted, however, that the absence of data on the metabolic stability of analogues of [^{3}H]-GW 4776 not containing fluorine in the ω and ω-1 positions of the decanamide moiety makes it impossible to judge the true contribution of perfluorination of these positions to the observed enhancement of metabolic stability.

5.4. *In Vivo* Activity of Inhibitors with Improved Metabolism and Pharmacokinetics

5.4.1. RAT CARRAGEENAN EDEMA

The anti-inflammatory activity of GW 4776 was assessed in the carrageenan paw edema model. Our standard testing regimen involved an initial evaluation of the activity of compounds after i.v. dosing, followed by evaluation of active compounds for oral activity. GW 4776 gave 23% inhibition of paw edema when given by i.v. administration at a dose of 30 mg/kg. This inhibition was statistically significant and was similar in magnitude to the inhibition observed for dexamethasone (28% inhibition at 1 mg/kg), the positive control that gives maximal inhibition of edema in this assay. In contrast to the results obtained on i.v. administration, GW 4776 was inactive when dosed orally in the same model at 50 mg/kg. The reasons for this lack of oral activity were not clear.

A second fluorinated compound, GW 8338, was tested for anti-inflammatory activity. The metabolism of this compound was not studied; however, it differs from GW 7027 by a single fluorine-for-hydrogen substitution at the ω position of the decanamide moiety, and it was therefore postulated that it might have similar bioavailability. Furthermore, the fluorine modification was expected to minimize metabolism via the fatty acid oxidation pathway. In contrast to the results for GW 4776, GW 8338 showed significant inhibition of paw edema after both i.v. and oral administration. The observed inhibition was 16% for the i.v. route (10 mg/kg) and 41% after oral administration (50 mg/kg).

5.4.2. PHORBOL ESTER-INDUCED MOUSE EAR EDEMA

In order to verify the anti-inflammatory activity observed for GW 4776 and GW 8338, an experiment was undertaken to assess activity in an alternative model of acute inflammation. The phorbol ester-induced mouse ear edema assay (De Young *et al.,* 1989) was chosen for this purpose. In addition to direct measurement of edema, this model allows one to assess activity of the enzyme myeloperoxidase as a specific marker of neutrophil infiltration. Ear plugs were taken 24 hr after the start of the experiment and edema was assessed by weighing the plugs. Myeloperoxidase activity in tissue homogenates was assessed spectrophotometrically. The presence of this enzyme in tissue is proportional to the degree of neutrophil infiltration (Bradley *et al.,* 1982; Krawisz *et al.,* 1984).

GW 4776 when applied topically to mouse ear at a total dose of 0.1 mg/ear gave 54% inhibition of edema as assessed by ear weight and 83% inhibition of myeloperoxidase activity. For GW 8338 tested at the same dose, the observed inhibitions of ear weight and myeloperoxidase activity were 43 and 82%, respectively. The inhibition of myeloperoxidase activity observed with both compounds was statistically significant ($p < 0.05$). The dexamethasone standard gave greater than 90% inhibition of both edema and myeloperoxidase activity when tested at the same dose.

6. SUMMARY AND CONCLUSIONS

Substrate-mimetic inhibitors of $sPLA_2$ with submicromolar *in vitro* potency were discovered by use of a novel dual substrate screening strategy. *In vivo* evaluation of selected inhibitors in the rat carrageenan paw edema model of inflammation, however, indicated that *in vitro* potency was not a good predictor of *in vivo* activity. Studies of the metabolic stability of early examples of these inhibitors suggested that the metabolic lability of these compounds was a major contributing factor to the observed weak *in vivo* activity. In an attempt to achieve improved *in vivo* activity, we prepared and tested compounds designed to overcome the observed metabolic instability. The design of the new compounds involved two types of changes in the inhibitor molecules. First, the C-2 ester moiety was replaced with an amide function so that direct cleavage by stomach acid and blood esterases at this site was minimized. Second, ω-oxidation of the decanamide moiety was eliminated by substitution of hydrogen with fluorine in this position.

Compounds containing fluorine in the terminal positions of the alkyl chain retained $sPLA_2$ inhibitory activity and also possessed improved *in vitro* metabolic stability and pharmacokinetic parameters relative to nonfluorinated inhibitors in this series. As exemplified by GW 4776, improvements in metabolic stability alone, however, were not sufficient to ensure oral activity. Thus, GW 4776 did not show oral activity in the carrageenan edema model and had only modest activity after i.v. dosing in the same model. In fact, the results for GW 9624 and GW 8219 suggested that factors in addition to potency of $sPLA_2$ inhibition and metabolism affect the observed *in vivo* activity. Despite the fact that these two compounds varied only by a single oxygen-to-sulfur substitution, one was active whereas the other was not. One possible explanation for the observed variability is a compound-dependent difference in the rate of equilibration into tissue. This possibility is relevant as both the carrageenan paw edema model and the phorbol ester edema model involve a localized inflammation. No measurements were made to assess differences in the distribution of the different inhibitors between the blood and the localized site of inflammation.

In summary, a series of bioavailable inhibitors of $sPLA_2$ was prepared using an iterative approach that combined medicinal chemistry, *in vitro* and *in vivo* evaluation of biological activity, and metabolic and pharmacokinetic studies. Although some compounds in the series showed *in vivo* activity, the anti-inflammatory effect observed in animal models was modest and a decision was made to abandon $sPLA_2$ as a molecular target for the development of anti-inflammatory agents.

Acknowledgments

The authors gratefully acknowledge C. O. Harris for PLA_2 assays, D. J. Parks for PLA_2 and myeloperoxidase assays, and K. Connolly, H. Sauls, J. Wakefield, and L. Sekut for *in vivo* work. J. Wiseman, D. Thakker, and D. Karenewsky are acknowledged for helpful discussions.

REFERENCES

Bennion, C., Connolly, S., Gensmantel, N. P., Hallam, C., Jackson, C. G., Primrose, W. U., Roberts, G. C. K., Robinson, D. H., and Slaich, P. K., 1992, Design and synthesis of some substrate analogue inhibitors of phospholipase A2 and investigations by NMR and molecular modeling into the binding interactions in the enzyme–inhibitor complex, *J. Med. Chem.* **35:**2939–2951.

Blanchard, S. G., Harris, C. O., and Parks, D. J., 1994, A fluorescence-based assay for human type II phospholipase A_2, *Anal. Biochem.* **222:**435–440.

Bradley, P. P., Priebat, D. A., Christensen, R. D., and Rothstein, G., 1982, Measurement of cutaneous inflammation: Estimation of neutrophil content with an enzyme marker, *J. Invest. Dermatol.* **78:**206–209.

Deems, R. A., Eaton, B. R., and Dennis, E. A., 1975, Kinetic analysis of phospholipase A2 activity toward mixed micelles and its implications for the study of lipolytic enzymes, *J. Biol. Chem.* **250:**9013–9020.

Dennis, E. A., 1973, Kinetic dependence of phospholipase A2 activity on the detergent Triton X-100, *J. Lipid Res.* **14:**152–159.

De Young, L. M., Kheifets, J. B., Ballaron, S. J., and Young, J. M., 1989, Edema and cell infiltration in the phorbol ester-treated mouse ear are temporally separate and can be modulated by different pharmacologic agents, *Agents Actions* **26:**335–341.

Jain, M. K., and Berg, O. G. , 1989, The kinetics of interfacial catalysis by phospholipase A2 and regulation of interfacial activation: Hopping versus scooting, *Biochim. Biophys. Acta* **1002:**127–156.

Jain, M. K., and Gelb, M. H., 1991, Phospholipase A2-catalyzed hydrolysis of vesicles: Uses of interfacial catalysis in the scooting mode, *Methods Enzymol.* **197:**112–125.

Jain, M. K., Tao, W., Rogers, J., Arenson, C., Eibl, H., and Yu, B. Z., 1991, Active-site directed specific competitive inhibitors of PLA_2: Novel transition state analogues, *Biochemistry* **30:** 10256–10268.

Krawisz, J. E., Sharon, P., and Stenson, W. F., 1984, Quantitative assay for acute intestinal inflammation based on myeloperoxidase activity: Assessment of inflammation in rat and hamster models, *Gastroenterology* **87:**1344–1350.

Murakami, M., Kudo, I., Nakamura, H., Yokoyama, Y., Mon, H., and Inoue, K., 1990, Exacerbation of rat adjuvant arthritis by intradermal injection of purified mammalian 14-kDa group II phospholipase A2, *FEBS Lett.* **268:**113–116.

Reynolds, L. J., Washburn, W. N., Deems, R. A., and Dennis, E. A., 1991, Assay strategies for phospholipases, *Methods Enzymol.* **197:**3–23.

Schultz, H., 1991, Oxidation of fatty acids, in: *Biochemistry of Lipids, Lipoproteins and Membranes* (D. E. Vance and J. Vance, eds.), pp. 87–110, Elsevier, Amsterdam.

Stryer, L., 1988, *Biochemistry,* Chapter 20, Freeman, San Francisco.

Wheeler, T. N., Blanchard, S. G., Andrews, R. C., Fang, F., Gray-Nunez,Y., Harris, C. O., Lambert, M. H., Mehrotra, M. M., Parks, D. J., Ray, J. A., and Smalley, T. L., Jr., 1994, Substrate specificity on short-chain phospholipid analogs at the active site of human synovial phospholipase A_2, *J. Med. Chem.* **24:**4118–4129.

Yuan, W., Quinn, D. M., Sigler, P. B., and Gelb, M. H., 1990, Kinetic and inhibition studies of phospholipase A2 with short-chain substrates and inhibitors, *Biochemistry* **29:**6082–6094.

Chapter 20

The Anxieties of Drug Discovery and Development

CCK-B Receptor Antagonists

Franco Lombardo, Steven M. Winter, Larry Tremaine, and John A. Lowe III

1. INTRODUCTION

The pharmacology of cholecystokinin (CCK), a polypeptide hormone originally discovered in 1929, has a rich history (Mutt, 1980). Its control of gallbladder function and digestive enzyme secretion has been extensively characterized, but its occurrence as a C-terminal octapeptide version in the brain was puzzling. The announcement in 1986 of the discovery of the first nonpeptide CCK antagonists promised a resolution of this conundrum, as compounds able to potently and selectively displace CCK binding in both pancreas and brain were reported (Evans and Bock, 1993). Subsequent confirmation of this result came with the cloning and sequencing of the gene for the CCK-A receptor from rat (Wank *et al.*, 1992) and human (de Weerth *et al.*, 1993) gut, and the CCK-B receptor from human brain (Lee *et al.*, 1992).

We were especially intrigued by the specific CCK-B receptor antagonist L-365,260 (Bock *et al.*, 1993), a benzodiazepine resulting from an extensive medicinal chemistry program, for its potential in examining the activity of CCK in the

Franco Lombardo, Steven M. Winter, Larry Tremaine, and John A. Lowe III • Central Research Division, Pfizer Inc., Groton, Connecticut 06340.

Integration of Pharmaceutical Discovery and Development: Case Studies, edited by Borchardt *et al.*, Plenum Press, New York, 1998.

central nervous system (CNS). This interest was heightened by a report that the selective CCK-B receptor agonist CCK-4 (the C-terminal tetrapeptide portion of CCK) induces a paniclike response in humans similar to the disease panic disorder (Harro *et al.,* 1993). The connection between panic and anxiety based on CCK that was subsequently discovered (Dooley and Klamt, 1993) gave an added commercial incentive to the discovery of a new CCK-B receptor antagonist. The importance of selectivity was underscored by the involvement of the CCK-A receptor in digestion. In addition, the CCK-B receptor has been suggested to play a role in pain (Noble *et al.,* 1993) and control of central dopaminergic function (Rasmussen *et al.,* 1993). These results provided the rationale for us to initiate a program aimed at finding a selective CCK-B receptor antagonist for the treatment of panic and anxiety.

2. CHEMISTRY

The medicinal chemistry of CCK-B antagonists from numerous structural classes has been recently reviewed (Makovec, 1993). One of the most thoroughly investigated classes is the benzodiazepine family, represented by the potent and selective CCK-B antagonist L-365,260 (Fig. 1) (Bock *et al.,* 1993). Using this structure as a starting point, we selected the 5-phenyl-3-ureidobenzazepin-2-one system to provide a mimic of each structural feature likely to be important for receptor recognition. The 5-phenyl-3-ureidobenzazepin-2-one system affords flexibility for incorporation of functionality in each key element of the structure, and provides an opportunity to explore the effect of stereochemistry through its two asymmetric centers.

The synthesis and structure–activity relationships (SAR) of the 5-phenyl-3-ureidobenzazepin-2-one series of CCK-B antagonists have been detailed by us

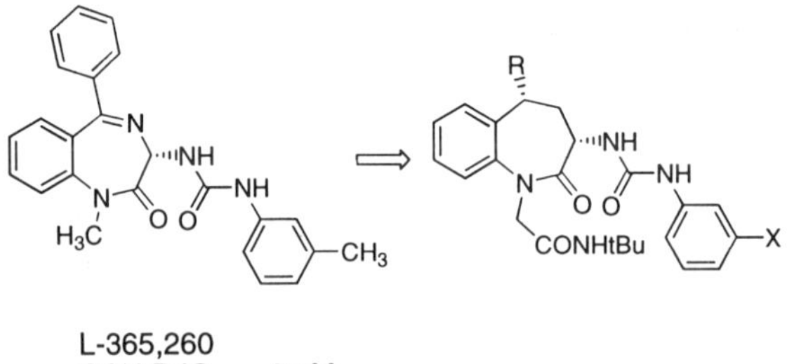

L-365,260
CCK-B IC_{50} = 8 nM
CCK-A IC_{50} = 120 nM

CP-212,454 R=Ph, X=Cl
CCK-B IC_{50} = 0.48 nM
CCK-A IC_{50} = 180 nM

CP-310,713 R=c-hexyl, X=CO_2H
CCK-B IC_{50} = 0.10 nM
CCK-A IC_{50} = 1,400 nM

Figure 1. Nonpeptide CCK-B receptor antagonists.

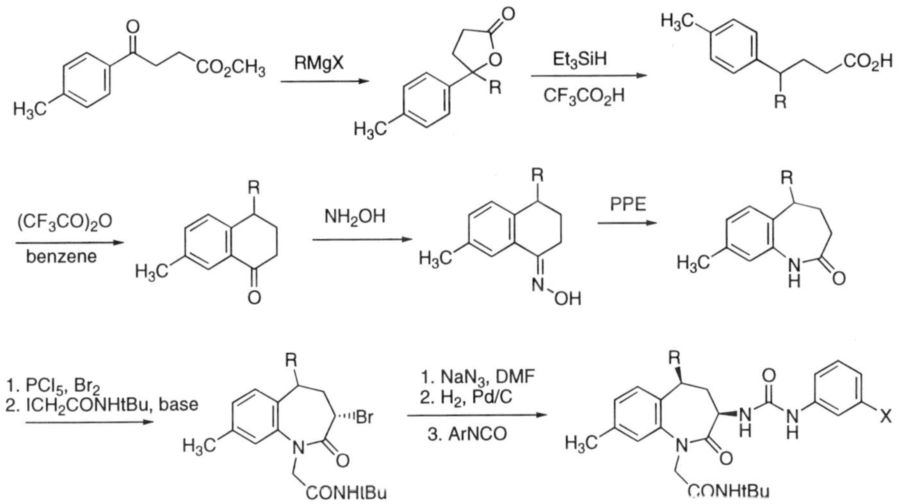

Figure 2. Synthesis of 5-phenyl-3-ureidobenzazepin-2-one series of CCK-B antagonists.

elsewhere (Lowe *et al.,* 1994) and will be summarized here. The synthetic route that ultimately proved the most flexible in determining the scope of SAR is outlined in Fig. 2.

This approach allowed incorporation of various substituents at the 5-position (R) and at the *meta* position of the phenylureido group (X), both of which play important roles in determining CCK-B receptor affinity. The stereochemistry was determined largely during the bromination step, whereas the absolute stereochemistry was set by a resolution process involving derivatization with phenylalanine and separation of diastereomers.

SAR studies indicated that the 8-methyl group on the benzazepinone nucleus, *t*-butylacetamide group at N-1, and *cis* stereochemistry are optimal. Several R groups provided potent and selective CCK-B receptor affinity, with both phenyl and cyclohexyl among the best (Table I). Although X = chloro was selected for its protection of the phenyl ureido group against metabolic derivatization, this position proved the most flexible for incorporation of polar functionality to improve solubility as the project progressed. SAR studies leading to the selection of X = CO_2H from among a number of acid and acid surrogate groups were reported (Lowe *et al.,* 1995). This work led to the selection of two compounds, CP-212,454 (CCK-B IC_{50} = 0.48 nM, CCK-A IC_{50} = 180 nM) and CP-310,713 (CCK-B IC_{50} = 0.10 nM and CCK-A IC_{50} = 1400 nM), for advanced evaluation. This selection process, however, depended on critical findings in both drug metabolism and pharmaceutical formulation, which are described in the following sections.

Table I

SAR Results for Benzazepine CCK-B Antagonists, with L-365,260 as Control[a]

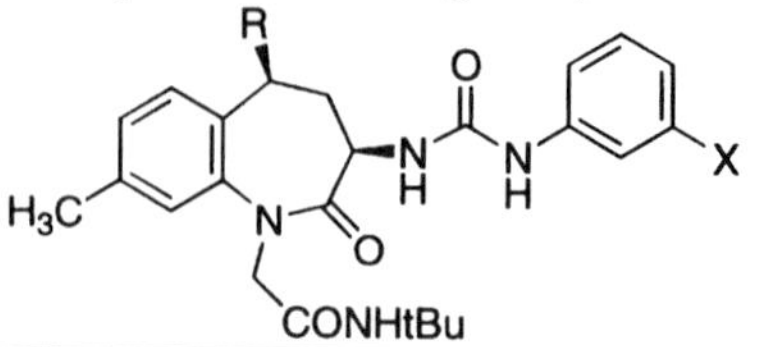

R	X	CCK-B IC_{50} (nM)	CCK-A IC_{50} (nM)
Phenyl(+)	Cl	0.48 ± 0.079	176 ± 46
Cyclohexyl	Cl	0.60 ± 0.25	560 ± 203
Benzyl	Cl	1.4 ± 0.06	>10,000
Isopropyl	Cl	3.4 ± 1.3	72 ± 14
Cyclohexyl	5-Tetrazolyl	0.22 ± 0.04	490 ± 130
Cyclohexyl	$CONHSO_2CH_3$	0.26 ± 0.075	520 ± 34
Cyclohexyl	CO_2CH_3	0.60 ± 0.15	370 ± 75
Cyclohexyl(+)	CO_2H	0.10 ± 0.015	1400 ± 240
L-365,260	—	8.1 ± 1.5	86 ± 27

[a]CCK-B binding to guinea pig cortex; CCK-A binding to guinea pig pancreas.

3. INITIAL DRUG METABOLISM STUDIES

CP-212,454 had been identified as a candidate for advanced evaluation, not because of its activity in an *in vivo* animal model, but rather based on its potent and selective receptor antagonism in *in vitro* studies and the demonstration of good CNS penetration via measurement and comparison of brain and plasma drug concentrations. In a close-in series of analogues including CP-212,454, brain to plasma concentration ratios in the mouse ranged from 0.02 to 0.17. Minimal ratios of 0.1 were desired, because a compound not penetrating the blood–brain barrier might exhibit a ratio of 0.05 simply as a result of the blood present within the cerebral vasculature. Penetration into the brain was thus viewed as an area requiring improvement for this series of compounds, and drug metabolism studies, rather than animal pharmacology, played a large role in the evaluation of the *in vivo* properties of these compounds.

In an initial pharmacokinetic study in male rats using the racemic form of the compound and administered in a formulation of DMSO/Emulphor/0.9% saline (5/5/90), the half-life and oral bioavailability of the compound were 1.5 hr and 17%, respectively. The intravenous and oral pharmacokinetics of CP-212,454 were next characterized in detail in rats as a matter of routine prior to conducting formal safety studies. Plasma samples from animals in the *in vivo* rodent model were also provided to relate drug concentration to effect. The pharmacokinetic parameters derived from the intravenous dose indicated that CP-212,454 had a high

Table II
Pharmacokinetics of CP-212,454 in Preclinical Animal Species Measured after Intravenous Administration at 5 mg/kg

Species	Clearance (ml/min/kg)	Volume of distribution (liters/kg)	$t_{1/2}$ (hr)
Male Sprague–Dawley rat	63	5.0	1.2
Beagle dog	3.0	1.4	5.8

clearance, a moderate to high volume of distribution, and short half-life in this species (Table II). The oral bioavailability was 13% at 5 to 10 mg/kg, with the range of 7.8 to 19% in nine animals studied, when formulated as an aqueous suspension in 0.5% methyl cellulose. In two additional animals receiving 25 mg/kg, the plasma concentration profile was dramatically changed, with a substantially longer apparent half-life, and plasma AUC increased supraproportionally in relation to dose. Altogether, the data suggested that the compound underwent first-pass metabolism that became saturable with dose. An underlying absorption issue was signaled by the fact that the oral bioavailability was lower in methyl cellulose than in the DMSO/Emulphor/saline formulation.

Pharmacokinetics in the dog was also determined at a 5 mg/kg i.v. dose, and oral doses of 5 and 50 mg/kg administered in aqueous, methyl cellulose, suspension, to assess both absolute bioavailability and the oral dose–exposure relationship. Unlike in the rat, CP-212,454 has a low clearance and low volume of distribution in the dog, resulting in a half-life of 6 hr after i.v. administration (Table II). Drug C_{max} after the 5 mg/kg dose was approximately 200 ng/ml, and oral bioavailability was estimated at 4%. With the 10-fold increase in oral dose, C_{max} and AUC increased by 1.7- and 2.2-fold, respectively, and the calculated oral bioavailability was 1%. The decrease in oral bioavailability with increasing dose indicated that oral absorption was poor when administered as an aqueous suspension. These results, coupled with those in the rat, suggested that formulation studies would be necessary to achieve suitable drug levels for toxicological evaluation and progression to clinical evaluation in humans.

Intestinal Absorption and the MAD Number

The poor oral bioavailability of CP-212,454 was first thought to be related to its very low aqueous solubility, as its equilibrium value was determined to be 0.2 μg/ml. This is generally an undesirable property as aqueous solubility is of paramount importance for oral absorption, reflecting the fact that the flux of passively absorbed drugs across the intestinal membrane(s) is a function of their solubility, in the aqueous intestinal fluids, and of their permeation properties across the membranes. Because the majority of drugs are passively absorbed by the small intestine, aqueous solubility is essential for oral absorption of drugs from conventional dosage forms. Tablets and capsules prepared from solid crystalline compounds would have to undergo disintegration and would have to release the active drug which should dissolve in a reasonably short time, if substantial absorption is to occur during the drug transit time through the small intestine. In the case of CP-212,454, preliminary pharmacokinetic studies, conducted using a well-dispersed methylcellulose suspension, showed that a conventional formulation could not be expected to produce, in humans, plasma concentrations associated with clinical efficacy. Following this specific example, it is thus worthwhile, in this section of the chapter, to illustrate one of the tools we use in our approach toward the general problem of absorption prediction and analysis.

Although lipophilic drugs may be poorly soluble, they likely permeate intestinal membranes quite well, whereas it is often observed that very soluble, hydrophilic compounds show the opposite behavior. Any dramatic shift in either direction (high lipophilicity or high hydrophilicity) may result in an overall diminished absorption, and a sensible balance of these factors needs to be considered during the drug design phases. In general, we treat the solubility and permeability parameters independently and we have found it convenient to define and use, in comparing drug candidates, the maximum absorbable dose (MAD) in humans (Johnson and Swindell, 1996):

$$\text{MAD (mg)} = S \times k_{\text{a}} \times \text{IFV} \times \text{RT}$$

S is the solubility in phosphate buffer at pH 6.5 (in mg/ml), k_{a} is the absorption rate constant in rats (in min^{-1}), IFV is the intestinal fluid volume (250 ml), and RT is an average residence time in the small intestine, taken to be equal to 270 min. It is of course arguable that some or all of these parameters may be arbitrarily chosen, and although their review and discussion is outside of the scope of this work, it will suffice, in this context, to say that these parameters may be modified, for example, if there are reasons to believe that the solubility may be higher (or lower) than the value determined at pH 6.5, as well as if there is the possibility of a longer residence time. We shall return to this point later, but the important factor to consider is that this equation provides a good comparison tool among drug candidates, when they are ranked in terms of their absorption properties, using the two intrinsic variables (solubility and absorption rate constant) and the two other parameters, which we typically keep constant. In typical clinical trials, we use a volume of 250 ml of fluids to administer drug candidates and the residence time of 270 min is probably quite appropriate.

The k_{a} value should also be briefly discussed in this context as it provides a physiological measure of intestinal permeability *in vivo.* This value is determined by single-pass intestinal perfusion (SPIP) of a diluted solution of the drug in PBS (30–50 μg/ml), perfused through a 10-cm section of the jejunum of anesthetized rats, with a typical flow rate of 0.2 ml/min. HPLC analysis of a sample of the perfusate collected at 15-min intervals, during a 90-min perfusion, yields the k_a value:

$$k_a = (1 - C_{out}/C_{in}) \times Q/V$$

where C_{in} and C_{out} are the initial and final concentration of the drug in the perfusate (before and after single pass), respectively, Q is the flow rate, and V is the intestinal volume, which is usually taken as the volume of a cylinder of radius 0.2 cm, yielding a value of 1.26 ml for a 10-cm section of jejunum. Reasonably good absorption rate constants measured with this method would range between 0.1 and 0.01 min^{-1} and values below the lower limit would be associated with poorly permeable drug. In the case of CP-212,454, we measured an absorption rate constant of 0.02 min^{-1}.

If the solubility is good (>1 mg/ml), the MAD could be on the order of several hundred milligrams even in the case of a poorly permeable compound, as a high value for S would yield a high MAD. This finding should in turn be reflected by a higher flux of drug because of the high concentration gradient established between the intestinal lumen and the portal circulation.

It is important to keep in mind that this number should not be used as an absolute value as it is best applied to a series of compounds to be compared. Furthermore, its application should be relative to the projected dose while attempting to determine the potential for absorption issues for any given candidate. If the MAD largely exceeds (>10-fold) the projected dose, the compound is likely to be well absorbed, whereas if the MAD is of comparable magnitude or much lower than the projected dose, incomplete to poor absorption may result if a conventional dosage form such as a tablet or capsule is desired.

As mentioned, the parameters used in the MAD equation can be adjusted to reflect, for example, data that suggest that the solubility might be significantly improved by intestinal surfactants (bile acid salts). Furthermore, a basic compound may retain supersaturation after transit from the stomach to the higher pH environment of the intestine. The solubility under those conditions might then be used to generate "limiting" MAD numbers. It should be emphasized, however, that the MAD is generally an upper limit value, using some approximations, and it should not be expected to exactly yield the actual dose that will be absorbed in each specific case.

4. FORMULATION STUDIES

Despite the *in vitro* potency of CP-212,454, data from several *in vivo* animal models including a pentagastrin-induced acid secretion model in rat, a CCK-4-induced cardiovascular effect model in dogs, and a CCK-4-induced panic attack model in primates suggested that relatively high plasma concentrations of CP-212,454 (200–500 ng/ml), as well as high parenteral doses (>1 mg/kg subcutaneously), were required for efficacy. Although initially perplexing, this was in part rationalized by the high plasma protein binding of CP-212,454. Plasma protein binding, as determined by equilibrium dialysis at a concentration of ~5 μg/ml, in rat, dog, and monkey was 98.27 ± 1.22, 99.97 ± 0.02, and 99.73 ± 0.11%, respectively. Another possible factor leading to the need for high plasma drug concentrations for efficacy was a lack of CNS penetration. Rats were dosed intravenously with CP-212,454 at 3 mg/kg, and terminal blood and cerebrospinal fluid (CSF) samples were obtained from individual animals at 0.25 and 1 hr postdose (n = 2/time point). The CSF/plasma concentration ratios were 0.008 and 0.035 at these respective time points. This study suggested that extracellular concentrations within the CNS might be equal to unbound concentrations in plasma, and distribution of drug across the blood–brain barrier was not a limiting factor in drug effect. Although the data from the plasma protein binding study provided a basis for the high plasma drug concentrations required for efficacy, the substantial differences in extrapolated unbound plasma concentrations needed for efficacy between species suggested that this analysis might not provide any better estimate of the plasma concentrations required for a pharmacological response in humans. Given the probability of similarly high plasma protein binding in humans, it was projected, quite simplistically perhaps, that plasma concentrations of CP-212,454 of approximately 200–500 ng/ml might also be required for clinical efficacy.

The MAD calculation, considering the poor solubility of CP-212,454, gave a value of 0.3 mg for the maximum amount of the compound that could be absorbed. With the clinically projected therapeutic concentration of 200–500 ng/ml, toxicology studies would ideally require plasma concentrations of drug on the order of 2–5 μg/ml. Because unformulated drug would be unsuitable for reaching these required plasma levels, formulation studies were undertaken to improve absorption. Initially a simple 0.5% carboxymethylcellulose vehicle and subsequently a 2% Tween 80 (Polysorbate 80) vehicle in water were tried, followed by neat Cremophor EL® (a polyoxyethylated castor oil derivative) and a vehicle consisting of Cremophor EL®/PEG 400/water (30/10/60 v/v/v). None of these efforts, however, afforded a maximal plasma concentration of CP-212,454 above 2 μg /ml in the rat (Fig. 3).

The latter vehicle was developed to try to overcome the potential problems of Cremophor EL®, namely, its viscosity and tendency to gelify in the presence of water, in addition to the possibility of gastrointestinal irritation. The ternary mixture described above seemed to alleviate these concerns although at the price of a significantly lower solubility of the drug candidate(s) dosed. As an example, the solubility of CP-212,454 was found to be 73 mg/ml in 70% Cremophor EL®, whereas it was 8 mg/ml in the ternary mixture containing only 30% of this surfactant.

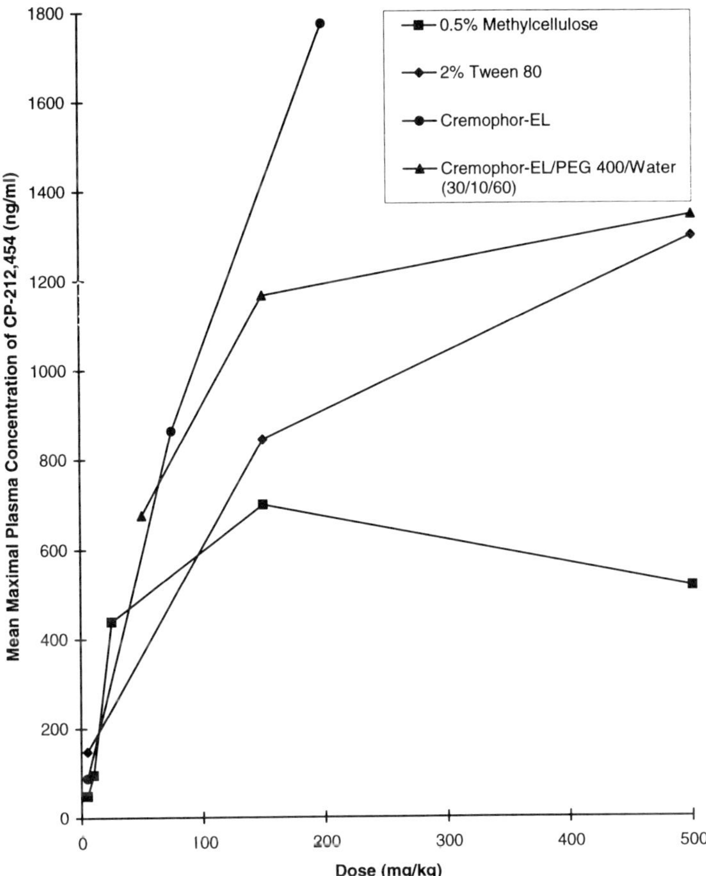

Figure 3. Relationship between oral dose and systemic exposure (mean maximal plasma concentration, C_{max}) for CP-212,454 in Sprague–Dawley rats administered single oral doses of CP-212,454 in various vehicles at 10 ml/kg.

Similar studies dosing CP-212,454 in various vehicles were also conducted in dog and a primate species. Because systemic exposure of the drug in primates was less than that observed in dog at comparable doses, formulation efforts were more extensively evaluated in the dog (Fig. 4). Although a dose-limited exposure was also observed in the dog, a satisfactory systemic exposure was achievable, using the Cremophor EL® /PEG 400/water vehicle described, at a volume of 4 ml/kg. Assuming that exposure to drug in rats and dogs was sufficient for toxicology studies and that both species could tolerate these vehicles for the duration of the stud-

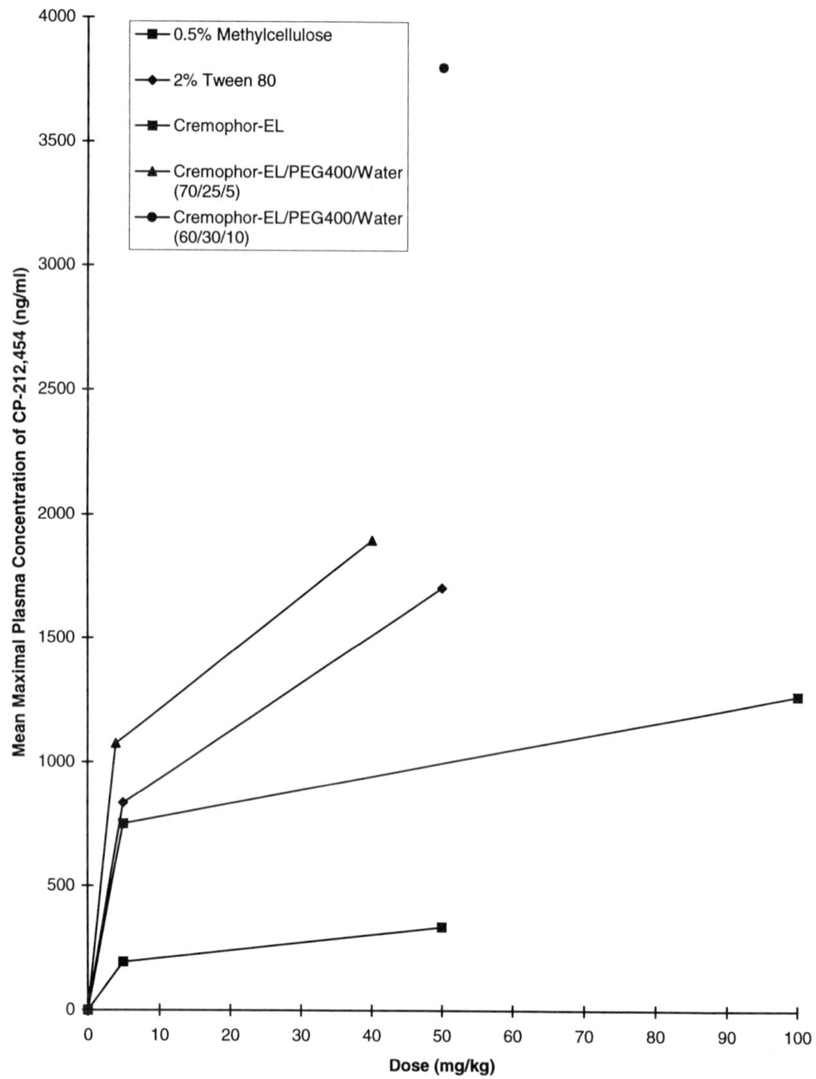

Figure 4. Relationship between oral dose and systemic exposure (mean maximal plasma concentration, C_{max}) for CP-212,454 in beagle dogs administered single oral doses of CP-212,454 in various vehicles.

ies, one potential obstacle to the development of CP-212,454 was successfully overcome through formulation research.

Clinical Vehicles

As it was unlikely that a solid dosage formulation would produce therapeutic concentrations of CP-212,454 in humans, a commercially viable formulation to enhance drug exposure was sought. One potential vehicle (PEG 400/glycerol/water, 92/5/3 v/v/v), identified earlier for incorporation into soft gelatin capsules, was evaluated in capsule form in the dog. Assuming exposure to drug observed in dogs was predictive of that in humans, studies with capsules containing 40–50 mg of drug suggested that exposure to drug could be enhanced with small volumes (0.6 ml) of this vehicle contained within a capsule. As the capsules contained the drug at the maximum solubility in this vehicle, multiple capsule administration may still have been needed to achieve therapeutic concentrations of CP-212,454 in humans. On the other hand, when CP-212,454 was administered to monkeys at 50 mg/kg in Cremophor EL® /PEG 400/water (75/25/10, 1 ml/kg), the mean peak plasma concentration was only 291 ng/ml. Furthermore, some stability concerns became apparent when the stability of CP-212,454 was studied in the PEG 400/glycerol/water formulation described above. Preliminary data generated on this formulation had shown that a substantial decay of CP-212,454 was occurring in relatively short amounts of time (1–3 weeks) at 50°C, and the addition of BHA (a mixture of 2- and 3-*tert*-butyl-4-hydroxyanisole) was able to slow down the decay but not stop it. The indication that an oxidative mechanism was likely to be responsible for the decay of CP-212,454 raised some concerns over the feasibility of a commercial softgel formulation for this compound and the efforts necessary to accomplish that goal. Furthermore, the program had steered toward the discovery of more soluble analogs and no further studies were pursued with this formulation.

Other vehicles were considered with the objective of increasing clinical exposure and also for their potential use in commercial dosage forms. Oleic acid, Capmul MCM® (a mixture of medium chain mono- and diglycerides), glycerol monooleate, triacetin, olive oil, and sesame oil were chosen as they represented a wide variety of distinct vehicles with potential for a commercial formulation. The equilibrium solubility determinations showed that the olive and sesame oils would be inadequate as softgel vehicles, with a solubility of 4 and 3 mg/ml, respectively, whereas all of the other vehicles showed a solubility of at least 90 mg/ml. Table III lists the observed solubility of CP-212,454 in each vehicle.

However, in consideration of stability concerns for a liquid formulation and in particular because a more conventional dosage form was desired, it had become increasingly apparent that a more soluble analogue able to yield a higher exposure from a conventional tablet or capsule, was needed.

Table III
Solubility of CP-212,454 in Various Vehicles

Vehicle	Solubility (mg/ml)
Oleic acid	>90
Capmul MCM®	>102
Glycerol monooleate	>104
Triacetin	>141
Olive oil	>3.9
Sesame oil	>2.8

5. A NEW ANALOGUE WITH IMPROVED AQUEOUS SOLUBILITY: CP-310,713

The development support for CP-212,454 had shown that a more soluble analogue was needed in order to develop a product that would not require a softgel capsule. Solubility and absorption rate constant determinations, on several analogues that were being synthesized, yielded a gloomy picture in terms of solubility and it was impossible, because of their low solubility, to determine the absorption rate constant (k_a) for many of them. For example, the analogues shown in Table I for R = cyclohexyl and X = 5-tetrazolyl and for R = cyclohexyl and X = $NHSO_2CH_3$ were found to have solubilities lower than 0.1 μg/ml in PBS at pH 6.5.

More extensive screening, however, revealed an analogue with greatly improved solubility in the potassium salt of CP-310,713 (R = cyclohexyl and X = CO_2H, Table I). This potassium salt had a solubility of 136 μg/ml in PBS, whereas in unbuffered water its solubility was 4.8 mg/ml. The solubility in PBS, coupled with an absorption rate constant of 0.02 min^{-1}, would yield a MAD of 184 mg. Furthermore, our interest in this compound had been heightened by the fact that it appeared to be a very potent compound in guinea pigs. The focus of our attention thus shifted from CP-212,454, a compound requiring extensive formulation support, to CP-310,713, a seemingly promising candidate for the development of a conventional dosage form.

In guinea pigs, CP-310,713 appeared to be a very potent compound. An approximately 50% blockade of CCK-4-induced cardiovascular changes (heart rate and blood pressure) was observed at 20 min following oral administration of the drug at 0.05 mg/kg. The plasma concentration of CP-310,713 at this time point was 26 ng/ml or less. However, in a CCK-4-induced monkey panic attack model, considerably higher doses leading to higher concentrations of the drug (approx. 500 ng/ml) appeared to be required for efficacy. If reversal of the CCK-4 challenge in the monkey model is indicative of central activity, it is likely that the large discrepancy between concentrations (and doses) of drug required for efficacy in monkey panic models and guinea pig cardiovascular models reflected a low CNS penetration for the compound. At this point in the project, it became apparent that problems with efficacy and bioavailability were beyond our ability to afford a timely solution, and the project was terminated. There were, however, important lessons learned from this failed attempt that have served as a guide for numerous subsequent projects.

6. LESSONS LEARNED

1. *In vitro* potency is not a guarantee of a successful drug candidate. Properties such as solubility and bioavailability can become central to progression of a project, even to the ability to carry out safety evaluation. Lead compounds should be evaluated more thoroughly for these properties, and the SAR program in chemistry guided more carefully, by consideration of these latter properties, in addition to potency and efficacy (Lipinski *et al.*, 1997).
2. Administration of the drug at two doses early in the drug evaluation process to ensure dose-proportional exposure at the higher dose is important for planning toxicology studies. When poor absorption limits drug exposure at higher doses, toxicology evaluation may not reveal the potential for side effects.
3. The use of less conventional formulations (primarily soft gelatin capsules) is not a panacea for insoluble and poorly absorbed compounds, even though it may help in the short term. They can lead to longer development times and present more challenging problems such as stability in hydrophilic or lipophilic vehicles. They may also require a high solubility of the drug candidate in the vehicles in order to dose relatively small volumes.
4. Where reliable *in vivo* animal pharmacological assays are unavailable, pharmacokinetics measurements can substitute by determining clearance, plasma protein binding, distribution to the tissue, nonspecific tissue binding, and oral absorption. But the extrapolation of these parameters to humans to predict clinical efficacy is hazardous, given the potential for variations in distribution to the target site, nonspecific binding to plasma and tissue macromolecules, and the rate of metabolism and the formation of active metabolites.
5. Tissue penetration and determination of drug concentration in the relevant "biophase" is particularly difficult for CNS drugs (Lombardo *et al.*, 1996). Because of the tight endothelial lining of the blood–brain barrier, compounds that reach the extracellular space of peripheral tissues may not permeate across endothelial cells and into the extracellular fluid

bathing neuronal cells. Measuring total concentrations in brain tissue and plasma can provide a qualitative indication of penetration across the blood–brain barrier, but provides little insight into the concentration of drug at the receptor. Given that drug appears to freely diffuse across the blood–brain barrier, the most reasonable assumption is that this concentration equals the unbound concentration in plasma. Brain concentrations of drug above this likely reflect additional nonspecific binding to tissue macromolecules.

In conclusion, while serendipity will continue to play a major role in successful drug development, project teams can benefit from a few of the lessons learned in the CCK-B project discussed here by starting with a lead that has favorable solubility and absorption properties and coordinating the optimization of *in vitro* potency and selectivity with *in vivo* activity and pharmacokinetics at each stage of the program. This approach ensures that each advance in potency and selectivity translates to a truly improved drug candidate. Luck can never be removed as a factor, but the above approach can minimize "the anxieties of drug discovery and development".

Acknowledgments

The authors wish to thank the following colleagues for their hard work and dedication to the CCK-B project: Troy A. Appleton, Mark G. Biron, Dianne K. Bryce, Anthony M. Campeta, Rosemary T. Crawford, Michael DePasquale, Susan E. Drozda, Eugene F. Fiese, Anthony Fossa, Weldon Horner, Deepa Johnson, Stafford McLean, Fred Nelson, and Nita Patel.

In addition, F.L. wishes to acknowledge William J. Curatolo, Michael J. Gumkowski, Kevin C. Johnson, and Steven C. Sutton for helpful discussions and their insight during the period this work was carried out.

REFERENCES

Bock, M., DiPardo, R. M., Evans, B. E., Rittle, K. E., Whitter, W. L., Garsky, V. M., Gilbert, K. F., Leighton, J. L., Carson, K. L., Mellin, E. C., Veber, D. F., Chang, R. S. L., Lotti, V. J., Freedman, S. B., Smith, A. J., Patel, S., Anderson, P. S., and Freidinger, R. M., 1993, Development of 1,4-benzodiazepine cholecystokinin type B antagonists, *J. Med. Chem.* **36:**4276–4292.

de Weerth, A., Pisegna, J. R., Huppi, K., and Wank, S. A., 1993, Molecular cloning, functional expression and chromosomal localization of the human cholecystokinin type A receptor, *Biochem. Biophys. Res. Commun.* **194:**811–818.

Dooley, D. J., and Klamt, I., 1993, Differential profile of the CCK-B receptor antagonist CI-988 and diazepam in the four-plate test, *Psychopharmacology* **112:**452–454.

Evans, B. E., and Bock, M. G., 1993, Promiscuity in receptor ligand research: Benzodiazepine-based cholecystokinin antagonists, *Adv. Med. Chem.* **2:**111–152.

Harro, J., Vasar, E., and Bradwejn, J., 1993, CCK in animal and human research on anxiety, *Trends Pharmacol. Sci.* **14:**244–249.

Johnson, K. C., and Swindell, A. C., 1996, Guidance in the setting of drug particle size specifications to minimize variability in absorption, *Pharm. Res.* **13:**1794–1797.

Lee, Y.-M., Beinborn, M., McBride, E. W., Lu, M., Kolakowski, L. F., and Kopin, A. S., 1992, The human brain cholecystokinin-B/gastrin receptor, *J. Biol. Chem.* **268:**8164–8169.

Lipinski, C. A., Lombardo, F., Dominy, B. W., and Feeney, P. J., 1997, Experimental and computational approaches to estimate solubility and permeability in drug discovery and development settings, *Adv. Drug Del. Rev.* **23:**3–25.

Lombardo, F., Blake, J. F., and Curatolo, W. J., 1996, Computation of brain-blood partitioning of organic solutes via free energy calculations, *J. Med. Chem.* **39:**4750–4755.

Lowe, J. A., III, Hageman, D. L., Drozda, S. E., McLean, S., Bryce, D. K., Crawford, R. T., Zorn, S., Morrone, J., and Bordner, J., 1994, 5-Phenyl-3-ureidobenzazepin-2-ones as cholecystokinin-B receptor antagonists, *J. Med. Chem.* **37:**3789–3811.

Lowe, J. A., III, Drozda, S. E., McLean, S., Bryce, D. K., Crawford, R. T., Zorn, S., Morrone, J., Appleton, T. A., and Lombardo, F., 1995, A water soluble benzazepine cholecystokinin-B receptor antagonist, *Bioorg. Med. Chem. Lett.* **5:**1933–1936.

Makovec, F., 1993, CCK-B/gastrin-receptor antagonists, *Drugs Future* **18:**919–931.

Mutt, V., 1980, Cholecystokinin: Isolation, structure, and functions, in: *Gastrointestinal Hormones* (G. B. J. Glass, ed.), pp. 169–221, Raven Press, New York.

Noble, F., Derrien, M., and Roques, B. P., 1993, Modulation of opioid antinociception by CCK at the supraspinal level: Evidence of regulatory mechanisms between CCK and enkephalin systems in the control of pain, *Br. J. Pharmacol.* **109:**1064–1070.

Rasmussen, K., Czachura, J. F., Stockton, M. E., and Howbert, J. J., 1993, Electrophysiological effects of diphenylpyrazolidinone cholecystokinin-B and cholecystokinin-A antagonists on midbrain dopamine neurons, *J. Pharmacol. Exp. Ther.* **264:**480–488.

Wank, S. A., Harkins, R., Jensen, R. T., Shapira, H., de Weerth, A., and Slattery, T., 1992, Purification, molecular cloning, and functional expression of the cholecystokinin receptor from rat pancreas, *Proc. Natl. Acad. Sci. U.S.A.* **89:**3125–3129.

Chapter 21

CI-1015

An Orally Active CCK-B Receptor Antagonist with an Improved Pharmacokinetic Profile

Bharat K. Trivedi and Joanna P. Hinton

1. INTRODUCTION

Cholecystokinin (CCK), a 33-amino-acid polypeptide, occurs in a variety of biologically active forms throughout the peripheral and central nervous systems (Larsson and Rehfeld, 1979; Rehfeld *et al.*, 1979; Rehfeld and Nielsen, 1995). It has been implicated as a putative neurotransmitter, and is involved in the modulation of other neurotransmitters such as dopamine and GABA (Beinfeld, 1983; Voigt *et al.*, 1986; Crawley, 1989). These effects of CCK are mediated by its interaction with the CCK receptors. CCK receptors have been divided into two receptor subtypes: CCK-A receptors present predominantly in the periphery, and CCK-B receptors present predominantly in the brain (Innis and Snyder, 1980; Dourish and Hill, 1987). Recently, cloning and expression of both the CCK receptors from rat (Wank *et al.*, 1992) and the CCK-A receptor from human (De Weerth *et al.*, 1993) have been achieved. Human CCK-B receptors have also been cloned (Lee *et al.*, 1993). Several lines of evidence suggest that modulation of various pharmacological actions of CCK may provide an opportunity to design and

Bharat K. Trivedi • Department of Medicinal Chemistry, Parke-Davis Pharmaceutical Research, Warner-Lambert Company, Ann Arbor, Michigan 48105. *Joanna P. Hinton* • Department of Pharmacokinetics and Drug Metabolism, Parke-Davis Pharmaceutical Research, Warner-Lambert Company, Ann Arbor, Michigan 48105.

Integration of Pharmaceutical Discovery and Development: Case Studies, edited by Borchardt *et al.*, Plenum Press, New York, 1998.

develop therapeutically useful agents. Thus, efforts have been made to develop CCK-A-receptor-selective agonists for the treatment of satiety (Holladay *et al.*, 1992), and potential clinical utility of CCK-A receptor antagonists has been reviewed (D'Amato *et al.*, 1994). More recently, clinical evidence has suggested the possibility that CCK-B receptors may be involved in the pathogenesis of panic attacks. CCK-4 has been shown to induce panic attacks in patients with panic disorder (Bradwejn *et al.*, 1990, 1991; Bradwejn and Koszycki, 1992). This has provided further impetus for the pharmaceutical industry to develop novel and possibly nonsedative anxiolytic agents.

1.1. First-Generation CCK-B Antagonists

Significant effort has been made regarding the identification of receptor-selective antagonists (reviewed by Trivedi, 1994a,b). In particular, L-365,260, a benzodiazepine-based CCK-B antagonist, was developed clinically and shown to block the anxiogenic effect of CCK-4 in panic patients when dosed orally (Bradwejn *et al.*, 1994). Similarly, CI-988, a peptoid analogue designed from the CCK-4 tetrapeptide, which will be discussed later, was also shown to modestly block CCK-4-induced panic symptoms in healthy individuals (Bradwejn *et al.*, 1995). These clinical observations as well as preclinical CNS effects of CCK have recently been summarized (Bourin *et al.*, 1996). Because of our continued interest in designing potent and selective CCK-B receptor-selective antagonists as potential novel therapeutic agents, efforts over the last several years at Parke-Davis have led to the identification of a series of peptoid derivatives as CCK-B antagonists (Horwell *et al.*, 1991; Boden *et al.*, 1993). From this class of compounds, CI-988 (Fig. 1) was identified as a potent and selective CCK-B receptor antagonist with anxiolytic-like activity in established *in vivo* paradigms such as the X-maze (Hughes *et al.*, 1991). However, during the preclinical and clinical development of this compound, its oral bioavailability was found to be very low in rat (Feng *et*

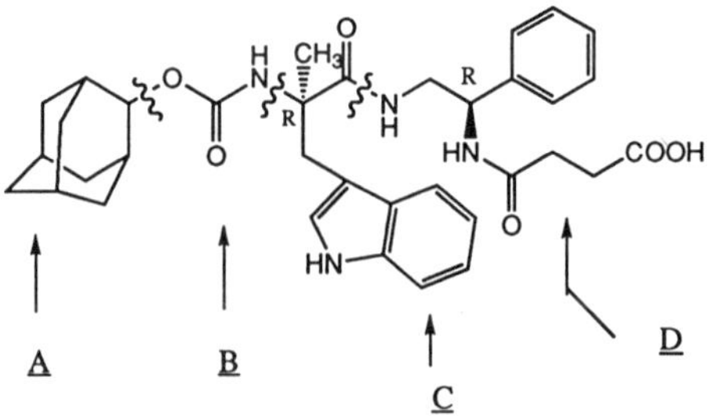

Figure 1. Structure of CI-988.

al., 1993), monkey (Hinton *et al.*, 1991, 1993) and human (Bradwejn *et al.*, 1995). The low bioavailability in preclinical studies was attributed to inefficient absorption as well as high biliary excretion in part related to the high molecular weight (MW = 614) of the compound.

It is of interest to note that similar issues faced the development of L-365,260. For this compound the oral bioavailability in nonrodent and rodent species ranged from 2 to 14% (Chen *et al.*, 1992). The lack of oral bioavailability was attributed in part to the poor aqueous solubility (<0.002 mg/ml) of the compound. Thus, over the past few years, Merck scientists have made attempts to improve solubility and absorption of the compound by incorporation of polar ionic functionalities into the molecule to enhance aqueous solubility (Bock *et al.*, 1994; Showell *et al.*, 1994). The results of this endeavor are summarized in Table I. Most of the ligands that bind to CCK receptors are hydrophobic and/or lipophilic in nature, which reduces the aqueous solubility of the ligands, and in turn renders these molecules less bioavailable. Furthermore, to achieve sufficient brain concentrations of these ligands essential for the pharmacological actions, a certain amount of lipophilicity is mandatory. Thus, it appears that for this class of compounds, a delicate balance of aqueous solubility and lipophilicity is required. This clearly is a challenging task as evident by recent observations of Bock *et al.* (1994). They were able to improve the aqueous solubility profile for their series of benzodiazepine CCK-B antagonists, although the ability to cross the blood–brain barrier was unaltered (Table I). Thus, for L-369,466 and L-368,935, the aqueous solubility was enhanced 200- to 700-fold. However, when assessed for their ability to cross the blood–brain barrier in an *ex vivo* binding experiment, these analogues showed no improvement over L-365,260. With CI-988, we faced essentially similar issues regarding to the low bioavailability and brain penetration. Thus, it was essential for us to identify an analogue with an overall improved profile.

1.2. CI-988 Pharmacokinetic Retrospective

Before discussing the strategy and design of CI-1015, it is important to review the pharmacokinetic development issues of CI-988. From this historical perspective, the pharmacokinetic challenges encountered in the CCK-B program will be highlighted.

CI-988's novel pharmacology, its extraordinary potency in preclinical models of anxiolytic potential, and the intense competitive interest by other pharmaceutical companies in the CCK arena resulted in an accelerated development program for CI-988. At the time of lead compound declaration (1989), the biodisposition of CI-988 had not been characterized. The evolution in the pharmaceutical industry toward greater and earlier involvement of pharmacokinetics in the drug discovery process has paralleled that of this CCK program. As such, we are

Table I
Benzodiazepine-Based CCK-B Antagonists

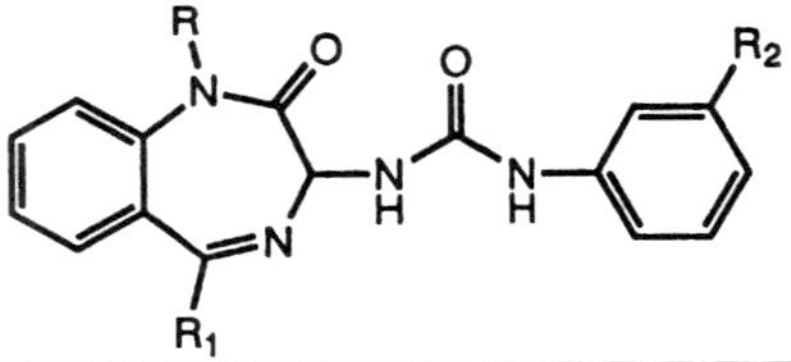

Compound	R	R_1	R_2	IC_{50} (nM) CCK-B	IC_{50} (nM) CCK-A	Solubility (mg/ml)	*Ex vivo* binding ED_{50} (mg/kg i.v.)
L-365,260[a]	CH_3	Ph	Ch_3	8.5	736	<0.002 [7.4]	13
L-368,935	i-Bu	Ph	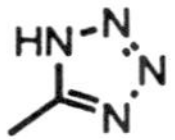	0.14	1434	1.4 [7.0] >11 [8.0]	5.6
L-369,466	CH_3	Ph	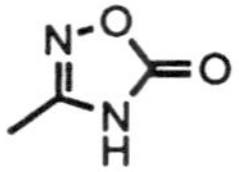	0.26	983	0.41 [7.4] 1.66 [8.0]	6.5
L-708,474	CH_3	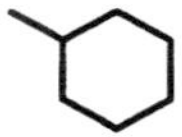	CH_3	0.28	6500	<10 ng/ml	
L-736,309[b]	CH_3	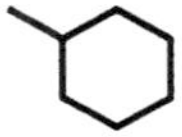	$CONHSO_2$-*o*-tolyl	0.27	5900	0.41	
L-740,093 HCl salt	CH_3	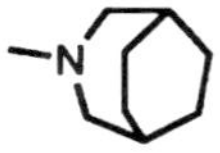	CH_3	0.1	1604	0.15	

[a]Bioavailability: 14% (rat), 9% (dog), 2% (monkey).
[b]Bioavailability: 14% (rat)

able to offer one example of how pharmacokinetic studies were useful in selecting a backup candidate to CI-988 and demonstrate the importance of integrating nonclinical biodisposition studies earlier into the discovery process.

1.2.1. CI-988 PRECLINICAL PHARMACOKINETIC PROFILE

The preclinical pharmacokinetics of CI-988 was studied extensively in both rat and monkey during its development (Table II). The initial pharmacokinetic studies examined the disposition of CI-988 in rat. Absolute oral bioavailability at 20 mg/kg p.o. was measured in fasted Wistar rats and found to be $\leq$3% and independent of delivery vehicle (Feng *et al.*, 1993; Trivedi *et al.*, 1998). This low value was initially cause for more curiosity than concern because the compound had exhibited superior *in vivo* anxiolytic-like activity. Because of the excellent aqueous solubility (>2 mg/ml), stability, and optimal octanol/water partition coefficient (Hansch *et al.*, 1987) at near-neutral pH ($\log P = 2$), limited gastrointestinal permeability and absorption of CI-988 were not immediate causes for concern. Hence, the low bioavailability seemed perhaps more related to the high systemic plasma clearance (34 ml/min per kg) observed, which suggested the potential for significant first-pass metabolism. One premise was that perhaps a metabolite of CI-988 was the active agent and not CI-988 itself.

Bioavailability studies proceeded into cynomolgus monkeys under both fasted and fed conditions. Although a sensitive and selective HPLC-fluorescence method for quantitation of CI-988 had been developed (Hinton *et al.*, 1995a), oral doses of 50 mg/kg, which were considerably higher than those projected to be therapeutic, were studied in order to ensure adequate pharmacokinetic characterization. Again the absolute oral bioavailability was quite low (ca. 2%) and was decreased further under fed conditions (ca. 1%) (Table II; Hinton *et al.*, 1991). Systemic plasma clearance was again high at 38 ml/min per kg, nearly approaching liver blood flow. Potential for first-pass metabolism was suspected. The high

Table II

Mean (% RSD)[a] Pharmacokinetic Parameters of CI-988 for Fasted Rats and Monkeys

Species	Formulation	Physical form	Dose (mg/kg p.o.)	C_{max} (ng/ml)	t_{max} (hr)	$t_{1/2}$ (hr)	%F^b
Rat	Saline	Solution	20	32 (19)	2.3 (140)	ND	3.2 (44)
	HPβCD[c]	Solution	20	5 (51)	0.25 (0)	ND	<1
Monkey	Saline	Solution	50	44 (80)	6.6 (62)	4.2 (100)	1.8 (61)
	Saline (Fed)	Solution	50	33 (118)	8.5 (167)	ND	0.9 (48)

[a]% RSD, percent relative standard deviation; C_{max}, maximum plasma concentration; t_{max}, time to C_{max}; $t_{1/2}$, terminal elimination half-life; %*F*, absolute oral bioavailability.

[b]Intravenous doses are provided in Table VI.

[c]1:2 molar ratio of CI-988 to HPβCD in water.

molecular weight of CI-988 at 614 did not go unnoticed; nor did the work by Doyle and others (Doyle *et al.,* 1984; Gores *et al.,* 1986a,b; Hunter *et al.,* 1990) that showed the extensive hepatic extraction and metabolism of CCK fragments. Thus, it became critical to investigate whether metabolism was the primary factor responsible for the low oral bioavailability.

With the availability of [^{14}C]-CI-988, we designed a monkey mass balance study. One advantage in using [^{14}C]-CI-988 was it permitted study of the p.o. disposition of CI-988 at a lower dose than our earlier monkey bioavailability study. The study was a three-way crossover conducted in bile duct-cannulated monkeys with a 2-week washout between treatments (Hinton *et al.,* 1993). The doses were 0.5 mg/kg i.v., 0.5 mg/kg p.o., and 10 mg/kg i.v. Bile, urine, and feces were collected out to 144 hr. After the i.v. dose, the majority of the radioactivity (>80%) was recovered within 2 hr postdose and nearly half of the radioactivity was recovered as unchanged CI-988. These results indicated efficient hepatic extraction as the main component of the high clearance. The principal monkey metabolite, which was extracted from bile, was identified as a hydroxylated adamantyl derivative (unpublished data). It had micromolar binding affinity to CCK-B receptor, and thus was considered an unlikely "active" moiety. After the 0.5 mg/kg p.o. dose, the majority of radioactivity (76%) was recovered in feces as unchanged CI-988. This result indicated that CI-988 was inefficiently absorbed regardless of its good aqueous solubility. Indeed, the low bioavailability of CI-988 in monkey was attributed at first to poor gastrointestinal absorption (ca. 25%) and then to efficient hepatic extraction ($E_H \sim 0.9$).

Whole-body autoradiography in rat was also studied with [^{14}C]-CI-988 after i.v. dosing. This study dramatically revealed the inefficient brain penetration of CI-988. Other studies (Dubroeucq *et al.,* 1994; Patel *et al.,* 1994) and some of our own *ex vivo* brain binding data corroborated these findings. Poor brain penetration for a potential anxiolytic seemed like a contradiction. This paradox was also apparent to the Merck scientists working on the benzodiazepine-based CCK-B receptor antagonists (Bock *et al.,* 1994; Freedman *et al.,* 1994; Showell *et al.,* 1994). Achieving an optimal balance of molecular weight, aqueous solubility, and lipophilicity was recognized as key to developing a compound with acceptable oral bioavailability and essential brain penetration.

1.2.2. CI-988 PRECLINICAL TOXICOLOGY AND BIODISPOSITION IN HUMAN

Phase I safety studies proceeded into human with little difficulty as the overall toxicity of CI-988 was rather limited. Mild to moderate gastric mucosal degeneration was observed in cynomolgus monkeys after p.o. dosing at ≥25 mg/kg CI-988. Toxicokinetic evaluations demonstrated that these gastrointestinal changes were not related to systemic plasma drug concentrations but rather were a result of local and possible pharmacologically related effects (Dethloff and Hinton,

1997). Therefore, safety margins based on dose (initially 100 mg/day and ultimately 100 mg t.i.d.) and not plasma concentrations were used to assess CI-988's potential gastrointestinal risks in humans. For CI-988 the occurrence of gastrointestinal changes in monkey and stimulation of gastric acid secretion in human were causes for concern, as these characteristics were deemed "nonideal" for a new anxiolytic.

As CI-988 progressed into the clinical program, we learned more about its pharmacokinetic disposition. Food dramatically reduced both the rate and overall extent of absorption of CI-988 (Hinton *et al.*, 1995b) as had been observed in monkeys. Relative bioavailability of CI-988 taken with food was only 30% of that for the fasted state (Fig. 2). Consequently, all subsequent clinical trials were conducted such that CI-988 was given on an empty stomach (no food or milk within ± 2 hr of dose) to ensure maximum systemic exposure. Maalox was also shown to markedly reduce systemic exposure of CI-988 (Cook *et al.*, 1995). Furthermore, a pilot i.v. safety study of CI-988 in humans allowed estimation of the absolute oral biovailability of CI-988 (unpublished data). Under fasted conditions, bioavailability of CI-988 was estimated to be less than 1%. The relatively modest activity of CI-988 in blocking CCK-4-induced panic symptoms in healthy volunteers (Bradwejn *et al.*, 1995) was attributed in part to its poor and variable systemic availability. Clinical development of CI-988 proceeded through a 6-week trial in generalized anxiety disorder. Its further development was terminated because of the absence of clinical efficacy, which was primarily attributed to poor pharmacokinetics.

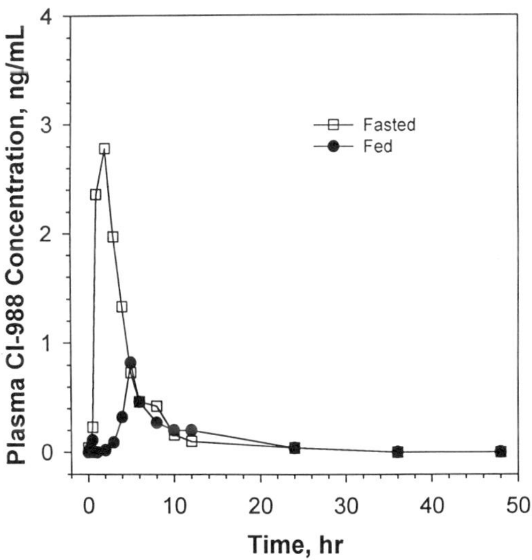

Figure 2. Mean plasma CI-988 concentration–time profiles following single oral dose of 100 mg CI-988 to healthy subjects (n = 12).

CI-988 is not alone in being discontinued because of its unacceptable pharmacokinetic profile. Nearly 30 to 40% of compounds withdrawn during clinical development are terminated for pharmacokinetic reasons (Prentis *et al.,* 1988; DiMassi, 1995). The challenge for the drug discovery scientists is to learn from these endeavors, improve the processes, and define clear objectives for identifying and developing a backup compound worthy of clinical development.

1.3. Objectives of the Discovery Team

The goal of the team was to discover an improved analogue of CI-988 with the following criteria: (1) it must have a molecular weight less than 500, (2) the overall pharmacological profile (*in vitro* binding and *in vivo* activity) should be equal to or better than that of CI-988, (3) the compound should have a reduced tendency to stimulate gastric acid secretion, (4) it must have improved oral bioavailability, and (5) the compound should have improved brain penetration.

2. DISCOVERY OF CI-1015

2.1. Design Strategy

With the above-mentioned objectives in mind, we chose to develop structure–activity relationships (SAR) with the intent to lower the molecular weight. We dissected CI-988 (**1**) in four different parts (Fig. 1). We analyzed each part and concluded that fragments A and B, required for high affinity, contributed minimally toward the molecular weight. Fragment C, although it contributed significantly toward the molecular weight (30%), was essential for the binding affinity. Thus, we were left with fragment D, which represented 38% of the molecular weight. Additionally, based on earlier SAR it was determined that this portion of the molecule was amenable to further manipulation without any significant impact on the binding affinity. Thus, we decided to develop an SAR study by inserting relatively smaller amines in this part of the molecule.

2.2. Structure–Activity Relationship Study

2.2.1. CYCLOALKYL AMINE AND HYDRAZINE DERIVATIVES

It was previously determined that a simple phenethyl moiety at the C-terminus (Table III) provided compound **2** with modest binding affinity. Interestingly, insertion of an acetic acid function onto the phenethyl amine side chain provided

Table III
Receptor Binding Affinities of Selected C-Terminus Amides

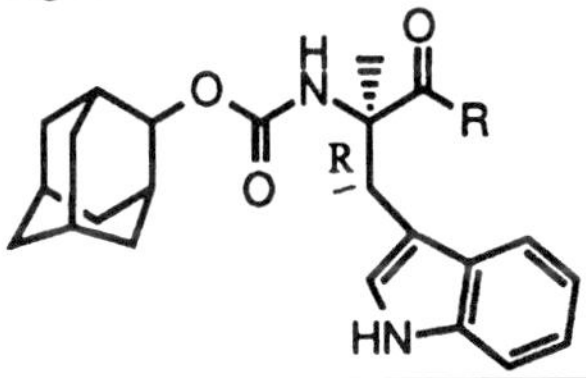

Example	R	IC_{50} (μM) CCK-B	IC_{50} (μM) CCK-A	Ratio
1 (CI-988)	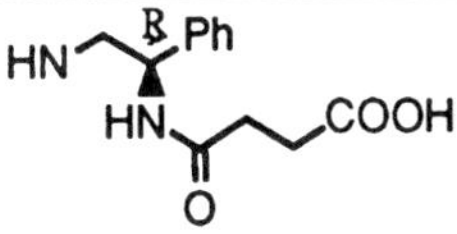	1.7	4300	2529
2		32	650	20
3	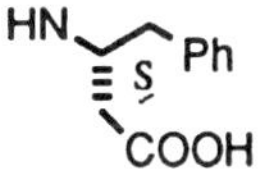	0.15	25.5	170
4	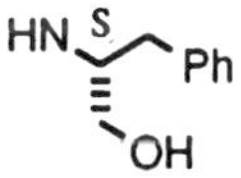	6.3	780	123
5	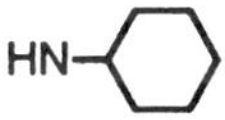	50	2400	48
6	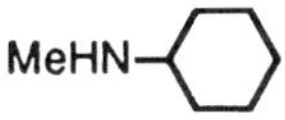	314	3334	10.6
7	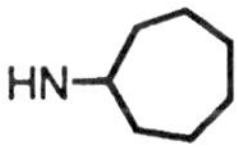	45.5	3234	71
8	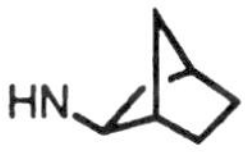	78.5	1464	18.6
9		166	1249	7.5

compound **3,** which is the most potent ligand for the CCK-B receptor for this class of compounds. However, the selectivity for the CCK-B receptor was only modest. Replacement of the acetic acid side chain with a hydroxy methyl function provided compound **4** with binding affinity of 6.3 nM for the CCK-B receptor. For both compounds **3** and **4,** the corresponding *R*-isomer showed significantly less binding affinity for the CCK-B receptor. We then chose to evaluate a series of compounds in which we incorporated various cycloalkyl functionalities. Insertion of cyclohexyl amine at the C-terminus provided compound **5** with modest binding affinity for the CCK-B receptor. *N*-Alkylation of **5** gave compound **6** with a six-fold loss in the binding affinity suggesting that the free NH was essential for high affinity at the CCK-B receptor. Increase in the size of the cycloalkyl moiety (**7**) did not affect the binding affinity. The corresponding bicyclic *exo*-(**8**) and *endo*-(**9**) norbornyl analogues showed stereoselective interactions at the CCK-B receptor, although without any improvement in the binding affinity.

We then prepared a series of hydrazide derivatives from readily available cycloalkyl hydrazines (Table IV). For this series of analogues, increase in the size and the lipophilicity showed incremental improvement in the receptor binding affinity and selectivity. Thus, the homopiperidine analogue (**12**) showed a better than 10-fold increase in binding affinity relative to the pyrrolidine analogue (**10**). The corresponding bicyclic analogue (**13**) showed further enhancement in binding affinity with an IC_{50} value of 6.5 nM for the CCK-B receptor. Interestingly, incorporation of a methoxymethyl functionality onto 1-amino pyrrolidine provided the corresponding *S*- (**14**) and the *R*- (**15**) methoxymethyl pyrrolidine derivatives. As anticipated, these analogues rendered stereospecific interactions at both receptors. The *S*-isomer (**12**) improved the binding affinity 2-fold over the unsubstituted pyrrolidine derivative (**8**) with a binding affinity of 65 nM. However, the corresponding *R*-isomer (**15**) was greater than 50-fold more potent with a binding affinity of 2.5 nM at the CCK-B receptor. Furthermore, this manipulation provided detrimental interactions at the CCK-A receptor and, thus, for the first time greater selectivity (>1370-fold) for the CCK-B receptor was achieved.

2.2.2. INCORPORATION OF POLAR FUNCTIONAL GROUPS

We continued our SAR study by synthesizing additional analogues in which we incorporated hydrophilicity by adding polar functionality such as an alcohol into the molecules (Table V). Insertion of a hydroxymethyl group onto the cyclopentyl and cyclohexyl amines provided compounds (**16** and **17**) with reduced binding affinity. The corresponding carboxylic acid analogue (**18**) maintained the binding affinity of compound **5,** the unsubstituted cyclohexyl analogue. Interestingly, vicinal substitution on the cycloalkyl amine provided analogues with enhanced binding affinity and selectivity for the CCK-B receptor. Thus, compound **19** showed binding affinity of 39 nM for the CCK-B receptor. However, replace-

Table IV
Receptor Binding Affinities of Hydrazide Analogues

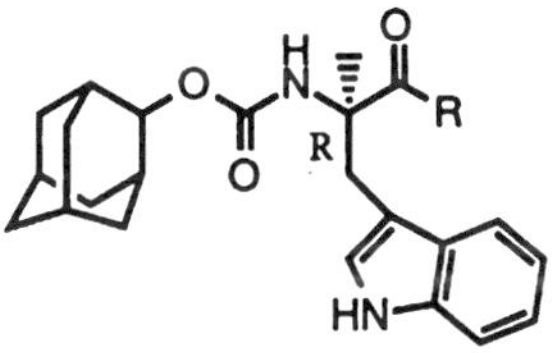

Example	R	IC_{50} (μM) CCK-B	CCK-A	Ratio
10	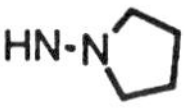 HN-N (pyrrolidine)	132	10,804	82
11	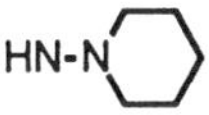 HN-N (piperidine)	19.7	3,829	194
12	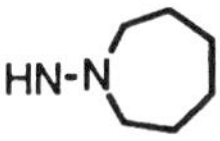 HN-N (azepane)	12.2	3,040	248
13	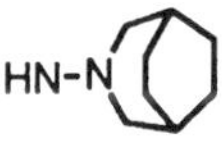 HN-N (bicyclic)	6.5	1,560	238
14	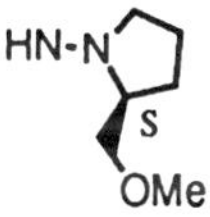 HN-N (S, OMe)	65.4	6,899	105
15	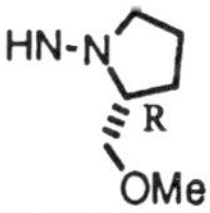 HN-N (R, OMe)	2.5	3,430	1372

ment of the methyl group with a cyano (**20**) or a carboxyl function (**21**) provided compounds with 10- and 100-fold increases in binding affinity. Thus, compound **21,** a mixture of diastereomers, showed an excellent binding affinity of 0.99 nM for the CCK-B receptor and was 700-fold selective.

Further exploration of the SAR revealed that incorporation of a hydroxyl moiety on the vicinal carbon provided compound **22** with 6.2 nM binding affinity for the CCK-B receptor. Encouraged by this result, we separated the individual diastereomers on HPLC. Compound **23** showed binding affinity of 3.0 and 2900 nM for the CCK-B and CCK-A receptors, respectively. The other diastereomer (**24**)

Table V
Receptor Binding Affinities of Selected Cycloalkyl Amides

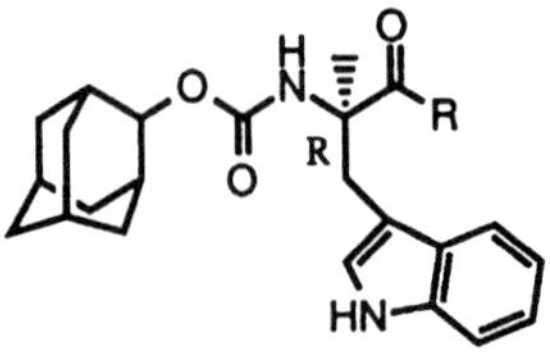

Example	R	IC_{50} (μM)		Ratio
		CCK-B	CCK-A	
16	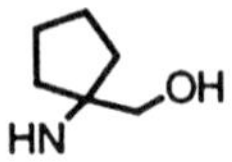	268	1419	5.3
17	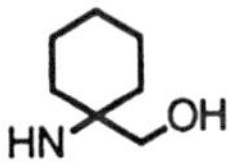	121	1509	12.5
18	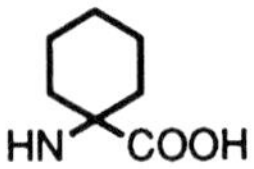	83	1525	18.4
19	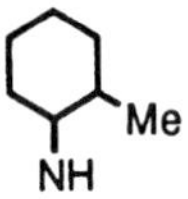	39	2784	71.4
20	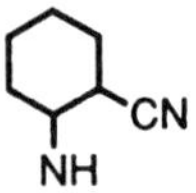	3.5	1500	429
21	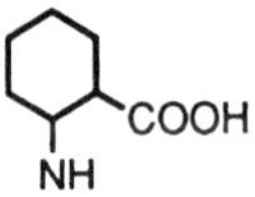	0.99	701	708
22	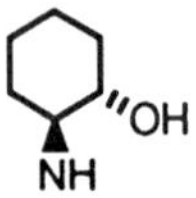	6.2	1380	223
23 (CI-1015)	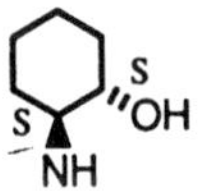	3.0	2900	967
24	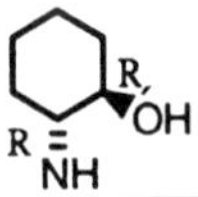	6.2	1380	223

was less active and less selective. The analogues were resynthesized using chiral amino alcohols as previously reported (Overman and Sugai, 1985; Aubé *et al.*, 1992). Accordingly, we prepared the chiral amino alcohols, and assigned the absolute stereochemistry based on the comparative physicochemical data (i.e., melting point, rotation). In order to reconfirm the absolute stereochemical assignment, we obtained an X-ray crystal structure for compound **23** (Fig. 3), which confirmed the relative (*trans*) and absolute stereochemistry at both of the chiral centers being *S, S*. Thus, we assigned the *R, R* stereochemistry for the other isomer (**24**). From this *in vitro* SAR study, a few compounds were further evaluated based on their affinity and selectivity.

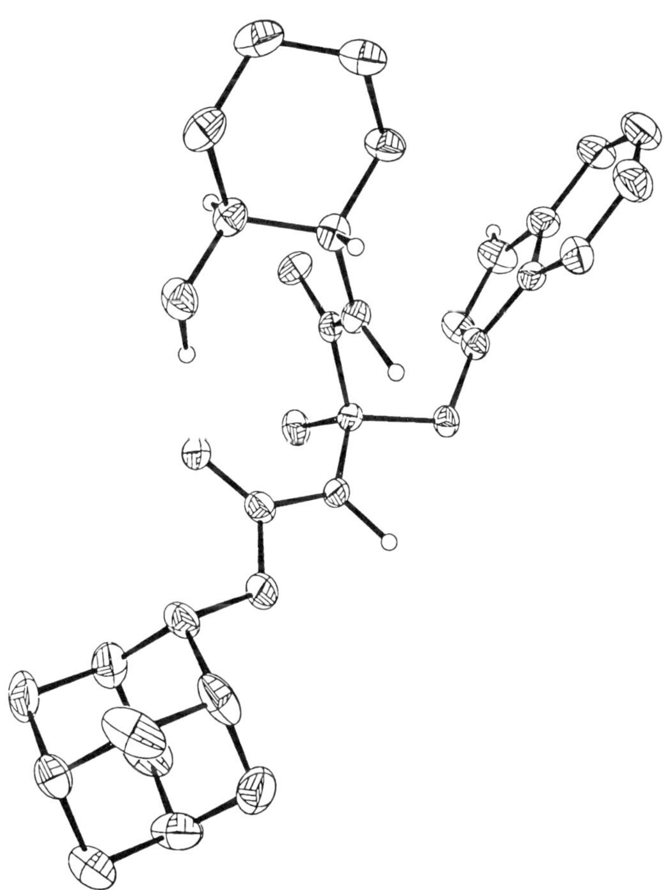

Figure 3. X-ray crystal structure of CI-1015.

3. PRECLINICAL CHARACTERIZATION OF BACKUP CANDIDATES

3.1. *In Vitro* and *in Vivo* Comparison

We chose to further evaluate compounds **4** and **23** in the secondary *in vitro* and *in vivo* assays. As shown in Table VI, the molecular weight of both compounds is around 500, and the log *P* values are significantly higher than that for CI-988. Although the solubility profile was not ideal, the increase in lipophilicity was considered a benefit for a CNS-active agent. It should be noted from our earlier observations on this class of compounds that analogues having a free carboxyl moiety were poorly absorbed. Thus, we chose not to pursue similar analogues (i.e., **18**) despite their high potency. Compound **15,** although potent and selective, showed relatively greater propensity toward the acid secretion in the mouse isolated stomach preparation and thus was not considered for further evaluation. Compounds **4** and **23** showed an antagonist profile in the ventromedial hypothalamus assay and in the Ghosh and Schild assay (Table VII). In these experiments, the Ke and ED_{50} values obtained compared well with those for CI-988. These compounds provided a reasonable profile both *in vitro* and *in vivo*. CI-988 had been shown to have an anxiolytic-like profile (Singh *et al.,* 1991) in the rat X-maze with a minimum effective dose (MED) of 10 μg/kg orally. Compound **23,** interestingly, showed a MED of 1 μg/kg, although the dose response was relatively flat. Compound **4** also showed increased potency with a MED of 0.1 μg/kg. The increase in the percent time and entries onto the end sections of the X-maze suggests that these compounds possess anxiolytic-like activity. Similarly, in the light–dark test in mice in which standard anxiolytics have been shown to be active, these compounds also showed an increase in time spent by mice in the illuminated side with a MED of 1–10 μg/kg (Trivedi *et al.,* 1998).

3.2. Pharmacokinetic Evaluations in Rat

The pharmacokinetics of these compounds was studied in rat (Table VIII). Because of the poor aqueous solubility (Table VI), different vehicles were employed. A marked improvement in the oral bioavailability (%*F*) of **23** relative to that for CI-988 was obtained with all four p.o. formulations studied. The best bioavailability (28%) was achieved with **23** dosed as a solution in hydroxypropyl-β-cyclodextrin [50% (w/v) HPβCD in saline : 0.1 N HCl in ethanol (7:3, v/v)]. This formulation offered a nearly 10-fold improvement in bioavailability over CI-988. The other formulations had lower yet similar bioavailabilities of about 10%, which represented about a 2- to 3-fold improvement relative to CI-988. Oral bioavailability of **4** was also better than that for CI-988 for three of the four formulations tested.

Table VI

Summary of Physicochemical Parameters and Systemic Plasma Clearance Values of Selected CCK-B Antagonists

			Solubility (mg/ml)			Clearance (ml/min/kg)	
Compound	MW	Log P (pH 7)	0.1 N HCl	pH 7.4 buffer	PEG 400	Rat	Monkey
CI-988	614	2.0	<0.001	>2	>200	34 (24)[b]	38 (18)[c]
23	494	4.3	<0.001	0.002	33	46 (24)[d]	39 (30)[c]
4	530	5.1[a]	0.02	0.01	>200	37 (18)[d]	ND

[a]clog P.
[b]i.v. dose = 40 mg/kg.
[c]i.v. dose = 10 mg/kg.
[d]i.v. dose = 20 mg/kg.

Table VII

In Vitro and *in Vivo* Profile of Selected CCK-B Receptor Antagonists

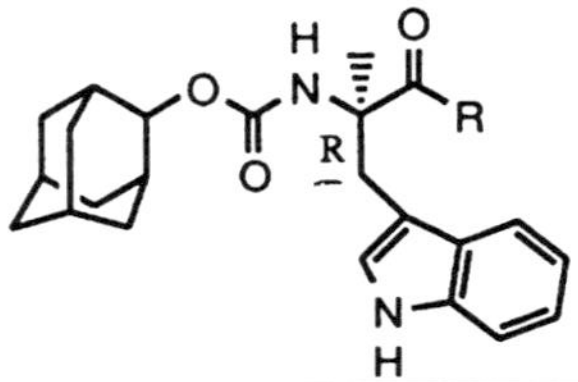

Compound	R	Binding IC_{50} (nM)		VMH[a] Ke (nM)	G & S[b] pA_2	Rat X-maze MED (μg/kg p.o.)	Mouse BWB[c] MED (μg/kg p.o.)
		CCK-B	CCK-A	CCK-B	CCK-B		
[CI-988]	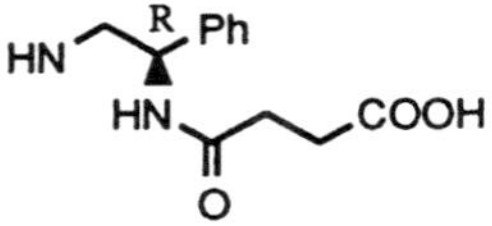 HN, R, Ph, HN, COOH, O	1.7	4300	2.1	9.4	10	10
4	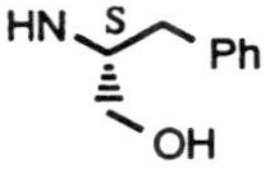 HN, S, Ph, OH	6.3	780	1.8	8.4	0.1	1
23	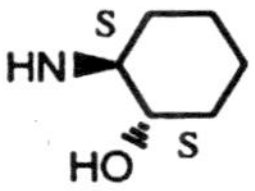 S, HN, HO, S	3.0	2900	34	7.8	1	10

[a]VMH, ventromedial hypothalamus assay.
[b]Ghosh and Schild test.
[c]BWB, black and white box assay.

Table VIII

Summary of Mean (% RSD)[a] Pharmacokinetic Parameters for Fasted Wistar Rats Following a Single Oral Dose of 20 mg/kg of CCK-B Receptor Antagonists

Compound	Formulation	Physical form	C_{max} (ng/ml)	t_{max} (hr)	$t_{1/2}$ (hr)	%F[h]
CI-988	Saline	Solution	32 (19)	2.3 (140)	ND	3.2 (44)
	HPβCD[b]	Solution	5 (51)	0.25 (0)	ND	<1
23	0.5% MC[c]	Suspension	185 (56)	3.0 (38)	1.0 (10)	8.9 (47)
	10% PEG[d]	Suspension	203 (35)	3.5 (75)	1.1 (41)	13 (35)
	100% PEG[e]	Solution	219 (14)	1.0 (71)	1.3 (22)	9.6 (25)
	HPβCD[f]	Solution	756 (35)	1.8 (89)	0.86 (24)	28 (12)
4	0.5% MC[c]	Suspension	42 (20)	2.9 (45)	ND	2.0 (35)
	DMA:H_2O[g]	Solution	212 (45)	1.2 (42)	2.3 (39)	9.6 (30)
	100% PEG[e]	Solution	167 (54)	2.8 (54)	1.1 (18)	9.6 (66)
	HPβCD[f]	Solution	416 (31)	0.9 (88)	1.5 (13)	16.0 (30)

[a]% RSD, percent relative standard deviation; C_{max}, maximum plasma concentration; t_{max}, time to C_{max}; $t_{1/2}$, terminal elimination half-life; %F, absolute oral bioavailability.
[b]1:2 molar ratio of CI-988 to HPβCD in water.
[c]0.5% methylcellulose in water.
[d]10% PEG400 in water.
[e]100% PEG400.
[f]50% (w/v) HPβCD in saline:ethanol (7:3, v/v).
[g]dimethylacetamide:water (3:7, v/v).
[h]Intravenous doses are given in Table VI.

For both of these compounds, solution formulations in general achieved better systemic availability than did suspension formulations. Solubility-limited absorption of these neutral compounds was anticipated based on their poor aqueous solubility. Nevertheless, the results suggested that with the right formulation, overall systemic exposure of these second-generation "dipeptoid" analogues could be markedly improved relative to CI-988.

3.3. Brain Penetration Studies

Two different methods were used to evaluate brain penetration of these analogues relative to CI-988. The first method compared the extent of brain penetration after an i.v. dose in mice by an *ex vivo* binding technique described previously (Trivedi *et al.,* 1998). The results are summarized in Table IX. For compounds that cross the blood–brain barrier by passive diffusion, their brain penetration and lipid solubility are well correlated. Thus, an increase in lipophilicity may improve brain uptake of drugs (Begley, 1996). Although the exact mechanism of blood-to-brain passage of these compounds is unknown, brain uptake for **23** appears dramatically improved relative to CI-988 and **4**. In mice, *ex vivo* binding data suggested nearly a 200-fold improvement for **23,** although this value may be slightly higher since as blood in brain capillary space was not removed by transcardiac perfusion.

The second method determined the brain:plasma ratios in rats after i.v. and p.o. administration. Blood samples were taken at various times postdose by cardiac puncture, and whole brains were harvested after a transcardiac perfusion with saline. Plasma and whole brain homogenate samples were assayed for drug using validated liquid chromatographic methods with fluorescence detection (Hinton *et al.,* 1996). In rats, 50-fold enhancement was seen after i.v. administration (Fig. 4). After PO administration, the brain:plasma ratio for **23** was lower than after IV administration (0.10) but remained constant (20% RSD) for 6 hr postdose. Similar

Table IX
Brain Levels of Selected Antagonists by *ex Vivo* Binding

Compound	Dose (i.v.)	Brain level (pmole/forebrain) Time after injection			Mice (*n*)
		5 min	10 min	20 min	
CI-988	10	10 ± 5	—[a]	8 ± 4	3–6
23	1	177 ± 46	154 ± 39	166 ± 35	9
4	1	40 ± 13	68 ± 19	43 ± 17	11–13

[a]Below level of detection (approx. 1 pmole/forebrain).

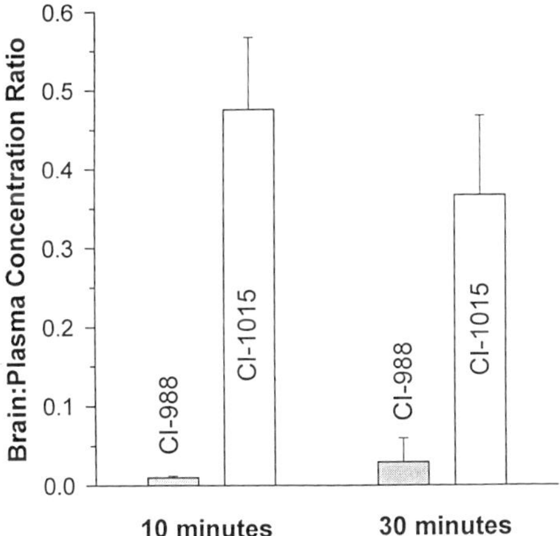

Figure 4. Comparison of brain:plasma concentration ratios of CI-988 and CI-1015 at 10 and 30 min postdose (5 mg/kg) after i.v. administration in 6:1:3 propylene glycol:ethanol:5% dextrose water. Transcardiac perfusion step was incorporated prior to harvesting brain.

ratios for CI-988 after p.o. dose could not be determined by this method because brain concentrations of CI-988 were below the limit of quantitation by our assays. Overall, compound **23** demonstrated improved blood–brain barrier penetration relative to CI-988.

3.4. Evaluation of Potential for Gastric Acid Secretion

The effect of compound **23** and other analogues on gastric acid secretion was examined in both *in vitro* and *in vivo* assays with the intent to identify compounds with less propensity for gastric acid secretion than CI-988. The mouse isolated lumen-perfused stomach model proved insufficient in distinguishing these compounds for stimulatory effects. Differences, however, were evident between **23** and CI-988 in a conscious rat acid secretion model. When dosed at 0.03 to 30 mg/kg (s.c.), compound **23** had no significant effect on the acid content of the stomach. However, CI-988 produced a significant increase in the acid content of the stomach 30 min postdose (10 mg/kg s.c.).

A similar study was also conducted in monkey. Under the conditions of the study, gastric secretory responses were quite variable, and there were no statistically significant differences. Nevertheless, rank order of acid output suggested that **23** showed less propensity for stimulation of acid secretion compared with CI-988.

3.5. Pharmacokinetic Evaluation in Monkey

The pharmacokinetics of compound **23** in monkey was investigated as the drug delivery group continued their efforts to develop a suitable solid formulation to obviate the solubility-limited absorption. Several simple suspension and solution formulations similar to those used in the rat studies were evaluated in monkey in a crossover design (Table X) (Wang *et al.,* 1995). As was observed in rats, plasma concentrations were distinctly higher for the solution formulations than for the suspension formulation; again consistent with solubility-limited absorption. The best bioavailability found in monkeys was 9.2%, which represented a nearly fivefold improvement over that of CI-988. Plasma concentration–time profiles for CI-988 and compound **23** in monkey are compared in Fig. 5.

Comparison of systemic plasma clearance values in rats (Table VI) reveals a rank order of **23** > **4** > CI-988. Observed perhaps as counterintuitive, absolute oral bioavailabilities also followed the same rank order. For monkeys, clearance values were similar for both compounds, yet oral bioavailability was distinctly higher for CI-1015. Although the absolute bioavailability for **23** was still low, significant improvements had been achieved over CI-988. Expectations are that formulation optimization will improve the fraction of drug absorbed across the gastrointestinal tract. However, it is unclear how significant an obstacle first-pass effects might be.

4. CONCLUSION

Our strategy of lowering the molecular weight to enhance absorption and bioavailability met with some success. Compound **23** has a molecular weight of 494, which represents a 20% decrease relative to CI-988. This is one of the few analogues of this class of compounds that lacks the free carboxylic function previously thought essential for high binding affinity to the CCK-B receptor. Compound **23,** identified by systematic SAR studies, showed excellent specificity both *in vitro* and *in vivo,* and afforded a better balance of lipophilicity and solubility than CI-988. Additional *in vivo* studies suggested that compound **23** has less propensity for inducing gastric acid secretion. Furthermore, marked improvements in both oral bioavailability and apparent brain penetration were achieved. Although ideally we would have preferred a compound with reduced liability for first pass effects (i.e., lower systemic clearance), it has provided us with an excellent tool to modulate CCK-mediated pharmacological actions in CNS. Based on this overall profile, compound **23** was selected as a development candidate (CI-1015) and is undergoing toxicological evaluation. Finally, we have demonstrated that incorporation of pharmacokinetic evaluation early in the discovery phase can have

Table X

Summary of Mean (% RSD)[a] Pharmacokinetic Parameters for Fasted Cynomolgus Monkeys Following a Single Oral Dose of CCK-B Antagonists

Compound	Formulation	Physical form	NC_{max} (ng/ml)	t_{max} (hr)	$t_{1/2}$ (hr)	%*F*[f]
CI-988	Saline	Solution	8.8 (80)	6.6 (62)	4.2 (100)	1.8 (61)
	Saline (fed)	Solution	3.3 (118)	8.5 (167)	ND	0.9 (48)
23	0.5% MC[b]	Suspension	19.9 (150)	4.4 (73)	9.3 (24)	4.9 (110)
	RTP[c]	Suspension	7.9 (140)	4.0 (130)	9.3 (47)	1.8 (100)
	100% PEG[d]	Solution	58.8 (72)	6.5 (58)	2.3 (32)	9.2 (37)
	HPβCD[e]	Solution	71.7 (71)	3.0 (27)	4.4 (110)	9.2 (59)

[a] % RSD, percent relative standard deviation; NC_{max}, maximum plasma concentration normalized to 10 mg/kg dose; t_{max}, time to C_{max}; $t_{1/2}$, terminal elimination half-life; %*F*, absolute oral bioavailability.

[b] 0.5% methylcellulose in water.

[c] Proprietary lecithin-coated suspension formulation prepared by Research Triangle Pharmaceuticals, Durham, NC.

[d] 100% PEG400.

[e] 50% (w/v) HPβCD in saline:ethanol (7:3, v/v).

[f] Intravenous doses are given in Table VI.

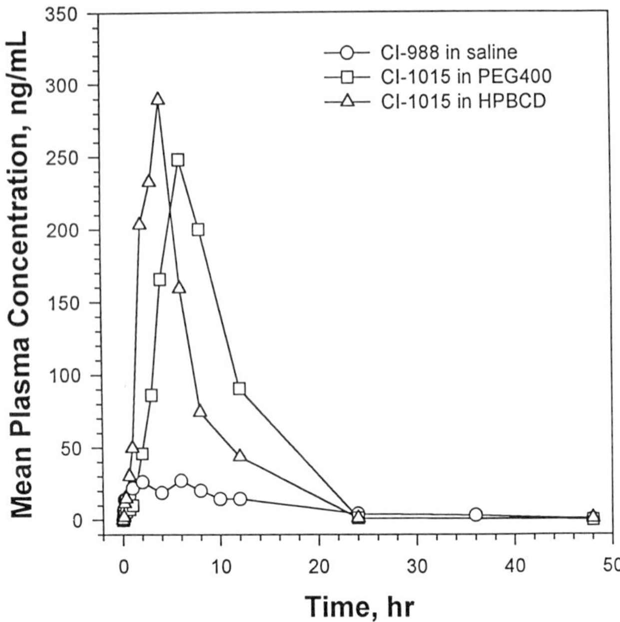

Figure 5. Comparison of dose-normalized (50 mg/kg) mean plasma concentration–time profiles following administration of CI-988 as solution in saline (○), CI-1015 as solution in PEG400 (□), and CI-1015 as solution in 50% (w/v) hydroxypropyl-β-cyclodextrin (HPβCD) in saline:ethanol (7:3, v/v) (△) to cynomolgus monkeys.

a significant impact on design and discovery of a viable clinical candidate. The future holds further promise for increased efficiency in the selection of promising clinical candidates as a result of exciting new drug discovery tools such as combinatorial chemistry and multiple compound cassette dosing *in vivo* for pharmacokinetic profiling using LC-MS/MS techniques.

Acknowledgments

The authors would like to thank the following individuals for their contribution toward the identification of CI-1015: Janak Padia, Ann Holmes, Martyn Pritchard, Clare Kneen, Jon Eden, Bruce Roth, David Horwell, Louise Wabdale, Nirmala Suman-Chauhan, Phil Boden, Lakhbir Singh, Geoffrey Woodruff, David Hill, John Hughes, Kathleen Jennings, Jim Atherton, Yow-Ming Wang, Gary Hudson, Steve Rose, D. Scott Wright, Al Kearney, Surendra Mehta, and Lloyd Dethloff.

REFERENCES

Aubé, J., Wolfe, M. S., Yantiss, R. K., Cook, S. M., and Takusagawa, F., 1992, Synthesis of enantiopure N-tert-butoxycarbonyl-2-aminocycloalkanones, *Synth. Commun.* **22**(20)**:**3003–3012.

Begley, D. J., 1996, The blood–brain barrier: Principles for targeting peptides and drugs to the central nervous system, *J. Pharm. Pharmacol.* **48:**136–146.

Beinfeld, M. C., 1983, Cholecystokinin in the central nervous system: A minireview, *Neuropeptides* **3:**411.

Bock, M. J., Mellin, E. C., Newton, R. C., Veber, D. F., Freedman, S. B., Smith, A. J., Patel, S., Kemp, J. A., Marshall, G. R., Fletcher, A. E., Chapman, K. L., Anderson, P. S., and Freidinger, R. M., 1994, Second-generation benzodiazepine CCK-B antagonists. Development of subnanomolar analogues with selectivity and water solubility, *J. Med. Chem.* **37:**722–724.

Boden, P. R., Higginbottom, M., Hill, D. R., Horwell, D. C., Hughes, J., Rees, D. C., Roberts, E., Singh, L., Suman-Chauhan, N., and Woodruff, G. N., 1993, Cholecystokinin dipeptoid antagonists: Design, synthesis and anxiolytic profile of some novel CCK-A and CCK-B selective and mixed antagonists, *J. Med. Chem.* **36:**552–565.

Bourin, M., Malinge, M., Vasar, E., and Bradwejn, J., 1996, Two faces of cholecystokinin: Anxiety and schizophrenia, *Fundam. Clin. Pharmacol.* **10:**116.

Bradwejn, J., and Koszycki, D., 1992, CCK receptors and panic attacks in man, in: *Multiple Cholecystokinin Receptors in the CNS* (C. T. Dourish, S. J. Cooper, S. D. Iversen, and L. I. Iverson, eds.), p. 121, Oxford University Press, London.

Bradwejn, J., Koszycki, D., and Meterissian, G., 1990, Cholecystokinin tetrapeptide in panic disorder, *Can. J. Psychiatry* **35:**83.

Bradwejn, J., Koszycki, D., and Shriqui, C., 1991, Enhanced sensitivity to cholecystokinin tetrapeptide in panic disorder, *Arch. Gen. Psychiatry* **48:**603.

Bradwejn, J., Koszycki, D., Couetoux, du T. A., van Megan, H., den Boer, J Westenbergh, and Annable, L., 1994, The panicogenic effects of cholecystokinin tetrapeptide are antagonized by L-365260, a central cholecystokinin receptor antagonist, in patients with panic disorder, *Arch. Gen. Psychiatry* **51:**486.

Bradwejn, J., Koszycki, D., Paradis, M., Reece, P., Hinton, J., and Sedman, A., 1995, Effect of CI-988 on cholecystokinin tetrapeptide-induced panic symptoms in healthy volunteers, *Biol. Psychiatry* **38:**742.

Chen, I.-W., Dorley, J. M., Ranjit, H. G., Pitzenberger, S. M., and Lin, J. H., 1992, Physiological disposition and metabolism of L-365,260, a potent antagonist of brain cholecystokinin receptor, in laboratory animals, *Drug Metab. Dispos.* **20:**390–395.

Cook, J., Siedlik, P., and Reece, P., 1995, Effect of Maalox TC on CI-988 pharmacokinetics, *Pharm. Res.* **12**(9)**:**S390.

Crawley, J. N., 1989, Micro-injection of cholecystokinin into the rat ventral tegmental area potentiates dopamine-induced hypolocomotion, *Synapse* **3:**34336.

D'Amato, M., Makovec, F., and Rovati, L., 1994, Potential clinical applications of CCK-A receptor antagonists in gastroenterology, *Drug News Perspect.* **7**(2)**:**87–95.

Dethloff, L. A., and Hinton, J. P. , 1997, Toxicokinetic comparison of a CCK-B / gastrin receptor antagonist given orally and intravenously, *Drug Dev. Res.* **40:**292–298.

De Weerth, A., Pisegna, J. R., Hupi, K., and Wank, S. A., 1993, Molecular cloning, functional expression and chromosomal localization of the human cholecystokinin type A receptor, *Biochem. Biophys. Res. Commun.* **194:**811.

DiMassi, J. A., 1995, Success rates for new drugs entering clinical testing in the United States, *Clin. Pharmacol. Ther.* **58:**1–14.

Dourish, C. T., and Hill, D. R., 1987, Classification and function of CCK receptors, *Trends Pharm. Sci.* **8:**207.

Doyle, J. W., Wolfe, M., and McGuigan, J. E., 1984, Hepatic clearance of gastrin and cholecystokinin peptides, *Gastroenterology* **87:**60–68.

Dubroeucq, M. C., Guyon, C., Manfre, F., Capet, M., Barreau, M., and Betrand, P., 1994, Evaluation of brain penetration of CCK-B antagonists, *Ann. N.Y. Acad. Sci.* **713:**377–379.

Feng, R., Hinton, J. P., Hoffman, K., Parker, T. D., and Wright, D. S., 1993, Pharmacokinetics and oral bioavailability of CI-988 ester prodrugs in Wistar rats, *Pharm. Res.* **10**(10)**:**S346.

Freedman, S. B., Patel, S., Smith, A. J., Chapman, K., Fletcher, A., Kemp, J. A., Marshall, G. R., Hargreaves, R. J., Scholey, K., Mellin, E. C., DiPardo, R. M., Bock, M. G., and Freidinger, R. M., 1994, A second generation of non-peptide cholecystokinin receptor antagonists and their therapeutic potential, *Ann. N.Y. Acad. Sci.* **713:**312–318.

Gores, G. J., LaRusso, N. F., and Miller, L. J., 1986a, Hepatic processing of cholecystokinin peptides. I. Structural specificity and mechanism of hepatic extraction, *Am. J. Physiol.* **250**(Gastrointest. Liver Physiol. 13)**:**G344–G349.

Gores, G. J., Miller, L. J., and LaRusso, N. F., 1986b, Hepatic processing of cholecystokinin peptides. II. Cellular metabolism, transport, and biliary excretion, *Am. J. Physiol.* **250**(Gastrointest. Liver Physiol. 13)**:**G350–G356.

Hansch, C., Bjorkroth, J. P., and Leo, A., 1987, Hydrophobicity and central nervous system agents: On the principle of minimal hydrophobicity in drug design, *J. Pharm. Sci.* **76:**663–687.

Hinton, J. P., Rutkowski, K., Johnson, E. L., and Wright, D. S., 1991, Single dose pharmacokinetics and absolute bioavailability of the anxiolytic CI-988 in fasted and fed cynomolgus monkeys, *Pharm. Res.* **8**(10)**:**S267.

Hinton, J., Hoffmann, G., Poisson, A., Klemisch, W., and Wright, D. S., 1993, Mass balance and disposition of [^{14}C]CI-988 in cynomolgus monkeys, *Pharm. Res.* **10**(10)**:**S330.

Hinton, J. P., Jennings, K., Johnson, E. L., and Wright, D. S., 1995a, A sensitive HPLC assay for the cholecystokinin-B antagonist, CI-988, in human and monkey plasma, *Biomed. Chrom.* **9:**94–97.

Hinton, J. P., Jennings, K., Wright, D. S., Reece, P. A., and Sedman, A. J., 1995b, A food-effect study of CI-988 capsules and solution in healthy volunteers, *Pharm. Res.* **12**(9)**:**S420.

Hinton, J. P., Pablo, J., Bjorge, S., Hoffman, K., Jennings, K., and Wright, D. S., 1996, Three complementary liquid chromatographic methods for determination of the peptoid cholecystokinin-B antagonist, CI-988, in rat plasma, *J. Pharm. Biomed. Anal.* **14:**815–824.

Holladay, M. W., Bennett, M. J., Tufano, M. D., Lin, C. W., Asin, K. E., Witte, D. G., Miller, T. R., Bianchi, B. R., Nikkel, A. L., Bednarz, L., and Nadzan, A. M., 1992, Synthesis and biological activity of CCK heptapeptide analogues. Effects of conformational constraints and standard modifications on receptor subtype selectivity, functional activity in vitro and appetite suppression in *vivo, J. Med. Chem.* **35:**2919.

Horwell, D. C., Hughes, J., Hunter, J. C., Pritchard, M. C., Richardson, R. S., Roberts, E., and Woodruff, G. N., 1991, Rationally designed "dipeptoid" analogueues of CCK. Methyltryptophan derivative as highly selective and orally active gastrin and CCK-B antagonists with potent anxiolytic properties, *J. Med. Chem.* **34:**404–414.

Hughes, J., Boden, P., Costall, B., Domeney, A., Kelly, E., Horwell, D. C., Hunter, J. C., Pinock, R. D., and Woodruff, G. N., 1990, Development of a class of selective cholecystokinin type B receptor antagonists having potent anxiolytic activity, *Proc. Natl. Acad. Sci. USA* **87:**6728.

Hunter, E. B., Powers, S. P., Kost, L. J., Pinon, D. I., Miller, L. J., and LaRusso, N. F., 1990, Physiochemical determinants in hepatic extraction of small peptides, *Hepatology* **12**(1)**:**76–82.

Innis, R. B., and Snyder, S. H., 1980, Distinct cholecystokinin receptors in brain and pancreas, *Proc. Natl. Acad. Sci. USA* **77:**6917.

Larsson, L.-I., and Rehfeld, J. F., 1979, Localization and molecular heterogeneity of cholecystokinin in the central and peripheral nervous system, *Brain Res.* **165:**201.

Lee, Y. M., Beinborn, M., McBride, E. W., Lu, M., Kolakowski, L. F., and Kopin, A. S., 1993, The human brain cholecystokinin-B / gastrin receptor, *J. Biol. Chem.* **268:**8164.

Overman, L. E., and Sugai, S., 1985, A convenient method for obtaining trans-2-amino cyclohexanol and trans-2-aminocyclopentanol in enantiomerically pure form, *J. Org. Chem.* **50:**4154–4155.

Patel, S., Chapman, K. L., Heald, A., Smith, A. J., and Freedman, S. B., 1994, Measurement of central nervous system activity of systemically administered CCK-B receptor antagonists by ex vivo binding, *Eur. J. Pharmacol.* **253:**237–244.

Prentis, R. A., Lis, Y., and Walker, S. R., 1988, Pharmaceutical innovation by the seven UK-owned pharmaceutical companies (1964–1985), *Br. J. Clin. Pharmacol.* **25:**387–396.

Rehfeld, J. F., and Nielsen, F. C., 1995, Molecular forms and regional distribution of cholecystokinin in the central nervous system, in: *Cholecystokinin and Anxiety: From Neuron to Behavior* (J. Bradwejn and E. Vasar, eds.), pp. 33–56, RG Landes Company, Austin, TX.

Rehfeld, J. F., Goltermann, N., Larsson, L.-I., Emson, P. M., and Lee, C. M., 1979, Gastrin and cholecystokinin in central and peripheral neurons, *Fed. Proc.* **38:**2325.

Showell, G. A., Bourrain, S., Neduvelil, J. G., Fletcher, S. R., Baker, R., Watt, A. P., Fletcher, A. E., Freedman, S. B., Kemp, J. A., Marshall, G. R., Patel, S., Smith, A. J., and Matassa, V. G., 1994, High affinity and potent, water soluble 5-amino-1,4-benzodiazepine CCK-B/gastrin receptor antagonists containing a cationic solubilizing group, *J. Med. Chem.* **37:**719–721.

Singh, L., Field, M. J., Hughes, J., Menzies, R., Oles, R. J., Vass, C. A., and Woodruff, G. N., 1991, The behavioral properties of CI-988, a selective cholecystokinin B receptor antagonist, *Br. J. Pharmacol.* **104**(1)**:**239–245.

Trivedi, B. K., 1994a, Ligands for cholecystokinin receptors: Recent developments, *Curr. Opin. Ther. Patents* **4**(1)**:**31–44.

Trivedi, B. K., 1994b, Cholecystokinin receptor antagonists: Current status, *Curr. Med. Chem.* **1:**313–327.

Trivedi, B. K., Padia, J. K., Holmes, A., Rose, S., Wright, D. S., Hinton, J. P., Pritchard, M. C., Eden, J. M., Kneen, C., Webdale, L., Suman-Chauhan, N., Boden, P., Singh, L., and Hill, D., 1998, Second generation "dipeptoid" CCK-B antagonists: Identification and development of CI-1015 with an improved pharmacokinetic profile, *J. Med. Chem.* **41**(1)**:**38–45.

Voigt, M., Wang, R. Y., and Westfall, T. C., 1986, Cholecystokinin octapeptides alter the release of endogenous dopamine neurons in vitro, *J. Pharmacol. Exp. Ther.* **237:**147.

Wang, Y.-M., Hinton, J. P., Atherton, J. P., and Wright, D. S., 1995, Pharmacokinetics and bioavailability of PD 145942 in cynomolgus monkeys, *Pharm. Res.* **12**(9)**:**S426.

Wank, S. A., Pisegna, J. R., and DeWeerth, A. 1992, Brain and gastrointestinal cholecystokinin receptor family and functional expression, *Proc. Natl. Acad. Sci. USA* **89:**8691.

Chapter 22

Orally Active Nonpeptide CCK-A Agonists

Elizabeth E. Sugg, Lawrence Birkemo, Liang-Shang L. Gan, and Timothy K. Tippin

1. INTRODUCTION

Cholecystokinin (CCK) is a gastrointestinal hormone and neurotransmitter involved in nutrient assimilation, including the secretion of bile and digestive enzymes and the regulation of enteric transit (Crawley and Corwin, 1994). Although a variety of endogenous molecular forms of CCK have been isolated, the C-terminal octapeptide [Asp-Tyr(SO_3H)-Met-Gly-Trp-Met-Asp-PheNH_2, CCK-8] appears to be the minimum sequence required for bioactivity. CCK-8 potently activates both peripheral (CCK-A) and central (CCK-B) receptor subtypes. The utility of a CCK receptor agonist for the treatment of obesity is suggested by studies demonstrating that exogenous CCK can shorten meal duration and reduce meal size in several species, including lean (Kissileff *et al.*, 1981) and obese (Pi-Sunyer *et al.*, 1982) humans. Chronic administration of CCK-8 to patients on total parenteral nutrition has also demonstrated a role for CCK in the prevention of gallstones (Sitzmann *et al.*, 1990). The relevant target for both effects is the CCK-A receptor (Dourish *et al.*, 1989).

A directed screen of compounds from company registry files for contractile activity on the isolated guinea pig gallbladder (GPGB) led to the identification of

---

Elizabeth E. Sugg, Lawrence Birkemo, Liang-Shang L. Gan, and Timothy K. Tippin • Glaxo Wellcome Research and Development, Research Triangle Park, North Carolina 27709.

Integration of Pharmaceutical Discovery and Development: Case Studies, edited by Borchardt *et al.*, Plenum Press, New York, 1998.

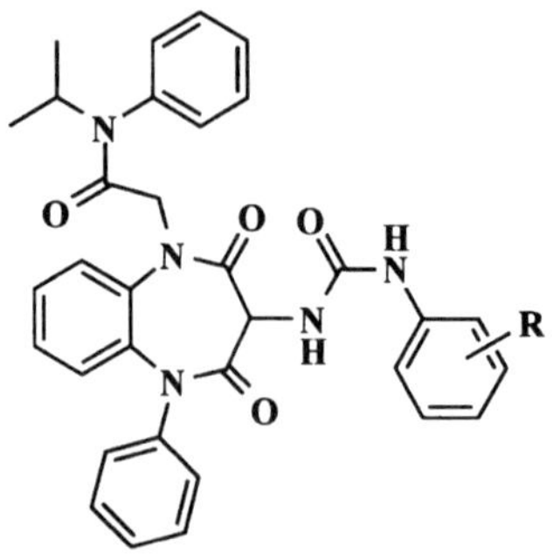

1 R = H, GW 4664

2 R = COOH, GW 7854

a series of 1,5-benzodiazepines, exemplified by **1** (GW3664), which were moderately potent CCK-A receptor agonists *in vitro* (GPGB) and *in vivo* (rat anorexia), but were not orally active in rat feeding models (Aquino *et al.,* 1996). In order to optimize this class of compounds, our initial strategy focused on chemical modifications to reduce molecular weight and/or increase aqueous solubility. Additionally, parallels with reported 1,4-benzodiazepine CCK-A or CCK-B receptor antagonists (Bock *et al.,* 1989, 1993; Evans *et al.,* 1986, 1988) directed a series of modifications at the C-3 position of the benzodiazepine ring (Henke *et al.,* 1996; Willson *et al.,* 1996; Hirst *et al.,* 1996). Optimal individual modifications were then combined (Henke *et al.,* 1996; Szewczyk *et al.,* in preparation).

Because the goal was identification of an analogue with oral activity in a suitable animal model, the screening strategy for new analogues was to confirm *in vitro* agonist efficacy on the GPGB following incubation (30–60 min) of a single concentration (30 or 1 μM) of test compound, then progress compounds with appropriate efficacy (≥40% of the CCK-8-induced contraction) to a suitable animal model.

Initially, analogues were evaluated in naive 18-hr food-deprived (18-hr FD) rats (8–10 animals per dose) and food intake was monitored for 30 min. In this assay, CCK-8 had an ED_{50} of 100 nmole/kg, following intraperitoneal dosing. Intraperitoneal (i.p.) dose–response curves (0.1, 1, 10 μmole/kg) and single-dose oral (p.o.) efficacy were evaluated for each new analogue. Of 40 compounds screened, only 1 (**2,** GW7854) was orally efficacious at the 10 μmole/kg dose.

2. *IN VIVO* PROFILE OF GW7854

GW7854 was more potent than CCK-8 following i.p. dosing (Fig. 1A) and anorectic activity was selectively reversed with the CCK-A receptor-selective an-

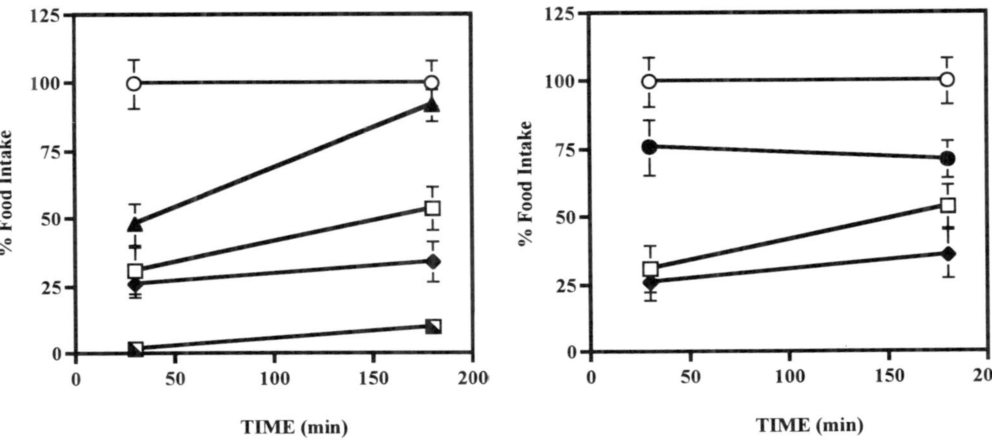

Figure 1. (A) Anorectic activity of GW7854 in 18-hr food-deprived rats following intraperitoneal dosing (n = 10/dose). ○, methylcellulose vehicle; ▲, CCK-8, 0.1 µmole/kg; □, GW7854, 0.1 µmole/kg; ◆, GW7854, 1 µmole/kg; ◪, GW7854, 10 µmole/kg. (B) Anorectic activity of GW7854 (0.1 µmole/kg) in the absence (□) or presence of L-365,260 (1 µmole/kg; ◆) or MK-329 (0.5 µmole/kg; ●).

tagonist MK-329 (Evans *et al.,* 1986), but not by a CCK-B receptor-selective antagonist (L-365,260; Bock *et al.,* 1989) (Fig. 1B). However, the oral anorectic activity of this compound was weak (30–40% reduction in feeding at 10 μmole/kg) and occasionally not reproducible. In order to understand the origin of this erratic *in vivo* response, a number of parameters were evaluated, involving efforts from Pharmacology, Pharmaceutics, and Drug Metabolism groups.

3. PHARMACEUTICAL STUDIES WITH GW 7854

3.1. Batch Variation

Both crystalline and amorphous (lyophilized) GW7854 were examined for *in vitro* and *in vivo* activity. Because amorphous material appeared to provide a more consistent agonist response, all subsequent compounds were prepared as lyophiles prior to biological assays.

3.2. Dosing Vehicle

No significant difference in bioactivity was observed when GW7854 was administered as a suspension (0.5% methylcellulose) or as a solution (propylene glycol or polyethylene glycol 400). Other vehicles, containing ethanol or dimethyl sulfoxide (DMSO), were found to adversely affect food intake when dosed alone.

4. PHARMACOLOGY STUDIES

4.1. The Mouse Gallbladder Emptying Assay

Because food intake is a behavioral response that can be sensitive to many external factors unrelated to drug effect (Sepinwall and Sullivan, 1991), other CCK-A receptor-mediated *in vivo* assays were sought that would provide a more physiological measurement of bioactivity. A mouse gallbladder emptying (MGBE) assay (Makovec *et al.,* 1987) was characterized in-house and found to be both extremely sensitive and reproducible. In addition, sample requirements were much lower (<5 mg) than those required for the rat anorexia studies (30–50 mg) and compounds could be screened with much higher throughput. Compounds that met the *in vitro* GPGB efficacy criteria were screened at a single dose (0.1 μmole/kg i.p. and 1.0 μmole/kg p.o., $n = 10$ mice/dose). Only compounds that were orally active in the mouse (>50% gallbladder emptying) were subsequently characterized in detail (full dose–response curves i.p. and p.o.) and advanced to food intake studies. Interestingly, although more than 20 compounds were identified that were potent, orally active CCK-A agonists in the mouse, none were orally active in the 18-hr FD rat model.

4.2. Alternate Species

Food intake studies were performed in 18-hr FD mice and guinea pigs. Mice were responsive to both CCK-8 and GW7854, but individual food intakes were so small that statistically significant data were difficult to obtain. Guinea pigs exhibited an adverse response to both CCK-8 and these benzodiazepine CCK-A agonists (prolapsed colon). Although this response was reversible, it was decided that this species was inappropriate for further studies.

4.3. The Conditioned Feeder Rat Model

As rats appeared to be the only species suitable for food intake studies, the decision was made to modify the protocol. It was suggested that the erratic response to GW7854 was related to the strong drive to eat induced by 18-hr food deprivation. Additionally, variable responses were observed between naive animals and animals that had experienced one or more feeding studies. A conditioned feeder rat (CF rat) model was developed in which rats were trained for 2 weeks to consume a palatable liquid diet following a 2-hr food deprivation (Aquino *et al.*, 1996). A saline preload was introduced by gastric lavage prior to dosing to further enhance the sensitivity of this model of CCK-A-mediated anorexia. Cumulative intake was evaluated at 30, 90, and 180 min. Potency (ED_{50}) was calculated from the 30-min dose–response curve. CCK-8 was threefold more potent in the CF rat model (ED_{50} = 30 nmole/kg) than in the 18-hr FD rat. The maximal reduction in food intake following oral dosing with GW7854 in the CF rat (58% at 10 μmole/ kg) was twice that observed in the 18-hr FD rat model (33% at 10 μmole/kg).

5. PHARMACOKINETIC PROFILE OF GW7854

The pharmacokinetic profile of GW7854 in rats was characterized by a moderate total body clearance (C_L = 22 ml/min per kg) and short half-life ($t_{1/2}$ = 1.2 hr) following i.v. administration (Fig. 2). Urinary excretion was insignificant

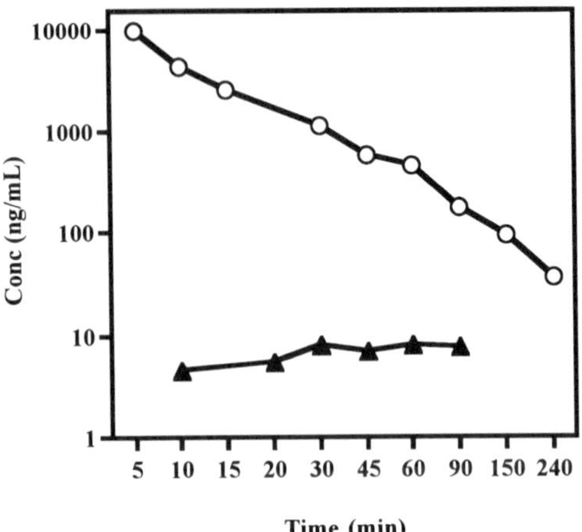

Figure 2. Blood concentration profile for GW7854 (10 μmole/kg; 6 mg/kg) following i.v. (○) or p.o. (▲) administration to conscious Long–Evans rats. Samples (0.1 ml) were taken at the indicated times and analyzed by analytical RP-HPLC using fluorescence detection.

(3–4%), whereas 67% of unchanged drug was excreted into bile. The oral bioavailability of a 10 μmole/kg (6 mg/kg) dose was 0.4%, with a C_{max} of only 10 ng/ml. Blood levels fell below the limits of detection after 90 min. The absence of liver metabolism was confirmed by a 3-hr incubation with rat liver microsomes (37°C, 5 mM GW7854, 1 mg microsomal protein). Because GW7854 was extremely potent following i.p. dosing, and i.p. dosing does deliver compound to the portal vein (Johnson, 1980), it was assumed that the low bioavailability of GW7854 was primarily related to poor absorption from the gastrointestinal tract.

6. THE Caco-2 MODEL FOR INTESTINAL ABSORPTION

Recently, a human colonic adenocarcinoma cell line (Caco-2) was proposed as a model to study passive drug absorption across intestinal epithelium (Hidalgo *et al.,* 1989; Artursson and Karlsson, 1991; Gan *et al.,* 1994). A good correlation was found between human intestinal absorption and the measured apparent permeability coefficients (P_{app}) for a series of drugs in the Caco-2 model (Artursson and Karlsson, 1991). This model was adapted in-house and used to characterize GW4664, GW7854, and a number of standard compounds (Table I). The P_{app} values for GW4664 and GW7854 suggested an extremely low potential for absorption.

Table I
Apparent Permeability Coefficients for Select Compounds

Compound	P_{app} ($\times 10^{-7}$ cm/sec)[a]
Chlordiazepoxide	114.8
MK-329[b]	32.1
L-365,260[c]	80.6
Amphetamine sulfate	120
Fenfluramine	>200
CCK-8	1.9
GW4664	<0.1
GW7854	0.9

[a]Caco-2 cells were cultured in minimum essential medium containing 10% fetal bovine serum and 1% nonessential amino acids on 1-cm^2 polycarbonate membranes of Costar Transwells™, seeded at a density of 71,000 cells/well. Cells were grown to late confluency (20–25 days), with the culture medium changed every 2 days. Culture medium was replaced with transport medium (Hanks' balanced salt solution containing 25 mM glucose and 25 mM Hepes buffer, pH 7.0) an hour before incubation with test compounds. A stock solution of each analogue was prepared in the transport medium (20 μM, pH 7, 0.5% DMSO). Compounds were added to the apical side (AP) of the transwell ($n = 3$ wells/compound) and incubated for 3 hr. Monolayer integrity was monitored by the measurement of [^{14}C]mannitol leakage or by transepithelial electrical resistance (Hidalgo *et al.*, 1989). Basolateral (BL) concentration for each analogue was evaluated by analytical HPLC (Vydac C-18 or BDS Hypersil) using gradients of acetonitrile in 0.1% aqueous trifluoroacetic acid. Calculation of the apparent permeability coefficient (P_{app}):

$$P_{app} = \frac{\text{Concn (BL)/hr}}{\text{Conc (AP)/cm}^3} \times \frac{1}{3600} \times \frac{1}{\text{SA(cm}^2\text{)}}$$

The surface area (SA) of the transwell was 1 cm^2 in all experiments. The P_{app} value is thus a measure of flux in units of centimeters per second normalized to the apical drug concentration, time, and membrane surface area.
[b]Evans *et al.* (1986).
[c]Bock *et al.* (1989).

6.1. Correlation with Rat Intestinal Absorption

In order to ascertain that P_{app} measurements were predictive of rat absorption, five compounds with P_{app}s ranging from 0.9 to 128 $\times 10^{-7}$ cm/sec were evaluated following intraduodenal dosing in anesthesized rats fitted with portal vein cannulas. As can be seen from Fig. 3, the correlation between P_{app} and rat intestinal absorption parallels that observed for human absorption. In this study, the P_{app} of 0.9 $\times 10^{-7}$ measured for GW7854 corresponds with 4% absorption.

6.2. Structure–Transport Relationships

More than 600 compounds have been prepared for the CCK-A agonist program, providing a large data base of compounds with which to study the correla-

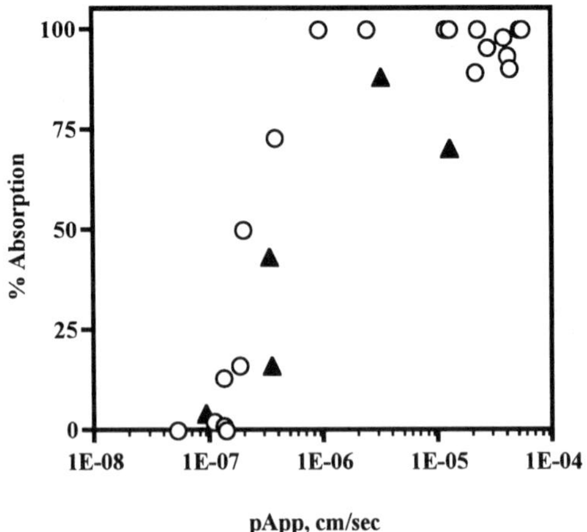

Figure 3. Human absorption data (○) from Artursson and Karlsson (1991). Rat data (▲) obtained for select 1,5-benzodiazepines following intraduodenal dosing in anesthesized rats fitted with portal vein cannulas. Portal vein blood samples were collected for 4 hr after dosing and analyzed by analytical RP-HPLC, using UV or fluorescence detection or mass spectrometry. % absorption = $(AUC_{p.o.}/AUC_{i.v.}) \times 100$.

tion of structure with transport potential. The Caco-2 *in vitro* assay was optimized as a rapid, single concentration screening tool and used to evaluate the structural requirements for good absorption of these benzodiazepines. Representative modifications are provided in Tables II–V. Compounds could be classified as poorly ($P_{app} < 2$), moderately ($2 < P_{app} < 10$), or well-transported ($P_{app} > 10$). Initial attempts to correlate transport rate coefficients with physicochemical parameters suggested that transport rate was moderately dependent on the number of potential hydrogen bonds, molecular size, and solvation energy. No correlation was found with either calculated (clog *P*) or measured (log *D*) lipophilicity.

Of the various template analogues within the C-3 phenyl urea derivatives (Table II), the 5-phenyl-1,5-benzodiazepine had the lowest transport potential. Substituting the 5-phenyl with 5-methyl, eliminating the benzo fusion, or replacing the 1,5-template with 1,4-benzodiazepine, benzolactam, or caprolactam templates greatly enhanced P_{app}. Unfortunately, most of these analogues had greatly reduced *in vitro* agonist efficacy.

Substitution of the N-1 anilido aromatic ring generally provided a 10- to 100-fold increase in P_{app} relative to GW4664 (Table III). Although these compounds retained good *in vitro* agonist efficacy, none were orally active in the rat (Aquino *et al.,* 1996). The 3- and 4-pyridyl C-3 amide derivatives had the highest P_{app} values observed within the 1,5-benzodiazepines (Table III). However, neither of these C-3 pyridyl amides had adequate *in vitro* efficacy (Hirst *et al.,* 1996). Interestingly, incorporation of a single carboxylic acid on the pyridine ring was sufficient to

Table II
Influence of Template on P_{app}

Structure	P_{app} ($\times 10^{-7}$ cm/sec)	Structure	P_{app} ($\times 10^{-7}$ cm/sec)
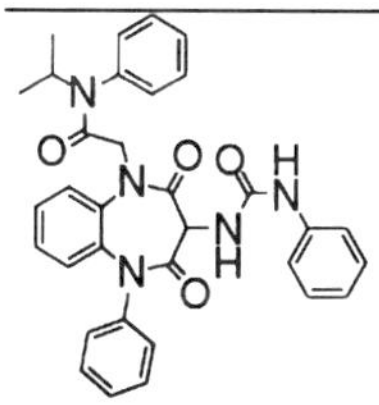	<0.1	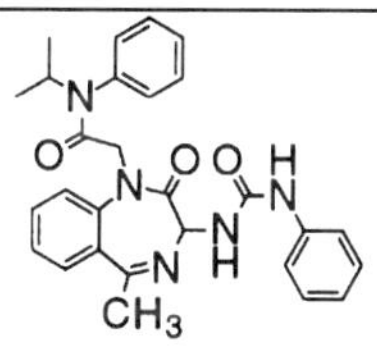	100
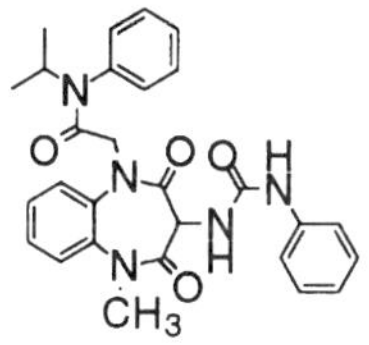	24	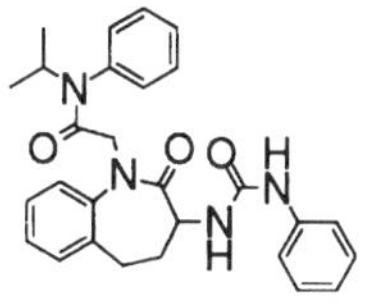	5
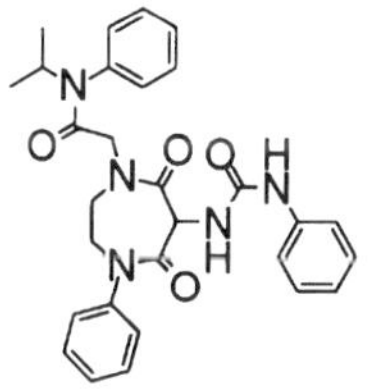	5.9	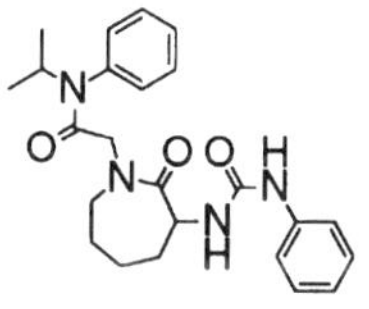	102
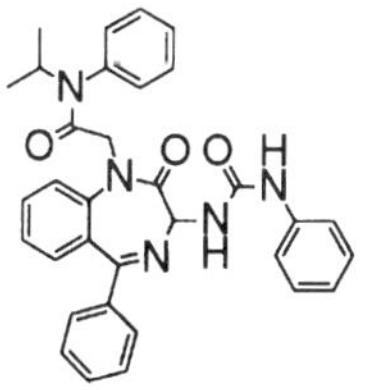	8.9		

reduce P_{app} 100-fold. Substitution of the C-3 phenyl urea gave compounds with up to a 50-fold improvement in P_{app} (Table III). All of these compounds retained good *in vitro* efficacy (Hirst *et al.*, 1996).

In general, modification or replacement of the C-3 phenyl urea moiety with a variety of substituents increased P_{app} (Table IV). Incorporation of the *m*-carboxylic acid group increased P_{app} 10-fold. Shifting (C-3 indolamide) or eliminating (C-3 methylene-linked phenylamide or 3-indazole) a single hydrogen bond resulted in a 70- to 260-fold increase in P_{app}. Quaternization of the C-3 center of these latter analogues further increased P_{app} 1.4 to 3-fold. Most of these analogues

Table III

Influence of Various Substituents on N-1 Anilido, C-3 Amide, or C-3 Urea Derivatives

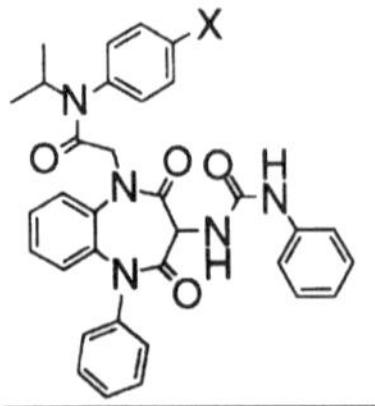

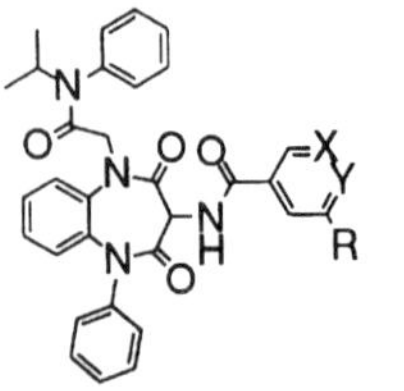

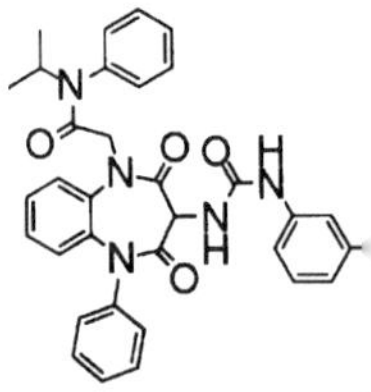

X	P_{app} ($\times 10^{-7}$ cm/sec)	X	Y	R	P_{app} ($\times 10^{-7}$ cm/sec)	Y	P_{app} ($\times 10^{-7}$ cm/s
-H	<0.1	N	CH	H	128	-H	<0.1
-F	5.0	CH	N	H	160	-F	5.0
-OH	1.4	N	CH	COOH	0.7	-OH	5.0
$-OCH_3$	9.4					-COOH	0.9
$-N(CH_3)_2$	2.2					$-COOCH_2CH_3$	1.3

retained good *in vitro* potency and efficacy (Henke *et al.,* 1996; Willson *et al.,* 1996; Hirst *et al.,* 1996).

Combination of the preferred N-1 anilido substitution (4-OCH_3, Table III) with these C-3 modifications did not produce a consistent change in P_{app} (Table V). The P_{app} value for the C-3 phenyl urea was increased 100-fold, whereas the values for the C-3 *m*-carboxyl phenyl urea or C-3 methylene-linked phenyl amide were increased only 2-fold. The P_{app} values for the remaining analogues were decreased from 2- to 10-fold. However, all of these analogues had P_{app} values still within the range of moderate transport potential (Fig. 3). More importantly, all of these combination analogues had greatly enhanced *in vitro* potency and efficacy (Henke *et al.,* 1996; Szewczyk *et al.,* in preparation).

7. BIOAVAILABILITY VERSUS BIOACTIVITY

Although more than 40 compounds were eventually identified that were orally active in the MGBE, only 7 were orally active in the CF rat model. These compounds had moderate to high total body clearance (C_L = 22–40 ml/min per kg) and short to moderate duration ($t_{1/2}$ = 0.3–4 hr) following i.v. administration. All had uniformly poor bioavailability in the rat ($F \leq 8\%$). The P_{app} values for these orally active compounds ranged from 0.9 to 91 $\times$ 10^{-7} cm/sec.

In contrast, two compounds (P_{app} = 3.4 and 32 $\times$ 10^{-7} cm/sec) that were not orally active in the CF rat were found to have good oral bioavailability (33–53%). Both compounds had very low total body clearance (C_L = 5–6 ml/min per kg) and moderate duration ($t_{1/2}$ = 3–4 hr) following i.v. administration.

Table IV
Influence of C-3 Substituent on Apparent Permeability

Structure	P_{app} ($\times 10^{-7}$ cm/sec)	Structure	P_{app} ($\times 10^{-7}$ cm/sec)
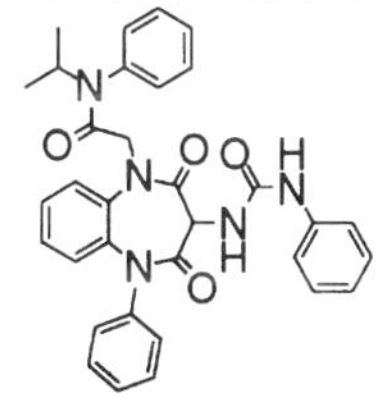	<0.1	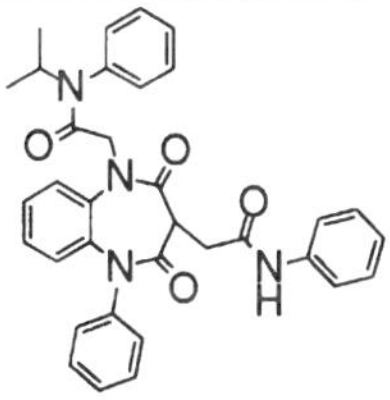	26
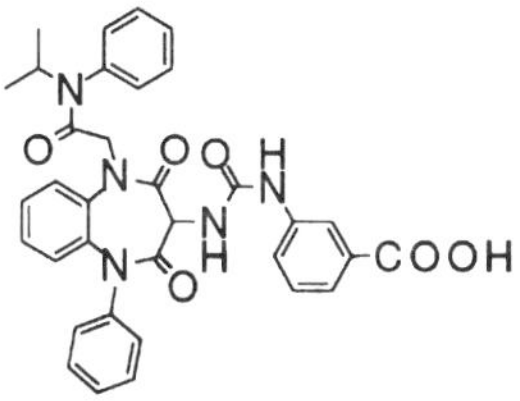	0.9	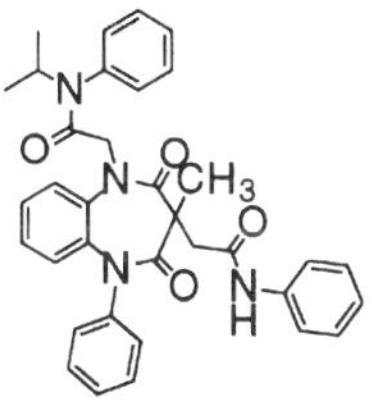	75.7
GW7854		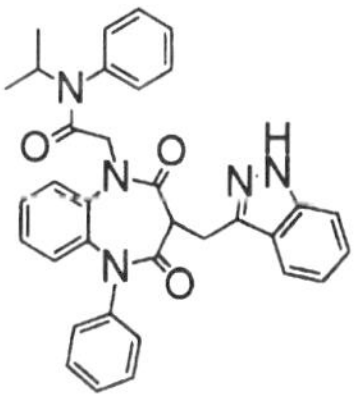	23
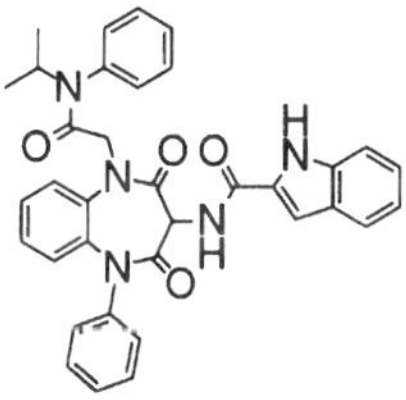	6.9		
		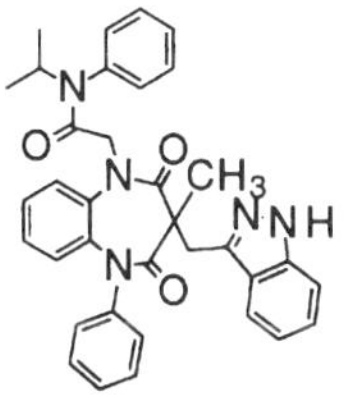	33

Thus, orally bioactive compounds were not bioavailable and bioavailable compounds were not orally bioactive. An almost linear inverse correlation was found between potent anorectic activity following i.p. administration and oral bioavailability (Fig. 4). The low, flat blood concentration versus time profiles observed with orally bioactive compounds were reminiscent of a sustained-release formulation, and suggested that absorption of these analogues was delayed at some point along the gastrointestinal tract.

Table V
Influence of C-3 Substituent on Apparent Permeability

Compound	P_{app} ($\times 10^{-7}$ cm/sec)	Compound	P_{app} ($\times 10^{-7}$ cm/sec)
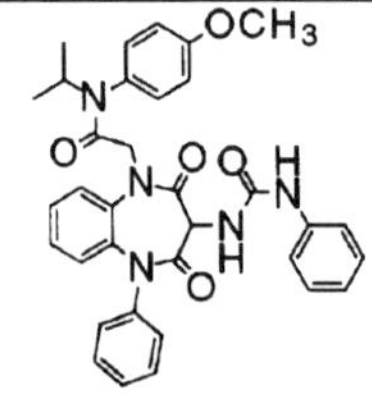	9.4	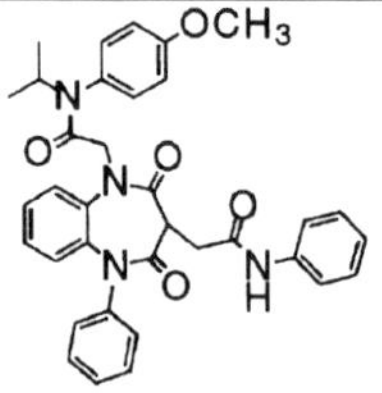	45.8
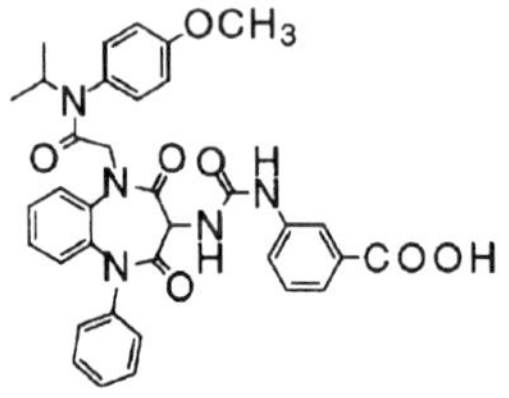 GW 6038	2.2	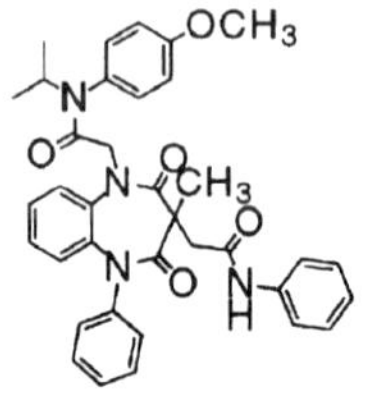	43
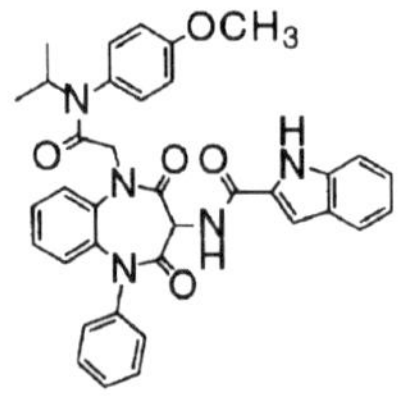 GW 0772	3.4	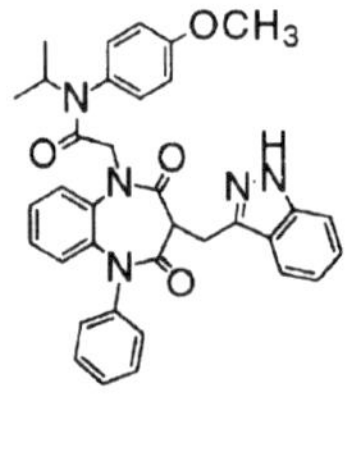 GW 5823	3.5
		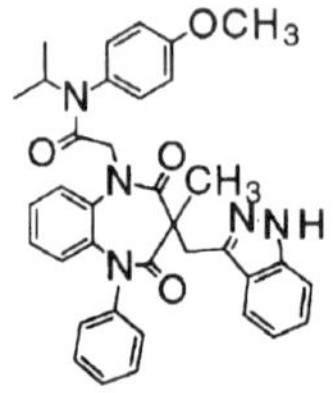	6.2

CCK-A Receptor-Mediated Gastric Emptying

Because CCK is reported to inhibit gastric emptying in a variety of species through CCK-A receptor-mediated contraction of the pyloric sphincter (Grider, 1994; Reidelberger, 1994), the low bioavailability of potent CCK-A receptor agonists could be related to gastric retention of drug. In order to better define the im-

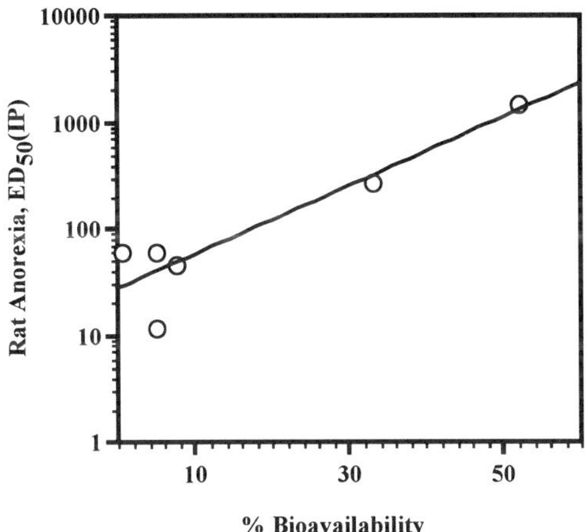

Figure 4. Inverse correlation between potency in the CF rat anorexia model (ED_{50}, i.p.) and bioavailability.

pact of delayed gastric emptying on oral bioavailability, two compounds were evaluated for their ability to inhibit gastric emptying in rats following i.p. or p.o. administration (Scarpignato *et al.*, 1980). These compounds had similar P_{app}s and *in vitro* GPGB potency, but GW5823 was fivefold more potent than GW0772 in the CF rat (Table VI) following i.p. administration. GW5823 was orally active in the CF rat (Henke *et al.*, 1996) whereas GW0772 was not (Szewczyk *et al.*, in preparation).

CCK-8, GW0772, and GW5823 were all able to completely inhibit gastric emptying following i.p. dosing (Fig. 5A). CCK-8 was threefold more potent than GW5823, which in turn was threefold more potent than GW0772. Following oral dosing, GW5823 was again threefold more potent than GW0772, and GW0772 was unable to completely inhibit gastric emptying in the rat, even at the highest dose tested (10 μmole/kg) (Fig. 5B). Thus, the higher bioavailability of GW0772

Table VI
In Vitro and *in Vivo* Profile of GW0772 and GW5823

Compound	P_{app}	F (%)[a]	GPGB pEC$_{50}$[b]	GPGB % max[c]	Anorexia ED$_{50}$ i.p.[d]
GW0772	3.4	33%	6.8 ± 0.6 (4)	47%	0.27 μmole/kg
GW5823	3.5	5%	7.0 ± 0.3 (3)	101%	0.05 μmole/kg

[a]Oral bioavailability in Long–Evans rats, = $(AUC_{p.o.})/(AUC_{i.v.}) \times 100$.
[b]-log of the concentration required for half-maximal contraction of the isolated GPGB, ± SEM (number of experiments).
[c]% maximal contraction at 10 μM, normalized the % contraction induced by 1 μM CCK-8.
[d]Dose required for half-maximal reduction in food intake in CF rats.

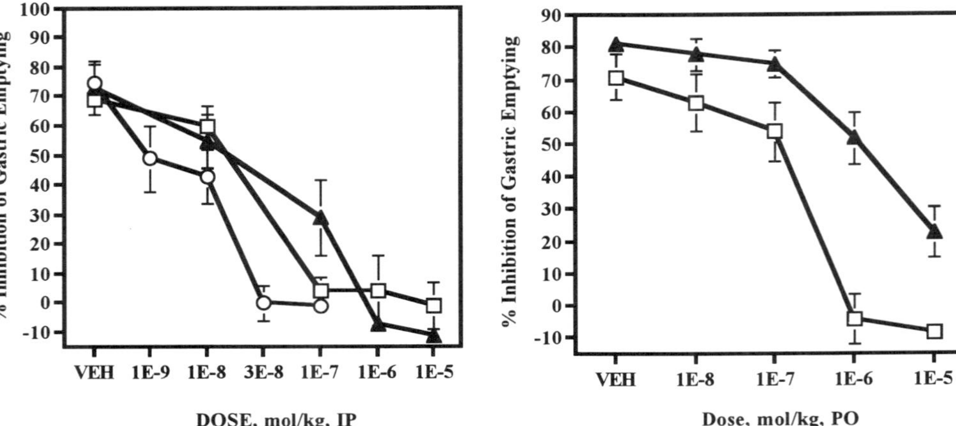

Figure 5. (A) Dose–response curves for % gastric emptying in rats (n = 10/dose) following i.p. dosing. ○, CCK-8; ▲, GW0772; □, GW5823. (B) Dose–response curves for % gastric emptying in rats following oral dosing of GW0772 (▲) or GW5823 (□). Vehicle (VEH) or compounds were dosed 20 min before gastric lavage of test meal (1.5% w/v methylcellulose colored with phenol red). Rats were sacrificed and stomachs removed 30 min after test meal.

Table VII
Intraduodenal versus Oral Anorectic Activity in Rats

	% anorexia[a]		
Compound	18-hr FD rat 10 μmole/kg p.o.	CF rat 10 μmole/kg p.o.	CF rat 10 μmole/kg i.d.
GW7854	33	58	59
GW6038	(6)	51	79
GW5823	(17)	40	79
GW8429	NT[b]	28	82
GW2126	NT	43	63

[a]% reduction in feeding at 30 min, versus control.
[b]NT, not tested.

appeared to be related to incomplete inhibition of gastric emptying. The P_{app} value of 3.4×10^{-7} cm/sec for GW0772 corresponded with 43% intestinal absorption in the anesthesized rat (Fig. 3). Additional compounds were evaluated for gastric emptying (data not shown) to confirm that incomplete inhibition of gastric emptying correlated with increased oral bioavailability and decreased oral anorectic activity.

8. ORAL VERSUS INTRADUODENAL DOSING

The pharmacodynamic and pharmacokinetic profiles of a select group of compounds were reevaluated using intraduodenally cannulated conscious rats. Whereas intraduodenal (i.d.) dosing did not improve the potency or efficacy of GW0772, other compounds were much more efficacious anorectic agents (Table VII). Full dose–response curves could be obtained with i.d. dosing, as almost maximal inhibition of food intake was achieved. The C_{max} after i.d. administration was generally fourfold higher than after oral administration and i.d. bioavailabilities were two- to threefold higher.

9. DISCUSSION

The primary goal of this program was to achieve consistent oral bioavailability and bioactivity in rats. Although screening for oral activity was a key step in the early stages of the project, recognition of the inverse correlation between oral bioactivity and oral bioavailability in rats eventually required modification of the compound progression strategy. A final project strategy evolved in which compounds were prioritized based on functional *in vitro* potency and efficacy, P_{app} values associated with good intestinal absorption (≥10), potent anorectic activity (ED_{50} ≤ 100 nmol/kg) in CF rats after i.p. dosing, and moderate pharmacokinetic duration ($t_{1/2}$ ≥ 2 hr) following i.v. dosing. No compound that met these criteria had better than 8% oral bioavailability in rats. Although we have focused on optimizing compounds for consistent systemic levels of drug, the relevance of systemic blood levels to anorectic activity is not clear. There is still much debate in the literature as to whether endogenous CCK induces satiety in rats through endocrine or paracrine mechanisms (Reidelberger *et al.*, 1994).

10. CLINICAL IMPLICATIONS

The utility of these orally active CCK-A agonists for the treatment of human obesity remains to be evaluated. CCK-A mechanism-based gastric retention may preclude oral dosing in humans, or may be beneficial, promoting satiety and limiting toxic exposure, or may not be relevant, as there is considerable species variation with respect to CCK-A receptor-mediated delayed gastric emptying. Humans are reported to be less sensitive than rats (Liddle *et al.*, 1989). Oral bioavailability over 30% in animal models is generally preferred for advancing compounds for human clinical trials in order to minimize unacceptable side effects resulting from variable absorption rates. The major side effects of CCK-8 in humans are nausea, vomiting, intestinal cramping, and diarrhea. Whether the extremely low bioavailability in rats will translate to humans, whether patients will experience wide variation in oral bioavailability related to variable inhibition of gastric emptying, and whether this variation will translate into unpredictable and/or unacceptable drug response remain to be determined.

Acknowledgments

The authors gratefully acknowledge Avis Bridgers and Souzan Yanni for Caco-2 P_{app} measurements.

REFERENCES

Aquino, C. J., Armour, D. R., Berman, J. M., Birkemo, L. S., Carr, R. A. E., Croom, D. K., Dezube, M., Dougherty, R. W., Ervin, G. N., Grizzle, M. K., Head, J. E., Hirst, G. C., James, M. K., Johnson, M. F., Long, J. E., Miller, L. J., Queen, K. L., Rimele, T. J., Smith, D. N., and Sugg, E. E., 1996, Discovery of 1,5-benzodiazepines with peripheral cholecystokinin (CCK-A) receptor agonist activity. I. Optimization of the agonist "trigger," *J. Med. Chem.* **39:**562–569.

Artursson, P., and Karlsson, J., 1991, Correlation between oral drug absorption in humans and apparent drug permeability coefficients in human intestinal epithelial (CACO-2) cells, *Biochem. Biophys. Res. Commun.* **175:**880–885.

Bock, M. G., DiPardo, R. M., Evans, B. E., Rittle, K. E., Whitter, W. L., Veber, D. L., Anderson, P. S., and Freidinger, R. M., 1989, Benzodiazepine gastrin and brain CCK receptor ligands L365,260, *J. Med. Chem.* **32:**13–16.

Bock, M. G., DiPardo, R. M., Evans, B. E., Rittle, K. E., Whitter, W. L., Garsky, V. M., Gilbert, K. F., Leighton, J. L., Carson, K. L., Mellin, E. C., Veber, D. F., Chang, R. S. L., Lotti, V. J., Freeman, S. B., Amith, A. J., Patel, S., Anderson, P. S., and Freidinger, R. M., 1993, Development of 1,4-benzodiazepine cholecystokinin type B antagonists, *J. Med. Chem.* **36:**4276–4292.

Crawley, J. N., and Corwin, R. L., 1994, Biological actions of cholecystokinin, *Peptides* **15:**731–755.

Dourish, C. T., Ruckert, A. C., Tattersall, F. D., and Iversen, S. D., 1989, Evidence that decreased feeding induced by systemic injection of cholecystokinin is mediated by CCK-A receptors, *Eur. J. Pharmacol.* **173:**233–234.

Evans, B. E., Bock, M. G., Rittle, K. E., DiPardo, R. M., Whitter, W. L, Veber, D. F., Anderson, P. S., and Freidinger, R. M., 1986, Design of potent, orally effective, nonpeptidyl antagonists of the peptide hormone cholecystokinin, *Proc. Natl. Acad. Sci. USA* **83:**4918–4922.

Evans, B. E., Rittle, K. E., Bock, M. G., DiPardo, R. M., Freidinger, R. M., Whitter, W. L., Lundell, G. F., Veber, D. F., Anderson, P. S., Chang, R. S. L., Lotti, V. J., Cerino, D. J., Chen, T. B., Kling, P. J., Kunkel, K. A., Springer, J. P., and Hirshfield, J., 1988, Methods for drug discovery: Development of potent, selective, orally effective, cholecystokinin antagonists, *J. Med. Chem.* **31:**2235–2246.

Gan, L. S., Eads, C., Niederer, T., Bridgers, A., Yanni, S. , Hsyu, P. H., Pritchard, F. J., and Thakker, D., 1994, Use of Caco-2 cells as an in vitro intestinal absorption and metabolism model, *Drug Dev. Ind. Pharm.* **20:**615–631.

Grider, J., 1994, Role of cholecystokinin in the regulation of gastrointestinal motility, *J. Nutr.* **124:**1334S–1339S.

Henke, B. R., Willson, T. M., Sugg, E. E., Croom, D. K., Dougherty, R. W., Jr., Queen, K. L., Birkemo, L. S., Ervin, G. N., Grizzle, M. K., Johnson, M. F., and James, M. K., 1996, 3-(1H-Indazol-3-ylmethyl)-1,5-benzodiazepines: CCK-A agonists that demonstrate oral activity as satiety agents, *J. Med. Chem.* **39:**2655–2658.

Hidalgo, I. J., Raub, T. J., and Borchardt, R. T., 1989, Characterization of the human colon carcinoma cell line (Caco-2) as a model system for intestinal epithelia permeability, *Gastroenterology* **96:**736–749.

Hirst, G. C., Aquino, C. J., Birkemo, L. S., Croom, D. K., Dezube, M., Dougherty, R. W., Jr., Ervin, G. N., Grizzle, M. K., Henke, B. R., James, M. K., Johnson, M. F., Momtahen, T. M., Queen, K. L., Sherrill, R. L., Szewczyk, J., Willson, T. M., and Sugg, E. E., 1996, Discovery of 1,5-benzodiazepines with peripheral cholecystokinin (CCK-A) receptor agonist activity (II): Optimization of the C3 amino substituent, *J. Med. Chem.* **39:**5236–5245.

Johnson, P., 1980, Pro-drugs and first-pass effects, *Chem. Ind.* **June:**443–447.

Kissileff, H. R., Pi-Sunyer, F. X., Thornton, J., and Smith, G. P., 1981, C-terminal octapeptide of cholecystokinin decreases food intake in man, *Am. J. Clin. Nutr.* **34:**154–160.

Liddle, R. A., Gertz, B. J., Kanayama, S., Beccaria, L., Coker, L. D., Turnbull, T. A., and Morita, E. T., 1989, Effects of a novel cholecystokinin (CCK) receptor antagonist, MK-329, on gallbladder contraction and gastric emptying in humans. Implications for the physiology of CCK, *J. Clin. Invest.* **84:**1220–1225.

Makovec, F., Bani, M., Cereda, R., Chiste, R., Pacini, M. A., Revel, L., and Rovati, L. C., 1987, Antispasmodic activity on the gallbladder of the mouse of cr1409 (lorglumide), a potent antagonist of peripheral cholecystokinin, *Pharmacol. Res. Commun.* **19:**41–51.

Pi-Sunyer, X., Kissileff, H. R., Thornton, J., and Smith, G. P., 1982, C-terminal octapeptide of cholecystokinin decreases food intake in obese men, *Physiol. Behav.* **29:**627–630.

Reidelberger, R. D., 1994, Cholecystokinin and control of food intake, *J. Nutr.* **124:**1327S–1333S.

Reidelberger, R. D., Varga, G., Liehr, R.-M., Castellanos, D. A., Rosenquist, G. L., Wong, H. C., and Walsh, J. H., 1994, Cholecystokinin suppresses food intake by a nonendocrine mechanism in rats, *Am. J. Physiol.* **267:**R901–R908.

Scarpignato, C., Capovilla, T., and Bertaccini, G., 1980, Action of caerulein on gastric emptying of the conscious rat, *Arch. Int. Pharmacodyn. Ther.* **246:**286–294.

Sepinwall, J., and Sullivan, A. C., 1991, Screening methods for anorectic, antiobesity and orectic agents, in: *Behavioral Models in Psychopharmacology: Theoretical, Industrial and Clinical Perspective* (P. Willner, ed.) pp. 215–236, Cambridge University Press, London.

Sitzmann, J. V., Pitt, H. A., Steinborn, P. A., Pasha, Z. R., and Sanders, R. C., 1990, Cholecystokinin prevents parenteral nutrition induced biliary sludge in humans, *Surg. Gynecol. Obstet.* **170:**25–31.

Szewczyk, J., Aquino, C. J., Birkemo, L. S., Croom, D. K., Dezube, M., Dougherty, R. W., Jr., Ervin, G. N., Grizzle, M. K., Henke, B. R., Hirst, G. C., James, M. K., Johnson, M. F., Momtahen, T. M., Queen, K. L., Sherrill, R. L., Szewczyk, J., Willson, T. M., and Sugg, E. E., in preparation.

Willson, T. M., Henke, B. R., Momtahen, T. M., Myers, P. L., Sugg, E. E., Unwalla, R. J., Croom, D. K., Grizzle, M. K., Johnson, M. F., Queen, K. L., Rimele, T. R., Yingling, J. D., and James, M. K., 1996, 3-[2-N-Phenylacetamide]-1,5-benzodiazepines: Orally active, binding selective CCK-A agonists, *J. Med. Chem.* **39:**3030–3034.

Chapter 23

Orally Active Growth Hormone Secretagogues

Arthur A. Patchett, Roy G. Smith, and Matthew J. Wyvratt

1. INTRODUCTION

Initially, clinical applications with growth hormone (GH) isolated from natural sources were limited to treatment of GH-deficient children. However, with the commercialization of recombinant human growth hormone (rhGH) in the mid-1980s, there has been an explosion in potential clinical uses of GH (Strobl and Thomas, 1994; Torosian, 1995). In addition to the treatment of GH-deficient children and adults, rhGH has shown promise in the treatment of patients with burns, wounds, bone fractures, and Turner's syndrome. More recently, rhGH has been shown to be beneficial in reversing the catabolic effects of glucocorticoids, chemotherapy, and AIDS and in improving body composition of individuals (Rudman *et al.,* 1990; Papadakis *et al.,* 1996; Welle *et al.,* 1996).

GH is synthesized and stored in the pituitary gland. Its release from the anterior lobe of the pituitary is regulated principally by two known hypothalamic peptides: growth hormone releasing hormone (GHRH) and the inhibitory hormone somatostatin (SRIF) (Fig. 1). In most cases, GH deficiency is related to a hypothalamic defect, not to a pituitary deficiency in GH. Thus, as an alternative to rhGH treatment, most GH-deficient patients could be treated with an agent that

Arthur A. Patchett, Roy G. Smith, and Matthew J. Wyvratt • Departments of Medicinal Chemistry and Biochemistry & Physiology, Merck Research Laboratories, Rahway, New Jersey 07065.

Integration of Pharmaceutical Discovery and Development: Case Studies, edited by Borchardt *et al.,* Plenum Press, New York, 1998.

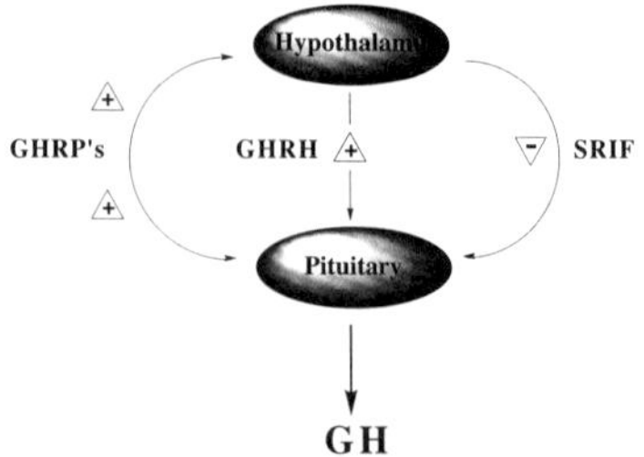

Figure 1. Regulation of growth hormone release.

would release endogenous GH from the pituitary gland (Schoen *et al.*, 1993). This can be achieved with GHRH and its analogues (Low, 1991); however, as with rhGH, their high cost and lack of oral bioavailability have restricted their clinical potential.

Inhibition of the cyclic inhibitory tetradecapeptide somatostatin as an alternative strategy to GHRH has received less attention primarily because of its diverse physiological properties (regulation of GH release and gastric acid secretion and modulation of glucagon, insulin, TSH, and prolactin levels) and the difficulty in identifying receptor antagonists. However, recently five distinct receptor subtypes for somatostatin have been identified and if one is uniquely associated with GH release, this could be a fruitful area for future research (Patel *et al.*, 1995; Reisine and Bell, 1995).

An additional regulatory pathway for GH release was identified by the pioneering work of C. Y. Bowers. In the late 1970s, Bowers *et al.* (1977) reported a series of peptide analogues of Leu and Met enkephalins that were devoid of opioid activity and that specifically released GH from the pituitary. These GH-releasing peptides (generally referred to as GHRPs) act directly on the pituitary and release GH via a unique mechanism distinct from GHRH (Fig. 1). In fact, GHRH and GHRPs act synergistically to release GH. A receptor for GHRPs has been identified that is present in both the pituitary and the hypothalamus. In the latter it appears also to be involved in the regulation of GHRH (Smith *et al.*, 1996a,b). Researchers at Merck (Howard *et al.*, 1996) have recently reported the cloning of the GHRP receptor [hereafter referred to as the GH secretagogue (GHS) receptor]. The endogenous ligand for this new orphan receptor has not been identified.

From Bowers's work, the hexapeptide GHRP-6 (His-D-Trp-Ala-Trp-D-Phe-Lys-NH_2) emerged as the early benchmark and was shown to be an extremely potent and relatively specific GH secretagogue in animals and in humans (Bowers *et al.*, 1990). Even though GHRP-6 exhibited only poor oral bioavailability ($<1\%$) in humans, it established an additional method of elevating GH as an alternative to subcutaneous treatment with rhGH or GHRH (Laron, 1995).

2. DISCOVERY OF GHRP-6 MIMICS: BENZOLACTAM L-692,429

In the late 1980s, Merck researchers became interested in potential uses of GHRP-6 and embarked on a research program to discover an orally active peptidomimetic of GHRP-6. Fortunately, extensive structure–activity relationships for the GHRPs had already been published (Bowers *et al.,* 1980; Momany *et al.,* 1981, 1984). Aromatic residues were known to be favored at positions 2, 4, and 5 and a basic amino terminus was critical for significant GH-releasing activity. This information along with the hypothesis that the GHS receptor may be G-protein linked was employed to select compounds from the Merck Sample Collection for screening in a GH-releasing rat pituitary cell culture assay (Cheng *et al.,* 1989). After screening only a few hundred samples mostly selected from other receptor programs, benzolactam **1** (Smith *et al.,* 1993) was discovered to release GH in a dose-dependent manner with an ED_{50} = 3 μM (GHRP-6, ED_{50} = 10 nM). This was truly a remarkable achievement at the time considering the rarity of nonpeptide agonists. The carboxylic acid moiety in **1** was initially replaced by a tetrazole, a well-established carboxylic acid bioisostere, to afford the more potent racemic analogue **2** (ED_{50} = 120 nM). Resolution of the C-3 chiral center in **2** led to the identification of the *R*-enantiomer **3** (L-692,429) as the biologically active isomer (ED_{50} = 60 nM). Mechanistically, **3** is identical to GHRP-6 *in vitro* (*vide infra*) and showed little or no activity in other receptor binding assays. In the rat pituitary membrane GHS receptor binding assay (data are presented here for critical compounds even though a receptor binding assay was not available in the early stages of this program), **3** exhibited a K_i = 63 nM compared with 6 nM for GHRP-6 and over 5000 nM for its inactive *S*-enantiomer (Pong *et al.,* 1996). Molecular modeling suggested that the benzolactam ring in **3** overlaid with the D-Trp residue in GHRP-6, its C-3 chiral center mapped onto the α-carbon of the D-Trp, and its basic amine occupied the same region as the N-terminal amino group in the hexapeptide (Schoen *et al.,* 1994a).

When administered intravenously, L-692,429 (**3**) was shown to release GH in rats, pigs, sheep, dogs, and rhesus monkeys. In dogs the release of GH occurred in a dose-dependent manner (≥0.1 mg/kg) with minimal effects on other hormones (Hickey *et al.,* 1994). However, L-692,429 exhibited poor oral activity in rats and dogs because of its low (2–8%) oral bioavailability (Leung *et al.,* 1996a). Extensive metabolic studies on L-692,429 showed that it was not metabolized *in vitro* or *in vivo* in rats, but was moderately metabolized in dogs to give the biologically active (ED_{50} = 200 nM) 7-hydroxylated derivative **4**. N-glucuronidation of the biphenyl tetrazole moiety in many angiotensin II antagonists is quite common; however, under similar in vitro conditions the biphenyl tetrazole unit in L-692,429 was not N-glucuronidated perhaps because of its zwitterionic nature. Radiolabeled experiments indicated that poor absorption of L-692,429, not metabolism, was pri-

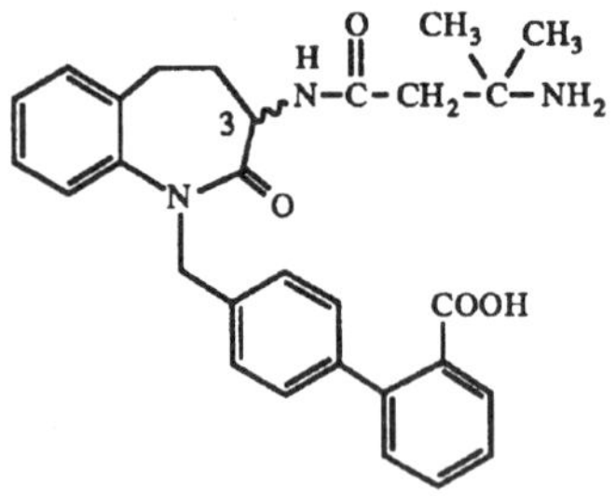

1

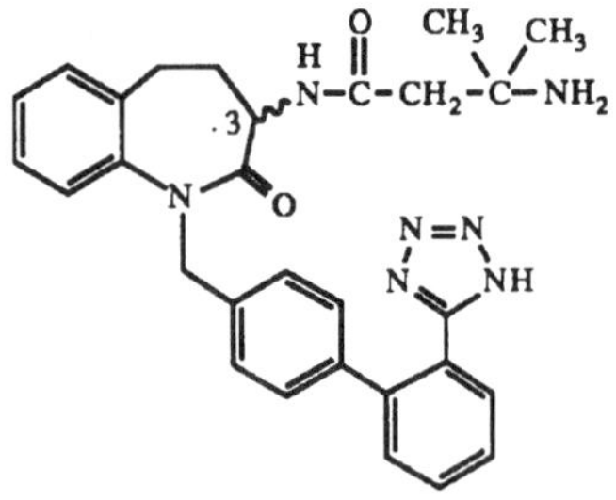

2

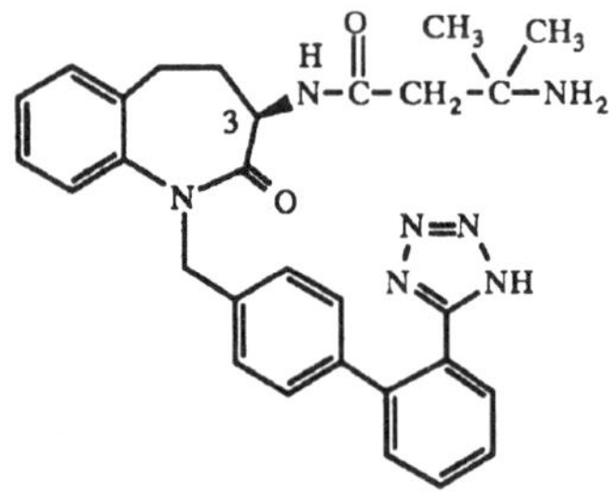

3
L-692,429

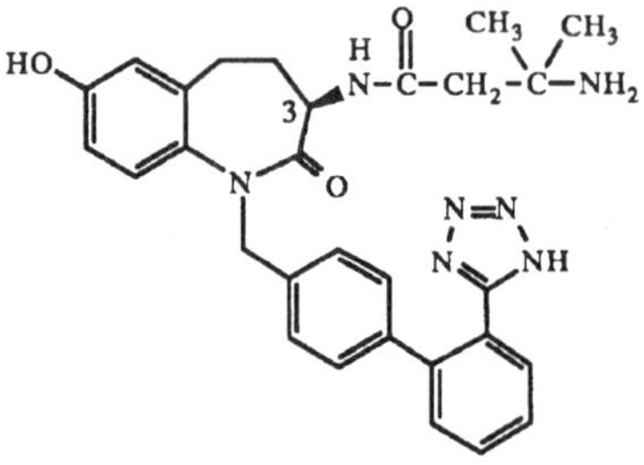

4

marily responsible for its poor oral bioavailability. This was subsequently confirmed in a human Caco-2 cell line (Artursson and Karlsson, 1991) in which very little transport of L-692,429 was observed.

2.1. Clinical Studies with L-692,429

As it was clear early on that the low oral bioavailability problem associated with this benzolactam lead would not be resolved quickly, L-692,429 was developed for intravenous administration in order to validate our approach to GH release in humans. When administered intravenously to healthy young males (Gertz *et al.*, 1993), L-692,429 ($t_{1/2}$ = 3.8 hr) was found to release GH in a dose-dependent fashion with all patients responding at 0.2 mg/kg (Table I), which was in excellent agreement with the dose required to release GH in dogs (Hickey *et al.*,

Table I
Evaluation of L-692,429 in Normal Males[a]

L-692,429 (mg/kg)	Mean peak GH (ng/ml)	No. of patients
0	1.95 ± 0.46	24
0.02	2.38 ± 1.25	6
0.05	8.95 ± 3.04	6
0.1	10.27 ± 6.04	6
0.2	41.03 ± 6.26	14
0.5	60.83 ± 7.34	8
1.0	82.51 ± 14.94	8

[a]Healthy 18- to 26-year-old males within 15% of ideal body weight; L-692,429 administered by i.v. infusion in saline (Gertz *et al.*, 1993).

1994). At this dose, the GH response to L-692,429 was greater than the standard intravenous dose of 1 μg/kg for GHRH, but somewhat less than the GH response for GHRP-6 at 1 μg/kg when administered intravenously. As observed with GHRP-6 and other GHRPs, there were small transient increases in cortisol and prolactin after L-692,429 administration. There were no significant changes in other pituitary hormones and no changes in IGF-1, glucose, or insulin levels. L-692,429 was well tolerated in these subjects with only a transient flushing or warm sensation being sporadically reported. In healthy elderly (71 ± 5 years) subjects, L-692,429 has been reported to release GH, although the response is somewhat diminished relative to healthy young men (Aloi *et al.*, 1994). This GH secretagogue has also been shown to partially reverse glucocorticoid suppression of GH secretion and therefore may be useful in reversing the catabolic effects of prednisolone and related steroids (Gertz *et al.*, 1994).

2.2. Structure–Activity–Bioavailability Relationships for the Benzolactams

With the validation that L-692,429 was a peptidomimetic of the GHRP class of GH secretagogues and that it appeared to be safe in humans, a major multidisciplinary effort at Merck was initiated with the objective to discover a more potent analogue of L-692,429 with good oral bioavailability suitable for development as an oral GH secretagogue. It was assumed that the zwitterionic character (pK_a, 4.6 and 9.2) of L-692,429 was probably responsible for its poor absorption in animal models. Consequently, much of the early medicinal chemistry on the benzolactam lead focused on modifying its structure to remove or attenuate either the protonated basic amine or the negatively charged tetrazole.

2.2.1. MODIFICATIONS OF THE C-3 AMINO ACID SIDE CHAIN

Extensive structure–activity relationships (Table II) were established for the C-3 dimethyl-β-alanine side chain in L-692,429 (Schoen *et al.*, 1994b; Ok *et al.*, 1994). Complete removal of the positive charge on this side chain (e.g., **5** and **6**) resulted in a loss of agonist activity confirming that, as was observed with GHRP-6, a basic amine is critical for GH-releasing activity. The spatial alignment for this basic amine is also important as its longer homologue **7** is less active, whereas its shortened analogue **8** is slightly more potent. The C-3 amide bond (-NHCO-) is critical for bioactivity as amide bond replacements (analogues **9–12**) are significantly less active (Ok *et al.*, 1996). In general, removal or attenuation of the positively charged character of this side chain led to improved plasma drug levels after oral administration in rats; however, these analogues were unfortunately accompanied by a significant loss in biological activity. The C-3 amide bond modifications did not have a positive effect on plasma drug levels after oral administration in rats, suggesting that the C-3 amide bond was not responsible for the low absorption observed for L-692,429.

Because the basic amino group in L-692,429 is critical for GH releasing activity and has a somewhat detrimental effect on oral absorption, a series of amino substituents with modulated basicity and lipophilicity were investigated (Table III). Simple alkyl substituents (e.g., propyl analogue **13**) were shown to be equivalent to L-692,429 in bioactivity. Although this alkyl substituent cannot bear a negatively charged group (e.g., acid **14**), hydroxy substitution leads to a potency improvement. For example, the 2(*R*)-hydroxypropyl derivative **16** (L-692,585) is 20-fold more potent than L-692,429 in the rat pituitary cell assay. The hydroxy substituent in **16** most likely forms an additional hydrogen bond to its receptor relative to L-692,429. This interpretation is in agreement with published structure–activity relationships (Schoen *et al.*, 1994b; Ok *et al.*, 1994) and recent binding data with the rat pituitary membrane GHS receptor (**16,** K_i = 0.8 nM; L-692,429, K_i = 63 nM) (Pong *et al.*, 1996). Unfortunately, attempts to add this potency-enhancing substituent to the slightly more potent and less basic α-methylalanine side chain in **8** resulted in a loss of bioactivity (**17**) even when additional spacers were incorporated into the side chain (e.g., **18**).

The 2(*R*)-hydroxypropyl analogue **16** represented a benchmark for this series as it was the first analogue that was more potent than the hexapeptide GHRP-6 (ED_{50}s 3 versus 10 nM, respectively) in the rat pituitary cell assay. Consequently, **16** was studied extensively in dogs where it was shown to be 20-fold more potent than L-692,429 and 2-fold more potent than GHRP-6 when administered intravenously (Jacks *et al.*, 1994). Although this fulfilled our objective of improving potency relative to L-692,429, the critical question of oral bioavailability remained. Benzolactam **16** was evaluated in rats, dogs, pigs, rhesus monkeys, and chimps for oral bioavailability. In all species, the oral bioavailability for **16** was between 1 and 8%. In rat and rhesus monkey liver microsomes, **16** was metabo-

Table II
Modifications of the C-3 Amino Acid Side Chain

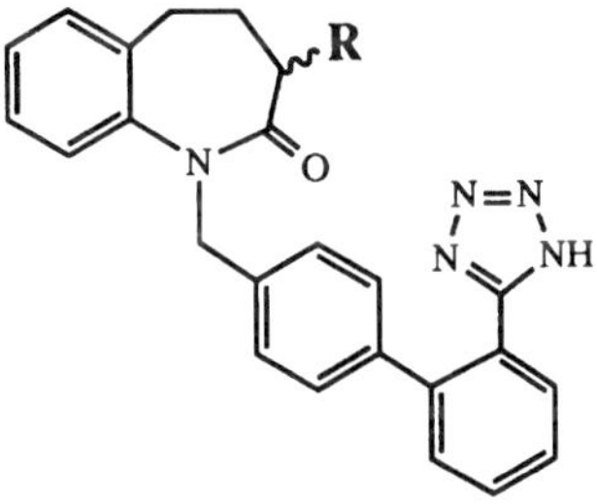

Compound	R	ED_{50} (μM)
3 L-692,429	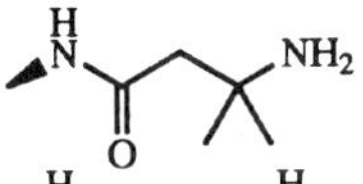	0.06
5	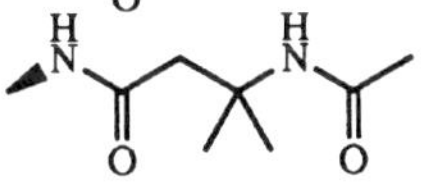	>20
6	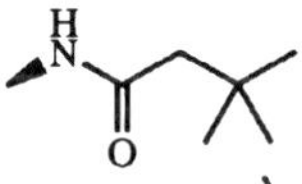	>20
7	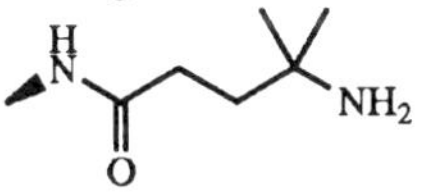	3
8	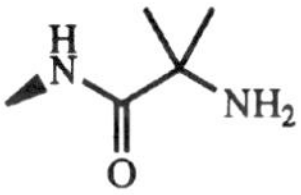	0.03
9	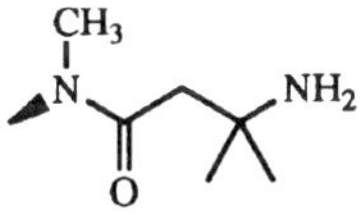	7
10	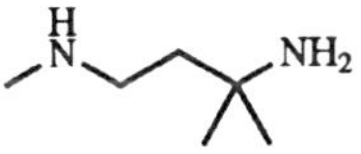	>1
11	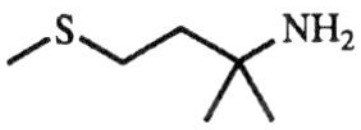	6
12	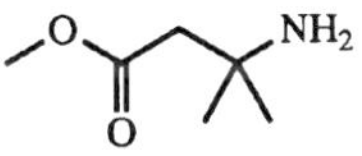	>1

Table III
Structure–Activity Profile of the Amino Substituent

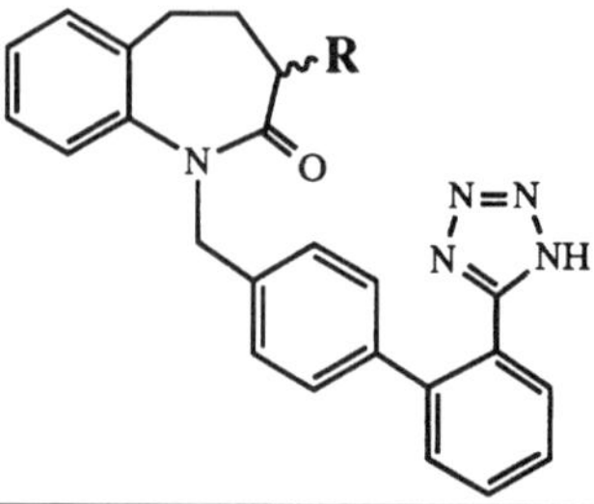

Compound	R	ED_{50} (μM)
3 L-692,429	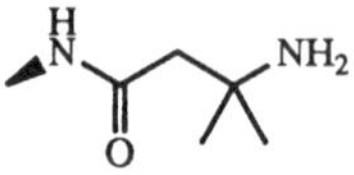	0.06
13	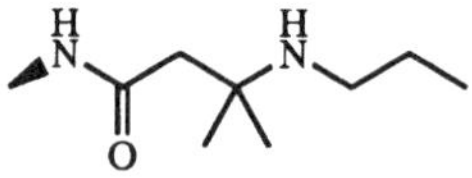	0.05
14	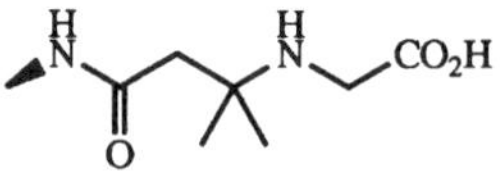	>1
15	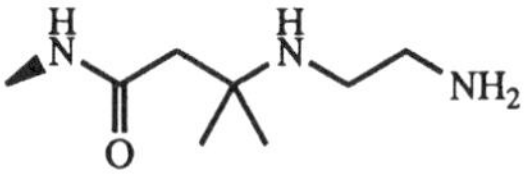	0.23
16	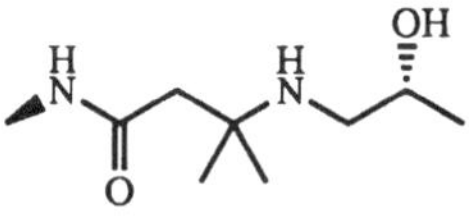	0.003
17	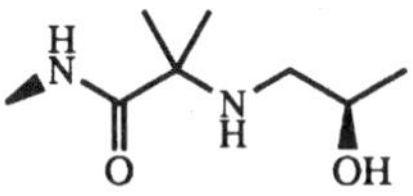	0.30
18	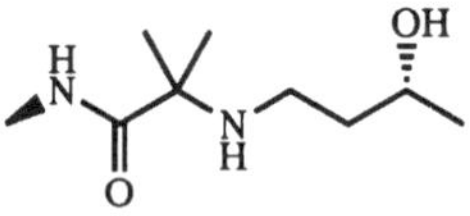	0.10

lized to the less biologically active L-692,429 via *N*-dealkylation of the amino substituent. Marginal improvements in oral bioavailability with **16** were possible via various formulations, but unfortunately they were unacceptable for clinical development. Although modifications of the C-3 amino acid side chain led to potency enhancement, the oral bioavailability issues remained.

2.2.2. REPLACEMENTS FOR THE 2′-BIPHENYL TETRAZOLE

Concurrent with these efforts to modify the amino acid side chain, attempts to replace the negatively charged tetrazole function were under way. As shown in Table IV, the tetrazole function in L-692,429 was 50-fold more potent *in vitro* than the corresponding carboxylic acid (**19**) or acylsulfonamide (**20**) derivatives despite their having similar pK_as (DeVita *et al.,* 1994a). Additionally, the 2′-methyl derivative **21** was slightly more potent than the carboxylic acid derivative **19,** thus suggesting that a negative charge is not required for biological activity. In fact, the tetrazole function can be replaced by other heterocycles (e.g., imidazole **22** and triazole **23**) (Bochis *et al.,* 1996) without loss of biological activity. As thiophene

Table IV
Replacements for the 2′-Biphenyl Tetrazole

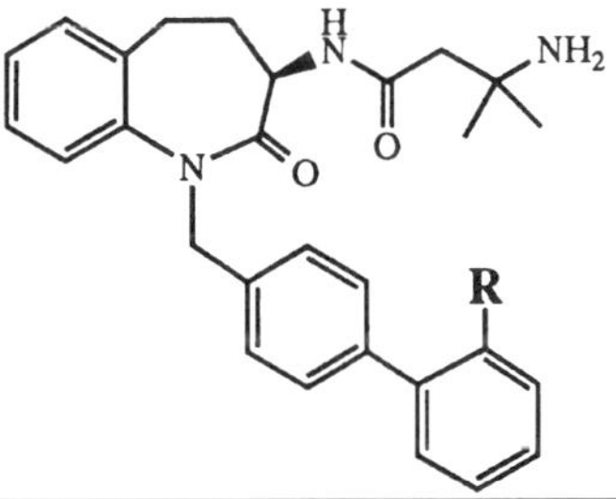

Compound	R	ED_{50} (μM)
3 L-692,429	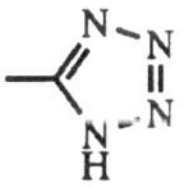 N N N N H	0.06
19	-COOH	3
20	$-SO_2NHCOPh$	3
21	$-CH_3$	1
22	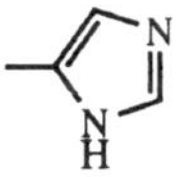 N N H	0.08
23	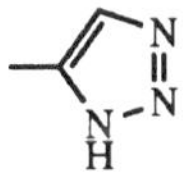 N N N H	0.04
24	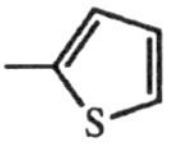 S	4
25	$-CONH_2$	0.08
26	$-NHCONHCH_3$	0.05
27	$-OCONHCH_3$	0.08

analogue **24** is not more active than the methylated derivative *21,* π–π stacking interactions of the 2′-heterocyclic substituent with the GHS receptor did not appear to play an important role. This suggested that the hydrogen bonding capabilities of these 2′-heterocycles may be important. To investigate this possibility, the 2′-carboxamide derivative **25** was prepared and found to be equipotent with tetrazole analogue L-692,429 (DeVita *et al.,* 1994b). Subsequently, urea **26** and carbamate **27** were also found to be equipotent to L-692,429, thus confirming a hydrogen bonding role for the 2′-substituent in L-692,429 (Bochis *et al.,* 1994).

With the identification of suitable neutral replacements for the anionic tetrazole function, the physicochemical properties of these molecules were changed dramatically and hopefully would have a profound positive effect on oral bioavailability. Because potency improvements were also required, the 2(*R*)-hydroxypropyl side chain was combined with 2′-neutral surrogates to afford, as expected, more potent analogues (e.g., **28,** ED_{50} = 4 nM, and **29,** ED_{50} = 2 nM). These analogues were evaluated in dogs for release of GH. Although many showed excellent oral activity at doses as low as 1 mg/kg, their oral bioavailability in dogs remained unacceptably low (<10%). It was very disheartening that such profound changes in the properties of the benzolactam lead did not improve its oral bioavailability. As efforts to solve the benzolactam oral bioavailability problem went on unabated, a strategic decision months earlier to continue screening for additional GH secretagogues was about to pay dividends.

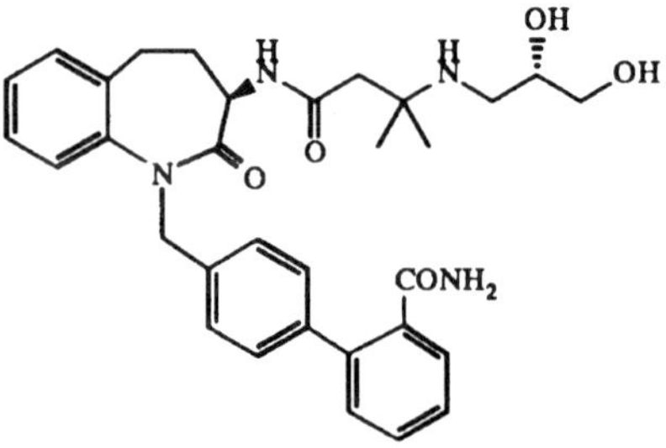

28 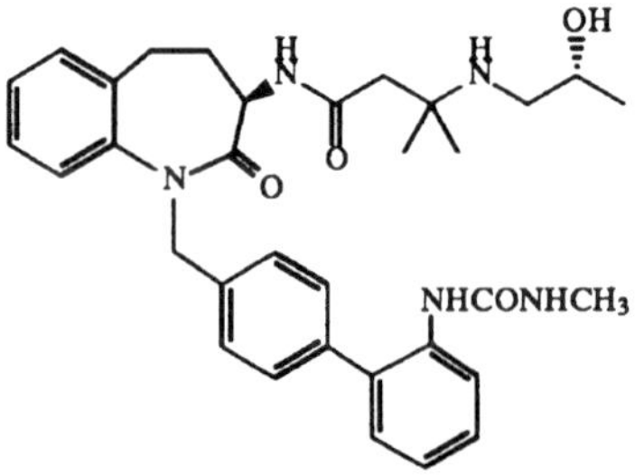 **29**

3. NEW STRUCTURAL LEADS

3.1. Privileged Structure Screening

As the benzolactam lead was being explored, Merck Sample Collection screening was also under way to discover additional peptidomimetic leads. Among the screening actives was **30** characterized by a modest ED_{50} = 300 nM in the rat pituitary growth hormone assay (Patchett *et al.,* 1995).

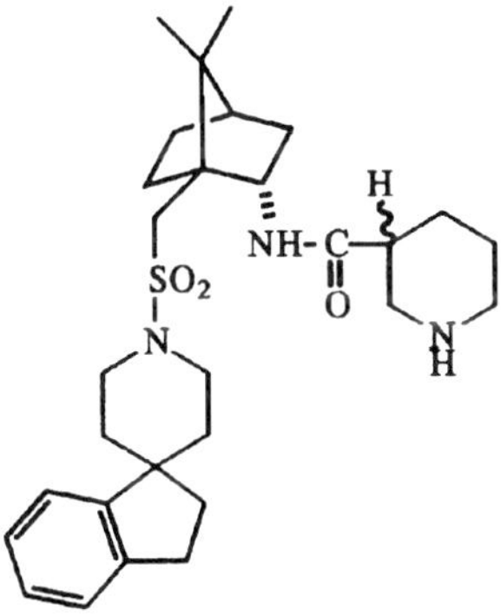

30

Attempts to improve the potency of **30** to a low nanomolar level were not successful (Nargund *et al.,* 1996). However, for broad screening and for the GHS target, it was decided to make some derivatives of the spiroindanylpiperidine component of this structure. This compound had come from an oxytocin antagonist project (IC_{50} = 68 nM) (Evans *et al.,* 1993) and spiroindanylpiperidines were known from other research at Merck to be sigma receptor antagonists (subsequently published by Chambers *et al.,* 1992). Therefore, we hypothesized that the spiroindanylpiperidine core was a "privileged structure" (Evans *et al.,* 1988) whose derivatization might afford potent ligands for a number of receptors in addition to the putative one involved in GH secretion. Subsequently, substituted spiropiperidines have also been reported to be antagonists of the NK1 receptor (Elliott *et al.,* 1996), of the NK2 receptor (Smith *et al.,* 1995), and as dual antagonists of the NK1 and NK2 receptors (Shah *et al.,* 1996), which today can be taken as additional support for their designation as privileged structures. From this spiroindanylpiperidine derivatization project, a highlight was the GHS lead **31** whose ED_{50} = 50 nM in the pituitary cell assay was remarkable for a mixture of four diastereomers (Patchett *et al.,* 1995). In retrospect, this excellent activity was as-

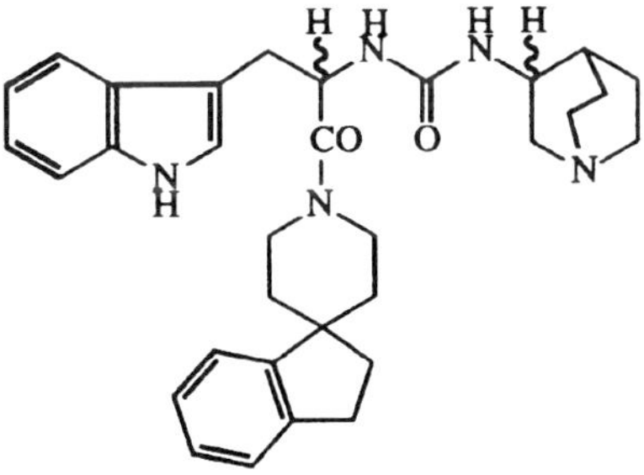

31

cribed to the presence of tryptophan which is a key amino acid in GHRP-6 and to the quinuclidine group which is a part-structure of another unpublished Merck GHS lead.

Evans *et al.* (1988) applied the term *privileged structures* to structural units that recur frequently in receptor ligands. However, the concept goes back to a classification of "multipotent competitive antagonists" that Ariens *et al.* (1979) proposed for many biogenic amine antagonists. The hydrophobic double-ring motif that they highlighted is frequently seen in CNS drugs (Andrews and Lloyd, 1982). What is unique in our design is the combination of amino acids with privileged structures and the agonist activity of these compounds.

3.2. Discovery of MK-0677

Unfortunately, the oral properties of **31** were disappointing. Despite an excellent elevation of GH in dogs when given intravenously at 0.1 mg/kg, there was no enhancement of GH levels even after oral administration at 5.0 mg/kg. Nonetheless, its structural novelty and high potency justified a major synthetic commitment. It was felt that different amines including some with reduced basicity and an elimination of the urea functionality might improve oral activity. We also hypothesized that there might be a common amine binding site in the GHS receptor for **31** and the benzolactams (e.g., L-692,429). Therefore, incorporation of amine side chains from the latter series was given a high priority. As shown in Table V, the best ones also afforded highly active spiroindanylpiperidine derivatives. Furthermore, the preferred stereochemistry of tryptophan in compound **32** as (*R*)- is identical to that at the benzolactam C-3 position, thereby suggesting a second link between the two series.

Compound **35,** an analogue of the highly active benzolactam **16,** was only poorly active orally in dogs at doses below 5.0 mg/kg. However, despite lesser intrinsic activity, compound **32** produced good GH elevations following oral administration at 2.0 mg/kg to dogs. Thus, it was chosen as the prototype spiroindanylpiperidine derivative for detailed study. We were particularly anxious to determine its activities on other receptors given its classification as a privileged structure derivative. The fact that no IC_{50} activities less than 10 μM were found for compound **32** in over 20 receptor assays was reassuring and, in many ways, unexpected so early in an analogue program. An oral bioavailability in rats greater than 40% (Chen *et al.,* 1996) further strengthened our confidence in the lead's potential.

The major goal in synthesizing analogues of compound **32** was simply to increase potency while retaining its specificity and oral bioavailability. When an alcohol or ketone was introduced in the indane benzylic position, ED_{50}s were lowered more than 10-fold in derivatives **37** and **38** (Table VI) (Patchett *et al.,* 1995).

Table V
Spiroindanylpiperidine Lead: Amino Acid Side Chain Modifications

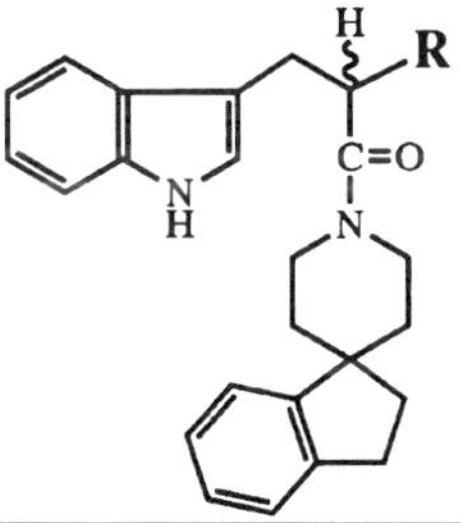

Compound	R	ED_{50} (nM)
32	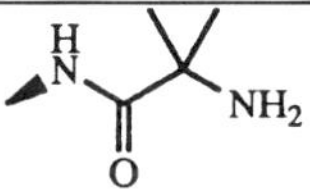	14
33	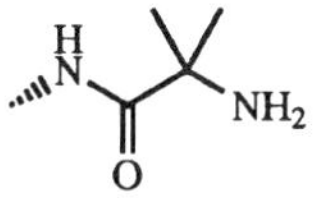	10,000
34	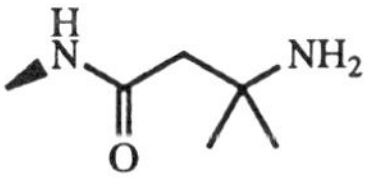	76
35	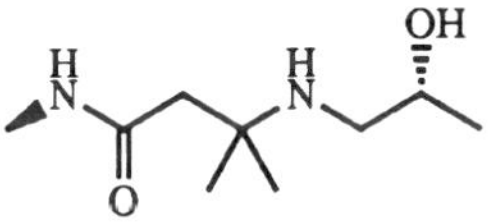	2.6
36	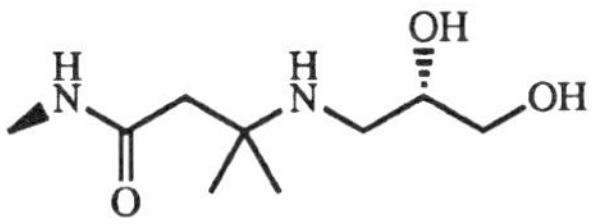	16

These compounds were orally active in dogs but not in proportion to their increased intrinsic activities. Ketone reduction and conjugation of the alcohol were possibly responsible for the disappointing results. Thus, other polar, hydrogen bond-accepting and -donating functional groups were tried in this position, which might be less subject to metabolism. The best of these analogues was the methanesulfonyl derivative **39**. Concurrently, studies with D-tryptophan replacements in the spiroindane series had uncovered particularly good oral activity using *O*-benzyl-D-serine. When this amino acid and the methanesulfonylspiroindoline partstructure were combined in one molecule, both high potency (K_i = 0.24 nM) and excellent oral activity were obtained. The resultant compound **40** (L-163,191) as

Table VI
Optimization of the Spiroindanylpiperidine Lead

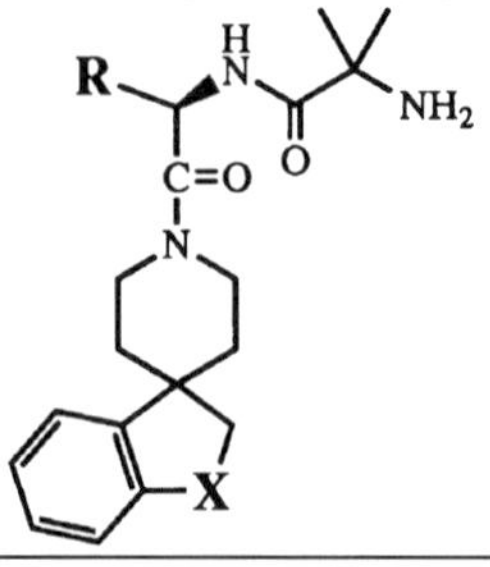

Compound	R	X	ED_{50} (nM)
32	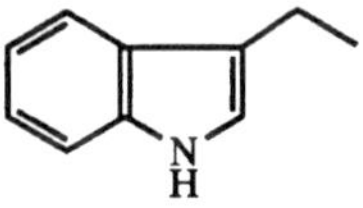	CH_2	14
37	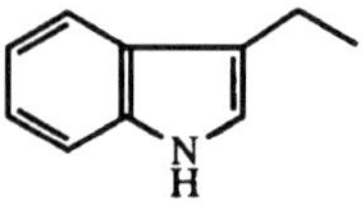	C=O	1.2
38	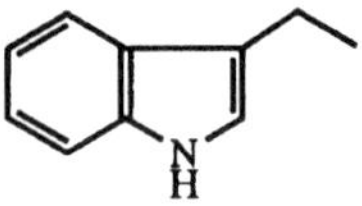	CHOH	0.6
39	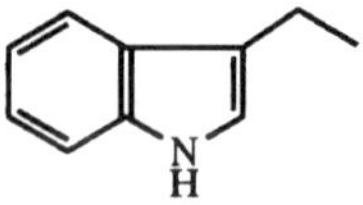	NSO_2CH_3	1.8
40 MK-0677	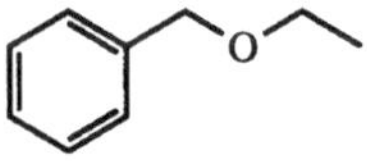	NSO_2CH_3	1.3
41	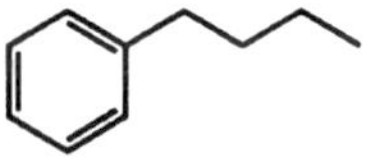	NSO_2CH_3	1.5

its mesylate salt was subsequently selected for safety assessment studies and it has since entered clinical trials as MK-0677 (Patchett *et al.*, 1995).

The properties of MK-0677 that led to its selection included an ED_{50} = 1.3 nM in the rat pituitary cell assay and an oral bioavailability in dogs of more than 60% (Patchett *et al.*, 1995). Contributing to this excellent bioavailability is peptidase stability presumably arising from the serine derivative's (D)-amino acid configuration and steric hindrance provided by the α-methylalanine (α-aminoisobutyric acid) component. Also, the compound's lipophilicity (log $P = 3.0$), water solubility (>250 mg/ml as its mesylate salt), and moderate basicity ($pK_a = 7.8$) are within generally accepted ranges for well-absorbed compounds. Furthermore, MK-0677 shows good permeability in Caco-2 cells, which frequently correlates with good drug absorption in man (Artursson and Karlsson, 1991).

In rats, the bioavailability of MK-0677 ranged between 6 and 22%. Absorption was good but clearance was rapid in this species (Leung *et al.,* 1996b). Fortunately, dogs rather than rats proved to be the better predictor of MK-0677's good oral activity in man. For example, in a study in elderly patients, Chapman *et al.* (1996b) reported IGF-1 elevations into the young adult range following once-a-day oral administration of only 25 mg of MK-0677. The excellent oral properties of MK-0677 are in marked contrast to those of GHRP-6, which was reported to be less than 1% bioavailable in man (Hartman *et al.,* 1992; Bowers *et al.,* 1992). Also, the oral absorption of the benzolactam L-692,429 was only about 3% in rats (Leung *et al.,* 1996a).

3.2.1. DURATION OF ACTION

MK-0677 is a long-acting GH secretagogue. Given orally to dogs at 1.0 mg/kg, it elevated GH for 6 hr (Fig. 2) (Jacks *et al.,* 1996a) and its terminal half-

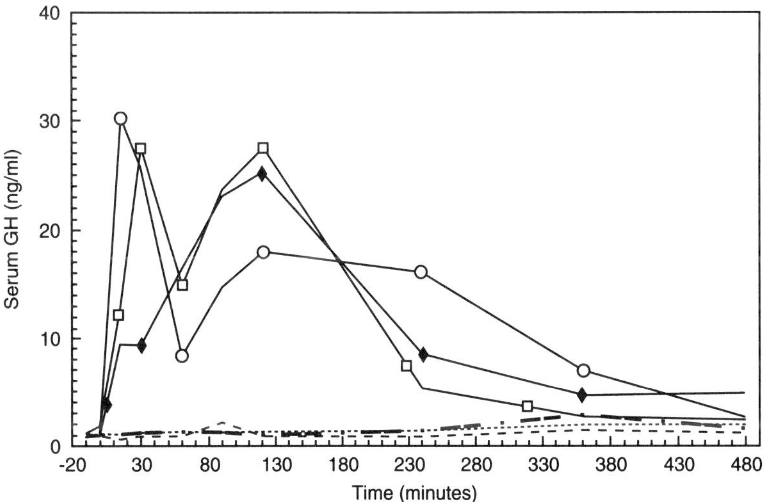

Figure 2. Serum growth hormone levels in beagles dosed with MK-0677 ($n = 3$, solid lines) or water ($n = 3$, broken lines). Reprinted with permission from Jacks *et al.,* 1996, MK-0677, a potent, novel orally-active growth hormone (GH) secretagogue—GH, IGF-1 and other hormonal responses in beagles, *Endocrinology* **137:**5284–5289. ©The Endocrine Society.

life in beagles following i.v. administration of 0.5 mg/kg was found to be between 5 and 6 hr (Leung *et al.,* 1996b). Most importantly, in clinical studies as noted above, MK-0677 has a duration of action adequate for once-a-day use as judged by its IGF-1 elevations.

MK-0677's extended duration of action is in marked contrast to that of the benzolactam secretagogue **16** (L-692,585) whose 21-fold peak elevation of GH in beagles following a 0.10 mg/kg i.v. dose returned to near baseline by 60 min (Jacks *et al.,* 1994). Similarly, the terminal half-life of the clinically studied benzolactam L-692,429 following i.v. administration of 0.9 mg/kg to dogs was determined to be only 0.88 ± 0.14 hr (Leung *et al.,* 1996a). As generally expected of peptides, a short serum half-life of approximately 20 min was reported for GHRP-6 in man following i.v. administration (Bowers *et al.,* 1992). The terminal half-life of the GHRP-6 analogue hexarelin in dogs was reported to be approximately 2 hr (Roumi *et al.,* 1995) and other studies suggest its duration of action is essentually the same as that of GHRP-6 (Bowers, 1996). GHRP-2 given intravenously to normal men has a terminal phase half-life of between 1 and 2 hr (Orczyk *et al.,* 1996). Thus, the extended duration of MK-0677 in man and animals contrasts with the earlier reported peptides and benzolactams.

3.2.2. DESENSITIZATION AND IGF-1 LEVELS

A once-a-day, orally active drug is preferred in medical practice. However, the relationship of an agonist's duration of action to desensitization and to IGF-1 levels must also be considered. Pituitary cells become rapidly desensitized to GHRPs (Bowers *et al.,* 1981; Badger *et al.,* 1984) and this is a general property of GHRPs and their peptidomimetics. Complete recovery requires at least 1 hr. However, desensitization *in vivo* is not a problem when the short-acting GHRPs are given several times a day as growth effects in animals and man have been observed with these dosing schedules over extended periods of time (Bowers *et al.,* 1984, 1991; Bowers, 1993). No desensitization of the GH peak or AUC responses was observed when 0.1 mg/kg of the potent, short-acting benzolactam **16** was administered to beagles by repeated once-a-day injection in a 15-day study. IGF-1 was elevated transiently, and to extend this elevation, the authors suggested that multiple injections or a compound with a longer duration of action might be more effective (Jacks *et al.,* 1994).

GHRP-6 and **16** have also been continuously infused in guinea pigs over a 6-hr period (Fairhall *et al.,* 1995) with resultant enhancement of GH secretion in its normal pulsatile pattern. Most importantly, continuous i.v. infusion studies of GHRP-6 have been conducted in man lasting 24 hr (Huhn *et al.,* 1993) and 36 hr (Jaffe *et al.,* 1993). Some desensitization was observed at the end of the 24-hr study as determined by administration of a bolus injection of GHRP-6. However, in both cases these infusions significantly augmented pulsatile GH secretion and increased mean plasma IGF-1 concentrations compared with saline controls. Pulsatile GH elevation was also observed during continuous infusion in man of the benzolactam L-692,429 (Chapman *et al.*, 1996a). From these studies it could be inferred that a long-duration, once-a-day orally active GH secretagogue should elevate GH in man in a pulsatile manner to mimic normal physiology and that IGF-1 would also be elevated. However, determining the levels of GH and IGF-1 that could be achieved in man over many days would need to be established in extended clinical trials.

As expected, a rapid desensitization of MK-0677's GH-releasing effect was seen in isolated rat pituitary cells (Patchett *et al.*, 1995). Downregulation of the GH response was also observed in a 14-day experiment in which MK-0677 was administered once a day to beagles orally at 1.0 mg/kg (Table VII) and at 0.25 and 0.50 mg/kg dose levels (data not shown) (Hickey *et al.*, 1997). This was in contrast to the benzolactam secretagogue **16** where no downregulation of the GH response was observed following once-a-day i.v. administration for 2 weeks (Jacks *et al.*, 1994). However, as shown in Table VII, the GH elevations produced by MK-0677 remained significantly higher than controls at 7 days and were still at that level on day 14 following the 1.0 mg/kg oral dose. IGF-1 levels were elevated up to 126% of control values. Numerically, there was no difference between the absolute values of the plasma IGF-1 levels on days 7 and 14. Importantly, predose levels of IGF-1 on days 7 and 14 were also significantly higher than control pretreatment levels.

To explain these observations, Hickey *et al.* (1997) suggested that sustained levels of IGF-1 produced by a long-duration GH secretagogue feed back on the pituitary and hypothalamus to reduce GH release. This suggestion is in accord with a reduction of GH levels that has been observed in man following injection of IGF-1 (Bermann *et al.*, 1994). In the experiment of Hickey *et al.* (1997), MK-0677 was

Table VII
Serum GH and IGF-1 Concentrations in Beagles Following Once-Daily Administration of MK-0677 at 1.0 mg/kg for 14 Days

		Growth hormone		Mean IGF-1 (ng/ml)	
Treatment	Day	Mean peak (ng/ml)	Mean AUC (ng/ml per hr)	0 hr	8 hr
MK-0677	1	70.2 ± 11.3*	189.7 ± 15.3*	63.9 ± 14.8	96.5 ± 9.8
(1.0 mg/kg)	7	11.4 ± 1.8*	43.2 ± 3.5*	111.9 ± 18.1*	121.8 ± 16.3*
	14	13.2 ± 2.1	44.5 ± 3.6	105.9 ± 21.1*	124.2 ± 22.5*
Vehicle	1	3.1 ± 0.7	19.4 ± 2.2	79.4 ± 6.8	72.2 ± 13.9
	7	3.8 ± 0.9	21.3 ± 2.4	84.5 ± 10.2	68.2 ± 13.7
	14	5.3 ± 0.2	25.3 ± 2.9	87.8 ± 12.0	81.3 ± 16.2

*Statistically different ($p < 0.05$) from the associated placebo mean. For IGF-1 analysis, statistics were based on % change in concentrations (Hickey *et al.*, 1997).

able to bring the GH and IGF-1 axis to a higher but plateaued level of activity. If there are mechanism-based side effects resulting from bolus injections of rhGH, they may be less likely with the self-limiting, pulsatile GH increase that MK-0677 produces.

3.2.3. SPECIFICITY

The specificity of MK-0677 was initially checked in over 50 *in vitro* assays including cholinergic, galanin, serotonin, histamine, sigma, opiate, vasopressin, and oxytocin receptor assays. In all of them, the IC_{50} values for MK-0677 were greater than 10 μM (Patchett *et al.,* 1995). Subsequently, it was shown that MK-0677 did not bind to somatostatin receptors nor would GHRH displace [^{35}S]-MK-0677 from the GHS receptor (Pong *et al.,* 1996).

The effects of MK-0677 on serum hormone levels *in vivo* were studied in dogs following an i.v. bolus injection of 0.25 mg/kg (Jacks *et al.,* 1996a). Total GH levels over 6 hr expressed as GH AUC were increased 6-fold whereas luteinizing hormone (LH) and prolactin (PRL) were unaltered and thyroxine (T_4) was slightly lowered. Insulin and glucose levels were higher although within normal ranges for dogs. However, cortisol AUC was increased 2.3-fold over the 6-hr period. Similar observations were made of the benzolactam L-692,429 in beagles (Hickey *et al.,* 1994). The effects of MK-0677 on cortisol levels were carefully monitored in the clinic although, in early clinical trials with measurements at 7 days and beyond, as noted below, elevations outside of the normal range were not observed. In dogs the responses of both cortisol and GH to MK-0677 are downregulated after several days of treatment (Hickey *et al.,* 1997). Perhaps as the authors suggest, both of these attenuated responses are mediated by the increased IGF-1 levels.

3.2.4. FUNCTIONAL STUDIES IN DOGS

The effects of an MK-0677 analogue on muscle formation and function were studied in a dog hind limb immobilization protocol. In this study the right hind leg of beagles was kept immobile by a fixation device for 10 weeks, and after its removal, the dogs were free to move about for an additional 5 weeks. Throughout the entire study, the beagles were treated once a day orally with either compound **41** at 5 mg/kg or a water placebo. IGF-1 levels were increased around 60% in the treated group whereas they were modestly depressed throughout the experiment in the control group. At week 15 the treated beagles had lost 0.21 ± 0.20 kg compared with a loss of 1.13 ± 0.19 kg in the placebo group. Muscle strength measured as isometric torque in the restrained limb declined in both groups during the first 10 weeks, but at 15 weeks, torque had increased 43% in the dogs receiving **41** compared with a 16% improvement in the untreated dogs. There was also a strong correlation between the diameter of the vastus lateralis muscle fiber and torque (Jacks *et al.,* 1996b).

These results with a secretagogue closely related to MK-0677 demonstrated sustained IGF-1 elevations over many weeks with accompanying anabolic improvements compared with controls, especially with combined exercise and drug treatment. The accelerated rate of muscle recovery that was observed in this experiment suggests possible utility in the treatment of frail elderly patients and in the rehabilitation therapy of individuals after extended periods of inactivity.

3.2.5. CLINICAL STUDIES

The first published clinical results using MK-0677 were of a 7-day study in normal young men, aged 18–30, who were given placebo, 5- or 25-mg doses of MK-0677 once a day at bedtime. On day 8, serum levels of IGF-1 had increased in a dose-responsive manner (placebo, 248 ± 101 μg/liter; low dose, 325 ± 110 μg/liter, and high dose, 360 ± 95 μg/liter). In this study, GH's AUC was similar in all three groups, although an increase especially in the number of low-amplitude GH pulses was observed. MK-0677 had no effect on 24-hr cortisol levels in plasma or in urine (Copinschi *et al.,* 1996).

The effects of MK-0677 in healthy elderly individuals were investigated in two separate study periods of 14 and 28 days involving 15 women and 17 men, aged 64–81, in randomized, double-blind, placebo-controlled studies. They were given 0, 2, 10, or 25 mg MK-0677 orally, and on days 14 and 28, blood was collected every 20 min for 24 hr and analyzed for changes in GH, IGF-1, cortisol, and other hormones as noted below. The 2-mg dose was ineffective in raising GH or IGF-1, whereas the two higher doses produced significant increases in both at 2 weeks. With the 10-mg dose, it was also demonstrated that MK-0677 raised GH and IGF-1 more effectively when given in the morning rather than at bedtime. Following oral, once-a-day administration of 25 mg MK-0677 for 2 weeks, mean 24-hr GH concentration had been increased by 97 ± 23%. Its secretion was pulsatile with an increase in AUC but no change in the number of GH peaks. Remarkably with this oral dose, IGF-1 concentrations had increased into the normal range for young adults: 219 ± 21 μg/liter at 2 weeks and 265 ± 29 μg/liter at 4 weeks. Mean serum cortisol levels were not significantly different from placebo controls at 14 days. PRL concentrations had increased 23% but were within the normal range and thyroid hormones were not affected significantly by MK-0677. Fasting blood glucose had increased 26.9 ± 6.8% above baseline at 4 weeks with an accompanying increase in insulin levels. The latter effects were possibly the result of increased insulin resistance produced by elevated GH. The authors suggested that with longer use, body fat may decrease along with a reduction in insulin resistance (Chapman *et al.,* 1996b).

Plotkin *et al.* (1996) reported a double-blind, placebo-controlled study involving 104 musculoskeletally impaired elderly patients who were given for the first 2 weeks doses of 0, 5, 10, or 25 mg of MK-0677. Thereafter, all patients on MK-0677 received 25 mg orally for a total of 9 weeks. At that time, hormonal changes were determined in a predefined subgroup of patients. In them, mean 24-hr serum GH levels had increased 250 $\pm$ 170% (n = 11) relative to baseline, there was no change in cortisol levels, PRL had increased 17%, and there was a small (ca. 8%) increase in mean fasting glucose levels (n = 12). When sampled at week 9, GH secretion remained pulsatile.

MK-0677's once-a-day oral efficacy as measured by sustained elevated levels of GH and IGF-1 over many weeks is supported by these studies. These initial trials provide a basis for the more extensive studies that will be required to establish important clinical benefits.

4. MECHANISM OF ACTION OF GH SECRETAGOGUES

4.1. Biochemistry

The early studies of Cheng *et al.* (1991) demonstrated that GHRP-6 in contrast to GHRH apparently signaled through the phospholipase C pathway. The effects of GHRP-6 are mimicked by activators of protein kinase C and antagonized by the protein kinase C inhibitor phloretin. Also, chronic exposure of rat anterior pituitary cells to a protein kinase C activator almost completely desensitized the cells to treatment with GHRP-6 without affecting their response to GHRH. Thus, the receptors for GHRH and GHRP-6 were distinct but, like GHRH, GHRP-6 appeared to interact with a G-protein-coupled receptor. The conclusions from these early studies suggesting that the GHRP-6 signal is transduced through phospholipase C are now supported by more direct evidence. Both GHRP-6 and the nonpeptide mimetic L-692,429 increase inositol triphosphate and translocation of protein kinase C (Smith *et al.*, 1993; Adams *et al.*, 1995; Mau *et al.*, 1995), and GHRP-6 stimulates Ca^{2+} release from intracellular stores (Bresson-Bepoldin and Dufy-Barbe, 1994; Herrington and Hille, 1994). These observations endorsed the notion that GHRP-6 and L-692,429 interact with a G-protein-coupled receptor that signals through phospholipase C to liberate the second messengers IP_3 and diacylglycerol.

Influx of extracellular Ca^{2+} is a common signal for the release of hormones stored in secretory granules. The role of Ca^{2+} in the signaling pathway involved in the action of GHRP-6, L-692,429, and MK-0677 was demonstrated using fura-2 to monitor changes in fluorescence in somatotrophs (Smith *et al.*, 1993, 1996a,b). The effects on fura-2 fluorescence and GH release were dependent on activation of L-type Ca^{2+} channels on somatotrophs. These results revealed that although the receptors, the signal transduction pathways, and the second messengers activated by the GH secretagogues were different from those activated by GHRH, the pathways converged such that GH release is caused by influx of Ca^{2+} through L-type Ca^{2+} channels (Smith *et al.,* 1993).

GHRP-6 and the peptidomimetics behave as functional antagonists of somatostatin (Cheng *et al.,* 1989; Blake and Smith, 1991) apparently by depolarizing somatotrophs (Smith *et al.,* 1993; Patchett *et al.,* 1995). The antagonism appears to be mediated by inhibiting K^+ channels and opening Ca^{2+} channels. Electrophysiology studies performed on somatotrophs showed that L-692,429 and GHRP-6 block K^+ currents resulting in depolarization and electrical spiking to enhance Ca^{2+} entry through voltage-gated channels. Depolarizing agents mimic the effects of the GHRP-6 and peptidomimetics by amplifying GHRH-induced GH release; however, the depolarizing agents are very weak GH secretagogues when used alone, illustrating the additional importance of activation of the phospholipase C pathway (Smith *et al.,* 1996a,b; Pong *et al.,* 1992, 1993; McGurk *et al.,* 1993; Leonard *et al.,* 1991).

4.2. Characterization of the GH Secretagogue Receptor (GHS-R)

The paucity of receptors in rat pituitary membranes and the high level of high-capacity low-affinity binding associated with radiolabeled peptide ligands frustrated early efforts to characterize a specific high-affinity site for GHRP-6 and the peptidomimetics. Following the identification of MK-0677, a radiolabeled analogue was synthesized by incorporating ^{35}S in place of ^{32}S to provide a high-specific-activity (700–1100 Ci/mole) ligand suitable for characterization of the receptor (Smith *et al.,* 1996b; Dean *et al.,* 1996; Pong *et al.,* 1996). It was shown by Scatchard analysis that [^{35}S]-MK-0677 bound to pig pituitary membranes with a K_d = 140 pM and B_{max} of 6 fmole/mg membrane protein. A similar binding site of lower concentration (2 fmole/mg protein) was measured in rat pituitary membranes. Competition binding studies with [^{35}S]-MK-0677 showed that other compounds in the MRL series such as L-692,429 and L-692,585 (**16**) bound to the same site as MK-0677. Most importantly, the peptides GHRP-6 and GHRP-2 also displaced [^{35}S]-MK-0677 from the binding site confirming that the MRL compounds were indeed mimetics of the GHRPs. Indeed, GHRP-6 was shown to be a competitive inhibitor of MK-0677 binding . When binding affinity was estimated from the ability to compete with [^{35}S]-MK-0677 binding, there was an excellent correlation with efficacy in stimulating GH release from rat pituitary cells. Similar high-affinity specific binding was also demonstrated in membranes isolated from rat hypothalamus but not in membranes from rat liver and posterior pituitary gland. Binding of [^{35}S]-MK-0677 was Mg^{2+} dependent and both GTP-γ-S and GDP-NHP accelerated dissociation of the ligand from pituitary and hypothalamic membranes consistent with binding to a G-protein-coupled receptor (Smith *et al.,* 1996a,b; Pong *et al.,* 1996; Chaung *et al.,* 1996).

4.3. Cloning the GH Secretagogue Receptor

Having demonstrated that a specific receptor was expressed in the anterior pituitary gland, poly-A^+ RNA was isolated from pig pituitary glands. The strategy for cloning was provided by the demonstration that the receptor for GHRP-6, L-692,429, and MK-0677 signaled through the phospholipase C pathway (Smith *et al.,* 1996a,b). Poly-A^+ RNA isolated from pig pituitaries was injected into xenopus oocytes and after incubation for 2–3 days the addition of MK-0677 caused activation of a Ca^{2+}-activated Cl^- current. However, because this signal was not reproducible enough to allow the cloning of a rare cDNA from a pituitary cDNA library efficiently, a new strategy for expression cloning had to be developed. To give a more robust signal for the detection of changes in Ca^{2+}, aequorin bioluminescence was used instead of electrophysiology, and to improve the efficiency of G-protein coupling; the $G_{\alpha 11}$ protein was coexpressed. An MK-0677-inducible aequorin bioluminescent signal was reproducibly observed when cRNA encoding aequorin and cRNA encoding $G_{\alpha 11}$ were coinjected with RNA derived from a pig pituitary cDNA library. Using this bioluminescence assay to fractionate pools from the cDNA library, a single cDNA clone encoding a protein that bound [^{35}S]-MK-0677 with high affinity was isolated. Displacement of [^{35}S]-MK-0677 binding correlated with the biological activity of MK-0677, GHRP-2, and GHRP-6 (Howard *et al.,* 1996).

The predicted amino acid sequence of GHS-R1a was consistent with that of a new G-protein-coupled receptor (GPC-R). Based on the nucleotide and predicted amino acid sequence, the GHS-R1a appears to be the first member of a new family of GPC-Rs. Cloning of the human and rat homologues show that the receptor is highly conserved across these species (Howard *et al.,* 1996; McKee *et al.,* 1997); therefore, it is likely that a natural ligand for the GHS-R exists.

4.4. GH Secretagogue Receptor and GH Pulsatility

In situ hybridization using nucleic acid probes shows that the GHS-R is expressed in the anterior pituitary, arcuate nucleus, ventromedial hypothalamus, and hippocampus (Howard *et al.,* 1996; Guan *et al.,* 1997). Localization of the receptor in the hypothalamus is consistent with observations that treatment of rats and mice with GHRP-6, L-692,429, L-692,585, and MK-0677 results in both increased electrical activity and increases in expression of c-*fos* in arcuate neurons (Dickson *et al.,* 1995; Sirinathsinghji *et al.,* 1996; Dickson, 1996; Bailey *et al.,*

1996). A significant proportion of these neurons contain GHRH (Dickson and Luckman, 1997). Activation can be prevented by administration of somatostatin or its longer-acting analogue sandostatin (Dickson *et al.,* 1996) suggesting a negative feedback role for somatostatin on GHRH neurons activated by GHRP-6 and the peptidomimetics.

Based on their properties as antagonists of somatostatin and amplifiers of GHRH activity, it is clear how ligands for the GHS-R increase GH secretion. Less obvious is how these molecules sustain pulsatile GH release. Single daily oral dosing with MK-0677 causes amplification of pulsatile GH release for 24 hr in humans (Chapman *et al.,* 1996b) as does L-692,429 and GHRP-6 when constantly infused (Chapman *et al.,* 1996a; Huhn *et al.,* 1993). Similarly, in guinea pigs L-692,585 (**16**) and GHRP-6 initiate pulsatility on commencement of infusion, suggesting that molecules interacting with the GHS-R reset the pulse generator regulating the episodic GH profile (Smith *et al.,* 1996b; Fairhall *et al.,* 1995). However, when pituitary cells are exposed to these secretagogues *in vitro,* GH release occurs almost instantaneously and the cells rapidly become refractory to repeated stimulation (Blake and Smith, 1991). To explain the apparent paradox between the *in vivo* and *in vitro* results, other factors such as somatostatin must play a regulatory role *in vivo.*

GH appears to self-entrain its pulsatility (Carlsson and Jansson, 1990). Somatostatin apparently plays a key role in GH-mediated negative feedback (Plotsky and Vale, 1985; Tannenbaum and Ling, 1984; Turner and Tannenbaum, 1995; Clark *et al.,* 1988; Bertherat *et al.,* 1995; Conway *et al.,* 1985; Frohman *et al.,* 1992) and is involved in resensitizing the GHS-R. A model, based on a series of observations from *in vivo* experiments, for the control of GH pulsatility is that GH sustains its own pulsatility through the coupling of three biological oscillators, somatostatin, GHRH, and the natural ligand for GHS-R. When the peptidomimetic GH secretagogues are administered acutely, they reset the coupled oscillators presumably by antagonizing somatostatin. They interrupt endogenous somatostatin tone, hence relieving repression on GHRH neurons and resulting in GHRH release. In response to GHRH, GH is released and then feeds back on the hypothalamus to entrain a new cycle.

5. CONCLUSION

The discovery of small orally active compounds that are able to cause the release of GH from the pituitary has stimulated considerable scientific and clinical interest. It is the latest milestone in a fascinating research story that began 20 years ago when Cyril Bowers of Tulane University announced the discovery of analogues of the enkephalins that specifically released GH from pituitary cells. Largely as a result of the dedicated efforts of Dr. Bowers, the potential of these peptides was nurtured, their potency and metabolic stability were improved, and their efficacy was demonstrated in animals and in the clinic. These relatively small peptides have activity after oral administration despite limited bioavailability. Most importantly, when given by infusion, they stimulate the pulsatile release of GH and, thereby, augment the normal, physiological pattern of GH release.

A research project was begun at Merck in the late 1980s with the goal of discovering a small molecule peptidomimetic of the GHRP hexapeptides. The discovery of the benzolactams is a noteworthy achievement, as at the time only the opiates were known to be peptidomimetic agonists.

The difficulty of achieving good oral bioavailability and once-a-day duration of action, however, is not automatically solved when nonpeptide compounds are designed from peptide leads. In addition to the obvious interplay between medicinal chemists and biologists, the GHS program at Merck was heavily dependent on *in vivo* testing for efficacy and oral bioavailability. Metabolism and absorption issues for the benzolactam lead were the primary focus of many researchers. Although the complete benzolactam story cannot be discussed at this time, the discovery of benzolactam L-692,429 and its "proof of concept" clinical trials were, at the same time, very rewarding and extremely frustrating.

Despite the limited capacity of the rat pituitary GH release assay, the strategic decision to continue screening for additional structural leads afforded the breakthrough that resulted in the oral clinical candidate MK-0677. This lead came from a project to derivatize "privileged structures" with capped amino acids and, in retrospect, its success was ascribed to the fact that the modular units of the lead were derived from three different compounds with GHS activity including GHRP-6.

The synthesis of the potent, selective agonist MK-0677 as an ^{35}S ligand led to important advances in our knowledge of how GH secretion is controlled. In particular, this ligand allowed the identification of a low-abundance, specific, high-affinity receptor of the GHRPs in pituitary and hypothalamic membranes. It was subsequently cloned and identified as the first member of a new family of G-protein coupled receptors. The presence of this unique receptor underscores the strong likelihood that a natural ligand interacts with it. If it were found, a new dimension in our understanding of GH secretion would have been established. Importantly, the study of its levels as a function of aging and of GH deficiency syndromes would then be of interest with implications for the most effective clinical use of GHRPs and their mimetics in replacement therapy. Even in our present stage of knowledge, the interaction of drug design research with basic science is well exemplified in the evolution of these orally active peptidomimetic GH secretagogues.

MK-0677 has been shown to release GH in healthy young males orally with once-a-day doses as low as 5 mg. In the elderly, a 25-mg oral dose has been reported to elevate IGF-1 to levels at the low end of normal for young adults. Based on these encouraging results, additional clinical trials have been conducted to establish the clinical benefits/endpoints for MK-0677 and will be reported at a later time.

REFERENCES

Adams, E. F., Petersen, B., Lei, T., Buchfelder, M., and Fahlbusch, R., 1995, The growth hormone secretagogue, L-692,429, induces phosphatidylinositol hydrolysis and hormone secretion by human pituitary tumors, *Biochem. Biophys. Res. Commun.* **208:**555–561.

Aloi, J. A., Gertz, B. J., Hartman, M. L., Huhn, W. C., Pezzoli, S. S., Wittreich, J. M., Krupa, D. A., and Thorner, M. O., 1994, Neuroendocrine responses to a novel growth hormone secretagogue, L-692,429, in healthy older subjects, *J. Clin. Endocrinol. Metab.* **79:**943–949.

Andrews, P. R., and Lloyd, E. J., 1982, Molecular conformation and biological activity of central nervous system active drugs, *Med. Res. Rev.* **2:**355–393.

Ariens, E. J., Beld, A. J., Rodrigues de Miranda, J. F., and Simonis, A. M., 1979, The pharmacon-receptor-effector concept, a basis for understanding the transmission of information in biological systems, in: *The Receptors: A Comprehensive Treatise,* Volume 1 (R. D. O'Brien, ed.), pp. 33–91, Plenum Press, New York.

Artursson, P., and Karlsson, J., 1991, Correlation between oral drug absorption in humans and apparent drug permeability coefficients in human intestinal epithelial (Caco-2) cells, *Biochem. Biophys. Res. Commun.* **175:**880–885.

Badger, R. M., Millard, W. J., McCormick, G. F., Bowers, C. Y., and Martin, J. B., 1984, The effects of growth hormone (GH) releasing peptides on GH secretion in perifused pituitary cells of adult male rats, *Endocrinology* **115:**1432–1438.

Bailey, A., Smith, R. G., and Leng, G., 1996, Activation of hypothalamic arcuate neurons in the urethane-anesthetized rat following systemic injection of growth-hormone secretagogues, *J. Physiol. London* **495:**115–116.

Bermann, M., Jaffe, C. A., Tsai, W., DeMott-Friberg, R., and Barkan, A. L., 1994, Negative feedback regulation of pulsatile growth hormone secretion by insulin-like growth factor-1, *J. Clin. Invest.* **94:**138–145.

Bertherat, J., Bluet-Pajot, M. T., and Epelbaum, J., 1995, Neuroendocrine regulation of growth hormone, *Eur. J. Endocrinol.* **132:**12–24.

Blake, A. D., and Smith, R. G., 1991, Desensitization studies using perifused rat pituitary cells show that growth hormone releasing hormone and His-D-Trp-Ala-Trp-D-Phe-Lys-NH_2 stimulates growth hormone release through distinct receptor sites, *J. Endocrinol.* **129:**11–19.

Bochis, R. J., Wyvratt, M., and Schoen, W. R., 1994, Benzo-fused lactams promote release of growth hormone, U.S. Patent No. 5,283,241.

Bochis, R. J., Fisher, M. H., DeVita, R. J., Schoen, W. R., and Wyvratt, M. J., 1996, Benzo-fused lactams promote release of growth hormone, U.S. Patent No. 5,583,130.

Bowers, C. Y., 1993, GH releasing peptides—Structure and kinetics, *J. Pediatr. Endocrinol.* **6:**21–31.

Bowers, C. Y., 1996, Xenobiotic growth hormone secretagogues: Growth hormone releasing peptides, in: *Growth Hormone Secretagogues* (B. B. Bercu and R. F. Walker, eds.), pp. 9–28, Springer-Verlag, Berlin.

Bowers, C. Y., Chang, J., Momany, F., and Folkers, K., 1977, Effects of the enkephalins and enkephalin analogues on release of pituitary hormones *in vitro,* in: *Molecular Endocrinology* (I. MacIntyne, ed.), pp. 287–292, Elsevier/North-Holland Biomedical Press, Amsterdam.

Bowers, C. Y., Momany, F., Reynolds, G. A., Chang, D., Hong, A., and Chang, K., 1980, Structure activity relationships of a synthetic pentapeptide that specifically releases growth hormone in vitro, *Endocrinology* **106:**663–667.

Bowers, C. Y., Reynolds, G. A., Chang, D., Hong, A., Chang, K., and Momany, F., 1981, A study on the regulation of growth hormone release from the pituitaries of rats in vitro, *Endocrinology* **108:**1071–1080.

Bowers, C. Y., Momany, F. A., Reynolds, G. A., and Hong, A., 1984, On the in vitro and in vivo activities of a new synthetic hexapeptide that acts on the pituitary to specifically release growth hormone, *Endocrinology* **114:**1537–1545.

Bowers, C. Y., Reynolds, G. A., Durham, D., Barrera, C. M., Pezzoli, S. S., and Thorner, M. O., 1990, Growth hormone (GH)-releasing peptide stimulates GH release in normal men and acts synergistically with GH-releasing hormone, *J. Clin. Endocrinol. Metab.* **70:**975–982.

Bowers, C. Y., Sartor, A. O., Reynolds, G. A., and Badger, T. M., 1991, On the actions of the growth hormone-releasing hexapeptide, GHRP, *Endocrinology* **128:**2027–2035.

Bowers, C. Y., Alster, D. K., and Frentz, J. M., 1992, The growth hormone-releasing activity of a synthetic hexapeptide in normal men and short statured children after oral administration, *J. Clin. Endocrinol. Metab.* **74:**292–298.

Bresson-Bepoldin, L., and Dufy-Barbe, L., 1994, GHRP-6 induces a biphasic calcium response in rat pituitary somatotrophs, *Cell Calcium* **15:**247–258.

Carlsson, L. M. S., and Jansson, J. O., 1990, Endogenous growth hormone (GH) secretion in male rats is synchronized to pulsatile GH infusions given at 3-hour intervals, *Endocrinology* **126:**6–10.

Chambers, M. S., Baker, R., Billington, D. C., Knight, A. K., Middlemiss, D. N., and Wong, E. H. F., 1992, Spiropiperidines as high-affinity, selective σ ligands, *J. Med. Chem.* **35:**2033–2039.

Chapman, I. M., Hartman, M. L., Pezzoli, S. S., and Thorner, M. O., 1996a, Enhancement of pulsatile growth hormone secretion by continuous infusion of a growth hormone-releasing peptide mimetic, L-692,429, in older adults—A clinical research center study, *J. Clin. Endocrinol. Metab.* **81:**2874–2880.

Chapman, I. M., Bach, M. A., Van Cauter, E., Farmer, M., Krupa, D., Taylor, A. M., Schilling, L. M., Cole, K. Y., Skiles, E. H., Pezzoli, S. S., Hartman, M. L., Veldhuis, J. D., Gormley, G. J., and Thorner, M. O., 1996b, Stimulation of the growth hormone (GH)/insulin-like growth factor-1 axis by daily oral administration of a GH secretagogue (MK-0677) in healthy elderly subjects, *J. Clin. Endocrinol. Metab.* **81:**4249–4257.

Chaung, L.-Y. P., Pong, S.-S., Dean, D., Schaeffer, J. M., and Smith, R. G., 1996, Characterization of a G-protein-linked receptor for peptidyl and nonpeptidyl growth hormone secretagogue in rat and porcine hypothalamic and pituitary membranes, *Proceedings of the 78th Annual Meeting of the Endocrine Society,* **288:**P1–613 (Abstract).

Chen, M.-H., Steiner, M. G., Patchett, A. A., Cheng, K., Wei, L., Chan, W. W.-S., Butler, B., Jacks, T. M., and Smith, R. G., 1996, Analogues of the orally active growth hormone secretagogue L-162,752, *BioMed. Chem. Lett.* **6:**2163–2168.

Cheng, K., Chan, W. W.-S., Barreto, A., Convey, E. M., and Smith, R. G., 1989, The synergistic effects of His-D-Trp-Ala-Trp-D-Phe-Lys-NH_2 on growth hormone (GH)-releasing factor-stimulated GH release and intracellular adenosine 3′,5′-monophosphate accumulation in rat primary pituitary cell culture, *Endocrinology* **124:**2791–2798.

Cheng, K., Chan, W. W.-S., Butler, B. S., Barreto, A., and Smith, R. G., 1991, Evidence for a role of protein kinase-C in His-D-Trp-Ala-Trp-D-Phe-Lys-NH_2 induced growth hormone release from rat pituitary cells, *Endocrinology* **129:** 3337–3342.

Clark, R. G., Carlsson, L. M. S., Rafferty, B., and Robinson, I. C. A. F., 1988, The rebound release of growth hormone (GH) following somatostatin infusion in rats involves hypothalamic GH-releasing factor release, *J. Endocrinol.* **119:**397–404.

Conway, S., McCann, S. M., and Krulich, L., 1985, On the mechanism of growth hormone autofeedback regulation: Possible role of somatostatin and growth hormone-releasing factor, *Endocrinology* **117:**2284–2292.

Copinschi, G., Van Onderbergen, A., L'Hermite-Baleriaux, M., Mendel, C. M., Caufriez, A., Leproult, R., Bolognese, J. A., De Smet, M., Thorner, M. O., and Van Cauter, E., 1996, Effects of a 7-day treatment with a novel, orally active growth hormone (GH) secretagogue, MK-0677, on 24-hour GH profiles, insulin-like growth factor I, and adrenocortical function in normal young men, *J. Clin. Endocrinol. Metab.* **81:**2776–2782.

Dean, D. C., Nargund, R. P., Pong, S.-S., Chaung, L.-Y. P., Griffin, P. R., Melillo, D. G., Ellsworth,

R. L., Van der Ploeg, L. H. T., Patchett, A. A., and Smith, R. G., 1996, Development of a high specific activity sulfur-35-labeled sulfonamide radioligand that allowed the identification of a new growth hormone secretagogue receptor, *J. Med. Chem.* **39:**1767–1770.

DeVita, R. J., Schoen, W. R., Ok, D., Barash, L., Brown, J. E., Fisher, M. H., Hodges, P., Wyvratt, M. J., Cheng, K., Chan, W. W.-S., Butler, B. S., and Smith, R. G., 1994a, Benzolactam growth hormone secretagogues: Replacements for the 2′-tetrazole moiety of L-692,429, *BioMed. Chem. Lett.* **4:**1807–1812.

DeVita, R. J., Schoen, W. R., Fisher, M. H., Frontier, A. J., Pisano, J. M., Wyvratt, M. J., Cheng, K., Chan, W. W.-S., Butler, B. S., Hickey, G. J., Jacks, T. M., and Smith, R. G., 1994b, Benzolactam growth hormone secretagogues: Carboxamides as replacements for the 2′-tetrazole moiety of L-692,429, *BioMed. Chem. Lett.* **4:**2249–2254.

Dickson, S. L., 1996, Evidence for a central site and mechanism of action of growth hormone releasing peptide (GHRP6), in: *Growth Hormone Secretagogues* (B. B. Bercu and R. F. Walker, eds.), pp. 237–251, Springer-Verlag, Berlin.

Dickson, S. L., and Luckman, S. D., 1997, Induction of c-*fos* messenger ribonucleic acid in neuropeptide Y and growth hormone (GH)-releasing factor neurones in the rat arcuate nucleus following systemic injection of the GH secretagogue, GH-releasing peptide-6, *Endocrinology* **138:**771–777.

Dickson, S. L., Leng, G., Dyball, R. E. J., and Smith, R. G., 1995, Central actions of peptide and non-peptide growth hormone secretagogues in the rat, *Neuroendocrinology* **61:**36–43.

Dickson, S. L., Doutrelant-Viltart, O., McKenzie, D. N., and Dyball, R. E. J., 1996, Somatostatin inhibits arcuate neurons excited by GH-releasing peptide (GHRP-6) in rat hypothalamic slices, *J. Physiol. London* **495:**109–110.

Elliott, J. M., Cascieri, M. A., Davies, S., Huscroft, I. T., Kelleher, F. J., Lewis, R. T., MacLeod, A. M., Merchant, K. J., Sadowski, S., and Stevenson, G. I., 1996, Serine derived NK1 antagonists, in: *Abstracts, 211th American Chemical Society Meeting,* New Orleans, MEDI 075.

Evans, B. E., Rittle, K. E., Bock, M. G., DiPardo, R. M., Freidinger, R. M., Whitter, W. L., Lundell, G. F., Veber, D. F., Anderson, P. S., Chang, R. S. L., Lotti, V. J., Cerino, D. J., Chen, T. B., Kling, P. J., Kunkel, K. A., Springer, J. P., and Hirshfield, J., 1988, Methods for drug discovery: Development of potent, selective, orally effective cholecystokinin antagonists, *J. Med. Chem.* **31:**2235–2246.

Evans, B. E., Lundell, G. F., Gilbert, K. F., Bock, M. G., Rittle, K. E., Carroll, L. A., Williams, P. D., Pawluczyk, J. M., Leighton, J. L., Young, M. B., Erb, J. M., Hobbs, D. W., Gould, N. P., DiPardo, R. M., Hoffman, J. B., Perlow, D. S., Whitter, W. L., Veber, D. F., Pettibone, D. J., Clineschmidt, B. V., Anderson, P. S., and Freidinger, R. M., 1993, Nanomolar-affinity, non-peptide oxytocin receptor antagonists, *J. Med. Chem.* **36:**3993–4005.

Fairhall, K. M., Mynett, A., Smith, R. G., and Robinson, I. C. A. F., 1995, Consistent GH responses to repeated injections of GH-releasing hexapeptide (GHRP-6) and the non-peptide GH secretagogue L-692,585, *J. Endocrinol.* **145:**417–426.

Frohman, L. A., Downs, T. R., and Chomczynski, P., 1992, Regulation of growth hormone secretion, *Front. Neuroendocrinol.* **13:**344–405.

Gertz, B. J., Barrett, J. S., Eisenhandler, R., Krupa, D. A., Wittreich, J. M., Seibold, J. R., and Schneider, S. H., 1993, Growth hormone response in man to L-692,429, a novel nonpeptide mimic of growth hormone releasing peptide, *J. Clin. Endocrinol. Metab.* **77:**1393–1397.

Gertz, B. J., Sciberras, D. G., Yogendran, L., Christie, K., Bador, K., Krupa, D., Wittreich, J. M., and James, I., 1994, L-692,429, a nonpeptide growth hormone (GH) secretagogue, reverses glucocorticoid suppression of GH secretion, *J. Clin. Endocrinol. Metab.* **79:**745–749.

Guan, X.-M., Yu, H., Palyha, O. C., McKee, K. K., Feighner, S. D., Sirinathsinghji, D. J. S., Smith, R. G., Van der Ploeg, L. H. T., and Howard, A. D., 1997, Distribution of mRNA encoding the growth hormone secretagogue receptor in brain and peripheral tissues, *Mol. Brain Res.* **48:**23–29.

Hartman, M. L., Farello, G., Pezzoli, S. S., and Thorner, M. O., 1992, Oral administration of growth hormone (GH)-releasing peptide stimulates GH secretion in normal men, *J. Clin. Endocrinol. Metab.* **74:**1378–1384.

Herrington, J., and Hille, B., 1994, Growth hormone-releasing hexapeptide elevates intracellular calcium in rat somatotrophs by two mechanisms, *Endocrinology* **135:**1100–1108.

Hickey, G., Jacks, T., Judith, F., Taylor, J., Schoen, W. R., Krupa, D., Cunningham, P., Clark, J., and Smith, R. G., 1994, Efficacy and specificity of L-692,429, a novel nonpeptidyl growth hormone secretagogue in beagles, *Endocrinology* **134:**695–701.

Hickey, G., Jacks, T., Schleim, K., Frazier, E., Chen, H., Krupa, D., Feeney, W., Nargund, R., Patchett, A., and Smith, R. G., 1997, Repeat administration of the growth hormone secretagogue MK-0677 increases and maintains elevated IGF-1 levels in beagles, *J. Endocrinol.* **152:**183–192.

Howard, A. D., Feighner, D. S., Cully, D. F., Arena, J. P., Liberator, P. A., Rosenblum, C. I., Hamelin, M., Hreniuk, D. L., Palyha, O. C., Anderson, J., Paress, P. S., Diaz, C., Chou, M., Liu, K. K., Pong, S.-S., Chaung, L.-Y., Elbrecht, A., Dashkevicz, M., Heavens, R., Rigby, M., Sirinathsinghji, D. J. S., Dean, D. C., Melillo, D. G., Patchett, A. A., Nargund, R., Griffin, P. R., DeMartino, J. A., Gupta, S. K., Schaeffer, J. M., Smith, R. G., and Van der Ploeg, L. H. T., 1996, A receptor in pituitary and hypothalamus that functions in growth hormone release, *Science* **273:**974–977.

Huhn, W. C., Hartman, M. L., Pezzoli, S. S., and Thorner, M. O., 1993, Twenty-four-hour growth hormone (GH)-releasing peptide (GHRP) infusion enhances pulsatile GH secretion and specifically attenuates the response to a subsequent GHRP bolus, *J. Clin. Endocrinol. Metab.* **76:**1202–1208.

Jacks, T., Hickey, G., Judith, F., Taylor, J., Chen, H., Krupa, D., Feeney, W., Schoen, W., Ok, D., Fisher, M., Wyvratt, M., and Smith, R., 1994, Effects of acute and repeated intravenous administration of L-692,585, a novel non-peptidyl growth hormone secretagogue, on plasma growth hormone, IGF-1, ACTH, cortisol, prolactin, insulin, and thyroxine levels in beagles, *J. Endocrinol.* **143:**399–406.

Jacks, T., Smith, R., Judith, F., Schleim, K., Frazier, E., Chen, H., Krupa, D., Hora, D., Nargund, R., Patchett, A., and Hickey, G., 1996a, MK-0677, a potent, novel orally-active growth hormone (GH) secretagogue—GH, IGF-1 and other hormonal responses in beagles, *Endocrinology* **137:**5284–5289.

Jacks, T., Lieber, R., Schleim, K.-D., Mohler, R., Haven, M., Feeney, W., Hora, D., and Hickey, G., 1996b, L-163,255, a GH secretagogue, increased muscle torque in the remobilization phase of a canine hind limb immobilization model, in: *Abstracts, 10th International Congress of Endocrinology,* San Francisco, P1-602.

Jaffe, C. A., Ho, P. J., Demott-Friberg, R., Bowers, C. Y., and Barkan, A. L., 1993, Effects of a prolonged growth hormone (GH)-releasing peptide infusion on pulsatile GH secretion in normal men, *J. Clin. Endocrinol. Metab.* **77:**1641–1647.

Laron, Z., 1995, Growth hormone secretagogues—clinical experience and therapeutic potential, *Drugs* **50:**595–601.

Leonard, R. J., Chaung, L.-Y. P., and Pong, S.-S., 1991, Ionic conductances of identified rat somatotroph cells studied by perforated patch recording are modulated by growth hormone secretagogues, *Biophys. J.* **59:**254a (Abstract).

Leung, K. H., Cohn, D. A., Miller, R. R., Doss, M. A., Stearns, R. A., Simpson, R. E., Feeney, W. P., and Chiu, S.-H. L., 1996a, Pharmacokinetics and disposition of L-692,429, a novel non-peptidyl growth hormone secretagogue, in preclinical species, *Drug. Metab. Dispos.* **24:**753–760.

Leung, K. H., Miller, R. R., Cohn, D., Colletti, A., McGowan, E., Feeney, W. P., Nargund, R. N., Rosegay, A., Wallace, M. A., and Chiu, S.-H. L., 1996b, Pharmacokinetics and disposition of MK-0677, a novel growth hormone secretagogue, in rats and dogs, in: *Program and Abstracts, International Society for the Study of Xenobiotics 7th International Meeting,* San Diego, p. 277.

Low, L. C. K., 1991, Growth hormone-releasing hormone: Clinical studies and therapeutic aspects, *Neuroendocrinology* **53**(Suppl. 1)**:**37–40.

Mau, S. E., Witt, M. R., Bjerrum, O. J., Saermark, T., and Vilhardt, H., 1995, Growth hormone releasing hexapeptide (GHRP-6) activates the inositol (1,4,5)-trisphosphate/diacylglycerol pathway in rat anterior pituitary cells, *J. Receptor Signal Transduction Res.* **15:**311–323.

McGurk, J. F., Pong, S.-S., Chaung, L.-Y. P., Gall, M., Butler, B. S., and Arena, J. P., 1993, Growth hormone secretagogues modulate potassium currents in rat somatotrophs, *Soc. Neurosci.* **19:**1559 (Abstract).

McKee, K. K., Palyha, O. C., Feighner, S. D., Hreniuk, D. L., Tan, C., Smith, R. G., Van der Ploeg, L. H. T., and Howard, A. D., 1997, Molecular analysis of growth hormone secretagogue receptors (GHS-Rs): Cloning of rat pituitary and hypothalamic GHS-R type 1a cDNAs, *Mol. Endocrinol.* **11:**415–423.

Momany, F. A., Bowers, C. Y., Reynolds, G. A., Chang, D., Hong, A., and Newlander, K., 1981, Design, synthesis, and biological activity of peptides which release growth hormone in vitro, *Endocrinology* **108:**31–39.

Momany, F. A., Bowers, C. Y., Reynolds, G. A., Hong, A., and Newlander, K., 1984, Conformational energy studies and in vitro and in vivo activity data on growth hormone-releasing peptides, *Endocrinology* **114:**1531–1536.

Nargund, R. P., Barakat, K. H., Cheng, K., Chan, W. W.-S., Butler, B. R., Smith, R. G., and Patchett, A. A., 1996, Synthesis and biological activities of camphor-based non-peptide growth hormone secretagogues, *BioMed. Chem. Lett.* **6:**1265–1270.

Ok, D., Schoen, W. R., Hodges, P., DeVita, R. J., Brown, J. E., Cheng, K., Chan, W. W.-S., Butler, B. S., Smith, R. G., Fisher, M. H., and Wyvratt, M. J., 1994, Structure activity relationships of the non-peptidyl growth hormone secretagogue L-692,429, *BioMed. Chem. Lett.* **4:**2709–2714.

Ok, H. O., Szumiloski, J. L., Doldouras, G. A., Schoen, W. R., Cheng, K., Chan, W. W.-S., Butler, B. S., Smith, R. G., Fisher, M. H., and Wyvratt, M. J., 1996, Benzolactam growth hormone secretagogues: Replacement of the C-3 amide bond in L-692,429, *BioMed. Chem. Lett.* **6:**3051–3056.

Orczyk, G. P., Meng, X., DiLea, C., French, D. C., Gonen, B., and Chiang, S. T., 1996, Growth hormone (GH) response to growth hormone releasing peptide-2 (GHRP-2) orally administered to normal men, in: *Program and Abstracts, 77th Meeting of the Endocrine Society,* P3-145.

Papadakis, M. A., Grady, D., Black, D., Tierney, M. J., Gooding, G. A. W., Schambelan, M., and Grunfeld, C., 1996, Growth hormone replacement in healthy older men improves body composition but not functional ability, *Ann. Intern. Med.* **124:**708–716.

Patchett, A. A., Nargund, R. P., Tata, J. R., Chen, M.-H., Barakat, K. J., Johnston, D. B. R., Cheng, K., Chan, W. W.-S., Butler, B., Hickey, G., Jacks, T., Schleim, K., Pong, S.-S., Chaung, L.-Y. P., Chen, H. Y., Frazier, E., Leung, K. H., Chiu, S.-H. L., and Smith, R. G., 1995, Design and biological activities of L-163,191 (MK-0677): A potent orally active growth hormone secretagogue, *Proc. Natl. Acad. Sci. USA* **92:**7001–7005.

Patel, Y. C., Greenwood, M. T., Panetta, R., Demchyshyn, L. Niznik, H., and Srikant, C. B., 1995, The somatostatin receptor family, *Life Sci.* **57:**1249–1265.

Plotkin, D., Ng, J., Farmer, M., Gelato, M., Kaiser, F., Kiel, D., Korenman, S., McKeever, C., Munoz, D., Schwartz, R., Bolognese, J., Gormley, G. J., and Bach, M. A., 1996, Use of MK-0677, an oral GH secretagogue, in frail elderly subjects, in: *Program and Abstracts, Growth Hormone Research Society Meeting,* London, November 13–16, 1996, O-068.

Plotsky, P. M., and Vale, W., 1985, Patterns of growth hormone-releasing factor and somatostatin into the hypophysial portal circulation of the rat, *Science* **230:**461–463.

Pong, S.-S., Chaung, L.-Y. P., Smith, R. G., Ertel, E. A., Smith, M. M., and Cohen, C. J., 1992, Role of calcium channels in growth hormone secretion induced by GHRP-s (His-D-Trp-Ala-Trp-D-Phe-Lys-NH_2) and other secretagogues in rat somatotrophs, in: *Proceedings of the 74th Annual Meeting of the Endocrine Society,* 255 (Abstract).

Pong, S.-S., Chaung, L.-Y. P., and Leonard, R. J., 1993, The involvement of ions in the activity of a novel growth hormone secretagogue L-692,429 in rat pituitary cell culture, in: *Proceedings of the 75th Annual Meeting of the Endocrine Society,* 172 (Abstract).

Pong, S.-S., Chaung, L.-Y. P., Dean, D. C., Nargund, R. P., Patchett, A. A., and Smith, R. G., 1996, Identification of a new G-protein coupled receptor for growth hormone secretagogues, *Mol. Endocrinol.* **10:**57–61.

Reisine, T., and Bell, G. I., 1995, Molecular biology of somatostatin receptors, *Endocr. Rev.* **16:**427–442.

Roumi, M., Lenaerts, V., Boutignon, F., Wuthrich, P., Deghenghi, R., Bellemare, M., Adam, A., and Ong, H., 1995, Radioimmunoassay for hexarelin, a peptidic growth hormone secretagogue, and its pharmacokinetic studies, *Peptides* **16:**1301–1306.

Rudman, D., Feller, A. G., Nagraj, H. S., Gergans, G. A., Lalitha, P. Y., Goldberg, A. F., Schlenker, R. A., Cohn, L., Rudman, I. G., and Mattson, D. E., 1990, Effects of human growth hormone in men over 60 years old, *N. Engl. J. Med.* **323:**1–6.

Schoen, W. R., Wyvratt, M. J., and Smith, R. G., 1993, Growth hormone secretagogues, in: *Annual Reports in Medicinal Chemistry,* Volume 28 (J. A. Bristol, ed.), pp. 177–186, Academic Press, San Diego.

Schoen, W. R., Pisano, J. M., Prendergast, K., Wyvratt, M. J., Fisher, M. H., Cheng, K., Chan, W. W.-S., Butler, B., Smith, R. G., and Ball, R. G., 1994a, A novel 3-substituted benzazepinone growth hormone secretagogue (L-692,429), *J. Med. Chem .* **37:**897–906.

Schoen, W. R., Ok, D., DeVita, R. J., Pisano, J. M., Hodges, P., Cheng, K., Chan, W. W.-S., Butler, B. S., Smith, R. G., Wyvratt, M. J., and Fisher, M. H., 1994b, Structure activity relationships in the amino acid sidechain of L-692,429, *BioMed. Chem. Lett.* **4:**1117–1122.

Shah, S. K., Hale, J. J., Qi, H., Miller, D. J., Dorn, C. P., Mills, S. G., Sadowski, S. J., Cascieri, M. A., Metzger, J. M., Eiermann, G. J., Forrest, M. J., MacIntyre, D. E., and MacCoss, M., 1996, Discovery of substituted spiroindolinepiperidines as orally active dual antagonists of NK1 and NK2 receptors, in: *Abstracts, 212th American Chemical Society Meeting,* Orlando, MEDI 136.

Sirinathsinghji, D. J. S., Chen, H. Y., Hopkins, R., Trumbauer, M., Heavens, R., Rigby, M., Smith, R. G., and Van der Ploeg, L. H. T., 1996, Induction of c-fos mRNA in the arcuate nucleus of normal and mutant growth hormone-deficient mice by a synthetic non-peptidyl hormone secretagogue, *Neuro. Rep.* **6:**1989–1992.

Smith, P. W., Cooper, A. W. J., Bell, R., Beresford, I. J. M., Gore, P. M., McElroy, A. B., Pritchard, J. M., Saez, V., Taylor, N. R., Sheldrick, R. L. G., and Ward, P., 1995, New spiropiperidines as potent and selective non-peptide tachykinin NK2 receptor antagonists, *J. Med. Chem.* **38:**3772–3779.

Smith, R. G., Cheng, K., Schoen, W. R., Pong, S.-S., Hickey, G., Jacks, T., Butler, B., Chan, W. S.-S., Chaung, L.-Y. P., Judith, F., Taylor, J., Wyvratt, M. J., and Fisher, M. H., 1993, A nonpeptidyl growth hormone secretagogue, *Science* **260:**1640–1643.

Smith, R. G., Cheng, K., Pong, S.-S., Leonard, R. J., Cohen, C. J., Arena, J. P., Hickey, G. J., Chang, C. H., Jacks, T. M., Drisko, J. E., Robinson, I. C. A. F., Dickson, S. L., and Leng, G., 1996a, Mechanism of action of GHRP-6 and nonpeptidyl growth hormone secretagogues, in: *Growth Hormone Secretagogues* (B. B. Bercu and R. F. Walker, eds.), pp. 147–163, Springer-Verlag, Berlin.

Smith, R. G., Pong, S.-S., Hickey, G., Jacks, T., Cheng, K., Leonard, R., Cohen, C. J., Arena, J. P., Chang, C. H., Drisko, J., Wyvratt, M., Fisher, M., Nargund, R., and Patchett, A., 1996b, Modulation of pulsatile GH release through a novel receptor in hypothalamus and pituitary gland, in: *Recent Progress in Hormone Research,* Volume 51 (P. M. Conn, ed.), pp. 261–286, The Endocrine Society, Bethesda, Maryland.

Strobl, J. S., and Thomas, M. J., 1994, Human growth hormone, *Pharmacol. Rev.* **46:**1–34.

Tannenbaum, G. S., and Ling, N., 1984, The interrelationship of growth hormone (GH)-releasing factor and somatostatin in the generation of the ultradian rhythm of GH secretion, *Endocrinology* **115:**1952–1957.

Torosian, M. H. (ed.), 1996, *Growth Hormone in Critical Illness—Research and Clinical Studies,* R. G. Landes Company, Austin, Texas.

Turner, J. P., and Tannenbaum, G. S., 1995, *In vivo* evidence of a positive role for somatostatin to optimize pulsatile growth hormone secretion, *Am. J. Physiol. (Endocrinol. Metab.)* **269:**E683–E690.

Welle, S., Thornton, C., Statt, M., and McHenry, B., 1996, Growth hormone increases muscle mass and strength but does not rejuvenate myofibrillar protein synthesis in healthy subjects over 60 years old, *J. Clin. Endocrinol. Metab.* **81:**3239–3243.

Chapter 24

Dorzolamide, a 40-Year Wait

From an Oral to a Topical Carbonic Anhydrase Inhibitor for the Treatment of Glaucoma

Gerald S. Ponticello, Michael F. Sugrue, Bernard Plazonnet, and Geneviève Durand-Cavagna

1. INTRODUCTION

Chronic open-angle glaucoma is by far the most prevalent form of glaucoma with it being the second most common form of blindness in the United States (Liesegang, 1996). Glaucoma is a chronic disease lacking a cure and, if left untreated, continues to progress. Currently, the only high-risk factor that can be modified in glaucoma is intraocular pressure, which is regulated by the rate at which aqueous humor is secreted and eliminated from the eye. The increase in intraocular pressure associated with glaucoma is related to an increased resistance to the outflow of aqueous humor from the eye through the trabecular meshwork. All drugs in current use to treat glaucoma are ocular hypotensive agents.

The carbonic anhydrase (CA) inhibitor, acetazolamide, was developed as a diuretic and its pharmacology was reported in 1954 (Maren *et al.*, 1954). In the

Gerald S. Ponticello and Michael F. Sugrue • Merck Research Laboratories, West Point, Pennsylvania 19486. *Bernard Plazonnet and Geneviève Durand-Cavagna* • Merck Sharp & Dohme-Chibret Research Center, Riom, 63203 France.

Integration of Pharmaceutical Discovery and Development: Case Studies, edited by Borchardt *et al.*, Plenum Press, New York, 1998.

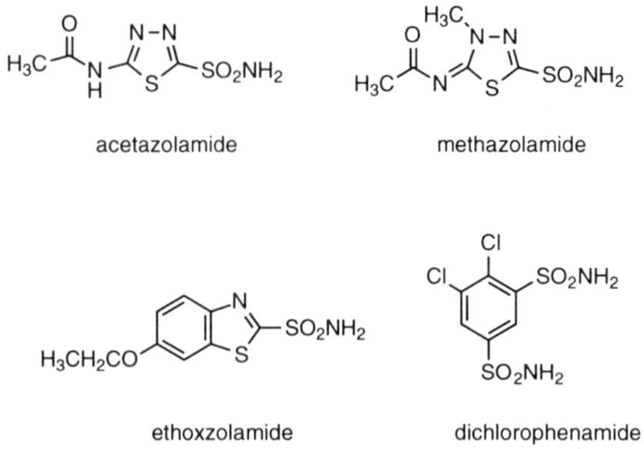

acetazolamide methazolamide

ethoxzolamide dichlorophenamide

Figure 1. Structure of acetazolamide, methazolamide, ethoxzolamide, and dichlorphenamide.

same year, orally administered acetazolamide was shown to lower the elevated intraocular pressure of glaucoma patients (Becker, 1954). CA was first observed in the anterior uvea of the rabbit eye (Wistrand, 1951) and subsequent studies confirmed its presence in human ciliary processes (Lütjen-Drecoll *et al.,* 1983). The enzyme is responsible for the generation of bicarbonate anions secreted from the ciliary process into the posterior chamber with sodium being the counter ion. Its inhibition decreases the rate of aqueous humor secretion.

The demonstrated effectiveness of acetazolamide was followed by the introduction of other oral agents, i.e., dichlorphenamide, ethoxzolamide, and methazolamide (Fig. 1). Although these drugs are very good ocular hypotensive agents, the extraocular inhibition of the enzyme results in a myriad of side effects and, as a consequence, patient compliance is very poor. It was rapidly realized that extraocular side effects could be dramatically reduced by the introduction of an agent that elicited ocular hypotension following local administration to the eye. Early attempts to develop a topically active agent were unsuccessful, as reviewed elsewhere (Maren, 1995). This was related to the fact that, in order to elicit a reduction in intraocular pressure, CA must be essentially inhibited 100% for 24 hr of the day and the agents tested for topical activity were incapable of achieving this. An intensive research program at Merck Research Laboratories (MRL) has been the quest for a topically active, ocular hypotensive CA inhibitor. This culminated in the discovery of dorzolamide (Trusopt®), which became available in a number of countries in 1995. Hence, there is a 40-year gap between the demonstrated oral effectiveness of acetazolamide and the introduction of topical dorzolamide.

This chapter will summarize the studies at MRL leading to the discovery of dorzolamide and will review the overall profile of the drug.

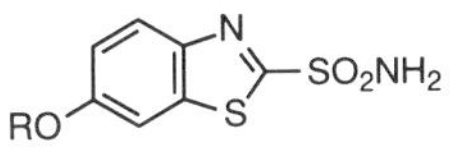

N
SO_2NH_2
RO
S

L-643,799 R = H
L-645,151 R = $(CH_3)_3CC(O)$-

Figure 2. Structure of benzothiazoles L-643,799 and L-645,151.

2. BENZOTHIAZOLES

Initial studies focused on modifications of the structure of ethoxzolamide and this effort resulted in the synthesis of the benzothiazole derivative L-645,151 (2-sulfamoyl-6-benzothiazolyl-2,2-dimethylpropionate), which is the *0*-pivaloyl ester of L-643,799 (6-hydroxybenzothiazole-2-sulfonamide). The former is a prodrug of the latter (Fig. 2), and corneal esterases are responsible for the generation of L-643,799, which is the active species (Schwam *et al.,* 1984). The importance of L-645,151 resides in the fact that the instillation of one 50-μl drop of doses as low as 0.25% significantly lowered the intraocular pressure of ocular hypertensive albino rabbits, and it was the first topical CA inhibitor to demonstrate ocular hypotensive activity following the administration of a single drop. Moreover, its site of action was localized within the eye as evidenced by a lack of effect following contralateral instillation (Sugrue *et al.,* 1985).

Allergic reactions to L-645,151 were observed during a 3-month ocular safety study in rabbits and dogs (Graham *et al.,* 1989; Durand-Cavagna *et al.,* 1996). These reactions appeared from week 5 and consisted of a persistent moderate redness of bulbar and palpebral conjuctivae and a slight discharge in half of the rabbits and a slight persistent redness in limbal conjunctivae and/or a slight discharge in half of the dogs. These changes were associated with cholesterol deposits in the cornea of dogs. In most of the rabbits, there was microscopically a very slight to moderate cellular infiltration, mainly lymphocytes, in the limbus corneae, in the eyelids, and in the nictitating membrane. In the most severe cases, keratitis was seen in which the cellular infiltration extended from the limbus corneae into the equatorial cornea. In the treated eyes of most dogs, a very slight to slight infiltration of lymphocytes and a few plasma cells were seen in the limbus corneae, the eyelids, and the nictitating membrane along with hyperplasia of the resident lymphoid tissue.

Subsequent evaluation in a guinea pig model for dermal sensitization (Magnusson and Kligman, 1969) revealed that L-645,151 was a potent allergen. Acetazolamide, methazolamide, and ethoxzolamide were also shown to elicit contact dermal sensitization in this test. A number of other benzothiazole sulfonamides were subsequently found to share this property. These compounds reacted easily with glutathione (GSH) and underwent displacement of the sulfamoyl moiety, a result suggesting that similar reactions with macromolecular nucleophiles might

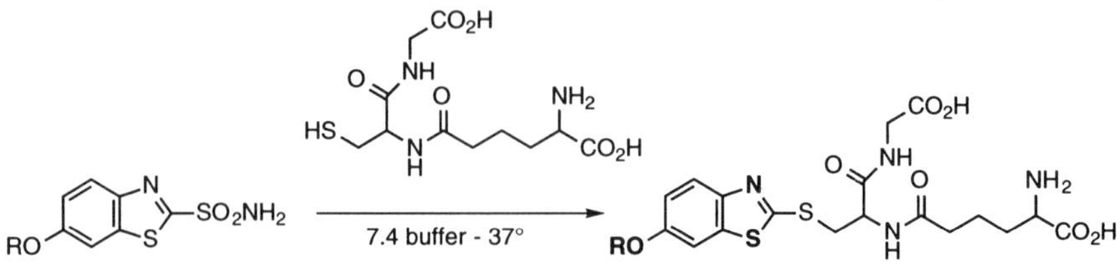

Figure 3. Reaction of GSH with benzothlazole-2-sulfonamides.

produce potent allergens (Fig. 3). In order to overcome the problems of electrophilicity and sensitizing potential, an *in vitro* test with reduced GSH under simulated physiological conditions was developed (Shepard *et al.,* 1991). It was found that CA inhibitors that reacted with excess GSH generally proved to be contact sensitizers in the guinea pig maximization test following dermal challenge.

3. BENZOTHIOPHENES

The reactivity of the benzothiazole class with sulfhydryl groups and the observation of the resulting sensitization reaction in guinea pigs prompted a search for a chemically more stable structure. The decision to pursue the benzo[*b*]thiophene-2-sulfonamide class (Graham *et al.,* 1989) was based on the premise that the electrophilic nature of the benzothiazole nucleus reflected the presence of the nitrogen atom in the ring system. De-aza analogues were considered to be less reactive toward nucleophiles. As a result, two benzo[*b*]thiophenes, L-650,719 and L-651,465, emerged that were free of sensitization potential in guinea pigs and received in-depth evaluations. L-650,719 (6-hydroxybenzo[*b*]thiophene-2-sulfonamide) and L-651,465, its acetate ester, displayed good ocular hypotensive activity in experimental animals and no problems were encountered in safety assessment studies (Fig. 4). However, both agents failed to sufficiently lower intraocular pressure in humans following topical dosing (Lippa, 1991).

Both L-650,719 and L-651,465 possessed limited water solubility and could only be administered as suspensions. Other drugs such as steroids are widely used as suspensions. The use of suspensions has specific prerequisites: availability of micronized sterile solid; sterile manufacturing and homogenizations; good sus-

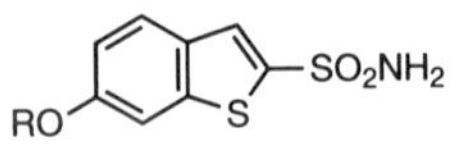

L-650,719 R = H
L-651,465 R = $CH_3C(O)$-

Figure 4. Structure of benzothiophenes L-650,719 and L-651,465.

pending properties and good resuspendability; good chemical and physical stability; and ease of dispensing for and acceptance by the patient. Among these characteristics, the physical properties, and in particular the crystal shape and size, are critical. Many of the "insoluble" CA inhibitors studied have had some solubility in water. Because water is a very good recrystallization solvent for sulfonamides, temperature changes induce variable solubility of the "insoluble" product resulting in changes in particle size and shape.

An alternative strategy for compounds such as L-650,719 and the oral CA inhibitors is their formulation in solution as an alkaline derivative at pH greater than 8.0 stemming from the presence of the acidic sulfamoyl moiety. However, this is not suitable for chronic administration to the human eye. In addition, the high pH is a potential source of instability for the pharmaceutical formulation.

4. THIENOTHIOPYRANS

Because of the potential disadvantage of suspensions as a pharmaceutical vehicle, an effort was initiated to discover a CA inhibitor with increased water solubility. The design strategy was based on combining structural features from the potent benzo[*b*]thiophene-2-sulfonamide series such as L-649,522 and L-650,719 and introducing functional groups capable of increasing water solubility. This approach was addressed by replacing the lipophilic benzene ring of L-650,719 and incorporating the electron-withdrawing sulfonyl moiety attached to the thiophene ring of L-649,522 to an annulated thiopyran-1,1-dioxide nucleus. To enhance water solubility, polar substituents were also introduced onto the thiopyran ring. This approach led to the synthesis of the thieno[2,3-*b*]thiopyran-7,7-dioxide class of heteroaryl sulfonamides. L-654,230 (5,6-dihydro-4*H*-4-hydroxythieno[2,3-*b*]thiopyran-7,7-dioxide-2-sulfonamide) was the first example that exhibited sufficient water solubility to be formulated as a 0.8% solution. In addition, the compound was devoid of contact sensitization potential and no adverse effects were encountered in safety assessment studies. However, the compound lacked sufficient efficacy in humans to justify development (Lippa, 1991). Replacement of the 4-hydroxy group of L-654,230 by amino groups led to the discovery of MK-927, the 4-(2,2-dimethylethyl) amino derivative. The presence of an amino group enabled the compound to be studied for water solubility over a wide pH range. The compound was formulated as a 2% solution via the protonated species at pH 5.2. Furthermore, resolution of MK-927 provided the more active *S*-enantiomer, MK-417, which was also evaluated in humans. However, MK-417 was less soluble than the racemate, MK-927, and could only be formulated as a 1.8% solution. The ability to achieve a proper balance between water solubility and lipophilicity for ocular penetration was a key feature of these compounds. The degree of binding to ocular pigment was also important and has been discussed elsewhere (Sugrue, 1996). Like L-654,230 and the

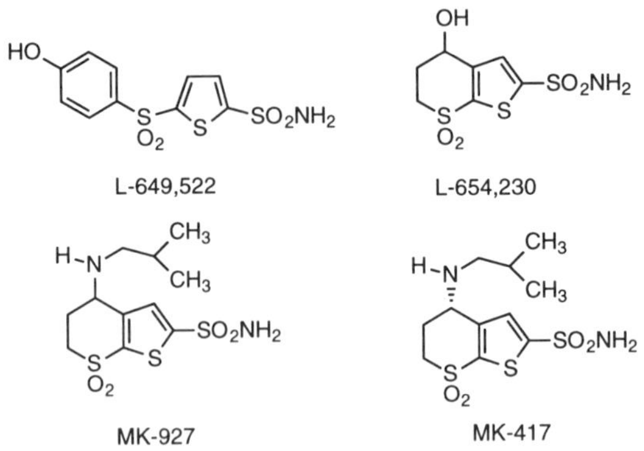

Figure 5. Structure of thiophene L-649,522 and thienothiopyrans L-654,230, MK-927, and MK-417.

alkylaminothienothiopyrans, MK-927 and MK-417 were negative in the guinea pig maximization test (Fig. 5). These compounds showed a good local tolerance in subsequent ocular irritation studies. No microscopic changes were observed in the examination of eyes and ocular adnexa. Both MK-927 (Lippa *et al.,* 1988) and MK-417 (Lippa *et al.,* 1991) were subsequently shown to effectively lower intraocular pressure in humans, and, in fact, MK-927 was the first CA inhibitor to display this property. Encouraged by these results, SAR studies were continued to develop a compound with an improved spectrum of activity and increased water solubility. This effort identified dorzolamide (MK-507), which met these goals.

5. DORZOLAMIDE

Dorzolamide contains two chiral centers and was prepared in overall 5–10% yield via the 10-step reaction sequence described in Fig. 6. The thieno[2,3-*b*] thiopyran structure **5** was prepared using 2-mercaptothiophene (**1**) and crotonic acid (**2**) as previously described (Ponticello *et al.,* 1988). Transformation of **5** to **12** was accomplished in a straightforward fashion (Ponticello *et al.,* 1987; Baldwin *et al.,* 1989). The mixture of *cis–trans* isomers was chromatographed to provide pure **13** and **14.** The *trans*-isomer **13** was resolved with (−)di-*p*-toluoyl-L-tartaric acid (DPT-L-TA) in ethanol to provide the *S,S*-isomer (**15**). The chromatographic separation of diastereomers followed by a tedious resolution was mainly responsible for the low overall yield obtained for dorzolamide in the process.

Subsequently, a more practical synthesis was developed for preparing dorzolamide for clinical testing as outlined in Fig. 7. The process provided dorzolamide

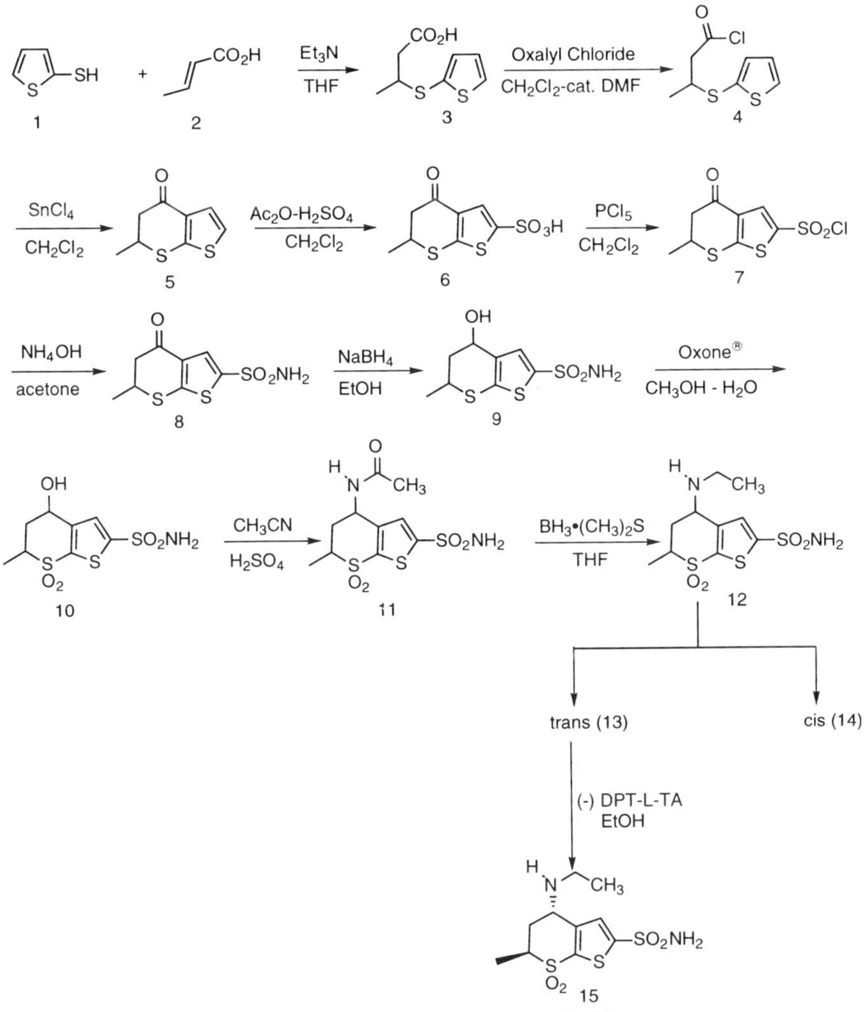

Figure 6. Chemical synthesis of MK-507.

in greater than 32% overall yield (Blacklock *et al.*, 1993). The key feature involved the introduction of the requisite two asymmetric centers in an efficient manner. The first was incorporated via an S_N2 inversion utilizing the *m*-chlorobenzenesulfonate of methyl (*R*)-3-hydroxybutyrate (**17**) to introduce the (*S*)-6-methyl group (**18** to **20**) in greater than 97% ee. The second chiral center was obtained through diastereomeric control by the classic Ritter reaction to provide the *trans*(*S*)-4-acetamide

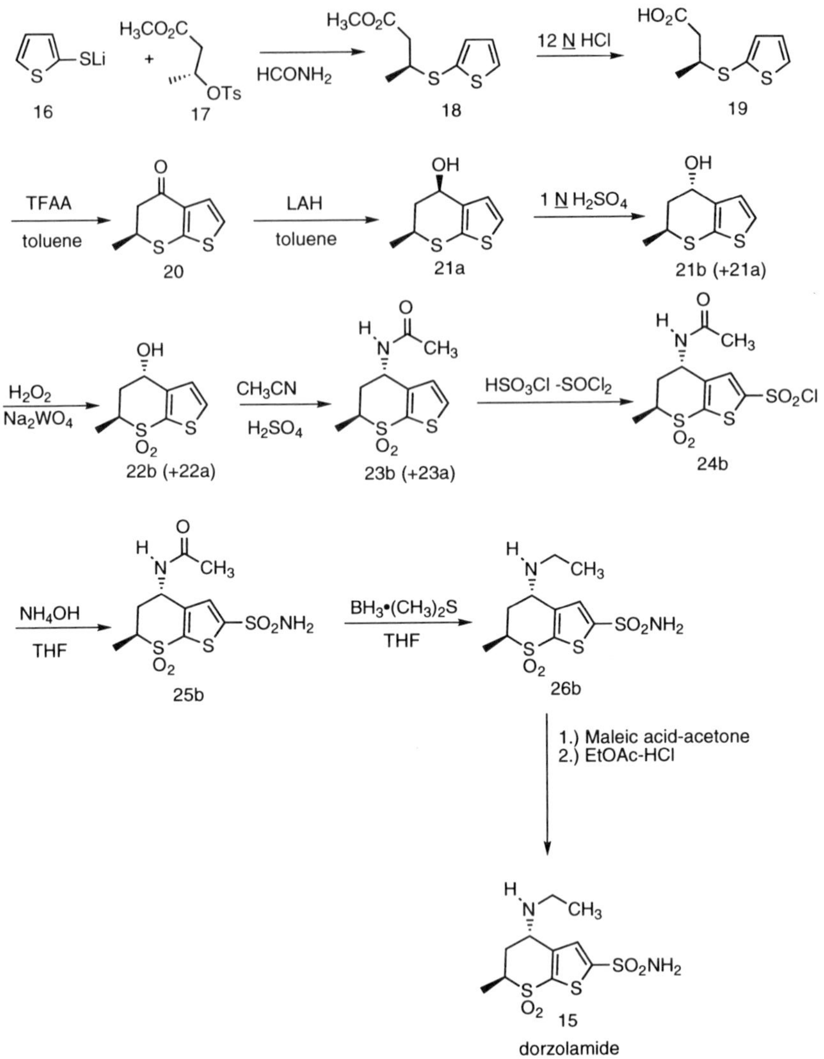

Figure 7. R & D process for MK-507.

intermediate (**23b**) in greater than 78% de, the precursor to the (*S*) 4-ethylamino moiety (**26b**). The asymmetric synthesis and the elimination of a low-yielding resolution were responsible for the preparation of *trans*-**23b** in high overall yield (a and b refer to *cis* and *trans* isomers, respectively).

Table I
In Vitro Inhibition of Human CA Isoenzymes I, II, and IV[a]

Compound	IC_{50} (nM)		
	CA-I	CA-II	CA-IV
Acetazolamide	13.9 ± 2.3	3.4 ± 0.3	14.7 ± 4.1
Dorzolamide	600.0 ± 13.6	0.18 ± 0.03	6.9 ± 0.7
Ethoxzolamide	6.9 ± 0.6	0.18 ± 0.02	15.7 ± 3.1
Methazolamide	4.7 ± 0.2	8.1 ± 2.0	80.3 ± 14.4
L-706,803	10.0	1.7	120

[a]Results are expressed as IC_{50}, and each is the mean ± SEM of at least three determinations. Isoenzymes I and II were obtained from red blood cells and isoenzyme IV from lung.

6. PHARMACOLOGY

6.1. *In Vitro*

Dorzolamide was observed to be a potent inhibitor of human CA isoenzyme II possessing an IC_{50} value of 0.18 nM (Table I). The drug fits very well into the active site of CA isoenzyme II and three-dimensional X-ray crystallography has revealed that its ethylamino group changes the position of histidine-64 in the active site of the enzyme with the result that water can no longer be bound (Smith *et al.*, 1994). In contrast, it was a very weak inhibitor of CA isoenzyme I, its IC_{50} value being 600 nM. Both isoenzymes I and II were isolated from red blood cells. L-706,803 (Fig. 8) is the primary metabolite of dorzolamide and its IC_{50} values against CA isoenzymes I and II were observed to be 10 and 1.7 nM, respectively.

The difference in the affinity of dorzolamide for isoenzymes I and II is important in the overall pharmacology of the drug. The human red blood cell contains approximately 150 μM of CA, 20 μM being isoenzyme II and 130 μM being isoenzyme I (Maren, 1967). Steady-state levels of dorzolamide and L-706,803 in human red blood cells following multiple topical dosing with 2% dorzolamide are approximately 20 and 8 μM, respectively (Strahlman *et al.*, 1996), and this is far from saturating the total red blood cell content of CA. This is also reflected in the 20% of total CA activity remaining in the human red blood cell at steady state.

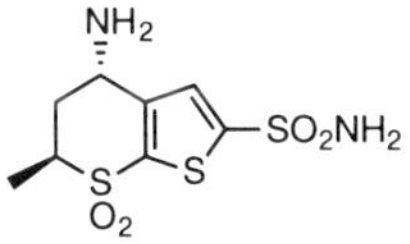

NH_2 … SO_2NH_2 … S … S … O_2

L-706,803

Figure 8. Structure of MK-507 metabolite, L-706,803.

In order to elicit a pharmacological response, CA must be blocked by essentially 100%. Hence, dorzolamide cannot elicit respiratory side effects stemming from the inhibition of red blood cell CA at steady state. Furthermore, the concentration of dorzolamide in the plasma of humans at steady state was observed to be less than the limit of detection (5 ng/ml) of the assay (Biollaz *et al.,* 1995). In contrast to dorzolamide, which is selective for CA isoenzyme II against isoenzyme I, the oral agents, acetazolamide and methazolamide, are nonselective, their IC_{50} values against CA I being 13.9 and 4.7 nM, respectively, and against CA II, 3.4 and 8.1 nM, respectively.

The cytosolic CA isoenzyme II has been traditionally viewed as the critical isoenzyme in the formation of aqueous humor. However, there is currently considerable speculation on the significance of the membrane-bound CA isoenzyme IV and a strong case for its importance has been presented (Maren, 1995). An argument against a role for isoenzyme IV in humans is the failure to detect its presence in the ciliary process (Hageman *et al.,* 1991). Dorzolamide is more potent than either acetazolamide or methazolamide at inhibiting isoenzyme IV isolated from human lung, the respective IC_{50} values being 6.9, 14.7, and 80.3 nM.

As stated previously, the modest binding of the amino-substituted thienothiopyran-2-sulfonamides to pigment is viewed as playing an important role in their ability to lower intraocular pressure. A 1-hr incubation at room temperature of 10 μM dorzolamide in phosphate buffer, pH 7.1, with pigment isolated from the bovine iris-ciliary body, resulted in 18.9% of the compound being bound, a value consistent for this class of compounds.

6.2. *In Vivo*

Dorzolamide is a much better ocular penetrator than either acetazolamide or methazolamide following topical dosing. This was illustrated in experiments in which CA activity in a homogenate of the iris-ciliary body of albino rabbits was measured 1 hr after topical dosing. Treatment with 0.1% dorzolamide achieved a 100% inhibition and lowering the dose to 0.02% resulted in a blockade of enzymatic activity of 87%. In contrast, the instillation of acetazolamide and methazolamide, both at 0.1%, blocked enzyme activity by 26.5 and 12.4%, respectively (Sugrue *et al.,* 1990). The instillation of 2% dorzolamide resulted in peak concentrations in the cornea, aqueous humor, and iris-ciliary body of pigmented rabbits of 24.0 μg/g, 7.8 μg/ml, and 27.0 μg/g, respectively (Table II). In addition to being present in the anterior segment of the eye, dorzolamide was also present in the retina with the peak retinal content of 5.3 μg/g being present at 4 hr postdosing.

Dorzolamide was studied for ocular hypotensive activity in ocular normotensive and hypertensive rabbits and monkeys. The topical instillation of a 2% solution of dorzolamide significantly lowered the intraocular pressure of ocular

Table II
Concentrations of Dorzolamide in Ocular Tissues and Fluids of Pigmented Rabbits at Selected Time Points Following 2% Dosing[a]

	Time after dosing			
Sample	1 hr	2 hr	4 hr	8 hr
Cornea (μg/g)	24.0 ± 2.7	20.4 ± 2.6	9.8 ± 1.8	3.3 ± 0.5
Aqueous humor (μg/ml)	4.6 ± 0.6	7.8 ± 1.7	2.2 ± 0.2	0.4 ± 0.1
Iris-ciliary body (μg/g)	9.9 ± 0.8	27.0 ± 4.4	17.6 ± 1.4	9.0 ± 1.7
Retina (μg/g)	4.3 ± 0.7	4.2 ± 0.4	5.3 ± 1.8	2.1 ± 0.1

[a]Concentrations were measured by HPLC and each result is the mean ± SEM of 18 determinations.

normotensive albino rabbits, the peak decrease of 3.3 mm Hg occurring at 1 hr postdosing. Dorzolamide was very effective in albino rabbits whose intraocular pressure had been experimentally elevated by the prior injection of α-chymotrypsin into the eye. Dosing with 0.01, 0.1, and 0.5% solutions of dorzolamide maximally lowered intraocular pressure by 3.1, 6.1, and 9.8 mm Hg, respectively, in a dose-dependent manner (Table III). In contrast, the unilateral instillation of 0.5% dorzolamide onto the contralateral eye did not significantly alter the intraocular pressure of the untreated ocular hypertensive eye. This clearly reveals that topically applied dorzolamide has a local action within the eye and that its activity is not related to systemic absorption followed by a subsequent redistribution. In addition to reducing the intraocular pressure of rabbits, topically applied dorzolamide was a very effective ocular hypotensive agent in monkeys. The instillation of 0.5, 1, and 2% solutions of dorzolamide maximally lowered in a dose-dependent manner the intraocular pressure of glaucomatous monkeys by 22, 30, and 37%, respectively. As in the case of α-chymotrypsin-treated rabbits, the unilateral instillation of 2% dorzolamide onto the contralateral eye had little effect on the intraocular pressure of the untreated, glaucomatous eye. This again confirms that the site of action of the drug is local (Sugrue, 1996).

Table III
Effect of Topically Applied Dorzolamide on the α-Chymotrypsin-Induced Elevation in the Intraocular Pressure (IOP) of Albino Rabbits[a]

	IOP decrease (mm Hg)					
	Time after treatment					
Dose	0.5 hr	1 hr	2 hr	3 hr	4 hr	5 hr
0.01%	−2.2 ± 0.6	−2.4 ± 0.7	−3.1 ± 0.8	−2.6 ± 0.8	−2.0 ± 0.9	−2.2 ± 0.9
0.1%	−4.9 ± 0.9	−5.9 ± 0.7	−6.1 ± 0.6	−5.9 ± 0.7	−5.3 ± 0.7	−4.8 ± 0.9
0.5%	−6.5 ± 0.7	−9.1 ± 0.7	−9.7 ± 1.0	−9.8 ± 1.0	−8.8 ± 1.1	−9.1 ± 1.2

[a]Each IOP value is the mean ± SEM of 12 observations.

The concentrations of dorzolamide and timolol in clinical use are 2 and 0.5%, respectively. When compared at these concentrations in glaucomatous monkeys, both drugs were comparable at peak, i.e., 4 hr postdosing. However, at 16 hr timolol retained greater intraocular pressure-lowering activity than dorzolamide, 59 versus 22% of peak decline, respectively (Sugrue, 1996).

The concurrent administration of dorzolamide and timolol was also studied in the glaucomatous monkey. However, 0.5% timolol and 2% dorzolamide could not be used because both doses are maximal in this paradigm. The instillation of a 0.005% solution of timolol was followed 10 min later by 0.5% dorzolamide. The reductions in intraocular pressure elicited by 0.005% timolol and 0.5% dorzolamide were comparable. The concurrent administration of both drugs was more effective than either agent alone in lowering intraocular pressure with significant differences being present from 1 hr onwards between 0.005% timolol alone and 0.005% timolol plus 0.5% dorzolamide (Sugrue, 1996). This study indicates that the ocular hypotensive effect of timolol in glaucomatous monkeys can be enhanced by the concurrent administration of dorzolamide.

The possible role of endogenous prostaglandins and/or prostanoids in the ocular hypotensive effect of dorzolamide in albino rabbits was studied following cyclooxygenase inhibition. The ability of 2% dorzolamide to lower the intraocular pressure of ocular normotensive albino rabbits was unaltered by either a 1-hr pretreatment with indomethacin (5 mg/kg, i.p.) or by topically administered 0.03% flurbiprofen. Epinephrine was included as a positive control and, in contrast to dorzolamide, the ocular hypotensive effect of 1% epinephrine was blunted by both indomethacin and flurbiprofen (Sugrue and O'Neill-Davis, 1991).

In terms of its mechanism of action, topically administered dorzolamide has been observed to reduce aqueous humor production in rabbits (Sugrue, 1996), monkeys (Wang *et al.*, 1991), and humans (Yamazaki *et al.*, 1994), and in this respect acts like oral CA inhibitors.

7. PHARMACEUTICAL RESEARCH AND DEVELOPMENT STUDIES

Dorzolamide was developed as a bifunctional drug with a sulfonamide moiety (pK_a 8.5) and a secondary amine (pK_a 6.35). Being an amphoteric compound, it displays reduced solubility close to neutrality and higher solubility in acidic or alkaline solution.

Thus, the selection of the pH of the pharmaceutical formulation was of paramount importance for the following reasons: to ensure the solubility and stability of dorzolamide, an adequate shelf life for the formulation; and to provide good tolerance of the formulation by the patient during chronic use. Because a 2% concentration was selected, a target pH of 5.65 was chosen to meet these criteria and was adjusted with a minimal amount of sodium citrate, so that the buffering capacity of the tear film was not exceeded (Moses, 1981). Studies conducted under more acidic (pH 4.5–5) or more alkaline (pH > 6) conditions did not demonstrate any advantages, as stability decreased at pH 6 and a low pH did not increase solubility (Grove *et al.*, 1995) and both high and low pH could also be a source of ocular irritation.

Although noncorneal routes of penetration of topical ocular drugs have been described, the main route of drug penetration into the eye is transcorneal and, for ionic compounds, is governed by the pH partition hypothesis. The partition coefficient expressed as the concentration in the organic phase versus the concentration in aqueous buffered phase was found to be 1.96 at 33°C for the *N*-octanol/pH 7.4 McIlvaine buffer system and 0.48 for the chloroform/pH 7.4 McIlvaine buffer system. Hence, dorzolamide adequately partitions between the aqueous and organic phases under these standard conditions. With the tear film/formulation mixture at a pH between 5.65 and 7.4, dorzolamide is more ionized and less prone to partition in the "lipid phase," i.e., the corneal epithelium.

Hydroxyethylcellulose, benzalkonium chloride, and mannitol were used as ancillary ingredients. Both preclinical and clinical studies indicated that an increase in the viscosity of the formulation elicited a better pharmacological response. Hydroxyethylcellulose was selected and its concentration adjusted to approximately 0.5% to accommodate tolerance by the patient and manufacturing feasibility. For use as multidose eye drops, the formulation had to be protected against microbial contamination during its use by the patient. Benzalkonium chloride is an antimicrobial preservative that has adequate activity against gram-negative and gram-positive bacteria, and against molds. However, as it may be irritating and/or sensitizing in some patients, its concentration was kept at a minimum level necessary to effectively protect the formulation from bacterial contamination. It was found that 2% dorzolamide eye drops could be preserved with 0.0075% of benzalkonium chloride and that such a formulation fulfilled the requirements of the European and U.S. pharmacopoeias. Mannitol was selected as the isotonizing agent because sodium chloride decreased the solubility of dorzolamide by the common ion effect.

8. SAFETY ASSESSMENT STUDIES

Topically administered CA inhibitors could conceivably elicit systemic effects and, therefore, the preclinical safety profile of dorzolamide included studies designed to address both ocular and systemic safety. The studies for the approval of dorzolamide included cutaneous hypersensitivity studies in guinea pigs; ocular tolerance studies for up to 1 year in dogs and/or monkeys, and up to 3 months in rabbits; acute toxicity studies in rats and mice; oral toxicity studies for up to 1 year in rats and dogs; several genetic toxicity studies; lifetime carcinogenicity assays in mice and rats; fertility and late gestation/lactation studies in rats; and developmental toxicity studies in rats and rabbits.

Dorzolamide was negative in the guinea pig maximization test at concentrations of 2% (intradermal induction) and 8% (topical induction and challenge).

Topically applied dorzolamide was studied for ocular tolerance at concentrations ranging from 2 to 4% in rabbits, dogs, and monkeys for periods of time ranging from 1 month to 1 year (Table IV). There were no significant ocular findings in any species studied at concentrations higher than that intended for clinical use (3 and 4% versus 2%). The only findings at the higher concentrations were slight increases in blinking in rabbits. There were no ocular changes in dogs or monkeys treated with 3% dorzolamide for up to 1 year. There was no increase in corneal thickness in rabbits given 4% for 3 months. Hence, the long-term topical administration of dorzolamide to animals was well tolerated and did not cause gross or microscopic changes in ocular tissues.

Systemic side effects of dorzolamide were entirely consistent with the inhibition of CA and represent a "class effect." Systemic side effects were urothelial hyperplasia, renal pelvic epithelial hyperplasia (RPEH) and mineralization, renal papillary cytoplasmic granularity (RPCG), gastric fundus mucosal hyperplasia, and bone changes.

Hyperplasia of the urinary bladder was seen with all tested CA inhibitors in rats and mice but not in rabbits, dogs, and monkeys and the overall incidence of these changes correlated with urinary changes (Table V). Based on data in the lit-

Table IV
Ocular Irritation Studies

	Rabbits				Dogs				Monkeys		
	1 mo		3 mo		3 mo		6 mo/1 yr		1 mo		6 mo/1 yr
	2%	4%	2%	4%	2%	4%	2%	4%	2%	3%	3%
One eye treated	•	•	•	•	•				•	•	•
Both eyes treated						•	•	•			
Vehicle control	•	•	•	•	•	•	•	•	•	•	•
Saline control	•	•	•	•							
General survey of eyes	•	•	•	•	•	•	•	•	•	•	•
Draize scoring	•	•	•	•	•	•	•	•	•	•	•
Ophthalmological exam[a]	•	•	•	•	•	•	•	•	•	•	•
Gross exam	•	•	•	•	•	•	•	•	•	•	•
Microscopic exam[b]	•	•	•	•	•	•	•	•	•	•	•
Corneal thickness	•	•	•	•							
Hematological exam					•	•	•	•			•
Biochemical exam					•	•	•	•			•

[a]Including ophthalmoscopic and slit lamp.
[b]Eyes, lacrimal glands, upper and lower lids, nictitating membranes in rabbits and dogs and harderian gland in rabbits.

Table V
Hyperplasia of Urinary Bladder in Rats and Mice

Compound	Study duration (weeks)	Dosage levels (mg/kg/day)	Urinalysis		Urothelial hyperplasia
			Na	pH	
MK-927					
Rats	5 to 53	0.1–1	—	—	−
		2 to 100	↑	↑	+
Mice	7 to 14	0.4	—	↑	−
		2 to 100	↑	↑	+
Dorzolamide					
Rats	5 to 53	0.05–0.1	—	—	−
		0.3 to 20	↑	↑	+
Mice	14	0.05–0.15	—	—	−
		1 to 100	↑	↑	+
Acetazolamide					
Rats	5	2	—	—	−
		20 to 400	↑	↑	+
Mice	14	4 to 400	↑	↑	+

erature, it was hypothesized that the hyperplasia was induced by high urinary pH and /or urinary sodium. In order to test this hypothesis, studies with systemic acidification and decreased dietary sodium were conducted in rats. The involvement of urinary pH was addressed in an oral study in rats given 15 mg/kg per day of MK-927 with 5% anhydrous monobasic potassium phosphate or 5% ammonium chloride in the diet. The group given the potassium phosphate meal had reductions in urinary pH and a decreased incidence of urothelial hyperplasia whereas the group fed the ammonium chloride meal had decreases in both urinary pH and sodium ion concentrations and the lowest incidence of urothelial hyperplasia when compared with MK-927 alone. The role of sodium was addressed in a study with MK-927 at 25 mg/kg per day in which rats were fed a low-sodium diet (110 mg/kg of sodium). Rats had very low urinary sodium and essentially no urothelial hyperplasia (Durand-Cavagna *et al.*, 1992). Furthermore, similar findings have been reported in rats given a number of sodium salts including sodium saccharin and sodium bicarbonate (Fukushima and Cohen, 1980; Hasegawa and Cohen, 1986; Shibata *et al.*, 1989) and in mice given 4-ethylsulfonylnaphthalene-1-sulfonamide (Sen Gupta, 1962). Because this rodent-specific change did not progress and even regressed despite continued treatment, and because a clear no-effect level was established for urine physiologic changes and urothelial hyperplasia, this is not regarded as a safety concern for humans.

RPEH and mineralization were noted in rats treated longer than 1 year with dorzolamide. These were dose-related increases greater than the spontaneous age-related incidences of RPEH. Minor increases were seen at the lowest dose in males

(0.05 mg/kg per day) but not in females. The RPEH was seen in rats given acetazolamide at the human therapeutic dose of 20 mg/kg per day as well as with hydrochlorothiazide (Bucher *et al.,* 1990). The mineralization was associated with, and was not seen without, RPEH in the 2-year study with dorzolamide. RPCG was seen with all CA inhibitors in mice and rats. The incidence tended to increase slowly with increasing duration of dosing. Reports from the literature support the hypothesis that this change is related to potassium loss in the urine and a decrease in renal medullary potassium (Hansen *et al.,* 1980; Owen *et al.,* 1993; Toback *et al.,* 1976). Because CA inhibitors cause hyperkaliuria, a similar mechanism may play a role in the development of RPCG. To test this hypothesis, a study in rats was conducted with acetazolamide in which the animals were supplemented with potassium chloride in their drinking water. The incidence of RPCG was significantly reduced by this protocol. Additionally, the syndrome was reversible when treatment was stopped. The RPCG was seen at a dosage level slightly below the maximum dose intended for humans (0.05 mg/kg per day for MK-0507) and this change is thought to be of no toxicological concern for humans. Indeed, acetazolamide also caused this change in rats at a dosage level (2 mg/kg per day) below the human therapeutic dose of 20 mg/kg per day.

Hyperplasia of the mucous neck cells of the gastric fundus mucosa was seen at doses that induced systemic acidosis with all CA inhibitors in dogs and/or monkeys. This change appears to be related to acidosis and this hypothesis was supported in a study in which the hyperplasia was ameliorated in dogs given MK-927 at 3 mg/kg per day supplemented with 1.5% sodium bicarbonate. It was also shown that this change regresses despite continued treatment with dorzolamide in dogs and was not present in long-term studies. Gastric mucous neck cell hyperplasia was seen in rats treated for longer than 6 months at 1 mg/kg per day with dorzolamide and at 20 mg/kg per day with acetazolamide and seems to be an exaggeration of a spontaneous age-related change. This change was seen in monkeys with dorzolamide at 50 mg/kg per day. In monkeys and dogs given 500 or 10 mg/kg per day of acetazolamide, respectively, the same lesion was seen. Based on these findings, this change is thought to be of little toxicological concern.

In ribs from young adult dogs and in femurs from juvenile monkeys given dorzolamide, there was a minor effect on endochondral bone formation seen as minimal retention of the primary spongiosa reflecting decreased remodeling of the bone at 1.5 mg/kg per day in female and 3 mg/kg per day in male dogs, and 50 mg/kg per day in monkeys. The hypothesis for the etiology of this change is the inhibition of CA in osteoclasts as this enzyme is required for acid production for bone resorption (Robbins *et al.,* 1984). The retention of the primary spongiosa is not considered an important toxicological finding because the change was of minimal degree at high doses and in dogs it disappeared despite continued treatment. Also, there is a clear no-effect dose that is at least 25 times the clinical dose. In addition, acetazolamide, at 500 mg/kg per day, caused the same kind of changes in the bone of monkeys.

A battery of genetic toxicology studies were undertaken and all were negative. These included the microbial mutagenesis assay, the V-79 assay for mammalian cell mutagenesis, the alkaline elution assay in rat hepatocytes for DNA strand breaks, the *in vitro* chromosomal aberration assay in Chinese ovary cells, and the *in vivo* chromosomal aberration assay in bone marrow from treated mice.

A 92-week carcinogenicity study in mice given dorzolamide at doses of up to 75 mg/kg per day showed no treatment-related tumors. In male rats given 20 mg/kg per day of dorzolamide for 2 years, an increased incidence of papillomas of the urinary bladder was seen. These papillomas were attributed to the pharmacological action of the drug: increases in urinary pH and sodium levels, and/or crystalluria/urolithiasis. Acetazolamide caused a non-dose-related increase in urinary bladder papillomas in rats treated for 2 years at dosage levels of 2, 20, and 200 mg/kg/day. There was a clear no-effect level (1 mg/kg per day) for this change and a good margin of safety (papillomas seen only at 250 times the maximum intended human dose). In rats, there were no fetal anomalies observed up to the highest dose tested (10 mg/kg per day). In rabbits, there were some malformations of vertebral bodies and ribs at 2.5 mg/kg per day. However, because of the wide clinical experience with acetazolamide and the observation that it caused specific limb defects in rats at 350 mg/kg per day (Layton and Hallesy, 1965), or about 17 times the human therapeutic dose, this finding was not considered to represent a risk for humans.

9. SUMMARY

Dorzolamide, on the basis of its pharmacological profile and lack of undesirable side effects in safety assessment studies together with the fact that it could be formulated in solution at 2%, underwent extensive clinical studies.

Early clinical studies in the development of dorzolamide have been described elsewhere (Maren, 1995; Serle and Podos, 1995). In a 1-year study in which a comparison was undertaken in patients for intraocular pressure lowering effects between 2% dorzolamide administered three times daily, 0.5% betaxolol twice daily, and 0.5% timolol twice daily, the peak reductions in intraocular pressure were 23, 21, and 25%, respectively. Tachyphylaxis did not develop to dorzolamide nor were electrolyte and/or systemic side effects encountered (Strahlman *et al.*, 1995). The latter is consistent with results of a pharmacokinetic study in humans in which plasma levels of dorzolamide were lower than the limit of detection (5 ng/ml) at a time when the red blood cell content of dorzolamide had reached steady state which was appreciably less than the red blood cell content of the enzyme (Biollaz *et al.*, 1995). Patients taking 0.5% timolol twice daily received either 2% dorzolamide twice daily or 2% pilocarpine four times daily for 6 months and the additional reductions in intraocular pressure elicited by dorzolamide and pilocarpine were very similar. However, pilocarpine usage resulted in a higher discontinuation rate (Strahlman *et al.,* 1996). In a separate study in which dorzolamide and pilocarpine were compared at these dosage schedules, patients preferred dorzolamide to pilocarpine by a ratio of over 7 to 1 in terms of quality of life (Laibovitz *et al.,* 1995).

In summary, the quest for a topical, ocular hypotensive, CA inhibitor, though time-consuming, was a successful one with the introduction of dorzolamide into general clinical practice.

ACKNOWLEDGMENT

The authors would like to acknowledge the secretarial assistance of Ms. Jo Hagan.

REFERENCES

Baldwin, J. J., Ponticello, G. S., Anderson, P. S., Christy, M. E., Murcko, M. A., Randall, W. C., Schwam, H., Sugrue, M. F., Springer, J. P., Gautheron, P., Grove, J., Mallorga, P., Viader, M.-P., McKeever, B. M., and Navia, M. A., 1989, Thienothiopyran-2 sulfonamides: Novel topically active carbonic anhydrase inhibitors for the treatment of glaucoma, *J. Med. Chem.* **32:**2510–2513.

Becker, B., 1954, Decrease in intraocular pressure in man by a carbonic anhydrase inhibitor, Diamox, *Am. J. Ophthalmol.* **37:**13–14.

Biollaz, J., Munafo, A., Buclin, T., Gervasoni, J.-P., Magnin, J. L., Jaquet, F., and Brunner-Ferber, F., 1995, Whole-blood pharmacokinetics and metabolic effects of the topical carbonic anhydrase inhibitor dorzolamide, *Eur. J. Clin. Pharmacol.* **47:**453–460.

Blacklock, T. J., Sohar, P., Butcher, J. W., Lamanec, T., and Grabowski, E. J. J., 1993, An enantioselective synthesis of the topically-active carbonic anhydrase inhibitor MK-507: 5,6-Dihydro-(S)-4-(ethylamino)-(S)-6-methyl-4H-thieno[2,3-b]thiopyran-2-sulfonamide-7,7-dioxide-hydrochloride, *J. Org. Chem.* **58:**1672–1679.

Bucher, J. R., Huff, J., Haseman, J. K., Eustis, S. L., Elwell, M. R., Davis, W. E., and Meierhenry, E. E., 1990, Toxicology and carcinogenicity studies of diuretics in F344 rats and B6C3F1 mice. 1. Hydrochlorothiazide, *J. Appl. Toxicol.* **10:**359–367.

Durand-Cavagna, G., Delort, P., Gordon, L. R., Peter, C. P., and Boussiquet-Leroux, C., 1992, Urothelial hyperplasia induced by carbonic anhydrase inhibitors (CAIs) in animals and its relationship to urinary Na and pH, *Fundam. Appl. Toxicol.* **18:**137–143.

Durand-Cavagna G., Gerin, G., and Gordon, L. R., 1996, Evaluating delayed contact hypersensitivity reactions from ocular medications, *J. Toxicol. Cut. Ocular Toxicol.* **15:**235–248.

Fukushima, S., and Cohen, S. M., 1980, Saccharin-induced hyperplasia of the rat urinary bladder, *Cancer Res.* **40:**734–736.

Graham, S. L., Shepard, K. L., Anderson, P. S., Baldwin, J. J., Best, D. B., Christy, M. E., Freedman, M. B., Gautheron, P., Habecker, C. N., Hoffman, J. M., Lyle, P. A., Michelson, S. R., Ponticello, G. S., Robb, C. M., Schwam, H., Smith, A. M., Smith, R. L., Sondey, J. M., Strohmaier, K. M., Sugrue, M. F., and Varga, S. L., 1989, Topically active carbonic anhydrase inhibitors. 2. Benzo(b) thiophenesulfonamide derivatives with ocular hypotensive activity, *J. Med. Chem.* **32:**2548–2554.

Grove, J., Quint, M. P., and Plazonnet, B., 1995, Influence of pH on the ocular penetration of the topical carbonic anhydrase inhibitor, MK-507, in the albino rabbit, *Invest. Ophthalmol. Vis. Sci.* **36:**S159.

Hageman, G. S., Zhu, X. L., Waheed, A., and Sly, W. S., 1991, Localization of carbonic anhydrase IV in a specific capillary bed of the human eye, *Proc. Natl. Acad. Sci. USA* **88:**2716–2720.

Hansen, G. P., Tisher, C. C., and Robinson, R. R., 1980, Response of the collecting duct to disturbances of acid–base and potassium balance, *Kidney Int.* **17:**326–337.

Hasegawa, R., and Cohen, S. M., 1986, The effect of different salts of saccharin on the rat urinary bladder, *Cancer Lett.* **30:**261–268.

Laibovitz, R., Strahlman, E. R., Barber, B. L., and Strohmaier, K. M., 1995, Comparison of quality of life and patient preference of dorzolamide and pilocarpine as adjunctive therapy to timolol in the treatment of glaucoma, *J. Glaucoma* **4:**306–313.

Layton, W. M., and Hallesy, D. W., 1965, Deformity of forelimbs in rats: Association with high doses of acetazolamide, *Science* **149:**306–308.

Liesegang, T. J., 1996, Glaucoma: Changing concepts and future directions, *Mayo Clin. Proc.* **71:**689–694.

Lippa, E. A., 1991, The eye: Topical carbonic anhydrase inhibitors, in: *The Carbonic Anhydrases* (R. E. Tashian, G. Gros, and N. D. Carter, eds.), pp. 171–181, Plenum Press, New York.

Lippa, E. A., von Denffer, H. A., Hofmann, H. M., and Brunner-Ferber, F. L., 1988, Local tolerance and activity of MK-927, a novel topical carbonic anhydrase inhibitor, *Arch. Ophthalmol.* **106:**1694–1696.

Lippa, E. A., Schuman, J. S., Higginbotham, E. S., Kass, M. A., Weinreb, R. N., Skuta, G. L., Epstein, D. L., Shaw, B., Holder, D. J., Deasy, D. A., and Wilensky, J. T., 1991, MK-507 versus Sezolamide, *Ophthalmology* **98:**308–313.

Lütjen-Drecoll, E., Lönnerholm, G., and Eichhorn, M., 1983, Carbonic anhydrase distribution in the human and monkey eye by light and electron microscopy, *Graefe's Arch. Clin. Exp. Ophthalmol.* **220:**285–291.

Magnusson, B., and Kligman, A. M., 1969, The identification of contact allergens by animal assay: The guinea pig maximization test, *J. Invest. Dermatol.* **52:**268–276.

Maren, T. H., 1967, Carbonic anhydrase: Chemistry, physiology, and inhibition, *Physiol. Rev.* **47:**595–781.

Maren, T. H., 1995, The development of topical carbonic anhydrase inhibitors, *J. Glaucoma* **4:**49–62.

Maren, T. H., Mayer, E., and Wadsworth, B. C., 1954, Carbonic anhydrase inhibition. I. The pharmacology of Diamox® 2-acetylamino-1,3,4-thiadiazole-5-sulfonamide, *Bull. Johns Hopkins Hosp.* **95:**199–243.

Moses, A. R. (ed.), 1981, *Adler's Physiology of the Eye,* 7th ed., p. 20, Mosby, St. Louis.

Owen, R. A., Durand-Cavagna, G., Molon-Noblot, S., Boussiquet-Leroux, C., Berry, P., Tonkonoh, N., Peter, C. P., and Gordon, L. R. , 1993, Renal papillary cytoplasmic granularity and potassium depletion induced by carbonic anhydrase inhibitors in rats, *Toxicol. Pathol.* **21:**449–455.

Ponticello, G. S., Freedman, M. B., Habecker, C. N., Lyle, P. A., Schwam, H., Varga, S. L., Christy, M. E., Randall, W. C., and Baldwin, J. J., 1987, Thienothiopyran-2-sulfonamides: A novel class of water-soluble carbonic anhydrase inhibitors, *J. Med. Chem.* **30:**591–597.

Ponticello, G. S., Freedman, M. B., Habecker, C. N., Holloway, M. K., Amato, J. S., Conn, R. S., and Baldwin, J. J., 1988, Utilization of α,β-unsaturated acids as Michael acceptors for the synthesis of thieno[2,3-b] thiopyrans, *J. Org. Chem.* **53:**9–13.

Robbins, S. L., Cotran, R. S., and Kumar, V., 1984, *Pathologic Basis of Disease,* p. 1319, Saunders, Philadelphia.

Schwam, H., Michelson, S. R., Sondey, J. M., and Smith, R. L., 1984, L-645,151, a topically effective ocular hypotensive carbonic anhydrase inhibitor: Part I. Biochemistry and metabolism, *Invest. Ophthalmol. Vis. Sci.* **25**(Suppl.)**:**180.

Sen Gupta, K. P., 1962, Hyperplasia of urinary tract epithelium induced by continuous administration of sulphonamide derivatives, *Br. J. Cancer* **16:**110–119.

Serle, J. B., and Podos, S. M., 1995, Topical carbonic anhydrase inhibitors in the treatment of glaucoma, *Ophthalmol. Clin. North Am.* **8:**315–325.

Shepard, K. L., Graham, S. L., Hudcosky, R. J., Michelson, S. R., Scholz, T. H., Schwam, H., Smith, A. M., Sondey, J. M., Strohmaier, K. M., Smith, R. L., and Sugrue, M. F., 1991, Topically active carbonic anhydrase inhibitors. 4. (Hydroxyalkyl)sulfonylbenzene and (hydroxyalkyl)sulfonylthiophenesulfonamides, *J. Med. Chem.* **34:**3098–3105.

Shibata, M. A., Tamano, S., Kurata, Y., Hagiwara, A., and Fukushima, S., 1989, Participation of urinary sodium, potassium, pH and L-ascorbic acid in the proliferative response of the bladder epithelium after the oral administration of various salts and/or ascorbic acid to rats, *Food Chem. Toxic.* **27:**403–413.

Smith, G. M., Alexander, R. S., Christianson, D. W., McKeever, B. M., Ponticello, G. S., Springer, J. P., Randall, W. C., Baldwin, J. J., and Habecker, C. N., 1994, Positions of His-64 and a bound water in human carbonic anhydrase II upon binding three structurally related inhibitors, *Protein Sci.* **3:**118–125.

Strahlman, E., Tipping, R., Vogel, R., and the International Dorzolamide Study Group, 1995, A double-masked, randomized 1-year study comparing dorzolamide (Trusopt), timolol and betaxolol, *Arch. Ophthalmol.* **113:**1009–1016.

Strahlman, E. R., Vogel, R., Tipping, R., Clineschmidt, C. M., and the Dorzolamide Additivity Study Group, 1996, The use of dorzolamide and pilocarpine as adjunctive therapy to timolol in patients with elevated intraocular pressure, *Ophthalmology* **103:**1283–1293.

Sugrue, M. F., 1996, Review: The preclinical pharmacology of dorzolamide hydrochloride, a topical carbonic anhydrase inhibitor, *J. Ocular Pharmacol. Ther.* **12:**363–376.

Sugrue, M. F., and O'Neill-Davis, 1991, The effect of cyclooxygenase inhibition on the ocular hypotensive action of topical carbonic anhydrase inhibitors in rabbits, *J. Ocular Pharmacol.* **7:**201–211.

Sugrue, M. F., Gautheron, P., Schmitt, C., Viader, M. P., Conquet, P., Smith, R. L., Share, N. N., and Stone, C. A., 1985, On the pharmacology of L-645,151: A topically effective ocular hypotensive carbonic anhydrase inhibitor, *J. Pharmacol. Exp. Ther.* **232:**534–540.

Sugrue, M. F., Mallorga, P., Schwam, H., Baldwin, J. J., and Ponticello, G. S., 1990, A comparison of L-671,152 and MK-927, two topically effective ocular hypotensive carbonic anhydrase inhibitors, in experimental animals, *Curr. Eye Res.* **9:**607–615.

Toback, F. G., Ordonez, N. G., Bortz, S. L., and Spargo, B. H., 1976, Zonal changes in renal structure and phospholipid metabolism in potassium-deficient rats, *Lab. Invest.* **34:**115–124.

Wang, R.-F., Serle, J. B., Podos, S. M., and Sugrue, M. F., 1991, MK-507 (L-671,152), a topically active carbonic anhydrase inhibitor, reduces aqueous humor production in monkeys, *Arch. Ophthalmol.* **109:**1297–1299.

Wistrand, P. J., 1951, Carbonic anhydrase in the anterior uvea of the rabbit, *Acta Physiol. Scand.* **24:**144–148.

Yamazaki, Y., Miyamoto, S., and Sawa, M., 1994, Effect of MK-507 on aqueous humor dynamics in normal human eyes, *Jpn. J. Ophthalmol.* **38:**92–96.

Chapter 25

Discovery and Development of Novel Melanogenic Drugs

Melanotan-I and -II

Mac E. Hadley, Victor J. Hruby, James Blanchard, Robert T. Dorr, Norman Levine, Brenda V. Dawson, Fahad Al-Obeidi, and Tomi K. Sawyer

1. INTRODUCTION

The melanocortins include the melanotropins [melanocyte-stimulating hormones (MSHs)] and corticotropin [adrenal cortical-stimulating hormone (ACTH)]. α-Melanotropin (α-MSH) is a tridecapeptide that in many vertebrates is derived from the pars intermedia of the pituitary gland. This peptide regulates pigmentation of the skin and hair (pelage) in many animals (Hadley, 1996). Until recently, the

---

Mac E. Hadley • Department of Cell Biology and Anatomy, University of Arizona, Tucson, Arizona 85724. *Victor J. Hruby* • Department of Chemistry, University of Arizona, Tucson, Arizona 85724. *James Blanchard* • Arizona Health Sciences Center, University of Arizona, Tucson, Arizona 85724. *Robert T. Dorr* • Arizona Cancer Center, University of Arizona, Tucson, Arizona 85724. *Norman Levine* • Department of Dermatology, University of Arizona, Tucson, Arizona 85724. *Brenda V. Dawson* • Health Sciences, The University of Auckland, 92019 Auckland, New Zealand. *Fahad Al-Obeidi* • Department of Chemistry, Selectide Research Center, Hoechst-Marion Roussel, Tucson, Arizona 85724. *Tomi K. Sawyer* • Ariad Pharmaceuticals, Cambridge, Massachusetts 02139.

Integration of Pharmaceutical Discovery and Development: Case Studies, edited by Borchardt *et al.*, Plenum Press, New York, 1998.

Table I
Possible Clinical Uses of Melanotan-I (MT-I) or Melanotan-II (MT-II)

Melanogenesis	MT-I and MT-II for use in the stimulation of skin melanogenesis and development as photoprotective or cosmetic drugs
Melanoma diagnosis	Multivalent MT-based peptide conjugates with fluorescent markers for use in the detection and localization of melanoma
Melanoma chemotherapy	Conjugates of MT-based peptides with anticancer drugs for use in site-specific delivery and chemotherapy of melanoma
Eating disorders	MT-based peptide or peptidomimetic agonists or antagonists for use in the treatment of eating disorders (obesity or anorexia)
Inflammation	MT-based peptide or peptidomimetic agonists for use in the treatment of inflammation (contact hypersensitivity) of the skin
Erectogenic dysfunction	MT-based peptide or peptidomimetic agonists for use in the diagnosis and/or treatment of erectogenic dysfunction

melanotropic peptides (α-MSH, β-MSH, γ-MSH) were of interest to only a limited scientific audience (e.g., comparative endocrinologists). However, with the recent discovery of a number of melanocortin receptor (MCR) types, including several localized to specific areas in the brain, interest has been aroused in the pharmaceutical industry that the melanocortins may be of important clinical relevance (Hadley *et al.,* 1996). Several possible clinical uses are summarized in Table I as related to two α-MSH superagonist analogues, Melanotan-I (MT-I) and Melanotan-II (MT-II) (*vide infra*).

Over the past 25 years, we have synthesized and biologically evaluated over 1000 analogues of α-MSH (Castrucci *et al.,* 1989; Hruby *et al.,* 1987). Several of these α-MSH analogues have exhibited superpotency and prolonged activity with respect to their melanogenic (skin tanning) properties, in addition to being resistant to degradation by proteolytic enzymes. Two α-MSH analogues, MT-I (Hadley *et al.,* 1993; Levine *et al.,* 1991) and MT-II (Dorr *et al.,* 1995), have been extensively studied and are currently in phase II clinical trials. The discovery and development of these two promising melanogenic drugs is described below, and this effort reflects the collaborative efforts of a team of academic scientists having diverse backgrounds in chemistry, endocrinology, pharmacology, pathology, toxicology, drug delivery, and dermatology.

2. THE MELANOCORTIN PEPTIDES AND RECEPTORS

2.1. Melanocortin Peptides

The melanocortin family of peptides are evolutionarily related and share an identical "active site" sequence, His-Phe-Arg-Trp (Fig. 1), as well as limited structural homology at their N- and/or C-termini. The melanocortins are derived from

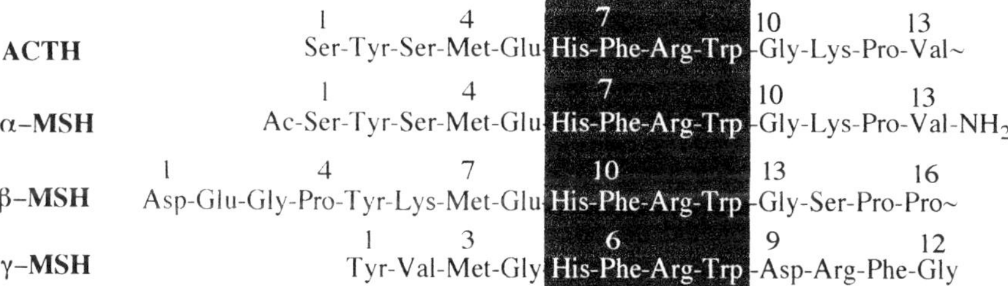

Figure 1. Amino acid sequences of the melanocortin peptides depicting their structural identity at the central His-Phe-Arg-Trp sequence.

a large precursor protein referred to as proopiomelanocortin (POMC). Messenger RNA for POMC has been localized to a variety of cells throughout the body. Depending on the specific enzymatic profile of the cell, the prohormone can be enzymatically cleaved to produce the peptide specific to that cell type (e.g., ACTH in corticotrophs and α-MSH in melanotrophs of the pituitary gland).

ACTH is synthesized and secreted by cells (corticotrophs) of the pars distalis (anterior lobe) of the pituitary gland. As a systemically acting hormone, ACTH stimulates steroidogenesis leading to secretion of cortisol from the adrenal glands. α-MSH is synthesized and released by melanotrophs of the pars intermedia of most vertebrate species. Acting as a systemic hormone, this tridecapeptide (Fig. 2) stimulates melanogenesis (melanin formation) within pigment cells or melanocytes of the epidermis of many animals. In humans, α-MSH or a related melanotropin is postulated to be produced by keratinocytes of the epidermis. By a local paracrine action on melanocytes, melanogenesis is enhanced leading to increased skin pigmentation (tanning).

In the brain, the localization of POMC to neurons (Gee *et al.*, 1983) suggests that one or more melanocortin peptides might function as a neurohormone (i.e., neurotransmitter or neuromodulator). In this regard, melanocortin peptides have been shown to enhance cognitive skills, short-term memory retention, induce satiety, sexual behavior, and other CNS activities (O'Donohue and Dorsa, 1982).

2.2. Melanocortin Receptors

Melanocortin peptides mediate their actions through MCRs that are restricted to certain cells (Hadley *et al.*, 1996). Presently, five distinct human MCR types have been cloned, expressed, and biochemically characterized (Barret *et al.*, 1994; Chhajlani *et al.*, 1993; Desarnaud *et al.*, 1994; Fathi *et al.*, 1995; Gantz *et al.*, 1993a,b, 1994; Griffon *et al.*, 1994; Labbé *et al.*, 1994; Mountjoy *et al.*, 1994; Roselli-Rehfuss *et al.*, 1993): MC1R, MC2R, MC3R, MC4R, and MC5R. The hu-

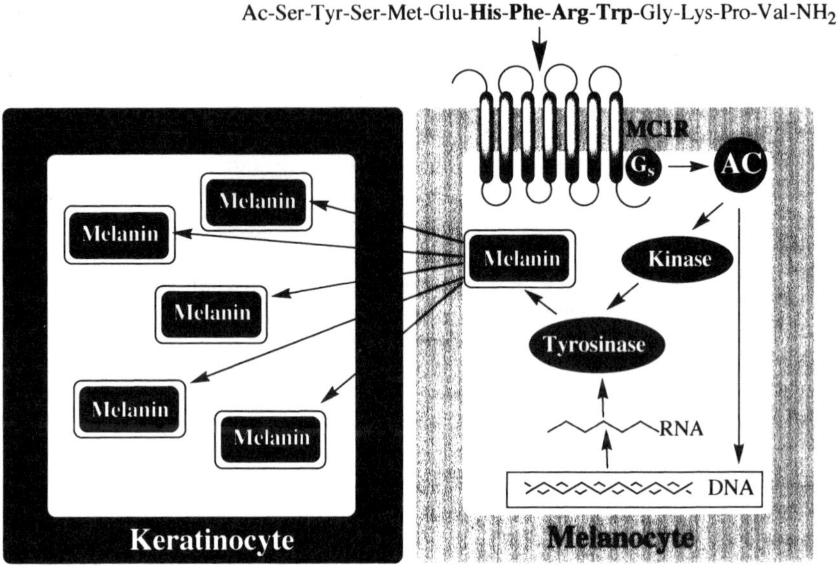

Figure 2. Schematic illustration of α-MSH-stimulated melanogenesis in melanocytes and delivery of melanin pigment into adjacent keratinocytes of the skin. In melanocytes, α-MSH binding to its receptor (the G-protein, G_s, coupled MC1R) activates adenylate cyclase (AC) to increase intracellular cAMP levels. This results in activated cAMP-dependent protein kinase and tyrosinase, and in enhanced biosynthesis of melanin.

man MC1R type is schematically illustrated in Fig. 3 to show its putative seven transmembrane (TM) a-helices, and extracellular and intracellular loops.

Constitutively activated MC1R mutants have been characterized (Robbins *et al.*, 1993), and site-directed mutagenesis studies of the MC1R have examined α-MSH binding (Frandberg *et al.*, 1994). The MC1R is localized to epidermal melanocytes and their malignant progeny, melanoma cells. The MC2R is primarily responsive to ACTH and is localized to the adrenocortical cells that produce cortisol. The MC3R and MC4R are primarily found in the brain, whereas the MC5R is widely distributed throughout the body, including various gut tissues. In contrast to both the MC1R and MC2R, the physiological roles of melanocortins interacting with the MC3R, MC4R, and MC5R remain to be unambiguously defined. Activation of the MC1R of normal human epidermal melanocytes (NHEMs) as well as human melanoma cells (HMCs) results in cAMP formation. Although enhanced cAMP levels lead to enhanced melanin formation in both normal and abnormal pigment cells, NHEMs proliferate whereas HMC growth is retarded (Hadley *et al.*, 1996; Jiang *et al.*, 1995). At the MC2R of adrenocortical cells, ACTH also increases cAMP formation. For the MC3R, MC4R, and MC5R, the

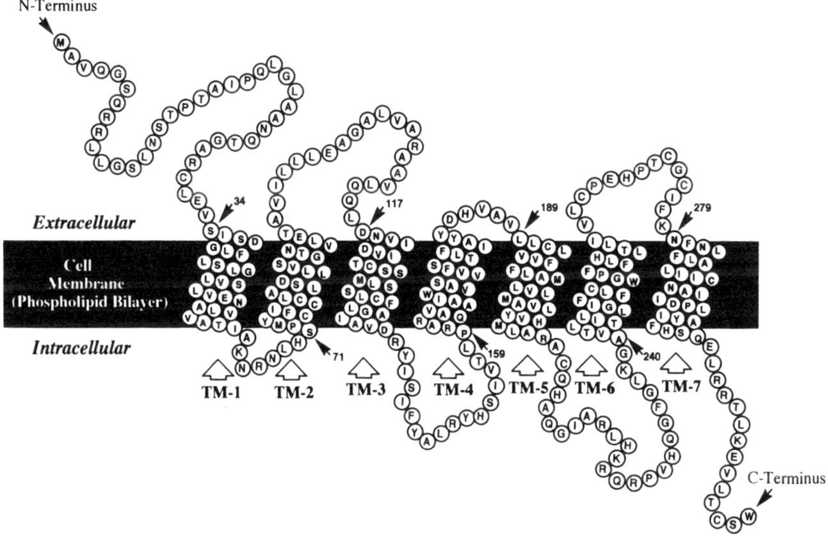

Figure 3. Human MC1R primary structure.

mechanisms of action of melanocortins remain to be fully characterized, but might be expected to involve G-protein-coupled adenylate cyclase stimulation and cAMP production.

3. DISCOVERY OF MT-I AND MT-II AS MSH SUPERAGONISTS

3.1. Structure–Activity Studies of α-MSH

We have designed and synthesized over 1000 analogues of α-MSH for structure–activity studies and have determined the minimal fragment of the native peptide needed for biological activity. Specifically, using the classical frog and lizard skin bioassays, it was determined that the minimal fragment of α-MSH required for agonist activity was its central tetrapeptide sequence, His^6-Phe^7-Arg^8-Trp^9, as exemplified by the analogue Ac-His-Phe-Arg-Trp-NH_2 (Hruby *et al.,* 1987). Both N-and C-terminal amino acid extension of the central tetrapeptide sequence effect increased potency to match that of α-MSH, as exemplified by Ac-α-MSH(4–11)-NH_2 on the lizard skin bioassay and by Ac-α-MSH(4–12)-NH_2 on the frog skin bioassay (Hruby *et al.,* 1984a). These results have also contributed to α-MSH-based drug design strategies with respect to key sites for chemical conjugation with

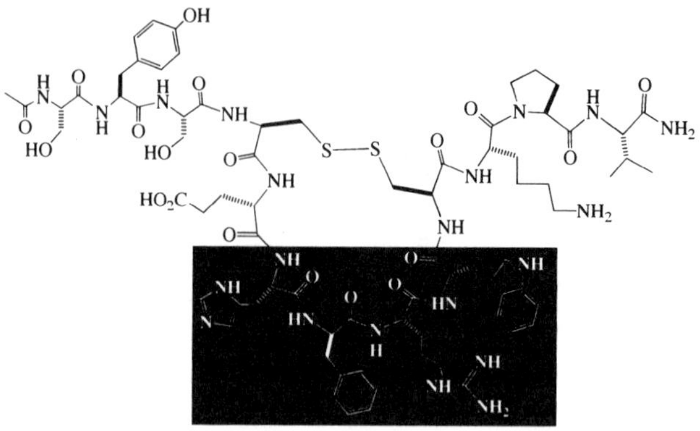

Ac-Ser-Tyr-Ser-Cys-Glu-His-Phe-Arg-Trp-Cys-Lys-Pro-Val-NH_2
1 4 7 10 13

Figure 4. Structure of c[Cys^4,Cys^{10}]-α-MSH.

macromolecules, diagnostic and/or cytotoxic agents to produce selectivity for melanoma cells (Hadley *et al.*, 1996).

The "bioactive conformation" of α-MSH was first shown to possibly exist as a reverse-turn-type conformation at the His-Phe-Arg-Trp sequence by the design of the macrocyclic analogue c[Cys^4,Cys^{10}]-α-MSH (Fig. 4), a superagonist in the frog skin bioassay (Hruby *et al.*, 1984a). A plethora of structure–conformation–activity studies of both linear and cyclic "second-generation" α-MSH analogues have subsequently been advanced to further explore the three-dimensional properties of α-MSH as related to its binding (molecular recognition) and/or activation (signal transduction) at α-MSH receptors on frog and lizard melanocytes (Al-Obeidi *et al.*, 1989a; Castrucci *et al.*, 1989; Cody *et al.*, 1988, Haskell-Luevano *et al.*, 1996b; Hruby *et al.*, 1987), human and mouse melanoma cells (Abdel-Malek *et al.*, 1985; Hadley *et al.*, 1985), and, more recently, cloned human MC1R (Haskell-Luevano *et al.*, 1996a, 1997).

3.2. Design and Chemistry of MT-I and MT-II

Two α-MSH superagonists, the tridecapeptide [Nle^4,D-Phe^7]α-MSH (MT-I; Sawyer *et al.*, 1980) and the cyclic heptapeptide Ac-c[Nle^4,Asp^5,D-Phe^7,Lys^{10}]α-MSH(4–10)-NH_2 (MT-II; Al-Obeidi *et al.*, 1989a), have been designed and shown to possess sustained-acting biological activity *(vide infra)*. The chemical structures of MT-I and MT-II (Fig. 5) share an identical central sequence, His-D-Phe-Arg-Trp, which has been proposed to exist as a "reverse-turn" conformation when the agonist binds to the MC1R.

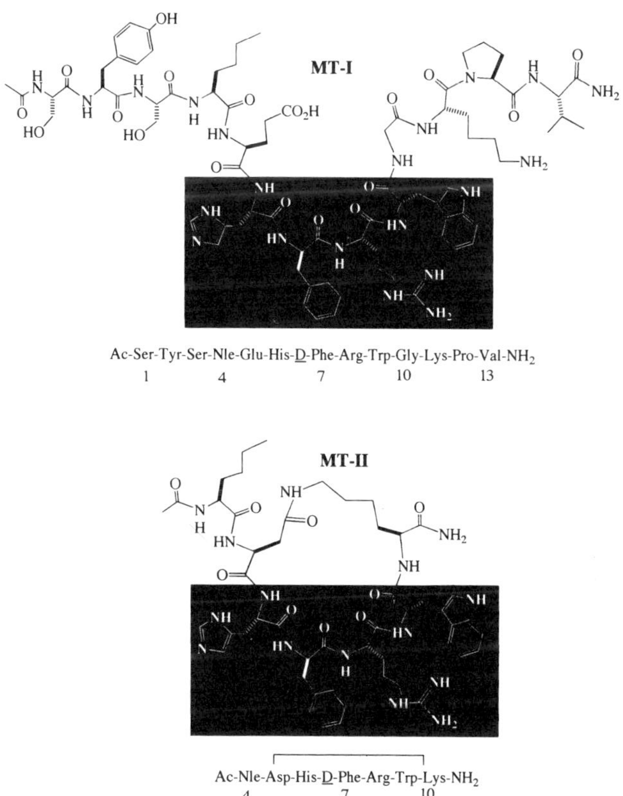

Figure 5. Structures of α-MSH superagonists MT-I and MT-II.

The substitution of D-Phe-7 in both MT-I and MT-II is critical for their superagonist activity. This particular modification was the result of previous studies involving partial racemization of α-MSH and [Nle^4]α-MSH by heat–alkali treatment (Bool *et al.,* 1981). This resulted in potentiation and sustained-acting melanotropic properties in terms of frog skin darkening. In fact, this study confirmed earlier studies (Smith and Graeser, 1924) showing that heat–alkali treatment of crude extracts of the posterior pituitary gland resulted in significantly enhanced melanotropic activity *in vivo.* To pinpoint and quantitate the extent of racemization (conversion of L-amino acids to their D-isomers) within α-MSH following heat–alkali treatment, we used high-resolution GC methodology (Engel *et al.,* 1981) and found Phe-7 to be racemized to a greater extent than expected. Conceptually, the mixture of possible diastereomers within the heat–alkali-treated α-MSH is 2^{12} (Gly excluded because it is achiral) or 4096 peptides, thus exemplifying a "combinatorial mixture" by virtue of chemically induced, partial racemization. The synthesis of MT-I provided proof-of-concept that the D-Phe-7 substitution accounted for the observed increased potency and sustained melanotropic activity as determined in several biological assays (Hruby *et al.*, 1984a; Sawyer *et al.*, 1980).

The design of MT-II was based on creating a macrocyclic analogue of Ac-[Nle^4, D-Phe^7]α-MSH(4–10)-NH_2 by virtue of two amino acid substitutions flanking the His-*D*-Phe-Arg-Trp sequence (namely, Asp-5 and Lys-10) and incorporating a lactam bridge between their side chains (Al-Obeidi *et al.*, 1989b). Both NMR spectroscopy and computer-assisted molecular modeling studies on MT-II have supported the proposed existence of a "reverse-turn" conformation within its active-site sequence. Structure–activity studies of both MT-I and MT-II have been reported (Haskell-Luevano *et al.*, 1994, 1996a,b; Sawyer *et al.*, 1982, 1993), and such work has been focused primarily on systematic modifications of the $MSH_{6–9}$ active-site sequence. Competitive antagonists have been discovered (Adan *et al.*, 1994; Al-Obeidi *et al.*, 1990; Hruby *et al.*, 1995) and have exemplified the effects of key structural modifications within the $MSH_{6–9}$ active-site sequence. Radiolabeled or fluorescently labeled derivatives of MT-I, as well as macromolecular conjugates thereof, have been developed to use for the visualization (diagnostic agents) and/or site-specific delivery of drugs (therapeutic agents) to melanoma cells based on the MC1R binding of such compounds (Hadley *et al.*, 1996). Very recently, a prototype series of peptidomimetic agonists have been reported (Haskell-Luevano *et al.*, 1996b, 1997) which are based on the $MSH_{6–9}$ active site sequence. Noteworthy among these tetra- and tripeptide agonists are Ac-His-D-Phe-Arg-Trp-NH_2 and Ac-D-Phe-Arg-D-Trp-NH_2 (*vide infra*). Such compounds may be significant leads for the discovery of orally bioavailable α-MSH analogues for therapeutic use.

3.3. *In Vitro* and *in Vivo* Pharmacology of MT-I and MT-II

As stated above, the two classical *in vitro* assays that have been used to determine the structure–activity of α-MSH analogues have been the frog and lizard skin bioassays. More recently, the use of cell lines expressing cloned human MCRs have provided the opportunity to determine the receptor specificity of MT-I, MT-II, and other α-MSH analogues (Haskell-Luevano *et al.*, 1996a,b, 1997). A comparative analysis of MSH, MT-I, and MT-II relative to the frog and lizard skin bioassays and the human MC1R (binding and cAMP activation) is shown in Table II.

The sustained-acting melanotropic activities of MT-I and MT-II have been shown using the frog and lizard skin bioassays (Al-Obeidi *et al.*, 1989a,b; Sawyer *et al.*, 1980), as well as in the frog *in vivo* (Hadley *et al.*, 1981). Relative to α-MSH, the biological effects of MT-I and MT-II resist "washout" in these skin preparations, and subcutaneous injection of MT-I into the frog results in several weeks of skin darkening *in vivo* versus only several hours for α-MSH.

Using transformed melanocytes, MT-I was first shown to exhibit superago-

Table II
Comparative Biological Properties of MT-I and MT-II

	Relative potency		Human MC1R	
	Frog skin	Lizard skin	Binding, IC_{50} (RP)[a]	cAMP, EC_{50} (RP)
α-MSH[b]	1.0	1.0	6.5 nM (1.0)	2.0 nM (1.0)
MT-I[c]	60.0	5.0	1.2 nM (5.4)	0.5 nM (4.0)
MT-II[d]	0.8	90.0	0.57 nM (11.4)	0.2 nM (10.0)

[a]RP, relative potency.
[b]α-MSH, Ac-Ser-Tyr-Ser-Met-Glu-His-Phe-Arg-Trp-Gly-Lys-Pro-Val-NH_2.
[c]MT-I, Ac-Ser-Tyr-Ser-Nle-Glu-His-D-Phe-Arg-Trp-Gly-Lys-Pro-Val-NH_2.
[d]MT-II, Ac-Nle-Asp-His-D-Phe-Arg-Trp-Lys-NH_2.

nist properties as observed with a mouse S91 melanoma tyrosinase assay (Abdel-Malek *et al.,* 1985). Specifically, following initial contact with MT-I, tyrosinase activity within these cells is enhanced for at least 7 days (after removal of the peptide by several washes) as based on transcriptional or translational readouts. These results are even more noteworthy when taking into account that the melanoma cells typically divide about every 24 hr.

Geschwind *et al.* (1972) found that injections of melanotropic peptides into certain strains of yellow-colored mice could result in a change in pelage color (i.e., from light yellow to a dark brown or even black color). We have confirmed such an *in vivo* melanogenic effect in mice using MT-I (Levine *et al.,* 1987). Specifically, MT-I was determined to be at least 100-fold more potent than α-MSH when injected subcutaneously.

4. DEVELOPMENT OF MT-I AND MT-II AS NOVEL MELANOGENIC DRUGS

The development of MT-I and MT-II as novel melanogenic drugs has been advanced in terms of stability, pharmacokinetic, toxicological, and drug delivery studies. Phase I pilot studies in humans have been successfully completed. A synopsis of these studies is described below.

4.1. Stability, Pharmacokinetic, and Toxicological Studies

Both MT-I and MT-II are resistant to metabolism by serum enzymes or by purified proteases. However, α-MSH is rapidly degraded by serum enzymes or purified proteases. Interestingly, the sustained-acting properties of either MT-I or MT-II, as demonstrated *in vitro,* are not related to their stability against proteolytic degradation. Nevertheless, *in vivo* efficacy of either MT-I or MT-II would be enhanced as a result of their metabolic stabilities.

4.1.1. MT-I STUDIES

The development of a sensitive, specific, and stability-indicating assay for the determination of MT-I in cell culture transport media and in human plasma has been accomplished (Surendran *et al.,* 1995a). A reversed-phase HPLC isocratic assay was developed with an analysis time per sample of less than 20 min. In another study (Surendran *et al.,* 1996), the partitioning properties and solution stability of MT-I were determined. The partitioning studies indicated that the absorption potential of MT-I was quite promising. The stability studies indicated that MT-I was relatively stable under acidic conditions, but was increasingly less stable as the pH was raised above 7. Both ionic strength and phosphate buffer concentration had no effect on the degradation kinetics of MT-I. MT-I exhibited apparent first-order degradation with an estimated shelf-life (t_{90}) at pH 7.4 of 40 days.

Toxicological studies were performed in mice given MT-I (Dorr *et al.,* 1988). MT-I is very slowly metabolized *in vivo* and is active at concentrations 1000-fold lower than α-MSH. Mice were administered up to 2 mg/kg of MT-I daily and weekly over 4–12 weeks by topical application (in 90% DMSO) or by intraperitoneal injections (in physiological saline). At the end of this period, no toxic effects were observed in various organs, hematological indices, or on weight gain. In a follow-up trial in rats, a slight (30%) increase in alkaline phosphatase levels was observed. There was no evidence in either species of a behavioral effect or any ACTH-like endocrine actions such as elevated serum cortisol levels. Similar results were observed in pigs. These studies demonstrated the nontoxicity of MT-I in both chronic and acute high dosage in rodent and larger species, and such results formed the basis of subsequent clinical trials on male volunteers.

To address the controversial issue as to whether or not α-MSH is trophic for fetal growth and if it affects fetal adrenal development, we then evaluated MT-I for its possible effects on gestation or embryonic fetal development in rats (Dawson *et al.,* 1993). The rat was used as a model to study such processes based on its similarities to the human (Moore, 1982; Wilson, 1965a,b; Witchi, 1962). MT-I was delivered directly to the conceptus *in utero* during organogenesis. No changes were found in the parameters examined (e.g., sex ratio, weight, morphology, or histology) between treated and control fetuses. Also, there was no evidence of premature parturition or pigmentation changes in the fetuses. These studies were considered especially relevant for the potential use of MT-I as a melanogenic drug by women of childbearing age.

4.1.2. MT-II STUDIES

Quantification of MT-II in biological fluids was determined using a reversed-phase HPLC method involving isocratic elution (Ugwu and Blanchard, 1992). This assay was used to ascertain the influence of pH, phosphate buffer concentration, temperature, and ionic strength on the rate of MT-II degradation (Ugwu *et al.*, 1994a). It was found that MT-II degradation followed apparent first-order kinetics, with a maximum stability at pH ~ 5. The degradation rate of MT-II was directly proportional to phosphate buffer concentration and temperature, but was independent of ionic strength of the buffer. The shelf-life (i.e., t_{90}, the time for 10% degradation) in aqueous phosphate buffer at 25°C was 27 hr. We then determined the dissociation constants of MT-II as well as its partition coefficients at three pH values (Lan *et al.*, 1994). The bioavailability of MT-II in the rat was then evaluated by comparing the area under the plasma concentration–time curve (AUC) following intrajejunal and intravenous doses. The calculated bioavailability of MT-II was 4.6%, which was significantly greater than the reported intrajejunal bioavailability of 0.3% for octeotride (Drewe *et al.*, 1993), a somatostatin agonist analogue of similar size to MT-II. These observations led us to conclude that it might be feasible to deliver MT-II orally.

The pharmacokinetic profile of MT-II was then determined in rats following a 0.3 mg/kg intravenous dose (Ugwu *et al.*, 1994b). The plasma concentration–time profile of MT-II was biphasic with an α-phase of about 15 min and a β-phase of about 1.5 hr. In this study we also compared the blood concentrations in rat plasma using our HPLC assay with the values obtained using the classic frog skin bioassay that had been developed by Hadley and co-workers. An excellent linear relationship between plasma concentrations determined by the two methods was observed.

4.2. Drug Delivery and Clinical Studies

The success of any promising new therapeutic entity depends, in large part, on the development of a suitable delivery system (i.e., route of administration and dosage form). This problem becomes magnified in the case of many peptide and protein therapeutics because of their short half-lives, metabolic instability, relatively high polarity, and larger molecular size than most traditional drug molecules which limits their transport across the gastrointestinal barrier.

Transdermal delivery of MT-I was of early interest as an alternative strategy of being noninvasive and avoiding the likelihood of poor absorption by the oral route. Four major facets of transdermal delivery of MT-I were examined. First, it was demonstrated that MT-I induced pigmentation in the hair follicles of the yellow C57BL/6JA mouse model after topical application (Levine *et al.*, 1987). MT-I was topically applied to an area of the back of these mice and within 24–48 hr eumelanin production was visible microscopically within hair bulb melanocytes in both treated and untreated areas. The presence of melanized organelles (eumelanosomes) within melanocytes was confirmed by electron microscopy. Thus, these results showed that MT-I was delivered through the skin and into systemic circulation. In another study (Dawson *et al.*, 1988) , MT-I was transdermally delivered to the extent of 0.002 or 0.05% of a 10^{-4} M preparation using a DMSO/water solution or PEG/alcohol cream base, respectively, through full-thickness mouse skin. However, similar studies in the rat showed that MT-I could not be transdermally delivered (Dawson *et al.*, 1988). Therefore, transdermal delivery using human skin was critical to accurately determine the possibility of using this route of administration. Previous *in vitro* transdermal penetration studies have accurately predicted the *in vivo* situation and have shown good correlation (Wester and Maibach, 1985; Shaw *et al.*, 1975; Bronaugh *et al.*, 1982). Accordingly, MT-I was applied to the surface of human skin samples using a standard permeation apparatus. Penetration of MT-I was examined for 24 hr at 37°C, and passage of MT-I was determined using both bioassay and radioimmunoassay for the collection fluid. Differences in the degree of transdermal penetration were regional as well, and skin thickness was a critical factor. Split thickness skin (i.e., only upper dermal tissue) allowed greater penetration, suggesting dermal binding of MT-I. Passage of MT-I from the topically applied vehicle (PEG) across the skin into a subcutaneous receiving vessel was demonstrated by standard frog skin bioassay, and transdermal delivery of MT-I through human skin *in vitro* was indicated. Methods for improved and more consistent delivery across skin remain to be developed, and the possibility of iontophoretic techniques (Bronaugh *et al.*, 1982) appears promising.

A study (Surendran *et al.*, 1995b) was designed to evaluate the potential of MT-I to be delivered orally. An *in vitro* cell monolayer (Caco-2) was used to screen the effects of several absorption enhancers on MT-I transport. The most promising enhancers also were evaluated using an *in situ* closed loop rat intestine model. In the Caco-2 cell monolayer model, the coadministration of aprotinin (a protease inhibitor) produced a 2.4-fold increase in the transport of MT-I. The transport data for the Caco-2 cell model and the rat model were in good agreement and indicated that inhibition of MT-I degradation by proteases was a promising approach to delivering MT-I orally.

Our next efforts involved a pilot study in human subjects to evaluate the melanogenic properties and pharmacokinetics of MT-I following oral, s.c., and i.v. dosing (Ugwu *et al.*, 1997). Although s.c. dosing had been utilized in nearly all of the previous MT studies, this was the first attempt to determine the pharmacokinetic profile and bioavailability of MT-I by the s.c. route. The s.c. dose was determined to be completely bioavailable as was the i.v. dose, but no detectable MT-I levels were observed following oral dosing. The plasma half-lives following s.c. dosing ranged from 0.7 to 0.79 hr for the adsorption phase and 0.8 to 1.7 hr for the beta-phase. Side effects were minimal and significant tanning of the forehead, arms, and neck was noted following i.v. and s.c. dosing. This effect peaked at 1 week following a 2-week MT-I dosing regimen, but it was still present 3 weeks after completing the 10-dose regimen given once daily, Monday–Friday, for 2 consecutive weeks.

Based on the short half-life observed for MT-I in the previous study, it was decided to focus our efforts on the development of a sustained-release injectable formulation. Our initial effort in this regard (Bhardwaj and Blanchard, 1996) involved the use of Poloxamer 407 (P407), a thermally reversible gel-forming agent. Various aqueous formulations containing MT-I and 25% w/v P407 alone or with one of several additives present were evaluated. The *in vivo* release kinetics of selected formulations was evaluated in guinea pigs following i.p. administration. The plasma concentration–time profiles demonstrated an extended release of MT-I formulated in the P407 gel compared with the i.p. administration of MT-I in solution.

Although the previous study had demonstrated that P407 gel formulations could provide prolonged plasma levels of MT-I, the prolongation was too short-lived to permit the peptide to be dosed less frequently than once daily. Therefore, we decided to develop a biodegradable polymeric implant dosage form that could be administered once a month and would provide a controlled release of MT-I over that time period (Bhardwaj and Blanchard, 1997). The implants were prepared by melt-extrusion method and utilized a poly(D,L-lactide-coglycolide) (PLGA) copolymer. The implants were characterized by evaluating the effects of viscosity and molecular weight and molecular weight distributions of the polymer on the factors controlling the release of MT-I from the polymer (i.e., degradation and erosion). The release rate of MT-I from the polymer implant was examined at different loading levels and in the presence of some hydrophilic additives. In addition, the effect of gamma radiation on the release kinetics of the peptide was analyzed to determine the optimal radiation dose for sterilization of the PLGA implants. The results indicated that the PLGA-based formulation has the potential to increase the therapeutic efficacy of MT-I by prolonging the release of the peptide into the circulation.

In a related study (Bhardwaj and Blanchard, 1998), the properties of PLGA implants were evaluated further. First, the surface morphology of the implants was assessed using scanning electron microscopy. The time-dependent changes in the molecular weight distribution of the polymer and its erosion were monitored in order to help characterize the hydrolytic degradation processes occurring *in vivo*. The time for the average molecular weight of PLGA in the implant to decrease to 50% of its initial value, determined by size-exclusion chromatography, was about 12 days compared with 5 weeks for 50% erosion of the copolymer mass to occur. The release of lactic acid from PLGA was also quantitated simultaneously in order to characterize the degradation. The point at which the lactic acid increased was found to coincide with the onset of the tertiary phase of the MT-I release profile in guinea pigs. The MT-I released from the depot implanted subcutaneously in guinea pigs exhibited a release profile that extended over 1 month, in agreement with data from *in vitro* studies.

Our most recent efforts have focused on the use of a very unique animal model to further study the ability of our PLGA implant formulations to stimulate melanogenesis while concurrently determining the pharmacokinetic profile of the delivery system in order to evaluate the controlled release of MT-I by the implants.

The melanotropic effects of MT-I were studied using a special breed of pigmented hairless and haired guinea pigs developed by Dr. John Pawelek. The pigmented guinea pigs combine the convenience of a hairless model with a pigmentary system that is similar to human skin in structure and in its response to various stimuli (Bolognia *et al.*, 1990). The guinea pig skin contains active interfollicular epidermal melanocytes as well as active follicular melanocytes. The former are located in the basal layer of the epidermis in a pattern similar to that observed in human skin. The hairless guinea pigs are very useful models as their hairless surface is convenient for testing the effect of UV irradiation as well as for assessing the changes in cutaneous pigmentation in response to external agents such as MT-I or MT-II.

The goal of this study was to evaluate the *in vitro* and *in vivo* melanotropic activity of PLGA implants designed for 1-month duration of action. The biological activity of the MT-I released *in vitro* from implants prepared with and without gamma irradiation was measured using frog skin bioassays (Castrucci et al., 1984). The effect of MT-I on skin pigmentation was measured with a Minolta Chromameter® (reflectometer) and the plasma levels of MT-I were measured using the α-MSH RIA (Kreutzfeld and Bagnara, 1989) following s.c. MT-I implants in guinea pigs. Eumelanin, the black/brown melanin pigment, was quantified in guinea pig skin biopsies via HPLC. The MT-I released *in vitro* after 1 day of incubation exhibited 100% melanotropic activity on the frog skins when compared with a standard, indicating that there was no degradation of MT-I during the fabrication of the implants and the gamma irradiation sterilization.

The plasma concentration versus time profile following the s.c. administration of 4 mg MT-I was similar to the triphasic profile for the *in vitro* release kinetics observed in earlier studies (Bhardwaj and Blanchard, 1998). The maximum MT-I concentration was observed in about 3 weeks after a slow release phase and the release of peptide continued for about 5 weeks. This peak observed at 3 weeks reflected the onset of erosion of the PLGA polymer. The melanotropic effect of MT-I continued during the slow release phase before the erosion of the polymer and persisted long after the MT-I levels were below the RIA detection limit.

Figure 6 illustrates the cutaneous and follicular effect of implantation of the MT-I depot in hairless and haired guinea pigs. The melanotropic activity of MT-I in hairless guinea pigs was observed as a cutaneous effect only, whereas in the haired animals darkening of the hair color from brown to black was observed as well as the skin darkening. To visualize the enhanced pigmentation in skin, histological sections prepared from guinea pig skin biopsy samples were stained with Fontana-Masson stain to highlight the melanin-positive cells. Figures 7 and 8 show the histology of the epidermis layers of the hairless and haired guinea pig skin, respectively. The increased pigmentation after implantation of the MT-I depot resulted in an increased number of melanin-containing cells shown as black granules in the epidermal region of the skin. The melanotropic effect peaked in 1 month and the melanin levels decreased after 3 months.

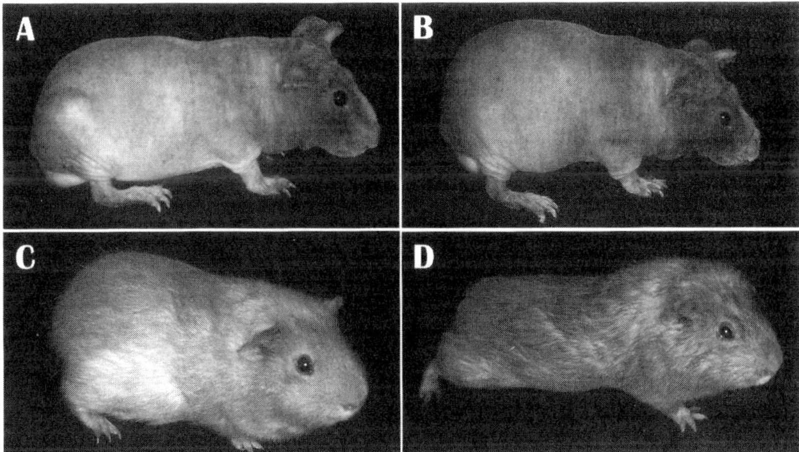

Figure 6. Photographs of pigmented guinea pigs after subcutaneous implantation of a depot containing 4 mg MT-I. Hairless guinea pigs at zero time (A) and at 3 months (B). Haired guinea pig at zero time (C) and after 3 months (D).

The reflectance reading showed a decrease in the luminance (L) of the guinea pigs, which measures the change in white to black hues. A negative L value signifies skin darkening and a decrease in L value by one unit indicates a visually perceptible skin darkening effect. The skin darkening was observed within a week following implantation and the maximum L value observed was −4.82. The pigmentation continued for 3 months even after the MT-I levels were undetectable in plasma, possibly reflecting an increased affinity and activity of MT-I for the melanocyte receptors in the epidermis.

The melanin pigments, eumelanin and pheomelanin, participate in skin pigmentation. Eumelanin and its precursor, 5,6-dihydroxyindole, appear to possess potent photoprotective (antioxidant) properties as opposed to the photodamaging effects of pheomelanin, on exposure to UV radiation. Hence, quantitation of eumelanin by HPLC (Ito and Wakamatsu, 1994) in guinea pigs after MT-I administration is another measure of MT-I's protective effect against UV rays of the sun. The measurement of eumelanin in skin biopsies revealed a concentration versus time profile similar to the skin reflectance (luminance) values. A 2.5-fold increase in eumelanin was observed in about 1 month and the effect persisted for 3 months.

The results indicate that the PLGA implant delivery system could provide a therapeutic tanning of the skin to lower the risk of UV-induced melanomas. Based on the prolonged release, enhanced biological activity of low, constant levels of MT-I were noted and the melanotropic action thus lasted for months. This reduces the frequency of administration from a once-a-month implant to once every 3

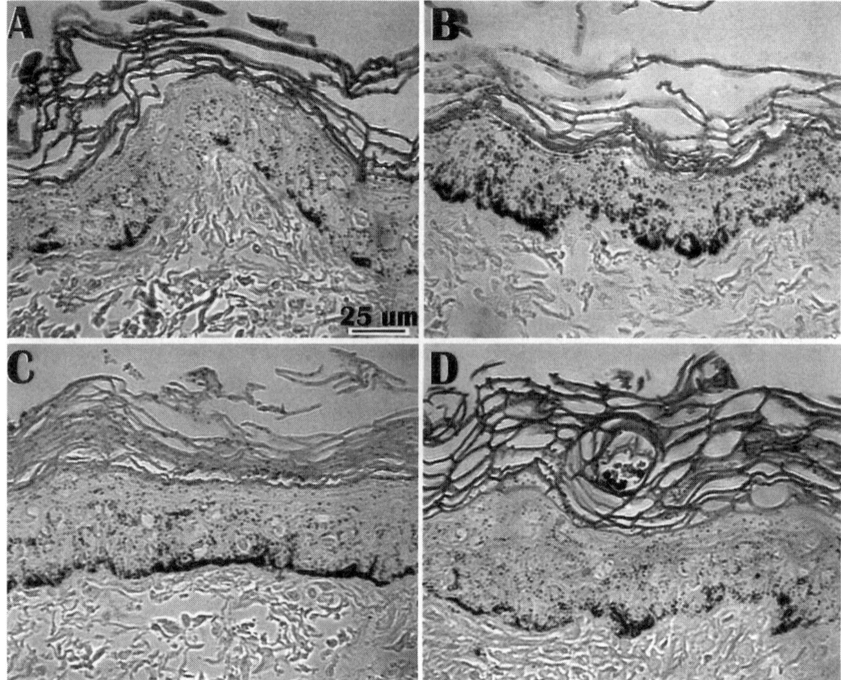

Figure 7. Histological sections of hairless guinea pigs at baseline level (A), 1 month (B), 2 months (C), and 3 months (D) after implantation of a 4-mg MT-I depot. The melanin pigment can be seen as dark granules at the junction of the basal epidermis (upper) and the dermis (lower).

months. In addition, the increase in melanin pigment, especially eumelanin, could provide protection against the photodamaging effects of UV radiation, thereby aiding in the prevention of skin cancers.

5. SUMMARY AND FUTURE DIRECTIONS

The discovery and development of MT-I and MT-II provides impetus to the future use of MT-based superagonists for a variety of MC1R-related applications, ranging from melanogenesis (skin tanning) to diagnostic or anticancer drug conjugates for melanoma chemotherapy. The recent discovery of the MC3R, MC4R, and MC5R types provides new possibilities for the discovery and development of novel MT-based agonists or antagonists for other MCR-targeted therapeutic uses, including eating disorders (obesity), inflammation, and erectogenic dysfunction.

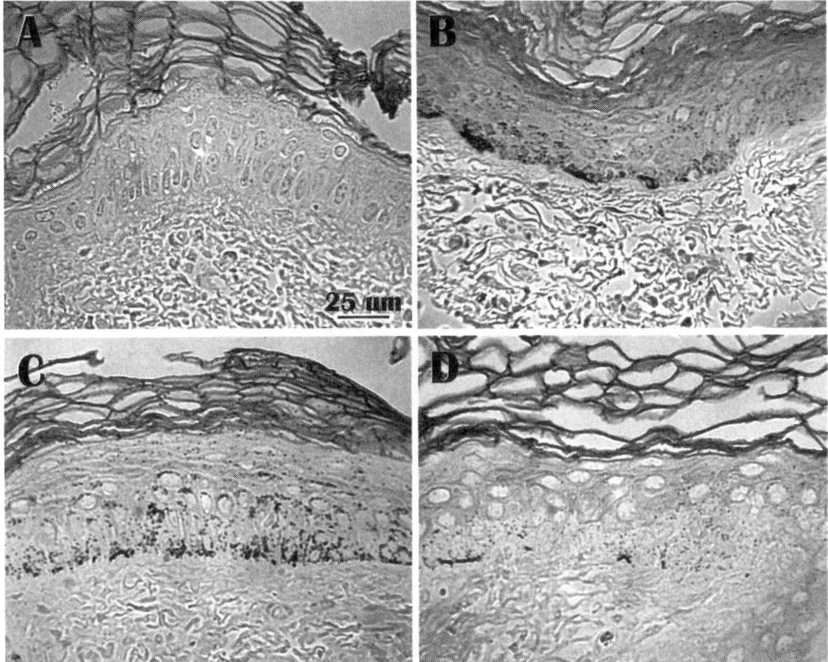

Figure 8. Histological sections of haired guinea pigs at baseline (A), 1 month (B), 2 months (C), and 3 months (D) after implantation of a 4-mg MT-I depot. The dark-stained melanin granules in the basal epidermis are less dense in the haired animals as a result of deposition of the melanin in the base of the hair follicles (not shown).

ACKNOWLEDGMENTS

This research has been the inspired and productive team effort that is credited to our colleagues, including many graduate students (Young Yang, Christopher Heward, Brian Fuller, Brian Wilkes, Paul Darman, Wayne Cody, Zalfa Abdel-Malek, Carrie Haskell-Luevano, Narayanan Surendran, Sidney Ugwu, Renu Bhadwaj), postdoctoral fellows (James Knittle, Elizabeth Sugg, Dhirendra Chatuverdi, Shubh Sharma), and a key collaborator (Ana Marie Castrucci).

This research was supported by Public Health Service Grant P01 CA 27502 (RTD), and DK 17420 (VJH).

REFERENCES

Abdel-Malek, Z., Kreutzfeld, K. L., Marwan, M. M., Hadley, M. E., Hruby, V. J., and Wilkes, B. C., 1985, Prolonged stimulation of S91 melanoma tyrosinase by [Nle^4, D-Phe^7]-substituted α-melanotropins, *Cancer Res.* **45:**4735–4740.

Adan, R. A. H., Oosterom, J., Ludvigsdolter, G., Brabbee, J. H., Beerbach, J. R. H., and Gispen, W. H., 1994, Identification of antagonists for MC3, MC4 and MC5 receptors, *Eur. J. Pharmacol.* **269:**331–338.

Al-Obeidi, F., Hruby, V. J., Pettitt, B. M., and Hadley, M. E., 1989a, Design of a new class of superpotent cyclic α-melanotropins based on quenched dynamic simulations, *J. Am. Chem. Soc.* **111:**3413–3416.

Al-Obeidi, F., Hruby, V. J., Castrucci, A. M. L., and Hadley, M. E., 1989b, Potent and prolonged acting cyclic lactam analogues of α-melanotropin: Design based on molecular dynamics, *J. Med. Chem.* **32:**2555–2561.

Al-Obeidi, F., Hruby, V. J., Hadley, M. E., Sawyer, T. K., and Castrucci, A. M. L., 1990, Design, synthesis, and biological activities of a potent and a selective α-melanotropin antagonist, *Int. J. Pept. Protein Res.* **35:**228–234

Barret, P., MacDonald, A., Helliwell, R., Davidson, G., and Morgan, P., 1994, Cloning and expression of a new member of the melanocyte-stimulating hormone receptor family, *J. Mol. Endocrinol.* **12:**203–213.

Bhardwaj, R., and Blanchard, J., 1996, Controlled-release delivery system for the α-MSH analog Melanotan-I using poloxamer 407, *J. Pharm. Sci.* **85:**915–919.

Bhardwaj, R., and Blanchard, J., 1997, *In vitro* evaluation of poly(D,L-lactide-co-glycolide) polymer-based implants containing the α-Melanocyte stimulating hormone analog, Melanotan-I, *J. Control. Rel.* **45:**49–55.

Bhardwaj, R., and Blanchard, J., 1998, *In vitro* characterization and *in vivo* release profile of a poly(D, L-lactide-co-glycolide)-based implant delivery system for the α-MSH analog, melanotan-I, *Int. J. Pharm.* **170.**

Bolognia, J. L., Murray, M. S., and Pawelek, J. M., 1990, Hairless pigmented guinea pigs: A new model for the study of mammalian pigmentation, *Pigment Cell Res.* **3:**150–156.

Bool, A. M., Gray, G. H., II, Hadley, M. E., Heward, C. B., Hruby, V. J., Sawyer, T. K., and Yang, Y. C. S., 1981, Racemization effects on melanocyte stimulating hormones and related peptides, *J. Endocrinol.* **88:**57–65.

Bronaugh, R. L., Stewart, R. F., and Coudgon, E. L., 1982, Methods for *in vitro* percutaneous absorption studies, *Toxicol. Appl. Pharmacol.* **62:**481–488.

Castrucci, A. M. L., Hadley, M. E., and Hruby, V. J., 1984, Melanotropin bioassays: *In vitro* and *in vivo* comparisons, *Gen. Comp. Endocrinol.* **55:**104–111.

Castrucci, A. M. L., Hadley, M. E., Sawyer, T. K., Wilkes, B. C., Al-Obeidi, F., Staples, D. J., DeVaux, A. E., Dym, O., Hintz, M. F., Riehm, J., Rao, K. R., and Hruby, V. J., 1989, α-Melanotropin: The minimal active sequence in the lizard skin bioassay, *Gen. Comp. Endocrinol.* **73:**157–163.

Chhajlani, V., Muceniece, R., and Wikberg, J. E. S., 1993, Molecular cloning of a novel human melanocortin receptor, *Biochem. Biophys. Res. Commun.* **195:**866–873.

Cody, W. L., Stevenson, J. W. S., Al-Obeidi, F., Sugg, E. E., and Hruby, V. J., 1988, Melanotropin three-dimensional structural studies by physical methods and computer-assisted molecular modeling, in: *The Melanotropic Peptides* (M. E. Hadley, ed.), Volume III, pp. 93–109, CRC Press, Boca Raton.

Dawson, B. V., Hadley, M. E., Kreutzfeld, K., Dorr, R. T., Hruby, V. J., Al-Obeidi, F., and Don, S., 1988, Transdermal delivery of a melanotropic peptide hormone analog, *Life Sci.* **43:**1111–1117.

Dawson, B. V., Ford, C. A., Holloway, H., Dorr, R. T., and Johnson, P., 1993, Administration of melanotropic peptides during gestation in the rodent, *Toxicology* **77:**91–101.

Desarnaud, F., Labbé, O., Eggerickx, D., Vassart, G., and Parmentier, M., 1994, Molecular cloning, functional expression and pharmacological characterization of a mouse melanocortin receptor gene, *Biochem. J.* **299:**367–373.

Dorr, R. T., Dawson, B. V., Hadley, M. E., Levine, N., and Hruby, V. J., 1988, Toxicological studies of a superpotent melanotropin, *Invest. New Drugs* **6:**251–258.

Dorr, R. T., Lines, T., Levine, N., Brooks, C., Xiang, L., Hruby, V. J., and Hadley, M. E., 1995, Pilot phase-1 trial of a superpotent melanotropic peptide in 3 normal volunteers, *Life Sci.* **58:**1777–1784.

Drewe, J., Fricker, G. W., Vonderscher, J. G., and Beglinger, C., 1993, Enteral absorption of octreotide absorption enhancement by polyoxyethylene-24-cholesterol ether, *Br. J. Pharmacol.* **108:**298–303.

Engel, M. H., Sawyer, T. K., Hadley, M. E., and Hruby, V. J., 1981, Quantitative determination of amino acid racemization in heat-alkali treated melanotropins: Implications for peptide hormone structure–function studies, *Anal. Biochem.* **116:**303–311.

Fathi, Z., Lawrence, G. I., and Parker, E. M., 1995, Cloning, expression, and tissue distribution of a fifth melanocortin receptor subtype, *Neurochem. Res.* **20:**107–113.

Frandberg, P.-A., Muceniece, R., Prusis, P., Wikberg, J., and Chhajlani, V., 1994. Evidence for alternate points of attachment for α-MSH and its stereoisomer [Nle^4, D-Phe^7]-α-MSH at the melanocortin-1 receptor, *Biochem. Biophys. Res. Commun.* **202:**1266–1271.

Gantz, I., Konda, Y., Tashiro, T., Shimoto, Y., Miwa, H., Munzert, G., Watson, S. J., DelValle, J., and Yamada, T., 1993a, Molecular cloning of a novel melanocortin receptor, *J. Biol. Chem.* **268:**8246–8250.

Gantz, I., Miwa, H., Konda, Y., Shimoto, Y., Tashiro, T., Watson, S. J., DelValle, J., and Yamada, T., 1993b, Molecular cloning, expression, and gene localization of a fourth melanocortin receptor, *J. Biol. Chem.* **268:**15174–15179.

Gantz, I., Shimoto, Y., Konda, Y., Miwa, H., Dickinson, C. J., and Yamada, T., 1994, Molecular cloning, expression, and characterization of a fifth melanocortin receptor, *Biochem. Biophys. Res. Commun.* **200:**1214–1220.

Gee, C. E., Chen, C. L. C., Roberts, J. L., Thompson, R., and Waston, S. J., 1983, Identification of proopiomelanocortin neurons in the rat hypothalamus by *in situ* cDNA–mRNA hybridization, *Nature* **306:**374–375.

Geschwind, I. I., Huseby, R. A., and Nishioka, R., 1972, The effect of melanocyte-stimulating hormone on coat color in the mouse, *Recent Prog. Horm. Res.* **28:**91–130.

Griffon, N., Mignon, V., Facchinetti, P., Diaz, J., Schwartz, J.-C., and Sokoloff, P., 1994, Molecular cloning and characterization of the rat fifth melanocortin receptor, *Biochem. Biophys. Res. Commun.* **200:**1007–1014.

Hadley, M. E., 1996, *Endocrinology,* 4th ed., Prentice–Hall, Englewood Cliffs, NJ.

Hadley, M. E., Anderson, B., Heward, C. B., Sawyer, T. K., and Hruby, V. J., 1981, Calcium-dependent, irreversible effects of [4-norleucine, 7-D-phenylalanine]melanotropin on melanophores, *Science* **213:**1025–1027.

Hadley, M. E., Abdel-Malek, Z., Marwan, M. M., Kreutzfeld, K. L., and Hruby, V. J., 1985, [Nle^4, D-Phe^7] α-MSH: A superpotent melanotropin that "irreversibly" activates melanoma tyrosinase, *Endocr. Res.* **11:**157–170.

Hadley, M. E., Sharma, S. D., Hruby, V. J., Levine, N., and Dorr, R. T., 1993, Melanotropic peptides for therapeutic and cosmetic tanning of the skin, *Ann. N.Y. Acad. Sci.* **680:**424–439.

Hadley, M. E., Hruby, V. J., Jiang, J., Sharma, S. D., Fink, J. L., Haskell-Luevano, C., Bentley, D. L., Al-Obeidi, F. A., and Sawyer, T. K., 1996, Melanocortin receptors: Identification and characterization by melanotropic agonists and antagonists, *Pigment Cell Res.* **9:**213–234.

Haskell-Luevano, C., Miwa, H., Dickinson, C., Hruby, V. J., Yamada, T., and Gantz, I., 1994, Binding and cAMP studies of melanotropin peptides with the cloned human peripheral melanocortin receptor, hMC1R, *Biochem. Biophys. Res. Commun.* **204:**1337–1342.

Haskell-Luevano, C., Miwa, H., Dickinson, C., Hadley, M. E., Hruby, V. J., Yamada, T., and Gantz, I., 1996a, Characterization of the unusual dissociation properties of melanotropin peptides from the melanocortin receptor, hMC1R, *J. Med. Chem.* **39:**432–435.

Haskell-Luevano, C., Sawyer, T. K., Hendrata, S., North, C., Panahinia, L., Stum, M., Staples, D. J., Castrucci, A. M. L., Hadley, M. E., and Hruby, V. J., 1996b, Truncation studies of melanotropin peptides identify tripeptide analogues exhibiting prolonged agonist bioactivity, *Peptides* **17:**995–1002.

Haskell-Luevano, C., Sawyer, T. K., Trumpp-Kallmeyer, S., Bikker, J. A., Humblet, C., Gantz, I., and Hruby, V. J., 1996c, Three-dimensional molecular models of the hMC1R melanocortin receptor: Complexes with melanotropin peptide agonists, *Drug Des. Disc.* **14:**197–211.

Haskell-Luevano, C., Nikiforovich, G., Sharma, S. D., Yang, Y.-K., Dickinson, C., Hruby, V. J., and Gantz, I., 1997, Biological and conformational examination of stereochemical modifications using the template melanotropin peptide, Ac-Nle-c[Asp-His-Phe-Arg-Trp-Ala-Lys]-NH_2, on human melanocortin receptors, *J. Med. Chem.* **40:**1738–1748.

Hruby, V. J., Wilkes, B. C., Cody, W. L., Sawyer, T. K., and Hadley, M. E., 1984a, Melanotropins: Structural, conformational and biological considerations in the development of superpotent and superprolonged analogs, *Pep. Protein Rev.* **3:**1–64.

Hruby, V. J., Wilkes, B. C., Hadley, M. E., Al-Obeidi, F., Sawyer, T. K., Staples, D. J., deVaux, A. E., Dym, O., Castrucci, A. M. L., Hintz, M. E., Riehm, J. P., and Rao, R., 1987, α-Melanotropin: The minimal active sequence. I. Frog skin bioassay, *J. Med. Chem.* **30:**2126–2130.

Hruby, V. J., Sharma, S. D., Lu, D., Castrucci, A. M. L., Al-Obeidi, F. A., Cone, R., and Hadley, M. E., 1995, Cyclic lactam α-Melanotropin analogues of Ac-c[Nle^4, Asp^5, D-Phe^7,Lys^{10}]α-MSH(4–10)-NH_2 with bulky aromatic amino acids at position 7 show high antagonist potency and selectivity at melanocortin receptors, *J. Med. Chem.* **38:**3454–3461.

Ito, S., and Wakamatsu, K., 1994, An improved modification of permanganate oxidation of eumelanin that gives a constant yield of pyrrole-2,3,5-tricarboxylic acid, *Pigment Cell Res.* **7:**141–144.

Jiang, J., Sharma, S. D., Nakamura, S., Lai, J.-Y., Fink, J., Hadley, M. E., Hruby, V. J., and Hendrix, M. J. C., 1995, The melanotropic peptide, [Nle^4, D-Phe^7]α-MSH, stimulates human melanoma tyrosinase activity and inhibits cell proliferation, *Pigment Cell Res.* **8:**301–323.

Kreutzfeld, K. L., and Bagnara, J. T., 1989, Application of α-Melanocyte stimulating hormone (α-MSH) radioimmunoassay to the detection of the superpotent analog, [Nle^4, D-Phe^7]α-MSH, *Pigment Cell Res.* **2:**65–69.

Labbé, O., Desarnaud, F., Eggerickx, D., Vassart, G., and Parmentier, M., 1994, Molecular cloning of a mouse melanocortin 5 receptor gene widely distributed in peripheral tissues, *Biochemistry* **33:**4543–4549.

Lan, E.-L., Ugwu, S. O., Blanchard J., Fang, X., Hruby, V. J., and Sharma, S., 1994, Preformulation studies with Melanotan-II: A potential skin cancer chemopreventative peptide, *J. Pharm. Sci.* **83:**1081–1084.

Levine, N., Lemus-Wilson, A., Wood, S. H., Abdel-Malek, A. Z., Hruby, V. J., and Hadley, M. E., 1987, Follicular melanogenesis in the mouse: Stimulation by injected and topical melanotropins, *J. Invest. Dermatol.* **89:**269–273.

Levine, N., Sheftel, S. N., Eytan, T., Dorr, R. T., Hadley, M. E., Weinrach, J. C., Ertl, G. A., Toth, K., McGee, D. L., and Hruby, V. J., 1991, Induction of skin tanning by the subcutaneous administration of a potent synthetic melanotropin, *J. Am. Med. Assoc.* **266:**2730–2736.

Moore, K. L., 1982, The circulatory system, in: *The Developing Human,* 3rd ed., pp. 298–322 and 333–343, Saunders, Philadelphia.

Mountjoy, K. G., Mortrud, M. T., Low, M. J., Simerly, R. B., and Cone, R. D., 1994, Localization of the melanocortin-4 receptor (MC4-R) in neuroendocrine and autonomic control circuits in the brain, *Mol. Endocrinol.* **8:**1298–1308.

O'Donohue, T. L., and Dorsa, D. M., 1982, The opiomelanotropinergic neuronal and endocrine system, *Peptides* **3:**353–395.

Robbins, L. S., Nadeau, J. H., Johnson, K. R., Kelly, M. A., Roselli-Rehfuss, L., Baack, E., Mountjoy, K. G., and Cone, R. D., 1993, Pigmentation phenotypes of variant extension locus alleles result from point mutations that alter MSH receptor function, *Cell* **72:**827–834.

Roselli-Rehfuss, L., Mountjoy, K. G., Robbins, L. S., Mortrud, M. T., Low, M. J., Tatro, J. B., Entwistle, M. L., Simedy, R. B., and Cone, R. D., 1993, Identification of a receptor for melanotropin and other proopiomelanocortin peptides in the hypothalamus and limbic system, *Proc. Natl. Acad. Sci. USA* **90:**8856–8860.

Sawyer, T. K., Sanfilippo, P. J., Hruby, V. J., Engel, M. H., Heward, C. B., Burnett, J. B., and Hadley, M. E., 1980, 4-Norleucine, 7-D-phenylalanine-α-Melanocyte stimulating hormone: A highly potent melanotropin with ultralong biological activity, *Proc. Natl. Acad. Sci. USA* **77:**5754–5758.

Sawyer, T. K., Hruby, V. J., Wilkes, B. C., Draelos, M. T., Hadley, M. E., and Bergsneider, M., 1982, Comparative biological activities of highly potent analogues of α-melanotropin(4–10), *J. Med. Chem.* **25:**1022–1027.

Sawyer, T. K., Castrucci, A. M. L., Staples, D. J., Affholter, J. A., deVaux, A. E., Hruby, V. J., and Hadley, M. E., 1993, Structure–activity relationships of [Nle4, D-Phe7] α-MSH: Discovery of a tripeptidyl agonist exhibiting sustained biological activity, *Ann. N.Y. Acad. Sci.* **680:**597–599.

Shaw, J. E., Chandrasekaran, S. K., Michaels, A. S., and Taskovict, L., 1975, Controlled transdermal delivery *in vitro* and *in vivo,* in: *Animal Models in Dermatology* (H. I. Maibach, ed.), pp. 138–146, Churchill–Livingstone, Edinburgh.

Smith, P. E., and Graeser, J. B., 1924, A differential response of the melanophore stimulant and oxytocic autocord of the posterior pituitary, *Anat. Rec.* **27:**187.

Surendran, N., Ugwu, S. O., Sterling, E. J., and Blanchard, J., 1995a, A HPLC assay for the determination of [Nle^4, D-Phe^7]α-MSH analog (MT-I) in biological matrices, *J. Chromatogr.* **670:**235–242.

Surendran, N., Ugwu, S. O., Nguyen, L. D., Sterling, E. J., Dorr, R. T., and Blanchard, J., 1995b, Absorption enhancement of Melanotan-I: Comparison of the Caco-2 and rat *in situ* models, *Drug Delivery* **2:**49–55.

Surendran, N., Bhardwaj, R., Ugwu, S. O., Sterling, E. J., and Blanchard, J., 1996, Partitioning properties and degradation kinctics of [Nle^4, *D*-Phe^7] α-MSH analog Melanotan-I (MT-I), *Int. J. Pharm.* **135:**81–89.

Ugwu, S. O., and Blanchard, J., 1992, High-performance liquid chromatographic assay for the α-Melanotropin(4–10) fragment analog (Melanotan-II) in rat plasma, *J. Chromatogr.* **584:**175–180.

Ugwu, S. O., Lan, E.-L., Sharma, S., Hruby, V., and Blanchard, J., 1994a, Kinetics of degradation of a cyclic lactam analog of α-Melanotropin (MT-II) in aqueous solution, *Int. J. Pharm.* **102:**193–199.

Ugwu, S. O., Blanchard, J., Nguyen, L. D., Hadley, M. E., and Dorr, R. T., 1994b, Comparison of HPLC and bioassay methods for plasma Melanotan-II (MT-II) determination: Application to pharmacokinetic study in rats, *Biopharm. Drug Dispos.* **15:**383–390.

Ugwu, S. O., Blanchard, J., Dorr, R. T., Levine, N., Brooks, C., Hadley, M., Aikin, M., and Hruby, V., 1997, Skin pigmentation and pharmacokinetics of Melanotan-I in humans, *Biopharm. Drug Dispos.* **18:**259–269.

Wilson, J. G., 1965, Methods for administering agents and detecting malformations in experimental animals, in: *Teratology Principles and Techniques* (J. G. Wilson and T. Warkan, eds.), pp. 262–277, University of Chicago Press, Chicago.

Wilson, J. G., 1965b, Developmental abnormalities, rats, in: *Pathology of Laboratory Animals* (V. Benirschke, F. M. Garner, and T. C. Jones, eds.), Volume 2, pp. 1840–1847, Springer-Verlag, Berlin.

Witchi, E., 1962, Development: Rat, in: *Growth, Including Reproduction and Morphologic Development* (P. L. Altman and D. S. Dittmer, eds.), pp. 304–314, FASEB, Washington, DC.

Index

A-65317, 15–17
A-74203, 19–22
A-75998 LHRH antagonist, 137
  aggregation and formation, 141
Absolute stereochemistry, 493
Absorption vs. P_{App}, 514
Absorption, 483
Acidolytic cleavage, 158
Acquired immune deficiency syndrome, 211
Aggregation, 165
Agonists, 154, 158–163, 174, 176–178
Allometric interspecies scaling, 413
Anaphylactic reaction, 176
Androgen-sterilized, 156
Angiotensin converting enzyme (ACE), 8, 9, 13
Angiotensin I (AI), 8
Angiotensin II
  antagonists, 29–56
    balanced, 49–51
  receptors, 29, 30, 48, 49
    subtypes, 30, 48, 49
Angiotensin II (AII), 8, 16
Antagonists, 158, 159
α_{1A} Antagonists, 424
  angiotensin, 29–56
Antihypertensives, 29, 47, 48
Antiviral activity, 212, 216
Anxiolytic-like activity, 485
Aqueous solubility, 13, 15, 16, 18, 19
AS 101 (Ossirene), 371, 372
Aspartic proteinase, 9, 10
Assays, 379
  colony stimulating, 379
  hematopoietic synergistic factor, 379
  HPLC, 382
Atevirdine, *see also* PNU-87201
  clinical studies, 292
  metabolism, 292
4-Azasteroid, 398, 399, 405, 408, 410–412
6-Azasteroid, 405–410

Balloon angioplasty, 121
Benhydrilamine resin, 158

Benign prostatic hyperplasia (BPH), 379–399, 416, 417, 423
Benzazepinone, 466, 467
Benzothiazoles, 557, 558
Benzothiophenes, 558, 559
Bestatin (Ubenimex), 372, 374
Betafectin, 371
BHAPs, 291
Biliary excretion, 483
Bioassays, 10, 154, 161
Bioavailability, 9, 13–15, 17, 21–24, 165, 166, 177, 351, 353, 357, 360–362, 407, 412, 415, 482
  oral, 345, 346, 350, 351, 353, 355, 357, 360, 362
Biochemical castration, 156, 165, 176, 178
Biodegradable microspheres, 381
Biodisposition, 483
Biologic half-life, 153, 156, 165
Biological evaluation, 90
Blood–brain barrier, 5, 483
Bosentan, 99, 101, 103
BQ 123, 119
Brain penetration, 486
Bulk drug manufacture, 161–163
Bulk drug synthesis, 158
Buserelin, 156

Caco-2 cells, 122, 453, 454, 460, 512, 513
Canine model of thrombosis, 67
Canine thrombogenesis model, 66
Carbapenem, 345, 348
Carbonic anhydrase inhibitors (CAI)
  acetazolamide, 556
  dichlorophenamide, 556
  ethoxzolamide, 556
  methazolamide, 556
  MK-417, 559, 560
  MK-507: *see* Dorzolamide
  MK-927, 559, 560
Carbonic anhydrases I, II and IV, 563
Cardiovascular disease
  coagulation cascade, 57
  enzymes, 58
  fibrinolytic cascade, 59
Carrageenan paw edema, 448, 449, 461
Cassette dosing, 502
CCK-A receptor, 481
CCK-B receptor, 481
Cephalosporin, 348, 350, 353
  cefaclor, 345, 353–355
  cefcamate pivoxil (S 1108), 347, 354
  cefdaloxime pentexil, 347, 354, 355
  cefetamet, 347, 355
    pivoxil, 347, 355
  cefixime, 353–355
  cefotaxime, 345
  cefotiam, 360
  cefpodoxime, 347, 353, 354
    proxetil ester, 347, 353–355
  ceftazidime, 345
  ceftriaxone, 345
  cefuroxime, 345, 347, 353, 354
    axetil ester, 347, 353–355
  cephalexin, 345

Cephalosporin (*cont.*)
  cephaloglycine, 347, 354, 355
    acetoxymethyl ester, 347, 354, 355
    pivaloyloxymethyl ester, 347, 354, 355
  E1101, 347, 354, 355
  parenteral, 360
  Ro 40-6890, 347, 355
    isobutoxycarbonyl-2-propylidene ethyl (Ro 41-3399), 347, 354–356
    pivaloyloxymethyl ester, 355, 356
CFU-GM, 374, 375, 379
CFU-S, 375
Chemical modification, 153
Chemistry/chemical development, 88–90
Chiral centers, 162
Cholecystokinin, 481
Chromatography, 158–160, 163, 164, 166, 170, 174
Chymotrypsin, 14
α-Chymotrypsin rabbit assay, 565
CI-1015, 488
CI-988, 482
Clearance, 215, 220
Clinical requirements, 153
Cloramine-T, 166
Cold sores: *see* Herpes labialis
Colony stimulating factors (CSFs)
  erythropoietin, 369, 370
  G-CSF, 368–370
  GM-CSF, 368–370
  M-CSF, 369, 370
  PIXY321, 369
  thrombopoietin, 369, 370
Combinatorial chemistry, 502
Condensing agents, 158
Congestive heart failure, 16
Convergent segment strategy, 160
Corticotropin, 575
  adrenal cortical-stimulating hormone (ACTH), 575–577
Crixivan®
  Indinavir, 241, 242
  L-735,524, 234, 240–242
    MK-639, 241, 242, 244, 247, 248
    metabolism, 248, 249
    oral absorption, 246, 247
    pH solubility, 241–243
    pH stability, 243–246
Cyclooctylpyrone, 213
Cynomolgus monkey, 12, 13
CYP-450 inhibitor, 251, 253
Cytokines, 368, 383
  IFN-γ,
  IL-1, 368–370
  IL-2, 369
  IL-3, 368–370
  IL-6, 368–370
  IL-11, 369, 370
  IL-12, 369, 370

Decolorization carbons, 162
Delavirdine, *see also* PNU-90152
  bioavailability, 296, 306
  clinical studies, 309, 310
  metabolism, 308
  protein binding, 307
  solid forms, 303, 304
  solubility, 296, 303
Dexamethasone, 460, 461
Diastereomers, 491
Dihydropyrone, 213
Dihydrotestosterone (DHT), 393–417

Dimethylbenzanthracene (DMBA), 156
DMP 811, 45–47
Dorzolamide
 clinical studies, 571
 formulation, 566, 567
 metabolite, 563
 pharmacology
  in vitro, 563, 564
  in vivo, 554–566
  pK_a, 566
  safety, 567–571
  stereochemistry, 561
  synthesis, 560–562
  systemic effects, 567–571
Dosing vehicle, 350, 361
Drug candidate selection, 153
Drug delivery, 585
 depot, 588
 transdermal (topical), 585
Drug discovery
 metabolic and pharmacokinetic issues, 2
 rational drug design, 1
 screening, 1
DuP 532, 45–47

Efegatran
 clinical data, 70
 HPLC of arginal, 69
Efficacy, 216
Elucidation of structure, 152
Enalkiren (A-64662), 15–17, 19–21, 24
Endometriosis, 156, 169
Endothelin
 biosynthesis, 116
 converting enzyme (ECE), 115
 ^{3}H-NMR structure, 120
Endothelin (*cont.*)
 peptide mimicry, 117
 peptides (ET-1, 2, 3), 114
 receptor mutagenesis, 119
 receptor subtypes (ETA, ETB), 113
Endothelin antagonists, 81–86, 97, 101, 103, 105
Endothelin B-type receptor gene, 83
Endothelin receptor antagonists: *see* Endothelin antagonists
Endothelin (ET) receptors, 83, 84, 86, 87, 91, 99, 100, 105
Endothelins, 81, 84, 85, 90–93, 96, 97, 99–101, 103
Enzyme immunoassay/ radioimmunoassay, 94, 95
 ET-A, 83–85, 87, 89, 90–93, 99–101, 105
 ET-B, 83–85, 87, 89, 90–92, 101, 103, 105
Epidermis, 577
 keratinocytes, 577, 578
 melanocytes, 577, 578, 585
Erectogenic dysfunction, 576
Esterase, 347, 358, 360
Esters
 acetoxymethyl, 353, 355
 acid-stable, 350
 acyloxymethyl, 351
 alkyl, 347, 350, 351, 357, 359
 2-(alkyloxycarbonyl)-2-alkylideneethyl, 349
 aryl, 350, 351, 357, 359
 arylalkyl, 347
 daloxate, 349
 dialkylaminoethyl, 348
 ethoxycarbonyloxyethyl, 349, 351

Esters (*cont.*)
hexetil, 351
indanyl, 351
isobutoxycarbonyl-2-propylidene ethyl (Ro 41-3399), 355
methylenedioxy diester, 348
phenyl, 351, 358
phthalide, 349, 351
pivaloyloxymethyl, 348, 351, 353, 355, 360
Ex vivo, 486
Excretion, 350
EXP3174, 44–47

Famciclovir
bioavailability, 325–327, 330–333
clinical efficacy, 333–337
metabolism, 326–328, 332, 333
structure, 314, 324
Fetal malformations, 176
Fetal mortality, 176
Finasteride, 393–417
First-pass effect, 500
First-pass metabolism, 485
Formulation, 500

Gastric acid secretion, 499
Gastric emptying
effect on absorption, 518, 519
effect on bioactivity, 521
Genital herpes, 313, 335, 336
GG745, 393, 410–417
Glaucoma, 555
Glaucomatous monkey assay, 566
Glutathione (GSH) test, 557
GnRH (Factrel®), 176
Goralatide (Seraspenide), 372, 373
G-protein coupled receptors, 115
ligand binding site, 120
Growth factors, 368
myelopoietic, 368
hematopoietic, 368
SCF, 369, 370
TCG-β, 369, 370
Growth hormone secretagogues
clinical studies
L-692,429, 528, 529
MK-0677, 543, 544
peptides
GHRP-6, 526
growth hormone releasing hormone (GHRH), 525, 526
growth hormone releasing peptides (GHRPs), 526
peptidomimetics
benzolactam L-692, 527, 528
spiroindanylpiperidine MK-0677, 536–539
receptors, 545, 546
Guinea pig maximization test: *see* Magnusson and Kligman

Half-life, 153, 177
Hansch π constant, 158
Heart failure, 99, 100
Hematopoiesis, 367
Hematopoietic synergistic factor (HSF), 379, 383
Hematoregulator peptide 5b (HP-5b), 374, 375
Hematoregulator, 371
low molecular-weight, 371
Hepatitis B virus, 313, 315, 316, 320, 321, 330, 337
Herpes labialis, 313, 336, 337

Herpes simplex virus, 313–319, 329, 330, 335–337
Herpes zoster, 313, 333–335
HIV protease inhibitors
  C_2 symmetric linear diols, 258
  Cyclic ureas
    Caco-cell permeability, 177
    conformational analysis, 262
    de novo design, 259
    displacement of structural water, 260
    HIVPR crystal structures, 265
    molecular recognition, 265
    preorganization, 266
  potency, 275
    resistance profile, 275
    structure–activity relationship, 263
  DMP, 323
    clinical study, 269
    design and discovery, 267
    pharmacokinetics; rat, dog, 268
    solubility, 269
    synthesis, 268
  DMP, 450
    clinical study, 273
    design and discovery, 267
    pharmacokinetics; rat, dog, 268
    protein binding, 273
    synthesis, 272
HIV protease, 234, 235
Hormonal drugs, 151
Human immunodeficiency virus, 211
Human, 82, 86, 87, 90–92, 99, 101
Hydrolysis, 348, 357, 358, 360
  enzymatic, 347,350
  nonenzymatic, 348
Hydrolysis (*cont.*)
  of pivampicillin, 360
  rates, 357, 358
3β-Hydroxy-Δ^5-steroid dehydrogenase/3-keto-Δ^5-steroid isomerase (3BHSD), 398, 405–408, 410, 412
Hydroxyethylene isostere, 10, 12
Hydroxypropyl-β-cyclodextrin, 494
Hypertension, 96–98
Hypertensive patients, 16, 24
Hypogastric nerve stimulation, 425
Hypophyseal portal circulation, 152
Hypotensive response, 13, 15, 16, 21, 22, 24
Hypothalamus, 152, 153

i.d. rat model, 14–17, 19, 22, 24
Implants, 587
  polymeric, 587
  prolonged release, 578
In vitro assays, 8–10, 13–15
In vitro metabolism, 428, 433
In vivo models, 9, 12–15
Indomethacin, 450
Inhibitor, 211, 213
Intestinal permeability, 122
Intraocular pressure, 564–566
Ischemic stroke, 124
Isohormone, 161
Isomerization, 347, 349, 360
Isosteres, of tetrazoles, 33–39

L-364,505, 235, 236
L-365, 260, 482
L-682,679, 235, 236
L-685,434, 235, 237, 238, 241
L-687,630, 238, 239

L-687,908, 235, 236
L-689,520, 237, 238
L-700,497, 238, 239
L-704,486, 240, 241
L-731,723, 238, 239
LC/MS, 434, 435
LC-MS/MS, 502
[D-Leu^{6}-desGly^{10}]LHRH ethylamide, 158–160, 163, 164, 166, 170, 174
Leuprolide, 154, 155–172, 174, 176–178
LHRH agonists, structures, 133
LHRH analogs, 131
  mechanism of action, 131
LHRH antagonists
  biological tests for, 137
  N-methyl substitution in, 136
  reduced size, 135
  structures of, 133
LHRH receptors, 156, 166
Ligand assays, 161
Limulus amebocyte lysate, 170
Lipophilicity, 347, 360, 483
Lisofylline, 372, 373
Liver microsomes, 408, 409, 428
Log *P*, 494
Losartan, 33–39
  activity in humans, 47, 48
  metabolism, 32, 33, 44, 45
  preparation, 39–44
Lupron Depot®, 167, 177
Luteinizing hormone (LH), 152, 165

4MA, 398, 401
Magnusson and Kligman, 557, 560
Male pseudohermaphroditism, 394
Maximum absorbable dose (MAD), 470–472
Medicinal chemistry, 84, 85
Melanin, 578
Melanocortin, 575, 576
  agonist, 582
  antagonist, 582
  receptor, 577–579
Melanocyte, 585
  stimulating hormone (MSH), 575–577
Melanogenesis, 576–578
Melanogenic drugs, 583, 586
Melanoma, 576, 578
Melanotropin, 575
Metabolism, 92, 93, 222, 347, 350, 361
  of losartin, 30, 32, 33, 44, 45
  of tetrazoles, 30, 32, 33
Methylcellulose, 469, 470, 472
  oral bioavailability, 468, 469
MK-639 back-ups, 249
  L-754,394, 250–253
  L-756,170, 250, 252
  L-758,825, 250, 252
MK996, 38
Molecular modeling, 117
Molecular structures, 88
Monkey, 9, 12, 13, 15–17, 21, 22, 24
Monobactams, 345
Multiple drug resistance, 4
Myeloperoxidase, 461

Neurological transmitters, 160
N-in-one dosing, 436
Nonsteroidal anti-inflammatory agents, 446, 449

Octreoscan: *see* Radiolabeled analogs

Octreotide, *see also* Sandostatin
antiproliferative effects, 197
in combination therapy, 199
as single-agent therapy, 197
synthesis, 189
Ocular hypertensive animal assay, 564
Ocular normotensive animal assay, 564
OncoLAR
clinical studies with Sandostatin pamoate LAR in oncology, 202
development strategy, 196
Optical integrity, 162
Oral bioavailability, 215, 224
desolvation energy, 3
Oral efficacy models, 73
Oxidative metabolism, 408

Panic, 465, 466, 471, 476
Partition coefficient, 15, 18, 485
PD 156707, 82, 84, 87–105
Penciclovir
antiviral activity, 314–316, 329, 330
bioavailability, 321, 327–329, 331
clinical efficacy, 336, 337
mechanism, 316–321
prodrugs, 321–327
structure, 314, 324
Penems, 345, 348, 350, 351
CP-65,207 pivaloyloxymethyl ester, 351
FCE 22101, 353
FCE 22891, 346
Penicillins, 346–348, 350
(Z)-alkyloxyimino, 346
amoxicillin, 345
Penicillins (*cont.*)
ampicillin, 346, 348, 349, 351, 353
bacampicillin, 346, 349, 351, 353, 358
5,6-dimethoxyphthalidyl ester, 351
lenampicillin, 346
pivampicillin, 346, 348, 349, 351, 353, 358, 360
talampicillin, 346, 349, 351, 353, 358
carbenicillin, 345, 346, 349–352, 357, 358, 359
carfecillin, 350, 357, 358
carindacillin, 350
phenyl ester, 358
cloxicillin, 345, 351
mecillinam, 346
methicillin
N,N-diethylaminoethyl ester, 360
nafcillin, 345
penicillin G, 347
penicillin V, 345, 351
temocillin, 345
ticarcillin, 345
ureido, 345
Pentoxyfylline, 373
Peptide, aldehyde, 60
Peptide growth factors, 160
Peptides, 3, 576
Peptidomimetics, 576
Peptoid, 482
P-glycoprotein, 4
Pharmacodynamics, 398, 399, 413, 414, 440
Pharmacokinetics, 73, 84, 89, 166, 168, 169, 215, 220, 399, 407, 408, 410–417, 433, 437

Pharmacologic marker, 165
Pharmacology, 84, 86
Phospholipase A2, substrates, 446–447
Phospholipase A2 inhibitors
acid stability, 453, 460
formulation, 449
metabolism, 455–457, 460
pharmacokinetics, 452, 460
stereoselectivity, 451
Plasma protein binding, 472, 477
Plasma renin activity (PRA), 12, 16
PNU-87201, *see also* Atevirdine
activity, 289
bioavailability, 291
structure, 289
PNU-90152, *see also* Delavirdine
activity, 296, 300
bioavailability, 296, 303
solubility, 296, 303
structure, 294
Potency, 215
Preclinical pharmacokinetics, 485
Precocious puberty, 156, 169
Pressor response, 92, 93
Proopiomelanocortin (POMC), 577
Prostate, 394–397, 399, 404, 407, 411
cancer, 177, 178, 397
Prostatic carcinoma, 156
Protease, 211
Protein binding, 215, 216
Pulmonary hypertension, 100–102, 125
Purification, 158–161, 163, 164
Pyrone, 213
Racemization, 158, 159, 163
Radiolabeled analogs
targeting sst_2 receptor expressing tumors, 203
tumor imaging with ^{111}In-DTPA analog Octreoscan, 203
tumor therapy with ^{90}Y-DOTA analog SMT 487, 204
Receptors
angiotensin, 29
subtypes, 30, 48, 49
G-protein coupled, 579
melanocortin, 576
5α-Reductase (5AR), 393–417
Renal failure, 121
Renin inhibitors, 8–28
Renin–angiotensin cascade (RAS), 7, 8, 12
Renin–angiotensin systems, 29
RIA, 166, 178
RO-31,8959, 235, 240
saquinavir, 238, 240

Salt-depleted models, 12, 13, 16, 21, 24
Sandostatin
pharmacodynamics, 190
pharmacokinetics, 191
preclinical safety assessment, 191
therapeutic potential, 193
Sandostatin LAR
biodegradation of the polymer, 195
clinical efficacy and safety, 195
SAR studies, 154, 156
SB 209670, 119

SB 217242, 123
Segment condensation, 158, 159
D-Ser(Bu^t)^{6}desGly^{10}, 154, 156
Shingles: *see* Herpes zoster
Single-pass intestinal perfusion (SPIP), 470
SK&F 107647, 375–383, 384
  drug disposition studies, 382
  mechanism of action, 376
  oral activity, 380
  parenteral dosage form, 381
  pharmacokinetic studies, 383
  pharmacophore of, 377
  preclinical studies, 380
  structure–activity relationships, 376–378
  synthesis, 380
SK&F 108636, 375
SK&F 66861, 115
Skin pigmentation, 575
  melanin and, 585–590
Solubility, 357, 360
  aqueous, 347
  lipid, 347
Somatostatin analogues, structure–activity relationships, 188
Somatostatin receptors
  gene family of sst receptor subtypes, 184
  tissue distribution, 185
Specific rotation $[\alpha]_D$, 162, 171
Stability, 347, 357
Statine, 10, 12
Stroke, 103, 104
Structure–activity relationship (SAR), 86, 154, 396, 406, 426, 428, 488
Superagonists, 156
Systemic exposure, 224

Testosterone, 393–417
Tetrazoles
  isosteres, 33–39
  metabolism, 30, 32, 33
  preparation, 30–32, 41, 42
  stability, 30–32
Therapeutic concentration, 216, 219, 224
Thienothiopyrans, 559, 560
Thrombin inhibitor
  clinical data, 77
  oral, 71
Time-dependent inhibition, 399–405
Topical administration, 556, 564
Toxicity, 160, 163
Toxicokinetic evaluations, 486
Toxicology, 486
  thrombin inhibitor, 74
  rat and dog after oral dosing, 75
Transcardiac perfusion, 498
Transition-state mimics, 10–12, 18, 22
Trinem, 345, 350
  GV 104326, 346, 351
  GV 118819, 346
  hexetil ester, 351
Trophic effect, 153, 160
Trusopt®: *see* Dorzolamide
Type II′ β-bend, 156

Unnatural amino acids, 62, 158
  conformationally constrained, 64
Urinary recovery, 350
Ussing chamber, 380
Uterine fibroids, 156, 169

Varicella zoster virus, 313, 315–320, 333–335
Vasoconstriction, 91, 92
Ventromedial hypothalamus assay, 494

Warfarin, 213, 216, 220
Whole-body autoradiography, 486

X-maze, 494
X-ray crystal structure, 493

Y-25510, 372, 373

Zankiren (A-72517), 22–24

Pharmaceutical Biotechnology

Chronological Listing of Volumes

Volume 1 PROTEIN PHARMACOKINETICS AND METABOLISM
Edited by Bobbe L. Ferraiolo, Marjorie A. Mohler, and Carol A. Gloff

Volume 2 STABILITY OF PROTEIN PHARMACEUTICALS, Part A: Chemical and Physical Pathways of Protein Degradation
Edited by Tim J. Ahern and Mark C. Manning

Volume 3 STABILITY OF PROTEIN PHARMACEUTICALS, Part B: *In Vivo* Pathways of Degradation and Strategies for Protein Stabilization
Edited by Tim J. Ahern and Mark C. Manning

Volume 4 BIOLOGICAL BARRIERS TO PROTEIN DELIVERY
Edited by Kenneth L. Audus and Thomas J. Raub

Volume 5 STABILITY AND CHARACTERIZATION OF PROTEIN AND PEPTIDE DRUGS: Case Histories
Edited by Y. John Wang and Rodney Pearlman

Volume 6 VACCINE DESIGN: The Subunit and Adjuvant Approach
Edited by Michael F. Powell and Mark J. Newman

Volume 7 PHYSICAL METHODS TO CHARACTERIZE PHARMACEUTICAL PROTEINS
Edited by James N. Herron, Wim Jiskoot, and Daan J. A. Crommelin

Volume 8 MODELS FOR ASSESSING DRUG ABSORPTION AND METABOLISM
Edited by Ronald T. Borchardt, Philip L. Smith, and Glynn Wilson

Volume 9 FORMULATION, CHARACTERIZATION, AND STABILITY OF PROTEIN DRUGS: Case Histories
Edited by Rodney Pearlman and Y. John Wang

Volume 10 PROTEIN DELIVERY: Physical Systems
Edited by Lynda M. Sanders and R. Wayne Hendren

Volume 11 INTEGRATION OF PHARMACEUTICAL DISCOVERY AND DEVELOPMENT: Case Histories
Edited by Ronald T. Borchardt, Roger M. Freidinger, Tomi K. Sawyer, and Philip L. Smith